T0156269

Handbook of Genetic Programming Applications

Amir H. Gandomi • Amir H. Alavi • Conor Ryan
Editors

Handbook of Genetic Programming Applications

Editors
Amir H. Gandomi
BEACON Center for the Study of Evolution in Action
Michigan State University
East Lansing, MI, USA

Amir H. Alavi
Department of Civil and Environmental Engineering
Michigan State University
East Lansing, MI, USA

Conor Ryan
Department of Computer Science and Information Systems
University of Limerick
Limerick, Ireland

Supplementary material and data can be found on link.springer.com

ISBN 978-3-319-36313-4 ISBN 978-3-319-20883-1 (eBook)
DOI 10.1007/978-3-319-20883-1

Springer Cham Heidelberg New York Dordrecht London

Softcover re-print of the Hardcover 1st edition 2015

Printed on acid-free paper

Springer International Publishing AG Switzerland is part of Springer Science+Business Media (www.springer.com)

To my love, Elnaz, meaning of my life . . .
Amir H. Gandomi

To my wife, Fariba, echo of my heart . . .
Amir H. Alavi

To Beautiful Heather, my own global optimum . . .
Conor Ryan

Foreword

In the past two decades, artificial intelligence algorithms have proved to be promising tools for solving a multitude of tough scientific problems. Their success is due, in part, to the elegant manner in which they avoid the sort of handicaps that often plague mathematical programming-based tools, such as smooth and continuous objective functions. Thus, globally optimal (or close approximations of) design can be achievable with a finite and reasonable number of search iterations.

One of the most exciting of these methods is Genetic Programming (GP), inspired by natural evolution and the Darwinian concept of "Survival of the Fittest". GP's ability to evolve computer programs has seen it enjoy a veritable explosion of use in the last 10 years in almost every area of science and engineering.

This handbook brings together some of the most exciting new developments in key applications of GP and its variants, presented in a hands-on manner to facilitate researchers tackle similar applications and even use the same data for their own experiments.

The handbook is divided into four parts, starting with review chapters to quickly get readers up to speed, before diving into specialized applications in Part II. Part III focuses on hybridized systems, which marry GP to other technologies, and Part IV wraps up the book with a detailed look at some recent GP software releases.

The handbook serves as an excellent reference providing all the details required for a successful application of GP and its branches to challenging real-world problems. Therefore, for most chapters, the used data are either available as supplementary materials or publicly accessible.

East Lansing, MI, USA — Amir H. Gandomi
East Lansing, MI, USA — Amir H. Alavi
Limerick, Ireland — Conor Ryan

Contents

Part III Hybrid Approaches

Part IV Tools

Part I
Overview of Genetic Programming Applications

Chapter 1
Graph-Based Evolutionary Art

Penousal Machado, João Correia, and Filipe Assunção

1.1 Introduction

The development of an evolutionary art system implies two main considerations: (1) the design of a generative system that creates individuals; (2) the evaluation of the fitness of such individuals (McCormack 2007). In the scope of this chapter we address both of these considerations.

Influenced by the seminal work of Sims (1991), the vast majority of evolutionary art systems follows an expression-based approach: the genotypes are trees encoding symbolic expressions and the phenotypes—i.e., images—are produced by executing the genotypes over a set of x, y values. While this approach has been proven fruitful, it has several shortcomings, most notably: (1) although it is theoretically possible to evolve any image (Machado and Cardoso 2002), in practice, expression-based evolutionary art tends to produce abstract, mathematical images; (2) due to the representation, the images lack graphic elements that are typically present in most forms of art, such as lines, strokes, clearly defined shapes and objects; (3) creating an appealing image by designing a symbolic expression by hand, or even understanding an evolved expression, is a hard endeavour.

Extending previous work (Machado et al. 2010; Machado and Nunes 2010), we describe an approach that overcomes these limitations and introduces new possibilities. Inspired on the work of Stiny and Gips (1971), who introduced the concept of *shape grammars*, we explore the evolution of context free design grammars (CFDGs) (Horigan and Lentczner 2009), which allow the definition of

Electronic supplementary material The online version of this chapter (doi: 10.1007/978-3-319-20883-1_1) contains supplementary material, which is available to authorized users.

P. Machado (✉) • J. Correia • F. Assunção
CISUC, Department of Informatics Engineering, University of Coimbra,
3030 Coimbra, Portugal
e-mail: machado@dei.uc.pt; jncor@dei.uc.pt; fga@student.dei.uc.pt

A.H. Gandomi et al. (eds.), *Handbook of Genetic Programming Applications*,
DOI 10.1007/978-3-319-20883-1_1

complex families of shapes through a compact set of production rules. As such, in our approach, each genotype is a well-constructed CFDG. Internally, and for the purposes of recombination and mutation, each genotype is represented as a hierarchical directed graph. Therefore, the evolutionary engine deviates from traditional tree-based Genetic Programming (GP) and adopts graph-based crossover and mutation operators. The details of the representation are presented in Sect. 1.3, while Sect. 1.4 describes the genetic operators.

In Sect. 1.5 we introduce several fitness assignment schemes based on evolutionary art literature. Then, in the same Section, we describe how we combine several of these measures in a single fitness function.

We conduct several tests to assess the adequacy of the system and determine reasonable experimental settings. In particular, we focus on the impact of unexpressed code in the evolutionary process, presenting and analyzing different options for handling these portions of code. Furthermore, we study how non-deterministic mapping between genotypes and phenotypes influences the robustness of the evolved individuals. These experiments are reported in Sect. 1.6. Based on the results of these tests, we conduct experiments using each of the previously defined fitness functions individually. The description and analysis of the experimental results is presented in Sect. 1.7. The analysis of the results highlights the type of images favored by each fitness function and the relations among them. We then proceed by presenting results obtained when using a combination of functions to guide fitness (Sect. 1.7.2). The analysis of these results is focused on the ability of the system to create imagery that simultaneously addresses the different components of the fitness functions. We finalize by drawing overall conclusions and identifying future work.

1.2 State of the Art

Although there are noteworthy expression-based evolutionary art systems (e.g. Sims (1991); World (1996); Unemi (1999); Machado and Cardoso (2002); Hart (2007)), systems that allow the evolution of images that are composed of a set of distinct graphic elements such as lines, shapes, colors and textures are extremely scarce.

Among the exceptions to the norm, we can cite the work of: Baker and Seltzer (1994), who uses a Genetic Algorithm (GA) operating on strings of variable size to evolve line drawings; den Heijer and Eiben (2011) who evolve Scalable Vector Graphics (SVG), manipulating directly SVG files through a set of specifically designed mutation and recombination operators. Unlike GP approaches, where the representation is procedural, the representations adopted in these works are, essentially, descriptive—in the sense that the genotypes describe the elements of the images in a relatively directed way instead of describing a procedure, i.e. program, that once executed or interpreted produces the image as output.

In addition to our early work on this topic (Machado et al. 2010; Machado and Nunes 2010), there are two examples of the use of CFDG for evolutionary art purposes. Saunders and Grace (2009) use a GA to evolve parameters of specific CFDG hand-built grammars. As the name indicates, *CFDG Mutate* (Borrell 2014) allows the application of mutation operators to CFDGs. Unfortunately the system only handles deterministic grammars (see Sect. 1.3) and does not provide recombination operators.

O'Neill et al. (2009) explore the evolution of shape grammars (Stiny and Gips 1971) using Grammatical Evolution (O'Neill and Ryan 2003) for design purposes, generating 2D shapes (O'Neill et al. 2009) and 3D structures (O'Neill et al. 2010). Although they do not use CFDGs, their work is, arguably, the one that is most similar in spirit to the described in this Chapter, due to the adoption of a procedural representation based on grammars and a GP approach.

1.3 Representation

Context Free (Horigan and Lentczner 2009) is a popular open-source application that renders images which are specified using a simple language entitled CFDG (for a full description of CFDG see Coyne (2014)). Although the notation is different from the one used in formal language theory, in essence, a CFDG program is an augmented context free grammar, i.e., a 4-tuple: (V, Σ, R, S) where:

1. V is a set of non-terminal symbols;
2. Σ is a set of terminal symbols;
3. R is a set of production rules that map from V to $(V \cup \Sigma)^*$;
4. S is the initial symbol.

Figure 1.1 depicts the CFDG used to illustrate our description. Programs are interpreted by starting with the S symbol (in this case $S = Edera$) and proceeding by the expansion of the production rules in breath-first fashion. Predefined Σ symbols call drawing primitives (e.g., SQUARE). CFDG is an *augmented* context free grammar: it takes parameters that produce semantic operations (e.g., s produces a scale change). Program interpretation is terminated when there are no V symbols left to expand, when a predetermined number of steps is reached, or when the rendering engine detects that further expansion does not induce changes to the image (Machado et al. 2010).

Like most CFDGs, the grammar depicted in Fig. 1.1 is non-deterministic: several production rules can be applied to expand the symbols *Ciglio* and *Ricciolo*. When several production rules are applicable one of them is selected randomly and the expansion proceeds. Furthermore, the probability of selecting a given production may be specified by indicating a weight (e.g., 0.08). If no weight is specified a default value of 1 is assumed. The non-deterministic nature of CFDGs has profound implications: each CFDG implicitly defines a language of images produced using the same set of rules (see Fig. 1.2). Frequently, these images share structural and aesthetic properties. One can specify the seed used by the random number generator of the grammar interpreter, which enables the replicability of the results.

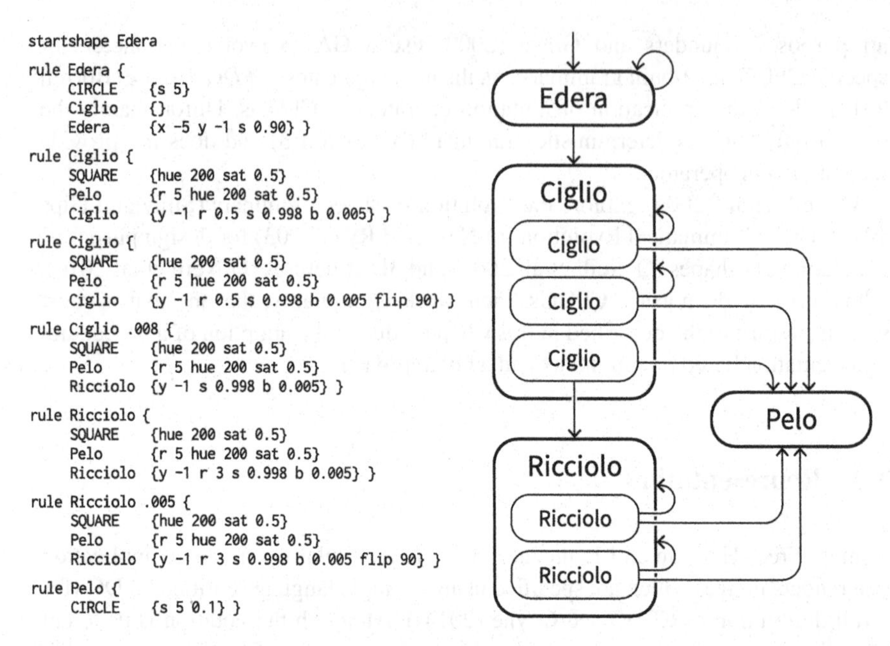
startshape Edera
rule Edera {
CIRCLE {s 5}
Ciglio {}
Edera {x -5 y -1 s 0.90} }
rule Ciglio {
SQUARE {hue 200 sat 0.5}
Pelo {r 5 hue 200 sat 0.5}
Ciglio {y -1 r 0.5 s 0.998 b 0.005} }
rule Ciglio {
SQUARE {hue 200 sat 0.5}
Pelo {r 5 hue 200 sat 0.5}
Ciglio {y -1 r 0.5 s 0.998 b 0.005 flip 90} }
rule Ciglio .008 {
SQUARE {hue 200 sat 0.5}
Pelo {r 5 hue 200 sat 0.5}
Ricciolo {y -1 s 0.998 b 0.005} }
rule Ricciolo {
SQUARE {hue 200 sat 0.5}
Pelo {r 5 hue 200 sat 0.5}
Ricciolo {y -1 r 3 s 0.998 b 0.005} }
rule Ricciolo .005 {
SQUARE {hue 200 sat 0.5}
Pelo {r 5 hue 200 sat 0.5}
Ricciolo {y -1 r 3 s 0.998 b 0.005 flip 90} }
rule Pelo {
CIRCLE {s 5 0.1} }
Edera
Ciglio
Ciglio
Ciglio
Ciglio
Pelo
Ricciolo
Ricciolo
Ricciolo

Fig. 1.1 On the *left*, a CFDG adapted from www.contextfreeart.org/gallery/view.php?id=165; On the *right*, the same CFDG represented as a graph (the labels of the edges were omitted for the sake of clarity)

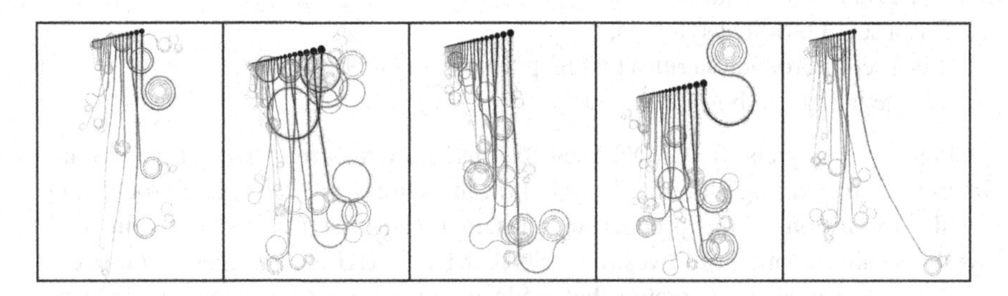

Fig. 1.2 Examples of images produced by the CFDG depicted in Fig. 1.1

In the context of our evolutionary approach each genotype is a well-constructed CFDG grammar. Phenotypes are rendered using Context Free. To deal with non-terminating programs a maximum number of expansion steps is set. The genotypes are represented by directed graphs created as follows:

1. Create a node for each non-terminal symbol. The node may represent a single production rule (e.g., symbol *Edera* of Fig. 1.1) or encapsulate the set of all production rules associated with the non-terminal symbol (e.g., symbols *Ciglio* and *Ricciolo* of Fig. 1.1);
2. Create edges between each node and the nodes corresponding to the non-terminals appearing in its production rules (see Fig. 1.1);
3. Annotate each edge with the corresponding parameters (e.g., in Fig. 1.1 the edges to *Pelo* possess the label '{r 5 hue 200 sat 0.5}').

1.4 Genetic Operators

In this Section we describe the genetic operators designed to manipulate the graph-based representation of CFDGs, namely: initialization, mutation and crossover.

1.4.1 Random Initialization

The creation of the initial population for the current evolutionary engine is of huge importance, being responsible for generating the first genetic material that will be evolved through time. In our previous works on the evolution of CFDGs the initial population was supplied to the evolutionary engine: the first population was either composed of human-created grammars (Machado and Nunes 2010) or of a single minimal grammar (Machado et al. 2010). Although both those options have merit, the lack of an initialization procedure for the creation of a random population of CFDGs was a limitation of the approach.

In simple terms, the procedure for creating a random CFDG can be described as follows: we begin by randomly determining the number of non-terminal symbols and the number of production rules for each of the symbols (i.e. the number of different options for its expansion). Since this defines the nodes of the graph, the next step is the random creation of connections among nodes and calls to non-terminal symbols. The parameters associated with the calls to terminal and non-terminal symbols are also established randomly. Finally, once all productions have been created, we randomly select a starting node and background color. Algorithm 1 details this process, which is repeated until the desired number of individuals is reached. Figure 1.3 depicts a sample of a random initial population created using this method.

1.4.2 Crossover Operator

The crossover operator used for the experiments described in this Chapter is similar to the one used in our previous work on the same topic (Machado et al. 2010; Machado and Nunes 2010). The rational was to develop a crossover operator that would promote the meaningful exchange of genetic material between individuals. Given the nature of the representation, this implied the development of a graph-based crossover operator that is aware of the structure of the graphs being manipulated. The proposed operator can be seen as an extension of the one presented by Pereira et al. (1999). In simple terms, this operator allows the exchange of subgraphs between individuals.

The crossover of the genetic code of two individuals, a and b, implies: (1) selecting one subgraph from each parent; (2) swapping the nodes and internal edges

Algorithm 1 Random initialization of an individual

```
procedure RandomInitialization
    terminal ← set of terminal symbols
    min_v, max_v ← minimum, maximum number of non-terminal symbols
    min_p, max_p ← minimum, maximum number of production rules per non-terminal
    min_c, max_c ← minimum, maximum number of calls per production
    nonterminal ← RandomlyCreateNonTerminalSet(min_v, max_v)
    for all V ∈ nonterminal do
        numberofproductions ← random(min_p, max_p)
        for i ← 1, numberofproductions do
            productionrule ← NewProductionRule(V)
            numberofcalls ← random(min_c, max_c)
            for j ← 1, numberofcalls do
                if random(0, 1) < prob_t then
                    productionrule.InsertCallTo(RandomlySelect(terminal))
                else
                    productionrule.InsertCallTo(RandomlySelect(nonterminal))
                end if
                productionrule.RandomlyInsertProductionRuleParameters()
            end for
        end for
    end for
    individual.setProductionRules(productionrules)
    individual.RandomlySelectStartShape(nonterminal)
    individual.RandomlyCreateBackgroundColor()
end procedure
```

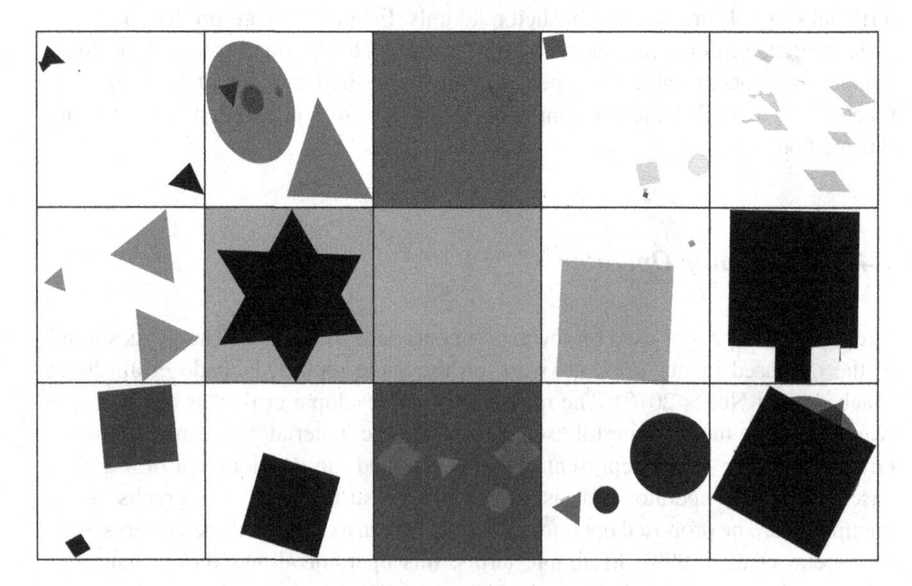

Fig. 1.3 Examples of phenotypes from a randomly created initial population

of the subgraphs, i.e., edges that connect two subgraph nodes; (3) establishing a correspondence between nodes; (4) restoring the outgoing and incoming edges, i.e., respectively, edges from nodes of the subgraph to non-subgraph nodes and edges from non-subgraph nodes to nodes of the subgraph.

Subgraph selection—randomly selects for each parent, a and b, one crossover node, v_a and v_b, and a subgraph radius, r_a and r_b. Subgraph s_{ra} is composed of all the nodes, and edges among them, that can be reached in a maximum of r_a steps starting from node v_a. Subgraph s_{rb} is defined analogously. Two methods were tested for choosing v_a and v_b, one assuring that both v_a and v_b are in the connected part of the graph and one without restrictions. The radius r_a and r_b were randomly chose being the maximum allowed value the maximum depth of the graph.

Swapping the subgraphs—swapping s_{ra} and s_{rb} consists in replacing s_{ra} by s_{rb} (and vice-versa). After this operation the outgoing and the incoming edges are destroyed. Establishing a correspondence between nodes repairs these connections.

Correspondence of Nodes—let s_{ra+1} and s_{rb+1} be the subgraphs that would be obtained by considering a subgraph radius of $r_a + 1$ and $r_b + 1$ while performing the subgraph selection. Let mst_a and mst_b be the minimum spanning trees (MSTs) with root nodes v_a and v_b connecting all s_{ra+1} and s_{rb+1} nodes, respectively. For determining the MSTs all edges are considered to have unitary cost. When several MSTs exist, the first one found is the one considered. The correspondence between the nodes of s_{ra+1} and s_{rb+1} is established by transversing mst_a and mst_b, starting from their roots, as described in Algorithm 2.

Restoring outgoing and incoming edges—the edges from a $\notin s_{ra}$ to s_{ra} are replaced by edges from a $\notin s_{rb}$ to s_{rb} using the correspondence between the nodes established in the previous step (e.g. the incoming edges to v_a are redirected to v_b, and so on). Considering a radius of $r_a + 1$ and $r_b + 1$ instead of r_a and r_b in the previous step allows the restoration of the outgoing edges. By definition, all outgoing edges from s_a and s_b link to nodes that are at a minimum distance of $r_a + 1$ and $r_b + 1$, respectively. This allows us to redirect the edges from s_b to b $\notin s_b$ to a $\notin s_a$ using the correspondence list.

1.4.3 Mutation Operators

The mutation operators were designed to attend two basic goals: allowing the introduction of new genetic material in the population and ensuring that the search space is fully connected, i.e., that all of its points are reachable from any starting point through the successive application of mutation operators. This resulted in the use of a total of ten operators, which are succinctly described on the following paragraphs.

Algorithm 2 Transversing the minimum spanning trees of two subgraphs

```
procedure TRANSVERSE(a, b)
    setcorrespondence(a, b)
    mark(a)
    mark(b)
    repeat
        if unmarked(a.descendants) ≠ NULL then
            next_a ← RandomlySelect(unmarked(a.descendants))
        else if a.descendants ≠ NULL then
            next_a ← RandomlySelect(a.descendants)
        else
            next_a ← a
        end if
        **** do the same for next_b ****
        transverse(next_a, next_b)
    until unmarked(a.descendants) = unmarked(b.descendants) = NULL
end procedure
```

Startshape mutate—randomly selects a non-terminal as starting symbol.

Replace, Remove or Add symbol—when applied to a given production rule, these operators: replace one of the present symbols with a randomly selected one; remove a symbol and associated parameters from the production rule; add a randomly selected symbol in a valid random position. Notice that these operators are applied to terminal and non-terminal symbols.

Duplicate, Remove or Copy & Rename rule—these operators: duplicate a production rule; remove a production rule, updating the remaining rules when necessary; copy a production rule, assigning a new randomly created name to the rule and thus introducing a new non-terminal.

Change, Remove or Add parameter—as the name indicates, these operators add, remove or change parameters and parameter values. The change of parameter values is accomplished using a Gaussian perturbation.

1.5 Fitness Assignment

Fitness assignment implies interpreting and rendering the CFDG. This is accomplished by calling the Context Free (Horigan and Lentczner 2009) application. Grammars with infinite recursive loops are quite common. As such, it was necessary to establish an upper bound to the number of steps that a CFDG is allowed to make before its expansion is considered complete. The original version of Context Free only allows the definition of an upper bound for the number of drawn shapes. This is insufficient for our goals, because it allows endless loops, provided that no shapes are drawn. As such, it was necessary to introduce several changes to the source code of Context Free (which is open source) to accommodate our needs. When calling Context Free we give as input (1) the CFDG to be interpreted and rendered,

(2) the rendering size, (3) the maximum number of steps (4) the rendering seed. We receive as output an image file. The maximum number of steps was set to 100,000 for all the experiments described in this Chapter. The "rendering seed" defines the seed of the random number generator used by Context Free during the expansion of the CFDGs. The rendering of the same CFDG using different rendering seeds can, and often does, result in different images (see Sect. 1.3). We performed tests using fixed and randomly generated rendering seeds. The results of those tests will be described in Sect. 1.6.

We use six different hardwired fitness functions based on evolutionary art literature and conduct tests using each of these functions to guide evolution. In a second stage, we perform runs using a combination of these measures to assign fitness. In the reminder of this Section we describe each of the functions and the procedure used to combine them.

1.5.1 JPEG Size

The image returned by Context Free is encoded in JPEG format using the maximum quality settings. The size of the JPEG file becomes the fitness of the individual. The rationale is that complex images, with abrupt transitions of color are harder to compress and hence result in larger file sizes, whereas simple images will result in small file sizes (Machado and Cardoso 2002; Machado et al. 2007). Although this assignment scheme is rather simplistic, it has the virtue of being straightforward to implement and yield results that are easily interpretable. As such, it was used to assess the ability of the evolutionary engine to complexify and to establish adequate experimental settings.

1.5.2 Number of Contrasting Colors

As the name indicates, the fitness of an individual is equal to the number of contrasting colors present in the image returned by Context Free. To calculate the number of contrasting colors we: (1) reduce the number of colors using a quantization algorithm; (2) sort all colors present in the image by descending order of occurrence; (3) for all the colors, starting from the most frequent ones, compute the Euclidean distance between the color and the next one in the ordered list, if it is lower than a certain threshold remove it from the group; (4) return as fitness the number of colors present on the list when the procedure is over. In these experiments, the Red, Green, Blue (RGB) color space was adopted. We quantize the image to 256 colors using the quantization algorithm from the graphics interchange format (GIF) format (Incorporated 1987). The threshold was set to 1 % of the maximum Euclidean distance between colors (255^3 for the RGB color space).

1.5.3 Fractal Dimension, Lacunarity

The use of fractal dimension estimates in the context of computational aesthetic has a significant tradition (Spehar et al. 2003; Mori et al. 1996). Although not as common, lacunarity measures have also been used (Bird et al. 2008; Bird and Stokes 2007). For the experiments described in this Chapter the fractal dimension is estimated using the box-counting method and the λ lacunarity value estimated by the Sliding Box method (Karperien 1999–2013). By definition, the estimation of the fractal dimension and lacunarity requires identifying the "object" that will be measured. Thus, the estimation methods take as input a binary image (i.e. black and white), where the white pixels define the shape that will be measured, while the black pixels represent the background. In our case, the conversion to black and white is based on the CFDG background primitive. All the pixels of the same color as the one specified by the CFDG background primitive are considered black, and hence part of the background, the ones that are of a different color are considered part of the foreground (see Fig. 1.4). Once the estimates are computed we assign fitness according to the proximity of the measure to a desired value, as follows:

$$\text{fitness} = \frac{1}{1 + |\text{target}_{\text{value}} - \text{observed}_{\text{value}}|} \tag{1.1}$$

We use the target values of 1.3 and 0.90 for fractal dimension and lacunarity, respectively. These values were established empirically by calculating the fractal dimension and lacunarity of images that we find to have desirable aesthetic qualities.

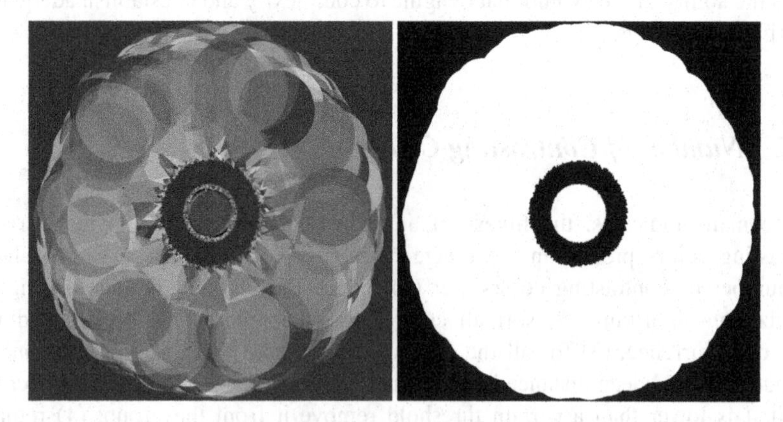

Fig. 1.4 Example of the transformation from the input color image (*left image*) to the background/foreground image (*right image*) used for the *Fractal Dimension* and *Lacunarity* estimates

1.5.4 Complexity

This fitness function, based on the work of Machado and Cardoso (2002); Machado et al. (2007, 2005), assesses several characteristics of the image related with complexity. In simple terms, the rationale is valuing images that constitute a complex visual stimulus but that are, nevertheless, easy to process. A thorough discussion of the virtues and limitations of this approach is beyond the scope of this Chapter, as such, we focus on practical issues pertaining its implementation. The approach relies on the notion of compression complexity, which is defined as calculated using the following formula:

$$C(i, \text{scheme}) = \text{RMSE}(i, \text{scheme}(i)) \times \frac{s(\text{scheme}(i))}{s(i)} \tag{1.2}$$

where i is the image being analysed, *scheme* is a lossy image compression scheme, *RMSE* stands for the root mean square error, and s is the file size function.

To estimate the complexity of the visual stimulus ($IC(i)$) they calculate the complexity of the JPEG encoding of the image (i.e. $IC(i) = C(i, JPEG)$). The processing complexity ($PC(i)$) is estimated using a fractal (quadratic tree based) encoding of the image (Fisher 1995). Considering that as time passes the level of detail in the perception of the image increases, the processing complexity is estimated for different moments in time ($PC(t0, i)$, $PC(t1, i)$) by using fractal image compression with different levels of detail. In addition to valuing images with high visual complexity and low processing complexity, the approach also values images where PC is stable for different levels of detail. In other words, according to this approach, an increase in description length should be accompanied by an increase in image fidelity. Taking all of these factors into consideration, Machado and Cardoso (2002); Machado et al. (2007, 2005) propose the following formula for fitness assignment:

$$\frac{IC(i)^a}{(PC(t0, i) \times PC(t1, i))^b \times (\frac{PC(t1,i)-PC(t0,i)}{PC(t1,i)})^c} \tag{1.3}$$

where a, b and c are parameters to adjust the importance of each component.

Based on previous work (Machado et al. 2005), the ability of the evolutionary engine to exploit the limitations of the complexity estimates was minimized by introducing limits to the different components of this formula, as follows:

$$\begin{cases} IC(i) & \rightarrow \max(0, \alpha - |IC(i) - \alpha|) \\ PC(t0, i) \times PC(t1, i) & \rightarrow \gamma + |(PC(t0, i) \times PC(t1, i)) - \gamma| \\ PC(t1, i) - PC(t0, i) & \rightarrow \delta + |(PC(t1, i) - PC(t0, i)) - \delta| \end{cases} \tag{1.4}$$

where α, γ and δ operate as target values for $IC(i)$, $(PC(t0, i) \times PC(t1, i)$ and $PC(t1, i) - PC(t0, i)$, which were set to 6, 24 and 1.1, respectively. These values were determined empirically through the analysis of images that we find to be desirable. Due to the limitations of the adopted fractal image compression scheme this approach only deals with greyscale images. Therefore, all images are converted to greyscale before being processed.

1.5.5 Bell

This fitness function is based on the work of Ross et al. (2006) and relies on the observation that many fine-art works exhibit a normal distribution of color gradients. Following Ross et al. (2006) the gradients of each color channel are calculated, one by one, in the following manner:

$$|\nabla r_{i,j}|^2 = \frac{(r_{i,j} - r_{i+1,j+1})^2 + (r_{i+1,j} - r_{i,j+1})^2}{d^2} \tag{1.5}$$

where $r_{i,j}$ is the image pixel intensity values for position (i, j) and d is a scaling factor that allows to compare images of different size; this value was set to 0.1 % of half the diagonal of the input image (based on Ross et al. (2006)). Then the overall gradient $S_{i,j}$ is computed as follows:

$$S_{i,j} = \sqrt{|\nabla r_{i,j}|^2 + |\nabla g_{i,j}|^2 + |\nabla b_{i,j}|^2} \tag{1.6}$$

Next, the response to each stimulus $R_{i,j}$ is calculated:

$$R_{i,j} = \log \frac{S_{i,j}}{S_0} \tag{1.7}$$

Where S_0 is a detection threshold (set to 2 as indicated in Ross et al. (2006)). Then the weighted mean (μ) and standard deviation (σ^2) of the stimuli are calculated as follows:

$$\mu = \frac{\sum_{i,j} R_{i,j}{}^2}{\sum_{i,j} R_{i,j}} \tag{1.8}$$

$$\sigma^2 = \frac{\sum_{i,j} R_{i,j}(R_{i,j} - \mu)^2}{\sum_{i,j} R_{i,j}} \tag{1.9}$$

At this step we introduce a subtle but important change to (Ross et al. 2006) work: we consider a lower bound for the σ^2, which was empirically set to 0.7. This prevents the evolutionary engine to converge to monochromatic images that, due to the use of a small number of colors, trivially match a normal distribution. This change has a profound impact in the experimental results, promoting the evolution of colorful images that match a normal distribution of gradients.

Using μ, σ^2 and the values of $R_{i,j}$ a frequency histogram with a bin size of $\sigma/100$ is created, which allows calculating the deviation from normality (DFN). The DFN is computed using q_i, which is the observed probability and p_i, the expected probability considering a normal distribution. Ross et al. (2006) uses:

$$\mathrm{DFN} = 1000 \cdot \sum p_i \log \frac{p_i}{q_i} \tag{1.10}$$

However, based on the results of preliminary runs using this formulation, we found that we consistently obtained better results using:

$$\mathrm{DFN}_s = 1000 \cdot \sum (p_i - q_i)^2 \tag{1.11}$$

Which measures the squares of the differences between expected and observed probabilities. Therefore, in the experiments described in this Chapter Bell fitness is assigned according to the following formula: $1/(1 + DFN_s)$.

1.5.6 Combining Different Functions

In addition to the tests where the fitness functions described above were used to guide evolution, we conducted several experiments where the goal was to simultaneously maximize several of these functions. This implied producing a fitness score from multiple functions, which was accomplished using the following formula:

$$\mathrm{combined}_{\mathrm{fitness}}(i) = \prod_j \log(1 + f_j(i)) \tag{1.12}$$

where i is the image being assessed and f_j refers to the functions being considered. Thus, to assign fitness based on the *Complexity* and *Bell* functions we compute: $\log(1 + \mathrm{Complexity}(i)) \times \log(1 + \mathrm{Bell}(i))$. By adopting logarithmic scaling and a multiplicative fitness function we wish to promote the discovery of images that maximize all the measures being considered in the experiment.

1.6 Configuring the Evolutionary Engine

The evolutionary engine has several novel characteristics that differentiate it from conventional GP approaches. Therefore, it was necessary to conduct a series of tests to assess the adequacy of the engine for the evolution of CFDGs and to determine a reasonable set of configuration parameters. These tests were conducted using *JPEG Size* as fitness function and allowed us to establish the experimental parameters

Table 1.1 Parameters used for the experiments described in this chapter

Initialization (see Algorithm 1)	Values
min, max number of symbols	(1,3)
min, max number of rules	(1,3)
min, max calls per production rule	(1,2)
Evolutionary Engine	Values
Number of runs	30
Number of generations	100
Population size	100
Crossover probability	0.6
Mutation probability	0.1
Tournament size	10
Elite size	Top 2 % of the population
CFDG Parameters	Values
Maximum number expansion steps	100,000
Limits of the geometric transformations	rotate ∈ [0,359], size ∈ [-5,5] x ∈ [-5,5], y ∈ [-5,5], z ∈ [-5,5] flip ∈ [-5,5], skew ∈ [-5,5]
Limits of the color transformations	hue ∈ [0,359], saturation ∈ [-1,1] brightness ∈ [-1,1], alpha ∈ [-1,1]
Terminal symbols	SQUARE, CIRCLE, TRIANGLE

summarized in Table 1.1, which are used throughout all the experiments described herein. In general, the results show that the engine is not overly sensitive to the configuration parameters, depicting an adequate behavior for a wide set of parameter configurations. Although the optimal parameters settings are likely to depend on the fitness function, a detailed parametric study is beyond the scope of this Chapter. Therefore, we did not attempt to find an optimal combination of parameters.

The use of a graph-based representation and genetic operators is one of the novel aspects of our approach. The use of such operators may introduce changes to the graph that may make some of the nodes (i.e. some production variables) unreachable from the starting node. For instance, a mutation of the node *Edera* of Fig. 1.1 may remove the call to node *Ciglio* making most of the graph unreachable. Although, unreachable nodes have no impact on the phenotype, their existence may influence the evolutionary process. On one hand they may provide space for neutral variations and promote evolvability (unreachable nodes may become reattached by subsequent genetic operators), on the other they may induce bloat since they allow protection from destructive crossover. To study the impact of unreachable nodes in the evolutionary process we considered three variations of the algorithm:

Unrestricted—the crossover points are chosen randomly;
Restricted—the crossover points are chosen randomly from the list of reachable nodes of each parent;

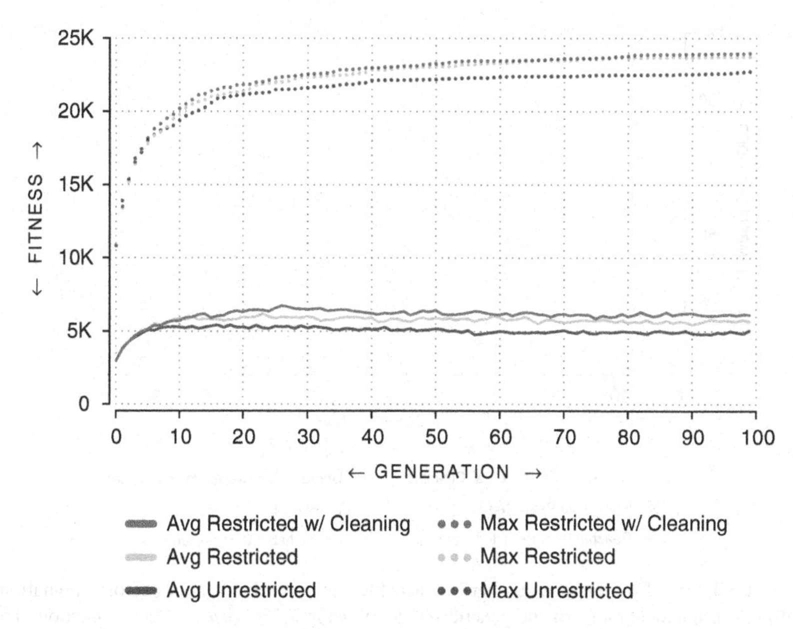

Fig. 1.5 Best and average fitness values for different implementations of the genetic operators using *JPEG Size* as fitness function. The results are averages of 30 independent runs

Restricted with Cleaning—in addition to enforcing the crossover to occur in a reachable region of the graph, after applying crossover and mutation all unreachable nodes are deleted.

Figure 1.5 summarizes the results of these tests depicting the best and average fitness for each population. As it can be observed, although the behaviors of the three different approaches are similar, the restricted versions consistently outperform the unrestricted implementation by a small, yet statistically significant, margin. The differences between the restricted approaches are not statistically significant.

The differences among the three approaches become more visible when we consider the evolution of the number of reachable and unreachable nodes through time. As it can be observed in Fig. 1.6, without cleaning, the number of unreachable nodes grows significantly, clearly outnumbering the number of reachable nodes. The number of reachable nodes of the restricted versions is similar, and smaller than the one resulting from the unrestricted version. Although cleaning does not significantly improve fitness in comparison with the restricted version, the reduction of the number of rules implies a reduction of the computational cost of interpreting the CFDGs and applying the crossover operators. As such, taking these experimental findings into consideration, we adopt the *Restricted with Cleaning* variant in all further tests.

The non-deterministic nature of the CFDGs implies that each genotype may be mapped into a multitude of phenotypes (see Sect. 1.3). The genotype to phenotype mapping of a non-deterministic grammar depends on a rendering seed, which is

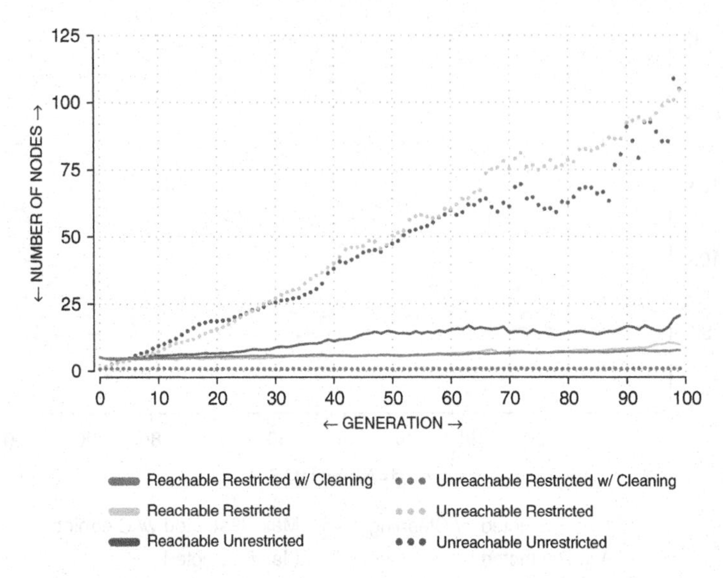

Fig. 1.6 Evolution of the average number of reachable and unreachable nodes across populations for different implementations of the genetic operators using *JPEG Size* as fitness function. The results are averages of 30 independent runs

passed to Context Free. We considered two scenarios: using a fixed rendering seed for all individuals; randomly generating the rendering seed whenever genotype to phenotype occurs. The second option implies that the fitness of a genotype may, and often does, vary from one evolution to the other, since the phenotype may change.

Figure 1.7 summarizes the results of these tests in terms of the evolution of fitness through time. As expected, using a fixed rendering seed yields better fitness, but the differences between the approaches are surprisingly small and decrease as the number of generations increases. To better understand this result we focused on the analysis of the characteristics of the CFGDs being evolved. Figure 1.8 depicts box plots of fitness values of the fittest individuals of each of the 30 evolutionary runs using different setups:

Fixed—individuals evolved and evaluated using fixed rendering seeds; Random—individuals evolved using random rendering seeds and evaluated using the same seeds as the ones picked randomly during evolution;
Fixed Random—individuals evolved using fixed rendering seeds and evaluated with 30 random seeds each;
Random Random—individuals evolved using random rendering seeds and evaluated with 30 random seeds each.

In other words, we take the genotypes evolved in a controlled static environment (fixed random seed) and place them in different environments, proceeding in the same way for the ones evolved in a changing environment. The analysis of the box plots shows that, in the considered experimental settings, the fitness of

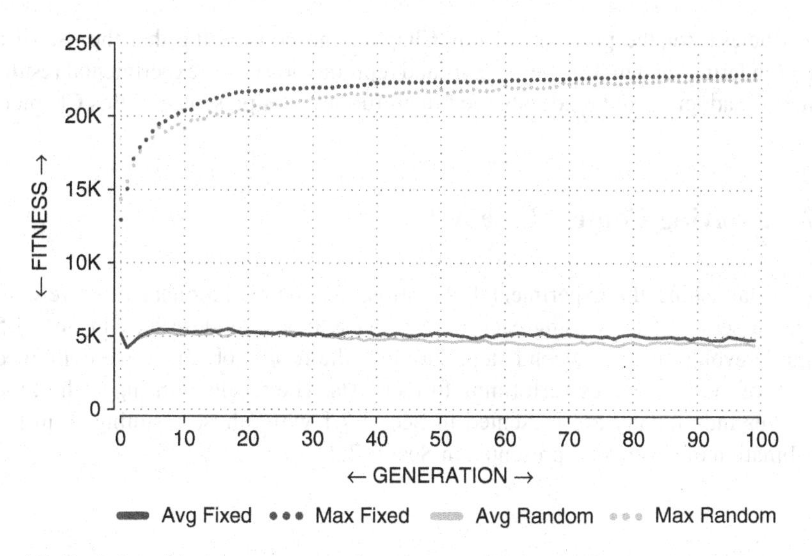

Fig. 1.7 Evolution of the best and average fitness across populations when using fixed and random rendering seeds using *JPEG Size* as the fitness function. The results are averages of 30 independent runs

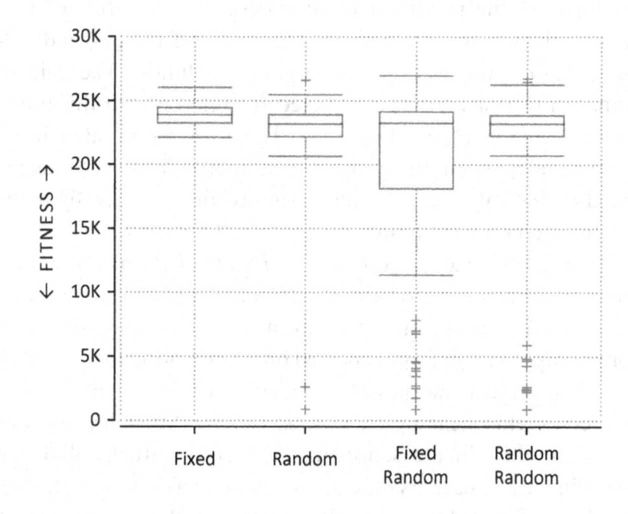

Fig. 1.8 *Box plots* of fitness values of the fittest individuals of each of the 30 evolutionary runs using different rendering seed setups

the individuals evolved in a fixed environment may change dramatically when the environmental conditions are different. Conversely, using a dynamic environment promotes the discovery of robust individuals that perform well under different conditions. Although this result is not unexpected, it was surprising to notice how fast the evolutionary algorithm was able to adapt to the changing conditions and find robust individuals. In future tests we wish to explore, and exploit, this ability.

Nevertheless, for the purposes of this Chapter, and considering that the use of a fixed rendering seed makes the analysis and reproduction of the experimental results easier, we adopt a fixed rendering seed in all further tests presented in this Chapter.

1.7 Evolving Context Free Art

After establishing the experimental conditions for the evolutionary runs we conducted a series of tests using each of the fitness functions described in Sect. 1.5 to guide evolution. In a second step, based on the results obtained, we combined several of these measures performing further tests. The results of using each of the measures individually are presented in Sect. 1.7.1 while those resulting from the combination of several are presented in Sect. 1.7.2.

1.7.1 Individual Fitness Functions

Figure 1.9 summarizes the results of these experiments in terms of evolution of fitness. Each chart depicts the evolution of the fitness of the best individual when using the corresponding fitness function to guide evolution. The values yield by the other 5 fitness functions are also depicted for reference to illustrate potential inter-dependencies among fitness functions. The values presented in each chart are averages of 30 independent runs (180 runs in total). To improve readability we have normalized all the values by dividing each raw fitness value by the maximum value for that fitness component found throughout all the runs.

The most striking observation pertains the *Fractal Dimension* and *Lacunarity* fitness functions. As it can be observed, the target values of 1.3 and 0.9 are easily approximated even when these measures are not used to guide fitness. Although this is a disappointing result, it is an expected one. Estimating the fractal dimension (or lacunarity) of an object that is not a fractal and that can be described using Euclidean geometry yields meaningless results. That is, although you obtain a value, this value is meaningless in the sense that there is no fractal dimension to be measured. As such, these measures may fail to capture any relevant characteristic of the images. In the considered experimental conditions, the evolutionary algorithm was always able to find, with little effort, non-fractal images that yield values close to the target ones. Most often than not, these images are rather simplistic. We conducted several tests using different target values, obtaining similar results.

An analysis of the results depicted in Fig. 1.9 reveals that maximizing *JPEG Size* promotes *Contrasting Colors* and *Complexity*, but does not promote a distributing of gradients approaching a normal distribution (*Bell*). Likewise, maximizing *Contrasting Colors* originates an improvement in *JPEG Size* and *Complexity* during the early stages of the evolutionary process; *Bell* is mostly unaffected. Using *Complexity* to guide evolution results in an increase of *JPEG Size* and *Contrasting Colors*

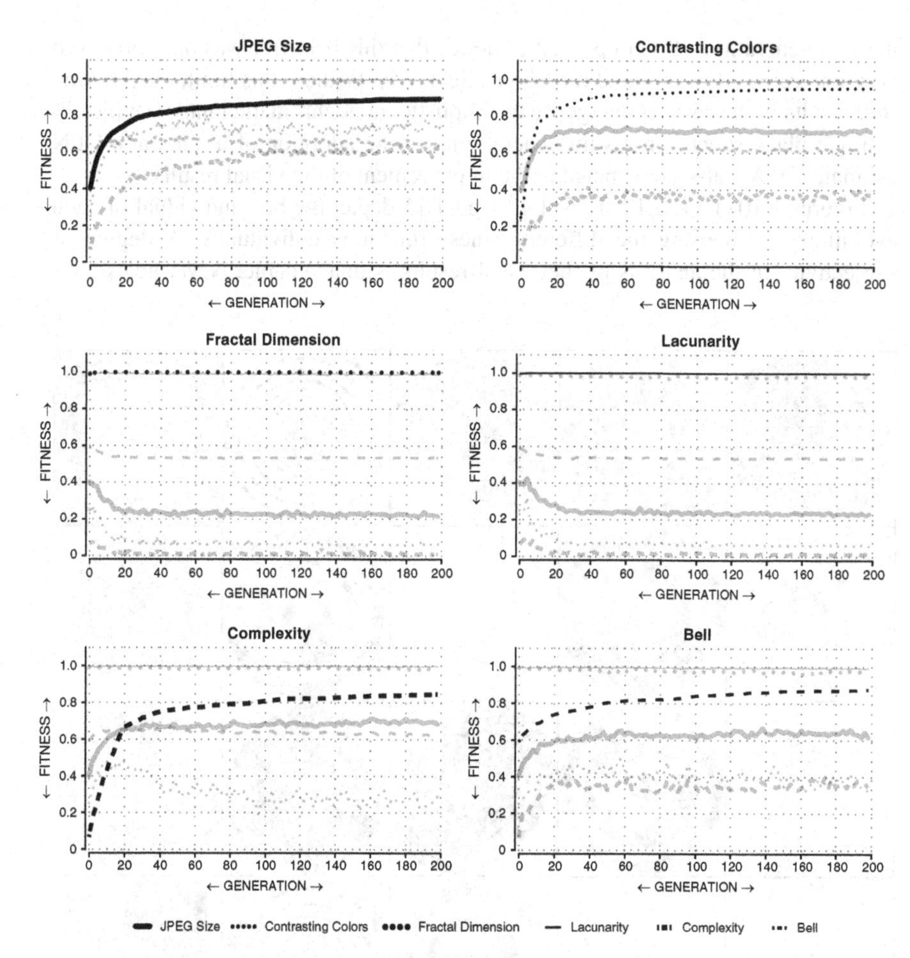

Fig. 1.9 Evolution of the fitness of the best individual across populations. The fitness function used to guide evolution is depicted in the title of each chart. The other values are presented for reference. The results are averages of 30 independent runs for each chart

during the early stages of the runs, but the number of *Contrasting Colors* tends to decrease as the number of generations progresses. The *Complexity* fitness function operates on a greyscale version of the images, as such it is not sensitive to changes of color. Furthermore, abrupt changes from black to white create artifacts that are hard to encode using JPEG compression, resulting in high IC estimates. Fractal image compression, which is used to estimate PC, is less sensitive to these abrupt changes. Therefore, since the approach values images with high IC and low PC, and since it does not take color information into consideration, the convergence to images using a reduced palette of contrasting colors is expected. Like for the other measures, *Complexity* and *Bell* appear to be unrelated. Finally, maximizing *Bell* promotes an increase of *JPEG Size*, *Contrasting Colors* and *Complexity* during the first generations. It is important to notice that this behavior was only observed after enforcing a lower bound for σ^2 (see Sect. 1.5). Without this limit, maximizing *Bell* results in the early convergence to simplistic monochromatic images (typically a single black square on a white background). The adoption of a quadratic DFN estimate (DFN_s) also contributed to the improvement of the visual results.

Figures 1.10, 1.11, 1.12, 1.13, 1.14, and 1.15 depict the best individual of each evolutionary run using the different fitness functions individually. A degree of subjectivity in the analysis of the visual results is unavoidable. Nevertheless, we

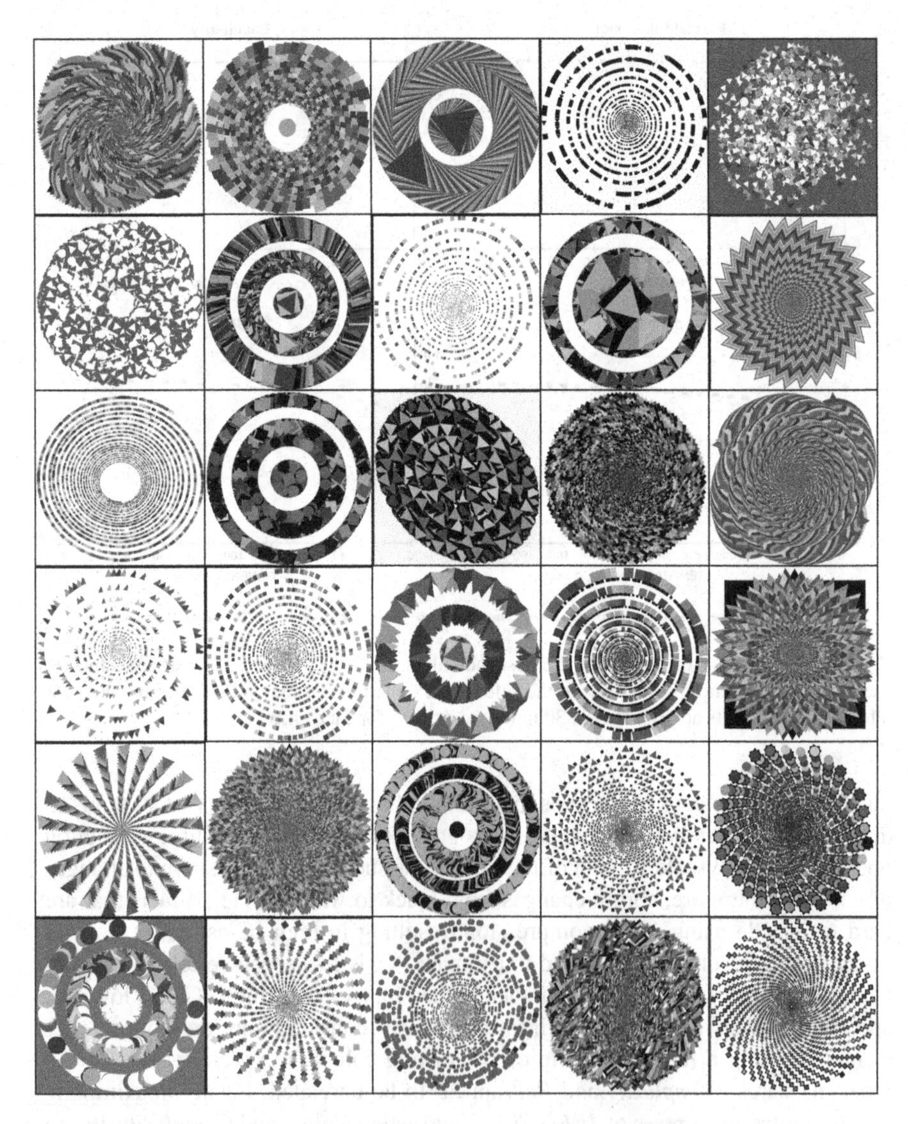

Fig. 1.10 Best individual of each of the 30 runs using *JPEG Size* as fitness function

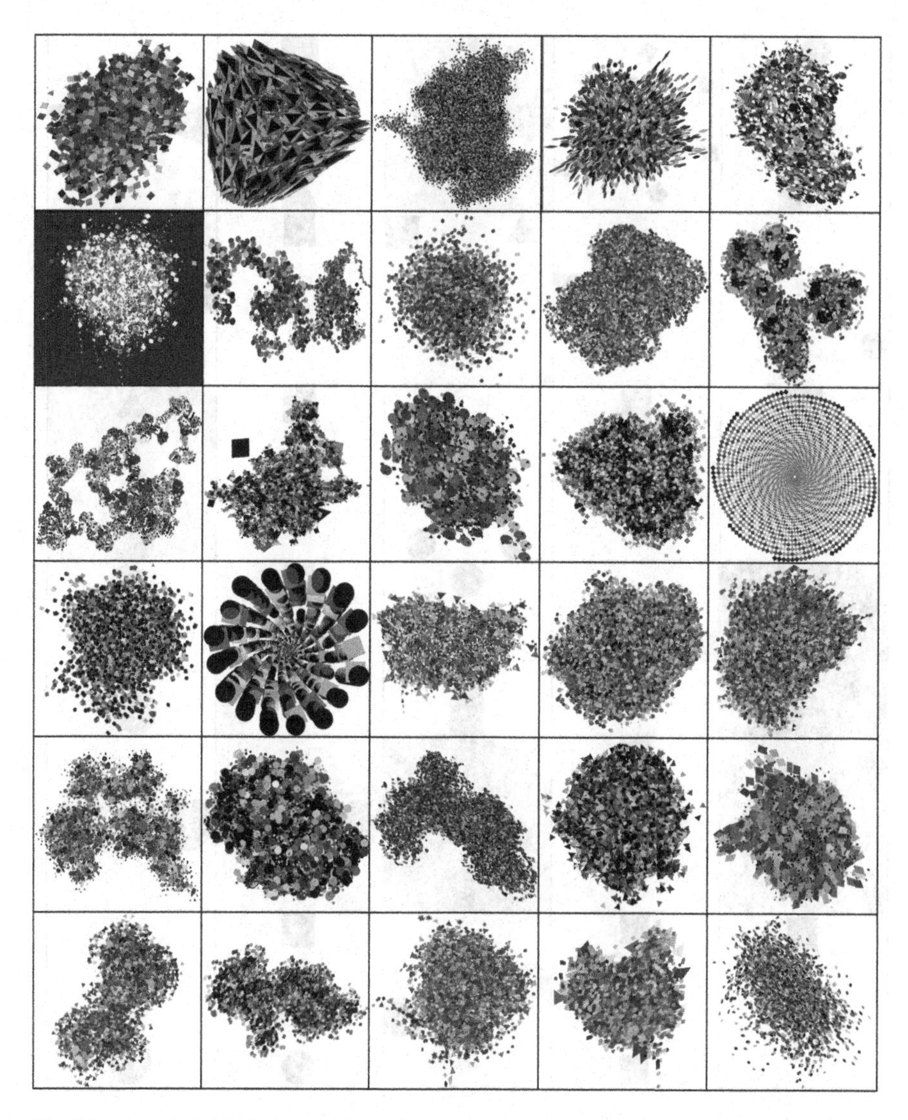

Fig. 1.11 Best individual of each of the 30 runs using *Contrasting Colors* as fitness function

believe that most of the findings tend to be consensual. When using *JPEG Size* to guide evolution, the evolutionary engine tended to converge to colorful circular patterns, with high contrasts of color (see Fig. 1.10). The tendency to converge to circular patterns, which is observed in several runs, is related with the recursive nature of the CFDGs and the particularities of the Context Free rendering engine. For instance, repeatedly drawing and rotating a square while changing its color will generate images that are hard to encode. Furthermore, the rendering engine automatically "zooms in" the shapes drawn cropping the empty regions of the

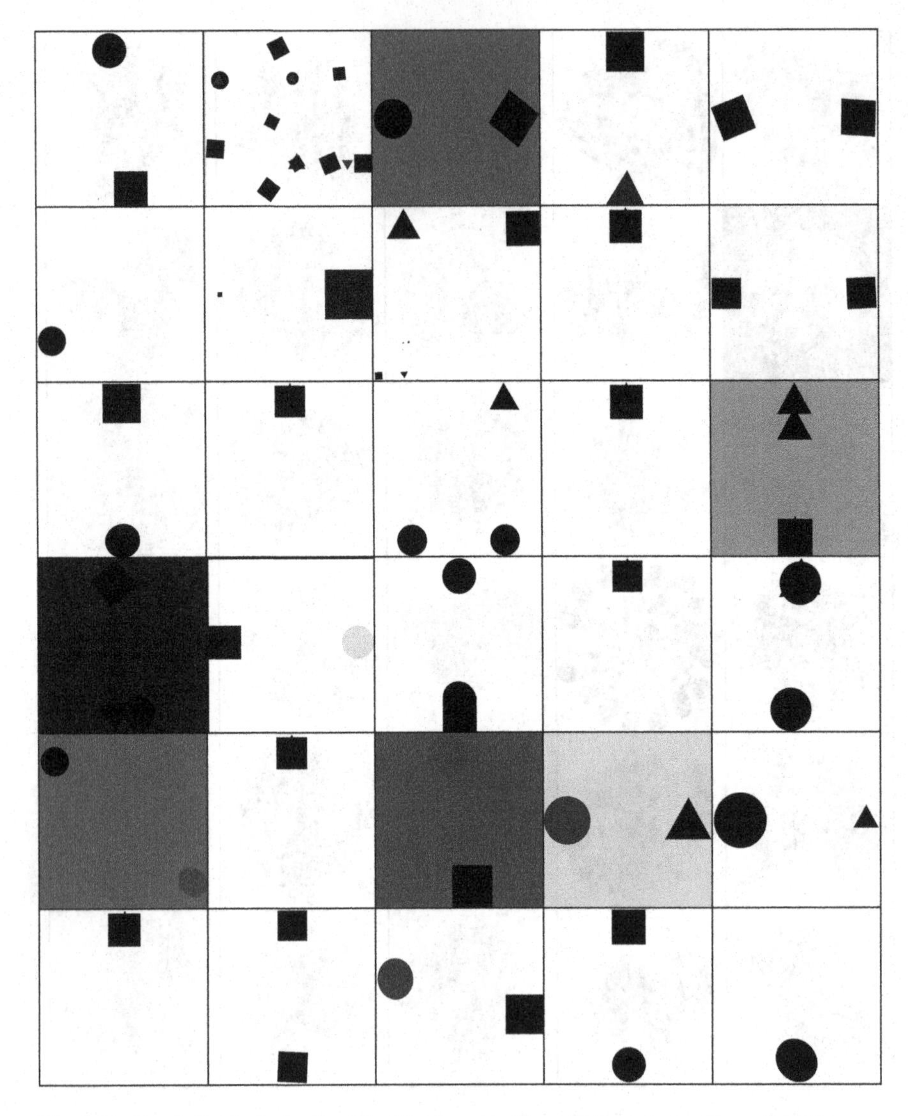

Fig. 1.12 Best individual of each of the 30 runs using *Fractal Dimension* as fitness function

canvas. As such, rotating about a fixed point in space tends to result in images that fill the entire canvas, maximizing the opportunities for introducing abrupt changes and, therefore, maximizing file size. Additionally, these CFDGs tend to be relatively stable and robust, which further promotes the convergence to this type of image.

Unsurprisingly, the results obtained when using *Contrasting Colors* are characterized by the convergence to images that are extremely colorful. Although some exceptions exist, most runs converged to amorphous unstructured shapes, which contrasts with circular patterns found when using *JPEG Size*. In our opinion

Fig. 1.13 Best individual of each of the 30 runs using *Lacunarity* as fitness function

this jeopardizes the aesthetic appeal of the images, that tend to have a random appearance, both in terms of shape and color.

As anticipated by the data pertaining the evolution of fitness, the visual results obtained using *Fractal Dimension* and *Lacunarity* (Figs. 1.12 and 1.13 are disappointing. None of the runs converged to images of fractal nature. These results reinforce earlier findings using expression based evolutionary art systems, indicating that these measures are not suitable for aesthetically driven evolution (Machado et al. 2007).

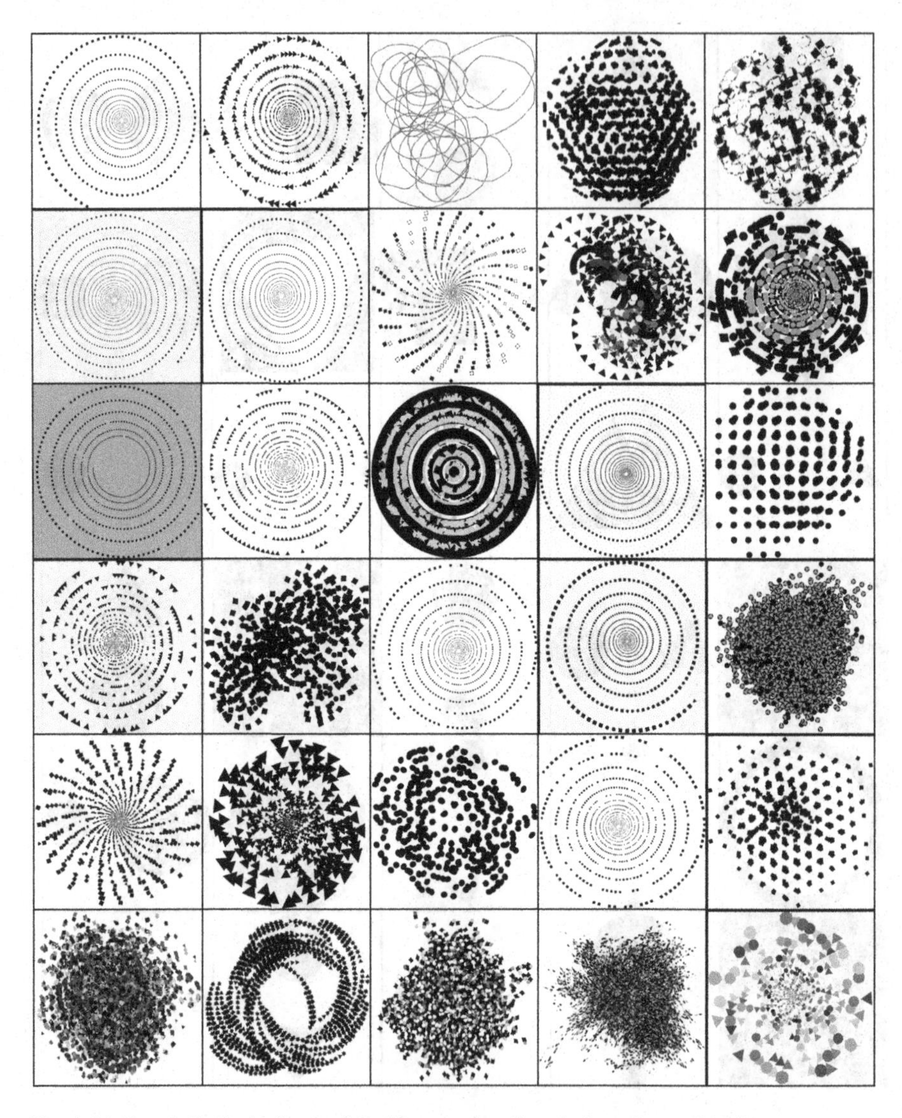

Fig. 1.14 Best individual of each of the 30 runs using *Complexity* as fitness function

As Fig. 1.14 illustrates, using *Complexity* tends to promote convergence to monochromatic and highly structured images. As previously, the tendency to converge to circular and spiral patterns is also observed in this case, and is explained by the same factors. Furthermore, since fractal image compression takes advantage of the self-similarities present in the image at multiple scales, the convergence to structured and self-similar structures that characterizes these runs was expected. As mentioned when analysing results pertaining the evolution of fitness, the

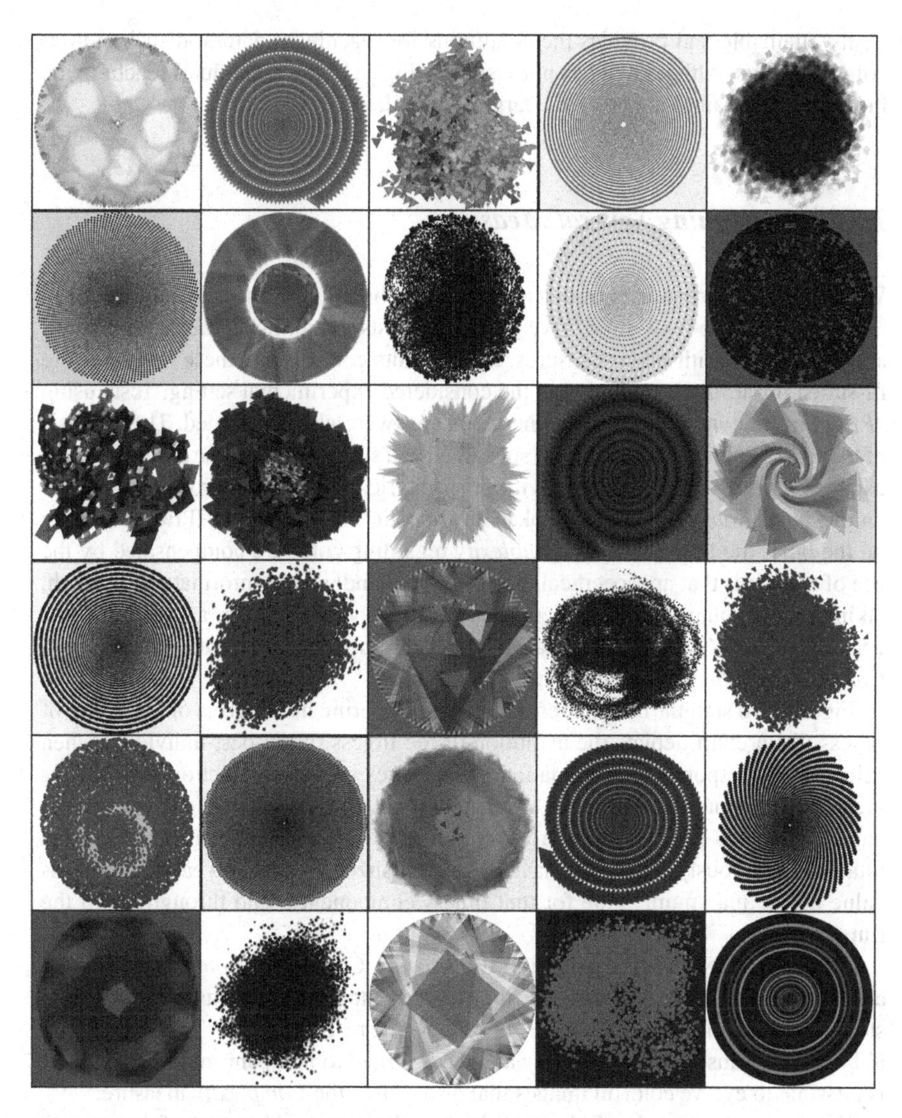

Fig. 1.15 Best individual of each of the 30 runs using *Bell* as fitness function

convergence to monochromatic images with high contrast is due to the different sensitivity of JPEG and fractal compression to the presence of abrupt transitions.

The most predominant feature of the images evolved using *Bell*, Fig. 1.15, is the structured variation of color, promoted by the need to match a natural distribution of color gradients. The shapes evolved result from an emergent property of the system. In other words, as previously explained, when using CFDG a circular pattern is easily attainable and provides the conditions for reaching a natural distribution of color gradients. Although this is not visible in Fig. 1.15 the individuals reaching the highest fitness values tend to use a large color palette.

1.7.2 Combining Several Measures

We performed several experiments where a combination of measures was used to assign fitness (see Sect. 1.5.6). We conducted tests combining *Fractal Dimension* and *Lacunarity* with other measures, these results confirm that these measures are ill-suited for aesthetic evolution in the considered experimental setting. Tests using *JPEG Size* in combination with other measures were also performed. The analysis of the results indicates that they are subsumed and surpassed by those obtained when using *Complexity* in conjunction with other metrics. This results from two factors: on one hand *Complexity* already takes into account the size of the JPEG encoding; on the other the limitations of *Complexity* regarding color are compensated by the use of measures that are specifically designed to handle color information. As such, taking into account the results described in the previous Section, as well as space constraints, we focus on the analysis of the results obtained when combining: *Contrasting Colors*, *Complexity* and *Bell*.

Figure 1.16 summarizes the results of these experiments in terms of evolution of fitness. Each chart depicts the evolution of the fitness of the best individual when using the corresponding combination of measures as fitness function. The values yield by the remaining measures are depicted but do not influence evolution. The values presented in each chart are averages of 30 independent runs (120 runs in total). As previously, the values have been normalized by dividing each raw fitness value by the maximum value for that fitness component found throughout all the runs.

As it can be observed, combining *Contrasting Colors* and *Complexity* leads to a fast increase of both measures during the early stages of the runs, followed by a steady increase of both components throughout the rest of the runs. This shows that, although the runs using *Complexity* alone converged to monochromatic imagery, it is possible to evolve colorful images that also satisfy the *Complexity* measure.

Combining *Contrasting Colors* and *Bell* results in a rapid increase of the number of contrasting colors during the first generations. Afterwards, increases in fitness are mainly accomplished through the improvements of the *Bell* component of the fitness function. This indicates that it is easier to maximize the number of contrasting colors than to attain a normal distribution of gradients. This observation is further attested by the analysis of the charts pertaining the evolution of fitness when using *Contrasting Colors*, *Complexity* and *Bell* individually, which indicate that *Bell* may be the hardest measure to address. The combination of *Complexity* and *Bell* is characterized by a rapid increase of complexity during the first populations,

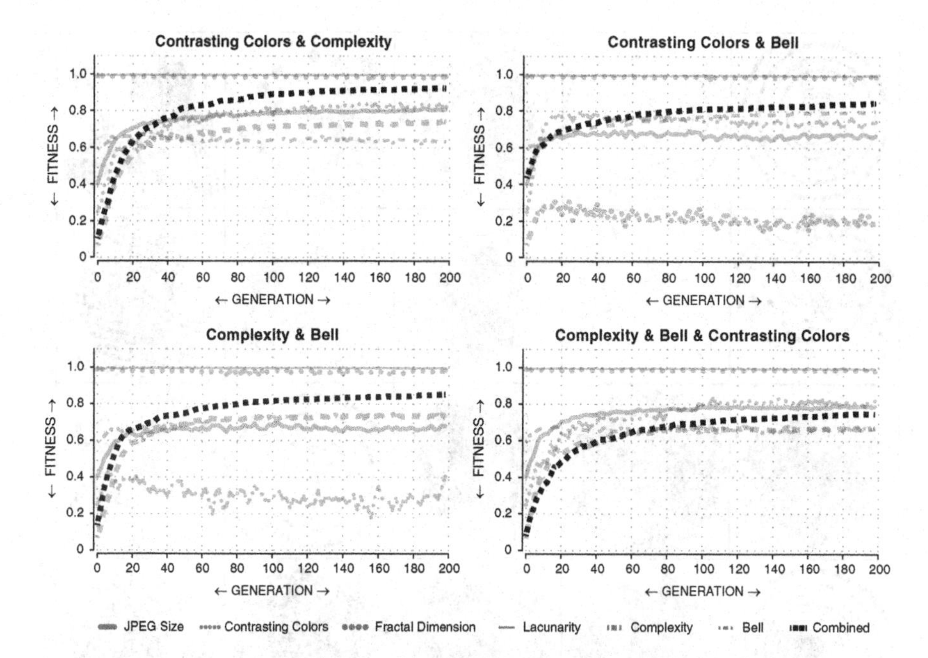

Fig. 1.16 Evolution of the fitness of the best individual across populations using a combination of measures. The combination used to guide evolution is depicted in the title of each chart. The other values are presented for reference, but have no influence in the evolutionary process. The results are averages of 30 independent runs for each chart and have been normalized to improve readability

followed by a slow, but steady, increase of both measures throughout the runs. The combination of the three measures further establishes *Bell* as the measure that is most difficult to address, since the improvements of fitness are mostly due to increases in the other two measures. Significantly longer runs would be necessary to obtain noteworthy improvements in *Bell*.

Figure 1.17 depicts the best individual of each evolutionary run using as fitness a combination of the *Contrasting Colors* and *Complexity* measures. As it can be observed, in most cases, the neat structures that characterize the runs using *Complexity* (see Fig. 1.14) continue to emerge. However, due to the influence of the *Contrasting Colors* measure, they tend to be colorful instead of monochromatic. Thus, the visual results appear to depict a good combination of both measures. The same can be stated for the images resulting from using *Contrasting Colors* and *Bell*. As can be observed in Fig. 1.18, they are more colorful than those evolved using *Bell* (see Fig. 1.15) but retain a natural distribution of color gradients, deviating from the "random" coloring schemes that characterize the images evolved using *Contrasting Colors* (see Fig. 1.11).

The images obtained when using *Complexity* and *Bell* simultaneously (Fig. 1.19) are less colorful than expected. Visually, the impact of *Complexity* appears to overshadow the impact of *Bell*. Nevertheless, a comparison between these images and those obtained using *Complexity* alone (Fig. 1.14) reveals the influence of *Bell*

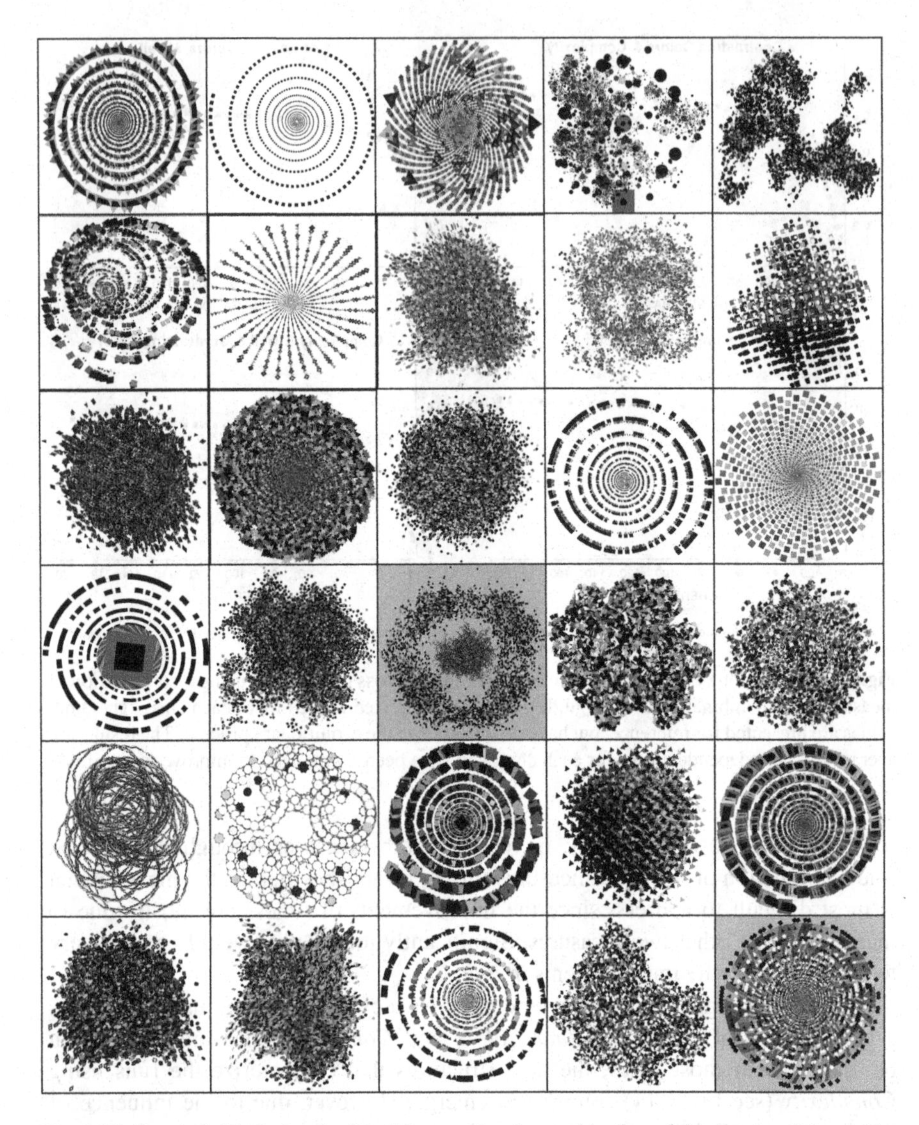

Fig. 1.17 Best individual of each of the 30 runs using the combination of *Contrasting Colors* with *Complexity* as fitness function

in the course of the runs: the monochromatic images are replaced by ones with a wider number of color gradients, and these color changes tend to be subtler.

Finally, as expected, the images obtained in the runs using the three measures (Fig. 1.20) often depict, simultaneously, the features associated with each of them. As previously, the influence of the *Bell* measure is less obvious than the others, but a comparison with the results depicted in Fig. 1.17 highlights the influence of this measure. Likewise, the structures that emerge from runs using *Complexity* and

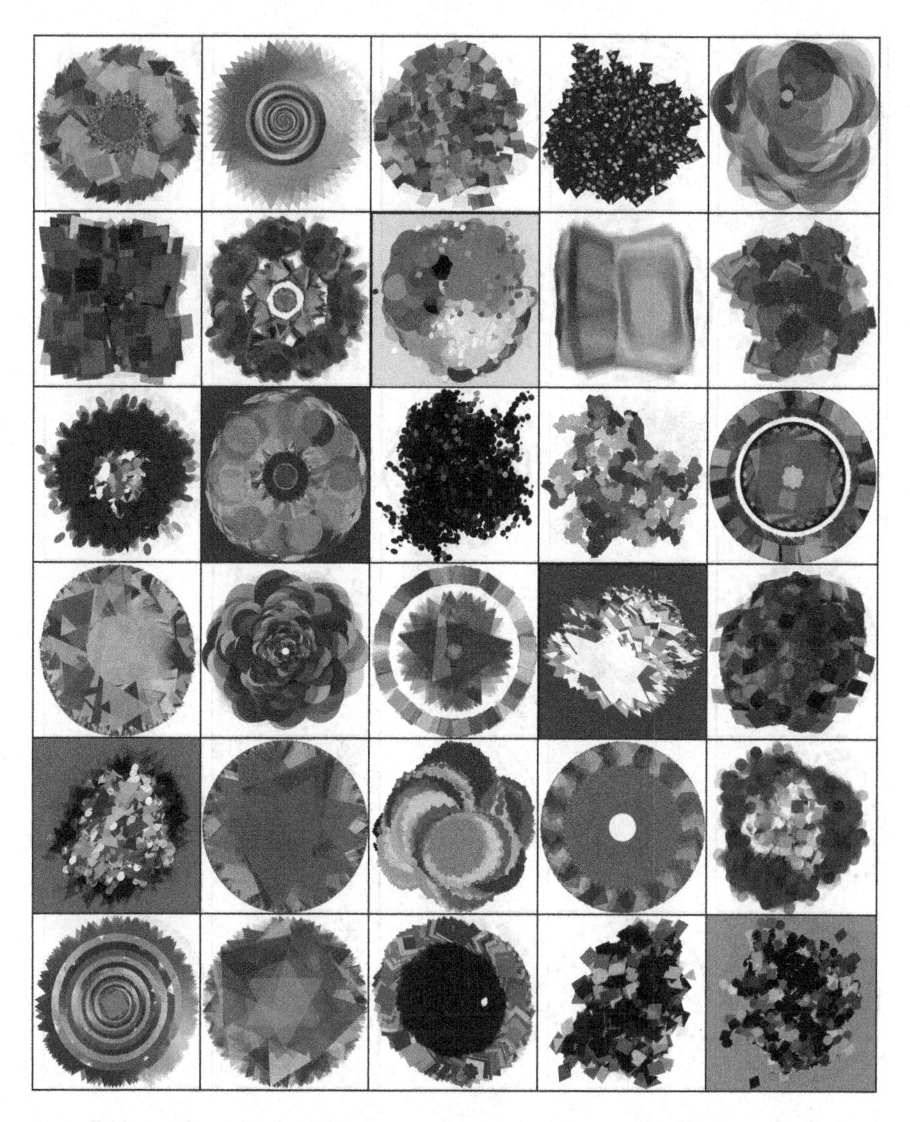

Fig. 1.18 Best individual of each of the 30 runs using the combination of *Contrasting Colors* with *Bell* as fitness function

the colorful images that characterize runs using *Contrasting Colors* are also less often. Thus, although the influence of each measure is observable, we consider that significantly longer runs would be necessary to enhance their visibility.

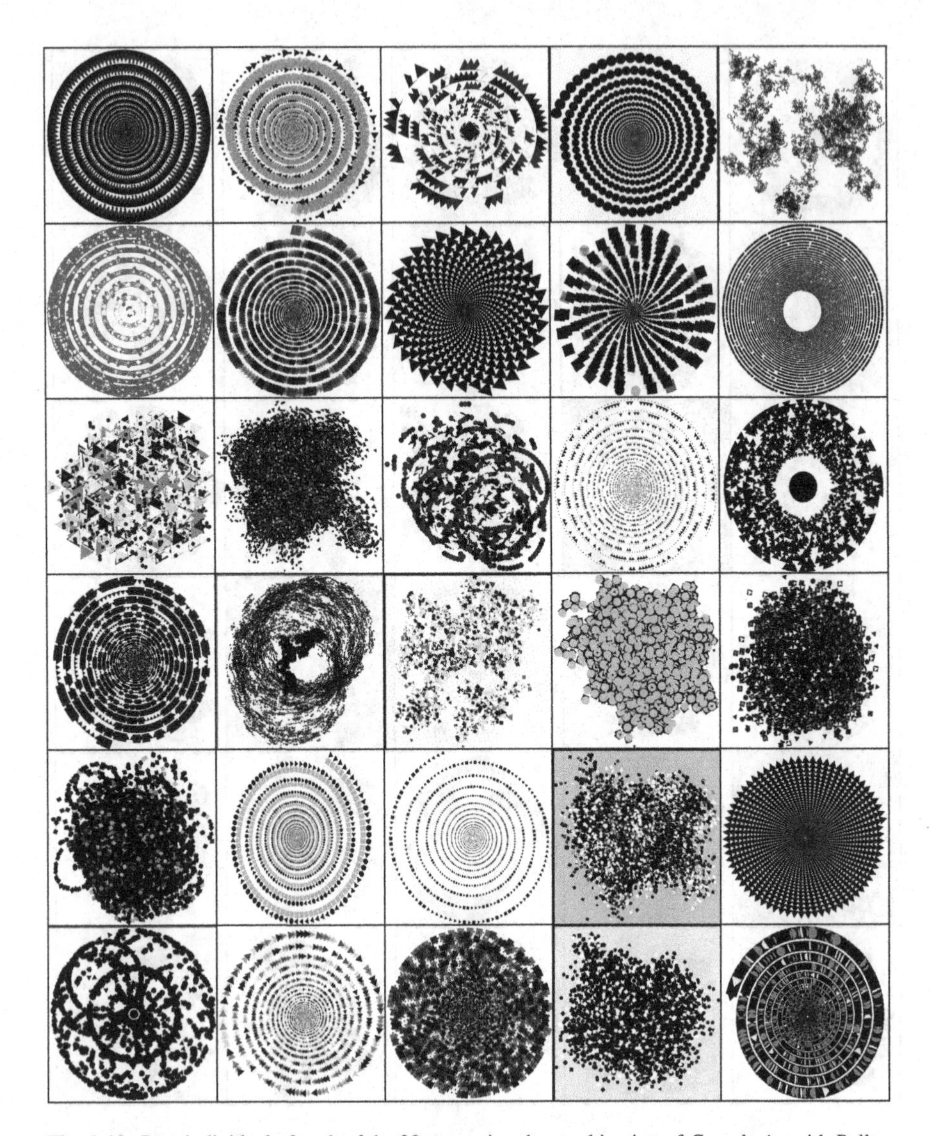

Fig. 1.19 Best individual of each of the 30 runs using the combination of *Complexity* with *Bell* as fitness function

1.8 Conclusions

We have presented a graph-based approach for the evolution of Context Free Design Grammars. This approach contrasts with the mainstream evolutionary art practices by abandoning expression-based evolution of images and embracing the evolution of images created through the combination of basic shapes. Nevertheless, the procedural nature of the representation, which characterizes Genetic Programming

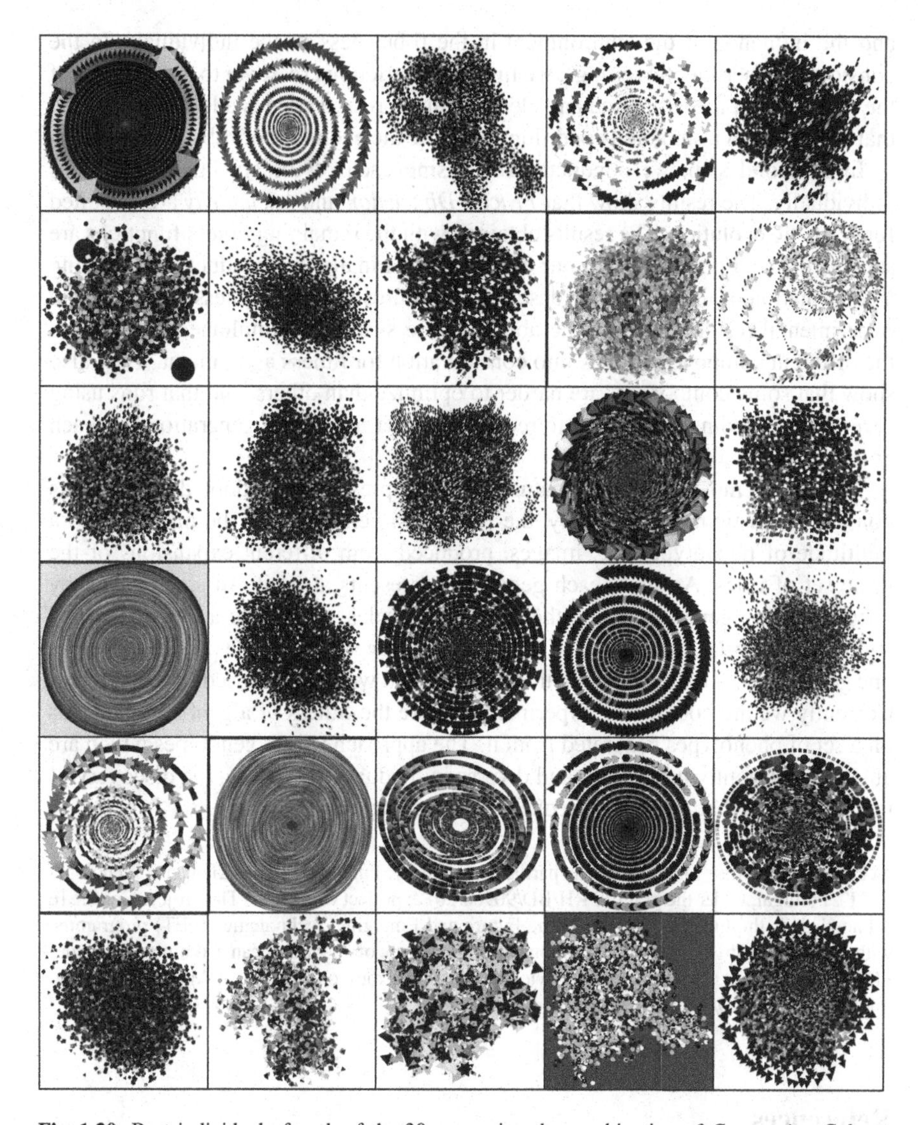

Fig. 1.20 Best individual of each of the 30 runs using the combination of *Contrasting Colors*, *Complexity* and *Bell* as fitness function

approaches, is retained. We describe the evolutionary engine, giving particular attention to its most discriminating features, namely: representation, graph-based crossover, mutation and initialization.

We introduce six different fitness functions based on evolutionary art literature and conduct a wide set of experiments. In a first step we assess the adequacy of the system and establish satisfactory experimental parameters. In this context, we study the influence of unexpressed genetic code in the evolutionary process and the influence of the environment in the robustness of the individuals. In the considered experimental settings, we find that restricting crossover to the portions of the genome that are expressed and cleaning unexpressed code is advantageous, and that dynamic environmental conditions promote the evolution of robust individuals.

In a second step, we conducted runs using each of the six fitness functions individually. The results show that *Fractal Dimension* and *Lacunarity* are ill-suited for aesthetic evolution. The results obtained with the remaining fitness functions are satisfactory and correspond to our expectations. Finally, we conducted runs using a combination of the previously described measures to assign fitness. Globally, the experimental results illustrate the ability of the system to simultaneously address the different components taken into consideration for fitness assignment. They also show that some components are harder to optimize than others, and that runs using several fitness components tend to require a higher number of generations to reach good results.

One of the most prominent features of the representation adopted herein is its non-deterministic nature. Namely, the fact that a genotype may be mapped into a multitude of phenotypes, i.e. images, produced from different expansions of the same set of rules. As such, each genotype represents a family of shapes that, by virtue of being generated using the same set of rules, tend to be aesthetically and stylistically similar. The ability of the system to generate multiple phenotypes from one genotype was not explored in this Chapter, and will be addressed in future work. Currently we are conducting experiments where the fitness of a genotype depends on a set of phenotypes generated from it. The approach values genotypes which are able to consistently produce fit and diverse individuals, promoting the discovery of image families that are simultaneously coherent and diverse.

Acknowledgements This research is partially funded by: Fundação para a Ciência e Tecnologia (FCT), Portugal, under the grant SFRH/BD/90968/2012; project ConCreTe. The project ConCreTe acknowledges the financial support of the Future and Emerging Technologies (FET) programme within the Seventh Framework Programme for Research of the European Commission, under FET grant number 611733. We acknowledge and thank the contribution of Manuel Levi who implemented the *Contrasting Colors* fitness function.

References

E. Baker and M. Seltzer. Evolving line drawings. In *Proceedings of the Fifth International Conference on Genetic Algorithms*, pages 91–100. Morgan Kaufmann Publishers, 1994.

J. Bird, P. Husbands, M. Perris, B. Bigge, and P. Brown. Implicit fitness functions for evolving a drawing robot. In *Applications of Evolutionary Computing*, pages 473–478. Springer, 2008.

J. Bird and D. Stokes. Minimal creativity, evaluation and fractal pattern discrimination. *Programme Committee and Reviewers*, page 121, 2007.

A. Borrell. CFDG Mutate. http://www.wickedbean.co.uk/cfdg/index.html, last accessed in November 2014.

C. Coyne. Context Free Design Grammar. http://www.chriscoyne.com/cfdg/, last accessed in November 2014.

E. den Heijer and A. E. Eiben. Evolving art with scalable vector graphics. In N. Krasnogor and P. L. Lanzi, editors, *GECCO*, pages 427–434. ACM, 2011.

Y. Fisher, editor. *Fractal Image Compression: Theory and Application*. Springer, London, 1995.

D. A. Hart. Toward greater artistic control for interactive evolution of images and animation. In *Proceedings of the 2007 EvoWorkshops 2007 on EvoCoMnet, EvoFIN, EvoIASP, EvoINTERACTION, EvoMUSART, EvoSTOC and EvoTransLog*, pages 527–536, Berlin, Heidelberg, 2007. Springer-Verlag.

J. Horigan and M. Lentczner. Context Free. http://www.contextfreeart.org/, last accessed in September 2009.

C. Incorporated. GIF Graphics Interchange Format: A standard defining a mechanism for the storage and transmission of bitmap-based graphics information. Columbus, OH, USA, 1987.

A. Karperien. Fraclac for imagej. In *http://rsb.info.nih.gov/ij/plugins/fraclac/FLHelp/Introduction.htm*, 1999–2013.

P. Machado and A. Cardoso. All the truth about NEvAr. *Applied Intelligence, Special Issue on Creative Systems*, 16(2):101–119, 2002.

P. Machado and H. Nunes. A step towards the evolution of visual languages. In *First International Conference on Computational Creativity*, Lisbon, Portugal, 2010.

P. Machado, H. Nunes, and J. Romero. Graph-based evolution of visual languages. In C. D. Chio, A. Brabazon, G. A. D. Caro, M. Ebner, M. Farooq, A. Fink, J. Grahl, G. Greenfield, P. Machado, M. ONeill, E. Tarantino, and N. Urquhart, editors, *Applications of Evolutionary Computation, EvoApplications 2010: EvoCOMNET, EvoENVIRONMENT, EvoFIN, EvoMUSART, and Evo-TRANSLOG, Istanbul, Turkey, April 7–9, 2010, Proceedings, Part II*, volume 6025 of *Lecture Notes in Computer Science*, pages 271–280. Springer, 2010.

P. Machado, J. Romero, A. Cardoso, and A. Santos. Partially interactive evolutionary artists. *New Generation Computing – Special Issue on Interactive Evolutionary Computation*, 23(42):143–155, 2005.

P. Machado, J. Romero, and B. Manaris. Experiments in computational aesthetics: an iterative approach to stylistic change in evolutionary art. In J. Romero and P. Machado, editors, *The Art of Artificial Evolution: A Handbook on Evolutionary Art and Music*, pages 381–415. Springer Berlin Heidelberg, 2007.

J. McCormack. Facing the future: Evolutionary possibilities for human-machine creativity. In J. Romero and P. Machado, editors, *The Art of Artificial Evolution: A Handbook on Evolutionary Art and Music*, pages 417–451. Springer Berlin Heidelberg, 2007.

T. Mori, Y. Endou, and A. Nakayama. Fractal analysis and aesthetic evaluation of geometrically overlapping patterns. *Textile research journal*, 66(9):581–586, 1996.

M. O'Neill, J. McDermott, J. M. Swafford, J. Byrne, E. Hemberg, A. Brabazon, E. Shotton, C. McNally, and M. Hemberg. Evolutionary design using grammatical evolution and shape grammars: Designing a shelter. *International Journal of Design Engineering*, 3(1):4–24, 2010.

M. O'Neill and C. Ryan. *Grammatical evolution: evolutionary automatic programming in an arbitrary language*, volume 4. Springer, 2003.

M. O'Neill, J. M. Swafford, J. McDermott, J. Byrne, A. Brabazon, E. Shotton, C. McNally, and M. Hemberg. Shape grammars and grammatical evolution for evolutionary design. In *Proceedings of the 11th Annual Conference on Genetic and Evolutionary Computation*, GECCO '09, pages 1035–1042, New York, NY, USA, 2009. ACM.

F. B. Pereira, P. Machado, E. Costa, and A. Cardoso. Graph based crossover – a case study with the busy beaver problem. In *Proceedings of the 1999 Genetic and Evolutionary Computation Conference*, 1999.

B. J. Ross, W. Ralph, and Z. Hai. Evolutionary image synthesis using a model of aesthetics. In G. G. Yen, S. M. Lucas, G. Fogel, G. Kendall, R. Salomon, B.-T. Zhang, C. A. C. Coello, and T. P. Runarsson, editors, *Proceedings of the 2006 IEEE Congress on Evolutionary Computation*, pages 1087–1094, Vancouver, BC, Canada, 16–21 July 2006. IEEE Press.

R. Saunders and K. Grace. Teaching evolutionary design systems by extending "Context Free". In *EvoWorkshops '09: Proceedings of the EvoWorkshops 2009 on Applications of Evolutionary Computing*, pages 591–596. Springer-Verlag, 2009.

K. Sims. Artificial evolution for computer graphics. *ACM Computer Graphics*, 25:319–328, 1991.

B. Spehar, C. W. G. Clifford, N. Newell, and R. P. Taylor. Universal aesthetic of fractals. *Computers and Graphics*, 27(5):813–820, Oct. 2003.

G. Stiny and J. Gips. Shape grammars and the generative specification of paintings and sculpture. In C. V. Freiman, editor, *Information Processing 71*, pages 1460–1465, Amsterdam, 1971. North Holland Publishing Co.

T. Unemi. SBART2.4: Breeding 2D CG images and movies, and creating a type of collage. In *The Third International Conference on Knowledge-based Intelligent Information Engineering Systems*, pages 288–291, Adelaide, Australia, 1999.

L. World. Aesthetic selection: The evolutionary art of Steven Rooke. *IEEE Computer Graphics and Applications*, 16(1), 1996.

Chapter 2
Genetic Programming for Modelling of Geotechnical Engineering Systems

Mohamed A. Shahin

2.1 Introduction

Geotechnical engineering deals with materials (e.g., soil and rock) that, by their very nature, exhibit varied and uncertain behaviour due to the imprecise physical processes associated with the formation of these materials. Modelling the behaviour of such materials is complex and usually beyond the ability of most traditional forms of physically-based engineering methods (e.g., analytical formulations and limit equilibrium methods). Artificial intelligence (AI) is becoming more popular and particularly amenable to modelling the complex behaviour of most geotechnical engineering materials as it has demonstrated superior predictive ability when compared to traditional methods. AI is a computational method that attempts to mimic, in a very simplistic way, the human cognition capability to solve engineering problems that have defied solution using conventional computational techniques (Flood 2008). The essence of AI techniques in solving any engineering problem is to learn by examples of data inputs and outputs presented to them so that the subtle functional relationships among the data are captured, even if the underlying relationships are unknown or the physical meaning is difficult to explain. Thus, AI models are data-driven approaches that rely on the data alone to determine the structure and parameters that govern a phenomenon (or system), without the need for making any assumptions about the physical behavior of the system. This is in contrast to most physically-based models that use the first principles (e.g., physical laws) to derive the underlying relationships of the system, which usually justifiably simplified with many assumptions and require prior knowledge about

Electronic supplementary material The online version of this chapter (doi: 10.1007/978-3-319-20883-1_2) contains supplementary material, which is available to authorized users.

M.A. Shahin (✉)
Department of Civil Engineering, Curtin University, Perth, WA 6845, Australia
e-mail: m.shahin@curtin.edu.au

A.H. Gandomi et al. (eds.), *Handbook of Genetic Programming Applications*,
DOI 10.1007/978-3-319-20883-1_2

the nature of the relationships among the data. This is one of the main benefits of AI techniques when compared to most physically-based empirical and statistical methods. Examples of the available AI techniques are artificial neural networks (ANNs), genetic programming (GP), support vector machines (SVM), M5 model trees, and k-nearest neighbors (Elshorbagy et al. 2010). Of these, ANNs are by far the most commonly used AI technique in geotechnical engineering and interested readers are referred to Shahin et al. (2001), where the pre-2001 ANN applications in geotechnical engineering are reviewed in some detail, and Shahin et al. (2009) and Shahin (2013), where the post-2001 papers of ANN applications in geotechnical engineering are briefly examined. More recently, GP has been frequently used in geotechnical engineering and has proved to be successful. The use of GP in geotechnical engineering is the main focus of this book chapter.

Despite the success of ANNs in the analysis and simulation of many geotechnical engineering applications, they have some drawbacks such as the lack of transparency and knowledge extraction, leading this technique to be criticised as being *black boxes* (Ahangar-Asr et al. 2011). Model transparency and knowledge extraction are the feasibility of interpreting AI models in a way that provides insights into how model inputs affect outputs. Figure 2.1 shows a representation of the classification of modelling techniques based on colours (Giustolisi et al. 2007) in which the higher the physical knowledge used during model development, the better the physical interpretation of the phenomenon that the model provides to the user. It can be seen that the colour coding of mathematical modelling can be classified into white-, black-, and grey-box models, each of which can be explained as follows (Giustolisi et al. 2007). White-box models are systems that are based on first principles (e.g., physical laws) where model variables and parameters are known and have physical meaning by which the underlying physical relationships of the system can be explained. Black-box models are data-driven or regressive systems in which the functional form of relationships between model variables are unknown and need to be estimated. Black-box models rely on data to map the relationships between model inputs and corresponding outputs rather than to

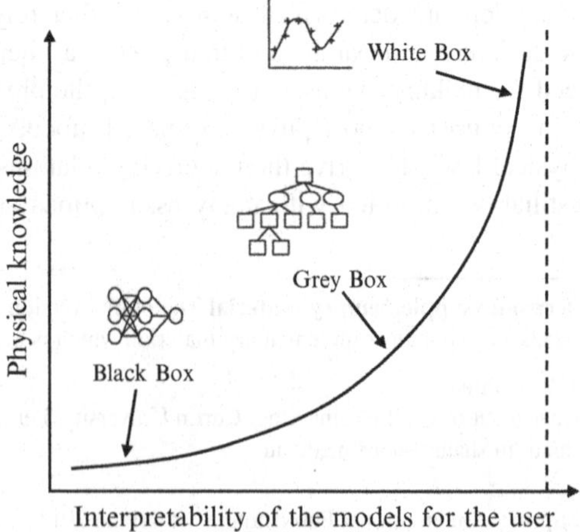

Fig. 2.1 Graphical classification of modelling techniques. *Source*: Adapted from Giustolisi et al. (2007)

find a feasible structure of the model input-output relationships. Grey-box models are conceptual systems in which the mathematical structure of the model can be derived, allowing further information of the system behaviour to be resolved. According to the abovementioned classification of modelling techniques based on colour, whereby meaning is related to three levels of prior information required, ANNs belong to the class of black-box models due to their lack of transparency and the fact that they do not consider nor explain the underlying physical processes explicitly. This is because the knowledge extracted by ANNs is stored in a set of weights that are difficult to interpret properly, and due to the large complexity of the network structure, ANNs fail to give a transparent function that relates the inputs to the corresponding outputs. Consequently, it is difficult to understand the nature of the derived input–output relationships (Shahin 2013). This urged many researchers to find alternative AI techniques that can overcome most shortcomings of ANNs; one of these techniques is the genetic programming.

GP is relatively new in geotechnical engineering but has proved to be successful. GP is based on evolutionary computing that aims to search for simple and optimal structures to represent a system through a combination of the genetic algorithm and natural selection. According to the classification of modelling techniques based on colour that is mentioned earlier, GP can be classified as "grey box" technique (conceptualisation of physical phenomena); despite the fact that GP is based on observed data, it returns a mathematical structure that is symbolic and usually uncomplicated. The nature of obtained GP models permits global exploration of expressions, which provides insights into the relationship between the model inputs and the corresponding outputs, i.e., it allows the user to gain additional knowledge of how the system performs. An additional advantage of GP over ANNs is that the structure and network parameters of ANNs should be identified a priori and are usually obtained using ad-hoc, trial-and-error approaches. However, the number and modelling parameters of GP are all evolved automatically during model calibration, as will be explained later. At the same time, the prior physical knowledge based on engineering judgment or other human knowledge can be used to make hypotheses about the elements of the objective functions and their structure, hence enabling refinement of final models. It should be noted that while white-box models provide maximum transparency, their construction may be difficult to obtain for many geotechnical engineering problems where the underlying mechanism is not entirely understood. In this chapter, the feasibility of utilising the GP technique to develop simple and transparent prediction models for solving some complex problems in geotechnical engineering will be explored and discussed.

2.2 Overview of Genetic Programming

Genetic programming (GP) is an extension of genetic algorithms (GA), which are evolutionary computing search (optimisation) methods that are based on the principles of genetics and natural selection. In GA, some of the natural evolutionary mechanisms, such as reproduction, cross-over, and mutation, are usually implemented to solve function identification problems. GA was first introduced by Holland (1975) and developed by Goldberg (1989), whereas GP was invented by Cramer (1985) and further developed by Koza (1992). The difference between GA and GP is that GA is generally used to evolve the best values for a given set of model parameters (i.e., parameters optimization), whereas GP generates a structured representation for a set of input variables and corresponding outputs (i.e., modelling or programming).

Genetic programming manipulates and optimises a population of computer models (or programs) proposed to solve a particular problem, so that the model that best fits the problem is obtained. A detailed description of GP can be found in many publications (e.g., Koza 1992), and an overview is given herein. The modelling steps by GP start with the creation of an initial population of computer models (also called *individuals* or *chromosomes*) that are composed of two sets (i.e., a set of functions and a set of terminals) that are defined by the user to suit a certain problem. The functions and terminals are selected randomly and arranged in a tree-like structure to form a computer model that contains a root node, branches of functional nodes, and terminals, as shown by the typical example of GP tree representation in Fig. 2.2. The functions can contain basic mathematical operators (e.g., $+$, $-$, $\times$, /), Boolean logic functions (e.g., AND, OR, NOT), trigonometric functions (e.g., sin, cos), or any other user-defined functions. The terminals, on the other hand, may consist of numerical constants, logical constants, or variables.

Once a population of computer models has been created, each model is executed using available data for the problem at hand, and the model fitness is evaluated depending on how well it is able to solve the problem. For many problems, the model fitness is measured by the error between the output provided by the model and the desired actual output. A generation of new population of computer models is then created to replace the existing population. The new population is created by applying the following three main operations: reproduction, cross-over, and mutation. These three operations are applied on certain proportions of the computer

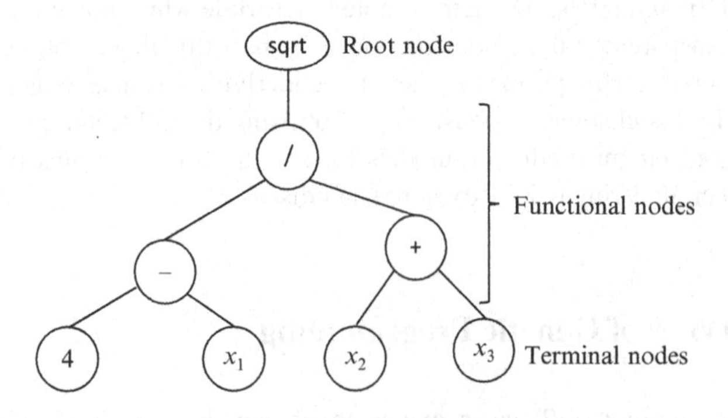

Fig. 2.2 Typical example of genetic programming (GP) tree representation for the function: $[(4 - x_1) / (x_2 + x_3)]^2$

models in the existing population, and the models are selected according to their fitness. Reproduction is copying a computer model from an existing population into the new population without alteration. Cross-over is genetically recombining (swapping) randomly chosen parts of two computer models. Mutation is replacing a randomly selected functional or terminal node with another node from the same function or terminal set, provided that a functional node replaces a functional node and a terminal node replaces a terminal node. The evolutionary process of evaluating the fitness of an existing population and producing new population is continued until a termination criterion is met, which can be either a particular acceptable error or a certain maximum number of generations. The best computer model that appears in any generation designates the result of the GP process. There are currently three variants of GP available in the literature including the linear genetic programming (LGP), gene expression programming (GEP), and multi-expression programming (MEP) (Alavi and Gandomi 2011). More recently, the multi-stage genetic programming (MSGP) (Gandomi and Alavi 2011) and multi-gene genetic programming (MGGP) (Gandomi and Alavi 2012) are also introduced. However, GEP is the most commonly used GP method in geotechnical engineering and is thus described in some detail below.

Gene expression programming was developed by Ferreira (2001) and utilises evolution of mathematical equations that are encoded linearly in chromosomes of fixed length and expressed non-linearly in the form of expression trees (ETs) of different sizes and shapes. The chromosomes are composed of multiple genes, each gene is encoded a smaller sub-program or sub-expression tree (Sub-ET). Every gene has a constant length and consists of a head and a tail. The head can contain functions and terminals (variables and constants) required to code any expression, whereas the tail solely contains terminals. The genetic code represents a one-to-one relationship between the symbols of the chromosome and the function or terminal. The process of information decoding from chromosomes to expression trees is called *translation*, which is based on sets of rules that determine the spatial organisation of the functions and terminals in the ETs and the type of interaction (link) between the Sub-ETs (Ferreira 2001). The main strength of GEP is that the creation of genetic diversity is extremely simplified as the genetic operators work at the chromosome level. Another strength is regarding the unique multi-genetic nature of GEP, which allows the evolution of more powerful models/programs composed of several sub-programs (Ferreira 2001).

The major steps in the GEP procedure are schematically represented in Fig. 2.3. The process begins with choosing sets of functions F and terminals T to randomly create an initial population of chromosomes of mathematical equations. One could choose, for example, the four basic arithmetic operators to form the set of functions, i.e., $F = \{+, -, \times, /\}$, and the set of terminals will obviously consist of the independent variables of a particular problem, for example, for a problem that has two independent variables, x_1 and x_2 would be $T = \{x_1, x_2\}$. Choosing the chromosomal architecture, i.e., the number and length of genes and linking functions (e.g., addition, subtraction, multiplication, and division), is also part of this step. The chromosomes are then expressed as expression trees of different sizes

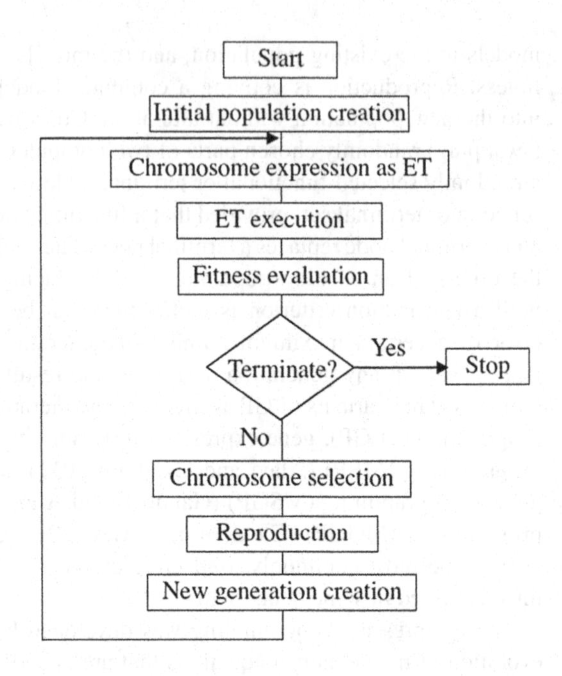

Fig. 2.3 Algorithm of gene expression programming (GEP) (Teodorescu and Sherwood 2008)

and shapes, and the performance of each individual chromosome is evaluated by comparing the predicted and actual values of presented data. One could measure the fitness f_i of an individual chromosome i using the following expression:

$$f_i = \sum_{j=1}^{C_t} \left(M - \left| C_{(i,j)} - T_j \right| \right) \tag{2.1}$$

where M is the range of selection, $C_{(i,j)}$ is the value returned by the individual chromosome i for fitness case j (out of C_t fitness cases), and T_j is the target value for the fitness case j. There are, of course, other fitness functions available that can be appropriate for different problems. If the desired results (according to the measured errors) are satisfactory, the GEP process is stopped, otherwise, some chromosomes are selected and mutated to reproduce new chromosomes, and the process is repeated for a certain number of generation or until the desired fitness score is obtained.

Figure 2.4 shows a typical example of a chromosome with one gene, and its ET and corresponding mathematical equation. It can be seen that, while the head of a gene contains arithmetic and trigonometric functions (e.g., +, −, ×, /, √, sin, cos), the tail includes constants and independent variables (e.g., 1, a, b, c). The ET is codified reading the ET from left to right in the top line of the tree and from top to bottom.

More recently, a genetic programming based technique called evolutionary polynomial regression (EPR) was developed and used in geotechnical engineering.

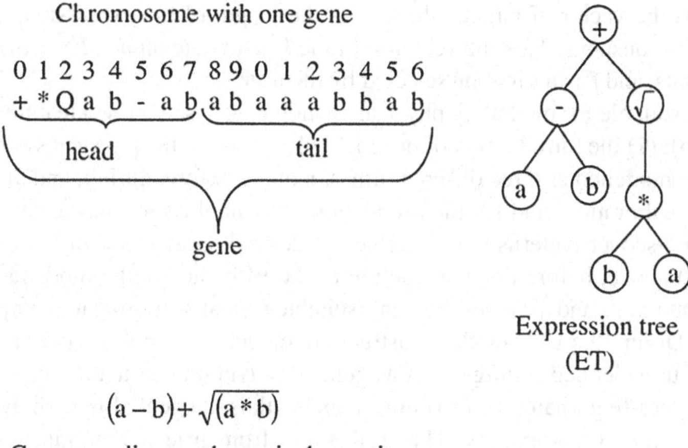

Fig. 2.4 Schematic representation of a chromosome with one gene and its expression tree (ET) and corresponding mathematical equation (Kayadelen 2011)

EPR is a hybrid regression technique that was developed by Giustolisi and Savic (2006). It constructs symbolic models by integrating the soundest features of numerical regression, with genetic programming and symbolic regression (Koza 1992). The following two steps roughly describe the underlying features of the EPR technique, aimed to search for polynomial structures representing a system. In the first step, the selection of exponents for polynomial expressions is carried out, employing an evolutionary searching strategy by means of GA (Goldberg 1989). In the second step, numerical regression using the least square method is conducted, aiming to compute the coefficients of the previously selected polynomial terms. The general form of expression in EPR can be presented as follows (Giustolisi and Savic 2006):

$$y = \sum_{j=i}^{m} F\big(X, f(X), a_j\big) + a_o \tag{2.2}$$

where y is the estimated vector of output of the process, m is the number of terms of the target expression, F is a function constructed by the process, X is the matrix of input variables, f is a function defined by the user, and a_j is a constant. A typical example of EPR pseudo-polynomial expression that belongs to the class of Eq. (2.2) is as follows (Giustolisi and Savic 2006):

$$\widehat{Y} = a_o + \sum_{j=i}^{m} a_j \;.\; (X_1)^{ES(j,1)} \ldots (X_k)^{ES(j,k)} . f\left[(X_1)^{ES(j,k+1)} \ldots (X_k)^{ES(j,2k)}\right] \tag{2.3}$$

where $\hat{Y}$ is the vector of target values, m is the length of the expression, a_j is the value of the constants, X_i is the vector(s) of the k candidate inputs, ES is the matrix of exponents, and f is a function selected by the user.

EPR is suitable for modelling physical phenomena, based on two features (Savic et al. 2006): (1) the introduction of prior knowledge about the physical system/process, to be modelled at three different times, namely before, during, and after EPR modelling calibration; and (2) the production of symbolic formulas, enabling data mining to discover patterns that describe the desired parameters. In the first EPR feature (1) above, before the construction of the EPR model, the modeller selects the relevant inputs and arranges them in a suitable format according to their physical meaning. During the EPR model construction, model structures are determined by following user-defined settings such as general polynomial structure, user-defined function types (e.g., natural logarithms, exponentials, tangential hyperbolics), and searching strategy parameters. The EPR starts from true polynomials and also allows for the development of non-polynomial expressions containing user-defined functions (e.g., natural logarithms). After EPR model calibration, an optimum model can be selected from among the series of models returned. The optimum model is selected based on the modeller's judgement, in addition to statistical performance indicators such as the coefficient of determination. A typical flow diagram of the EPR procedure is shown in Fig. 2.5, and a detailed description of the technique can be found in Giustolisi and Savic (2006).

2.3 Genetic Programming Applications in Geotechnical Engineering

In this section, the applications of GP techniques (including linear genetic programming, LGP; gene expression programming, GEP; multi-expression programming, MEP; multi-stage genetic programming, MSGP; multi-gene genetic programming, MGGP; and evolutionary polynomial regression, EPR) in geotechnical engineering are presented. The section provides a general view of GP applications that have appeared in the literature to date in the field of geotechnical engineering. Some of these applications are selected to be described in some detail, while others are acknowledged for reference purposes. The section starts with the overview of GP applications, followed by detailed description of some selected applications.

The behaviour of foundations (deep and shallow) in soils is complex, uncertain and not yet entirely understood. This fact has encouraged researchers to apply the GP techniques to predict the behaviour of foundations. The GP applications in foundations include the bearing capacity of piles (Gandomi and Alavi 2012; Alkroosh and Nikraz 2011, 2012, 2014; Shahin 2015), settlement and bearing capacity of shallow foundations (Rezania and Javadi 2007; Shahin 2015; Shahnazari et al. 2014; Pan et al. 2013; Tsai et al. 2013; Adarsh et al. 2012; Shahnazari and Tutunchian 2012), uplift capacity of suction caissons (Gandomi et al. 2011; Rezania et al. 2008), and pull-out capacity of ground anchors (Shahin 2015).

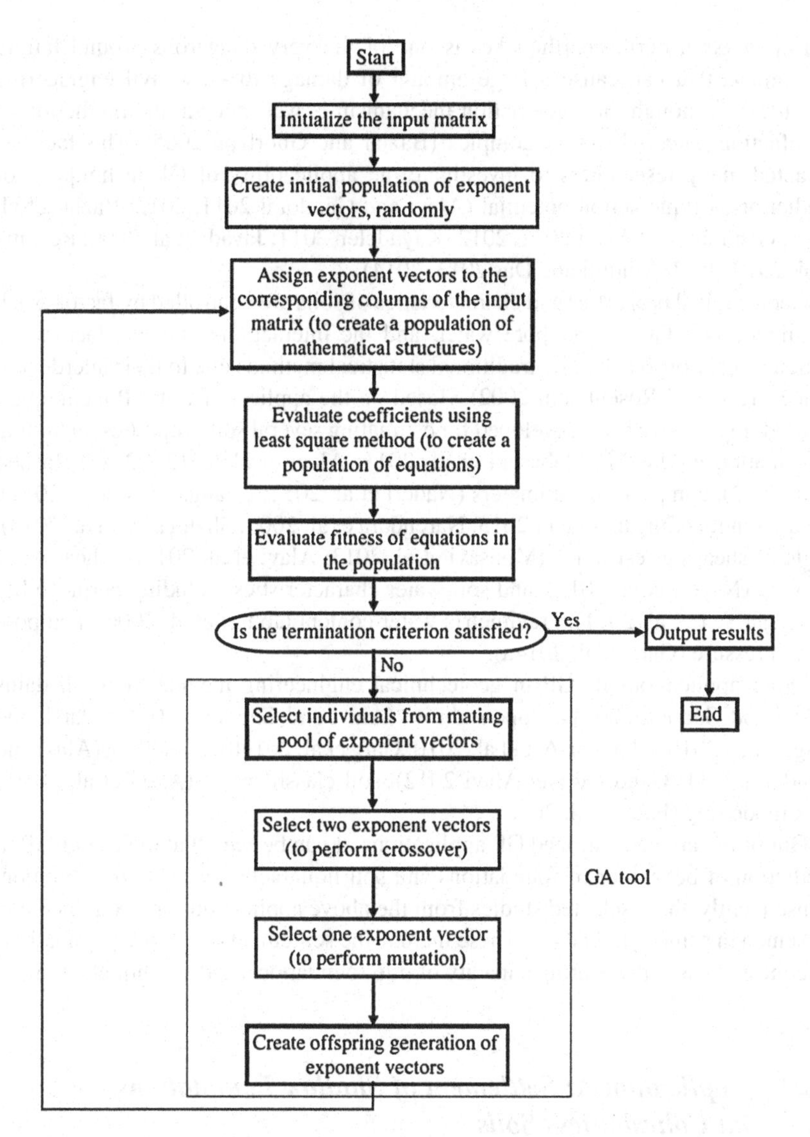

Fig. 2.5 Typical flow diagram of the evolutionary polynomial regression (EPR) procedure (Rezania et al. 2011)

Classical constitutive modelling based on elasticity and plasticity theories has limited capability to properly simulate the behaviour of geomaterials. This is attributed to reasons associated with the formulation complexity, idealization of material behaviour and excessive empirical parameters (Adeli 2001). In this regard, GP techniques have been proposed as a reliable and practical alternative to modelling the constitutive behaviour of geomaterials (Cabalar et al. 2009; Javadi and Rezania 2009; Shahnazari et al. 2010; Javadi et al. 2012a, b; Faramarzi et al. 2012; Feng et al. 2006).

Liquefaction during earthquakes is one of the very dangerous ground failure phenomena that can cause a large amount of damage to most civil engineering structures. Although the liquefaction mechanism is well known, the prediction of liquefaction potential is very complex (Baziar and Ghorbani 2005). This fact has attracted many researchers to investigate the applicability of GP techniques for prediction of liquefaction potential (Alavi and Gandomi 2011, 2012; Baziar et al. 2011; Gandomi and Alavi 2011, 2012; Kayadelen 2011; Javadi et al. 2006; Rezania et al. 2010, 2011; Muduli and Das 2013, 2014).

Geotechnical properties and characteristics of soils are controlled by factors such as mineralogy; fabric; and pore water, and the interactions of these factors are difficult to establish solely by traditional statistical methods due to their interdependence (Yang and Rosenbaum 2002). Based on the applications of GP techniques, methodologies have been developed for estimating several soil properties, including deformation moduli (Mollahasani et al. 2011; Alavi et al. 2012a, 2013; Rashed et al. 2012), compaction parameters (Naderi et al. 2012; Ahangar-Asr et al. 2011), shear strength (Cuisinier et al. 2013; Narendara et al. 2006; Shahnazari et al. 2013), angle of shearing resistance (Mousavi et al. 2013; Alavi et al. 2012b), shear wave velocity (Nayeri et al. 2013), and soil-water characteristics including permeability (Ahangar-Asr et al. 2011), gravimetric water content (Johari et al. 2006), and pore water pressure (Garg et al. 2014a).

Other applications of GP in geotechnical engineering include: rock-fill dams (Alavi and Gandomi 2011), slope stability (Alavi and Gandomi 2011; Adarsh and Jangareddy 2010; Ahangar-Asr et al. 2010; Garg et al. 2014b), tunnelling (Alavi and Gandomi 2011; Gandomi and Alavi 2012), soil classification (Alavi et al. 2010), rock modelling (Feng et al. 2006).

Out of the abovementioned GP applications, it can be seen that the use of GP in prediction of behaviour of foundations and soil liquefaction is the most common. Consequently, three selected studies from the above applications are examined and presented in some detail below. These include the settlement of shallow foundations on cohesionless soils, bearing capacity of pile foundations, and soil liquefaction.

2.3.1 Application A: Settlement of Shallow Foundations on Cohesionless Soils

The design of foundations is generally controlled by the criteria of bearing capacity and settlement, the latter often being the governing factor in design of shallow foundations, especially when the breadth of footing exceeds 1 m (Schmertmann 1970). The estimation of settlement of shallow foundations on cohesionless soils is complex, uncertain, and not yet entirely understood. This fact has encouraged a number of researchers to apply the GP techniques to the settlement of shallow foundations on cohesionless soils. For example, Shahin (2015) carried out a comprehensive study to predict the settlement of shallow foundations on cohesionless soils utilizing EPR technique. Using a large database that contains 187 data records of field measurements of settlement of shallow foundations as well as the corresponding information regarding the footings and soil, Shahin (2015) developed an EPR model that was found to outperform the most commonly used traditional methods. The data were obtained from the literature and cover a wide range of variation in footing dimensions and cohesionless soil types and properties. Details of the references from which the data were obtained can be found in Shahin et al. (2002a). The model was trained using five inputs representing the footing width, net applied footing pressure, average blow count obtained from the standard penetration test (SPT) over the depth of influence of the foundations as a measure of soil compressibility, footing length, and footing embedment depth. The single model output was the foundation settlement. The EPR returned several different models and the one selected to be optimal is as follows (Shahin 2015):

$$S_p^{EPR} = -8.327\frac{q}{N^2L} + 8.849\frac{q}{N^2} + 2.993\frac{B\sqrt{q}}{N} - 0.651\frac{B\sqrt{qD_f}}{N} + 2.883 \quad (2.4)$$

where S_p (mm) is the predicted settlement, B (m) is the footing width, q (kPa) is the net applied footing pressure, N is the average SPT blow count, L (m) is the footing length, and D_f (m) is the footing embedment depth.

The results between the predicted and measured settlements obtained by utilising GP model were compared with those obtained from an artificial neural networks (ANN) model previously developed by the author (Shahin et al. 2002b), and three traditional methods, namely, Meyerhof (1965), Schultze and Sherif (1973), and Schmertmann (1978). Comparisons of the results obtained using the GP model and the methods used for comparison in the validation set are given in Table 2.1. It can be seen that the EPR model performs better than the other methods, including the ANN model, in all performance measures used including the coefficient of correlation, r, coefficient of determination, R^2, root mean squared error, $RMSE$, mean absolute error, MAE, and ratio of average measured to predicted outputs, μ.

Using the same database of Shahin et al. (2002a) and similar model inputs and outputs used above, Rezania and Javadi (2007) and Shahnazari et al. (2014)

Table 2.1 Comparison of EPR model and other methods in the validation set for settlement of shallow foundations on cohesionless soils (Shahin 2015)

Performance measure	Method				
	EPR (Shahin 2015)	ANNs (Shahin et al. 2002a, b)	Meyerhof (1965)	Schultze and Sherif (1973)	Schmertmann (1978)
r	0.923	0.905	0.440	0.729	0.838
R^2	0.844	0.803	0.014	0.185	0.153
$RMSE$ (mm)	9.83	11.04	24.71	22.48	22.91
MAE (mm)	6.99	8.78	16.91	11.29	16.23
μ	1.03	1.10	0.91	1.73	0.79

developed two different genetic programming models (2014). The formulation of the GP model developed by Rezania and Javadi (2007) is as follows:

$$S_p^{GP} = \frac{q\,(1.80B + 4.62) - 346.15D_f}{N^2} + \frac{11.22L - 11.11}{L} \tag{2.5}$$

The formulation of the GP model developed by Shahnazari et al. (2014) is as follows:

$$S_p^{GP} = \frac{2.5B\left(\frac{N}{B} - 1 + \frac{B+1}{D_f+0.16B} + \frac{2B-N}{L} + \frac{q}{N}\right)}{\left(N + \frac{D_f}{B}\left(B - \frac{L}{B}\right) + \frac{B}{N}\right)} \tag{2.6}$$

The above GP models represented by Eqs. (2.5) and (2.7) were compared with the traditional methods and found to outperform most available methods.

2.3.2 *Application B: Bearing Capacity of Pile Foundations*

In contrast to design of shallow foundations, the load carrying capacity is often being the governing factor in design of pile foundations rather than settlement; hence, has been examined by several AI researchers. For example, Shahin (2015) developed EPR models for driven piles and drilled shafts that found to perform well. The data used to calibrate and validate the EPR models include a series of 79 in-situ driven pile load tests and 94 in-situ drilled shaft load tests, as well as cone penetration test (CPT) results. The conducted tests were located on sites of different soil types and geotechnical conditions, ranging from cohesive clays to cohesionless sands. The driven pile load tests include compression and tension loading conducted on steel and concrete piles. The driven piles used have different shapes (i.e., circular, square, and hexagonal) and range in diameter between 250 and 900 mm and embedment lengths between 5.5 and 41.8 m. The drilled shaft load tests were conducted on straight and belled concrete piles and include compression and tension loading but no tension loading for belled shafts. The drilled shafts used have stem diameters ranging from 305 to 1798 mm and embedment lengths from 4.5 to 27.4 m. The statistics of the data used can be found in Shahin (2015). The formulations of the developed EPR models yielded pile capacity, Q_u (kN), as follows (Shahin 2015):

For driven (steel) piles:

$$Q_{u(steel-driven)}^{EPR} = -2.277\frac{D\overline{q}_{c-tip}}{\sqrt{\overline{q}_{c-shaft}\overline{f}_{s-shaft}}} + 0.096DL + 1.714 \times 10^{-4}D^2\overline{q}_{c-tip}\sqrt{L} - 6.279 \times 10^{-9}D^2L^2\sqrt{\overline{q}_{c-tip}\overline{f}_{s-tip}} + 243.39 \tag{2.7}$$

Alternatively, for driven (concrete) piles:

$$Q_{u(concrete-driven)} = -2.277\frac{D\overline{q}_{c-tip}}{\sqrt{\overline{q}_{c-shaft}\overline{f}_{s-shaft}}} + 0.096DL + 1.714 \times 10^{-4}D^2\overline{q}_{c-tip}\sqrt{L} - 6.279 \times 10^{-9}D^2L^2\sqrt{\overline{q}_{c-tip}\overline{f}_{s-tip}} + 486.78 \tag{2.8}$$

For drilled shafts:

$$Q_{u(drilled-shafts)} = 0.6878L^2\sqrt{\overline{f}_{s-shaft}} + 1.581 \times 10^{-4}B^2\sqrt{\overline{f}_{s-shaft}} + 1.294 \times 10^{-4}L^2\overline{q}^2_{c-tip}\sqrt{D} + 7.8 \times 10^{-5}D\overline{q}_{c-shaft}\overline{f}_{s-shaft}\sqrt{\overline{f}_{s-tip}} \tag{2.9}$$

where D (mm) is the pile perimeter/π (for driven piles) or pile stem diameter (for drilled shafts), L (m) is the pile embedment length, B (mm) is the drilled shaft base diameter, $\overline{q}_{c-tip}$ (MPa) is the weighted average cone point resistance over pile tip failure zone, $\overline{f}_{s-tip}$ (kPa) is the weighted average cone sleeve friction over pile tip failure zone, $\overline{q}_{c-shaft}$ (MPa) is the weighted average cone point resistance over pile embedment length, and $\overline{f}_{s-shahft}$ (kPa) is the weighted average cone sleeve friction over pile embedment length.

The performance of the above EPR models, represented by Eqs. (2.7)–(2.9), was compared with four other models in the validation set and the results are given in Table 2.2. For driven piles, the methods considered for comparison include an ANN model developed by Shahin (2010), the European method (de Ruiter and Beringen 1979), LCPC method (Bustamante and Gianeselli 1982), and Eslami and Fellenius (1997) method. For drilled shafts, the methods considered for comparison include an ANN model (Shahin 2010), Schmertmann (1978) method, LCPC method (Bustamante and Gianeselli 1982), and Alsamman (1995) method. It can been seen from Table 2.2 that the performance of the EPR models is as good as the ANN model, or better, and outperforms the other available methods with the possible exception of Alsamman (1995).

The application of GP in estimating the capacity of pile foundations was carried out by Alkroosh and Nikraz (2011, 2012). Correlation models for predicting the relationship between pile axial capacity and CPT data using gene expression programming (GEP) technique were developed. The GEP models were developed for bored piles as well as driven piles (a model for each of concrete and steel piles). The performance of the GEP models was evaluated by comparing their results with experimental data as well as the results of a number of currently used CPT-based methods. The results indicated the potential ability of GEP models in predicting the bearing capacity of pile foundations and outperformance of the developed models over existing methods. More recently, Alkroosh and Nikraz (2014) developed GEP

Table 2.2 Comparison of EPR model and other methods in the validation set for bearing capacity of pile foundations (Shahin 2015)

	Methods for driven piles				
Performance measure	EPR (Shahin 2015)	ANNs (Shahin 2010)	de Ruiter and Beringen (1979)	Bustamante and Gianeselli (1982)	Eslami and Fellenius (1997)
r	0.848	0.837	0.799	0.809	0.907
R^2	0.745	0.753	0.219	0.722	0.681
RMSE (kN)	249.0	244.0	435.0	260.0	278.0
MAE (kN)	185.0	203.0	382.0	219.0	186.0
μ	1.00	0.97	1.36	1.11	0.94
	Methods for drilled shafts				
Performance measure	EPR (Shahin 2015)	ANNs (Shahin 2010)	Schmertmann (1978)	Bustamante and Gianeselli (1982)	Alsamman (1995)
r	0.990	0.970	0.901	0.951	0.984
R^2	0.944	0.939	0.578	0.901	0.939
RMSE (kN)	511.0	533.0	1404.0	681.0	534.0
MAE (kN)	347.0	374.0	702.0	426.0	312.0
μ	1.03	1.02	1.33	0.97	1.03

model that correlates the pile capacity with the dynamic input and SPT data. The performance of the model was assessed by comparing its predictions with those calculated using two commonly used traditional methods and an ANN model. It was found that the GEP model performed well with a coefficient of determination of 0.94 and 0.96 in the training and testing sets, respectively. The results of comparison with other available methods showed that the GEP model predicted the pile capacity more accurately than existing traditional methods and ANN model. Another successful application of genetic programming in pile capacity prediction was carried out by Gandomi and Alavi (2012), who used a multi-gene genetic programming (MGGP) method for the assessment of the undrained lateral load capacity of driven piles and undrained side resistance alpha factor of drilled shafts.

2.3.3 Application C: Soil Liquefaction

Soil liquefaction induced by earthquakes is one of the most complex problems in geotechnical engineering, and is an essential design criterion for many civil engineering structures. Many buildings, highways, embankments and other engineering structures have been damaged or destroyed as a result of liquefaction induced by strong earthquakes that have recently occurred around the world (Kayadelen 2011). Consequently, accurate determination of soil liquefaction potential is an essential part of geotechnical engineering investigation as it provides fairly significant and necessary tool for design of civil engineering structures located on active zones of earthquakes. Hence, many researchers used GP to develop new models for prediction of liquefaction potential of soils and induced deformation.

Alavi and Gandomi (2011) and Gandomi and Alavi (2012) developed generalized GP models including LGP, GEP, MEP, and MGGP for the classification of several liquefied and non-liquefied case records. Soil and seismic parameters governing the soil liquefaction potential were used for model development including the CPT cone tip resistance, q_c (MPa), sleeve friction ratio, R_f (%), effective stress at the depth of interest, σ_v' (kPa), total stress at the same depth, σ_v (kPa), maximum horizontal ground surface acceleration, a_{max} (g), and earthquake moment magnitude, M_w. the existence of the liquefaction (LC) was represented by binary variables, non-liquefied and liquefied cases were represented by 0 and 1, respectively. The GP models were developed based on CPT database that contains 226 case records, with 133 liquefied cases and 93 non-liquefied cases. Out of the available data, 170 case records were used for model training and 56 case records were used for model validation. The LGP, GEP, MEP, and MGGP models used to classify the non-liquefied and liquefied cases, LC, are given as follows (Alavi and Gandomi 2011; Gandomi and Alavi 2012):

$$LC_{LGP} = \frac{1}{\sigma_v'^2}\left(a_{\max}\sigma_v'^2 - 4q_c\sigma_v' - 9R_f\sigma_v' + 54\sigma_v' + 9\sigma_v - 54M_w - 378\right) \quad (2.10)$$

$$LC_{GEP} = a_{\max} - \frac{1}{\sigma_v'}\left(\frac{R_f}{a_{\max}} - 5a_{\max}\right) + \frac{q_c - (M_w - q_c)R_f}{2q_c - \sigma_v - 3} + \frac{4 - R_f}{q_c + 2} \quad (2.11)$$

$$\begin{aligned} LC_{MEP} = a_{\max} + \frac{1}{\sigma_v'}\Bigg(&4a_{\max} + 4M_w - 4q_c + \frac{9}{4\sigma_v'} - \frac{9R_f}{4} + \frac{\sigma_v'}{4} \\ &-4R_f\frac{2(q_c - a_{\max} - M_w)^2 + M_w}{\sigma_v'}\Bigg) \end{aligned} \quad (2.12)$$

$$\begin{aligned} LC_{MGGP} = {} & 0.5491 + 0.9634 \times 10^{-5} R_f q_c^4 \ln(q_c) - 0.6553 q_c \\ & + 0.6553 \ln(\tanh(a_{\max})) + 0.1288 R_f + 0.2576 M_w + 0.1288 \ln(\sigma_v') q_c \\ & + 0.2058 \ln(|\ln(a_{\max})|) + 0.2058 \ln(a_{\max}) R_f - 0.2861 \\ & \times 10^{-6} \sigma_v \sigma_v' (\sigma_v + q_c) - 0.2861 \times 10^{-6} q_c^2 \sigma_v'^2 \end{aligned} \quad (2.13)$$

When the return of Eqs (2.10)–(2.13) is greater than or equal to 0.5, the case is marked as "liquefied", otherwise, it is marked as "non-liquefied". The accuracy of the GP models in the training and validation sets were, respectively, LGP (training = 90 % and validation = 94.64 %), GEP (training = 88.82 % and Validation = 92.86 %), MEP (training = 86.47 % and validation = 85.71 %), MGGP (training = 90 % and validation = 96.4 %). These results clearly indicate that the GP models are efficiently capable of classifying the liquefied and non-liquefied cases. The best classification results are obtained by both LGP and MGGP models, which yielded similar performance, followed by the GEP and MEP models.

Javadi et al. (2006) introduced GP models for determination of liquefaction induced lateral spreading. The models were trained and validated using SPT-based case histories. Separate models were presented to estimate the lateral displacements for free face as well as gently sloping ground conditions. It was shown that the GP models are capable of learning the complex relationship between lateral displacement and its contributing factors, in the form of a high accuracy prediction function. It was also shown that the attained function can be used to generalize the learning to predict liquefaction induced lateral spreading for new cases that have not been used in model calibration. The results of the developed GP models were also compared with one of the most commonly used available methods in the literature, i.e., multi linear regression (MLR) model (Youd et al. 2002), and the advantages of the proposed GP models were highlighted. It was shown that the GP models offer an improved determination of the lateral spreading over the most commonly used MLR method.

Another successful application of genetic programming in soil liquefaction potential was carried out by Rezania et al. (2010), who used CPT results and EPR method for determination of liquefaction potential in sands. Furthermore, Kayadelen (2011) used GEP method to forecast the safety factor of soil liquefaction using standard penetration test (SPT) results. Both of the above GP models were found to provide more accurate results compared to the conventional available methods. More recently, Gandomi and Alavi (2013) developed a robust GP model, coupled with orthogonal east squares, for predicting the soil capacity energy required to trigger soil liquefaction, and Gandomi (2014) presented a short review for use of soft computing, including GP, in earthquake engineering.

2.4 Discussion and Conclusion

In the field of geotechnical engineering, it is possible to encounter some types of problems that are very complex and not well understood. In this regard, artificial intelligence (AI) techniques such as genetic programming (GP) provide several advantages over more conventional computing methods. For most traditional mathematical models, the lack of physical understanding is usually supplemented by either simplifying the problem or incorporating several assumptions into the models. Mathematical models also rely on assuming the structure of the model in advance, which may be less than optimal. Consequently, many mathematical models fail to simulate the complex behaviour of most geotechnical engineering problems. In contrast, AI techniques are a data-driven approach in which the model can be trained on input-output data pairs to determine the structure and parameters of the model. In this case, there is no need to either simplify the problem or incorporate any assumptions. Moreover, AI models can always be updated to obtain better results by presenting new training examples as new data become available. These factors combine to make AI techniques a powerful modelling tool in geotechnical engineering.

In contrast to most AI techniques, GP does not suffer from the problem of lack of transparency and knowledge extraction. GP has the ability to generate transparent, compact, optimum and well-structured mathematical formulations of the system being studied, directly from raw experimental or field data. Furthermore, prior knowledge about the underlying physical process based on engineering judgement or human expertise can also be incorporated into the learning formulation, which greatly enhances the usefulness of GP over other AI techniques. It was evident from the review presented in this chapter that GP has been applied successfully to several applications in geotechnical engineering. Based on the results of the studies reviewed, it can be concluded that genetic programming models provide high level of prediction capability and outperform most traditional methods.

References

ADARSH, S., DHANYA, R., KRISHNA, G., MERLIN, R. & TINA, J. 2012. Prediction of ultimate bearing capacity of cohesionless soils using soft computing techniques. *ISRN Artificial Intelligence,* 2012, 10pp.

ADARSH, S. A. & JANGAREDDY, M. 2010. Slope stability modeling using genetic programming. *International Journal of Earth Sciences and Engineering,* 3, 1–8.

ADELI, H. 2001. Neural networks in civil engineering: 1989-2000. *Computer-Aided Civil and Infrastructure Engineering,* 16, 126–142.

AHANGAR-ASR, A., FARAMARZI, A. & JAVADI, A. 2010. A new approach for prediction of the stability of soil and rock slopes. *Engineering Computations: International Journal of Computer-Aided Engineering and Software,* 27, 878–893.

AHANGAR-ASR, A., FARAMARZI, A., MOTTAGHIFARD, N. & JAVADI, A. A. 2011. Modeling of permeability and compaction characteristics of soils using evolutionary polynomial regression. *Computers and Geosciences,* 37, 1860–1869.

ALAVI, A. H. & GANDOMI, A. H. 2011. A robust data mining approach for formulation of geotechnical engineering systems. *Engineering Computations: International Journal of Computer-Aided Engineering and Software,* 28, 242–274.

ALAVI, A. H. & GANDOMI, A. H. 2012. Energy-based models for assessment of soil liquefaction. *Geoscience Frontiers.*

ALAVI, A. H., GANDOMI, A. H., NEJAD, H. C., MOLLAHASANI, A. & RASHED, A. 2013. Design equations for prediction of pressuremeter soil deformation moduli utilizing expression programming systems. *Neural Computing and Applications,* 23, 1771–1786.

ALAVI, A. H., GANDOMI, A. H., SAHAB, M. G. & GANDOMI, M. 2010. Multi expression programming: a new approach to formulation of soil classification. *Engineering with Computers,* 26, 111–118.

ALAVI, A. H., MOLLAHASANI, A., GANDOMI, A. H. & BAZA, J. B. 2012a. Formulation of secant and reloading soil deformation moduli using multi expression programming. *Engineering Computations: International Journal of Computer-Aided Engineering and Software,* 29, 173–197.

ALAVI, A. M., GANDOMI, A. H., BOLURY, J. & MOLLAHASANI, A. 2012b. Linear and tree-based genetic programming for solving geotechnical engineering problems. *In:* YANG,

X.-S., GANDOMI, H., TALATAHARI, S. & ALAVI, A. H. (eds.) *Metaheuristics in Water, Geotechnical and Transport Engineering*. London: Elsevier

ALKROOSH, I. & NIKRAZ, H. 2011. Correlation of pile axial capacity and CPT data using gene expression programming. *Geotechnical and Geological Engineering,* 29, 725–748.

ALKROOSH, I. & NIKRAZ, H. 2012. Predicting axial capacity of driven piles in cohesive soils using intelligent computing. *Engineering Applications of Artificial Intelligence,* 25, 618–627.

ALKROOSH, I. & NIKRAZ, H. 2014. Predicting pile dynamic capacity via application of an evolutionary algorithm. *Soils and Foundations,* 54, 233–242.

ALSAMMAN, O. M. 1995. *The use of CPT for calculating axial capacity of drilled shafts.* PhD Thesis, University of Illinois-Champaign.

BAZIAR, M. H. & GHORBANI, A. 2005. Evaluation of lateral spreading using artificial neural networks. *Soil Dynamics and Earthquake Engineering,* 25, 1–9.

BAZIAR, M. H., JAFARIAN, Y., SHAHNAZARI, H., MOVAHED, V. & TUTUNCHIAN, M. A. 2011. Prediction of strain energy-based liquefaction resistance of sand-silt mixtures: an evolutionary approach. *Computers and Geotechnics,* 37, 1883–1893.

BUSTAMANTE, M. & GIANESELLI, L. 1982 Published. Pile bearing capacity prediction by means of static penetrometer CPT. Proceedings of the 2nd European Symposium on Penetration Testing, 1982 Amsterdam. 493–500.

CABALAR, A. F., CEVIK, A. & GUZELBEY, I. H. 2009. Constitutive modeling of Leighton Buzzard sands using genetic programming. *Neural Computing and Applications,* 19, 657–665.

CRAMER, N. L. 1985 Published. A representation for the adaptive generation of simple sequential programs. Proceedings of the international conference on genetic algorithms and their applications, 1985 Carnegie-Mellon University, Pittsburgh, PA. 183–187.

CUISINIER, O., JAVADI, A., AHANGAR-ASR, A. & FARIMAH, M. 2013. Identification of coupling parameters between shear strength behaviour of compacted soils and chemical's effects with an evolutionary-based data mining technique. *Computers and Geotechnics,* 48, 107–116.

DE RUITER, J. & BERINGEN, F. L. 1979. Pile foundation for large North Sea structures. *Marine Geotechnology,* 3, 267–314.

ELSHORBAGY, A., CORZO, G., SRINIVASULU, S. & SOLOMATINE, D. P. 2010. Experimental investigation of the predictive capabilities of data driven modeling techniques in hydrology-part 1: concepts and methodology. *Hydrology and Earth System Science* 14, 1931–1941.

ESLAMI, A. & FELLENIUS, B. H. 1997. Pile capacity by direct CPT and CPTu methods applied to 102 case histories. *Canadian Geotechnical Journal,* 34, 886–904.

FARAMARZI, A., JAVADI, A. & ALANI, A. M. 2012. EPR-based material modelling of soils considering volume changes. *Computers and Geosciences,* 48, 73–85.

FENG, X. T., CHEN, B., YANG, C., ZHOU, H. & DING, X. 2006. Identification of visco-elastic models for rocks using genetic programming coupled with the modified particle swarm optimization algorithm. *International Journal of Rock Mechanics and Mining Sciences,* 43, 789–801.

FERREIRA, C. 2001. Gene expression programming: a new adaptive algorithm for solving problems. *Complex Systems,* 13, 87–129.

FLOOD, I. 2008. Towards the next generation of artificial neural networks for civil engineering. *Advanced Engineering Informatics,* 22, 4–14.

GANDOMI, A. H. 2014. Soft computing in earthquake engineering: a short review. *International Journal of Earthquake Engineering and Hazard Mitigation,* 2, 42–48.

GANDOMI, A. H. & ALAVI, A. H. 2011. Multi-stage genetic programming: a new strategy to nonlinear system modeling. *Information Sciences,* 181, 5227–5239.

GANDOMI, A. H. & ALAVI, A. H. 2012. A new multi-gene genetic programming approach to non-linear system modeling. Part II: geotechnical and earthquake engineering problems. *Neural Computing Applications,* 21, 189–201.

GANDOMI, A. H. & ALAVI, A. H. 2013. Hybridizing genetic programming with orthogonal least squares for modeling of soil liquefaction. *International Journal of Earthquake Engineering and Hazard Mitigation,* 1, 2–8.

GANDOMI, A. H., ALAVI, A. H. & GUN, J. Y. 2011. Formulation of uplift capacity of suction caissons using multi expression programming. *KSCE Journal of Civil Engineering,* 15, 363–373.

GARG, A., GARG, A., TAI, K. & SREEDEEP, S. 2014a. Estimation of pore water pressure of soil using genetic programming. *Geotechnical and Geological Engineering,* 32, 765–772.

GARG, A., GARG, A., TAI, K. & SREEDEEP, S. 2014b. An integrated SRP-multi-gene genetic programming approach for prediction of factor of safety of 3-D soil nailed slopes. *Engineering Applications of Artificial Intelligence,* 30, 30–40.

GIUSTOLISI, O., DOGLIONI, A., SAVIC, D. A. & WEBB, B. W. 2007. A multi-model approach to analysis of environmental phenomena. *Environmental Modelling and Software,* 22, 674–682.

GIUSTOLISI, O. & SAVIC, D. A. 2006. A symbolic data-driven technique based on evolutionary polynomial regression. *Journal of Hydroinformatics,* 8, 207–222.

GOLDBERG, D. E. 1989. *Genetic Algorithms in Search Optimization and Machine Learning,* Mass, Addison - Wesley.

HOLLAND JH. 1975 Published. Adaptation in natural and artificial systems. 1975 University of Michigan

JAVADI, A., AHANGAR-ASR, A., JOHARI, A., FARAMARZI, A. & TOLL, D. 2012a. Modelling stress-strain and volume change behaviour of unsaturated soils using an evolutionary based data mining technique, and incremental approach. *Engineering Applications of Artificial Intelligence,* 25, 926–933.

JAVADI, A., FARAMARZI, A. & AHANGAR-ASR, A. 2012b. Analysis of behaviour of soils under cyclic loading using EPR-based finite element method. *Finite Elements in Analysis and Design,* 58, 53–65.

JAVADI, A. & REZANIA, M. 2009. Intelligent finite element method: An evolutionary approach to constitutive modelling. *Advanced Engineering Informatics,* 23, 442–451.

JAVADI, A., REZANIA, M. & MOUSAVI, N. M. 2006. Evaluation of liquefaction induced lateral displacements using genetic programming. *Computers and Geotechnics,* 33, 222–233.

JOHARI, A., HABIBAGAHI, G. & GHAHRAMANI, A. 2006. Prediction of soil-water characteristic curve using genetic programming. *Journal of Geotechnical and Geoenvironmental Engineering,* 132, 661–665.

KAYADELEN, C. 2011. Soil liquefaction modeling by genetic expression programming and neuro-fuzzy. *Expert Systems with Applications,* 38, 4080–4087.

KOZA, J. R. 1992. *Genetic programming: on the programming of computers by natural selection,* Cambridge (MA), MIT Press.

MEYERHOF, G. G. 1965. Shallow foundations. *Journal of Soil Mechanics & Foundation Engineering Division,* 91, 21–31

MOLLAHASANI, A., ALAVI, A. H. & GANDOMI, A. H. 2011. Empirical modeling of plate load test moduli of soil via gene expression programming. *Computers and Geotechnics,* 38, 281–286.

MOUSAVI, S. M., ALAVI, A. H., MOLLAHASANI, A., GANDOMI, A. H. & ESMAEILI, M. A. 2013. Formulation of soil angle of resistance using a hybrid GP and OLS method. *Engineering with Computers,* 29, 37–53.

MUDULI, P. K. & DAS, S. K. 2013. SPT-based probabilistic method for evaluation of liquefaction potential of soil using multi-gene genetic programming. *International Journal of Geotechnical Earthquake Engineering,* 4, 42–60.

MUDULI, P. K. & DAS, S. K. 2014. CPT-based seismic liquefaction potential evaluation using multi-gene genetic programming approach. *Indian Geotechnical Journal,* 44, 86–93.

NADERI, N., ROSHANI, P., SAMANI, M. Z. & TUTUNCHIAN, M. A. 2012. Application of genetic programming for estimation of soil compaction parameters. *Applied Mechanics and Materials,* 147, 70–74.

NARENDARA, B. S., SIVAPULLAIAH, P. V., SURESH, S. & OMKAR, S. N. 2006. Prediction of unconfined compressive strength of soft grounds using computational intelligence techniques: A comparative study. *Computers and Geotechnics,* 33, 196–208.

NAYERI, G. D., NAYERI, D. D. & BARKHORDARI, K. 2013. A new statistical correlation between shear wave velocity and penetration resistance of soils using genetic programming. *Electronic Journal of Geotechnical Engineering,* 18K, 2071–2078.

PAN, C.-P., TSAI, H.-C. & LIN, Y.-H. 2013. Improving semi-empirical equations of ultimate bearing capacity of shallow foundations using soft computing polynomials. *Engineering Applications of Artificial Intelligence* 26, 478–487.

RASHED, A., BAZA, J. B. & ALAVI, A. H. 2012. Nonlinear modeling of soil deformation modulus through LGP-based interpretation of pressuremeter test results. *Engineering Applications of Artificial Intelligence,* 25, 1437–1449.

REZANIA, M., FARAMARZI, A. & JAVADI, A. 2011. An evolutionary based approach for assessment of earthquake-induced soil liquefaction and lateral displacement. *Engineering Applications of Artificial Intelligence,* 24, 142–153.

REZANIA, M. & JAVADI, A. 2007. A new genetic programming model for predicting settlement of shallow foundations. *Canadian Geotechnical Journal,* 44, 1462–1472.

REZANIA, M., JAVADI, A. & GIUSTOLISI, O. 2008. An evolutionary-based data mining technique for assessment of civil engineering systems. *Engineering Computations: International Journal of Computer-Aided Engineering and Software,* 25, 500–517.

REZANIA, M., JAVADI, A. & GIUSTOLISI, O. 2010. Evaluation of liquefaction potential based on CPT results using evolutionary polynomial regression. *Computers and Geotechnics,* 37, 82–92.

SAVIC, D. A., GIUTOLISI, O., BERARDI, L., SHEPHERD, W., DJORDJEVIC, S. & SAUL, A. 2006. Modelling sewer failure by evolutionary computing. *Proceedings of the Institution of Engineers, Water Management,* 159, 111–118.

SCHULTZE, E. & SHERIF, G. 1973 Published. Prediction of settlements from evaluated settlement observations for sand. Proceedings of the 8th International Conference on Soil Mechanics & Foundation Engineering, 1973 Moscow. 225–230.

SCHMERTMANN, J. H. 1970. Static cone to compute static settlement over sand. *Journal of Soil Mechanics & Foundation Engineering Division,* 96, 1011–1043.

SCHMERTMANN, J. H. 1978. Guidelines for cone penetration test, performance and design. Washington, D. C.: U. S. Department of Transportation.

SHAHIN, M. A. 2010. Intelligent computing for modelling axial capacity of pile foundations. *Canadian Geotechnical Journal,* 47, 230–243.

SHAHIN, M. A. 2013. Artificial intelligence in geotechnical engineering: applications, modeling aspects, and future directions. *In:* YANG, X., GANDOMI, A. H., TALATAHARI, S. & ALAVI, A. H. (eds.) *Metaheuristics in Water, Geotechnical and Transport Engineering.* London: Elsevier Inc.

SHAHIN, M. A. 2015. Use of evolutionary computing for modelling some complex problems in geotechnical engineering. *Geomechanics and Geoengineering: An International Journal,* 10(2), 109–125.

SHAHIN, M. A., JAKSA, M. B. & MAIER, H. R. 2001. Artificial neural network applications in geotechnical engineering. *Australian Geomechanics,* 36, 49–62.

SHAHIN, M. A., JAKSA, M. B. & MAIER, H. R. 2002a. Artificial neural network-based settlement prediction formula for shallow foundations on granular soils. *Australian Geomechanics,* 37, 45–52.

SHAHIN, M. A., JAKSA, M. B. & MAIER, H. R. 2009. Recent advances and future challenges for artificial neural systems in geotechnical engineering applications. *Journal of Advances in Artificial Neural Systems,* 2009, doi: 10.1155/2009/308239.

SHAHIN, M. A., MAIER, H. R. & JAKSA, M. B. 2002b. Predicting settlement of shallow foundations using neural networks. *Journal of Geotechnical & Geoenvironmental Engineering,* 128, 785–793.

SHAHNAZARI, H., DEHNAVI, Y. & ALAVI, A. H. 2010. Numerical modeling of stress-strain behavior of sand under cyclic loading. *Engineering Geology,* 116, 53–72.

SHAHNAZARI, H., SHAHIN, M. A. & TUTUNCHIAN, M. A. 2014. Evolutionary-based approaches for settlement prediction of shallow foundations on cohesionless soils. *International Journal of Civil Engineering,* 12, 55–64.

SHAHNAZARI, H. & TUTUNCHIAN, M. A. 2012. Prediction of ultimate bearing capacity of shallow foundations on cohesionless soils: An evolutionary approach. *KSCE Journal of Civil Engineering,* 16, 950–957.

SHAHNAZARI, H., TUTUNCHIAN, M. A., REZVANI, R. & VALIZADEH, F. 2013. Evolutionary-based approaches for determining the deviatoric stress of calcareous sands *Computers and Geosciences,* 50, 84–94.

TEODORESCU, L. & SHERWOOD, D. 2008. High energy physics event selection with gene expression programming. *Computer Physics Communications,* 178, 409–419.

TSAI, H.-C., TYAN, Y.-Y., WU, Y.-W. & LIN, Y.-H. 2013. Determining ultimate bearing capacity of shallow foundations using a genetic programming system. *Neural Computing and Applications,* 23, 2073–2084.

YANG, Y. & ROSENBAUM, M. S. 2002. The artificial neural network as a tool for assessing geotechnical properties. *Geotechnical Engineering Journal,* 20, 149–168.

YOUD, T. L., HANSEN, C. M. & BARLETT, S. F. 2002. Revised multilinear regression equations for prediction of lateral spread displacement *Journal of Geotechnical and Geoenvironmental Engineering,* 128, 1007–1017.

Chapter 3
Application of Genetic Programming in Hydrology

E. Fallah-Mehdipour and O. Bozorg Haddad

3.1 Introduction

In the real world, there are several natural and artificial phenomenons which follow some rules. These rules can model in a mathematical and/or logical form considering simple or complex equation/s may be difficult in some systems. Moreover, it is sometimes necessary to model just some parts of system without considering whole system information. Data-driven models are a programming paradigm that employs a sequence of steps to achieve best connection between data sets.

The contributions from artificial intelligence, data mining, knowledge discovery in databases, computational intelligence, machine learning, intelligent data analysis, soft computing, and pattern recognition are main cores of data-driven models with a large overlap in the disciplines mentioned.

GP is a data-driven tool which applies computational programming to achieve the best relation in a system. This tool can set in the inner or outer of system modeling which makes it more flexible to adapt different system states.

Electronic supplementary material The online version of this chapter (doi: 10.1007/978-3-319-20883-1_3) contains supplementary material, which is available to authorized users.

E. Fallah-Mehdipour
Department of Irrigation & Reclamation Engineering, Faculty of Agricultural Engineering & Technology, College of Agriculture & Natural Resources, University of Tehran, Karaj, Tehran, Iran

National Elites Foundation, Tehran, Tehran, Iran
e-mail: Falah@ut.ac.ir

O. Bozorg Haddad (✉)
Department of Irrigation & Reclamation Engineering, Faculty of Agricultural Engineering & Technology, College of Agriculture & Natural Resources, University of Tehran, Karaj, Tehran, Iran
e-mail: OBHaddad@ut.ac.ir

A.H. Gandomi et al. (eds.), *Handbook of Genetic Programming Applications*,
DOI 10.1007/978-3-319-20883-1_3

In the water engineering, there are several successful metaheuristic algorithm applications in general (e.g. Yang et al. 2013a, b; Gandomi et al. 2013) and GP in particular. Sivapragasam et al. (2009), Izadifar and Elshorbagy (2010), Guven and Kisi (2011), and Traore and Guven (2012, 2013) applied different GP versions to find best evaporation or evapotranspiration values with minimum difference from real values. Urban water management is other GP application field in which monthly water demand has forecasted by lags of observed water demand. Nasseri et al. (2011) applied GP for achieving an explicit optimum formula. These results can help decision makers of water resources to reduce their risks of online water demand forecasting and optimal operation of urban water systems (Nasseri et al. 2011). Li et al. (2014) extracted operational rules for multi-reservoir system by GP out of mathematical model. They used following steps to find operational rules: (1) determining the optimal operation trajectory of the multi-reservoir system using the dynamic programming to solve a deterministic long-term operation model, (2) selecting the input variables of operating rules using GP based on the optimal operation trajectory, (3) identifying the formulation of operating rules using GP again to fit the optimal operation trajectory, (4) refining the key parameters of operating rules using the parameterization-simulation-optimization method (Li et al. 2014). Results showed the derived operating rules were easier to implement for practical use and more efficient and reliable than the conventional operating rule curves and ANN rules.

Hydrology is a field of water engineering that focuses on the quantity and quality of water on Earth and other planets. In the scientific hydrologic studies, formation, movement and distribution of water are considered in hydrologic cycle, water resources and environmental watershed sustainability. The Earth is often called "blue planet" because of water distribution on its surface that appears blue from space. The total volume of water on Earth is estimated at 1.386 billion km^3 (333 million cubic miles), with 97.5 % and 2.5 % being salt and fresh water, respectively. Of the fresh water, only 0.3 % is in liquid form on the surface (Eakins and Sharman 2010). Due to, the key role of freshwater in life and different limitations of available water on the Earth, appropriate accuracy on hydrology models is necessary. On the other hand, increasing accuracy needs more data and application of expand conceptual methods in the hydrology models. Thus, GP have been applied as a popular, simple and user-friendly tool. This tool can summarize complex methods in a black-box process without modeling all system details. The purpose of this chapter is to assess the state of the art in GP application in hydrology problems.

3.2 Genetic Programming

GP is a data-driven model which borrows a random iterative searching base from evolutionary algorithms and move toward optimal solution (optimal relation) using advantage of these algorithms. Evolutionary algorithm is a subfield of artificial intelligence that involves combinatorial optimization and uses in the different fields

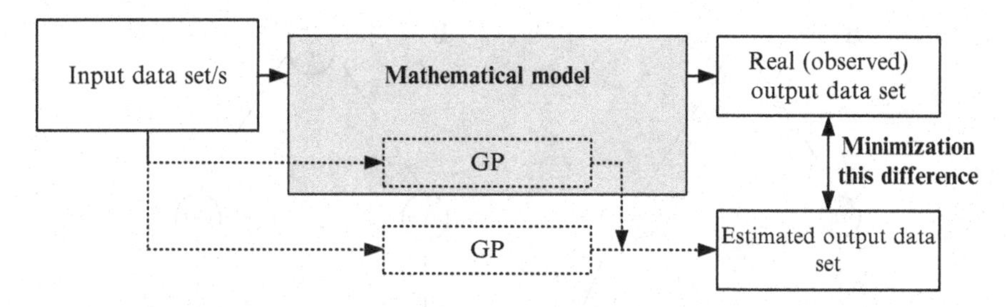

Fig. 3.1 GP presentation in the mathematical models

of water management considering single- and multi-objective. In the recent decades, there is a considerable growth in the development and improvement of evolutionary algorithms and application of hybrid algorithms to increase convergence velocity and find near-optimal solution.

Although, some new developed hybrid algorithms are capable to derive optimal solution, the decision variables have been considered only among the numerical variables. Thus, these algorithms present optimal value and not optimal equations. GP is one of the evolutionary algorithms, in which mathematical operators and functions are added to the numerical values as decision variables.

As shown in Fig. 3.1, GP equation can stand in or out of mathematical model to minimize difference between real (observed) and estimated output data set.

If GP equation presents in mathematical model, it will determine a constraint. In contrast, if GP equation is out of mathematical model, it will play a black-box role which can replace with mathematical model.

In evolutionary algorithms, each decision variable is called a gene, particle, frog and bee in the genetic algorithm (GA), particle swarm optimization (PSO), shuffled frog leaping algorithm (SFLA) and honey bees mating optimization (HBMO) algorithm and a set of aforementioned points with a fixed length is identified as solutions. However, in GP, the solutions have a tree structure which can include different numbers of decision variables and can produce a mathematical expression. Every tree node has an operator function and every terminal node has an operand, necessitating the evaluation of mathematical and logical expressions (Fallah-Mehdipour et al. 2012).

Figure 3.2a, b present two trees in the GP. As it is shown, in a tree structure, all the variables and operators are assumed to be the terminal and function sets, respectively.

Thus, $\{x, y, 47\}$ and $\{x, y\}$ are the terminal sets and $\{\sin, +, /\}$ and $\{\exp, \cos, /\}$ are the function sets of Fig. 3.2a, b, respectively. In the GP structure, the length of the tree creates the formula called depth of tree. The larger number of depth of tree, the more accuracy of the GP relation (Orouji et al. 2014). The GP searching process starts generating a random set of trees in the first iteration as same as other evolutionary algorithms. An error performance which is commonly assumed such as root mean squared error (RMSE) or mean absolute error (MAE) is then calculated. Thus, the error performance corresponds obtained objective function.

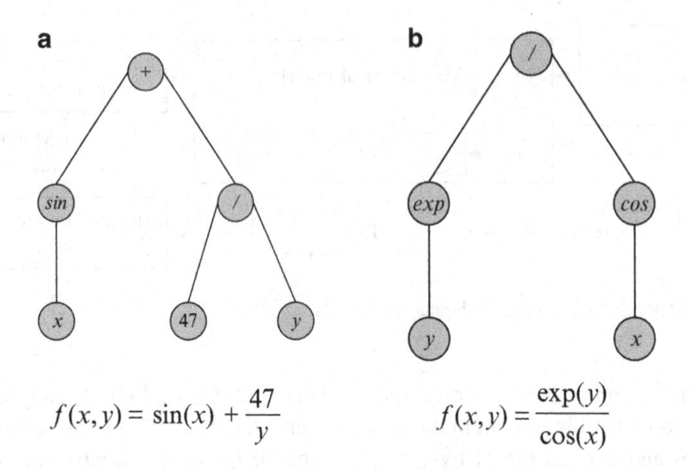

Fig. 3.2 Two GP expressions in the tree structure

To generate the next tree set, trees with the better fitness values are selected using techniques such as roulette wheel, tournament, or ranking methods (Orouji et al. 2014). In following, crossover and mutation as the two genetic operators as same as GA operators create new trees using the selected trees. In the crossover operator, two trees are selected and sub-tree crossover randomly (and independently) selects a crossover point (a node) in each parent tree. Then, two new trees are produced by replacing the sub-tree rooted at the crossover point in a copy of the first parent with a copy of the sub-tree rooted at the crossover point in the second parent, as illustrated in Fig. 3.3 (Fallah-Mehdipour et al. 2012).

In the mutation operator, point mutation is applied on a per node basis. That is, some node/s are randomly selected, it is exchanged by another random terminal or function, as it is presented in Fig. 3.4. The produced trees using genetic operators are the input trees for the next iteration and the GP process continues up to a maximum number of iterations or minimum of error performance.

3.3 GP Application in Hydrology Problems

GP is a data-driven model based on a tree-structured approach presented by Cramer (1985) and Koza (1992, 1994). This method belongs to a branch of evolutionary algorithm, based on the GA, which presents the natural process of struggle for existence. There are two approaches to apply GP in water problems: (1) outer and (2) inner mathematical model. In the first approach, GP extracts system behavior by using some or all characteristics without focus on the system modeling. In contrast, in the second approach, the derived equation by GP uses in system modeling as same as other basic equations. In this section, some applications of aforementioned approaches have been considered.

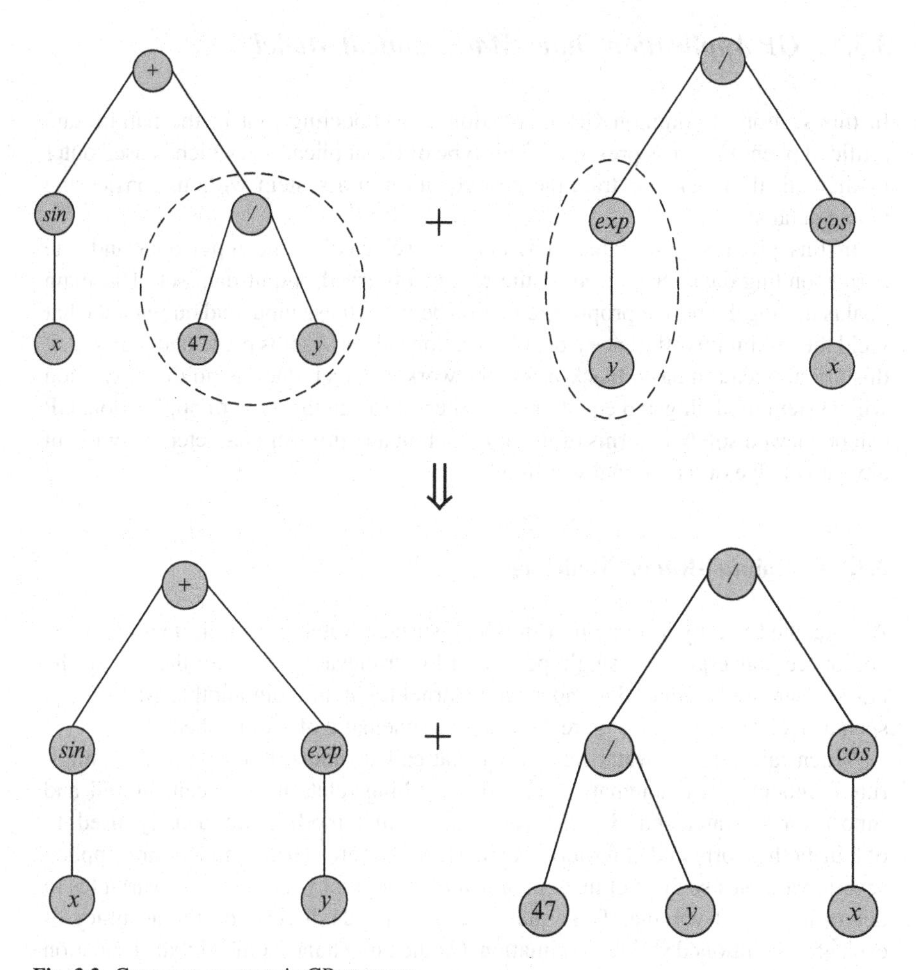

Fig. 3.3 Crossover operator in GP structure

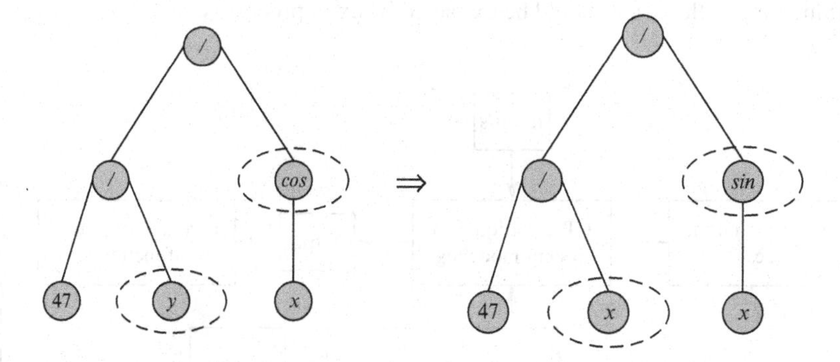

Fig. 3.4 Mutation operation in GP structure

3.3.1 GP Application Outer Mathematical Model

In this section, a common GP application as a modeling tool in the natural and artificial phenomenon is presented. This type of GP applications which is used outer mathematical model to extract the best equation in a system without considering whole details.

In this process, some characteristic/s are selected as the input data and one corresponding data set is used as the real or observed output data set. The main goal is finding the best appropriate equation between these input and output data that yield the minimum difference from observation values. As it is presented in Fig. 3.5, this GP application has a black-box framework in which there is no direct relation with system modeling and equations. In other words, in this type of application, GP can be viewed solely in terms of its input, output and transfer characteristic without any knowledge of its internal working.

3.3.1.1 Rainfall-Runoff Modeling

A watershed is a hydrologic unit in which surface water from rain, melting snow and/or ice converges to a single point at a lower elevation, usually the exit of the basin. Commonly, water that moves to external point and join another water body, such as river, lake or sea. Figure 3.6 presents schematic of a watershed.

When rain falls on watershed, water that called runoff, flows on it. A rainfall-runoff model is a mathematical model describing relations between rainfall and runoff for a watershed. In this case, conceptual models are usually used to obtain both short- and long-term forecasts of runoff. These models are applied several variables such as climate parameters, topography and land use variables to determine runoff volume. Thus, that volume depends directly on the accuracy of each aforementioned variable estimation. On the other hand, some global circulation model (GCM) that is used for runoff calculation apply for large scale and runoff volume for smaller scale should be extracted by extra processes.

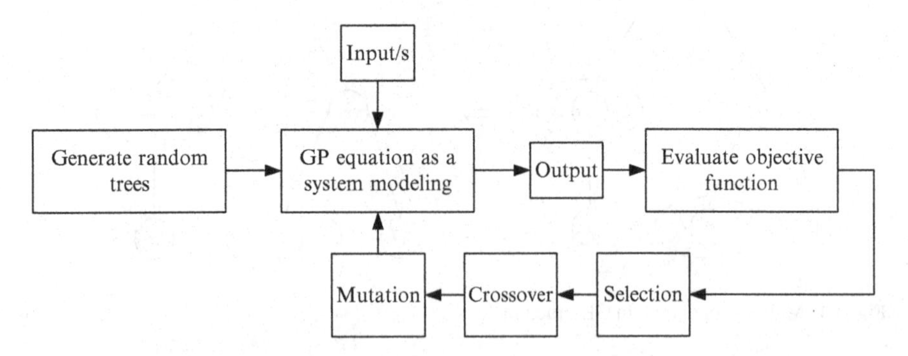

Fig. 3.5 GP framework in the outer mathematical model

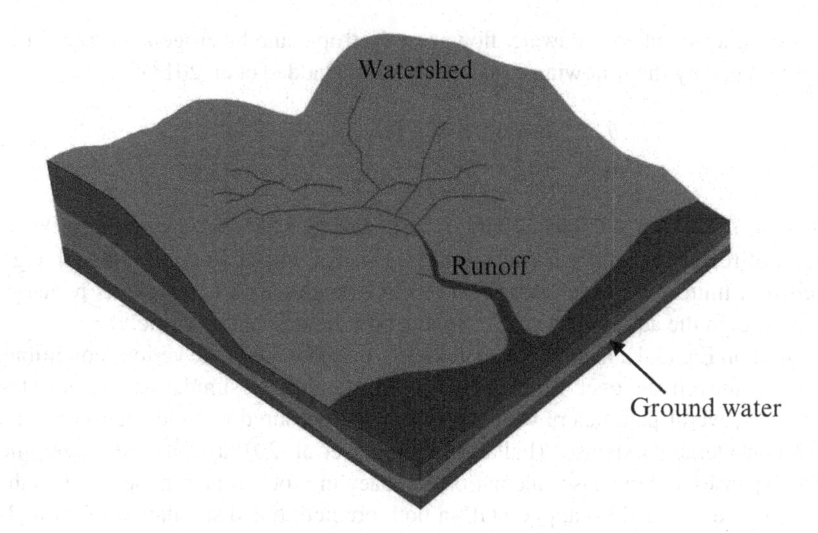

Fig. 3.6 Schematic of a watershed

Although conceptual models can calculate runoff for a watershed, their processes are long and expensive. Therefore, to overcome these problems, Savic et al. (1999) applied GP to estimate runoff volume for Kirkton catchment in Scotland.

Rainfall on the Kirkton catchment is estimated using a network of 11 period gauges and 3 automatic weather stations at different altitudes. The daily average rainfall is calculated from weighted domain areas for each gauge. Stream flow is measured by a weir for which the rating has been adjusted after intensive current metering (Savic et al. 1999). They compared obtained results with HYRROM, one conceptual model by Eeles (1994) that applied 9 and 35 parameters for runoff estimation considering different land use variables. Moreover, GP employed different combinations rainfall, runoff and evaporation for one, two and three previous periods and rainfall at current period as the input data to estimate runoff of current period as the output data. Results showed that GP can present better solution even by fewer input data sets than other conceptual models by Eeles (1994).

3.3.1.2 Groundwater Levels Modeling

When rain falls, extra surface water and runoff moves under earth and forms groundwater. In groundwater, soil pore spaces and fractures of rock formations fill from water and called an aquifer. The depth at which soil pores and/or fractures become completely saturated with water is water table or groundwater level.

Groundwater contained in aquifer systems is affected by various processes, such as precipitation, evaporation, recharge, and discharge. Groundwater level is typically measured as the elevation that the water rises in, for example, a test well.

Two-dimensional groundwater flow in an isotropic and heterogeneous aquifer is approximated by the following equation (Bozorg Haddad et al. 2013):

$$\frac{\partial}{\partial x}\left(T\frac{\partial h}{\partial x}\right)+\frac{\partial}{\partial y}\left(T\frac{\partial h}{\partial y}\right)\pm W=Sy\frac{\partial h}{\partial t} \tag{3.1}$$

in which, T = aquifer transmissivity; h = hydraulic head; Sy = storativity; W = the net of recharge and discharge within each a real unit of an aquifer model, e.g., a cell in a finite-difference grid; W is positive (negative) if it represents recharge (discharge) in the aquifer; and x, y = spatial coordinates, and t = time.

Based on Eq. (3.1), mathematical models are used to simulate various conditions of water movement over time. However, mathematical simulation necessitates values of several parameters which may not be measured or their measurements incur considerable expenses (Fallah-Mehdipour et al. 2013a). Thus, to overcome those expenses and increase calculation accuracy in groundwater modeling, Fallah-Mehdipour et al. (2013a) applied GP in both prediction and simulation of groundwater levels. Results of the prediction and simulation process respectively help determining unknown and missed data in a time series. In order to modeling, three observation well of Karaj aquifer with water level variation in a 7-year (84-month) period have been considered. This aquifer is recharged from precipitation and recharging wells. To judge fairly about GP capabilities in groundwater modeling, results of the GP have been compared with adaptive neural fuzzy inference system (ANFIS). Results showed that GP yields more appropriate results than ANFIS when different combinations of input data sets have been employed in both prediction and simulation processes.

3.3.2 GP Application in Inner Mathematical Model

In this section, reservoir presents as an example of hydro systems in which GP is applied in mathematical model. In this model, GP is extracted operational rule as a constraint that illustrates when and how release water from reservoir.

Reservoirs are one of the main water structures which operate for several purposes, such as supplying downstream demands, generating hydropower energy, and flood control. There are several investigations in the short, long, and integrating short and long term (e.g., Batista Celeste et al. 2008) reservoir operation without considering any operational decision rules (Fallah-Mehdipour et al. 2013b). In these investigations, released water from reservoir is commonly identified as the decision variable.

The result of this type of operation is only determined for the applied time series. In order to operate a reservoir system in real-time, an operational decision rule can be used in reservoir modeling which helps the operator to make an appropriate decision to calculate how much (amount) and when (time) to release water from the reservoir.

To determine a decision rule, a general mathematical equation is usually embedded in the simulation model:

$$R_t = F_1 (S_t, Q_t) \tag{3.2}$$

in which, R_t, S_t and Q_t are release, storage and inflow at t^{th} period. Moreover, F_1 is linear or nonlinear function for transferring storage volume and inflow to the released water from the reservoir at each period.

The common pattern of aforementioned decision rule which is a linear decision rule that a, b and c are the decision variables (e.g., Mousavi et al. 2007; Bolouri-Yazdeli et al. 2014):

$$R_t = a \times Q_t + b \times S_t + c \tag{3.3}$$

Although, application of Eq. (3.3) as a decision rule is useful in real-time operation, this rule has a pre-defined linear pattern. It is possible to exist some decision rules with other mathematical frame (not just linear). GP can extract an embed equation in this reservoir model without any assumed pattern which is adapted with storage and inflow and their fluctuations at each period.

Moreover, the aforementioned rule involves Q_t needs commonly a prediction model may be coupled with decision rule to estimate inflow as a stochastic variable. Inappropriate selection of this prediction model increases calculations and impacts the reservoir operation efficiency (Fallah-Mehdipour et al. 2012). To overcome this inappropriate selection, GP can find a flexible decision rule which develops a reservoir operation policy simultaneously with inflow prediction. In this state, GP which presented its capability in inflow prediction, has been used as the reservoir simulation tool and two operational rule curves including water release, storage volume, and previous inflow/s (not in the current period (t)) are extracted.

Fallah-Mehdipour et al. (2012, 2013b) applied the GP application considering inflow of the current and previous periods. In these investigations, GP tries to close released water from reservoir to the demand by using different functions and terminals in the decision rule. Thus, GP rules presented a considerable improvement compare to the common linear decision rule.

Figure 3.7 presents GP framework in the real-time operation of reservoir. As it is shown, the random trees are generated in the first iteration. These trees are decision rules which explain a mathematical function including inflow, storage and release.

Accordingly, decision rule is embedded in the reservoir operation model and the released water from reservoir is calculated using continuity equation and limited constraint storage volume between minimum and maximum allowable storage ($S_{Min} < S_t < S_{Max}$). Then, the objective function yields considering minimization of deficit and maximization of generated energy in the supplying downstream demand and hydropower energy generation purpose, respectively. To find released water and storage in a feasible range, the constraints are considered in the optimization

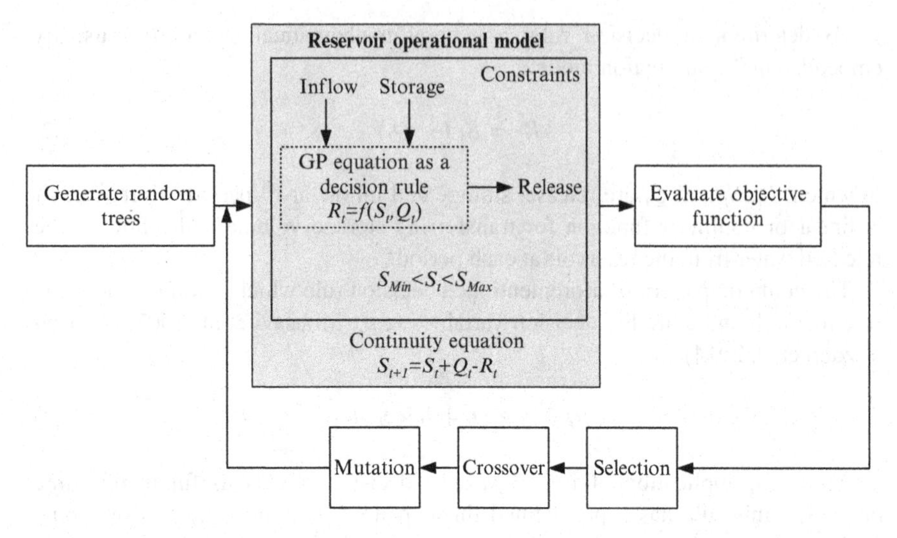

Fig. 3.7 GP framework in the real-time operation of reservoir

process by penalty. This penalty is added and subtracted in the minimization and maximization objective for each violation unit from feasible bound. The other GP process (selection, crossover and mutation) are continues to satisfy stopping criteria.

3.4 Concluding Remarks

There are many investigations that present successful application, development and adaptation of GP in the water engineering and hydrology. This chapter reviewed these investigations considering different aspects of GP application in the mathematical models that can be inner and outer of system modeling. Inner system modeling such as decision operational rule uses GP equation in the modeling process as same as other system equations. Thus, the output which is released water in reservoir system is adapted to the GP equation. In contrast, the outer mathematical model is widely used for developing an optimal existing relation between input and output data in water resources in a black-box method. In both aforementioned methods, GP illustrated appropriate solution and can be recommended for the future studies, because some highlight reasons:

- Appropriate capability to use in and out of models.
- Predict and simulate some phenomenon with a considerable fluctuation especially in the extreme bounds.
- Easy link with other models, softwares, and optimization techniques.

Acknowledgement Authors thank Iran's National Elites Foundation for financial support of this research.

References

Batista Celeste, A., Suzuki, K., and Kadota, A. (2008). "Integrating long-and short-term reservoir operation models via stochastic and deterministic optimization: case study in Japan." Journal of Water Resources Planning and Management (ASCE), 134(5), 440–448.

Bolouri-Yazdeli, Y., Bozorg Haddad, O., Fallah-Mehdipour, E., and Mariño, M.A. (2014). "Evaluation of real-time operation rules in reservoir systems operation." Water Resources Management, 28(3), 715–729.

Bozorg Haddad, O., Rezapour Tabari, M.M., Fallah-Mehdipour, E., and Mariño, M.A. (2013). "Groundwater model calibration by meta-heuristic algorithms." Water Resources Management, 27(7), 2515–2529.

Cramer, N.L. (1985). "A representation for the adaptive generation of simple sequential programs." In Proceedings of an International Conference on Genetic Algorithms and the Applications, Grefenstette, John J. (ed.), Carnegie Mellon University. 24-26 July, 183–187.

Eakins, B.W. and Sharman, G.F. (2010). Volumes of the World's Oceans from ETOPO1, NOAA National Geophysical Data Center, Boulder, CO, 2010.

Eeles CWO: (1994) Parameter optimization of conceptual hydrological models, PhD Thesis, Open University, Milton Keynes, U.K.

Fallah-Mehdipour, E., Bozorg Haddad, O., and Mariño, M. A. (2012). "Real-time operation of reservoir system by genetic programming." Water Resources Management, 26(14), 4091–4103.

Fallah-Mehdipour, E., Bozorg Haddad, O., and Mariño, M. A. (2013a). "Prediction and Simulation of Monthly Groundwater level by Genetic Programming." Journal of Hydro-environment Research, 7(4), 253–260.

Fallah-Mehdipour, E., Bozorg Haddad, O., and Mariño, M. A. (2013b). "Developing reservoir operational decision rule by genetic programming." Journal of Hydroinformatics, 15(1), 103–119.

Gandomi, A.H., Yang, X.S., Talatahari, S., and Alavi, A. H. (2013). "Metaheuristic Applications in Structures and Infrastructures" Elsevier. 568 pages.

Guven, A., and Kisi, O. (2011). "Daily pan evaporation modeling using linear genetic programming technique." Irrigation Science, 29(2), 135–145.

Izadifar, Z., and Elshorbagy, A. (2010). "Prediction of hourly actual evapotranspiration using neural network, genetic programming, and statistical models." Hydrological Processes, 24(23), 3413–3425.

Koza, J. R. (1992). Genetic programming: on the programming of computers by means of natural selection. MIT Press, Cambridge, MA.

Koza, J. R. (1994). Genetic Programming II: Automatic Discovery of Reusable Programs. MIT 293 Press. Cambridge, MA.

Li, L., Liu, P., Rheinheimer, D.E., Deng, C., and Zhou, Y. (2014). "Identifying explicit formulation of operating rules for multi-reservoir systems using genetic programming." Water Resources Management, 28(6), 1545–1565.

Mousavi, S. J., Ponnambalam, K., and Karray, F. (2007). "Inferring operating rules for reservoir operations using fuzzy regression and ANFIS." Fuzzy Sets and Systems, 158(10), 1064–1082.

Nasseri, M., Moeini, A., and Tabesh, M. (2011). "Forecasting monthly urban water demand using extended Kalman filter and genetic programming." Expert Systems with Applications, 38(6), 7387–7395.

Orouji, H., Bozorg Haddad, O., Fallah-Mehdipour, E., and Mariño, M.A. (2014). "Flood routing in branched river by genetic programming." Proceedings of the Institution of Civil Engineers: Water Management, 167(2), 115–123.

Savic, D. A., Walters, G. A., and Davidson, J. W. (1999). "A genetic programming approach to rainfall-runoff modeling." Water Resources Management, 13(3), 219–231.

Sivapragasam, C., Vasudevan, G., Maran, J., Bose, C., Kaza, S., and Ganesh, N. (2009). "Modeling evaporation-seepage losses for reservoir water balance in semi-arid regions." Water Resources Management, 23(5), 853–867.

Traore, S., and Guven, A. (2012). "Regional-specific numerical models of evapotranspiration using gene-expression programming interface in Sahel." Water Resources Management, 26(15), 4367–4380.

Traore, S., and Guven, A. (2013). "New algebraic formulations of evapotranspiration extracted from gene-expression programming in the tropical seasonally dry regions of West Africa." Irrigation Science, 31(1), 1–.10.

Yang, X.S., Gandomi, A.H., Talatahari, S., and Alavi, A. H. (2013a). "Metaheuristis in Water, Geotechnical and Transportation Engineering" Elsevier. 496 pages

Yang, X.S., Cui, Z., Xiao, R., Gandomi, A.H., and Karamanoglu, M. (2013b). "Swarm Intelligence and Bio-Inspired Computation: Theory and Applications", Elsevier. 450 pages.

Chapter 4
Application of Gene-Expression Programming in Hydraulic Engineering

A. Zahiri, A.A. Dehghani, and H.Md. Azamathulla

4.1 Introduction

Hydraulic engineering as a sub-discipline of civil engineering is the application of fluid mechanics principles to problems dealing with the collection, storage, control, transport, regulation, measurement, operation, and use of water (Prasuhn 1987). In other words, hydraulic engineering is the application of fluid mechanics and other science and engineering disciplines in the design of structures, and the development of projects and systems involving water resources (Roberson et al. 1998). An interesting believe for hydraulic engineering is from Liggett (2002) who defines this term as clearly a field for those who love nature and who are comfortable in applying the laws of fluid mechanics for the betterment of mankind while preserving nature. Familiar applications of hydraulic engineering are water supply and distribution systems, flood protection, flood hazard mapping, erosion protection, transport modeling of pollutants in surface water, irrigation, navigation, water quality modeling and environmental evaluation of projects. This broad field covers many aspects ranges from closed conduit (pipe, pump) to open channels (river, canal, lake, estuary, and ocean). However, civil engineers are primarily concerned with open channel flow, and especially natural rivers.

Electronic supplementary material The online version of this chapter (doi: 10.1007/978-3-319-20883-1_4) contains supplementary material, which is available to authorized users.

A. Zahiri (✉) • A.A. Dehghani
Department of Water Engineering, Gorgan University of Agricultural Sciences and Natural Resources, Gorgan, Iran
e-mail: zahiri_reza@yahoo.com; zahiri@gau.ac.ir; a.dehghani@gau.ac.ir

H.Md. Azamathulla
Associate Professor of Civil Engineering, Faculty of Engineering, University of Tabuk, Tabuk, Saudi Arabia
e-mail: mdazmath@gmail.com

A.H. Gandomi et al. (eds.), *Handbook of Genetic Programming Applications*,
DOI 10.1007/978-3-319-20883-1_4

The areas of theoretical, experimental and computational hydraulics have much progress in various field of application. However, each of these sub-groups has some individual or in many cases inter-dependent difficulties. Theoretical hydraulics generally is beyond of the scientific extent and trades of hydraulic and river engineers. Experimental works need large space and facilities. Finally, computational hydraulics' progress depends on the growth of theoretical aspects in hydraulic field. These features have caused limiting progress in these fields. On the other hands, evolutionary algorithms such as genetic algorithm, genetic programming and gene-expression programming have considerable progress and development through recent years. These algorithms known as soft computing techniques, with less complexity and cost, have received much attention by researchers in many fields of science and engineering (Guven and Gunal 2008; Azamathulla et al. 2010; Azamathulla and Zahiri 2012; Guven and Azamathulla 2012; Azamathulla and Jarrett 2013; Najafzadeh et al. 2013; Sattar 2014; Onen 2014). The soft computing includes the concepts and techniques to solve or overcome the difficulties in the real world especially in engineering sciences (Gandomi and Alavi 2011, 2012).

Guven and Azamathulla (2012) presented the following relations for estimation of maximum scour depth (d_s), width (w_s) and location (l_s) at the downstream of the flip bucket spillway by using GEP, respectively:

$$\frac{d_s}{d_w} = \left[\frac{d_{50}+q}{\phi} - (H_1 - q - 1.199)\left(5.616\frac{d_{50}}{\phi}\right)\right] \times \left[0.309\phi(R + (d_{50}+0.185)(H_1 d_{50}))^{-0.5}\right] \tag{4.1}$$

$$\frac{w_s}{d_w} = \left[\frac{q - 0.006^{d50} + 1.168^{R+H_1}}{2.336\phi}\right]^{0.5}\left[\frac{7.521d_{50} + 3.955H_1 - 2q}{0.428\phi^{-1}} + 15.42\phi\right]^{0.5} \tag{4.2}$$

$$\frac{l_s}{d_w} = \left[\frac{e^{\phi}}{R}\left(H_1 - d_{50} + 0.495 + 2.878(q)^{-1}\right)\right] \left[R(q + H_1 - 9.948d_{50})^{0.5}(2H_1 + q + \phi)\right] \tag{4.3}$$

In which q is unit discharge over the spillway, H_1 is total head, R is radius of the bucket, ϕ is lip angle of the bucket, d_w is tail water depth, d_{50} is median sediment size and g is acceleration due to gravity. Results of these equations were compared with the regression equation formulae and neural network approach. The comparison revealed that the GEP models (Eqs. 4.1–4.3) have higher accuracy.

Mujahid et al. (2012) used GEP for estimation of bridge pier scour. The following explicit relation was obtained and its performance was compared with artificial neural networks (ANNs) and conventional regression-based techniques. The results showed that GEP gives more accurate results than the other models.

$$\frac{d_s}{Y} = \frac{b}{Y}\left(0.595 - F_r - \left(\frac{d_{50}}{Y}\frac{b}{Y}\right)^2\right) + F_r\left(F_r + 0.063 - \frac{d_{50}}{Y}\right) - \left(\frac{d_{50}}{Y}\right)^2(\sigma - 1) + F_r\left(\frac{b}{Y}\left(F_r - \left(\frac{3.24^{d_{50}}/Y}{(F_r - 1)}\right)\right)\right) \tag{4.4}$$

In the above equation, Y is approach flow depth, Fr is Froude number, b is pier width and σ is standard deviation of particle grain size distribution.

Wang et al. (2013) using GEP model, presented Eq. (4.5) for estimating pier scours depth based on available experimental data from various researches. Four main dimensionless parameters such as pier width (D/d_{50}), approaching flow depth (Y/d_{50}), threshold flow velocity ($(V^2-V_c^2)/(\Delta g d_{50})$), and pier scour depth (d_s/D) were used as independent variables in Eq. (4.5).

$$\frac{d_s}{D} = \left(\left(\frac{V^2 - V_c^2}{\Delta g d} \Big/ \frac{D}{d_{50}}\right)\left(\frac{Y}{d_{50}} \Big/ \alpha\right)\right)^{\log\left(\frac{D}{d_{50}} \Big/ \frac{V^2 - V_c^2}{\Delta g d_{50}}\right)} \times \left(\left(\frac{V^2 - V_c^2}{\Delta g d_{50}}\right)^{2.05}\left(\frac{D}{d_{50}} \Big/ 4.94\right) \Big/ \left(\frac{D}{d_{50}}\alpha\right)^{\log\left(\frac{D}{d_{50}} \Big/ 1.44\right)}\right)^{0.156} \tag{4.5}$$

where D is pier width or diameter, Y is approaching flow depth, V is average approaching flow velocity, $\Delta = (\rho_s/\rho_w)$-1 is the relative submerged density of sediment which ρ_w and ρ_s are density of water and sediment, respectively, and α is channel open-ratio. Analysis of the above equation results showed the high capability of GEP model.

Moussa (2013) used GEP for estimation of scour depth downstream of stilling basin through a trapezoidal channel. The performance of GEP approach was compared with other modeling techniques such as artificial neural networks (ANNs) and multiple linear regression (MLR). The results showed that GEP gives significantly more accurate results than the ANN and MLR models.

Azamathulla et al. (2011) used GEP and developed stage-discharge (S-Q) relationship for the River Pahang as follows:

$$Q = 9.84S^2 - 64.391S - 4033.296 \tag{4.6}$$

The results showed that GEP as an effective tool can be used for estimating of daily discharge data in flood events. For developing flow rating curves, also Guven and Aytek (2009) used GEP technique in two stations of Schuylkill River (Pennsylvania). The performance of GEP approach was compared with more conventional methods, common stage rating curve (SRC) and multiple linear regression (MLR) techniques. The results showed that GEP gives more accuracy results than SRC and MLR models. Equations (4.7) and (4.8) were obtained for discharge as a function of stage for upstream (Berne) and downstream stations (Philadelphia), respectively:

$$Q = 10.313S^{1.5} + 4.738S^{-6} - 27.743 \tag{4.7}$$

$$Q = 2S - 4.925S^2 + 54.421(2S - 4.715/S)^2 - 8.349 \tag{4.8}$$

Zakaria et al. (2010) used GEP model to predict total load transport in three rivers (i.e., Kurau, Langat, and Muda). The explicit formulation of GEP for total bed material load is presented in Eq. (4.9):

$$\begin{aligned} Q_s = {} & \left[\left(-0.39R_hY_0\sqrt{S_0}\right) / \left(-0.72 + S_0\right)\right] + \left(R_h + e^{Sin(QVR_h)}\right. \\ & \left. + Tan^{-1}\left(-0.16R_hB\right) + R_h\sqrt{Q}\right) + \left(d_{50} - 3.39\right) d_{50}^3 S_0 + Tan^{-1}(V)\, e^V \\ & - \log\left(6.93 - Y_0\right)\left((\omega B) / (-2.075)\right) \end{aligned} \tag{4.9}$$

In which, Q_s is total load, B is river width, Y_0 is flow depth, R is hydraulic radius, Q is the flow discharge, S_0 is river slope and ω is particle fall velocity.

Azamathulla and Jarrett (2013) using field measurements data presented the following Equation for estimation of the manning's roughness coefficient for high gradient streams:

$$n = 3S_f - \left[1.87S_f\left(d_{84} - R_h + d_{84}^2\right)\right] - \left(9.13\frac{S_f^{\,2}}{R_h}\right) + \frac{\left(d_{84} - S_f\right)^{0.25}}{\left(26.2 - 4.68d_{84}\right)} \tag{4.10}$$

in which n is the Manning's roughness coefficient, S_f is the energy gradient or friction slope, and d_{84} is streambed particle size. The results showed that GEP presents more accurate results than the Jarrett's (1984) equation.

From the literatures it is found that GEP technique has recently received much attention by researchers in the field of hydraulic engineering. In this book chapter, some of soft computing techniques' applications on hydraulic engineering have been presented. In this regards, we have mainly focused on gene-expression programming.

4.2 Material and Methods

4.2.1 Gene-Expression Programming

Inspired by Darwin's theory of natural evolution and motivated by the development of computer technologies, Evolutionary Computation (EC) was introduced in the 1960s as a robust and adaptive search method. This technique is capable of solving complex problems that the traditional algorithms have been unable to conquer

(Sharifi 2009). The branch of Genetic based algorithms states that the survival of an organism is affected by the rule of "survival of the strongest species". This family of algorithms is often viewed as function optimizers, although the range of problems to which genetic algorithms have been applied is quite broad (Whitley 1991). Among the different methods belong to this family, the Gene expression programming (GEP) is the up-to-date technique. GEP was invented by Ferreira in 1999, and is the natural development of EAs. The great insight of GEP was the invention of chromosomes capable of representing any expression tree; GEP surpasses the genetic programming (GP) system by a factor of 60,000 (Ferreira 2001). In GEP, complex relations are encoded in simpler, linear structures of a fixed length called chromosomes. The chromosomes consist of a linear symbolic string of a fixed length composed of one or more genes (Sattar 2014). To express the genetic information encoded in the gene, Ferreira (2001) used expression tree (ET) representations. Due to the simple rules that determine the structure of the ET, it is possible to infer the gene composition given the ET and vice versa using the unequivocal Karva language. The Karva language represents genes in a sequence that begins with a start codon, continues with amino acid codons, and ends with a termination codon (Sattar 2014). Consider, for example, the following mathematical expression:

$$z = e^{\pi \cos(x) + \frac{\sin(x)}{\cos(y)}} \tag{4.11}$$

This mathematical form can be represented as an ET (Fig. 4.1):

In this example x, y and π are the set of terminals or the variables used in the example definition; and the basic mathematical operators of +, ×, /, sin, cos and exp are the rules (functions) that determine the spatial organization of the terminals.

The characteristics of the best chromosomes and the evolutionary strategy of the GEP have been explained in detail by many researchers. For complete details on GEP and the related genetic operations, interested readers can refer to Ferreira

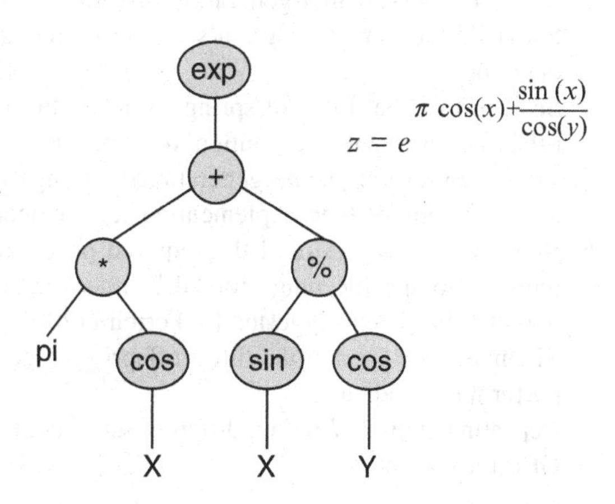

Fig. 4.1 Expression tree for a mathematical example (Eq. 4.11)

(2001) and Sattar (2014). GEP fitting computations for experimental or field data is can be performed using some commercial nonlinear data-mining softwares such as GeneXProTools (www.gepsoft.com). The fitness function of a program f_i in GEP is:

$$f_i = 1000\frac{1}{1 + RRSE_i} \tag{4.12}$$

The fitness function ranges from 0 to 1000, with 1000 corresponding to a perfect fit. In the above equation, RRSE is root relative square error of an individual program i (i-th offspring) and is defined by the following equation:

$$RRSE_i = \sqrt{\frac{\sum_{j=1}^{n} |Y_{ij} - X_j|^2}{\sum_{j=1}^{n} |\overline{X} - X_j|^2}} \tag{4.13}$$

where X and Y, respectively, are the actual and predicted targets, Y_{ij} is the value predicted by the program i for fitness case j, X_j is the target value for fitness case j, $\overline{X}$ is the average of the measured outputs (X) and n is the number of samples (Sattar 2014).

4.2.2 Analysis Procedure for GEP Model Development

The following procedure has been used to develop the final GEP equation (Sattar 2014);

1. Choosing an initial set of control variables as terminals for GEP.
2. Defining the chromosome architecture (number of genes, head size, functions) and mutation rates for the initial work environment of GEP.
3. Producing several first-generation offspring by GEP through randomly formulation of the parent program's chromosomes and implementation of genetic operators.
4. Selection of the fittest offspring by using the fitness criteria (Eq. 4.12). This offspring represents the solution to the problem in the first generation.
5. Producing several second-generation offspring by GEP using the fittest offspring as new parent the then implementing genetic operators.
6. Repeating steps 3–5 until the required program fitness is met. Unfortunately, there is no specific range for GEP fitness indicator f_i, however, a domain of 600–800 has been suggested by Ferreira (2001) for suitable model predictions. The final GEP-model (the fittest offspring of generation i) is scored on a set of performance indicators.
7. Repeating steps 1–7 with a different set of control variables to produce another GEP-model.

4.2.3 GEP Model Evaluation by Statistical Measures

To evaluate the accuracy of the final explicit GEP equation in both training and testing phases, some common statistical measures including correlation coefficient (R), the mean squared error (MSE), the mean absolute error (MAE) and performance index (ρ), are used in this study as follows:

$$R = \frac{\sum xy}{\sqrt{\sum x^2 \sum y^2}} \tag{4.14}$$

$$MSE = \frac{\sum (X - Y)^2}{N} \tag{4.15}$$

$$MAE = \frac{\sum \frac{|X - Y|}{X}}{N} \tag{4.16}$$

$$\rho = \frac{\sqrt{MSE}}{\overline{X}} \frac{1}{1 + R} \tag{4.17}$$

where $x{=}(X - \overline{X})$, $y{=}(Y - \overline{Y})$, $\overline{X}$ is the mean of X (measured outputs), $\overline{Y}$ is the mean of Y (predicted outputs) and N is the data point's number for GEP evaluation (experimental data). The last statistical parameter (ρ) is a new criterion proposed by Gandomi and Roke (2013) which combines both correlation and error functions.

4.3 Applications

4.3.1 Main Channel and Floodplain Discharges in Compound Channels

Rivers are vital carriers of water and sediments. At extreme discharge conditions floods may occur that could damage nearby infrastructure and also cause casualties (Huthoff 2007). Over more than three decades, hydraulics of compound channels has been extensively investigated by many researchers.

Flow hydraulic characteristics are completely different in main channel and floodplains, and hence, it's necessary to treat the channel into subsections for any analysis and computation (Lambert and Myers 1998). For dividing of compound

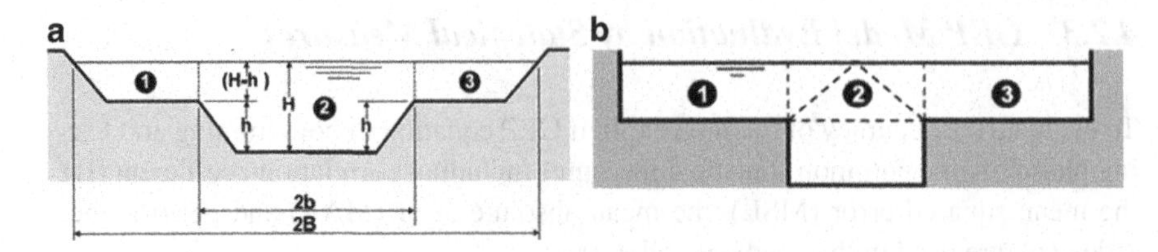

Fig. 4.2 A typical compound channel (**a**) with common dividing channel methods (**b**) (Bousmar 2002)

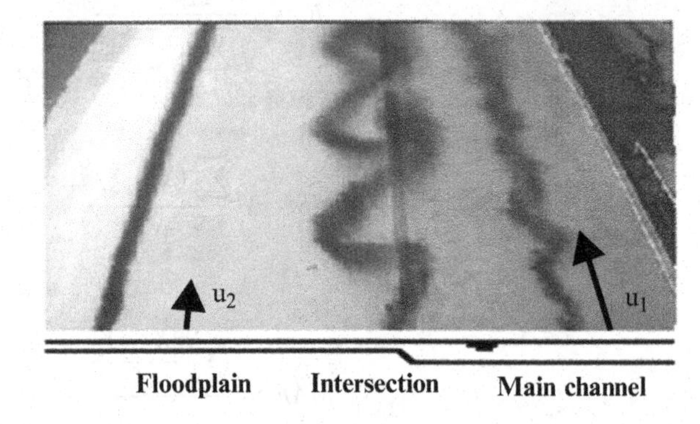

Fig. 4.3 Development of Strong lateral shear stress at the main channel/floodplain interface (Van Prooijen et al. 2005)

channels, there are three common approaches, vertical, horizontal and diagonal planes. For hydraulic modeling of river compound channels, the first approach has the most applications in one-dimensional commercial mathematical packages such as MIKE11, HEC-RAS, ISIS and SOBEC (Huthoff et al. 2008). However, this method has great over-prediction error for discharge estimation in field and laboratory compound sections.

In Fig. 4.2a, a typical compound channel with associated important parameters is shown. Also, in Fig. 4.2b, the division ways for vertical, horizontal and diagonal dividing channel methods have been illustrated. As mentioned by many researchers, it's assumed, in all these simple methods, that there is no shear stress and momentum transfer at the division lines between main channel and floodplains. Results of experimental works carried out in compound channels, have revealed that this assumption isn't correct and therefore, these methods are maybe very erroneous (Martin and Myers 1991; Ackers 1992).

Unreliability of dividing channel methods' assumption was demonstrated through Van Prooijen et al. (2005) experimental work. It's seen from Fig. 4.3, that due to lateral shear stress, a fully turbulent flow with high momentum transfer is induced at the main channel/floodplain interface. This shear stress is maybe comparable, in magnitude, with the bed shear stress. Furthermore, it's interesting to note that the flow velocity in the floodplain is considerably less than the main channel and it's insufficient to cause major movement of suspended sediment.

In accordance to large error of traditional divided channel methods, several modified approaches have been provided by many researchers (Wormleaton and

Merrett 1990; Ackers 1992; Bousmar and Zech 1999; Atabay 2001; Huthoff et al. 2008; Yang et al. 2014). Even though these methods perform well, but in most of these studies, the main aim is precise prediction of total flow discharge or average velocity. However, in many practical situations, distribution of flow rate in the main channel and the flood plains is also important. Ackers (1992) states that any suitable method should has such ability to predict flow discharge in subsections, especially in the main channel, with sufficient accuracy.

In overbank flows the river system not only behaves as a conveyance but also as a storage or pond. It is recognized that for sediment transport, only the flow discharge in main channel is effective and floodplain's discharge is nearly negligible (Ackers 1992). In fact the floodplains, due to their high capacities, play an important role in flood water level reduction, water retention and sediment deposition. These features are essential for wetlands restoration and preserve of river ecology as well as for success of flood mitigation works. The main channel flow discharge determination also covers the main input data for several hydraulic and morphologic computations such as pollutant dispersion, sediment transport and bed shear stress distribution in river compound channels.

The bank-full level is defined as the level at which the water has its maximum power to move sediment. In flood event and when the water rises above the bank-full level, flow spills onto the floodplain, which the average flow velocity and consequently stream power dramatically reduce. As the stream power is reduced, so too is its capacity sediment transport. Thus, for better monitoring of river behavior during flood events, accurate computation of flow velocity and hence sediment transport capacity of both main channel and floodplains are needed.

It should be noted that for computation of sediment transport capacity in flooded rivers, one initially should separate the main channel flow discharge from the total flow rate and then put it into a suitable empirical sediment transport equation.

Towards the finding suitable methods for prediction of main channel and floodplain discharges, first, the traditional methods are reviewed.

4.3.1.1 Divided Channel Methods

In Fig. 4.2b, main channel and floodplains sections separated by three dividing methods (e.g., horizontal, vertical, and diagonal) are shown. Total flow discharge is the sum of discharges calculated separately in each subsection using an appropriate conventional friction formula, for example, Manning's equation (Chow 1959):

$$Q_{DCM} = \sum_{i=1}^{3} Q_i = \sum_{i=1}^{3} \frac{A_i R_i^{2/3} S_0^{1/2}}{n_i} \tag{4.18}$$

where Q_{DCM} is total flow discharge in compound channel, A is area. In this equation, i refers to each subsection (main channel or floodplains).

4.3.1.2 Coherence Parameter

In a compound channel, the degree of interaction between main channel and floodplains depends on many factors, including relative depth of floodplain flow to main channel flow, width ratio between main channel and floodplain and relative roughness of floodplain to main channel and channel geometry (Ackers 1992). On this basis, Ackers (1992, 1993) introduced an important dimensionless parameter, named coherence parameter (COH):

$$COH = \frac{(1 + A_*)^{1.5} \Big/ \sqrt{\left(1 + P_*{}^{1.33} n_*{}^2 / A_*{}^{0.33}\right)}}{1 + A_*{}^{1.67} / n_* P_*{}^{0.33}} \tag{4.19}$$

where P is the wetted perimeter and * denotes the ratio of floodplain to main channel's value.

4.3.1.3 Data Used for Modeling

In this study, 102 laboratory stage-discharge data from 14 different compound channel sections were used among 72 were training data and the remaining 30 were taken as testing data. This data set include bank-full depth, bed slope, and main channel and floodplain characteristics such as width, side slope, flow discharge, flow depth and Manning roughness coefficient. These data are collected form experimental works carried out by HR Wallingford (FCF) in compound channel flumes with large-scale facility (Knight and Sellin 1987, www.flowdata.bham.ac.uk; Lambert and Myers 1998; Bousmar and Zech 1999; Bousmar et al. 2004; Fernandez et al. 2012). The ranges of geometric and hydraulic characteristics of compound channels used in this study are listed in Table 4.1.

Table 4.1 Overview of data sets used for development and assessment of GEP model

Variable definition	Variable range	Mean
Bank-full height, h (m)	0.05–0.2	0.103
Flow depth, H (m)	0.058–0.32	0.1482
Main channel width, $b_c(m)$	0.05–1.6	0.89
Floodplain width, $b_f(m)$	0.16–6	1.49
Bank-full discharge, $Q_b(m^3/s)$	0.0023–0.2162	0.096
Total flow discharge, $Q_t(m^3/s)$	0.003–1.1142	0.2145
Main channel flow discharge, $Q_{mc}(m^3/s)$	0.00233–0.6271	0.1499
Floodplain flow discharge, $Q_f(m^3/s)$	0.00046–0.6340	0.064
Bed slope	0.00099–0.013	0.0021

4.3.1.4 Selection of Input and Output Variables

For GEP model development, it is assumed, somewhat similar to Ackers' approach, that subsection's flow discharges are dependent on three input dimensionless parameters including depth ratio (floodplain depth to main channel depth, *Dr*), coherence parameter, and calculated flow discharge using vertical divided channel method. Accordingly, the following functions are proposed to predict the flow discharge both in main channel and floodplain:

$$Q_{mc} = f\,(Dr, COH, Q_{mc-VDCM}) \tag{4.20}$$

$$Q_{fp} = f\left(Dr, COH, Q_{fp-VDCM}\right) \tag{4.21}$$

where Q_{mc} and Q_{fp} are flow discharges in main channel and floodplain, respectively.

4.3.1.5 GEP Results

In Fig. 4.4a, b, results of three methods of dividing compound channels are shown for flow discharges in main channel and floodplains. As can be seen, for main channel discharge (Fig. 4.4a), vertical and horizontal approaches have still over and under predictions, respectively. Errors of these methods are growing with increasing flow discharges, especially for vertical method. Among these approaches, the diagonal planes, has a suitable result, although for large main channel's discharges, the errors are increasing. For floodplains (Fig. 4.4b), both vertical and diagonal dividing planes produce considerably better predictions than the horizontal case. It is interesting to note, that even for large discharges, these two methods have good results. Furthermore, for floodplains, the vertical divided method under-predicts the flow discharge, opposite to the main channels case. This is due to the interaction effect that causes the actual discharge to decrease in the main channel and increase in the floodplains. This flow exchange isn't considered in the vertical divided method, as well as for both horizontal and diagonal methods.

The formulations of GEP model for main channel and floodplains flow discharges, as a function of *Dr*, *COH* and vertical divided discharges, were obtained as following:

$$Q_{mc} = \frac{Q_{mc-VDCM}}{\sqrt{Dr} - 8.181} + \frac{e^{COH} + Q_{mc-VDCM} - e^{Dr}}{-5.103^3} + Q_{mc-VDCM} \tag{4.22}$$

$$Q_{fp} = \frac{COH^3\left(Q_{fp-VDCM}\left(Q_{fp-VDCM} + COH\right)\right)^3}{Dr - 1.963} + Q_{fp-VDCM} + \left[\frac{Q_{fp-VDCM}\,(1 - Dr)}{COH + Q_{fp-VDCM}}\right]^3 + \frac{Dr^2}{5.222}\,(COH + 5.222)\,\frac{Q_{fp-VDCM}}{9.495} \tag{4.23}$$

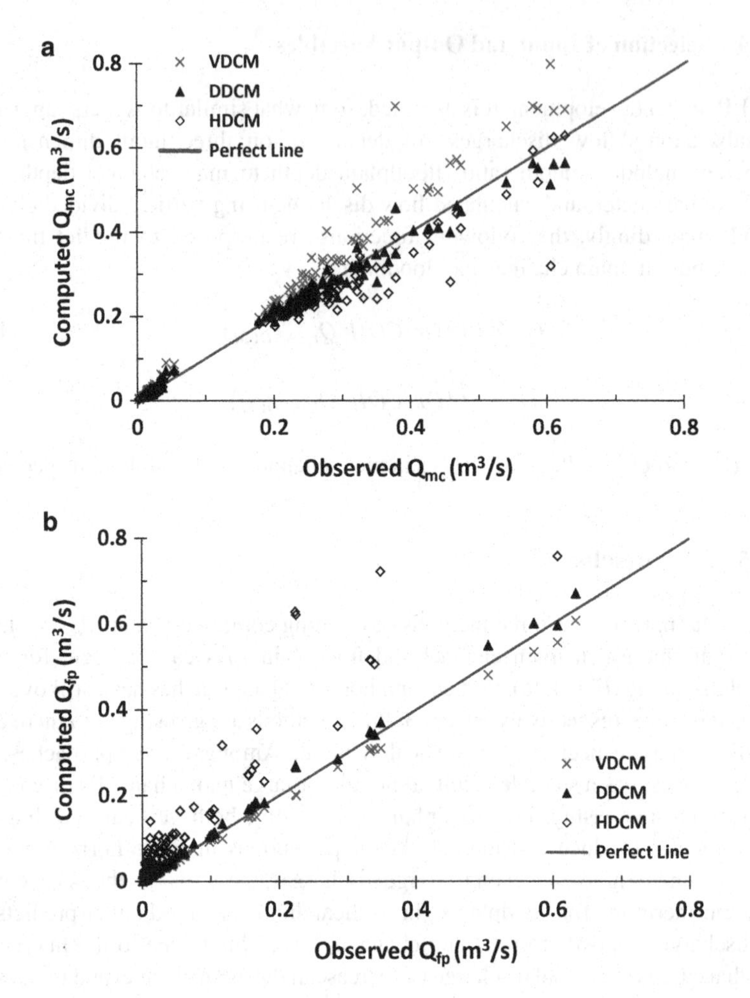

Fig. 4.4 Flow discharge calculation of traditional divided channel methods for main channels (**a**) and floodplains (**b**)

The performance of the GEP model was compared with the traditional vertical divided method. Figure 4.5a, b show the observed and estimated main channel and floodplains flow discharges of the all used data. As can be seen, the GEP model produced much enhanced results, especially for floodplaindischarges.

Table 4.2 presentsa comparison of R, MSE, MAE and ρ for predicted main channel flow discharges obtained from different models. It can be concluded that according to the error functions, especially the mean absolute error, the GEP model gives much better results than the other approaches. The GEP model produces the least errors (MSE = 0.0003 and MAE = 2.1 %). Among traditional methods, the

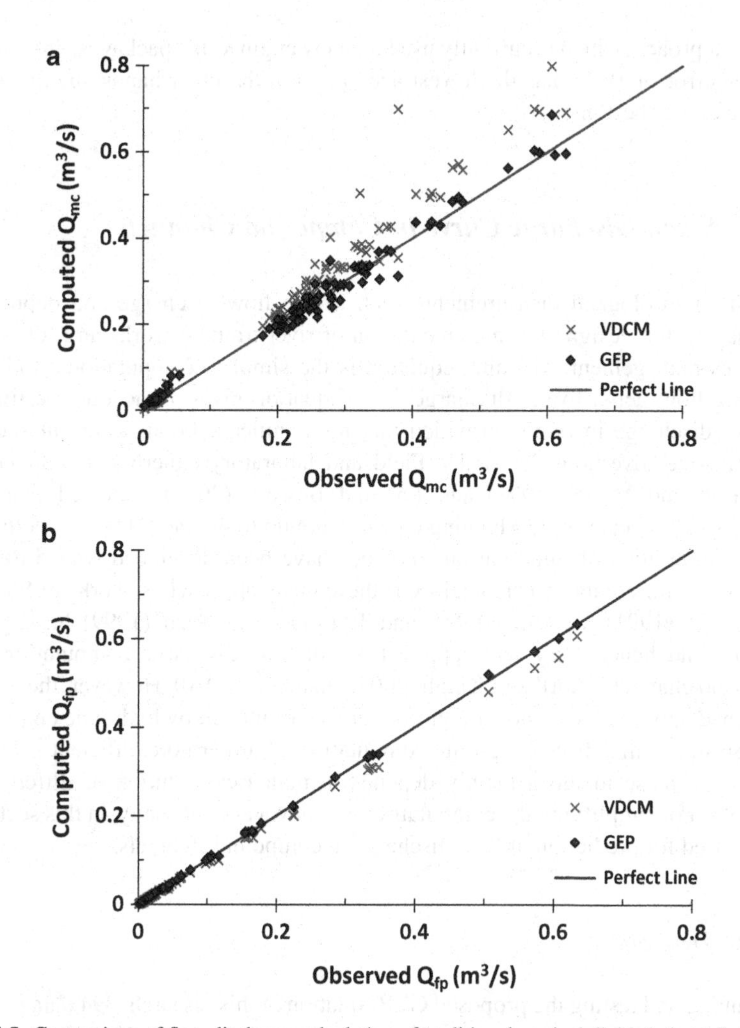

Fig. 4.5 Comparison of flow discharge calculation of traditional vertical divided channel method and GEP model for main channels (**a**) and floodplains (**b**)

Table 4.2 Correlation and error measures for different traditional predictors (divided channel methods) and GEP model for flow discharge in main channel

Models	R	MSE	MAE (%)	ρ
GEP model	0.994	0.0003	2.1	0.055
Vertical divided method	0.986	0.0026	18.98	0.152
Diagonal divided method	0.994	0.0003	11.05	0.056
Horizontal divided method	0.980	0.0015	15.91	0.117

vertical approach, which is currently used in many engineering packages, with mean absolute error of 19 %, has the lowest accuracy. On the other hands, the diagonal approach is the best one.

4.3.2 Stage-Discharge Curve in Compound Channels

In rivers, hydrological measurements such as the flow discharge and depth are essential for the design and implementation of river training works and for water resources management. Manning equation is the simplest computational tool for changing flow depth to the discharge. This equation gives an adequate estimate for flow discharge in rivers, provided that no significant flood occurs in such a way that river overflows its banks. Field and laboratory experiments conducted by Martin and Myers (1991) and Lai and Bessaih (2004) indicated that the maximum errors caused by Manning equation are up to 40 and 60 %, respectively. To overcome this difficulty, various methods have been developed with different assumptions for compound channels. Off these many approaches, works of Shiono and Knight (1991), Ackers (1992), and Bousmar and Zech (1999) have good accuracy and hence, very wide applications in flow discharge computations of compound channels (Abril and Knight 2004; Unal et al. 2010). However, the above mentioned approaches are not straightforward to be applied by hydraulic engineers and also may suffer from long-time computations. Furthermore, efficient solution of some of these methods mainly depends to numerical solution of differential equations. For simplifying the computations of conveyance capacity, in this section, GEP is used for prediction of flow discharge in compound channels.

4.3.2.1 Data Set

For training and testing the proposed GEP equation in this research, 394 data sets of flow hydraulic parameters from 30 different straight laboratory and river compound sections were selected. Most of these data are gathered form an experimental program undertaken by HR Wallingford (FCF) in large scale compound channel flumes (Knight and Sellin 1987). In addition, some extra laboratory data from other studies were used (Blalock and Sturm 1981; Knight and Demetriou 1983; Lambert and Sellin 1996; Myers and Lyness 1997; Lambert and Myers 1998; Bousmar and Zech 1999; Haidera and Valentine 2002; Guan 2003; Lai and Bessaih 2004; Bousmar et al. 2004). Field data were collected from natural compound rivers of River Severn at Montford Bridge (Ackers 1992; Knight et al. 1989), River Main (Martin and Myers 1991) and Rio Colorado (Tarrab and Weber 2004). A typical geometry for natural compound section having inclined berms is seen in Fig. 4.6. The domains of main parameters of flow hydraulics and cross section geometry of compound channels used in this book chapter are mentioned in Table 4.3.

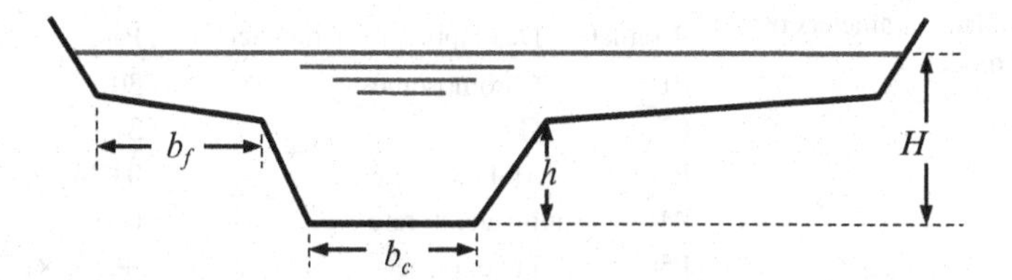

Fig. 4.6 Typical river compound channel cross section with berm inclination

Table 4.3 Range of geometric and hydraulic variables of compound channels

Variable definition	Variable range	Mean value
Bank-full height, h(m)	0.031–6	0.811
Main channel width, b_c(m)	0.152–21.4	3.2
Floodplain width, b_f(m)	0–63	6.5
Manning's n for main channel, n_c	0.01–0.036	0.0133
Manning's n for floodplains, n_f	0.01–0.05	0.0166
Bed slope, S_0	0.000185–0.005	0.0011
Flow depth, H(m)	0.036–7.81	0.985
Bank-full discharge, Q_b(m^3/s)	0.00268–172.048	20.99
Total flow discharge, Q_t(m^3/s)	0.003–560	30.486

4.3.2.2 Input and Output Variables

For developing a precise explicit equation to obtain total flow discharge in compound channels, GEP has been used. Through following equation, it is assumed that total flow discharge in compound channels is proportional to three dimensionless parameters:

$$\frac{Q_t}{Q_b} = f\left(Dr, COH, \frac{Q_{VDCM}}{Q_b}\right) \tag{4.24}$$

Where Q_t is total flow discharge and Q_b is bank-full discharge. Of the total data set, approximately 70 % (272 sets) were selected randomly and used for training. The remaining 30 % (112 sets) were considered for testing.

Using optimization procedure, following relationship has been obtained for training data:

$$\frac{Q_t}{Q_b} = 3.954Dr \times 0.457^{Dr+(1-COH)^{Dr}} + \frac{Q_{VDCM}}{Q_b}^{(1-COH)^{(1-COH)^{2^{Dr}}}} - Dr^{(1-COH)^{\frac{Q_{VDCM}}{Q_b}}}(1-COH)^{Dr^{2.462}} \tag{4.25}$$

In GEP, to avoid overgrowing programs, the maximum size of the program is generally restricted (Brameier and Banzhaf 2001). This configuration was tested for the

Table 4.4 The parameters of final GEP model

Parameter	Description of parameter	Parameter amount
P1	Chromosomes	30
P2	Genes	3
P3	Mutation rate	0.044
P4	Inversion rate	0.1
P5	Function set	+, −, ×, /, power
P6	One-point recombination rate	0.3
P7	Two-point recombination rate	0.3
P8	Gene recombination rate	0.1
P9	Gene transposition rate	0.1
P10	Program size	33

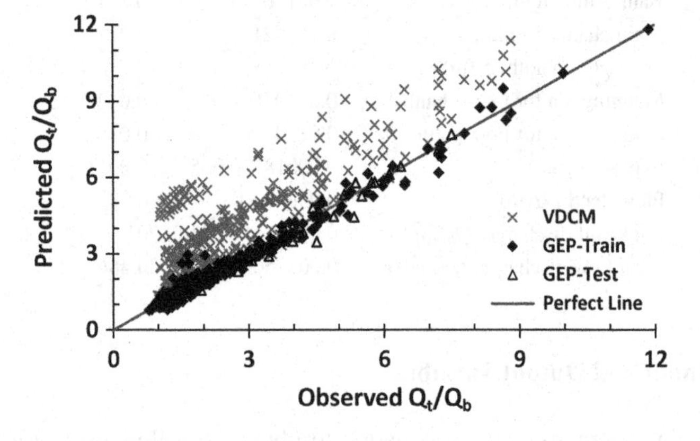

Fig. 4.7 Comparison of traditional vertical divided channel method and GEP model for discharge ratios

proposed GEP model and was found to be sufficient. The best individual (program) of a trained GEP can be converted into a functional representation by successive replacements of variables starting with the last effective instruction (Oltean and Grosan 2003). For developing Eq. (4.25), beside to the basic arithmetic operators and mathematical functions (+, −, ×, power), a large number of generations (5000) were used for testing. First, the maximum size of each program was specified as 256, starting with 64 instructions for the initial program. Table 4.4 shows the operational parameters and functional set used in the GEP modelling.

The computed discharge ratios (Q_t/Q_b) resulted from the GEP model for both training and testing data as well as the vertical divided method are presented in Fig. 4.7. It is clearly seen that GEP model in all variable ranges of selected data (laboratory and field compound sections), has very promised accuracy. Based on these prediction results, the mean absolute errors of discharge ratios for VDCM and GEP model have been calculated as 55.2 and 8.5 %, respectively. It indicates that the GEP model (Eq. 4.25) is highly satisfactory for total flow discharge in compound open channels.

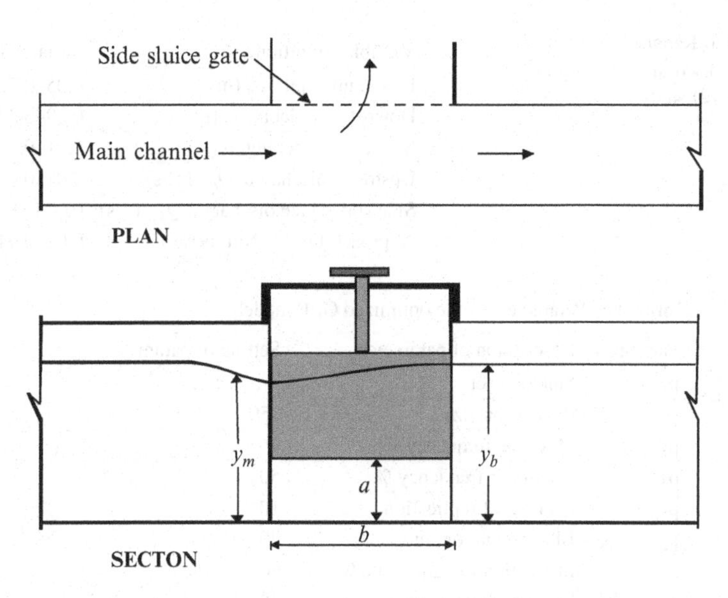

Fig. 4.8 Flow through side sluice gate

4.3.3 Flow Discharge Through Side Sluice Gates

Side sluice gates are underflow diversion structures placed along channels for spilling part of the liquid through it (Fig. 4.8). These structures are mainly used in irrigation, land drainage, urban sewage system, sanitary engineering, storm relief and as head regulators of distributaries (Ghodsian 2003).

The flow through a side sluice gate is a typical case known as spatially varied flow with decreasing discharge. By considering flow through side sluice gate as an orifice flow, the flow discharge through side sluice gate under free flow condition may be written as (Mostkow 1957):

$$Q_s = C_d ab\sqrt{2gy_m} \tag{4.26}$$

where C_d is discharge coefficient, a is opening height of the side gate, b is the gate length and y_m is upstream flow depth in the main channel.

Review of the literature shows that in spite of the importance of the side sluice gates, relatively little attention has been given to studying the behavior of flow through this structure (Panda 1981; Swamee et al. 1993; Ojah and Damireddy 1997; Ghodsian 2003; Azamathulla et al. 2012). They related the discharge coefficient of side sluice gates to the approach Froude number and ratio of flow depth to gate opening. In the recent work, Azamathulla et al. (2012) used GEP technique for developing a relationship for the discharge coefficient for the computation of discharge through sluice gates.

Table 4.5 Range of variables used in Azamathulla et al. (2012) study

Variable definition	Variable range
Upstream depth, y_m (m)	0.05–0.78
Downstream depth, y_b (m)	0.09–0.39
Sluice gate opening, a (m)	0.01–0.1
Upstream discharge, Q_m (m^3/s)	0.01–0.098
Side sluice gate discharge, Q_s (m^3/s)	0.005–0.099
Approach Froude Number, Fr	0.02–0.94

Table 4.6 Parameters of the optimized GEP model

Parameter	Description of parameter	Setting of parameter
p_1	Function set	+, −, ×, /
p_2	Population size	250
p_3	Mutation frequency %	96
p_4	Crossover frequency %	50
p_5	Number of replication	10
p_6	Block mutation rate %	30
p_7	Instruction mutation rate %	30
p_8	Instruction data mutation rate %	40
p_9	Homologous crossover %	95
p_{10}	Program size	initial 64, maximum 256

The experimental data of Ghodsian (2003) were used in Azamathulla et al. (2012) study. The experiments were restricted to subcritical flow in main and side channel. The range of various parameters used in this study is given in Table 4.5.

Basic arithmetic operators (+, −, ×, /) as well as main basic trigonometric and mathematical functions (sin, cos, tan, log, *power*) were used for GEP equation development. Furthermore, a large number of generations (5000) were tested. The functional set and operational parameters used in side slice gate flow hydraulic modelling with GEP during this study are listed in Table 4.6.

The GEP model presented by Azamathulla et al. (2012) for estimating the discharge coefficient of side sluice gates for free flow conditions is as follows:

$$\begin{aligned} C_d = {} & \left[-0.12574\tan^{-1}\left\{\tan^{-1}\sin\left(\log\frac{y_m}{a}\right)+Fr\right\}\right] \\ & - 0.1293\left[\tan^{-1}\left\{\sin\left(\cos\left(\sin\left(\frac{y_m}{a}\right)\right)\right)\right\}\right] \\ & + \left[\tan^{-1}\cos\left(\frac{Fr}{\frac{-8.54}{Fr}+\left(\frac{y_m}{a}\right)^{1/3}}\right)\right] \end{aligned} \tag{4.27}$$

The correlation coefficient (R) and mean square error (MSE) for model training (60 data) are 0.976 and 0.0012, respectively, while for testing phase (14 data) are 0.967 and 0.0043, respectively. The performance of the GEP model is shown in

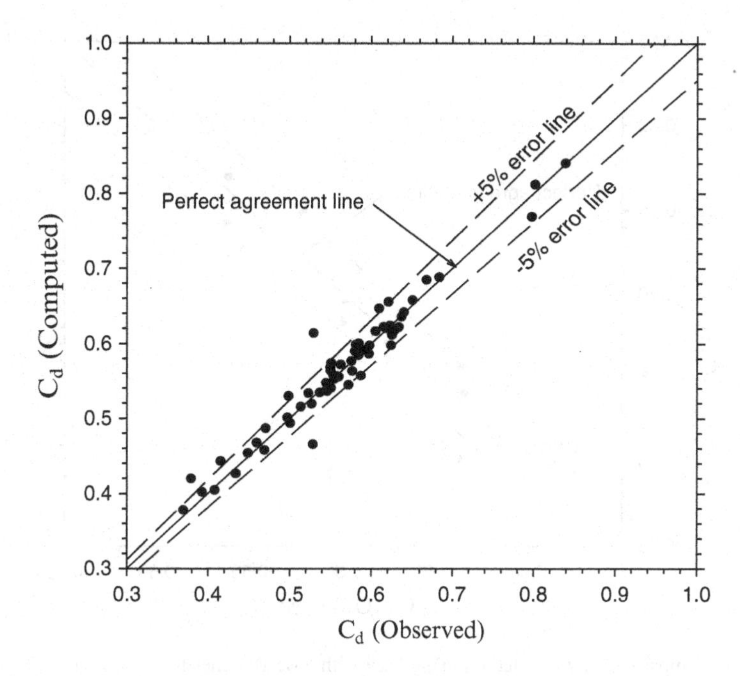

Fig. 4.9 Comparison of computed C_d using GEP with observed ones for training data

Figs. 4.9 and 4.10 for training and testing data, respectively. As can be seen, for both phases, the computed C_d is within ±5 % of the observed ones. The mean absolute percentage error of the computed discharge coefficient by proposed GEP model is about 2.15. It should be noted that although GEP models are somewhat complicated in the mathematical form, but are easy to use practically by engineers at the field by aid of available tools (e.g. spreadsheets).

4.3.4 Local Scour Depth Downstream of Bed Sills

Bed sills are a common solution to stabilize degrading bed rivers and channels. They are aimed at preventing excessive channel-bed degradation in alluvial channels by dividing them into partitions (Zahiri et al. 2014). For practical purposes, designers and civil engineers are often interested in a short-term local scouring and its extent in downstream of grade control structures. By this local scour, the structure itself (and many times other structures in vicinity of it, like bridge piers or abutments, or bank revetments) might be undermined (Bormann and Julien 1991; Gaudio and Marion 2003). Therefore, most researchers have focused on local scouring at isolated or series bed sill structures. Summaries of research for the bed sills can be found in Lenzi et al. (2002). Most of the studies on scouring at bed sills have been conducted through experimental works (Bormann and Julien 1991; Gaudio et al.

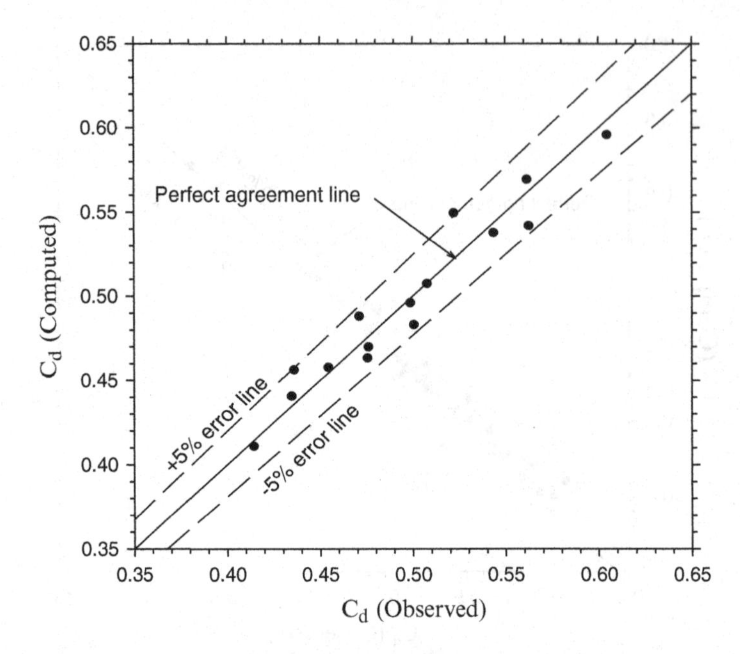

Fig. 4.10 Comparison of computed C_d using GEP with observed ones for testing data

Table 4.7 Empirical equations for maximum scour depth prediction

Empirical equation	Investigator	Eq. number
$\frac{y_s}{H_s} = 1.45\left(\frac{a}{H_s}\right)^{0.86} + 0.06\left(\frac{a}{\Delta d_{50}}\right)^{1.49} + 0.44$	Lenzi et al. (2004)	(4.28)
$\frac{y_s}{H_s} = 1.6\left(\frac{a}{H_s}\right)^{0.61} + 1.89\left(\frac{a}{\Delta d_{50}}\right)^{0.21} - 2.03$	Chinnarasri and Kositgittiwong (2008)	(4.29)
$\frac{y_s}{H_s} = 3\left(\frac{a}{H_s}\right)^{0.6} SI^{-0.19}\left(1 - e^{-0.25\frac{L}{H_s}}\right)$	Tregnaghi (2008)	(4.30)

2000; Lenzi et al. 2002, 2003; Lenzi and Comiti 2003; Marion et al. 2004; Tregnaghi 2008; Chinnarasri and Kositgittiwong 2008). In general, in laboratory works a non-linear regression equation is proposed based on curve fitting of experimental scour depth data and hydraulic quantities and sediment properties. Some well-known empirical equations based on regression analysis of experimental data have been presented in Table 4.7.

These regression equations have one key limitation which mainly originate from the wide ranges of hydraulic and sediment characteristics of flow in rivers. Owing to rapid increase in successful applications of artificial intelligence techniques, it is interesting to explore the applicability of the GEP in prediction of maximum scour depth at bed sills.

In this section, using the 226 experimental data set of maximum scour depth at bed sills from literatures in different canal bed slopes, applicability of GEP has been examined in prediction of relative maximum scour depth at bed sills. These data

Table 4.8 Range of geometric and hydraulic parameters for scouring at bed sills

Variable definition	Variable range	Mean value
Sills spacing, L (m)	0.4–2.5	1.07
Initial bed slope, S_0	0.0059–0.268	0.1099
Flow discharge, Q (l/s)	0.68–30.6	16.5
Sediment median diameter, d_{50} (mm)	0.6–9.0	6.17
Maximum scour depth, y_s (cm)	2.4–29.8	14.45

are collected from Lenzi et al. (2002), Gaudio and Marion (2003), Marion et al. (2004), Tregnaghi (2008) and Chinnarasri and Kositgittiwong (2006, 2008). Range of variations as well as the mean values of important flow hydraulic and sediment characteristics of experimental data are shown in Table 4.8.

4.3.4.1 Physical Definition of Scouring

Chinnarasri and Kositgittiwong (2008) by considering most effective parameters of flow and sediment characteristics on bed sill scouring (see Fig. 4.11) and using Buckingham's π-theorem, presented the following dimensionless groups:

$$\frac{y_s}{H_s} = f_2\left(\frac{a}{H_s}, \frac{a}{\Delta d_{50}}, \frac{L}{H_s}, \frac{d_{50}}{H_s}, S_0\right) \tag{4.31}$$

where y_s is equilibrium maximum scour depth, $a = \left(S_0 - S_{eq}\right) L$ is morphological jump which S_0 and S_{eq} are initial and equilibrium bed slopes, respectively, L is horizontal spacing between sills, $\Delta = \left(\rho_s - \rho_w\right) / \rho_w$ is the relative submerged density of sediment and $H_s = 1.5\sqrt[3]{q^2/g}$ is critical specific energy on the sills where q is water discharge per unit width.

4.3.4.2 Selection of Input and Output Parameters

Based on dimensional analysis of scour depth downstream bed sills, one can select the parameters of a/H_s, $a/\Delta d_{50}$, L/H_s, d_{50}/H_s and S_0 as input variables and y_s/H_s as output variable. Table 4.9 reports the ranges of input and output parameters, used in this section.

4.3.4.3 GEP Results

According to training data (174 data), an explicit equation has been developed GEP technique as following:

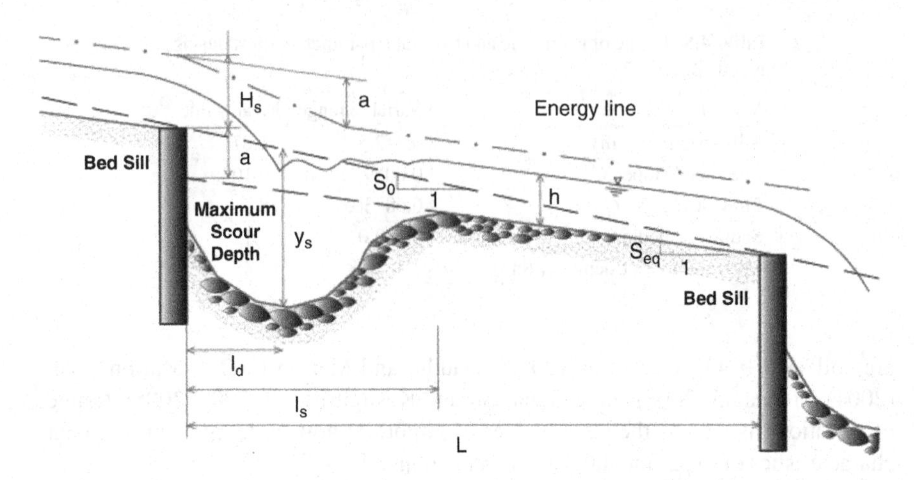

Fig. 4.11 Schematic of scour depth and length downstream of a bed sill (Tregnaghi 2008)

Table 4.9 Range of input and output parameters used in this study

Input/output parameter	Range	Mean value
a/H_s	0.096–9.703	2.12
$a/\Delta d_{50}$	0.494–164.62	23.906
L/H_s	0.1531–55.74	17.736
d_{50}/H_s	0.0136–0.4615	0.106
S_0	0.0059–0.268	0.1099
y_s/H_s	0.261–10.617	2.12

$$\frac{y_s}{H_s} = Ln\left(\frac{a}{\Delta d_{50}} + \frac{a}{H_s} + 9.8561\frac{d_{50}}{H_s}\right) + \frac{d_{50}}{H_s}\left(\frac{a}{\Delta d_{50}} - \frac{S_0}{d_{50}/H_s}\right) + \frac{d_{50}/H_s}{Log\left(a/H_s\right) - \left(a/\left(\Delta d_{50}\right)\right)^{1/3}} \tag{4.32}$$

The prediction results of bed sill scour depth for training and testing (52 data) GEP model have been showed in Fig. 4.12. Comparison of the GEP model with the empirical equations of scour depth at bed sills (Eqs. 4.28–4.30) are presented in Fig. 4.13.

The detailed information of GEP model as well as the empirical equations is indicated in Table 4.10. Based on this statistical analysis, it is indicated that among different models considered in this study, Eq. (4.28) (Lenzi et al. 2004) has the highest errors and therefore, doesn't recommended for application. On the other hand, GEP model can be proposed as an option for prediction of maximum scour depth at bed sills. In addition, the simple equation of Chinnarasri and Kositgittiwong (2008), with requiring to only two parameters and also having good accuracy, is may be considered as a suitable approach.

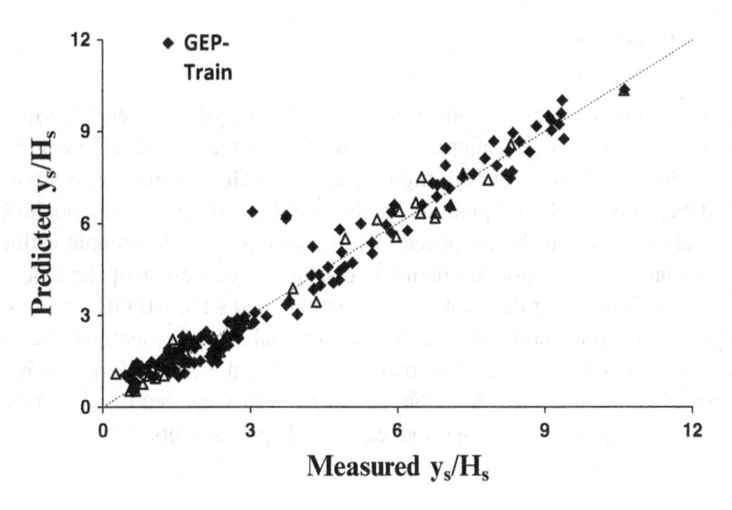

Fig. 4.12 GEP model results of relative maximum scour depth for training and testing data

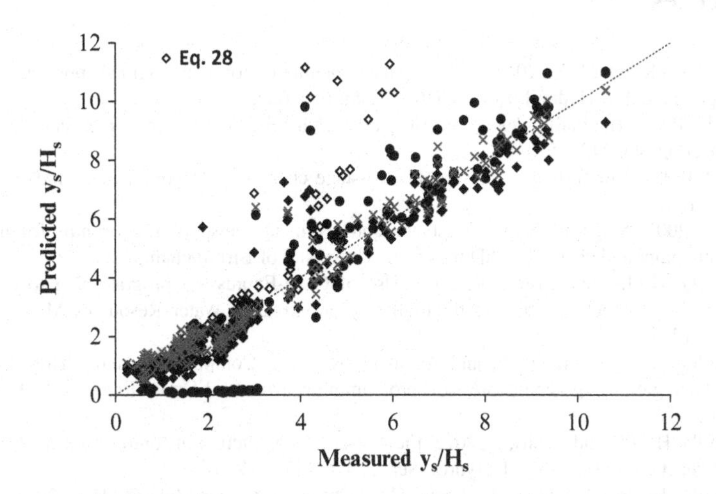

Fig. 4.13 Comparison of GEP model and well-known empirical equations for prediction of relative maximum scour depth at bed sills

Table 4.10 Evaluation of empirical equations and GEP model for bed sill scour depth prediction

	Training			Testing			All data		
Method	R	MSE	ρ	R	MSE	ρ	R	MSE	ρ
Empirical eqs.									
Eq. (4.28)	–	–	–	–	–	–	0.780	402	3.35
Eq. (4.29)	–	–	–	–	–	–	0.956	0.559	0.118
Eq. (4.30)	–	–	–	–	–	–	0.925	1.667	0.200
GEP model	0.976	0.203	0.067	0.986	0.308	0.013	0.979	0.286	0.081

4.4 Conclusions

In this book chapter, some applications of GEP in hydraulic engineering field have been presented. These examples cover a broad range of hydraulic engineering problems. Through these examples, high capability of GEP technique, as a powerful tool for developing explicit equations has been indicated. The main conclusion of this book chapter is that the proposed GEP equations provide reliable estimation of flow discharge in compound channels, discharge coefficient of the side sluice gates and maximum scour depth at the downstream bed sills. All GEP models have high degree of accuracy and are better than traditional or basic methods proposed in the literature. It is interesting to note that although the developed formulas by GEP have generally complex form in mathematical point of view, but they can be easily calculated by using available programs (e.g. Excel spreadsheets).

References

Abril, J.B., and Knight, D.W. 2004. Stage-discharge prediction for rivers in flood applying a depth-averaged model. J. Hydraul. Res., IAHR, 122(6): 616–629.

Ackers, P. 1992. Hydraulic design of two-stage channels. Journal of Water and Maritime Engineering, 96, 247–257.

Ackers, P. 1993. Flow formulae for straight two-stage channels. J. Hydraul. Res., IAHR, 31(4): 509–531.

Atabay, S. 2001. Stage-discharge, resistance and sediment transport relationships for flow in straight compound channels. PhD thesis, the University of Birmingham, UK.

Azamathulla, H.Md., and Jarrett, R. 2013. Use of Gene-Expression programming to estimate manning's roughness coefficient for high gradient streams. Water Resources Management, 27(3): 715–729.

Azamathulla, H.Md., Ahmad, Z., and AbGhani, A. 2012. Computation of discharge through side sluice gate using gene-expression programming. Irrig. and Drain. Engrg., ASCE, 62(1): 115–119.

Azamathulla, H.Md., and Zahiri, A. 2012. Flow discharge prediction in compound channels using linear genetic programming. J. Hydrol. Ser. C, 454(455):203–207.

Azamathulla, H.Md., AbGhani, A., Leow, C.S., Chang, C.K., and Zakaria, N.A. 2011. Gene-expression programming for the development of a stage-discharge curve of the Pahang River. Water Resources Management, 25(11): 2901–2916.

Azamathulla, H.Md., AbGhani, A., Zakaria, N.A., and Guven, A. 2010. Genetic programming to predict bridge pier scour. J. Hydraul. Eng., ASCE, 136(3): 165–169.

Blalock, M.E. and Sturm, T.W. 1981. Minimum specific energy in compound channel. J. Hydraulics Division, ASCE, 107: 699–717.

Bormann, N., and Julien, P.Y. 1991. Scour downstream of grade-control structures, J. Hydraul. Eng., ASCE, 117(5): 579–594.

Bousmar, D. 2002. Flow modelling in compound channels: momentum transfer between main channel and prismatic or non-prismatic floodplains. PhD Dissertation, Université catholique de Louvain, Belgium. http://dial.academielouvain.be/handle/boreal:4996

Bousmar, D. and Zech, Y. 1999. Momentum transfer for practical flow computation in compound channels. J. Hydraul. Eng., ASCE, 125(7), 696–70.

Bousmar, D., Wilkin, N., Jacquemart, H. and Zech, Y. 2004. Overbank flow in symmetrically narrowing floodplains. J. Hydraul. Eng., ASCE, 130(4), 305–312.

Brameier, M., and Banzhaf, W. 2001. A comparison of linear genetic programming and neural networks in medical data mining. IEEE Trans. Evol. Comput., 5: 17–26.
Chinnarasri, C., and Kositgittiwong, D. 2006. Experimental investigation of local scour downstream of bed sills. J. Research in Engineering and Technology, Kasetsart University, 3(2):131–143.
Chinnarasri, C., and Kositgittiwong, D. 2008. Laboratory study of maximum scour depth downstream of sills. ICE Water Manage., 161(5): 267–275.
Chow, V. T. 1959. Open channel hydraulics, McGraw-Hill, London.
Fernandes, J.N., Leal, J.B. and Cardoso, A.H. 2012. Analysis of flow characteristics in a compound channel: comparison between experimental data and 1-D numerical simulations. Proceedings of the 10th Urban Environment Symposium, 19: 249–262.
Ferreira, C. 2001. Gene expression programming: a new adaptive algorithm for solving problems. Complex Systems, 13(2): 87–129.
Gandomi A.H., and Alavi, A.H., 2011. Multi-stage genetic programming: A new strategy to nonlinear system modeling. Information Sciences, 181(23): 5227–5239.
Gandomi A.H., and Alavi, A.H., 2012. A new multi-gene genetic programming approach to nonlinear system modeling. Part II: Geotechnical and Earthquake Engineering Problems. Neural Computing and Applications, 21(1): 189–201.
Gandomi, A.H., and Roke, D.A. 2013. Intelligent formulation of structural engineering systems. In: Seventh M.I.T. Conference on Computational Fluid and Solid Mechanics, Massachusetts Institute of Technology, Cambridge, MA.
Gaudio, R., and Marion, A. 2003. Time evolution of scouring downstream of bed sills. J. Hydraul. Res., IAHR, 41(3): 271–284.
Gaudio, R., Marion, A., and Bovolin, V. 2000. Morphological effects of bed sills in degrading rivers. J. Hydraul. Res., IAHR, 38(2): 89–96.
Ghodsian, M. 2003. Flow through side sluice gate. J. Irrig. and Drain. Engrg., ASCE, 129(6): 458–463.
Guven, A., and Azamathulla, H.Md. 2012. Gene-Expression programming for flip-bucket spillway scour. Water Science & Technology, 65(11): 1982–1987.
Guven, A., and Gunal, M. 2008. Genetic programming approach for prediçtion of local scour downstream of hydraulic structures. J. Irrigation and Drainage Engineering, 134(2): 241–249.
Guan, Y. 2003. Simulation of dispersion in compound channels. M.Sc. Thesis in Civil Engineering, Ecole Polytechnique Federale de Lausanne (EPFL), Switzerland.
Guven, A., Aytek A., 2009. New approach for stage discharge relationship: gene expression programming. J. Hydrol. Eng., ASCE, 14: 812–820.
Haidera, M.A. and Valentine, E.M. 2002. A practical method for predicting the total discharge in mobile and rigid boundary compound channels. International Conference on Fluvial Hydraulics, Belgium, 153–160.
Huthoff, F. 2007. Modeling hydraulic resistance of floodplain vegetation. PhD Thesis, Twente University, the Netherland.
Huthoff, F., Roose, P.C., Augustijn, D.C.M., Hulscher, S.J.M.H. 2008. Interacting divided channel method for compound channel flow. J. Hydraul. Eng., ASCE, 134(8): 1158–1165.
Jarrett, R.D. 1984. Hydraulics of high gradient streams. J. Hydraul. Eng., ASCE, 110(1):1519–1539.
Knight, D.W. and Sellin, R.H.J. 1987. The SERC flood channel facility. Journal of Institution of Water and Environment Management, 1(2), 198–204.
Knight, D.W. and Demetriou. J.D. 1983. Flood plain and main channel flow interaction. J. Hydraulic Division, ASCE, 109(8): 1073–1092.
Knight, D.W., Shiono, K., and Pirt, J. 1989. Predictions of depth mean velocity and discharge in natural rivers with overbank flow. International Conference on Hydraulics and Environmental Modeling of Coastal, Estuarine and River Waters, England, 419–428.
Lai, S.H. and Bessaih, N. 2004. Flow in compound channels. 1st International Conference on Managing Rivers in the 21st Century, Malaysia, 275–280.

Lambert, M.F. and Myers, R.C. 1998. Estimating the discharge capacity in straight compound channels. Water, Maritime and Energy, 130, 84–94.
Lambert, M.F. and Sellin, R.H.J. 1996. Discharge prediction in straight compound channels using the mixing length concept. Journal of Hydraulic Research, IAHR, 34, 381–394.
Lenzi, M. A., Marion, A., Comiti, F., and Gaudio, R. 2002. Local scouring in low and high gradient streams at bed sills. J. Hydraul. Res., IAHR, 40(6): 731–739.
Lenzi, M.A., and Comiti, F. 2003. Local scouring and morphological adjustments in steep channels with check-dams sequences. Geomorphology, 55: 97–109.
Lenzi, M.A., Marion, A., and Comiti, F. 2003. Interference processes on scouring at bed sills. Earth Surface Processes and Landforms, 28(1): 99–110.
Lenzi, M.A., Comiti, F., and Marion, A. 2004. Local scouring at bed sills in a mountain river: Plima river, Italian alps. J. Hydraul. Eng., ASCE, 130(3):267–269.
Liggett, J.A. 2002. What is hydraulic engineering? J. Hydraul. Eng., ASCE, 128(1): 10–19.
Marion, A., Lenzi, M.A., and Comiti, F. 2004. Effect of sill spacing and sediment size grading on scouring at grade-control structures. Earth Surface Processes and Landforms, 29(8): 983–993.
Martin, L. A. and Myers, R. C. 1991. Measurement of overbank flow in a compound river channel. Journal of Institution of Water and Environment Management, 645–657.
Mostkow, M. A. 1957. A theoretical study of bottom type water intake. La Houille Blanche, 4: 570–580.
Moussa, Y.A.M. 2013. Modeling of local scour depth downstream hydraulic structures in trapezoidal channel using GEP and ANNs. Ain Shams Engineering Journal, 4(4): 717–722.
Mujahid, K., Azamathulla, H.Md., Tufail, M. and AbGhani, A. 2012. Bridge pier scour prediction by gene expression programming. Proceedings of the Institution of Civil Engineers Water Management, 165 (WM9):481–493.
Myers, R.C. and Lyness. J.F. 1997. Discharge ratios in smooth and rough compound channels. J. Hydraul. Eng., ASCE, 123(3): 182–188.
Najafzadeh, M., Barani, Gh-A., Hessami Kermani, M.R., 2013. GMDH Network Based Back Propagation Algorithm to Predict Abutment Scour in Cohesive Soils. Ocean Engineering, 59: 100–106.
Ojah, C.S.P., Damireddy, S. 1997. Analysis of flow through lateral slot. J. Irrig. and Drain. Engrg., ASCE, 123(5): 402–405.
Oltean, M., and Grosan, C. 2003. A comparison of several linear genetic programming techniques. Complex Syst., 14(1):1–29.
Onen, F. 2014. Prediction of scour at a side-weir with GEP, ANN and regression models. Arab J. Sci Eng, 39: 6031–6041.
Panda, S. 1981. Characteristics of side sluice flow. ME thesis, University of Roorkee, Roorkee, India.
Prasuhn, A. 1987. Fundamentals of hydraulic engineering. Holt, Rinehart, and Winston, New York, USA.
Roberson, J.A., Cassidy, J.J., and Chaudhry, M.H. 1998. Hydraulic Engineering. John Wiley and Sons, Inc., New York, USA.
Sattar, M.A. 2014. Gene expression models for the prediction of longitudinal dispersion coefficients in transitional and turbulent pipe flow. J. Pipeline Systems Engineering and Practice, ASCE, 5(1):.
Sharifi, S. 2009. Application of evolutionary computation to open channel flow modeling. PhD Thesis in Civil Engineering, University of Birmingham, 330 p.
Shiono, K. and Knight, D.W. 1991. Turbulent open-channel flows with variable depth across the channel. J. Fluid Mechanics, 222: 617–646.
Swamee, P.K., Pathak, S.K., Sabzeh Ali, M. 1993. Analysis of rectangular side sluice gate. J. Irrig. and Drain. Engrg., ASCE, 119(6): 1026–1035.
Tarrab, L., and Weber, J.F. 2004. Transverse mixing coefficient prediction in natural channels. Computational Mechanics, 13: 1343–1355.
Tregnaghi, M. 2008. Local scouring at bed sills under steady and unsteady conditions. PhD Thesis, University of Padova, Italy.

Unal, B., Mamak, M., Seckin, G., and Cobaner, M. 2010. Comparison of an ANN approach with 1-D and 2-D methods for estimating discharge capacity of straight compound channels. Advances in Engineering Software, 41: 120–129.

Van Prooijen, B.C., Battjes, J.A. And Uijttewaal, W.S.J. 2005. Momentum exchange in straight uniform compound channel flow. J. Hydraul. Eng., ASCE, 131:175–183.

Wang, C.Y., Shih, H.P, Hong, J.H. and Rajkumar, V.R. 2013. Prediction of bridge pier scour using genetic programming. J. Marine Science and Technology, 21(4), 483–492.

Whitley, D. 1991. Fundamental principles of description in genetic search. Foundations of genetic algorithms. G. Rawlings, ed., Morgan Kaufmann.

Wormleaton, P.R. and Merrett, D.J. 1990. An improved method of calculation for steady uniform flow in prismatic main channel/floodplain sections. J. Hydraul. Res., IAHR, 28: 157–174.

Yang, K., Liu, X., Cao, S., and Huang, E. 2014. Stage-discharge prediction in compound channels. J. Hydraul. Eng., ASCE, 140(4).

Zahiri, A., Azamathulla, H.Md., and Ghorbani, Kh. 2014. Prediction of local scour depth downstream of bed sills using soft computing models. T. Islam et al. (eds.), Computational Intelligence Techniques in Earth and Environmental Sciences, Springer, Chapter 11, 197–208.

Zakaria, N.A, Azamathulla, H.Md, Chang, C.K and AbGhani, A. 2010. Gene-Expression programming for total bed material load estimation – A Case Study. Science of the Total Environment, 408(21): 5078–5085.

Chapter 5
Genetic Programming Applications in Chemical Sciences and Engineering

Renu Vyas, Purva Goel, and Sanjeev S. Tambe

5.1 Introduction

With ever-increasing amounts of monitored, recorded and archived data in establishments such as manufacturing and service industries, and R&D institutions, the need for making sense of the collected data is also growing exponentially. Analyzing and interpreting data regarding structures, properties and reactions of chemicals as also plant operations, due to their sheer size, have become a challenging task. More often than not, systems encountered in chemical sciences and engineering/technology exhibit nonlinear behavior and analyzing data emanating from them using, for example, traditional classification and modeling techniques often leads to difficulties. The modern day chemical processes also comprise multiple equipment wherein a plethora of reactions and physical and chemical transformations take place. This characteristic together with their commonly encountered nonlinear behavior makes the "first principles" modeling (also termed *phenomenological* modeling) of such systems a complex, time-consuming, tedious and costly task. In this context, data-mining methods including data-driven modeling have assumed a great importance.

In the last two and half decades, *artificial intelligence* (AI), *machine intelligence* (ML), and *computational intelligence* (CI) based formalisms have found increasing data-mining applications in chemical sciences and engineering/technology. These computer science sub-fields are linked by a major common theme in that they attempt to meet one of the main challenges narrated by Samuel (1983)—"to get machines to exhibit behavior, which if done by humans, would be assumed to involve the use of intelligence." The principal methods employed by AI, ML, and CI, in getting the machines to exhibit an intelligent behavior are *artificial neural*

R. Vyas • P. Goel • S.S. Tambe (✉)
Artificial Intelligence Systems Group, Chemical Engineering and Process Development Division, CSIR-National Chemical Laboratory, Pune 41100, India
e-mail: ss.tambe@ncl.res.in

A.H. Gandomi et al. (eds.), *Handbook of Genetic Programming Applications*,
DOI 10.1007/978-3-319-20883-1_5

networks (ANNs), *fuzzy logic* (FL), *evolutionary algorithms* (EA) and *support vector machines/regression* (SVM/SVR).

There exists a novel member of the evolutionary algorithms family, namely *genetic programming* (GP) (Koza 1992), that addresses the above-stated challenge by providing a method for automatically creating a computer program that performs a prespecified task simply from a high-level statement of the problem. Genetic programming follows Darwin's theory of biological evolution comprising "survival of the fittest" and "genetic propagation of characteristics" principles. It addresses the goal of automatic generation of computer programs by: (i) genetically breeding a random population of computer programs, and (ii) iteratively transforming the population into a new generation of computer programs by applying analogs of nature-inspired genetic operations, namely, *selection*, *crossover* and *mutation*.

The operating mechanisms of GP are similar to that of the genetic algorithms (GA) (Goldberg 1989; Holland 1975). Though both these formalisms use the same evolutionary principles, their application domains are very different; while GA searches and optimizes the decision variables that would maximize/minimize a specified objective function, GP automatically generates computer codes performing prespecified tasks. In addition to generating computer programs automatically, there exist two important data-mining applications, namely, classification and symbolic regression, for which GP has been found to be a suitable methodology. Unlike the "divide and conquer" approach employed by machine learning algorithms to perform classification, an evolutionary algorithm such as GP does not directly construct a solution to a problem (e.g., a decision tree) but rather searches for a solution in a space of possible solutions (Eggermont et al. 2004). The GP-based symbolic regression (GPSR) is an extension of the genetic model of learning into the space of function identification. Here, members of the population are not computer programs but they represent mathematical models/expressions coded appropriately using symbols.

As compared to classification, GP has been used extensively to conduct symbolic regression in chemical sciences and engineering. GPSR possesses several advantages over the two widely employed strategies namely *artificial neural networks* (ANNs) and *support vector regression* (SVR) in developing exclusively data-driven models. ANNs and SVR construct models in terms of a non-linear transfer function and a kernel function, respectively. Depending upon the specific application for which an ANN (SVR) model is being developed and the nature of the nonlinearities between the corresponding input and output data, the complexity of the model differs. However, owing to the use of the transfer (kernel) function, the basic building blocks of the data-driven models fitted by the ANN (SVR) strategy remain the same irrespective of their application domains. In contrast, GPSR provides a system-specific linear or a nonlinear model that fits the given input–output data and that too without making any assumptions regarding the form of the fitting function (Kotanchek 2006). This is a remarkable feature of GP, which makes it a novel, ingenious and an effective data-driven modeling formalism. The GP models are also more compact and utilize less number of parameters than the existing classical statistical techniques. Some of the comparative studies have indicated GP to be superior in terms of accuracy of prediction than ANNs (Can and Heavy 2012).

Moreover, GP models may enable the user to gain an insight into the fundamental mechanisms (first principles) underlying the data. In a noteworthy study by Schmidt and Lipson (2009) GP has been demonstrated to yield a phenomenological model (natural law) governing the dynamics of a pendulum.

Despite its novelty and potential, GP—unlike ANNs and SVR—has not witnessed an explosive growth for data-driven modeling applications. One possible reason behind this scenario is that for a long time feature-rich, user-friendly and efficient GP software packages were not available. The situation has changed in recent years and a few software packages, both commercial and open source, have become available for performing GP-based classification and symbolic regression. These packages have definitely assisted in the development of a large number of diverse GP applications in various science, engineering and technology disciplines. In this chapter, GP-based classification and regression applications in chemical sciences including biochemical sciences and chemical engineering/technology are reviewed. Owing to the predominance of GPSR over GP-based classification, the implementation details of the former are presented in greater depth. For an in-depth generic treatment of the GP-based classification the reader is referred to, for example, Koza (1991), Bonet and Geffner (1999), Cantú-Paz and Kamath (2003), Eggermont et al. (2004), and Espejo et al. (2010).

Hereafter, this chapter is structured as follows. Section 5.2 provides a detailed discussion of symbolic regression, issues involved in conducting GPSR, the stepwise procedure of GPSR and a short list of GP software packages. In Sect. 5.3, a review of classification and regression applications of GP in various sub-areas of chemistry is provided. The specific GP application areas covered in this section include drug design, environmental chemistry, green technologies, analytical chemistry, polymer chemistry, biological chemistry, and proteomics. Section 5.4, presents GP applications in chemical engineering and technology. Here, the specific GP application areas that are considered comprise process modeling, energy and fuels, membrane technology, petroleum processes and heat transfer. Finally, Sect. 5.5 provides concluding remarks.

5.2 Symbolic Regression

Conventional regression analysis involves finding the parameters of a predefined function such that it best fits a given sample of input–output data. The principal difficulty with this approach is that if the data fit is poor then the model builder has to explore other functional forms until a well-fitting model is secured. This approach is time-consuming, tedious, and requires a skilled model builder to guess and evaluate various potential linear/nonlinear functional forms. In this type of search for an optimal data-fitting model, even domain experts tend to have strong mental biases that limit wider exploration of the function space. For instance, in many application areas traditionally only linear or quadratic models are used, even when the data might be fitted better by a more complex model (Poli et al. 2008). In the traditional nonlinear regression analysis, even after expending a major effort in exploring the function space there is no guarantee that a well-fitting model can indeed be secured in a finite number of trials.

Most of the above-stated difficulties are overcome by the symbolic regression (SR). It essentially involves *function identification* wherein a mathematical model/-expression coded in a symbolic form is found in a manner such that the model and the associated parameters provide a good, best, or a perfect fit between a given finite sampling of values of the independent variables (model inputs) and the corresponding values of the dependent variable (model output). Notably, SR does this without making any assumptions about the structure of that model. That is, it finds an appropriate linear or nonlinear form of the function that fits the data. A similarity between the original "automatic development of a computer program doing a specified job" and "symbolic regression" applications of GP is that both methods take the values of the independent variables as input and produce the values of the dependent variables as output (Koza 1990). Symbolic regression was one of the earliest applications of GP (Koza 1992), and continues to be widely studied (see, for example, Koza and Poli 2005; Poli et al. 2008; Iba et al. 2010; Keedwell and Narayanan 2005; Sumathi and Surekha 2010; Cartwright 2008; Cai et al. 2006; Gustafson et al. 2005; Lew et al. 2006). The major drawback of GPSR, however, is that it is computationally intensive since it searches wide function and associated parameter spaces. This however does not pose a major difficulty since GPSR procedure is amenable to parallel processing.

Consider a multiple input—single output (MISO) example data set, $D = \{(\mathbf{x}_1, y_1), (\mathbf{x}_2, y_2), \ldots, (\mathbf{x}_N, y_N)\}$, consisting of N patterns, where $\mathbf{x}_n$ $(n = 1, 2, \ldots, N)$ denotes an M-dimensional vector of inputs ($\mathbf{x}_n = [x_{n1}, x_{n2,\ldots,} x_{nM}]^T$), and y_n denotes the corresponding scalar output. Using the data set D, the task of GPSR is to search and optimize the exact form and the associated parameters of that unknown MISO linear/nonlinear function (f), which for the given set of inputs produces the corresponding outputs as closely as possible. The general form of the function/model to be fitted by GPSR is given as:

$$y = f\ (\mathbf{x}, \boldsymbol{\alpha}) \tag{5.1}$$

where $\alpha = [\alpha_1, \alpha_1, \ldots, \alpha_K]^T$ represents a K-dimensional parameter vector.

5.2.1 GPSR Implementation

In GPSR, to begin with a random population of probable (candidate) solutions to the function identification problem is generated. Each candidate solution is coded in the form of a "parse tree," which when decoded forms a candidate model for producing the desired outputs $\{y_n\}$ (Iba 1996). The tree structure emanates from a *root* node and consists of *operator* ("function") and *operand* ("terminal") nodes. The former class of nodes define mathematical operators while operands define model inputs

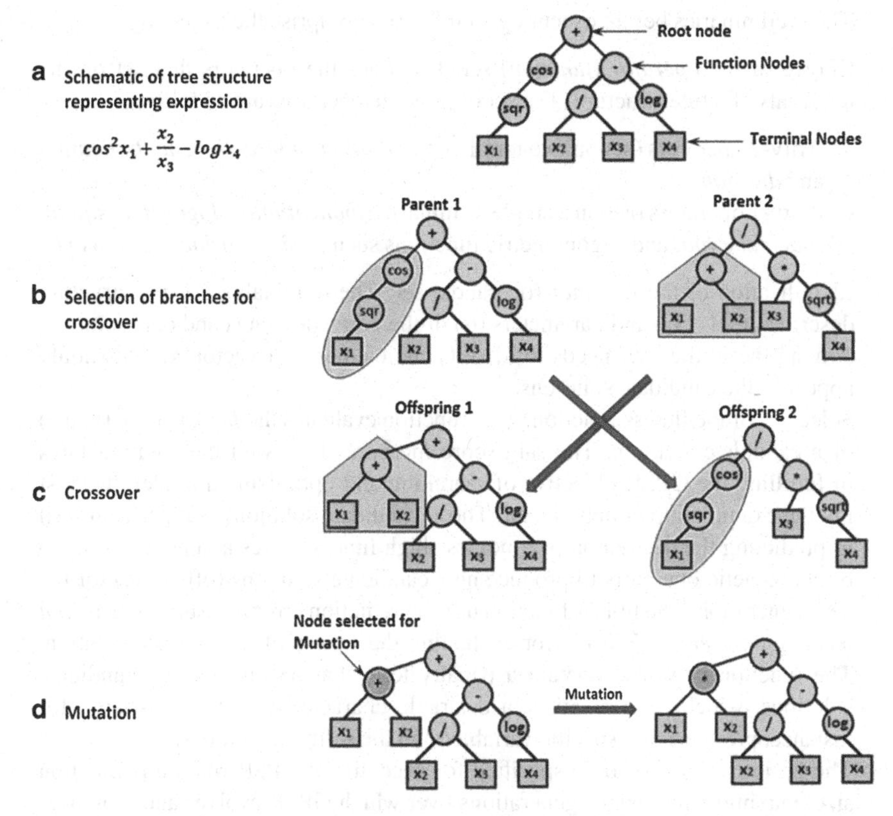

Fig. 5.1 Schematic of genetic programming: (**a**) basic tree structure, (**b**) random selection of branches for reproduction, (**c**) crossover operation, and (**d**) mutation operation. Symbols in the figure denote following operators (function nodes): ("+") addition, ("−") subtraction, ("*") multiplication, ("÷") division; x_n, $n = 1, 2, \ldots, N$, and numeric values define operands (terminal nodes)

(**x**) and parameters (α). The trees in a population are of different sizes and their maximum size is predefined. An illustrative tree structure representing an expression "$cos^2x_1 + \frac{x_2}{x_3} - logx_4$" is depicted in Fig. 5.1. Upon forming the initial population of candidate solutions, following steps are performed: evaluation of fitness scores of candidate solutions, formation of a mating pool of parents, and actions of the genetic operations, namely *crossover* and *mutation*. An iteration of these steps produces a new generation of offspring candidate solutions. Several such iterations are needed before convergence is achieved. The candidate solution possessing the highest fitness score encountered during the iterative process is chosen as the best-fitting model.

The preliminaries before executing a GPSR run comprise the following:

- Choice of the *operator (function)* set: It defines the operators that act on the terminals of a tree structure. The set of possible operators are as follows.
  - Arity-2 operators (act on two terminals): *addition*, *subtraction*, *multiplication* and *division*.
  - Arity-1 operators (act on a single terminal): *exponentiation, logarithm, square root, cube root* and trigonometric functions such as *sine, cosine, tan* and *cot*.
- Identification of the *terminal* (operands) set: The terminals of a tree structure describe inputs ($\mathbf{x}_n$) and parameters (α) of the corresponding candidate solution. Among these, the user needs to identify the elements of vector, $\mathbf{x}$, that should appear in the candidate solutions.
- Selection of the fitness function: This function evaluates the *fitness score* (*value*) of a candidate solution. The said score measures how well the solution fares in fulfilling the GPSR objective of searching and optimizing a model that best fits the example input–output data. Those candidate solutions that perform well in predicting the desired outputs possess high fitness scores and are acted upon by the genetic operators to produce new candidate solutions (offspring) for the next generation. The fitness function uses a prediction error measure such as *root mean square error* (RMSE) for evaluating the fitness of a candidate solution. The function may also contain a penalty term that penalizes those candidate solutions, which do not satisfy a desirable characteristic or a constraint, for instance, presence of a specific variable(s) in the solution's input space.
- The basic parameters to be specified for executing a GPSR run are population size, maximum number of generations over which GPSR evolves and crossover and mutation probabilities. More parameters are possible depending upon the specific software package used in the GPSR implementation. It is necessary to vary all these parameters systematically to obtain an overall optimal solution.
- Using prior knowledge: If some prior knowledge about the data-fitting function is available then it should be utilized while creating the initial population of candidate solutions as also choosing the members of an operator set. For example, if there exists a periodic relationship between the inputs (predictor variables) and the desired outputs, then inclusion of the operators, such as, *sine* and *cosine,* is recommended. Similarly, in the case of data emanating from exothermic reactions, choice of the exponentiation operator is suggested. It may be noted that some GP software packages allow even user-defined expressions in the initial population of candidate solutions.

A generic step-wise implementation of the GPSR is presented below.

Step 1: Generate an initial population of N_p number of candidate solutions randomly; each candidate solution is represented using a tree structure.

Step 2: Iteratively perform the following four sub-steps until a termination criterion is satisfied. The commonly used termination criteria are: (1) a pre-specified number of generations have been evolved, and (2) the fitness value of the best candidate solution in a population no longer increases significantly or remains constant over a large number of successive generations.

1. *Fitness score evaluation:* Using the example set inputs $\{\mathbf{x}_n\}$, $n = 1,2,\ldots N$, compute the output of each candidate solution in the current population. The computed outputs are utilized to calculate the magnitude of the pre-selected error metric (e.g., RMSE), which is then used to determine the fitness score of that solution. This procedure is repeated to calculate fitness scores of all candidate solutions in the current population.
2. *Selection*: Create a mating pool of candidate solutions termed "parents" to undergo crossover operation described in step 3. The members of the mating pool are selected in a manner such that only those candidate solutions possessing relatively high fitness scores can enter the pool. There exist a number of methods for selecting the candidate solutions in a mating pool, such as *Roulette-wheel selection*, *greedy over-selection*, *ranking selection*, *tournament selection* and *elite strategy* (Iba et al. 2010). Each one of these strategies possesses certain advantages and limitations.
3. *Crossover*: This step can be performed multiple ways, for example, *single-* and *two-point* crossover. In the former (see Fig. 5.1b), a pair of parent candidate solutions is selected randomly from the mating pool and two new candidate solutions (offspring) are created by slicing each parent tree at a random point along the tree length and mutually exchanging and recombining the sliced parts between the parents (see Fig. 5.1c). This crossover operation is conducted with a pre-specified probability value (termed *crossover probability*) and repeated with other randomly chosen parent pairs until N_p offspring candidate solutions are formed. The trees in GP do not have a fixed length and these can grow or shrink due to the crossover operation.
4. *Mutation*: In this step, small changes are applied to the operator and operand nodes of the offspring solutions to produce a new generation of candidate solutions (see Fig. 5.1d). This step is performed with a small magnitude of the probability termed "mutation probability."

Avoiding over-fitting of models An important issue that needs to be addressed during GPSR is *over-fitting* of the constructed models. Over-fitting can occur in two ways, that is, when a model is trained over a large number of iterations (termed *overtraining*), and/or the model contains more terms and parameters than necessary (*over-parameterization*). Over-parameterization tends to increase the complexity of the fitted model. Over-fitting results in a model that has learnt even the noise in the data at the cost of capturing a smooth trend therein. Such a model performs poorly at generalization, which refers to the model's ability to accurately predict outputs corresponding to a new set of inputs. A model incapable of generalization is of no practical use. To overcome the problem of over-fitting, the available input–output example set is partitioned into two sets namely *training* and *test* sets. While the former is used to train the model, the test set is used for evaluating the generalization capability of the model. After each training iteration or convergence the candidate solutions are assessed for their generalization capability using the test set data and those predicting the training and test set outputs with high and comparable accuracies are accepted. Sometimes, a third set known as *validation set* is formed from the available example data and a model performing well on all the three i.e. training, test and validation sets is selected.

GP implementation is computationally very intensive and often the evolved solution is anything but ideal thus requiring even greater numerical processing to secure an acceptable solution. The component of the GP algorithm that is computationally most expensive is fitness evaluation. It is therefore at most important to use fitness functions that are computationally economical yet efficient.

The most attractive feature of the GP-based symbolic regression is that it searches and optimizes the "structure" (form) of a suitable linear or nonlinear data-fitting function. It also obtains values of all the parameters of that function although relatively this is a less important GP characteristic since several deterministic and stochastic linear/nonlinear parameter estimation strategies are already available. Moreover, optimality of the parameter values searched by the GP cannot be guaranteed. It is therefore advisable that the parameters of the GP-searched function are optimized using an appropriate parameter optimization strategy such as Marquardt's method (Marquardt 1963).

5.2.2 Software Packages for Implementing GP

A non-exhaustive list of software packages for implementing GP algorithm is given below. It may be noted that the list is meant only for providing an idea of what is available and should not be construed as a recommendation.

- *Disciplus* ™ (commercial software) performs predictive modeling and utilized in data mining tasks requiring predictive analytics, classification, ROC curve and regression analysis (Register Machine Learning Technologies Inc. 2002)
- *Eureqa*® (Schmidt and Lipson 2009, 2014) uses symbolic regression to unravel the intrinsic relationships in data and explain them as simple mathematics. It uses GP heavily in its functioning and is optimized to provide parsimonious solutions. A number of modeling studies in various science and engineering disciplines, such as astronomy, biology, chemistry, chemical engineering, and computer, material and environmental sciences, have been conducted using *Eureqa.*
- *GPTIPS* (Searson et al. 2010) is a free, open source MATLAB toolbox for performing GPSR. It is specifically designed to evolve mathematical models of predictor-response data that are "multigene" in nature, i.e. linear combinations of the low order nonlinear transformations of the input variables.
- *HeuristicLab* (GNU general public license) (Wagner 2009) is an open-source environment for heuristic optimization. This software provides a number of well-known standard algorithms for classification and regression tasks and additionally includes an extensive implementation of GPSR.

- *RGP* (Flasch 2014) is a GP system based on, as well as fully integrated into, the *R* environment. It implements classical tree-based GP as well as other variants including, for example, strongly typed GP and Pareto GP. This package is flexible enough to be applied in nearly all possible GP application areas, such as symbolic regression, feature selection, automatic programming and general expression search.
- *ECJ* toolkit (ECJ 22 2014) is one of the popular computational tool with full support for GP. It is a evolutionary computation research system written in Java and developed at George Mason University's evolutionary computation Laboratory. ECJ 22 is the latest version of the toolkit provided with a GUI and various features such as flexible breeding architecture, differential evolution and multiple tree forest representation. This toolkit is reviewed by White (2012).

5.3 Applications of Genetic Programming (GP) in Chemical Sciences

Genetic programming has been used in chemistry with a great success for providing potential solutions to a variety of classification and data-driven modeling problems as well as to create new knowledge. It has been also established that GP models can approximate to a first principles models (Anderson et al. 2000). While GA has been used extensively for optimization in chemical sciences and chemical engineering/technology, GP-based applications in these disciplines are relatively fewer (Aguiar-Pulido et al. 2013). The applications of GP in chemical sciences have focused mainly on data mining, which can be further broadly categorized into *rule-based classification* and *symbolic regression* based model development (see Fig. 5.2).

GP is an apt tool for data-mining in chemical sciences. Formally, data mining is defined as "identification of patterns in large chunks of information" (Cabena et al. 1997). The data could be physicochemical data from small molecule based assays or spectral data emanating from the analytical instruments used for characterization of chemical or biological moieties. Several representations of the tree-based GP have been used exhaustively for the rule-based data classification (Li and Wong 2004). GP-based intelligent methodologies have been used for developing the rule-based systems in chemistry and biochemistry domains such as, chemical networks and reactivity, wherein the computer programs are all functional models of chemical or biochemical properties (Tsakonas et al. 2004). GP-based regression has been mainly employed for building quantitative structure—property relationship (QSPR) models (Barmpalexis et al. 2011). The flexibility of GPSR is at the core of the development of free form mathematical models from the observed data (Kotanchek 2006). Both the approaches persist and consequently the subsequent sections are devoted to the applications of the GPSR and GP-based classification methods, illustrated by using copious examples from the literature.

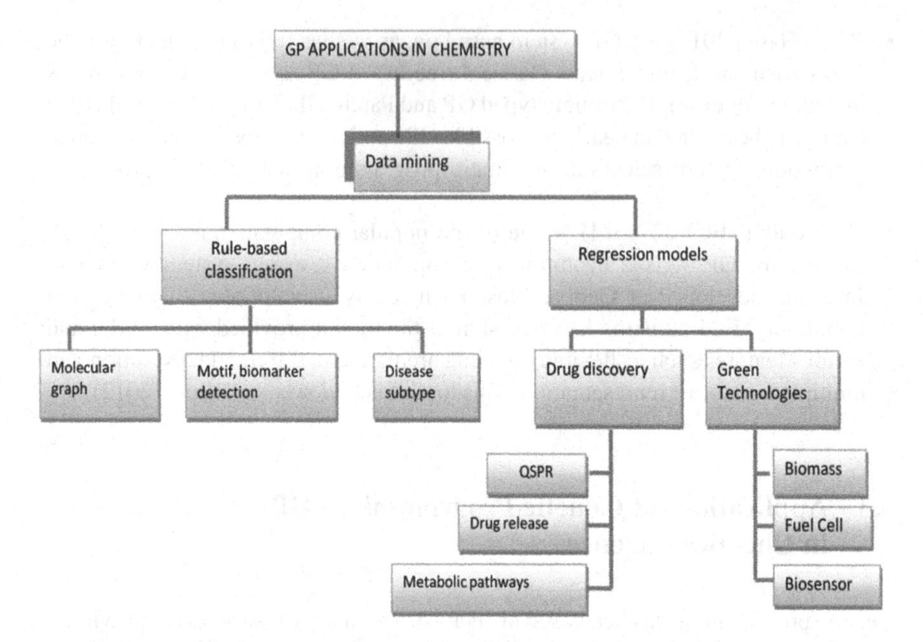

Fig. 5.2 An overview of the major applications of GP in chemical sciences

5.3.1 Applications of GP in Drug Design

Drug design and development is an essential component of pharmaceutical industry (Guido et al. 2008; Anderson 2003). In this tedious, time consuming and expensive endeavor, computational methods are utilized in every stage of the development, mainly for the prediction of properties of small molecules (Venkatraman et al. 2004) and their affinities towards the respective biological targets. Due to the *fail early* paradigm prevalent in the pharmaceutical industry, even an approximate computational method applied before the clinical stage, which can assist in eliminating molecules with undesirable properties is highly welcome (Atkinson et al. 2012). The GP-based methods have been employed in the field of drug design in conjunction with the machine learning methods to address the interactions between potential drugs/lead molecules or between drugs and large bio-molecules such as proteins (Garcia et al. 2008). These interactions are otherwise difficult to assess via experiments largely due to the involved ethical issues. As depicted in Fig. 5.3, GP approaches have been mainly applied in the four principal stages of the drug discovery pipeline viz. *lead selection*, *lead optimization*, *preclinical trials* and *clinical trials* stages. Wherever applicable, related examples have been cited in the text that follows.

Genetic programming has been compared with a few other advanced computational methods for predicting the critical ADME properties—such as the oral bioavailability (OB) of the drug molecules—during preclinical trials (Langdon

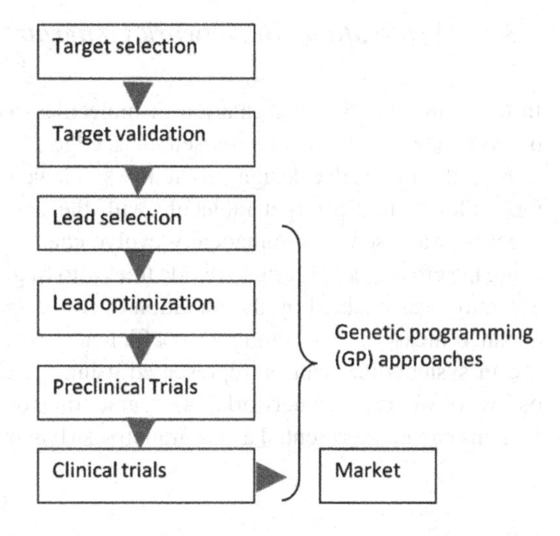

Fig. 5.3 A schematic of the drug discovery pipeline showing the steps wherein the GP method can be applied

and Barrett 2005). The results showed that classification of drugs into 'high' and 'low' OB classes could be performed on the basis of their molecular structure and the output given by the developed models would be useful in the pharmaceutical research. Moreover, the results indicated that the quantitative prediction of the oral bioavailability is also possible. In a study involving structure-property relationships, GP was utilized for the prediction of Caco-2 cell permeability (Vyas et al. 2014), which is an important ADMET parameter; the said GP model yielded high coefficient of correlation ($\approx$0.85) between the desired and model predicted values of the permeability and a low RMSE value of 0.4.

Mathematical models have been developed to predict drug release profiles (Ghosal et al. 2012; Güres et al. 2012). In a related study, GP-based models were built for the prediction of drug release from the solid-lipid matrices (Costa and Lobo 2001). Here, GP was used specifically for determining the parameters of the model—a modified Weibull equation—that is commonly used in the reliability engineering defined as:

$$f(T) = \frac{\beta}{\eta}\left(\frac{T-\gamma}{\eta}\right)^{\beta-1} e^{-\left(\frac{T-\gamma}{\eta}\right)^{\beta}} \tag{5.2}$$

where $f(T) \geq 0,\ T \geq 0$ or $\gamma, \beta > 0,\ \eta > 0$, and $-\infty < \gamma < \infty$. Here, β represents the shape parameter, also known as the *Weibull slope*, η is the scale parameter and γ denotes the location parameter. In this study, the calculated release profiles of the solid-lipid extrudates of varying dimensions compared well with the experimentally determined dissolution curves.

5.3.2 Applications in Network Generation

In the graph theory representation of molecules, atoms are denoted as nodes and bonds connecting them are represented as edges. In a classic study by Globus et al. (1999), the molecular design problem was viewed as a search of the space of all molecules to find a target molecule with the desired properties. The GP fitness function was used to automatically evolve chains and rings in new molecules by using the crossover operator to divide trees into fragments (Globus et al. 1999). Ring evolution was enabled by the mutation operator and the fitness function defined a distance measure, i.e., Tanimoto coefficient. Likewise, GP has found a potential use in systems that can be represented using the graph theory such as metabolic pathways wherein the networks of organic transformations occurring in biological systems can be represented as program trees (Ivanova and Lykidis 2009).

5.3.3 Applications in Environmental Chemistry

Whole cell biosensors have become an integral part of the environment monitoring (Gu et al. 2004). Here, the main task is to detect the substance specific patterns from the huge biosensor data being monitored continuously. GP has been found to be a suitable classification technique to handle the stated task. For example, GP has been employed in the classification of herbicide chemical classes and herbicides with high sensitivity albeit with a low selectivity (Podola and Melkonian 2012). Electronic noses are being employed as vapor sensors since they provide rich information regarding the analyte binding (Persaud and Dodd 1982). GP-based approaches were able to detect the airborne analytes in real time with a good sensitivity as also selectivity (Wedge et al. 1999).

Gene expression programming (GEP) is an extension of the genetic programming (GP) and genetic algorithms (GAs). It is a population-based evolutionary algorithm (Ferreira 2001) wherein a mathematical function defined as a *chromosome* consisting of multi-genes is developed using the data presented to it. In GEP, a mathematical expressions are encoded as simple linear strings of a fixed-length, which are subsequently expressed as nonlinear entities of different sizes and shapes (i.e. simple diagram representations or expression trees) (Cevik 2007). Singh and Gupta (2012) employed GEP for forecasting the formation trihalomethanes (THMs)—which are toxic to human health—in waters subjected to chlorination. In this study, five parameters namely dissolved organic carbon normalized chlorine dose, water pH, temperature, bromide concentration, and contact time, were used as model inputs. Similar to the GP-based model, ANN and SVM based models were developed for comparison purposes. The results of this comparison revealed that the ANN, SVM, and GEP models are capable of capturing the complex nonlinear relationship between the water disinfection conditions and the corresponding THM formation in the chlorinated water.

5.3.4 Applications in Green Technologies

Multigene genetic programming has been employed to gauge the performance of microbial fuel cells (Garg et al. 2014). The method provided a correlation between the output voltage and input factors of microbial fuel cell and was found to be superior to ANN and SVR in terms of the generalization ability.

Biomass is one of the upcoming important and renewable sources of green energy (Martin 2010). It is thus crucial—from the viewpoint of designing, fabricating, operating and optimizing biomass-based energy generating systems—to accurately know the amount of energy contained in a biomass fuel (biofuel). The higher heating value (HHV) is an important property defining the energy content of a biomass fuel. Experimental estimation of the energy content of a biofuel in terms of HHV is a slow and time-consuming laboratory procedure. Thus, a number of proximate and/or ultimate analysis based predominantly linear models have been proposed for predicting HHV magnitudes of biomass fuels. The basic assumption of linear dependence (Parikh et al. 2005) between the constituents of the proximate/ultimate analyses of biofuels and the respective HHVs is not unambiguously supported by the corresponding experimental data. Accordingly, Ghugare et al. (2014a) employed GP for developing two biomass HHV prediction models, respectively using the constituents of the proximate and ultimate analyses as the model inputs. In the development of the proximate (ultimate) analysis based model, data pertaining to 382 (536) different biomass samples were utilized. The GP-based two models developed using *Eureqa Formulize* software package (Schmidt and Lipson 2009) are as follows:

- Proximate analysis based optimal model:

$$HHV = 0.365 \times FC + 0.131 \times VM + \frac{1.397}{FC} + \frac{328.568 \times VM}{10283.138 + 0.531 \times FC^3 \times ASH - 6.893 \times FC^2 \times ASH} \tag{5.3}$$

  where FC, VM, and ASH are the weight percentages (dry basis) of fixed carbon, volatile matter, and ash respectively.
- Ultimate analysis based optimal model:

$$HHV = 0.367 \times C + \frac{53.883 \times O}{2.131 \times C^2 - 93.299} + \frac{C \times H - 115.971}{10.472 \times H + 0.129 \times C \times O} - \frac{91.531}{(35.299 + N)} + \frac{232.698}{77.545 + S} \tag{5.4}$$

  where, C, H, O, N and S are the weight percentages (dry basis) of carbon, hydrogen, oxygen, nitrogen, and sulfur, respectively. The *coefficient of correlation* (CC) magnitudes in respect of the experimental and GP model-predicted

HHVs were high (>0.95) while the corresponding magnitudes of *mean absolute percentage error* (MAPE) were low (<4.5 %). The HHV prediction accuracy and generalization performance of these models were rigorously compared with the corresponding multilayer perceptron (MLP) neural network based as also previously available high-performing linear and nonlinear HHV models. This comparison showed that the HHV prediction and generalization performance of the GP as also MLP-based models to be consistently better than that of their linear and/or nonlinear counterparts proposed earlier. The biofuel HHV prediction models proposed by Ghugare et al. (2014a), due to their excellent performance, possess a potential of replacing the models proposed earlier. Also, their GP-based strategy can be extended for developing HHV prediction models for other types of fuels.

Among the two commonly employed analyses for characterizing biomass fuels, proximate analysis is relatively easy to perform than the ultimate analysis. Accordingly, GPSR was employed for building non-linear models for the accurate prediction of *C*, *H* and *O* fractions of the solid biomass fuels from the constituents of the corresponding proximate analysis (Ghugare et al. 2014b). These models were constructed using a large data set of 830 fuels. For comparison purposes, *C*, *H* and *O* prediction models were developed using ANN and SVR approaches also. The results of the comparison of the prediction accuracy and generalization performance of GP, ANN and SVR based nonlinear models with that of the currently available linear models indicated that the nonlinear models have consistently and significantly outperformed their linear counterparts.

5.3.5 Applications in Analytical Chemistry

GP has been applied for conducting multivariate analysis of the nonlinear dielectric spectroscopy (NLDS) data of a yeast fermentation process (Woodward et al. 1999). In this study, GP was found to outperform the conventional methods like partial least squares (PLS) and ANNs. Genetic programming was also used for recognizing the bonds taking part in increasing or decreasing the dominant excitation wavelength by identifying the conjugated Π systems for lowest UV transition for a system of 18 anthocyanidins (Alsberg et al. 2000). The model stressed upon the important role of bond critical point (BCP) characterized by the electron density, the Laplacian operator and the ellipticity.

5.3.6 Applications in Polymer Chemistry

The reactivity ratios in free radical copolymerization are routinely estimated using Alfrey-Price (AP) model (Alfrey and Price 1947). However, the accuracy of predictions made by this model is sub-optimal. Accordingly, exclusively data-driven, GP-based nonlinear models have been developed for the reactivity ratio prediction in free radical copolymerization (Shrinivas et al. 2015). These models use the same Q and e parameters as utilized by the Alfrey-price model for characterizing the monomers. The GP-based models were further fine-tuned using Levenberg-Marquardt (LM) nonlinear regression method (Marquardt 1963). A comparison of the Alfrey-Price, GP, GP-LM and artificial neural network (ANN) based models indicated that the GP and GP-LM models exhibit superior reactivity ratio prediction accuracy and generalization performance (with correlation coefficient magnitudes close to or greater than 0.9) when compared with the AP and ANN models. The GP-based reactivity ratio prediction models possess the potential of replacing the widely used AP models mainly due to their higher accuracy and generalization capability. In the area of designing of new polymeric materials Porter et al. (1996) employed GP to perform a structural optimization of a monomer in order to achieve desired polymer properties.

5.3.7 *Applications in Biological Chemistry*

By itself GP is a biology inspired computational technique and finds several applications in this field. The ever increasing amounts of data being generated by today's sophisticated technologies such as microarray and single nucleotide polymorphism (SNP) make the usage of suitable methods for feature extraction and data analysis essential. These data mainly emanate from the fields of genomics, proteomics and clinical time series studies and are amenable to processing by GP (Schneider and Orchard 2011). An exhaustive review on GP applications in genomics has been recently published (Khan and Alam 2012). The applications essentially include genetic network inference (Lanza et al. 2000), gene expression data classification (Paul et al. 2006), SNP (Poli et al. 2008), epistasis (Estrada-Gil et al. 2007) and gene annotation (Stein 2001). Genetic programming neural networks (GPNN) have begun to be recently employed in the identification of the hidden relationships of gene–gene and gene-environment interactions in the context of disease of interest (Motsinger et al. 2006). The GP-based classification method has been used for automatically locating the property motif candidates in peptide sequences (Tomita et al. 2014). The discriminant nature of the GP-based rules was ascertained by the precise identification of twofold MHC class II binding peptides. Another important GP application lies in identifying the signal peptides and discerning their cleavage sites. In a report by Lennartsson and Nordin (2004), GP was used for the automatic evolution of classification programs and it compared favorably with ANNs. The best evolved motif could detect the h region composed of the hydrophobic amino acids in the signal peptide.

5.3.8 Applications in Proteomics

MS/MS spectroscopy technique plays an important role in proteomics for identifying proteins, peptides and metabolic data. Feature selection methods have been often used in clinical proteomics (Christin et al. 2013). In a study, GP-based approach was employed for biomarker detection and classification of MS data (Ahmed et al. 2014). Specifically, GP was employed for the feature ranking in mass spectroscopic data; this is considered a herculean task due to the presence of a large number of features. The inherent feature selection ability of GP was exploited to identify the selected features from the best evolved program. Here, GP not only proved to be superior to "Information GAIN" and "RELIEF" feature selection methods but outperformed the GA-based approach also. In the same study, the GP-based classifier was found to be superior to *J48*, *Naive Bayes* and *SVM* classifiers. The reduced set of features for a biomarker as selected by GP brings down the clinical cost of validating them in laboratories.

5.3.9 Applications of GP in Chemical Biology

Cancer is a major disease for which GP has found numerous applications ranging from the classification models of cancer tumors to mechanistic understanding of the underlying pathogenesis. Easy interpretability of these GP models greatly enhances our understanding of the underlying cellular and disease pathway dynamics at the systems biology level (Finley et al. 2014). Interested readers are referred to a comprehensive review of GP applications in cancer research (Worzel et al. 2009). As of today, for most of the neurodegenerative diseases there is no cure; however an early detection can provide a better life for the patients suffering from them. An example of the use of GP in the clinical time series data involves inducing classifiers capable of recognizing the movements characteristic of patients afflicted with a disease like Parkinson's wherein a diagnostic accuracy of 97 % was achieved (Castelli et al. 2014). Here, GP was used to identify patterns in the slow motor movements (Bradykinesia) related clinical data. Similar applications are found in the context of visuo-spatial diseases also where a graph based GP system termed *Implicit Context representation Cartesian Genetic Programming* (IRCGP), which functions similarly to the well known crossover operator was devised (Smith and Lones 2009).

5.3.10 Applications in Reaction Modeling

Genetic programming was used to generate a network of chemical reactions from the observed time domain data (Koza et al. 2001). Here, the concentration of the

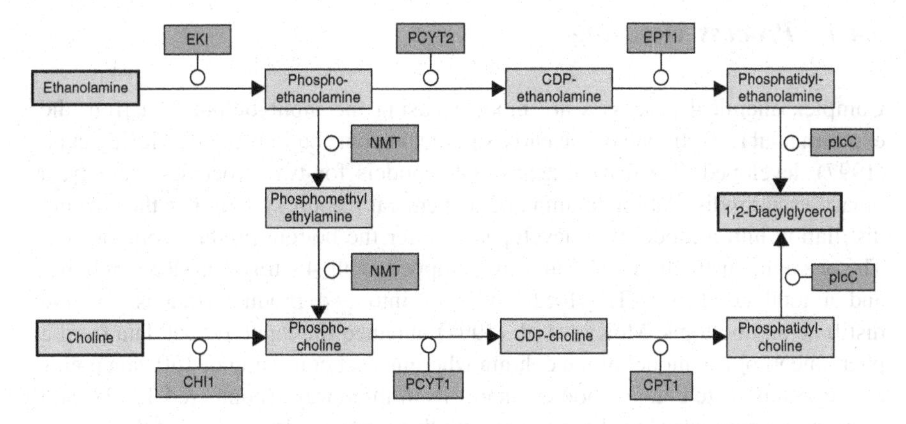

Fig. 5.4 A schematic representation of reactions involved in the human phospholipid pathway

last product of the predicted network model matched with a high accuracy with the experimental data; GP could successfully construct two metabolic pathways viz. phospholipids cycle and degradation of ketone bodies. The eukaryotic phospholipids biosynthetic pathway and its role in the cellular biology has been well studied (Vamce and Vance 2008) (see Fig. 5.4). Here, the researchers chose four enzymatic reactions from the said pathway with glycerol and fatty acid as inputs and diacyl-glycerol as the end product. A tree was constructed programmatically to represent the chemical reaction functions and selector functions as nodes, and reaction rates, substrates, products and enzymes, as leaves. The results of the GP run could be corroborated with the observed experimental data. Thus, GP could create metabolic pathways that included topological features such as an internal feedback loop, bifurcation point, accumulation point and rates for all reactions using the time domain concentration values.

5.4 GP Applications in Chemical Engineering/Technology

In chemical engineering and technology, GP formalism has been used in a wide variety of applications. Depending upon their domain these applications have been divided in the following eight major categories: process modeling, energy and fuels, water desalination and wastewater treatment (membrane technology), petroleum systems, heat transfer, unit operations, process identification, and miscellaneous.

5.4.1 Process Modeling

Complex chemical processes are modeled using the input–output data from the experimental tests. In one of the early significant contributions of GP, McKay et al. (1997) developed data driven steady-state models for two processes, namely, a binary vacuum distillation column and a chemical reactor system. For the vacuum distillation unit a model was developed to infer the bottom product composition. The vacuum distillation column was equipped with 48 trays, a steam reboiler and a total condenser. The feed was split into two product streams i.e., the distillate and bottoms. McKay et al. (1997) obtained the input–output data from a phenomenological model of the column wherein a set consisting of 150 data points of the steady-state composition estimates from three trays (numbered 12, 27, and 42 from the top) was considered along with the corresponding values of the bottom composition. A set of 50 data points was used in the model validation. The models were accepted only if the validation set *RMSE* was less than 0.02. An F-test was then performed to find the best model. The overall best model had an *RMSE* of 0.011 on the training set data and 0.015 on the validation set data. McKay et al. (1997) applied a similar method to obtain a functional relationship, the data for which was generated from an assumed relationship with three inputs (u_1, u_2, *and* u_3) and a single output, y:

$$y = 1000\, u_1 \times \exp(-5/u_2) + u_3 \tag{5.5}$$

They also modeled a continuous stirred tank reactor (CSTR) system for the prediction of the product composition.

In chemical processes, operating conditions need to be optimized for a variety of reasons such as maximization of conversion, profit and selectivity of desirable products, and minimization of cost and selectivity of undesirable products. For conducting such an optimization, it is necessary that a representative and accurate process model is available. Often, process behavior is nonlinear and complex and therefore developing phenomenological (also termed "first principles" or "mechanistic") process models becomes tedious, costly and difficult. In such instances, data-driven process models can be constructed. Cheema et al. (2001) presented GP-assisted stochastic optimization strategies for the optimization of glucose to gluconic acid bioprocess wherein *Aspergillus niger* strain was used for producing gluconic acid. Their study utilized two hybrid process modeling-optimization approaches wherein a GP-based model was first developed from the process data, following which the input space of the GP model was separately optimized using two stochastic optimization (SO) formalisms, namely, *genetic algorithms* (GA) and *simultaneous perturbation stochastic approximation* (SPSA) (Spall 1998). A schematic of the GP-based process modeling and GA-based optimization strategy is shown in Fig. 5.5. Cheema et al. (2002) used process data from 46 batch fermentation experiments conducted by them in building the GP-based model. The gluconic acid concentration (y)(g/L) which formed the output of the GP model (see Eq. 5.6) was predicted as a function of three process parameters, namely,

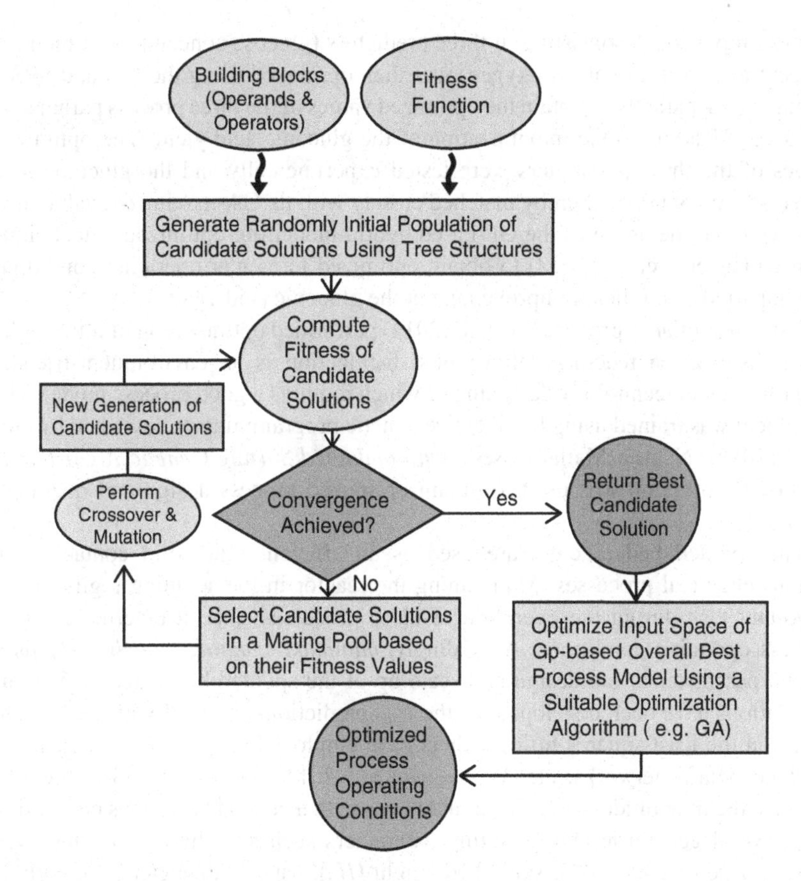

Fig. 5.5 Schematic of GP-model based process optimization

glucose concentration (x_1)(g/L), biomass concentration (x_2)(g/L), and dissolved oxygen (x_3) (mg/L).

$$y = \left[\frac{\beta_1 x_1}{(x_1 - \beta_2)^4 + \beta_3}\right]\left[\frac{1}{x_2{}^2 - \beta_4 x_2 + \beta_5}\right]\left[\frac{1}{\beta_6 x_3{}^2 - \beta_7 x_3 + \beta_8}\right] \tag{5.6}$$

The values of the eight model parameters fitted by the GPSR are: $\beta_1 = 3.1911 \times 10^{10}$, $\beta_2 = 158.219$, $\beta_3 = 2.974 \times 10^6$, $\beta_4 = 5.421$, $\beta_5 = 107.15$, $\beta_6 = 0.116$, $\beta_7 = 12.752$, and $\beta_8 = 448.112$. The magnitude of the variance (R^2) pertaining to the training (test) set output predictions made using Eq. (5.6) was 0.987 (0.986). Since the form of the model was known (determined by GP), Cheema et al. (2002) subjected the GP-based model to Marquardt's non-linear regression analysis (Marquardt 1963) to explore whether the model's eight parameters could be fine-tuned further to improve its prediction accuracy. This parameter fine-tuning indeed led to a better R^2 value of 0.9984 (0. 9979) for the training (test) set. The GP based model's input space consisting of three predictors (glucose concentration, biomass concentration and dissolved oxygen) was then optimized using the GA and SPSA formalisms separately to obtain the optimized values of the three process parameters (x_1, x_2, x_3) leading to the maximization of the gluconic acid yield. The optimized values of the three parameters were tested experimentally and the gluconic acid concentration obtained thereby matched closely with its GA-maximized value. It is thus seen that the usage of the GP-based hybrid modelling-optimization technique allowed Cheema et al. (2002) to obtain optimized fermenter operating conditions that imparted a significant improvement in the gluconic acid yield.

Using a similar approach, Xu et al. (2014) performed optimization of ultra-violet water disinfection reactors. Ultra-violet disinfection is an environment-friendly water treatment technology designing of which requires a good process model. The GP model was trained using bi-objective genetic programming as described by Giri et al. (2012). Next, a Matlab-based *Non-dominated Sorting Genetic Algorithm II* (NSGA II) program was used to obtain optimized process design and operating conditions.

The spouted bed reactors are used as an efficient fluid-solid contactors in various chemical processes. Maintaining the reactor in the spouting regime is an important task during the operation of these processes since it determines other process operating conditions. Accordingly, *minimum spouting velocity* (U_{ms})is a crucial parameter in the design and scale up of the spouted bed reactors. Various correlations have been developed for the U_{ms} prediction—most of which while are based on the least-squares fitting, others have employed support vector machines, artificial neural networks, etc. Hosseini et al. (2014) developed a GP model to estimate the magnitude of U_{ms} in spouted beds with a conical base. This correlation uses several geometric and operating parameters such as column diameter (D_c), spout nozzle diameter (D_i), static bed height (H_0), particle diameter (d_p), particle density (ρ_p), gas density (ρ_g) and gravitational force (g). It is given as:

$$\frac{U_{ms}}{\sqrt{2H_0 g}} = \left[\left(\frac{D_i}{D_c}\right)^{0.8} + \left(\frac{d_p}{D_c} + \left(\frac{D_i}{D_c}\right)^{0.8}\right) A\right](A - B) \tag{5.7}$$

where,

$$A = \frac{\left(\frac{d_p}{D_c}\right) log_2 \left(\frac{d_p}{D_c}\right)}{\left(\frac{H_0}{D_c}\right)} + \sqrt{\frac{d_p}{D_c}\left(\frac{\rho_p - \rho_g}{\rho_g}\right)^{0.3}} \tag{5.8}$$

$$B = \left(\frac{d_p}{D_c} + \sqrt{\frac{d_p}{D_c}} + \left(\frac{D_i}{D_c}\right)^{0.8}\right)^2 \frac{H_0}{D_c} \tag{5.9}$$

The results obtained from the above GP-based correlation are in good agreement with the experimental values.

Grosman and Lewin (2002) described the use of GP to generate empirical dynamic models of two processes. These GP models were derived to implement the nonlinear model predictive control (NMPC) strategy. The first process for which Grosman and Lewin (2002) formulated the GP model was a mixing tank consisting of two feed flows (fresh water and saturated salt water) for which the control objective was to maintain the level of fluid and concentration of the effluent salt. Since the GP models obtained for both the stated parameters were of linear type, an inference could be drawn regarding the almost linear character of the actual process. The second system considered by Grosman and Lewin (2002) is a Karr liquid–liquid extraction column with the controlled outputs being the dispersed phase effluent concentration and column hold-up. Here, both linear and nonlinear models were developed although the performance of the nonlinear model was found to be better than that of the linear model. Subsequently, Grosman and Lewin (2004) modified the GP approach to generate steady-state nonlinear empirical models for process analysis and optimization. The key feature was to improve the efficiency and accuracy of the algorithm, via incorporation of a novel fitness calculation, optimal creation of new generations, and parameter allocation. The first case study by Grosman and Lewin (2004) was similar to that of McKay et al. (1997) wherein the performance of GP was tested on the data generated using Eq. (5.5). The developed GP model was compared with the model presented by McKay et al. (1997). The GP model obtained by Grosman and Lewin (2004) yielded an RMSE of 0.017 whereas the GP model of McKay et al. (1997) resulted in an RMSE value of 0.47. As can be seen, the Grosman and Lewin (2004) model showed an improvement in the predictive capability by a good order of magnitude. Their second case study involved a catalytic reaction of hydrogen and toluene to produce methane and benzene.

$$C_6H_5CH_3 + H_2 \rightarrow C_6H_6 + CH_4 \tag{5.10}$$

Grosman and Lewin (2004) developed a GP model to predict the reaction rate of toluene. The developed model showed that the GP-based model possesses a good prediction accuracy (RMSE = 0.0038).

In semiconductor manufacturing, *rapid thermal processing* (RTP) has gained importance in recent years. In the processes using RTP, the principal issue is temperature regulation. Dassau et al. (2006) presented GP for the development of steady-state and dynamic temperature control models. To improve RTP an NMPC system was developed using GP, as done by Grosman and Lewin (2002). The models were based on the mathematical representation of the Steag RTP system and these were subsequently used in controlling the temperature at three different locations on the wafer, namely, centre of the wafer, 5 cm from the centre, and edge of the wafer (9.5 cm from the centre). The advantage of this NMPC system is that the process approaches the set point easily.

Hinchcliffe and Willis (2003) developed GP-based dynamic process models in two case studies—a system with time delay (Narendra and Parthasarathy 1990) and a cooking extruder (Elsey et al. 1997)—which were used to compare the performance of the GP algorithm with the filter based neural networks (FBNNs). It was observed that the two approaches exhibit comparable performances although GP has a potential advantage over FBNNs in that it can measure the performance of the model during its development. In another study involving development of GP based steady-state and dynamic input–output models, Willis et al. (1997) constructed models for a vacuum distillation column and a twin screw cooking extruder.

Coal gasification is more environment-friendly and efficient process for energy generation than coal combustion. The performance of a coal gasification process is significantly dependent on the quality of coal. For instance, process efficiency is adversely affected by the high ash content in a coal. Such coals are found in many countries, and form an important raw material for the coal gasification and combustion. Despite the wide-spread availability of high ash coals, modeling studies on gasification using these coals are much less in number when compared with the studies performed using low ash coals. Coal gasification is a nonlinear and complex process and its phenomenological modeling is a difficult, tedious and expensive task. In such circumstances, data-driven modeling provides a low-cost and relatively easier alternative to conduct process modeling if representative, statistically well-distributed and sufficient process data are available. Accordingly, Patil-Shinde et al. (2014) developed data-driven steady-state models for a pilot plant scale fluidized bed coal gasifier (FBCG) utilizing Indian coals with a high ash content. These models were constructed using process data from 36 experiments conducted in the FBCG. Specifically, four models predicting gasification related performance variables, namely, *$CO+H_2$ generation rate, syngas production rate, carbon conversion* and *heating value of syngas,* were developed using GP and multi-layer perceptron (MLP) neural network formalisms, separately. The input space of these models consisted of eight coal and gasifier process related parameters, namely *fuel ratio, ash content of coal, specific surface area of coal, activation energy of gasification, coal feed rate, gasifier bed temperature, ash discharge rate* and *air/coal ratio*. A comparison of the GP and MLP-based models revealed that their output prediction accuracies and generalization performance vary from good to excellent as indicated by the high training and test set correlation coefficient magnitudes lying between 0.920 and 0.996.

Gandomi and Alavi (2011) developed a new strategy for non-linear system modeling. They proposed a multistage genetic programming (MSGP) formulation to provide accurate predictions by incorporating the individual effects of predictor variables and interactions among them. The initial stage of MSGP formulates the output variable in terms of an influencing variable. Thereafter, a new variable is defined by considering the difference between the actual and predicted value. Finally, an interaction term is derived by considering the difference between the desired output values and those predicted by the individually developed terms. Gandomi and Alavi (2011) applied this strategy to various engineering problems such as simulation of pH neutralization process. The results yielded by the MSGP strategy were observed to be more accurate than that by the standard GP.

5.4.2 Energy and Fuels

The quality of fossil fuels is comprehended by their physical, chemical and thermodynamic properties. These properties are measured using various chemical, physical and instrumental methods. Often, the procedures involved in the property determination are tedious, time consuming and expensive. In such cases, the role of empirical models for property prediction becomes vital. A number of studies have been performed wherein, GP has been utilized for developing models for the prediction of a property.

Shokir (2008) developed a dew-point pressure (DPP) model to successfully predict the future performance of gas condensate reservoirs. This study used GP and the orthogonal least squares (GP-OLS) algorithm to generate a new DPP model as a function of the reservoir fluid composition consisting of mole fractions of methane to heptane+, nitrogen, carbon dioxide, hydrogen sulfide, molecular weight of heptane+ fraction, and reservoir temperature. The GP-OLS model was developed using a training dataset of 245 gas condensate samples and a test dataset of 135 samples. The prediction accuracy and generalization performance of the model were tested and validated by comparing the GP-OLS based DPP predictions with those from the correlations of Nemeth and Kennedy (1967), Elsharkawy (2002), and Peng—Robinson (1976); the said comparison indicated the accuracy of the GP-OLS based model to be better than the other correlations. Additionally, the impact of the independent variables on the predicted DPP was assessed by performing a sensitivity analysis, results of which were found to be comparable with that given by the equation of state (EoS) calculations. Shokir and Dmour (2009) employed similar strategy to develop a model for the prediction of viscosities of pure hydrocarbon gases (methane to pentane) and hydrocarbon gas mixtures that also contain minuscule amounts of non-hydrocarbon gases. Their GP-OLS based viscosity model covers wide ranges of temperatures (0–238 °C) and pressures (1–890 bar) and uses gas density, pseudo reduced pressure, pseudo reduced temperature, and molecular weight of pure and mixed hydrocarbons, as inputs. In the model development, Shokir and Dmour (2009) used training and test sets consisting of 6330 and 2870 data points, respectively. The results showed a good agreement between the model predicted and experimental gas viscosities, with only a 5.6 % average absolute relative error in respect of the test set outputs. This viscosity prediction model has an advantage that it does not require the measurement of gas viscosity at an atmospheric pressure, which is necessary in most of the previously developed correlations. Subsequently, Shokir et al. (2012) developed a model for the prediction of compressibility factor (*z*-factor) of sweet, sour and condensate gases. There exist over twenty complex correlations for calculating the *z*-factor. Firstly, the GP-OLS technique was applied to develop models for psuedo-critical pressure and temperature as a function of the gas composition (mole percent of C_1-C_{7+} , H_2S, CO_2 and N_2) and specific gravity of C_{7+}. A data set of 1150 gas samples was considered in the model development. These pseudo-critical pressure and temperature models were used to calculate the pseudo-reduced pressure and temperature. Next, using GP, a *z*-factor predicting model was developed as a function of the pseudo reduced pressure and pseudo reduced temperature. The *z*-factor model by Shokir et al. (2012) possesses higher prediction accuracy than that possessed by other empirical correlations and the EoS with an average absolute relative error of only 0.58 % and coefficient of correlation equal to 0.999.

With hydrogen gaining importance as an alternate to fossil fuels, it has become necessary to have the knowledge of its thermophysical properties. Muzny et al. (2013) developed a correlation for the prediction viscosity of normal hydrogen using GP. The correlation was developed for a wide range of temperatures—from the triple point to 1000 K—and pressures up to 200 MPa. This model agrees well with the experimentally determined viscosities over the temperature range of 200–400 K and for pressures up to 0.11 MPa with uncertainty magnitude less than 0.1 %. Outside this region, the model has an estimated uncertainty of 4 % for the saturated liquid and supercritical fluid phases. The uncertainty is larger along the saturated liquid boundary above 31 K and near the critical region.

The GP has also been utilized for the prediction of crude oil properties. For example, Fattah (2012, 2014) and AlQuraishi (2009) developed GP-based models for the prediction of *K*-value of crude oil components the gas-oil ratio of gas condensate and crude oil saturation pressure, respectively.

Pandey et al. (2015) proposed a multi gene genetic programming technique to predict the syngas yield and lower heating value for the municipal solid waste gasification in a fluidised bed gasifier. The predicted outputs were in good agreement with the experimental data.

5.4.3 Water Desalination and Wastewater Treatment (Membrane Technology)

Oils, fuels, solvents, paints, detergents, organic matter, and rusts, are a few typical contaminants present in the wastewater. With ever growing need for a high quality water, the need for treating the waste-water has also increased. In recent years, membrane technology has assumed an important role in the wastewater treatment. A significant difficulty with this technology is fouling of the membrane, which leads to a decline in the permeation flux. Accordingly, Lee et al. (2009) utilized genetic programming for the prediction of membrane fouling in a microfiltration (MF) system. The model was developed to predict the membrane fouling rate in a pilot scale drinking water production system consisting of a hollow fiber membrane of polyvinylidene fluoride (PVDF). The model was developed using the following input variables (predictors): operating conditions (*flow rate* and *filtration time*) and feed water quality (*turbidity*, *temperature* and *algae pH*). Lee et al. (2009) collected data from a membrane filtration system for 470 days, during which chemical washing was done three times. The operating conditions and water quality data used in the model development were analyzed separately during the three system runs.

The resultant GP model predicted accurately the membrane resistance and yielded very low RMSE values (ranging from 0.06 to 0.12). Shokrkar et al. (2012) also developed a GP model for an accurate quantification of the permeation flux decline during cross-flow membrane filtration of oily wastewater. A total of 327 data points covering the effects of individual variations in the five predictor variables, namely, temperature, cross flow velocity, trans-membrane pressure, oil concentration, and filtration time, were used in the model development. Simulations conducted using the GP model suggested that by increasing the filtration time, oil layer on the membrane surface thickens and permeation flux decreases. The permeation flux falls rapidly in the beginning of the filtration. The GP-model predicted results agreed within 95 % of the experimental data. This study clearly gives an idea of how each parameter affects the flux.

Reverse osmosis (RO) has been proved to be a promising technology for desalination. It has been observed that the performance of RO is negatively affected by the formation of scales of soluble salts. Cho et al. (2010) developed a GP model to study the effect of $CaSO_4$ scale formation on the RO membrane. The extent of RO fouling (permeation flux decline) is dependent on the applied pressure, time, volumetric concentration factor (VCF), stirring speed, and humic acid concentration. After training and validation, the correlation coefficient for the model predictions was 0.832. It was observed that the applied pressure and VCF have higher impact on the RO fouling. Park et al. (2012) have also developed a GP model for the analysis of the performance of an RO process. The input parameters considered by them are pH, oxidation reduction potential (ORP), conductivity, temperature, flux, TMP, and recovery time. The GP models were developed for trans-membrane pressure and membrane permeability separately for an early stage data (0–5 days) and the late stage data (20–24 days). The models matched the trend of the pilot plant data well. The sensitivity analysis of the GP models showed that the model-fitted early stage data exhibit higher sensitivity towards conductivity, flux and ORP, whereas the late stage data show higher sensitivity to temperature, recovery time, and ORP. It was also observed that the GP model predictions of membrane permeability were more accurate than the predictions of transmembrane pressure. In a recent study, Meighani et al. (2013) have conducted a thorough comparative analysis of the three modeling techniques (pore-blocking model, ANN and GP) for the prediction of permeate flux decline. The permeate flux is modeled as a function of the transmembrane pressure, feed temperature, cross flow velocity, pH and filtration time. Eight sets of experimental data were compiled from the literature to investigate the accuracy of the models. The correlation coefficients for the pore blocking, ANN, and GP models were found to be 0.9799, 0.9999 and 0.9723, respectively.

Okhovat and Mousavi (2012) predicted the performance of a nanofiltration process for the removal of heavy metals such as arsenic, chromium and cadmium. GP-based models were developed for studying the membrane rejection of arsenic, chromium and cadmium ions. Specifically, ions rejection (%) was considered as the model output while feed concentration and transmembrane pressure, formed the model inputs. The models showed satisfactory prediction accuracies with RMSE magnitudes ranging between 0.005 and 0.02. Suh et al. (2011) utilized GP for estimating the membrane damage during the membrane integrity test of a silica fluorescent nanoparticle microfiltration membrane. This model predicts the area of membrane damage for the experimental input parameters (concentration of fluorescent nanoparticles, permeate water flux, and transmembrane pressure). The GP model yielded good prediction results with mean absolute error (MAE) of 0.83.

5.4.4 Petroleum Systems

Wax or asphaltene precipitation is a common issue encountered by the petroleum industry, which leads to serious problems in crude oil production. This can occur due to recovery processes or a natural depletion in reservoir condition. Manshad et al. (2012a) utilized genetic programming neural network (GPNN) approach for the prediction of wax precipitation in crude oil systems. In GPNN, genetic programming is employed to choose an optimal architecture for the feedforward neural network (Ritchie et al. 2003, 2007). This methodology was proposed by Koza and Rice (1991). Manshad et al. (2012a) used a set of 87 experimental data points in the development of a GPNN model for the prediction of wax precipitation. Model's input parameters were compositions of C_1-C_3, C_4-C_7, C_8-C_{15}, C_{16}-C_{22}, C_{23}-C_{29} and C_{30+} fractions, specific gravity, system pressure and temperature. The CC magnitude pertaining to the training (test) set output predictions was 0.973 (0.930). Prediction performance of this GPNN model was compared with that of the multi-solid model and its prediction accuracy was found to be better than the latter model. Manshad et al. (2012b) utilized a similar strategy for the modeling of permeability reduction by asphaltene precipitation in Iranian crude oil reservoirs.

In petroleum production, the inflow performance relationship (IPR) is used for evaluating the reservoir deliverability. It is a graphical representation of the relationship that exists between the oil flow rate and bottom-hole flowing pressure. There are various empirical models for IPR modeling. Sajedian et al. (2012) developed a GP model for the prediction of the inflow performance of the vertical oil wells experiencing two phase flow and compared this model with the multi-layer perceptron model and empirical correlations. Their study investigated the ability of GP and MLP in establishing and predicting the well-inflow performance for the solution-gas-drive reservoirs. Though, for an IPR only the bottom-hole pressure and its corresponding oil flow rate are required, additional parameters such as recovery factor, average reservoir pressure, bubble point pressure, oil formation volume factor at bubble point pressure, solution gas-oil ratio at bubble point pressure and gas viscosity at bubble point pressure, become necessary for defining a specific IPR. For the GP-based model, Sajedian et al. (2012) considered data from sixteen different simulated reservoir models. Data from fourteen reservoirs were used in the model training while data from the remaining two reservoirs were utilized for testing and validation. The prediction performance of the developed GP and MLP models was compared with that of the existing empirical models. This comparison clearly indicated that the GP model produced the smallest error for the unseen data.

5.4.5 Heat Transfer

In designing thermal systems for industrial processes, it is necessary to predict the performance of system components. Theoretically, such a performance calculation can be carried out with the help of "first principles" based governing equations. However, complexities arising from the factors like turbulence, temperature dependence of properties, and the geometry, make the first principles based modeling a difficult exercise. To overcome the said difficulty, Cai et al. (2006) presented a GP-based methodology for developing heat transfer correlations predicting the performance of a thermal system component. They demonstrated the idea with two case studies—heat transfer in compact heat exchangers and, heating and cooling of liquids in pipes. To prevent complex heat transfer correlation functions, Cai et al. (2006) modified the GP method by imposing a penalty on such functions. The procedure was applied to the heat exchanger data reported by McQuiston (1978) for a compact multi-row multi-column heat exchanger, with air as the over-tube fluid and water as the in-tube fluid. Performance of the air side heat transfer was indicated by the Colburn j-factor, for which a model was developed in terms of the Reynolds number and a non-dimensional geometric parameter representing an air side area ratio. The GP-based models yielded a smaller prediction error when compared with that reported by McQuiston (1978). A comparison of the experimental and model predicted j-factors showed a minor scatter. Earlier, Lee et al. (1997) developed a GP model for the critical heat flux (CHF) prediction for an upward water flow in vertical round tubes, under low pressure and flow conditions. The data for modeling were obtained from the KAIST CHF data bank (414 and 314 CHF data). These models were developed for predicting CHF at the inlet (upstream condition) and local conditions (CHF point conditions). The errors pertaining to the GP-model predictions were small when compared with the predictions of other existing correlations. Pacheco-Vega et al. (2003) also used GP to construct heat transfer correlations for a compact heat exchanger. Two datasets were used in testing the capability of the GP-based models–first being the artificial data from a one dimensional function and the second from the previously determined correlations for a single phase air-water heat exchanger. In both cases, the GP-based heat transfer correlations showed good prediction capability.

5.4.6 Unit Operations

In one of its early applications, Greeff and Aldrich (1998) employed GP in various leaching experiments as described below.

1. *Acid pressure leaching of nickeliferous chromites*: Based on the data by Das et al. (1995) GP models were developed for the dissolution of nickel, cobalt and iron from the beneficiated lateritic chromite samples as a function of temperature, ammonium sulfate concentration, and acid concentration at different time intervals.

2. *Leaching of uranium and radium*: Here, models were developed for the co-extraction of uranium and radium from a high grade arseniferous uranium ore (Kondos and Demopoulos 1993), where the percentage of radium and uranium was modeled as a function of pulp density, concentration of hydrochloric leaching acid, concentration of the calcium chloride additive and time. Cross-validation (using *leave-k out* method, $k = 3$) was performed to evaluate the prediction performance of the models. All the GP-based models were found to be comparable and significantly more accurate than those developed by means of the standard least-squares methods. In another leaching related study, Biswas et al. (2011) analyzed the leaching of manganese from low grade sources, using genetic algorithms, GP and other strategies where they made an extensive comparison of the data-driven modeling techniques.

Wang et al. (2008a) applied GP to a complex heat-integrated distillation system to synthesize a flow-sheet for separating a multicomponent mixture into pure components at a minimum total annual cost. Both sharp and non-sharp distillations were considered in the modeling. Based on the knowledge of chemical engineering, unique solution encoding methods and solution strategies were proposed in this study. The GP-based synthesis algorithm automatically optimizes the problem of complex distillation systems. In related studies, Wang et al. (2008b) and Wang and Li (2008, 2010) made use of GP for the synthesis of non-sharp distillation sequences, synthesis of multicomponent products separation sequences, and synthesis of heat integrated non-sharp distillation.

5.4.7 System Identification

In a situation when a chemical process is too complex to be understood and modeled at a fundamental (first-principles) level, system identification is used for its modeling. It refers to the development of an empirical (often *black-box*) model for a dynamic system/process from the experimental data. Here also exists a scope for incorporating into the model any process knowledge available *a priori*. When compared with the first principles models, it is easier and time-saving to build data-driven dynamic process models since industries routinely collect large amounts of data via distributed control systems. Another advantage of the system identification is that it can handle unmeasured process dynamics and uncertainties, which are difficult to take care of using first-principles based modeling approaches.

Consider the dynamics of a single input—single output (SISO) system represented as:

$$y_{k+1} = f\left(y_k, y_{k-1}, y_{k-2}, \ldots, y_{k-m+1}; u_k, u_{k-1}, u_{k-2}, \ldots, u_{k-n+1}\right) \quad (5.11)$$

where k denotes the discrete time, y_{k+1} refers to the *one-time-step-ahead* process output, u is the manipulated variable, f refers to the linear/nonlinear functional relationship (to be identified) between y_{k+1} and the current (kth) and the lagged values of the input–output variables, and m and n refer to the number of lags in the output and input variables, respectively. Traditionally, system identification employs statistical methods for constructing models of dynamical systems from their experimental (measured) data consisting of y_k and u_k values. In one of the early studies on the applications of GP technique in chemical engineering, Kulkarni et al. (1999) utilized the methodology for system identification by conducting two case studies involving nonlinear pH and heat exchanger control systems. The objective in both case studies was to obtain an appropriate non-linear form of f given the time series values of the process input (u) and the corresponding output (y). To derive GP-based models, Kulkarni et al. (1999) used synthetic process data. In actual practice, these data are collected by conducting open-loop tests wherein manipulated variable (u) is varied randomly and its effect on y is monitored. The specific systems considered in two case studies were: (1) *continuous stirred tank reactor* (CSTR) wherein hydrochloric acid and sodium hydroxide streams are mixed and effluent stream's pH is measured and controlled using a model-based control strategy, and (2) a nonlinear heat exchanger control system wherein heater voltage and exchanger outlet temperature are the manipulated and controlled variables, respectively. The CC values in respect of the training and test set output predictions by both the models were greater than 0.99 indicating an excellent prediction and generalization performance by the models identified by GP. In a similar work, Nandi et al. (2000) performed GP-based system identification of a fluidized catalytic cracking (FCC) unit, wherein an exothermic reaction $\left(A \rightarrow B \rightarrow C\right)$ takes place. Here, two GP models each possessing an excellent prediction accuracy and generalization capability were developed for the prediction of *one-time-step-ahead* and *three-time-steps-ahead* outlet concentrations of species B.

Sankpal et al. (2001) utilized a GP-based model for the monitoring of a process involving continuous production of gluconic acid by the fermentation of sucrose and glucose solution in the presence of aspergillus niger immobilized on cellulose fabric. During the continuous conversion of glucose and sucrose the rate of gluconic acid formation drops as fermentation progresses. To compensate for this loss in efficiency the residence time needs a suitable adjustment. As online determination of the reaction rates and substrate concentration is cumbersome, a GP-based model was developed to predict the conversion (z) as a function of the time. For a given time series $\left\{z_t, z_{t-1}, \ldots, z_T\right\}$, of length T, a model was developed to compute z_{t+1}, where $z_{t+1} = f\left(z_t, z_{t-1}, \ldots, z_{t-\alpha}\right);\ \alpha \le t \le (T-1);0 \le \alpha \le L$; where t refers to the discrete time, α refers to the number of lags and L denotes the maximum permissible lags. The expression for the one-time-step-ahead prediction gave high prediction accuracies with correlation coefficient magnitudes ≈ 1.

Timely and efficient process fault detection and diagnosis (FDD) is of critical importance since it helps in, for example, energy savings, reduction in operating and maintenance costs, curbing damage to the equipment, avoiding economic losses due to process down-time, and most importantly preventing mishaps and injuries to plant personnel. Process identification and FDD are related since a good process model is needed for conducting the latter. Witczak et al. (2002) used a GP-based approach for process identification and fault diagnosis of non-linear dynamical systems. They proposed a new fault detection observer and also demonstrated the use of GP for increasing the convergence rate of the observer. The reliability and effectiveness of the identification network proposed by Witczak et al. (2002) were checked by constructing models for a few individual parts of the evaporation section at Lublin Sugar Factory S.A. and for an induction motor.

Madar et al. (2005) applied GP to develop nonlinear input–output models for dynamic systems. They hybridized GP and the orthogonal least squares (OLS) method for the selection of a model structure using a tree representation based symbolic optimization. The strategy was implemented using MATLAB GP-OLS Toolbox—a rapid prototyping system—for predicting (a) the structure of a known model, (b) the model order for a continuous polymerization reaction, and (c) both order and structure of the model for Van der Vusse reaction. The results of this modeling study indicated that the proposed strategy provides an efficient strategy for the selection of the model order and identification of the model structure. Recently, Faris and Sheta (2013) also adopted GP for the system identification of Tennessee Eastman Chemical process reactor.

5.4.8 Miscellaneous Applications

Other than the main areas of GP applications covered in the preceding subsections, there exist a number of studies wherein the formalism has been employed to address diverse problems in chemical engineering.

Genetic programming based methods are frequently employed in the multi-scale modeling of process and product data (Seavey et al. 2010). A number of properties such as critical flux heat prediction, viscosity, dew point pressure, compressibility, permeation flux, solubility and gas consumption have been modeled using the GP approach. It is often applied in conjunction with the genetic algorithms for automatically building the kinetic models in terms of ordinary differential equations (ODEs) to model complex systems of chemical reactions (Cao et al. 1999).

The proportional-integral-derivative (PID) controllers are the most commonly used industrial process control strategies. Implementation of a PID controller requires knowledge of three parameters, namely, the proportional gain (K_p), the integral time (T_i) and the derivative time (T_d). Ziegler-Nichols (ZN) proposed a method to determine these parameters. However, parameters evaluated in this manner usually have an overshoot of $\approx 25\,\%$ thus making their fine tuning essential. Almeida et al. (2005) used GP for fine tuning PID controller parameters designed via ZN technique. The GP algorithm was programmed to create an initial population of 500 individuals (candidate solutions), which evolved over 30 generations. The GP-based fine tuning of PID parameters is a simple and an efficient method and it improved the settling time of the system with a minimum overshoot and with a null steady-state error. This performance was clearly seen in the three case studies that Almeida et al. (2005) performed, namely, a high order process, a process with a large time delay, and a highly non-minimum phase process. Their GP-based approach when compared with four other fine tuning techniques, was found to exhibit superior performance. GP has been also used in the implementation of nonlinear model predictive control strategies for the rapid acquisition of efficient models to accurately predict the process trajectories (Tun and Lakshminarayanan 2004).

Often a situation arises, in which an appropriate hardware-based sensor for measuring a process variable is either unavailable or the alternative analytical procedure for its determination is time-consuming, expensive and tedious. In such cases, a suitably developed *soft-sensor* can be employed for estimating the magnitude of the "tricky-to-measure" process variable/parameter. Soft-sensor is a software module consisting of a mathematical model that utilizes the available quantitative information of other process variables and parameters for estimating the magnitude of the chosen variable/parameter. Recently, Sharma and Tambe (2014) demonstrated that GP can be effectively used to develop soft-sensors models for biochemical systems. Specifically, they developed the GP-based soft-sensors possessing excellent prediction accuracy and generalization capability for two biochemical processes, namely, extracellular production of lipase enzyme and bacterial production of poly(3-hydroxybutyrate-co-3-hydroxyvalerate) copolymer. The strategy developed by Sharma and Tambe (2014) is generic and can be extended to develop soft-sensors for various other types of processes.

Despite the fact that most industrial reactions employ heterogeneous catalysis, GP has received little attention in this area. Baumes et al. (2009), however, used GP for an advanced performance assessment of industrially relevant heterogeneous catalysts. Epoxides are cyclic ethers with three ring atoms. Their structure is highly strained, which makes them more reactive than other ethers. Epoxidation of double bonds to obtain epoxides, is carried out with micro- and meso-porous titanosilicates (Ti-MCM-41 and Ti-ITQ-2) as catalysts. The catalytic activity of these materials can be improved by controlling their surface properties. Baumes et al. (2009) achieved this control by anchoring alkyl-silylated agents onto the catalyst surface, which modifies the hydrophilic nature of the catalyst. In the absence of a rigorous kinetic study of the synthesized catalysts, they used GP for obtaining a model for the conversion of reactant in presence of a catalyst as a function of the reaction time. The catalyst activity was assessed via the conversion versus reaction time curve. Catalyst performance evaluation by this method is based on the total reaction time. The catalyst activity was monitored during 16 h of reaction in a batch reactor and the GP model constructed thereby resulted in an adjusted fitness of 0.93. Baumes et al. (2009) also presented a GP algorithm with the *context aware crossover* (CAX) operator, which did not perform better than the ordinary crossover operator.

The chemical industry requires reliable and accurate thermodynamic data for different fluids, covering a wide range of temperature, pressure and composition (Hendriks et al. 2010). The knowledge of thermodynamic properties of fluids plays a critical role in the design and operation of chemical processes. A large number of phenomenological and empirical models have been developed for the prediction of thermodynamic properties of fluids although application of GP in this area is not wide-spread. Bagheri et al. (2014) developed a linear genetic programming (LGP) based quantitative structure-property relationship (QSPR) model for the prediction of standard state real gas entropy of pure materials. The LGP was utilized for 1727 diverse chemicals comprising 82 material classes obtained from *Design Institute for Physical Properties* (DIPPR) database. The model yielding the best prediction accuracy contained four input parameters describing two topological features and a single 3D-MoRSE and a molecular property descriptor; the model is given by,

$$S^{\circ}_{298} = \Big(11.15 \pm 0.12 * R^{1.1}_{WW} - (1.54 \pm 0.04) \times BAC - (32.35 \pm 1.12) \times Mor11u - (1.91 \pm 0.08) \times TPSA(NO)^{0.9} + (188.80 \pm 2.00) \quad (5.12)$$

where S°_{298} refers to the standard state real gas entropy, R_{WW} represents the reciprocal hyper-detour index, *BAC* is the Balaban centric index, *TPSA*(*NO*) refers to the topological polar surface area, and *Mor11u* describes the 3D molecular representation of the structures based on electron diffraction (3D-MoRSE); all these parameters can be derived from the chemical structure. The predictions of the above QSPR model resulted in the RMSE and coefficient of determination (r^2) magnitudes of 52.24 J/(mol K) and 0.885, respectively. This model by Bagheri et al. (2014) is helpful in the design of materials and exergy analysis.

The vapor-liquid equilibrium (VLE) models are used for the estimation of vapor and liquid compositions under thermodynamic equilibrium conditions. Seavey et al. (2010) employed GP for modeling VLE as also polymer viscosity. Their VLE model relates the temperature (T), pressure (P) and overall molar composition (z_i) to the overall vapor mole fraction (Ψ) and composition of liquid and vapor phases (x_i and y_i, respectively).

$$\Psi,\ x_i, y_i = f_1\,(T, P, z_i) \quad (5.13)$$

In this study the unknown function f_1 is characterized using a combination of the fundamental and empirical modeling techniques to fit the data. The VLE model development was initiated using the well-known Rachford–Rice equation,

$$x_i = \frac{z_i}{1 + \Psi\,(K_i - 1)} \quad (5.14)$$

$$y_i = K_i x_i \quad (5.15)$$

$$\sum_i \frac{z_i\,(K_i - 1)}{1 + \Psi\,(K_i - 1)} = 0 \quad (5.16)$$

$$K_i = \frac{\gamma_i P_i^{sat}}{P} \quad (5.17)$$

where the overall vapor mole fraction (Ψ), K_i values, activity coefficients (γ_i), vapor pressure (P_i^{sat}), liquid mole fraction (x_i), and vapor mole fraction (y_i), are unknown; the vapor pressure can be calculated by using Antoine equation. The activity coefficient, γ_i, is a function of the temperature and the excess Gibbs energy of each component g_i (J/mol) is given by

$$RT \ln \gamma_i = g_i \tag{5.18}$$

where R is the ideal gas law constant (8.314 J/mol K). The excess Gibbs energy is then modeled using GP following which the overall vapor mole fraction and compositions of the vapor and liquid phases are evaluated. In the same study, Seavey et al. (2010) developed a GP-based model for predicting the polymer viscosity, wherein the fundamental Williams-Landel-Ferry (WLF) equation was used along with GP to capture the effect of temperature on the polymer viscosity. Integrating GP with the fundamental equations led to the models that are compact and containing fewer parameters.

Sugimoto et al. (2005) employed GP for obtaining dynamic models for two enzyme-catalyzed reactions involving adenylate kinase and phosphofructokinase. Data for developing the GP models were obtained by simulating the respective kinetic models. The topology (structure) and the corresponding parameters of the GP-based models obtained by Sugimoto et al. (2005) matched closely with the respective phenomenological models. Their study indicates that the GP-based modeling approach presented by them can be applied to identify metabolic reactions from the observable reaction data.

Principal component analysis (PCA) is a standard statistical technique, which is commonly employed in the dimensionality reduction of large highly correlated data sets. PCA's main limitation is that it is a linear technique and therefore it finds restricted utility in analyzing data from nonlinearly behaving chemical processes. Hiden et al. (1999) proposed a GP-based technique for non-linear PCA and demonstrated its applicability using two simple non-linear systems and data collected from an industrial distillation column.

Marref et al. (2013) studied the use of GP and GA for the derivation of corrosion rate expressions for steel and zinc. Here, GP-based corrosion rate (μm) predicting model was obtained using the major influential environmental factors as inputs and GA was used to estimate the parameters of the engineered GP-based model. The five inputs used in the modeling were temperature, time of wetness, contaminant (SO_2) content, contaminant (chloride) content, and exposure time. The corrosion rate expressions yielded by GP and GA exhibited good accuracy in predicting the corrosion rate.

Bagheri et al. (2012) predicted the sublimation enthalpies of organic contaminants using their 3D molecular structure. Gene expression programming was integrated with the quantitative structure-property relationship (QSPR), which produced promising results with the coefficient of determination magnitude of 0.931 and RMSE of 9.87 kJ/mol. A dataset of 1586 organic contaminants from 73 diverse material classes was used in the model development.

Bagheri et al. (2013) also developed a model for the prediction of the formation enthalpies of nitro-energetic materials based on multi expression programming (MEP). Multi expression programming is a sub-area of GP. A dataset of 35 nitro-energetic materials with formation enthalpies ranging between 115.4 and 387.3 kJ/mol were used by Bagheri et al. (2013). The MEP based model developed thereby considers three molecular descriptors—*Kier flexibility index, the mean information index* and *R maximal autocorrelation of lag* 2, as inputs. This model yielded an acceptable accuracy for the prediction of formation enthalpy.

5.5 Conclusion

Genetic programming is one of the most intellectually appealing computational intelligence formalisms. Its attractiveness stems from the following features: (a) unlike ANNs and SVR, genetic programming does not make any assumptions about the form of the data-fitting model and thus GP-based models exhibit far greater diversity, (b) depending upon the relationship between the dependent and predictor variables in the data set, it arrives at an appropriate linear or nonlinear data-fitting function and all its parameters, and (c) since it provides system-specific closed-form explicit linear/nonlinear data-fitting functions, GP-based models are easier to grasp, deploy, and use.

As this chapter has revealed, GP applications in chemical sciences and engineering/technology have spanned a very wide problem space. GP is a young field of research and efforts are directed at understanding its functioning in greater details and devising methodologies to make it more efficient and faster. On the other hand, practitioners in chemical sciences and engineering are finding ever increasing applications of GP. In general, the phenomenal increase in the CPU speeds in the last 30 years and the emergence of parallel computing have definitely assisted in developing GP-based solutions for some real-life modeling and classification problems. Despite its novelty and potential, genetic programming has not been explored in chemical sciences and engineering for classification and modeling applications as widely as ANNs and SVM/SVR. Thus, there is still a lot of work to be done in the context of the GP-based applications in chemical sciences and engineering. In what follows, some guidelines as also an outlook for the future developments involving GP applications in chemical sciences and engineering are provided.

There exist a number of studies wherein prediction and generalization performance of GP has been compared with other data-driven modeling/classification methodologies such as ANNs and SVR/SVM. Often it has been found that for non-linear systems, no single modeling/classification formalism consistently outperforms the other methods. Accordingly, it is advisable to explore multiple approaches such as GP, ANN and SVR/SVM for conducting modeling/classification of nonlinear systems and choose the best performing one.

Although there are a few studies demonstrating its capability to obtain phenomenological models (see for example, Schmidt and Lipson 2014) this fascinating and un-matched feature of GP has been largely ignored. It is thus necessary to exploit the stated GP characteristic extensively for developing first principles models in chemistry and chemical engineering.

A large number of semi-quantitative and purely empirical correlations are routinely used in chemistry and chemical engineering/technology. Prediction accuracies of many of these correlations are far from satisfactory. Also, in a number of instances, linear correlations—since being easy to develop—are utilized although the underlying phenomena being modelled are nonlinear. It is possible to construct these correlations freshly using GP to improve their prediction accuracy and generalization performance. The notable feature of GP that it is by itself capable of arriving at an appropriate linear or a nonlinear model, can be gainfully exploited for the development of the correlations alluded to above.

GP possesses certain limitations such as it is computationally demanding, the solutions provided by it may be over fitted and an extensive heuristics is involved in obtaining the best possible solution. Despite these limitations it is envisaged that owing to its several attractive and unique features together with the advent of user-friendly software such as *Eureqa Formulize*, *Discipilus* and *ECJ* tool kit, GP will be extensively and fruitfully employed for providing meaningful relationships and insights into the vast data available in the domain of chemical sciences and engineering.

References

Aguiar-Pulido V, Gestal M, Cruz-Monteagudo M et al (2013) Evolutionary computation and QSAR research. Curr Comput Aided Drug Des 9 (2):206-25

Ahmed S, Zhang M, Peng L (2014) Improving feature ranking for biomarker discovery in proteomics mass spectrometry data using genetic programming. Connect Sci 26(3): 215-243

Alfrey T Jr., Price CC, (1947) Relative reactivities in vinyl copolymerization. J Polym Sci 2: 101-106

Almeida GM, Silva VVR, Nepomuceno EG et al (2005) Application of genetic programming for fine tuning PID controller parameters designed through Ziegler-Nichols technique. In: Wang L, Chen K, Ong YS (eds) Proceedings of the first international conference on advances in natural computation (ICNC'05), vol part III. Springer, Heidelberg, 2005

AlQuraishi AA (2009) Determination of crude oil saturation pressure using linear genetic programming. Energy Fuels 23:884-887

Alsberg BK, Marchand-Geneste N, King RD (2000) A new 3D molecular structure representation using quantum topology with application to structure–property relationships. Chemometr Intell Lab 54(2): 75–91

Anderson AC (2003) The process of structure-based drug design. Chem Biol 10:787–797
Anderson B, Svensson P, Nordahl M et al (2000) On-line evolution of control for a four-legged robot using genetic programming. In: Cagnoni S et al (eds) Real World Applications of Evolutionary Computing, Springer, Berlin, p 319-326
Atkinson AJ Jr., Huang S-M, Lertora J, Markey SP (eds) (2012) Principles of Clinical Pharmacology. Elsevier, San Diego, USA
Bagheri M, Bagheri M, Gandomi AH, Golbraikh A (2012) Simple yet accurate prediction method for sublimation enthalpies of organic contaminants using their molecular structure. Thermochimica Acta 543: 96-106
Bagheri M, Gandomi AH, Bagheri M, Shahbaznezhad M (2013) Multi-expression programming based model for prediction of formation enthalpies of nitro-energetic materials. Expert Systems 30: 66–78
Bagheri M, Borhani TNG, Gandomi AH et al (2014) A simple modelling approach for prediction of standard state real gas entropy of pure materials. SAR QSAR Environ Res 25:695-710
Barmpalexis P, Kachrimanis K, Tsakonas A et al (2011) Symbolic regression via genetic programming in the optimization of a controlled release pharmaceutical formulation. Chemom Intell Lab Syst 107:75-82
Baumes LA, Blansché A, Serna P et al (2009) Using genetic programming for an advanced performance assessment of industrially relevant heterogeneous catalysts. Mater Manuf Processes 24:282-292
Biswas A, Maitre O, Mondal DN et al (2011) Data-driven multiobjective analysis of manganese leaching from low grade sources using genetic algorithms, genetic programming, and other allied strategies. Mater Manuf Processes 26:415-430
Bonet B, Geffner H (1999) Planning as heuristic search: New results. In: Proceedings of the 5th European Conference on Planning (ECP-99). Springer-Verlag, Heidelberg, Germany, p 360–372
Cabena P, Hadjnian P, Stadler R et al (1997) Discovering data mining: from concept to implementation, Prentice Hall, USA
Cai W, Pacheco-Vega A, Sen M et al (2006) Heat transfer correlations by symbolic regression. Int J Heat Mass Transfer 49(23-24):4352–4359
Can B, Heavy C (2012) A comparison of genetic programming and artificial neural networks in metamodeling of discrete event simulation models. Comput Oper Res 39(2):424-436
Cantú-Paz E, Kamath C (2003) Inducing oblique decision trees with evolutionary algorithms. IEEE Trans Evol Comput 7(1):54-68
Cao H, Yu J, Kang L et al (1999) The kinetic evolutionary modeling of complex systems of chemical reactions. Comput Chem 23(2):143-151
Cartwright H (2008) Using artificial intelligence in chemistry and biology. Taylor & Francis, Boca Raton, FL
Castelli M, Vanneschi L, Silva S (2014) Prediction of the unified Parkinson's disease rating scale assessment using a genetic programming system with geometric semantic genetic operators. Expert Syst Appl 41(10): 4608–4616
Cevik A (2007) Genetic programming based formulation of rotation capacity of wide flange beams. J Constr Steel Res 63: 884–893
Cheema JJS, Sankpal NV, Tambe SS et al (2002) Genetic programming assisted stochastic optimization strategies for optimization of glucose to gluconic acid fermentation. Biotechnol Progr 18:1356-1365
Cho J-S, Kim H, Choi J-S et al (2010) Prediction of reverse osmosis membrane fouling due to scale formation in the presence of dissolved organic matters using genetic programming. Desalin Water Treat 15:121-128
Christin C, Hoefsloot HC, Smilde AK et al (2013) A critical assessment of feature selection methods for biomarker discovery in clinical proteomics. Mol Cell Proteomics 12(1):263-76
Costa P, Lobo JMS (2001) Modeling and comparison of dissolution profile. Eur J Pharm Sci 13:123–133

Das GK, Acharya S, Anand S et al (1995) Acid pressure leaching of nickel-containing chromite overburden in the presence of additives. Hydrometallurgy 39:117-128

Dassau E, Grosman B, Lewin DR (2006) Modeling and temperature control of rapid thermal processing. Comput Chem Eng 30:686-697

ECJ 22- A java-based evolutionary computation research system (2014) http://cs.gmu.edu/~eclab/projects/ecj/. (Accessed 3 Dec 2014)

Eggermont J, Kok JN, Kosters WA (2004) Genetic programming for data classification: partitioning the search space. Presented at the 19th Annual ACM Symposium on Applied Computing (SAC'04), Nicosia, Cyprus, 14-17 March 2004, p 1001-1005

Elsey J, Riepenhausen J, McKay B et al (1997) Modelling and control of a food extrusion process. Comput Chem Eng 21 (supplementary): 5361-5366. Proceedings of the European Symposium on Computer Aided Process Engineering , ESCAPE-7 , Trondheim, Norway

Elsharkawy A (2002) Predicting the dewpoint pressure for gas condensate reservoir: Empirical models and equations of state. Fluid Phase Equilib 193:147–165

Espejo P, Ventura S, Herrera F (2010) A Survey on the Application of Genetic Programming to Classification. IEEE Trans Syst Man and Cybern 40(2): 121-144

Estrada-Gil JK, Fernández-López JC, Hernández-Lemus E, Silva-Zolezzi I, Hidalgo-Miranda A, Jiménez-Sánchez G, Vallejo-Clemente EE (2007) GPDTI: A Genetic Programming Decision Tree Induction method to find epistatic effects in common complex diseases. Bioinformatics 23(13):i167-i174

Faris H, Sheta AF (2013) Identification of the Tennessee Eastman chemical process reactor using genetic programming. Int J Adv Sci Technol 50:121-139

Fattah KA (2012) K-value program for crude oil components at high pressures based on PVT laboratory data and genetic programming. J King Saud University—Eng Sci 24:141-149

Fattah KA (2014) Gas-oil ratio correlation (RS) for gas condensate using genetic programming. J Petrol Explor Prod Technol 4:291-299

Ferreira C (2001) Gene Expression Programming in Problem Solving. In: Roy R, Köppen M, Ovaska S, Furuhashi T, Hoffmann F (eds) Soft computing and industry-recent applications, Springer-Verlag, Heidelberg, p 635-654

Finley SD, Chu LH, Popel AS (2014) Computational systems biology approaches to anti-angiogenic cancer therapeutics. Drug Discov Today. doi: 10.1016/j.drudis.2014.09.026.

Flasch O (2014) A Friendly Introduction to RGP, RGP release 0.4-1. Available online: http://cran.r-project.org/web/packages/rgp/vignettes/rgp_introduction.pdf. Accessed 6 Nov 2014

Gandomi AH, Alavi AH (2011) Multi-stage genetic programming: A new strategy to nonlinear system modeling. Information Sciences 181: 5227–5239

Garcia B, Aler R, Ledeezma A, Sanchis (2008) Genetic Programming for Predicting Protein Networks. In: Geffner H et al (eds) Proceedings of the 11th Ibero-American Conference on AI, IBERAMIA 2008, Lisbon, Portugal. Lecture notes in computer science, vol 5290. Springer-Verlag, Germany, p 432-441

Garg A, Vijayaraghavan V, Mahapatra SS, Tai K, Wong CH (2014) Performance evaluation of microbial fuel cell by artificial intelligence methods. Expert Syst Appl 41(4):1389-1399

Ghosal K, Chandra A, Rajabalaya R et al (2012) Mathematical modeling of drug release profiles for modified hydrophobic HPMC based gels. Pharmazie 67(2):147-55

Ghugare SB, Tiwary S, Elangovan V et al (2014a) Prediction of higher heating value of solid biomass fuels using artificial intelligence formalisms. Bioenergy Res 7:681-692

Ghugare SB, Tiwary S, Tambe SS (2014b) Computational intelligence based models for prediction of elemental composition of solid biomass fuels from proximate Analysis. Int J Syst Assur Eng Manag DOI 10.1007/s13198-014-0324-4 Published online December 7, 2014

Giri BK, Hakanen J, Miettinen K et al (2012) Genetic programming through bi-objective genetic algorithms with a study of a simulated moving bed process involving multiple objectives. Appl Soft Comput 13:2613-2623

Globus AI, Lawton J, Wipke T (1999) Automatic molecular design using evolutionary techniques. Nanotechnology 10: 290-299

Goldberg DE (1989) Genetic algorithms in search, optimization, and machine learning. Addison-Wesley, Reading, MA
Greeff DJ, Aldrich C (1998) Empirical modelling of chemical process systems with evolutionary programming. Comput Chem Eng 22:995-1005
Grosman B, Lewin DR (2002) Automated nonlinear model predictive control using genetic programming. Comput Chem Eng 26:631-640
Grosman B, Lewin DR (2004) Adaptive genetic programming for steady-state process modeling. Comput Chem Eng 28:2779-2790
Gu MB, Mitchell RJ, Kin BC (2004) Whole cell based biosensors for environmental monitoring and application. Adv Biochem Eng/Biotechnol 87: 269-305
Guido RV, Oliva G, Andricopulo AD (2008) Virtual screening and its integration with modern drug design technologies. Curr Med Chem 15(1):37-46
Güres S, Mendyk A, Jachowicz R et al (2012) Application of artificial neural networks (ANNs) and genetic programming (GP) for prediction of drug release from solid lipid matrices. Int. J. Pharm. 436(1–2): 877–879
Gustafson SM, Burke EK, Krasnogor N (2005) On improving genetic programming for symbolic regression. In: Proceedings of the 2005 IEEE Congress on Evolutionary Computation, Edinburgh, UK, 2-5 September 2005. Volume 1, IEEE Press, Los Alamitos, p 912–919
Hendriks E, Kontogeorgis GM, Dohrn R et al (2010) Industrial Requirements for Thermodynamics and Transport Properties. Ind Eng Chem Res 49:11131–11141
Hiden HG, Willis MJ, Tham MT et al (1999) Non-linear principal components analysis using genetic programming. Comput Chem Eng 23: 413-425
Hinchcliffe MP, Willis MJ (2003) Dynamic systems modeling using genetic programming. Comput Chem Eng 27:1841-1854
Holland JH (1975) Adaptation in natural and artificial systems. University of Michigan Press, Ann Arbor
Hosseini SH, Karami M, Olazar M et al (2014) Prediction of the minimum spouting velocity by genetic programming approach. Ind Eng Chem Res 53:12639-12643
Iba H (1996) Random tree generation for genetic programming. In: Voigt H-M, Ebeling W, Rechenberg I, Schwefel, H.-P (eds) 4th International Conference on Parallel Problem Solving from Nature (PPSN IV), Berlin, Germany, 22 - 26 September 1996. Lecture Notes in Computer Science, vol 1411. Springer-Verlag, Germany, p 144–153
Iba H, Paul TK, Hasegawa Y (2010) Applied genetic programming and machine learning. Taylor and Francis, Boca Raton, FL
Ivanova N, Lykidis A (2009) Metabolic Reconstruction. Encyclopedia of Microbiology (Third Edition) p 607–621
Keedwell E, Narayanan A (2005) Intelligent bioinformatics. John Wiley & Sons, Chichester, England
Khan MW, Alam M (2012) A survey of application: Genomics and genetic programming, a new frontier. Genomics 100(2): 65-71
Kondos PD, Demopoulos GP (1993) Statistical modelling of O_2-$CaCl_2$-HCl leaching of a complex U/Ra/Ni/As ore. Hydrometallurgy 32: 287-315
Kotanchek M (2006) Symbolic regression via genetic programming for nonlinear data modeling. In: Abstracts, 38th Central Regional Meeting of the American Chemical Society, Frankenmuth, MI, United States, 16-20 May 2006. CRM-160
Koza JR (1990) Genetic programming: a paradigm for genetically breeding populations of computer programs to solve problems. Stanford University, Stanford
Koza JR (1992) Genetic Programming: On the programming of computers by means of natural selection. MIT Press, Cambridge, MA, USA
Koza JR (1991) Concept formation and decision tree induction using the genetic programming paradigm. In: Proceedings of the 1st Workshop on Parallel Problem Solving from Nature (PPSN-1), London, UK. Springer-Verlag, Heidelberg, Germany, p 124–128
Koza JR, Mydlowee W, Lanza G, Yu J, Keane MA (2001) Reverse engineering of metabolic pathways from observed data using genetic programming. Pac Symp Biocomput 6: 434-445

Koza JR, Poli R (2005) Genetic programming. In: Burke EK, Kendall G (eds), Search Methodologies: Introductory Tutorials in Optimization and Decision Support Techniques, Springer, Boston, p 127-164

Koza JR, Rice JP (1991). Genetic generation of both the weights and architecture for a neural network. Neural Networks 2:397-404

Kulkarni BD, Tambe SS, Dahule RK et al (1999) Consider genetic programming for process identification. Hydrocarbon Process 78:89-97

Langdon WB, Barrett SJ (2005) Evolutionary computing in data mining In: Ghosh A, Jain LC (eds) Studies in Fuzziness and Soft Computing, vol 163. Springer-Verlag, Heidelberg, p 211-235

Lanza G, Mydlowec W, Koza JR (2000) Automatic creation of a genetic network for the lac operon from observed data by means of genetic programming. Paper presented at the First International Conference on Systems Biology (ICSB), Tokyo, 14-16 November 2000

Lee DG, Kim H-G, Baek W-P et al (1997) Critical heat flux prediction using genetic programming for water flow in vertical round tubes. Int Commun Heat Mass Transfer 24: 919–929

Lee TM, Oh H, Choung YK, et al (2009) Prediction of membrane fouling in the pilot-scale microfiltration system using genetic programming. Desalination 247:285–294

Lennartsson D, Nordin P (2004) A genetic programming method for the identification of signal peptides and prediction of their cleavage sites. EURASIP Journal on Applied Signal Processing 1:138-145

Lew TL, Spencer AB, Scarpa F et al (2006) Identification of response surface models using genetic programming. Mech Syst Sig Process 20(8):1819–1831

Li J, Wong L (2004) Rule-Based Data Mining Methods for Classification Problems in Biomedical Domains In: A tutorial note for the 15 the European conference on machine learning (ECML) and the 8th European conference on principles and practice of knowledge discovery in databases (PKDD), Pisa, Italy, September 2004

Madar J, Abonyi J, Szeifert F (2004) Genetic programming for system identification. In: 4th International conference on intelligent systems design and application. Available via CiteSeerx. http://conf.uni-obuda.hu/isda2004/7_ISDA2004.pdf. Accessed 18 Nov 2014

Madár J, Abonyi J, Szeifert F (2005) Genetic programming for the identification of nonlinear input output models. Ind Eng Chem Res 44:3178-3186

Madar J, Abonyi J, Szeifert F (2005) Genetic programming for system identification. Available online: http://conf.uni-obuda.hu/isda2004/7_ISDA2004.pdf. Accessed 3 Dec 2014

Manshad AK, Ashoori S, Manshad MK et al (2012a) The prediction of wax precipitation by neural network and genetic algorithm and comparison with a multisolid model in crude oil systems. Pet Sci Technol 30:1369-1378

Manshad AK, Manshad MK, Ashoori S (2012b) The application of an artificial neural network (ANN) and a genetic programming neural network (GPNN) for the modeling of experimental data of slim tube permeability reduction by asphaltene precipitation in Iranian crude oil reservoirs. Pet Sci Technol 30:2450-2459

Marquardt DW (1963) An algorithm for least-squares estimation of nonlinear parameters. J Soc Ind Appl Math 11:431-441

Marref A, Basalamah S, Al-Ghamdi R (2013) Evolutionary computation techniques for predicting atmospheric corrosion. Int J Corros. doi: http://dx.doi.org/10.1155/2013/805167

Martin MA (2010) First generation biofuels compete. N Biotechnol 27 (5): 596–608

McKay B, Willis M, Barton G (1997) Steady-state modeling of chemical systems using genetic programming. Comput Chem Eng 21:981-996

McQuiston FC (1978) Heat, mass and momentum transfer data for five plate-fin-tube heat transfer surfaces. ASHRAE Trans 84:266–293

Meighani HM, Dehghani A, Rekabdar F et al (2013) Artificial intelligence vs. classical approaches : a new look at the prediction of flux decline in wastewater treatment. Desalin Water Treat 51:7476-7489

Motsinger AA, Lee SL, Mellick G, Ritchie MD (2006) GPNN: power studies and applications of a neural network method for detecting gene-gene interactions in studies of human disease. BMC Bioinformatics 7:39

Muzny CD, Huber ML, Kazakov AF (2013) Correlation for the viscosity of normal hydrogen obtained from symbolic regression. J Chem Eng Data 58:969-979
Nandi S, Rahman I, Tambe SS, Sonolikar RL, Kulkarni BD (2000) Process identification using genetic programming: A case study involving Fluidized catalytic cracking (FCC) unit. In: Saha RK, Maity BR, Bhattacharyya D, Ray S, Ganguly S, Chakraborty SL (eds) Petroleum refining and petrochemicals based industries in Eastern India, Allied Publishing Ltd., New Delhi, p 195-201
Narendra K, Parthasarathy K (1990) Identification and control of dynamic systems using neural networks. IEEE Trans Neural Network 1: 4-27
Nemeth LK, Kennedy HTA (1967) Correlation of dewpoint pressure with fluid composition and temperature. SPE J 7:99–104
Okhovat A, Mousavi SM (2012) Modeling of arsenic, chromium and cadmium removal by nanofiltration process using genetic programming. Appl Soft Comput 12:793-799
Pacheco-Vega A, Cai W, Sen M et al (2003) Heat transfer correlations in an air–water fin-tube compact heat exchanger by symbolic regression. In: Proceedings of the 2003 ASME International Mechanical Engineering Congress and Exposition, Washington, DC, 15-21 November 2003
Pandey DS, Pan I, Das S, Leahy JJ, Kwapinski W (2015) Multi gene genetic programming based predictive models for municipal solid waste gasification in a fluidized bed gasifier. Bioresource Technology 179:524-533
Parikh J, Channiwala SA, Ghosal GK (2005) A correlation for calculating HHV from proximate analysis of solid fuels. Fuel 84:487–494
Park S-M, Han J, Lee S et al (2012) Analysis of reverse osmosis system performance using a genetic programming technique. Desalin Water Treat 43:281-290
Patil-Shinde V, Kulkarni T, Kulkarni R, Chavan PD, Sharma T, Sharma BK, Tambe SS, Kulkarni BD (2014) Artificial intelligence-based modeling of high ash coal gasification in a pilot plant scale fluidized bed gasifier. Ind Eng Chem Res. 53:18678–18689; doi: 10.1021/ie500593j
Paul TK, Hasegawa Y, Iba H (2006) Classification of gene expression data by majority voting genetic programming classifier. In: 2006 IEEE congress on evolutionary computation, Vancouver, 16-21 July 2006
Peng DY, Robinson DB (1976) A new two constant equation of state. Ind Eng Chem Fundam 15:59–64
Persaud K, Dodd G (1982) Analysis of discrimination mechanisms in the mammalian olfactory system using a model nose. Nature 299 (5881): 352–355
Podola B, Melkonian M (2012) Genetic Programming as a tool for identification of analyte-specificity from complex response patterns using a non-specific whole-cell biosensor. Biosens Bioelectron 33(1):254-259
Poli R, Langdon WB, McPhee NF (2008) A field guide to genetic programming. Available via lulu: http://lulu.com, http://www.gp-field-guide.org.uk . Accessed on 3 Dec 2014
Porter M, Willis M, Hiden HG (1996) Computer-aided polymer design using genetic programming. M Eng. Research Project, Dept. Chemical and Process Engng, University of Newcastle, UK.
Register Machine Learning Technologies Inc (2002) Discipulus 5 genetic programming predictive modelling. http://www.rmltech.com/. Accessed 7 Nov 2014
Ritchie MD, Motsinger AA, Bush WS (2007) Genetic programming neural networks: A powerful bioinformatics tool for human genetics. Appl Soft Comput 7:471–479
Ritchie MD, White BC, Parker JS et al (2003) Optimization of neural network architecture using genetic programming improves detection of gene–gene interactions in studies of human diseases. BMC Bioinf 4:28
Sajedian A, Ebrahimi M, Jamialahmadi M (2012) Two-phase inflow performance relationship prediction using two artificial intelligence techniques: Multi-layer perceptron versus genetic programming. Pet Sci Technol 30:1725-1736
Samuel, AL (1983) AI, Where it has been and where it is going. In: Proceedings of the 8th International Joint Conference on AI (IJCAI-83), Karlsruhe, Germany. p 1152–1157

Sankpal NV, Cheema JJS, Tambe SS et al (2001) An artificial intelligence tool for bioprocess monitoring: application to continuous production of gluconic acid by immobilized aspergillus niger. Biotechnol Lett 23: 911-916

Schmidt M, Lipson H (2009) Distilling free-form natural laws from experimental data. Science 324:81-85

Schmidt M, Lipson H (2014) Eureqa (Version 0.98 beta) [Software]. Available online: www.nutonian.com. Accessed 3 Dec 2014

Schneider MV, Orchard S (2011) Omics technologies, data and bioinformatics principles. Methods Mol Biol 719: 3-30

Searson DP, Leahy DE, Willis MJ (2010) GPTIPS: an open source genetic programming toolbox for multi-gene symbolic regression. In: International Multi-Conference of Engineers and Computer Scientists 2010 (IMECS 2010), Kowloon, Hong Kong, 17-19 March 2010. Volume 1. Available online: http://www.iaeng.org/publication/IMECS2010/IMECS2010_pp77-80.pdf

Seavey KC, Jones AT, Kordon AK (2010) Hybrid genetic programming - first principles approach to process and product modelling. Ind Eng Chem Res 49(5):2273-2285

Sharma S, Tambe SS (2014) Soft-sensor development for biochemical systems using genetic programming. Biochem Eng J 85:89-100

Shokir EMEl-M (2008) Dewpoint pressure model for gas condensate reservoirs based on genetic programming. Energy Fuels 22:3194-3200

Shokir EMEl-M, Dmour HN (2009) Genetic programming (GP)-based model for the viscosity of pure and hydrocarbon gas mixtures. Energy Fuels 23:3632-3636

Shokir EMEl-M, El-Awad MN, Al-Quraishi AA et al (2012) Compressibility factor model of sweet, sour, and condensate gases using genetic programming. Chem Eng Res Des 90:785-792

Shokrkar H, Salahi A, Kasiri N et al (2012) Prediction of permeation flux decline during MF of oily wastewater using genetic programming. Chem Eng Res Des 90:846-853

Shrinivas K, Kulkarni RP, Ghorpade RV et al (2015) Prediction of reactivity ratios in free-radical copolymerization from monomer resonance-polarity (Q-e) parameters: Genetic programming based models. International Journal of Chemical Reactor Engineering Published Online on 04/03/2015, DOI 10.1515/ijcre-2014-0039.

Singh KP, Gupta S (2012) Artificial intelligence based modeling for predicting the disinfection by-products in water. Chemom Intell Lab Syst 114:122–131

Smith SL, Lones MA (2009) Implicit Context Representation Cartesian Genetic Programming for the Assessment of Visuo-spatial Ability Evolutionary Computation. In: 2009 IEEE Congress on Evolutionary Computation, Trondheim, Norway, 18-21 May 2009

Spall JC (1998) Implementation of the Simultaneous Perturbation Algorithm for Stochastic Optimization. IEEE Trans Aerosp Electron Syst 34:817-822

Stein L (2001) Genome annotation: from sequence to biology. Nature Reviews Genetics 2:493-503

Sugimoto M, Kikuchi S, Tomita M (2005) Reverse engineering of biochemical equations from time-course data by means of genetic programming. Biosystems 80(2):155–164

Suh C, Choi B, Lee S et al (2011) Application of genetic programming to develop the model for estimating membrane damage in the membrane integrity test using fluorescent nanoparticle. Desalination 281:80-87

Sumathi S, Surekha P (2010) Computational intelligence paradigms theory and applications using MATLAB. Taylor and Francis, Boca Raton, FL

Tomita Y, Kato R, Okochi M et al (2014) A motif detection and classification method for peptide sequences using genetic programming. J Biosci Bioeng 85: 89–100

Tsakonas A, Dounias G, Jabtzen J et al (2004) Evolving rule-based systems in two medical domain using genetic programming. Artif Intell Med 32(3):195-216

Tun K, Lakshminarayanan S (2004) Identification of algebraic and static space models using genetic programming. Dyn Control Process Syst 1:311-328

Vamce DE, Vance JE (eds) (2008) Biochemistry of lipids, lipoproteins and membranes. Elsevier, Amsterdam

Venkatraman V, Dalby AR, Yang ZR (2004) Evaluation of mutual information and genetic programming for feature selection in QSAR. J Chem Inf Comput Sci 44(5):1686–1692. doi:10.1021/ci049933v

Vyas R, Goel P, Karthikeyan M, Tambe SS, Kulkarni BD (2014) Pharmacokinetic modeling of Caco-2 cell permeability using genetic programming (GP) method. Lett Drug Des Discovery 11(9): 1112-1118

Wagner S (2009) Heuristic optimization software systems - modeling of heuristic optimization algorithms in the HeuristicLab software environment. PhD Dissertation, Johannes Kepler University, Linz, Austria

Wang X-H, Hu Y-D, Li Y-G (2008b) Synthesis of nonsharp distillation sequences via genetic programming. Korean J Chem Eng 25:402-408

Wang X-H, Li Y-G (2008) Synthesis of multicomponent products separation sequences via stochastic GP method. Ind Eng Chem Res 47:8815-8822

Wang X-H, Li Y-G (2010) Stochastic GP synthesis of heat integrated nonsharp distillation sequences. Chem Eng Res Des 88:45-54

Wang X-H, Li Y-G, Hu Y-D et al (2008a) Synthesis of heat-integrated complex distillation systems via genetic programming. Comput Chem Eng 32:1908-1917

Wedge DC, Das A, Dost R et al (1999) Real-time vapour sensing using an OFET-based electronic nose and genetic programming. Bioelectrochem Bioenerg 48(2):389-96

White DR (2012) Software review: the ECJ toolkit. Genet. Prog. Evol. Mach 13: 65-67

Willis M, Hiden H, Hinchcliffe M et al (1997) Systems modeling using genetic programming. Comput Chem Eng 21:1161-1166

Witczak M, Obuchowicz A, Korbics J (2002) Genetic programming based approaches to identification and fault diagnosis of non linear dynamic systems. Int J Control 75:1012-1031

Woodward AM, Gilbert RJ, Kell DB (1999) Genetic programming as an analytical tool for non-linear dielectric spectroscopy. Bioelectrochem Bioenerg 48(2):389-396

Worzel WP, Yu J, Almal AA et al (2009) Applications of Genetic Programming in Cancer Research. Int J Biochem Cell Biol 41(2): 405–413

Xu C, Rangaiah GP, Zhao XS (2014) Application of neural network and genetic programming in modeling and optimization of ultra violet water disinfection reactors, Chem Eng Commun doi: 10.1080/00986445.2014.952813

Chapter 6
Application of Genetic Programming for Electrical Engineering Predictive Modeling: A Review

Seyyed Soheil Sadat Hosseini and Alireza Nemati

6.1 Introduction

Over the last decade, GP has received the interest of streams of researchers around the globe. First, we wanted to provide an outline of the basics of GP, to sum up valuable tasks that gave impetus and direction to research in GP as well as to discuss some interesting applications and directions. Things change fast in this area, as researchers discover new paths of doing things, and new things to do with GP. It is not possible to cover all phases of this field, even within the generous page limits of this chapter.

GP produces computer models to solve a problem utilizing the principle of Darwinian natural selection. GP results are computer programs that are represented as tree structures and shown in a functional programming language (such as LISP) (Koza 1992; Alavi et al. 2011). In other words, programs evolved by genetic programming are parse trees whose length can change throughout the run (Hosseini et al. 2012; Gandomi et al. 2012). GP provides the architecture of the approximation model together with the values of its parameters (Zhang et al. 2011; Gandomi et al. 2011). It optimizes a population of programs based on a fitness landscape specified by a program capability to perform a given task. The fitness of each program is assessed utilized an objective function. Therefore, a fitness function is the objective function that GP optimizes (Gandomi et al. 2010; Javadi and Rezania 2009; Torres et al. 2009). GP and other evolutionary methods have been successfully applied to different supervised learning work like regression (Oltean and Dioan 2009), and unsupervised learning work like clustering (Bezdek et al. 1994; Jie et al. 2004; Falco et al. 2006; Liu et al. 2005; Alhajj and Kaya 2008) and association discovery (Lyman and Lewandowski 2005).

S.S. Sadat Hosseini (✉) • A. Nemati
Department of Electrical Engineering and Computer Science,
University of Toledo, Toledo, OH 43606, USA
e-mail: s.sadathosseini@gmail.com; nemati.alireza@gmail.com

A.H. Gandomi et al. (eds.), *Handbook of Genetic Programming Applications*,
DOI 10.1007/978-3-319-20883-1_6

Our review on the application of GP is focused on electrical engineering, control, optimization and scheduling, signal processing, classification and power system operation. The distinctive features of GP create it a very convenient method in regard to optimization. The application of GP to these areas gives some interesting advantages, the principal one being its flexibility, which lets the algorithm be modified to the needs of each particular problem. In fact, GP typically performs an implicit process of feature selection and extraction. Interpretability can be quickly favored by the utilization of GP since it can utilize more interpretable representation formalism, like rules. It requires to be mentioned that GP approach is an evolutionary method that bears a strong resemblance to genetic algorithm's (GA's). The main differences between GA's and GP can be summed up as follows:

- GP codes solutions as tree structured, variable length chromosomes, but GAs make utilization of chromosomes of fixed length and structure.
- GP usually includes a domain specific syntax that governs meaningful arrangements of information on the chromosome. The chromosomes are syntax free for GAs.
- GP maintains the syntax of its tree-structured chromosomes during 're-production'.
- GP solutions are frequently coded in a way that lets the chromosomes be directly executed utilizing a suitable interpreter. GAs are hardly coded in a directly executable form.

The utilization of this flexible coding system permits the method to carry out structural optimization. This technique can be helpful to the solution of many engineering problems. For instance, GP may be utilized to implement symbolic regression. While conventional regression seeks to optimize the parameters for a pre-specified model architecture with symbolic regression, while the model design and parameters are specified simultaneously. Similarly, the evolution of control methods, Structural design, scheduling programs and signal processing algorithms can be seen as structural optimization problems appropriate for GP. Cramer created one of the first tree structured GAs for primary symbolic regression. Another early development was the BEAGLE technique of Forsyth, which produced classification rules utilizing a tree structured GA. However, it was Koza (1992) who was largely responsible for the popularization of GP within the area of computer science. His GP method (coded in LISP) was applied to a broad range of problems involving symbolic regression, control, robotics, games, classification and power system operation. Engineering applications have started to appear while still dominated by computer scientists. Thus, the objective of this paper is to discuss these recent engineering applications and give an entry point to this quickly expanding areas.

6.2 Genetic Programming

GP is a symbolic optimization method that produces computer programs to solve a problem using the principle of Darwinian natural selection. GP was introduced by Koza as an extension of genetic algorithms (GAs). In GP, a random population of individuals (trees) is created to achieve high diversity. While common optimization techniques represent the potential solutions as numbers (vectors of real numbers), the symbolic optimization algorithms present the potential solutions by structural ordering of several symbols. A population member in GP is a hierarchically structured tree comprising functions and terminals. The functions and terminals are selected from a set of functions and a set of terminals. For example, function set F can contain the basic arithmetic operations ($+$, $-$, $*$, /, etc), Boolean logic functions (AND, OR, NOT, etc.), or any other mathematical functions. The terminal set T contains the arguments for the functions and can consist of numerical constants, logical constants, variables, etc. The functions and terminals are chosen at random and constructed together to form a computer model in a tree-like structure with a root point with branches extending from each function and ending in a terminal. An example of a simple tree representation of a GP model is illustrated in Fig. 6.1.

The creation of the initial population is a blind random search for solutions in the large space of possible solutions. Once a population of models has been created at random, the GP algorithm evaluates the individuals, selects individuals for reproduction, generates new individuals by mutation, crossover, and direct reproduction, and finally creates the new generation in all iterations. During the crossover procedure, a point on a branch of each solution (program) is selected at random and the set of terminals and/or functions from each program are then swapped to create two new programs as can be seen in Fig. 6.2.

The evolutionary process continues by evaluating the fitness of the new population and starting a new round of reproduction and crossover. During this process, the GP algorithm occasionally selects a function or terminal from a model at random

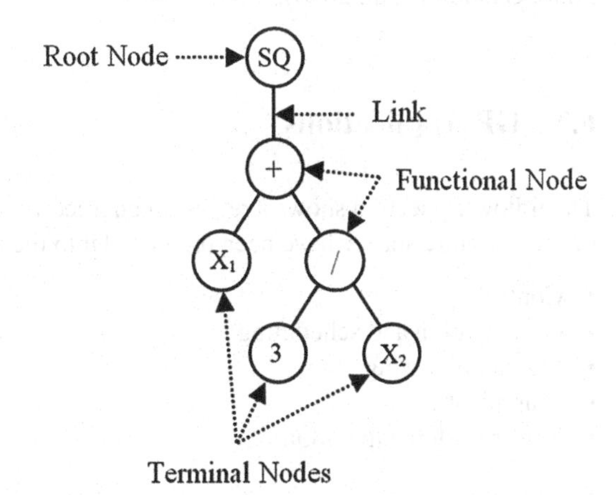

Fig. 6.1 The tree representation of a GP model $(X1 + 3/X2)^2$

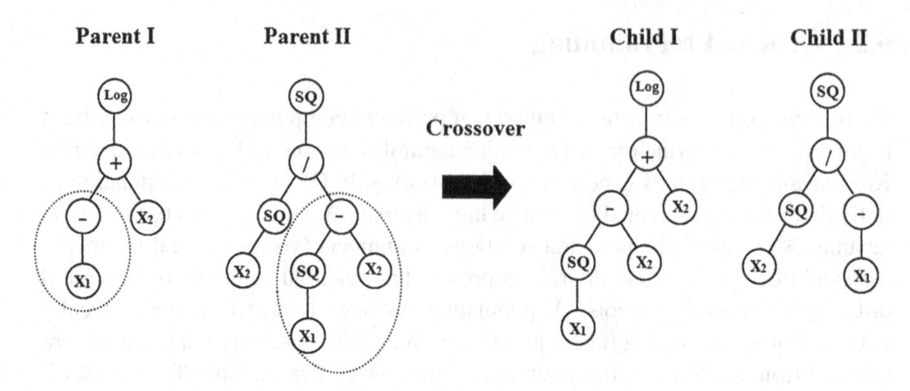

Fig. 6.2 Typical crossover operation in genetic programming

Fig. 6.3 Typical mutation operation in genetic programming

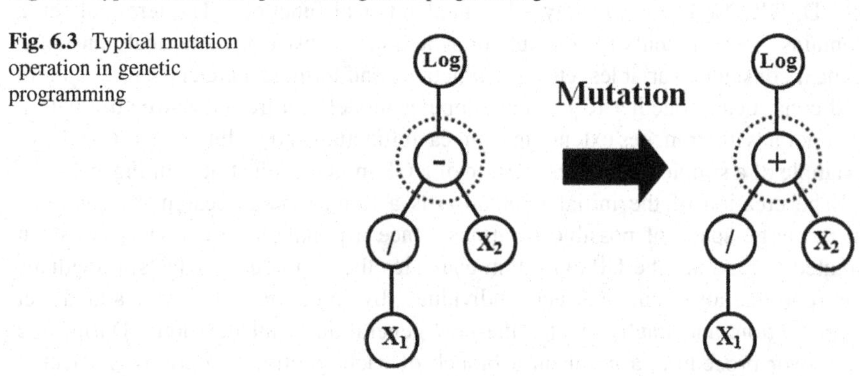

and mutates it (see Fig. 6.3). GEP is a linear variant of GP. The linear variants of GP make a clear distinction between the genotype and the phenotype of an individual. Thus, the individuals are represented as linear strings that are decoded and expressed like nonlinear entities (trees) (Yaghouby et al. 2010; Baykasoglu et al. 2008; Gandomi et al. 2008).

6.3 GP Applications

The following section shows a review of engineering applications of GP. The results of the literature survey have been organized into the following broad groups:

- Control
- Optimization and scheduling
- Signal processing
- Classification
- Power System Operation

6.3.1 Control

Mwaura and Keedwell (2010) used evolutionary algorithms (EAs) to automatically develop robot controllers and occasionally, robot morphology. This field of research is introduced as evolutionary robotics (ER). Through the utilizations of evolutionary methods such as genetic algorithms and genetic programming, ER has proved to be a promising approach through which robust robot controllers can be developed. Ebner (1999) explored the utilization of genetic programming for robot localization to evolve an inverse function mapping sensor readings. This inverse function is defined as an internal model of the environment. Environment is sensed utilizing dense distance information acquired from a laser range finder. An inverse function is developed to localize a robot in a simulated office environment.

Alfaro-Cid et al. (2008) assessed the implementation of genetic programming to design a controller structure. GP is utilized to evolve control strategies that provided the current and desired state of the propulsion and heading dynamics of a supply ship as inputs, produce the commanded forces needed to maneuver the ship. The controllers built utilizing GP are analyzed through real maneuverability tests and computer simulations in a laboratory water basin facility. The robustness of each controller is analyzed through the simulation of environmental disturbances.

Dracopoulos and Kent (1997) emphasized on the application of genetic programming to prediction and control. Results were shown for an oral cancer prediction task and a satellite attitude control problem. Using bulk synchronous model parallelization, the paper explained how the convergence of genetic programming can be significantly speeded up. Nordin and Banzhaf (1997) evaluated the utilization of genetic programming to a direct control a miniature robot. The GP system is employed to evolve real-time obstacle avoiding behavior. Genetic programming enables real-time learning with a real robot. A speed-up of the approach by a factor of more than 2000 was achieved by learning from past. Genetic programming was used in Zell (1999) to search the space of possible programs automatically. First a behavior-based control architecture utilizing computer simulations is evolved. Then one of the experiments with a service robot is replicated, displaying that Kozas classic experiment of evolving a control structure can be transferred to the real world with adjustment to representation. Suwannik and Chongstitvatana (2001) generated the control program by genetic programming to enhance the robustness of a robot arm. The robustness is measured in the real world. To enhance the robustness, multiple robot arm configurations used to evolve the control program. The result showed that the robustness of a control program is enhanced by 10 % in comparison to a control program evolved with a single configuration. Another control related application of GP has been done by Nordin et al. (1997). they have tried to control the khepera robot using GP. Their objective of using GP to control this miniature robot is to evolve real-time obstacle avoiding behavior. Their technique enables real time learning with actual robot. Figures 6.4 and 6.5 show the actual robot and its sensor placement, respectively. The learning that applied to experimental result, had papulation size of 50 individuals. the individuals used values from the sensors as an

Fig. 6.4 The Khepera robot

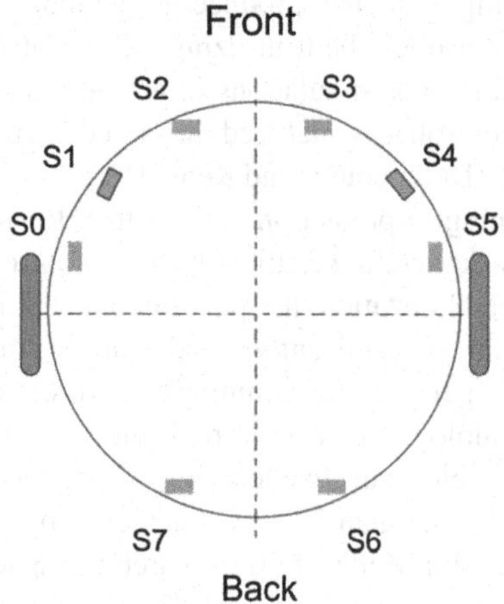

Fig. 6.5 Position of the IR proximity sensors

input and create two outputs values. The output values has been transmitted to the robot as motor speeds. the population of each individuals is processed by the GP system.

Figure 6.6 shows a schematic view of the system. This schematic has been captured from Nordin et al. (1997).

6.3.2 *Optimization and Scheduling*

Grimes (1995) were used genetic algorithm (GA) and genetic programming (GP) methods for track maintenance work with profit as the optimization criteria. The results were compared with an existing method. It was shown that the GP algorithm

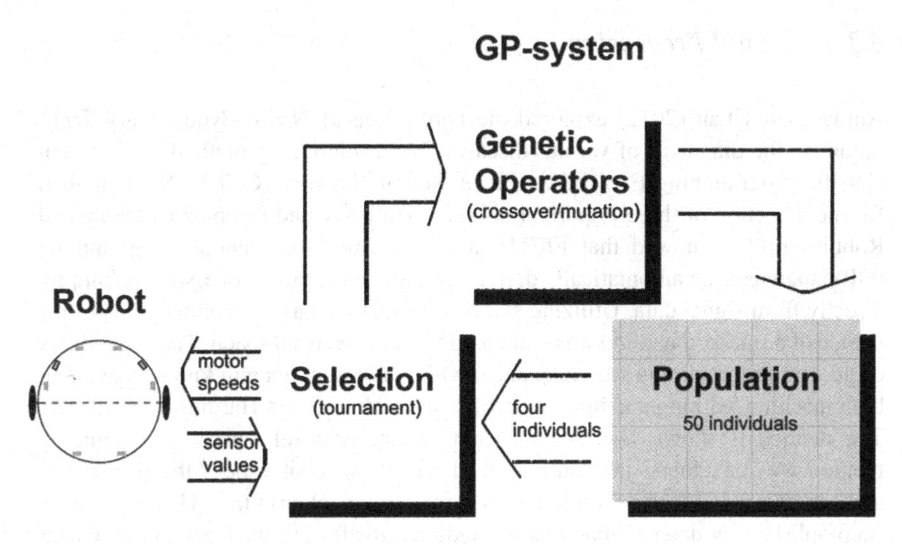

Fig. 6.6 Schematic view of the control system

provided the best results, with the GA approach providing good results for a short section and poor results for a long section of track. Genetic programming was used in Stephenson et al. (2003) to optimize the priority functions associated with two well-known compiler heuristics: predicted hyperblock formation and register allocation. Their system achieved remarkable speedups over a standard baseline for both problems. Vanneschi and Cuccu (2009) presented a new model of genetic programming with variable size population in this paper and applied to the reconstruction of target functions in dynamic environments. This models suitability was tested on a set of benchmarks based on some well-known symbolic regression problems.

Experimental results confirmed that their variable size population model found solutions of similar quality to the ones found by genetic programming, but with a smaller amount of computational effort. Ho et al. (2009) developed an algorithm to derive a distributed method automatically dynamically to optimize the coverage of a femtocell group utilizing genetic programming. The resulting evolved method showed the capability to optimize the coverage well. Also, this algorithm was able to offer increased overall network capacity compared with a fixed coverage femtocell deployment. The evolution of the best-known schedule illustrated in Langdon and Treleaven (1997) for the base South Wales problem utilizing genetic programming starting from the hand coded heuristics. Montana and Czerwinski (1996) applied a hybrid of a genetic algorithm and strongly typed genetic programming (STGP) to the problem of controlling the timings of traffic signals that optimize aggregate performance. STGP learns the single basic decision tree to be executed by all the intersections when determining whether to change the phase of the traffic signal.

6.3.3 Signal Processing

Ahmad and Khan (2012) explored the application of Neuro-Evolutionary Techniques to the diagnosis of various diseases. The evolutionary method of Cartesian Genetic programming Evolved Artificial Neural Network (CGPANN) is applied for the detection of three important diseases. Holladay and Robbins Holladay and Robbins (2007) showed that FIFTH, a new vector-based genetic programming (GP) language, can automatically derive very efficient signal processing techniques directly from signal data. Utilizing symbol rate estimate as an example, the performance of a standard method was compared to an evolved approach. The capabilities of genetic programming were expanded with combining domain knowledge about both machine learning and imaging processing techniques in Harding et al. (2013). The method is shown fast, scalable and robust. A novel genetic programming method was developed in Sharman et al. (1995) to evolve both the parameters and structure of adaptive digital signal processing algorithms. This process is accomplished by determining a set of node terminals and functions to implement the necessary operations commonly utilized in a broad class of DSP techniques. Also, simulated annealing was used to assist the GP in optimizing the numerical parameters of expression trees.

Esparcia Alczar (1998) presented a novel GP approach in the equalization of nonlinear channels. A new way of handling numerical parameters in GP, node gains, was defined. A node gain is a numerical parameter assigned to a node that multiplies its output value. Esparcia-Alczar and Sharman (1999) investigated the application of a combined genetic programming—simulated annealing (GP-SA) solution to a classical signal processing problem. This problem is called channel equalization where the goal is to build a system which adaptively compensates for imperfections in the path from the transmitter to the receiver. Authors were examined the reconstruction of binary data sequences transmitted through distorting channels. Alczar et al. (1996) are also have worked on some application of GP in signal processing in discrete-time manner. They have presented special tree nodes that maintain time recursion, sigmoidal nonlinear transfer functions and internal recursion, which are frequently used operations in signal processing. Table 6.1 which has capture from the same paper, describe these with nodes which implement frequently used algebraic operations in signal processing.

6.3.4 Classification

An algorithm to the utilization of genetic programming was proposed for multi-class image recognition problems in Smart and Zhang (2003). In their method, the terminal set is made with image pixel statistics, the function set includes arithmetic and conditional operators, and the objective function is based on classification precision in the training set. Instead of utilizing xed static thresholds as boundaries

Table 6.1 The function and terminal node set for evolving discrete-time systems

Symbol	Arty	Description	Symbol	Arty	Description
+,−	2	Addition, subtraction	**xN**	0	System input data. N indicates the delay (e.g. x2 returns xn−2)
*,/	2	Multiplication, division—if second argument is 0, then the node output is set to a large maximum value	**yN**	0	Previous output from the expression tree. The index N indicates 1+ the delay factor (e.g. y2 returns yn−3)
+1,−1	1	Increment, decrement	**z**	1	Unit sample time delay
*2,/2	1	Multiply/divide by two	**fN**	Variable	Execute the **N**th function tree
c**N**	0	Constant value. **N** is an index to a table of constants whose values may be predefined or chosen at random	**argN**	0	The Nth argument to a function tree
nl**N**	1	Non-linear transfer function. **N** indicates the amount of nonlinearity	**psh**	1	Push the argument value onto the stack
avg**N**	N	The average of its **N** arguments	**stkN**	0	Retrieve the Nth item from the stack

to distinguish between different classes, this technique proposed two dynamic algorithms of classification. These methods are centered dynamic range selection and slotted dynamic range selection, based on the returned value of an evolved genetic program where the boundaries between different classes can be dynamically decided during the evolutionary process. GP was applied to solve cost-sensitive classification by means of two techniques through a) manipulating training data and b) adapting the learning method in Li et al. (2005). A constrained genetic programming (CGP), a GP based the cost-sensitive classifier, has been proposed in this paper. CGP is capable of making decision trees to minimize not only the expected number of errors, but also the expected misclassification costs through a novel constraint objective function. The ensemble classification paradigm is an efficient way to enhance the accomplishment and stability of individual predictors. Evolutionary algorithms (EAs) also have been widely utilized to produced ensembles. In the context of heterogeneous ensembles, EAs have been successfully employed to modify weights of base classifiers or to select ensemble members. A novel genetic program was developed in Escalante et al. (2009) that learned a fusion function for integrating heterogeneous-classifiers outputs. It evolves a population of fusion functions to maximize the classification precision. A GP-based method was developed in Liu and Xu (2009) to evaluate multi-class micro-array data sets. In contrast to the standard GP, the individual formulated in this paper includes a set of small-scale ensembles, named as sub-ensemble (indicated by SE). Each SE includes a set of trees. In application, a multi-class problem is split into a set of two-class problems, each of which is addressed by an SE first. The SEs tackling the respective two-class problems are integrated to make a GP individual, so each can address a multi-class problem directly. Efficient algorithms are developed to address the problems arising in the fusion of SEs, and a greedy method is developed to keep high diversity in SEs. Three GP-based methods were proposed in Zhang and Nandi (2007) for addressing multi-class classification problems in roller bearing fault detection. The First method maps all the classes onto the one-dimensional GP output. The second algorithm singles out each class individually by evolving a binary GP for each class independently. The third technique also has one binary GP for each class, but these GPs are evolved together with the goal of choosing as few features as possible. It can also be mentioned that an application of GP in classifiers also could be dividing in three different data set which consist of several subsets. Each of the subsets is also used in an independent run of GP to build up each classifiers. Figure 6.7 shows the classification tasks where GP can be used. This figure has been captured from Zhang and Nandi (2007).

6.3.5 Power System Operation

One of the fundamental power systems planning responsibility of an electrical utility is to precisely anticipate load requirement for all time. The achieved results from load forecasting operation are utilized in various fields like planning and

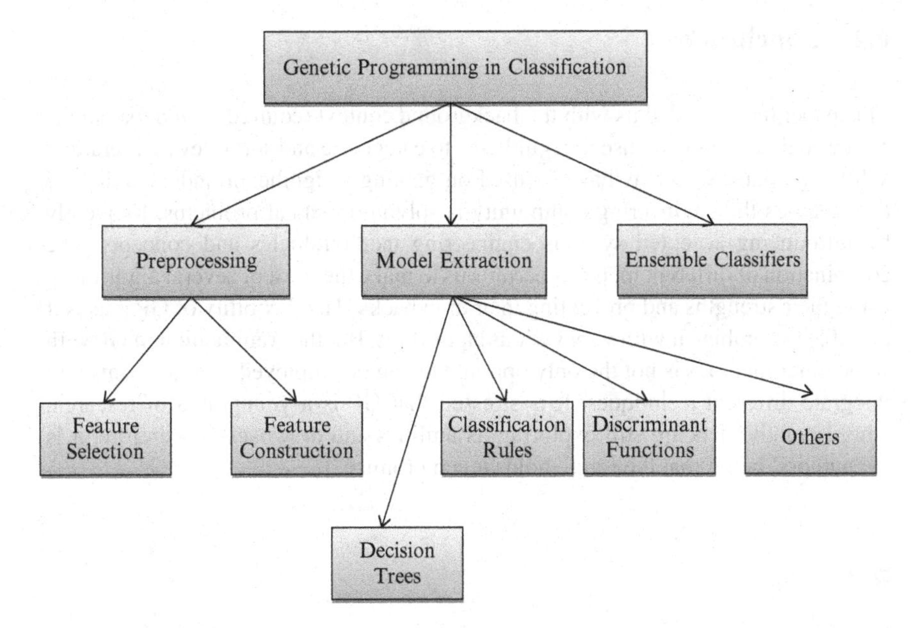

Fig. 6.7 Applications of GP in classification tasks

operation. Preparation of future expenses on construction, rely on the certainty of the long term load foretelling significantly, therefore, various estimation procedure have been tested for short and long term forecasting. Traditional load forecasting approaches are planted on statistical scheme. It needs to be mentioned that the evaluation of load aforetime is usually called as a Load forecasting. This estimation could be demand and energy which is essentially required to improve the system planning effectiveness. These forecasting is also utilized to organized approaches for construction and energy forecast which are essential for future fuel requirements determination. So a good forecast affecting the trend of power planning of present and future. Chaturvedi et al. (1995) are given the GP approach for long term load forecasting.

GP claims to support an optimal solution for the computational problem like power planning. Dr. Kamal (2002) worked on methods of calculation of problems of curve fitting by using GP. He showed that this problem can be carried out without use of equation shape. Farahat (2010) are also applied GP to forecast short term demand by using a new method. Some other researchers are also specified the comparison of different estimation algorithms for power system load forecasting. Genetic pronging, least absolute value filtering and least error squares are different approaches in their experiments. They have considered different forecasting models.

6.4 Conclusions

This paper has provided us with the background context required to understand the reviewed documents and use as a guideline to categorize and sort relevant literature. While computer scientists have focused on gaining a significant understanding of the methods the engineering community is solving practical problems, frequently by introducing accepted systems engineering methodologies and concepts. The combination of different methods permits us to make the most of several algorithms, using their strengths and preventing their drawbacks. The flexibility of GP makes it possible to combine it with very various algorithms. But the combination of GP with some other methods is not the only option; GP can be employed as a mechanism to integrate different techniques. It is stressed that GP is a young area of research, whose practitioners are still exploring its abilities and drawbacks. Therefore, it is the authors' belief that the future holds much promise.

References

Koza JR (1992) Genetic programming: on the programming of computers by means of natural selection. (3rd ed.) MIT press, Cambridge, MA

Alavi AH, Ameri M, Gandomi AH, Mirzahosseini MR (2011) Formulation of ow number of asphalt mixes using a hybrid computational method. Constr Build Mater 25(3):1338–1355

Sadat Hosseini SS, Gandomi AH (2012) Short-term load forecasting of power systems by gene expression programming. Neural Comput Appl 21(2):377–389

Gandomi AH, Alavi AH (2012) A new multi-gene genetic programming approach to non-linear system modeling. Part II: geotechnical and earthquake engineering problems. Neural Comput Appl 21(1):189–201

Zhang Q, Fang J, Wang Z, Shi M (2011) Hybrid genetic simulated annealing algorithm with its application in vehicle routing problem with time windows. Adv Mater Res 148–149:395–398

Gandomi AH, Yang X-S (2011) Benchmark problems in structural optimization. In: Computational Optimization, Methods and Algorithms. Springer, Berlin/Heidelberg, 259–281

Gandomi AH, Alavi AH, Arjmandi P, Aghaeifar A, Seyednour R (2010). Genetic programming and orthogonal least squares: a hybrid approach to modeling the compressive strength of CFRP-confined concrete cylinders. J Mech Mater Struct 5(5):735–753

Javadi A, Rezania M (2009) Applications of artificial intelligence and data mining techniques in soil modeling. Geomech Eng 1(1): 53–74

Torres RS, Falcão AX, Gonçalves MA, Papa JP, Zhang B, Fan W, Fox EA (2009) A genetic programming framework for content-based image retrieval. Pattern Recogn 42:283–92

Oltean M, Diosan L (2009) An autonomous GP-based system for regression and classification problems. Appl Soft Comput 9(1):49–60

Bezdek JC, Boggavarapu S, Hall LO, Bensaid A (1994) Genetic algorithm guided clustering.In: IEEE World Congress on Computational Intelligence

Jie L, Xinbo G, Li-Cheng J (2004) A CSA-based clustering algorithm for large data sets with mixed numeric and categorical values. Fifth World Congress on Intelligent Control and Automation, WCICA, 2303–2307

Falco ID, Tarantino E, Cioppa AD, Fontanella F (2006) An innovative approach to genetic programming-based clustering. In: Proc. 9th Online World Conf. Soft Comput. Ind.

Appl.(Advances in Soft Computing Series,34)., Berlin, Germany: Springer-Verlag, Sep./Oct, 55–64

Liu Y, Ozyer T, Alhajj R, Barker K (2005) Cluster validity analysis of alternative results from multi-objective optimization. In: Proc. 5th SIAM Int Conf Data Mining, Newport Beach, CA, 496–500

Alhajj R, Kaya M (2008) Multi-objective genetic algorithms based automated clustering for fuzzy association rules mining. J Intell Inf Syst 31(3): 243–264

Lyman M, Lewandowski G (2005) Genetic programming for association rules on card sorting data. In: Proc Genet Evol Comput Conf, Washington, DC: ACM, 1551–1552

Yaghouby F, Ayatollahi A, Yaghouby M, Alavi AH (2010) Towards automatic detection of atrial fibrillation: a hybrid computational approach. Comput in Biol Med 40(11–12):919–930

Baykasoglu A, Gullub H, Çanakç H, Özbakir L (2008) Prediction of compressive and tensile strength of limestone via genetic programming. Expert Syst Appl 35(1): 111–123

Gandomi A, Alavi A, Sadat Hosseini S (2008) A Discussion on Genetic programming for retrieving missing information in wave records along the west coast of Indian Applied Ocean Research 2007; 29 (3): 99–111.

Mwaura J, Keedwell E (2010) Evolution of robotic behaviours using Gene Expression Programming. In: IEEE Congress on Evolutionary Computation (CEC), 1–8

Ebner M (1999) Evolving an environment model for robot localization, Euro GP, Ebenhard-Karls-Universitat Tubingen, Germany, Springer Verlag, 184–192

Alfaro-Cid E, McGookin EW, Murray-Smith DJ, Fossen TI (2008) Genetic programming for the automatic design of controllers for a surface ship. IEEE Trans Intell Transp Syst 9(2):311–321

Dracopoulos DC, Kent S (1997) Genetic programming for prediction and control. Neural Comput Appl 6(4):214–228

Nordin P, Banzhaf W (1997) Real time control of a Khepera robot using genetic programmmg. Control Cybern 26(3)

Ebner M, Zell (1999) A Evolving a behavior-based control architecture-From simulations to the real world. In: Proceedings of the Genetic and Evolutionary Computation Conference, 1009–1014

Suwannik W, Chongstitvatana P (2001) Improving the robustness of evolved robot arm control programs with multiple configurations. In: 2nd Asian Symposium on Industrial Automation and Robotics, Bangkok, Thailand

Nordin, Peter, and Wolfgang Banzhaf. "Real time control of a Khepe. ra robot using genetic programmmg." Control and Cybernetics 26, no. 3 (1997).

Grimes CA, (1995) Application of genetic techniques to the planning of railway track maintenance work. in First International Conference on Genetic Algorithms in Engineering Systems: Innovations and Applications, GALESIA, IEE: Sheffield, UK, 414, 467–472

Stephenson M, OReilly UM, Martin MC, Amarasinghe S (2003) Genetic programming applied to compiler heuristic optimization, In: Proceedings of the European Conference on Genetic Programming, (Essex, UK), Springer, 238–253

Vanneschi L, Cuccu G (2009) A Study of Genetic Programming Variable Population Size for Dynamic Optimization Problems. In: Proceedings of the International Conference on Evolutionary Computation, part of the International Joint Conference on Computational Intelligence (IJCCI), ed. by A. Rosa et al

Ho, LTW, Ashraf I, Claussen H (2009) Evolving femtocell coverage optimization algorithms using genetic programming. In Personal, Indoor and Mobile Radio Communications, IEEE 20th International Symposium on, 2132–2136

Langdon WB, Treleaven P (1997) Scheduling maintenance of electrical power transmission networks using genetic programming. In KevinWarwick, Arthur Ekwue, and Raj Aggarwal, editors, Artificial Intelligence Techniques in Power Systems, chapter 10, 220–237

Montana DJ, Czerwinski S (1996) Evolving control laws for a network of traffic signals. In: Genetic Programming 1996: Proceedings of the First Annual Conference, Stanford University, CA, USA, 333–338. MIT Press, Cambridge

Ahmad AM, Khan GM (2012) Bio-signal processing using cartesian genetic programming evolved artificial neural network (cgpann). In: Proceedings of the 10th International Conference on Frontiers of Information Technology, 261–268

Holladay K, Robbins K (2007) Evolution of signal processing algorithms using vector based genetic programming. 15th International Conference in Digital Signal Processing, Cardiff, Wales, UK, 503–506

Harding S, Leitner J, Schmidhuber J (2013) Cartesian genetic programming for image processing. In Riolo, R., Vladislavleva, E., Ritchie, M. D., and Moore, J. H., editors, Genetic Programming Theory and Practice X, Genetic and Evolutionary Computation, 31–44. Springer New York

Sharman KC, Alcazar AIE, Li Y (1995) Evolving signal processing algorithms by genetic programming. First International Conference on Genetic Algorithms in Engineering Systems: Innovations and Applications, GALESIA, IEE, 414, 473–480

Esparcia Alcázar AI (1998) Genetic programming for adaptive digital signal processing. PhD thesis, University of Glasgow, Scotland, UK

Esparcia-Alcázar A, Sharman K (1999) Genetic Programming for channel equalisation. In R. Poli, H. M. Voigt, S. Cagnoni, D. Corne, G. D. Smith, and T. C. Fogarty, editors, Evolutionary Image Analysis, Signal Processing and Telecommunications: First European Workshop, 1596, 126–137, Goteborg, Sweden, Springer-Verlag

Alcázar, Anna I. Esparcia, and Ken C. Sharman. "Some applications of genetic programming in digital signal processing." In Late Breaking Papers at the Genetic Programming 1996 Conference Stanford University, pp. 24–31. 1996

Smart W, Zhang M (2003) Classification strategies for image classification in genetic programming. In: Proceeding of Image and Vision Computing Conference, 402–407, New Zealand

Li J, Li X, Yao X (2005) Cost-sensitive classification with genetic programming. The IEEE Congress on.Evolutionary Computation

Escalante HJ, Acosta-Mendoza N, Morales-Reyes A, Gago-Alonso A (2009) Genetic Programming of Heterogeneous Ensembles for Classification. in Progress in Pattern Recognition, Image Analysis, Computer Vision and Applications, Springer, 9–16

Liu KH, Xu CG (2009) A genetic programming-based approach to the classification of multiclass microarray datasets. Bioinformatics 25(3):331–337

Zhang L, Nandi AK (2007) Fault classification using genetic programming. Mech Syst Signal Pr 21(3):1273–1284

Chaturvedi DK, Mishra RK, Agarwal A (1995) Load Forecasting Using Genetic Algorithms Journal of The Institution of Engineers (India), EL 76, 161–165

Dr. Hanan Ahmad Kamal (2002) Solving Curve Fitting problems using Genetic Programming IEEE MELECON May, 7–9

Farahat MA (2010) A New Approach for Short-Term Load Forecasting Using Curve Fitting Prediction Optimized by Genetic Algorithms 14th International Middle East Power Systems Conference (MEPCON10)19–21

Chapter 7
Mate Choice in Evolutionary Computation

António Leitão and Penousal Machado

7.1 Introduction

Darwin's theory of Natural Selection (Darwin 1859) has been widely accepted and endorsed by the scientific community since its early years. Described as the result of competition within or between species affecting its individuals rate of survival, it has had a deep impact on multiple research field and is at the source of the ideas behind EC. The theory of Sexual Selection (Darwin 1906) was later developed by Darwin to account for a number of traits that were observed in various species, which seemed to have no place in his Natural Selection theory. Darwin described Sexual Selection as the result of the competition between individuals of the same species affecting their relative rate of reproduction, a force capable of shaping traits to high degrees of complexity and responsible for the emergence of rich ornamentation and complex courtship behaviour.

Despite having been discredited by the scientific community at the time, it is now widely regarded as a major influence on evolution theory. Interest arose in the 1970s through the works of Fisher (1915, 1930) and Zahavi (1975) and since then the community as gradually embraced it, having found its place in various research fields. While it has come a long way, Sexual Selection is still far from understood in EC, both regarding possible benefits and behaviour.

Electronic supplementary material: The online version of this chapter (doi: 10.1007/978-3-319-20883-1_7) contains supplementary material, which is available to authorized users.

A. Leitão (✉) • P. Machado
CISUC, Department of Informatics Engineering, University of Coimbra, 3030-290 Coimbra, Portugal
e-mail: apleitao@dei.uc.pt; machado@dei.uc.pt

A.H. Gandomi et al. (eds.), *Handbook of Genetic Programming Applications*,
DOI 10.1007/978-3-319-20883-1_7

Mate Choice was one of the processes of Sexual Selection described by Darwin and that mostly attracted his followers. This chapter describes a nature-inspired self-adaptive Mate Choice setup and covers the design steps necessary for applying it. A two-chromosome scheme where the first chromosome represents a candidate solution and the second chromosome represents mating preferences is proposed. Two approaches for encoding mating preferences are presented and differences discussed. Details on how to apply each of them to problems with different characteristics are given and design choices are discussed. The application of both approaches on different problems is reviewed and the observed behaviour discussed.

Section 7.2 introduces Mate Choice as well as its background, Sect. 7.3 gives a general overview of Mate Choice in Evolutionary Computation and covers the state of the art through a classification based on adaptation of mating preferences, popular preference choices and the role of genders. The section finally introduces the proposed setup, giving specific details on the ideas behind it and how to apply it. Section 7.4 describes the application of both the proposed approaches to multiple problems and discusses the obtained results and behaviours. Finally, Sect. 7.5 presents a summary.

7.2 Sexual Selection Through Mate Choice

Since his journey on the Beagle, Darwin has thoroughly studied the forces responsible for the evolution of species. The result of competition within or between species affecting their individuals relative rate of survival was named Natural Selection. Since the publication of Darwin's *On the Origin of Species by Means of Natural Selection, or the Preservation of Favoured Races in the Struggle for Life* in 1859 (Darwin 1859), the theory has become widely accepted by the scientific community. This was achieved thanks to the evidence gathered by Darwin, its co-discoverer Alfred Russel Wallace (1858) as well as multiple following researchers, ultimately overcoming other competitive ideas (Cronin 1993).

Despite such a success, Darwin battled with gaps in its theory. For instance, Darwin questioned how was it that Natural Selection could account for animal ornamentation or courtship behaviour. He observed a large number of species, where individuals displayed rich and costly ornamentations or complex and risky courtship behaviours that seemed to serve no purpose in survival, sometimes even risking it. These characteristics challenged the theory of Natural Selection and the idea that traits adapted to the environment in a purposeful way. Individuals carrying aimless and costly features should be unfavored in competition, making such features bound to face extinction.

However, as Darwin observed, that was not the case. Ornamentation and courtship behaviour were spread across populations and species, although Natural Selection could not explain their origin. He figured, however, that for these features to emerge they had to bring some kind of competitive advantage. As they didn't fit in Natural Selection, he envisioned the existence of another trait-shaping selection force in nature, one capable of shaping species in complex and diverse ways, by causing traits that help in competing for mates to spread through future generations. These traits were linked to reproduction, as he observed in nature, and brought evolutionary advantages, even when risking survivability. Darwin developed the theory of Sexual Selection (Darwin 1906) to explain this phenomena and described it as the result of competition within species affecting its individuals relative rate of reproduction.

Darwin therefore saw evolution as the interplay between two major forces, Natural Selection as the adaptation of species to their environment, and Sexual Selection as the adaptation of each sex in relation to the other, in a struggle of individuals of one sex for the possession of individuals of the other in order to maximize their reproductive advantages. While the outcome of failing in Natural Selection would be low survivability, the outcome of failing in Sexual Selection would be a low number or no offspring. From an evolutionary perspective, they reach the same outcome with competition in reproductive rates between individuals leading to evolutionary changes across populations.

Unlike his theory of Natural Selection, which easily found support on the scientific community, his theory of Sexual Selection was mostly rebuffed. The scientific community was not keen on Darwin's ideas regarding Sexual Selection, specially his ideas on Female Mate Choice and the impact it could have on evolution. It was clear for them that Natural Selection was the only force capable of adapting species and so a number of theories emerged in order to explain rich displays or courtship behaviour. One of the most avid opponents of Darwin's ideas was Wallace who came up with various reasons for the emergence of traits such as protection through dull colors in females, recognition of individuals of the same species, usage of surplus energy on courtship behaviour or non-selective side effects (Cronin 1993).

The community was better prepared to understand such ideas, which were embraced by various renowned researchers such as Huxley (Cronin 1993). This lead to a time where Darwin's ideas on Sexual Selection and specially Mate Choice were dismissed as non-important, and its impact to be regarded as a small part of Natural Selection. These ideas remained for over a century, with the exception of a few works by a select few researchers who made important contributions, such as Fisher (1915) who explored the origin of mating preferences and runaway sexual selection, Williams and Burt (1997) who discussed how ornaments should be considered as important as other adaptations or Zahavi (1975) who expanded on the role of displays as fitness indicators.

Overtime, the work of these researchers was able to gain some space in the community, eventually reaching more open-minded generations who were also better equipped to test and understand the workings of Sexual Selection and Mate Choice. The resulting discussion has for the past few decades attracted experimental biologists, psychologists and anthropologists that since then have put Darwin's ideas as well as those promoted by his followers to the test, contributing with increasing evidence to back the ideas behind Sexual Selection through Mate Choice. Nowadays, there is active research on various fields and the theory has been widely accepted by the community. Two extensive reviews on Sexual Selection have been published by Helena Cronin (1993) and Malte Andersson (1994).

Mate Choice is one of the main Sexual Selection processes described by Darwin and where he put much of his effort, as did most of his followers. They aimed to explain the emergence of aesthetic features such as ornamentation or courtship behaviour. One of the pillars of Darwin's theories on evolution was that species went through adaptations because they brought some kind of advantage over time. In this case, these traits emerged because they brought reproductive advantage as a result of preferences when selecting mating partners (Darwin 1906). Fisher helped explain the relation between mating preferences and traits, and the genetic link between them through his theory of runaway sexual selection. He addressed how displays may arise as a result of positive reinforcement between mental mating preferences and physical traits, through a feedback loop that can lead to extravagant adaptations such as the peacock's tail, colorful appearance or complex courtship behaviour (Fisher 1915, 1930).

Fisher's work suggests the inheritance of mating preferences much like any other trait, therefore adapting throughout the generations. This process can be better understood if mate choice is thought of like any other adaptive choice such as food choice (Miller 1994). Still, criticism of theses ideas remained, since the evolution of traits through such a runaway process with increasing speed could drastically risk the survival ability of individuals. Zahavi later expanded on this subject, suggesting that aesthetic displays can act as indicators of fitness, health, energy, reproductive potential etc. He argued through his handicap principal (Zahavi 1975) that even in the case of costly displays and behaviour, which seemed to have no purpose in Natural Selection, it was in fact their high cost that made them good fitness indicators. As these traits were handicaps, they couldn't be maintained by weak, unfit individuals and that only strong healthy individuals would be able to maintain them and survive. Therefore reinforcement of mating preferences for these traits would be beneficial for females, which would in turn reinforce such physical traits in males as suggested by Fisher (1915).

These ideas were explored and discussed by many other researchers, who finally brought Sexual Selection through Mate Choice into the spotlight. Their work corroborated Darwin's ideas and brought new evidence allowing Sexual Selection to be seen as an important force in Evolutionary Theory. The interplay between Natural Selection and Sexual Selection was found to have a deep impact on various traits on many different species, especially among those equipped with complex sensory systems (Cronin 1993). During the last few decades, Sexual Selection has found its place on various research fields such as Evolutionary Biology, Evolutionary Psychology and Evolutionary Anthropology. On the other hand, it is yet to attract the full attention of the Evolutionary Computation community, despite the publication of several papers over the last couple of decades.

The possible advantages that Sexual Selection, particularly through Mate Choice can bring to the field of Evolutionary Computation have been previously discussed by several researchers, and an extensive discussion on arguably the most relevant ones has been published by Miller and Todd (1993). They find that the addition of Mate Choice to Natural Selection can bring advantages such as (1) increased accuracy when mapping from phenotype to fitness, therefore reducing the "error" caused by different forms of Natural Selection; (2) increasing the reproductive variance of populations by distinguishing between individuals with no survival-relevant (fitness) differences; (3) help populations escape from local optima through a directional stochastic process; (4) contribute to the emergence of complex innovations which may eventually contribute to fitness increasing; (5) promote sympatric speciation, diversity and parallel evolutionary searches.

7.3 Mate Choice in Evolutionary Computation

Mate Choice has been modeled in Evolutionary Computation by applying more or less the same mechanism. Algorithm 1 succinctly describes the approach. First, parent1 is selected from the population using fitness-based traditional operators. Secondly, a pool of potential mating partners is determined. These could be the whole remaining population, a random subset or a group of individuals selected based on a given characteristic. Thirdly, the mating candidates are evaluated according to a given set of mating preferences. Finally, the candidate that according to the evaluation best matches the first parent is selected as parent2.

To further understand how Mate Choice works, Figs. 7.1 and 7.2 show how traditional approaches and mate choice approaches work respectively. As seen in Fig. 7.1, traditional approaches select each parent independently, based on their fitness alone, meaning that each individual should have reproductive success according to their fitness value. However, these individuals are paired randomly, without any knowledge about their mating partners. In mate choice approaches,

Algorithm 1 Parents Selection using Sexual Selection through Mate Choice

```
proc MateChoice(population) ≡
  parent1 ← parentSelection(population)
  candidates ← candidatesSelection(population)
  evaluateCandidates(parent1,candidates)
  parent2 ← selectBest(candidates)
end
```

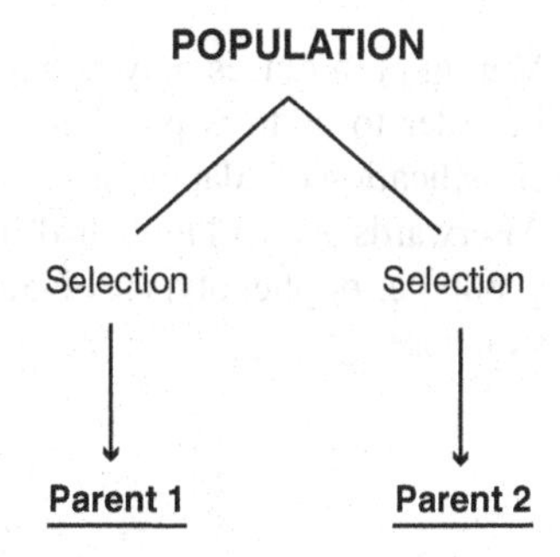

Fig. 7.1 Parents Selection using traditional approaches

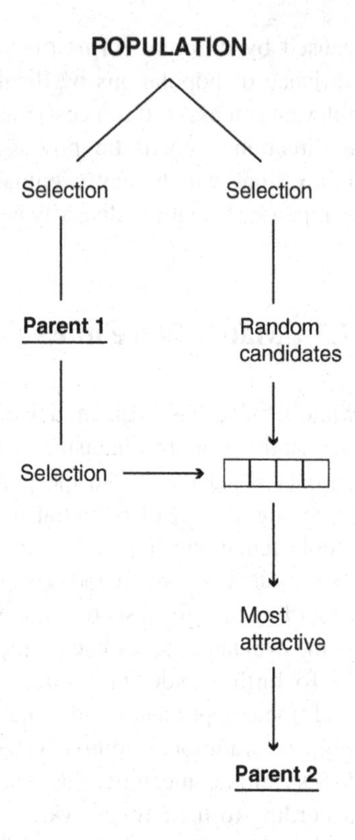

Fig. 7.2 Parents Selection using a Mate Choice approach

as seen in Fig. 7.2, individuals selected by traditional approaches are allowed to choose a mating partner based on their own criteria, meaning the pairing is no longer random but happens with a given characteristic in mind. This Mate Choice process is therefore ruled by the mating preferences as they will determine which individuals are good matches and are more likely to achieve reproductive success by producing fit, attractive offspring.

7.3.1 State of the Art

Mating preferences may remain static over the generations or undergo adaptation. In order to address preferences and adaptation mechanisms we will rely on the classification of adaptation of parameters and operators by Hinterding et al. (1997). Afterwards, we will look on different preferences and how they can be used to assess genotypes or phenotypes. Finally we will discuss the use of genders in Mate Choice strategies.

7.3.1.1 Adaptation of Mate Choice

Various authors rely on guiding the choice of mating partners using pre-established preferences, which remain static over the evolution process. The most widely known strategies are probably those that mate individuals based on similarity measures. Examples found in the literature include using Hamming distances (De et al. 1998; Fernandes and Rosa 2001; Galan et al. 2013; Ochoa et al. 2005; Ratford et al. 1996, 1997; Varnamkhasti and Lee 2012), Euclidean distances (Galan et al. 2013; Ratford et al. 1996), number of common building blocks (Ratford et al. 1996) or even simply using the fitness value as a distance measure (De et al. 1998; Galan et al. 2013; Goh et al. 2003; Ratford et al. 1996; Varnamkhasti and Lee 2012). In some implementations, the first selected parent chooses the candidate that maximizes similarity (Fernandes et al. 2001; Hinterding and Michalewicz 1998) while in other cases the candidate that minimises the measure is considered the best (Varnamkhasti and Lee 2012). Other approaches attribute a probability of selection proportional or inversely proportional to the distance measures (Ratford et al. 1996).

Moreover, some authors don't want to maximize or minimize distances but rather consider an ideal distance and favour mating candidates that have distances closer to that pre-established value. In this case the attractiveness of a candidate is established using bell curves or other functions with a predefined center and width parameters (Ratford et al. 1997).

Other metrics have been applied such as in Hinterding and Michaelwicz's study on constrained optimization (Hinterding and Michalewicz 1998). They suggest having the first parent select the mating candidate that in conjunction with itself maximizes the number of constraints satisfied. A second example is the study by Fernandes et al. on vector quantization problems where a problem specific metric is used (Fernandes et al. 2001). Sometimes different metrics are combined in a seduction function. This can be accomplished through the use of rules such as choosing the fittest candidate if two candidates both maximize or minimize the similarity measure or choosing between them randomly if they both also share the same fitness value (Varnamkhasti and Lee 2012). A different approach is to combine different metrics using different functions (such as weighted functions) which has been done for instance by Ratford et al. (1996).

While fixed parameters and mating preferences can often achieve competitive results and reproduce desired behaviours in Mate Choice algorithms, allowing their online control may be extremely valuable. Such approaches, which allow the Mate Choice strategy to change online without external control, therefore turning it into a dynamic process, can be subdivided into three groups: deterministic, adaptative and self-adaptative.

Deterministic approaches are the least common among the literature. Still, Ratford et al. give a good example (Ratford et al. 1997). On an aforementioned study they use a function to calculate a candidate's attractiveness which has two variables, centre and depth. It values individuals whose hamming distance from the first parent is closer to the centre of the function. However they complement this study by testing an approach where the centre of the function is adjusted at each generation, so that dissimilar individuals are favoured at the beginning of the run but the opposite happens at the end.

Adaptative approaches are more common and rely on information about the evolution process to adapt parameters or preferences. Fry et al. (2005) have applied such a strategy to control which operator is used to select the second parent, either a regular tournament selection or a Mate Choice operator. In their study they choose mating partners by combining fitness with a penalization for similar candidates (different similarity measures are applied). However they use this operator with a given probability which is increased or reduced depending on its relative success in producing enhanced offspring in previous generations. A second example can be found on studies by Sánchez-Velazco and Bullinaria (2003, 2013). Mating candidates in this case are evaluated based on a weighted function that combines three metrics: fitness, likelihood of producing enhanced offspring, and age. While age is adapted deterministically at each generation, the second factor represents a feedback on each individual's ability to produce fit offspring in the past.

Self-adaptive approaches better resemble the workings of Mate Choice in nature. By allowing preferences and parameters to be encoded in each individual, self-adaptation allows them to take part in the evolution process and to impact not only the individuals that encode them but the population as a whole. Possibly the simplest example of such an approach relies on encoding an index as an extra gene in each individual. When evaluating its mating candidates, each individual will order them from best to worst according to a given metric and select the candidate at the encoded position. Galan et al. (2013) have experimented with this approach using Euclidean distance and fitness to order mating candidates. On the aforementioned study by Fry et al. (2005), a second approach was tested, where each individual encodes its own probability of selecting a mating partner using a mate choice operator rather than regular tournament selection. The probability is inherited by the offspring and adapted by comparing their fitness with that of their parents.

More complex mating preferences can be found on self-adaptive approaches. Miller and Todd (1993) and Todd and Miller (1997) suggest encoding a reference position on the phenotype space marking each individual's ideal position for a mating candidate. When assessing potential mating partners, the probability of mating varies according to their distance to the reference position. New offspring inherit genetic material from both parents through two-point crossover thus allowing for its self-adaptation throughout the evolutionary process.

Holdener and Tauritz (2008) relied on an extra chromosome to encode a list of desired features to look for on mating candidates. They tackle a problem using a binary representation on the first chromosome but rely on a real value representation on the preferences chromosome. This chromosome has the same size as the first chromosome with each gene representing how much an individual wants the corresponding gene to be set to 1. This information is used to evaluate mating candidates by comparing the preferences chromosome with each candidate's potential solution, favouring desired genes. Preference genes are inherited from parents to offspring so that they match the genes they influence and adapt to match the offspring's relative success. On a related study, Guntly and Tauritz (2011) proposed a centralized approach in addition to an approach similar to the one described above. The centralized approach relies on two preference vectors common to the whole population: one relative to genes set to 0 and one relative to genes set to 1. These vectors are accessed by individuals when evaluating mating candidates in a similar fashion as in the previous approach. The value of each gene is adapted at each selection step according to the relative success of the offspring.

Smorodkina and Tauritz (2007) proposed a different approach where each individual encodes a mate choice function in addition to its own candidate solution to the problem at hand. This function is represented as a tree which is used to select a mating partner. The tree in each individual is initialized using only one terminal node which corresponds to the remaining of the population. As a non-terminal set, a number of selection operators can be used, which compare different metrics between mating candidates. Eventually the tree returns the preferred mating candidate. If the produced offspring shows enhancements then it inherits the tree used for evaluation from its parents, otherwise it inherits the product of recombination between the evaluation trees of both parents.

7.3.1.2 Mating Preferences

As described above, many possible mating preferences have been applied by different researchers. Some of them focus on the similarity between the first parent and each mating candidate. A measure of similarity can be assessed either using genotypic (De et al. 1998; Fernandes and Rosa 2001; Galan et al. 2013; Ochoa et al. 2005; Ratford et al. 1996, 1997; Varnamkhasti and Lee 2012) or phenotypic (Galan et al. 2013; Ratford et al. 1996) information.

Other metrics focus on characteristics of mating candidates such as previous reproductive success, fitness or age (Ratford et al. 1996). Characteristics can be compared to those of the first parent in an attempt to find a partner that complements it (Hinterding and Michalewicz 1998). Often, when multiple metrics are applied they are combined through rules or functions. In some cases similarity measures are also combined with such metrics (Ratford et al. 1996). More interestingly are perhaps approaches where the first parent is able to perceive certain genotypic or phenotypic traits on mating candidates and selects the one that best matches its preferences. This is often accomplished by encoding mating preferences in each individual and comparing those preferences with traits displayed by each candidate (Holdener and Tauritz 2008; Miller and Todd 1993; Smorodkina and Tauritz 2007; Todd and Miller 1997).

7.3.1.3 Gender Roles

In nature, Mate Choice is almost absolutely on the side of females. Due to their high reproductive investment, they are picky when selecting a mating partner, looking for a fit male that can provide good genes. On the other hand males are more willing to mate with as many females as possible in an attempt to increase their number of offspring and benefit the presence of their genes in following generations. Looking back at Algorithm 1, *parent1* takes the female role while the mating candidates and therefore *parent2* take the male role.

There are different approaches in the literature to establish which individuals will take a female role or a male role. For instance, on some approaches a gender is attributed randomly to each individual at the beginning of each generation (Goh et al. 2003; Sánchez-Velazco and Bullinaria 2013) while in other cases gender is attributed alternatively when offspring are produced (Varnamkhasti and Lee 2012). In such cases where each individual has a fixed role, females can be selected from their subpopulation using different strategies. In some cases all females are selected and produce offspring once (Goh et al. 2003), in other cases traditional selection operators are applied (Varnamkhasti and Lee 2012). An alternative approach has each individual participating once as a female and once as a male in the parent selection process (Holdener and Tauritz 2008). Mating candidates are selected from the males pool, often randomly. In other cases, all individuals in the population have the chance to play either role. In these cases, any individual can be selected as *parent1*, therefore for the role of female, and all the remaining can be selected as mating candidates, or for the role of male. In these cases it is popular to select females through traditional operators and males randomly. If an individual takes the role of female at a selection step, it could be selected as male on the next one and vice-versa.

7.3.2 Designing a Nature-Inspired Mate Choice Approach

When designing Mate Choice approaches we feel that in order to best resemble the natural process, models should follow three nature-inspired rules:

1. individuals must choose who they mate with based on their own mating preferences
2. mating preferences, as mental traits, should be inherited the same way as physical ones
3. mate selection introduces its own selection pressure but is subject to selection pressure itself

We see the evaluation of mating candidates as a complex process, where the relation between observed traits, or their weight on each individual's mate choice mechanism is difficult to establish beforehand. While some traits could be seen as valuable on a mating candidate, others could turn out to be irrelevant or even harmful. The relation between them is also certainly not straightforward as they could be connected in unforeseeable ways. Moreover, certain displayed traits may be very important for survival purposes but hold little value for mate choice, or the other way around. This value can also vary on each selection step, depending on the characteristics of *parent1*, its mating preferences and the mating candidates involved.

This discussion is particularly relevant when we recall the following aspects of Sexual Selection through Mate Choice:

1. each individual has its own characteristics and may benefit differently from reproducing with different mates. Each individual also has its own distinct mating preferences that may value different characteristics in mating candidates. The reproductive success of individuals depends on choosing appropriate mating partners the same way that it depends on how attractive they are to others.
2. the paradigm may result on cases where individuals with poor survival abilities attain a high reproductive success because they display characteristics that are favoured by mating preferences. Their offspring may achieve low fitness values but may contribute to exploration and the emergence of innovation which may eventually turn into ecological opportunities.
3. the handicap principle shows that certain traits may risk the survival ability of individuals while in fact being indicative of good gene quality, thus reducing the accuracy of fitness values. Mate Choice mechanisms may be able to help increase the accuracy of mapping between phenotypes and fitness values and translate that into reproductive success.
4. mating preferences and evolved physical traits have an intrinsic and deep dependence between them which results from the feedback loop described in the theory of runaway sexual selection. The resulting arms race causes mating preferences to evolve in relation to displayed traits and physical traits to adapt in relation to enforced mating preferences. This process can lead traits to a high degree of elaboration.

With these ideas in mind and with the goal of designing mate choice mechanisms that best resemble the natural process, we refrain from establishing what are good or bad mating preferences. Also, we avoid linking each individual's candidate solution with its mating preferences using any pre-established method, such as inheritance rules. We therefore treat genetic material regarding physical traits and mating preferences equally and leave the responsibility of adapting individuals up to the evolutionary process. Inheritance and selection pressure should be able to bring reproductive and survival advantages to individuals carrying genes linked to appropriate phenotypes and mating preferences through the intrinsic relation between Natural Selection and Sexual Selection through Mate Choice.

The following subsections detail how we design Mate Choice approaches that meet the presented rules and aspects.

7.3.2.1 Representation

We propose a setup where each individual is composed by two chromosomes. The first one encodes a candidate solution to the problem at hand while the second chromosome encodes an individual's mating preferences, which it will use to assess potential mating partners. The first chromosome could use any representation wanted for the problem at hand, be it a Genetic Algorithm (GA) vector, a GP tree or others. We propose representing mating preferences, therefore the second chromosome, as a GP tree.

Mating preferences can be represented using two possible approaches: (1) representing an ideal mating partner; (2) representing an evaluation function. The first approach requires that we are able to map GP representations to the phenotype space of the problem at hand. In this case, the terminal and non-terminal sets should be established accordingly. For instance, if we are using a GP representation for the candidate solutions in the first chromosomes, then we can use the same terminal and non-terminal sets. This approach can also be used if we rely on other representations in the first chromosome but there are know GP representations, as long as both representations map to the same phenotype space. The second approach requires that the individuals can extract characteristics from their mating candidates and evaluate them. The terminal set of the GP representation will be the evaluation of such characteristics. The non-terminal set will provide a number of operators that allows the creation of relations between characteristics into complex functions. At the end, in this approach, the second chromosome encodes a GP function that evaluates a number of characteristics on a mating candidate and through a number of operations linking different characteristics, produces an attractiveness value.

7.3.2.2 Evaluation

The two representation approaches rely on different evaluation mechanisms, however Fig. 7.3 shows a general view of the process. On the first approach, when an individual is assessing a mating candidate, an ideal mating partner according to that individual's preferences is mapped to the phenotype space and compared to the phenotype of the mating candidate. In order to do so, a similarity measure has to be established and used. The mating candidate that best resembles an ideal partner and therefore minimizes the metric used is selected as a mating partner. Notice that this is conceptually different from selecting mating partners so that the distance between parents is minimized or maximized. The behaviour resulting from this approach will be much different but the design effort required is actually quite similar. As long as there is a possible GP representation for the problem at hand, that can be used for the second chromosome and for evaluation to take place, only a similarity measure is required.

The second approach evaluates mating candidates in a different way. Instead of comparing each mating candidate to an ideal mating partner, it will evaluate them according to a number of displayed traits. Therefore, in order to design such an approach, we have to determine a perceptive system, or in other words, what characteristics can each individual observe on others. To do so, researchers have to rely on their own knowledge of the problem and determine what may or may not be relevant characteristics. This process often requires the deconstruction of the problem at hand and subdividing it on a set of simpler problems, or assessing simpler objectives that are implicit on the global fitness function. These characteristics will make up the terminal set, so that when an individual is evaluating a candidate, the terminal nodes on the GP tree representing its mating preferences will assume a numeric value representing how the candidate performs on the present characteristics. The operators included on the non-terminal set will determine how these values relate with each other and what weight they have on the mating preferences. Finally, the GP function returns a numeric value which represents the attractiveness of a mating candidate (Fig. 7.4 shows an example of a possible GP tree). The candidate that achieves the highest attractiveness is selected as a mating partner. This approach is more difficult to setup as it requires more knowledge from the person designing the system. However, the choice of relevant characteristics does not have to be perfect, the evolution process will be in charge of determining which are valuable, harmful or irrelevant. It has the advantage of not requiring a GP representation of the problem to be applicable.

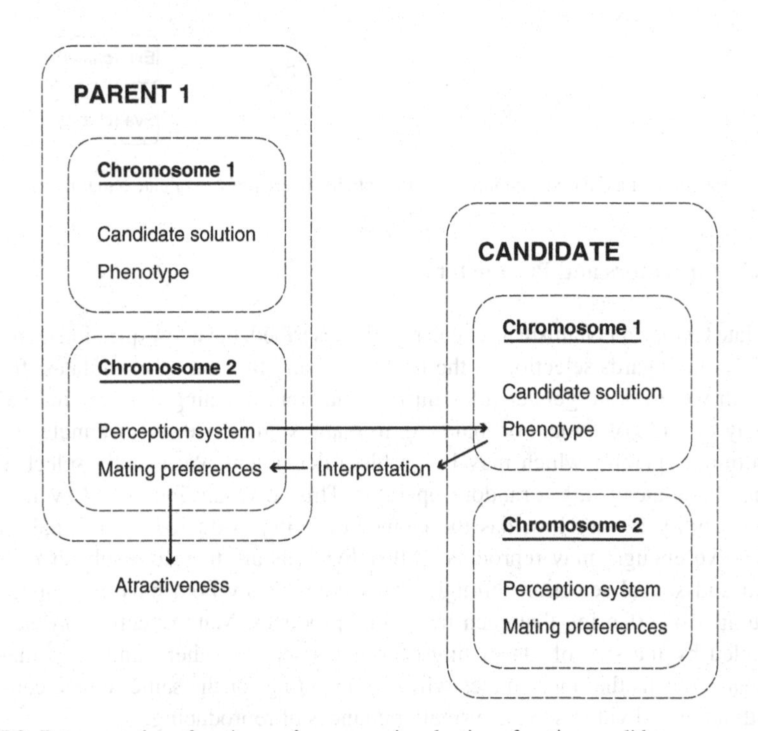

Fig. 7.3 Representation of mating preferences and evaluation of mating candidates

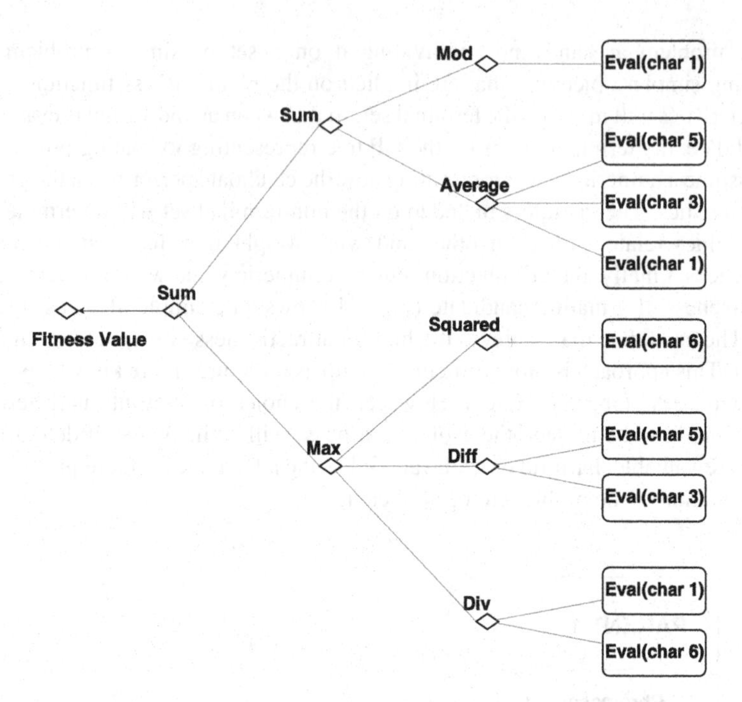

Fig. 7.4 Example of a GP tree combining a different characteristics to evaluate a mating candidate

7.3.2.3 Operators and Parameters

The Mate Choice mechanism has been mostly described, but a few questions remain. The first one regards selection of the first parent and the mating candidates. In our approach we attribute gender roles during selection, meaning that any individual can play the role of female or male. As in many studies, we select females using a traditional operator which may be problem dependent. We usually select male mating candidates using a random operator. This way, any individual even if its survival ability is very low, has the chance of being evaluated by a female and, if attractive enough, may reproduce. All individuals are therefore subject to both natural and sexual selection through mate choice. They both have an impact on the number of offspring that each individual produces. Mate selection pressure is controlled by the size of the mating candidates set. A higher number of mating candidates means that more males will be competing for the same female causing less attractive individuals to have smaller chances of reproducing.

When two parents have been selected, two new offspring are generated by means of reproductive operators. In our setup we apply these operators independently at each chromosome. This decision not only allows different strategies to be applied to each of them, for instance if different representations are used, but also allows operators to be applied with different probabilities and parameters. This is beneficial since the two chromosomes may impact behaviour differently with the same operators and parameters. Regarding the mating preferences chromosome, we suggest applying either crossover or mutation exclusively and with a given probability. This means that new offspring may inherit a combination of the mating preferences of both its parents, a mutated version of the mating preferences of one of its parents, or in case neither crossover or mutation occurs the offspring will inherit the exact mating preferences of one of its parents (Koza 1994). Unlike previous studies (Holdener and Tauritz 2008; Smorodkina and Tauritz 2007), we don't check the feasibility of new offspring to decide on how their mating preferences will be inherited. Finally, new offspring are inserted into the population of the following generation.

7.4 Applications

The presented Mate Choice setup has been previously applied to different problems, including real-world applications. We have had the chance to test both preference representation approaches and have also compared to a GA representation on a specific problem. The section discusses the obtained results and observed behaviour.

Applying Mate Choice to symbolic regression has probably been the easiest to setup (Leitão et al. 2013). Using a GP representation to tackle the problem allows us to rely on the same terminal and non-terminal sets on both chromosomes. This way, the first chromosome represents a candidate solution to the target function while the second chromosome represents an ideal mating partner, as described previously. In a nutshell, evaluation of mating candidates is quite similar to the standard evaluation of the phenotype of each individual, but instead of comparing the phenotype with the target function, it's compared to the ideal mating partner represented. The similarity measure applied was the same. The approach was tested on six setups with different characteristics, using a Tournament of five individuals to select the first parent which chooses its mate from a pool of five candidates.

Table 7.1 shows the obtained results. The instances where the proposed approach performed statistically better than a standard one or the other way around are shown in bold (A Wilcoxon Mann Whitney test with a significance level of 0.01 was used). The Mate Choice approach was able to outperform the standard approach on all instances of the problem except for the Koza-1 function. However, the Koza-1

Table 7.1 Mean Best fitness obtained over 50 runs obtained with a standard, mate choice and random approaches on the symbolic regression of six functions

Function	Standard	Mate Choice	Random
Keijzer-1	0.008005462	**0.0059473756**	0.0072442644
Keijzer-2	0.0063776454	**0.0052139161**	0.0062104645
Keijzer-3	0.0071500245	**0.0056003145**	0.0067438776
Keijzer-4	0.0890397335	**0.0833904122**	0.0840754187
Koza-1	**0.0006384168**	0.0014386396	0.0006481816
Nguyen-5	0.0014892713	**0.0004783439**	0.0025763115

approach is regarded as a particularly easy one (McDermott et al. 2012), where the standard approach is able to convert faster and therefore gain advantage. Also, the Nguyen-5 instance relies on a large population, which may explain why both approaches were able to largely outperform the random approach, with the Mate Choice approach still achieving statistical differences from the standard one.

It is clear by observing the results of the random approach, where the mating candidate is selected randomly, that the evolution process benefits from a lower selection pressure on the problem set. An analysis of the behaviour of each approach, focusing on the Mean Cumulative Destructive Crossovers (MCDC) and Mean Cumulative Neutral Crossovers (MCNC) shows that they behave differently. The Mate Choice approach achieved consistently a smaller number of neutral crossovers (with statistical differences except for the Koza-1 function), which don't contribute to fitness enhancements. At the same time it performed a larger number of destructive crossovers, which reduce the fitness of the offspring when compared with their parents, with statistical differences. This behaviour suggests that Mate Choice focus more on exploration than exploitation in this particular scenario, whereas the standard approach promotes mating between the fittest individuals. This exploration seems to be important for the enhanced results as it helps avoid convergence and turns explored traits into ecological opportunities thus achieving better Mean Best Fitness (MBF) values. The approach also obtains lower MCNC and higher MCDC values consistently when compared with the random approach suggesting that the promoted behaviour is not similar to randomly selecting mating partners. A larger analysis of the behaviour is available (Leitão et al. 2013).

Cluster Geometry Optimization (CGO) consists on finding the geometry of a cluster so that its potential energy is minimized. The problem is a NP-hard task (Cheng et al. 2009) with important applications in Nanoscience, Physics, Chemistry and Biochemistry (Zhao and Xie 2004). The problem provides a number of difficult test instances, from which we have tested the applicability of Mate Choice on Morse clusters ranging from 41 to 80 atoms (Leitão and Machado 2013). We have coupled our setup with an evolutionary algorithm that has previously achieved state of the art results (Pereira and Marques 2009). The initial approach relies on a steady-state population with a substitution mechanism that controls which offspring are allowed into the population. This mechanism is focused on maintaining diversity in the population as this is seen as a key factor when tackling CGO problems.

Each individual encodes the Cartesian coordinates of each particle in the cluster in the 3D space and evolve using GA operators. The Mate Choice setup adds a GP chromosome that allows each individual to evaluate mating candidates on a number of phenotypic features. When tackling the optimization of a Morse Cluster, each individual not only encodes a candidate solution to a N sized cluster but also to all $N-i$ instances. How an individual performs on each of these smaller instances may indicate good, bad or neutral genes, depending on the instance being optimized. It's up to the self-adaptive system to use this information in an appropriate way.

Table 7.2 shows the success rate on finding the putative optima on 30 runs for each instance using a tournament of five individuals to select the first parent and five mating candidates. A pairwise proportions test was used to test for significant

Table 7.2 Success rate over 30 runs obtained by a standard and mate choice approaches on the cluster geometry optimization of morse clusters of size N

N	41	42	43	44	45	46	47	48	49	50	51	52	53	54	55	56	57	58	59	60
Mate Choice	**24**	**21**	18	**18**	12	14	6	16	16	2	2	9	4	7	7	11	9	5	3	**7**
Standard	15	12	14	7	5	9	2	14	18	5	7	6	5	12	6	11	8	4	2	0
N	61	62	63	64	65	66	67	68	69	70	71	72	73	74	75	76	77	78	79	80
Mate Choice	1	6	8	8	9	10	5	3	6	6	6	3	6	5	5	3	3	4	2	5
Standard	1	12	6	11	8	8	5	4	6	8	6	7	6	1	2	4	7	3	6	5

differences (Taillard et al. 2008) with a significance level of 0.01. Overall, out of 40 instances, the Mate Choice approach was able to achieve higher success rates on 20 instances, 4 of them with significant differences. Ties were found on seven instances as well. A study on the impact of the Mate Choice approach on the behaviour of the algorithm focusing on population diversity as well as on the acceptance rate of the replacement operator has shown relevant differences from a standard approach.

The replacement operator has a powerful role on this problem as it determines which individuals are allowed in the population. If they are structurally similar (according to a distance measure) to one individual of the population, the fittest one is kept. If they are structurally dissimilar from all individuals, the offspring replaces the worst one as long as it has a better fitness. Otherwise the offspring is discarded. Therefore it is foreseeable that mating preferences will adapt so that new offspring can overcome these restrictions.

Results suggest that the algorithm has successfully done so, with an increase in the number of accepted individuals, both similar and dissimilar. Together, there has been an increase of roughly 50 % on the cumulative average number of substitutions. While an increase in the acceptance rate of new offspring doesn't necessarily translate into a better performance it does show an adaptation to the replacement strategy. The results in Table 7.2 suggests that such a behaviour contributes to competitive and sometimes better success rates, which is corroborated by average fitness values along the generations.

An analysis of the average population diversity also shows behavioural differences. While the standard approach is able to reach a higher average diversity, the Mate Choice approach is able to maintain it steadier while at the same time conducting a much larger number of substitutions in the population. This effect may indicate that individuals produced by Mate Choice in this scenario have smaller but steadier impact, which seems beneficial on the population level. A more complete analysis on the behaviour of Mate Choice on the optimization of Morse Clusters is available (Leitão and Machado 2013).

A third application of our Mate Choice mechanism has been done on the problem of Packing Circles in Squares (CPS). This problem consists on finding the configuration of a set of circles of fixed radius so that they minimize the area of a containing square (Machado and Leitão 2011). In order to represent candidate solutions to this problem, each individual encodes a vector of Cartesian coordinates representing the position of each circle, which is then mapped so that the area of the enclosing square is calculated. Our approach adds a GP chromosome that is used to assess characteristics on other individuals. Similarly to the CGO problem, we have divided the problem into several subproblems, in this case, for the instance of packing N circles, we assess how an individual performs on each $N - i$ instance.

Therefore, the GP functions are initialized using a terminal set composed by such characteristics and a number of arithmetic operators to combine them. Once again, performing well on a given instance may be an indication of either good, bad or neutral genes. The approach was compared to a standard approach, an approach where mating partners are selected randomly from the candidates set and a third approach where the mate evaluation function was encoded using GAs. The GA

Table 7.3 Mean Best Fitness over 30 runs obtained with a standard, random, mate choice with GA representation and mate choice with GP representation on the packing of N circles in squares

				Mate Choice	
N	Optimal	Standard	Random	GA	GP
2	3.4142	3.4142	3.4142	3.4142	3.4142
3	3.9319	3.9320	3.9320	3.9319	3.9319
4	4.0000	4.0266	4.0001	4.0255	4.0001
5	4.8284	5.0056	4.9911	**4.9250**	**4.9475**
6	5.3282	5.3669	5.3674	5.3685	5.3804
7	5.7321	5.8227	5.8081	5.8296	5.8098
8	5.8637	6.0212	5.9615	5.9913	5.9898
9	6.0000	6.5184	6.4907	6.5401	6.5154
10	6.7474	6.8936	6.8854	6.9110	**6.8536**
11	7.0225	7.1619	7.1764	7.2232	7.1564
12	7.1450	7.3966	7.3565	7.4809	7.3438
13	7.4630	7.8088	7.8167	7.8355	**7.7147**
14	7.7305	8.0705	8.0950	8.1509	**8.0048**
15	7.8637	8.3324	8.4173	8.4345	**8.2581**
16	8.0000	8.7014	8.8632	8.8153	**8.6012**
17	8.5327	8.8765	9.2345	9.0836	8.8665
18	8.6564	9.0996	9.4966	9.2724	9.0984
19	8.9075	9.4442	9.9422	9.6036	**9.3511**
20	8.9781	9.7212	10.2839	9.7641	**9.6030**
21	9.3580	9.9788	10.7402	10.1307	9.9425
22	9.4638	10.2610	11.0512	10.3705	10.2693
23	9.7274	10.5201	11.5476	10.6498	10.5892
24	9.8637	10.7725	11.8382	10.8163	10.8034

approach assesses all $N - i$ characteristics on mating candidates and combines them on a weighted sum. Each individual's second chromosome encodes therefore a set of weights, one for each characteristic and representing that characteristic's impact on the Mate Choice process. We expect the algorithm to evolve appropriate weights, either negative, close to zero or positive, according to the value of each characteristic.

Experiments were conducted on problem instances using 2–24 circles. The first parent was selected using a tournament of five individuals and sets of five mating candidates. Table 7.3 shows the MBF over 30 runs obtained by each approach. The GP based approach was able to achieve better results on 18 out of 23 instances, 8 of which with statistical differences (using a Wilcoxon Mann Whitney test with a significance level of 0.05) while the GA based approach performed better on 4 instance, 1 of which with statistical significant differences. The results achieved by the random approach suggest that the algorithm may benefit from a smaller selection pressure on most of the instances. The results obtained by the GA approach were disappointing as it was expected that it would be able to make a better use of the information provided or at least evolve into the original fitness function and therefore achieve results closer to the standard approach.

The results obtained by the GP approach suggest that the approach was able to make good use of the information provided about mating candidates and build appropriate functions to select mating candidates, which are likely to be radically different from the weighted sum designed by us using our knowledge of the problem. We believe that we would unlikely be able to design mating evaluation functions capable of matching the ones evolved by the GP approach. A larger analysis on the application of Mate Choice on the CPS problem is available (Machado and Leitão 2011).

Apart from the presented analysis on performance and behaviour of the Mate Choice setup on the three addressed scenarios, there are a two observations that are transversal and should be discussed. The first one regards the overhead of the setup, which results from the need to evaluate mating candidates using preferences specific to each individual. While on the first approach, based on ideal mating partners, this overhead depends on the metric being used, on the second approach it depends on the characteristics being evaluated. On the first approach, each candidate is evaluated once at each selection step as it is being compared to an ideal mating partner. On the second approach, individuals share what characteristics they can see in mating partners, therefore when a candidate is evaluated on a given characteristics, that process doesn't have to be repeated. The effort also depends on the mating preferences present in the population at each generation. This makes estimating an overhead for the Mate Choice operator a hard task. The second observation regards the complexity of the evolved GP functions, which makes it extremely difficult to assess exactly how individuals are evaluating others, which characteristics they value or not. Figure 7.4 shows what a GP tree representing a mating evaluation function may look like, however they can be much larger in size. Still, the example is enough to show that while we can see what characteristics are present, it is extremely difficult to assess exactly what role they play in the function. Some of them may be extremely relevant while others may be approximately neutral. Moreover, these functions are different from individual to individual and may vary drastically over the evolution process so that certain characteristics can be important at the beginning and others at the end, or may have different importance to each individual. What we can assess however is that we would unlikely be able to design such evaluation functions by hand and that they are radically different from previously used functions. The results suggest that despite how they work, these functions are making a good use of the information that they assess on mating candidates.

7.5 Synthesis

Sexual Selection through Mate Choice has been first proposed by Darwin to explain animal characteristics that didn't seem to fit in his theory of Natural Selection such as extravagant ornamentation and complex courtship behaviour. The theory, while being mostly rejected by the active research community at the time, eventually was able to gain the attention of researchers in multiple fields that have contributed with supporting evidence over the years. Nowadays Sexual Selection through Mate Choice is highly regarded by the scientific community as an important player in the evolution of species.

Despite the success of the theory on other fields, the impact of Mate Choice in the field of Evolutionary Computation is still very small, and a large number of questions remain unanswered regarding its design, implementation, behaviour and potential benefits. Several authors have proposed various approaches inspired by Mate Choice and have tackled multiple problems with more or less success. We have addressed a framework common for Mate Choice approaches and have reviewed important contributions through a classification study based on their adaptation mechanisms. We have also covered preference choices and gender role attribution mechanisms.

We finally propose a nature-inspired Mate Choice setup. First we discuss relevant aspects of Mate Choice in nature, propose three nature-inspired rules and follow up by covering design choices such as representation, evaluation, operators and parameters. The use of an extra chromosome to encode a GP tree representing mating preferences following two possible approaches was discussed as well as how it can be used to evaluate mating partners and take part of the evolution process.

We present a discussion on the application of Mate Choice on three problems and assess the behaviour of the algorithm. The discussed approaches differ drastically in behaviour from standard selection approaches as well as from an approach where the mating partner is selected randomly. The differences in behaviour impact performance as well as diversity, exploration and exploitation. A comparison with a self-adaptive Mate Choice approach based on a GA representation is also included and the differences discussed. Finally it's argued that assessing the overhead caused by this selection process is a difficult task and that the inner-workings of the evolved Mate Choice functions are very complex, making it extremely difficult to see which are relevant or irrelevant mating preferences. Still, the reported results suggest that the evolved GP functions are able to use the provided information in meaningful and beneficial ways. It is however unlikely that we would be able to design them by hand.

Acknowledgements The authors acknowledge the financial support from the Future and Emerging Technologies (FET) programme within the Seventh Framework Programme for Research of the European Commission, under the ConCreTe FET-Open project (grant number 611733).

References

Malte B Andersson. *Sexual selection*. Princeton University Press, 1994.

Longjiu Cheng, Yan Feng, Jie Yang, and Jinlong Yang. Funnel hopping: Searching the cluster potential energy surface over the funnels. *The Journal of chemical physics*, 130(21):214112, 2009.

Helena Cronin. *The ant and the peacock: Altruism and sexual selection from Darwin to today*. Cambridge University Press, 1993.

C. Darwin. The origin of species, 1859.

C. Darwin. *The descent of man and selection in relation to sex*. John Murray London, 1906.

Charles Darwin and Alfred Wallace. On the tendency of species to form varieties; and on the perpetuation of varieties and species by natural means of selection. *Journal of the proceedings of the Linnean Society of London. Zoology*, 3(9):45–62, 1858.

Susmita De, Sankar K Pal, and Ashish Ghosh. Genotypic and phenotypic assortative mating in genetic algorithm. *Information Sciences*, 105(1):209–226, 1998.

Carlos Fernandes and Agostinho Rosa. A study on non-random mating and varying population size in genetic algorithms using a royal road function. In *Evolutionary Computation, 2001. Proceedings of the 2001 Congress on*, volume 1, pages 60–66. IEEE, 2001.

Carlos Fernandes, Rui Tavares, Cristian Munteanu, and Agostinho Rosa. Using assortative mating in genetic algorithms for vector quantization problems. In *Proceedings of the 2001 ACM symposium on Applied computing*, pages 361–365. ACM, 2001.

R.A. Fisher. The evolution of sexual preference. *The Eugenics Review*, 7(3):184, 1915.

R.A. Fisher. The genetical theory of natural selection. 1930.

Rodney Fry, Stephen L Smith, and Andy M Tyrrell. A self-adaptive mate selection model for genetic programming. In *Evolutionary Computation, 2005. The 2005 IEEE Congress on*, volume 3, pages 2707–2714. IEEE, 2005.

Severino Galan, O Mengshoel, and Rafael Pinter. A novel mating approach for genetic algorithms. *Evolutionary computation*, 21(2):197–229, 2013.

Kai Song Goh, Andrew Lim, and Brian Rodrigues. Sexual selection for genetic algorithms. *Artificial Intelligence Review*, 19(2):123–152, 2003.

Lisa M Guntly and Daniel R Tauritz. Learning individual mating preferences. In *Proceedings of the 13th annual conference on Genetic and evolutionary computation*, pages 1069–1076. ACM, 2011.

Robert Hinterding and Zbigniew Michalewicz. Your brains and my beauty: parent matching for constrained optimisation. In *Evolutionary Computation Proceedings, 1998. IEEE World Congress on Computational Intelligence., The 1998 IEEE International Conference on*, pages 810–815. IEEE, 1998.

Robert Hinterding, Zbigniew Michalewicz, and Agoston E Eiben. Adaptation in evolutionary computation: A survey. In *Evolutionary Computation, 1997., IEEE International Conference on*, pages 65–69. IEEE, 1997.

Ekaterina A Holdener and Daniel R Tauritz. Learning offspring optimizing mate selection. In *Proceedings of the 10th annual conference on Genetic and evolutionary computation*, pages 1109–1110. ACM, 2008.

John R. Koza. Genetic programming as a means for programming computers by natural selection. *Statistics and Computing*, 4(2):87–112, 1994.

António Leitão and Penousal Machado. Self-adaptive mate choice for cluster geometry optimization. In Christian Blum and Enrique Alba, editors, *Genetic and Evolutionary Computation Conference, GECCO '13, Amsterdam, The Netherlands, July 6–10, 2013*, pages 957–964. ACM, 2013.

António Leitão, Jose Carlos Neves, and Penousal Machado. A self-adaptive mate choice model for symbolic regression. In *IEEE Congress on Evolutionary Computation*, pages 8–15, 2013.

Penousal Machado and António Leitão. Evolving fitness functions for mating selection. In Sara Silva, James A. Foster, Miguel Nicolau, Penousal Machado, and Mario Giacobini, editors, *Genetic Programming – 14th European Conference, EuroGP 2011, Torino, Italy, April 27–29, 2011. Proceedings*, volume 6621 of *Lecture Notes in Computer Science*, pages 227–238. Springer, 2011.

James McDermott, David R White, Sean Luke, Luca Manzoni, Mauro Castelli, Leonardo Vanneschi, Wojciech Jaskowski, Krzysztof Krawiec, Robin Harper, Kenneth De Jong, et al. Genetic programming needs better benchmarks. In *Proceedings of the fourteenth international conference on Genetic and evolutionary computation conference*, pages 791–798. ACM, 2012.

Geoffrey F Miller. Exploiting mate choice in evolutionary computation: Sexual selection as a process of search, optimization, and diversification. In *Evolutionary Computing*, pages 65–79. Springer, 1994.

Geoffrey F Miller and Peter M Todd. Evolutionary wanderlust: Sexual selection with directional mate preferences. In *From Animals to Animats 2: Proceedings of the Second International Conference on Simulation of Adaptive Behavior*, volume 2, page 21. MIT Press, 1993.

Gabriela Ochoa, Christian Mädler-Kron, Ricardo Rodriguez, and Klaus Jaffe. Assortative mating in genetic algorithms for dynamic problems. In *Applications of Evolutionary Computing*, pages 617–622. Springer, 2005.

Francisco B Pereira and Jorge MC Marques. A study on diversity for cluster geometry optimization. *Evolutionary Intelligence*, 2(3):121–140, 2009.

Michael Ratford, AL Tuson, and Henry Thompson. An investigation of sexual selection as a mechanism for obtaining multiple distinct solutions. *Emerg. Technol*, 1997.

Michael Ratford, Andrew Tuson, and Henry Thompson. The single chromosome's guide to dating. In *In Proceedings of the International Conference on Artificial Neural Networks and Genetic Algorithms*. Citeseer, 1996.

José Sánchez-Velazco and John A Bullinaria. Gendered selection strategies in genetic algorithms for optimization. *survival*, 8(6):11, 2003.

Josīe Sīanchez-Velazco and John A Bullinaria. Sexual selection with competitive/co-operative operators for genetic algorithms. 2003.

Ekaterina Smorodkina and Daniel Tauritz. Toward automating ea configuration: the parent selection stage. In *Evolutionary Computation, 2007. CEC 2007. IEEE Congress on*, pages 63–70. IEEE, 2007.

Éric D Taillard, Philippe Waelti, and Jacques Zuber. Few statistical tests for proportions comparison. *European Journal of Operational Research*, 185(3):1336–1350, 2008.

Peter M Todd and Geoffrey F Miller. Biodiversity through sexual selection. In *Artificial life V: Proceedings of the Fifth International Workshop on the Synthesis and Simulation of living Systems*, volume 5, page 289. MIT Press, 1997.

Mohammad Jalali Varnamkhasti and Lai Soon Lee. A genetic algorithm based on sexual selection for the multidimensional 0/1 knapsack problems. In *International Journal of Modern Physics: Conference Series*, volume 9, pages 422–431. World Scientific, 2012.

George C Williams and Austin Burt. *Adaptation and natural selection*. na, 1997.

A. Zahavi. Mate selection–a selection for a handicap. *Journal of theoretical Biology*, 53(1):205–214, 1975.

Jijun Zhao and Rui-Hua Xie. Genetic algorithms for the geometry optimization of atomic and molecular clusters. *Journal of Computational and Theoretical Nanoscience*, 1(2):117–131, 2004.

Part II
Specialized Applications

Chapter 8
Genetically Improved Software

William B. Langdon

Sources and data sets are available on line.

8.1 Introduction

As other chapters in the book show, genetic programming (Koza 1992; Banzhaf et al. 1998; Poli et al. 2008) has been very widely applied.[1] For example in modelling (Kordon 2010), prediction (Langdon and Barrett 2004; Podgornik et al. 2011; Kovacic and Sarler 2014), classification (Freitas 1997), design (Lohn and Hornby 2006) (including algorithm design Haraldsson and Woodward 2014), and creating art (Reynolds 2011; Jacob 2001; Langdon 2004; Romero et al. 2013). Here we concentrate upon application of genetic programming to software itself (Arcuri and Yao 2014). We start by briefly summarising research which evolved complete software but mostly we will concentration on newer work which has very effectively side stepped, what John Koza referred to as the *S-word* in artificial intelligence, the *scaling* problem, by using genetic programming not to create complete software but rather to enhance existing (human written) software.

The next section describes early successes with using GP to evolve real, albeit small, code and for automatically fixing bugs and then Sects. 8.3–8.6 describe recent success in which GP improved substantial (human written) C or C++ programs. The last part of the chapter (Sect. 8.7 onwards) describes in detail one of these.

[1]Genetic programming bibliography http://www.cs.bham.ac.uk/~wbl/biblio/ gives details of more than nine thousand articles, papers, books, etc.

W.B. Langdon (✉)
Computer Science, University College, London, UK
e-mail: W.Langdon@cs.ucl.ac.uk

A.H. Gandomi et al. (eds.), *Handbook of Genetic Programming Applications*,
DOI 10.1007/978-3-319-20883-1_8

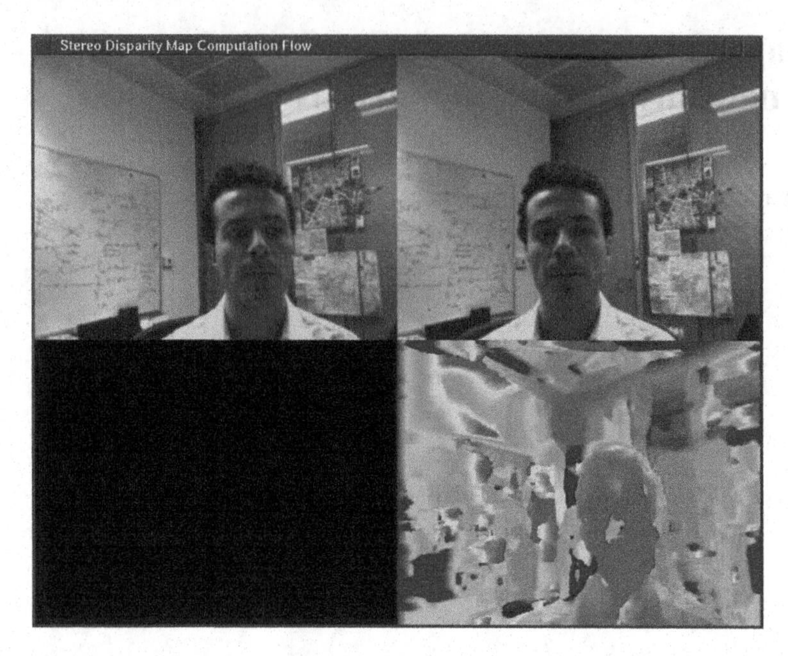

Fig. 8.1 *Top*: *left* and *right* stereo images. *Bottom*: Discrepancy between images, which can be used to infer distances

It shows how genetic programming was used to automatically evolve an almost seven fold speedup in parallel graphics code for extracting depth from stereoscopic image pairs. (See Fig. 8.1.)

8.2 Background

8.2.1 Hashes, Caches and Garbage Collection

Three early examples of real software being evolved using genetic programming are: hashing, caching and garbage collection. Each has the advantages of being small, potentially of high value and difficult to do either by hand or by theoretically universal principles. In fact there is no universally correct optimal answer. Any implementation which is good in one circumstance may be bettered in another use case by software deliberately designed for that use case. Thus there are several examples where not only can GP generate code but for particular circumstances, it has exceeded the state-of-the art human written code. Whilst this is not to say a human could not do better. Indeed they may take inspiration, or even code, from the evolved solution. It is that to do so, requires a programmer skilled in the art, for each new circumstance. Whereas, at least in principle, the GP can be re-run for each new use case and so automatically generate an implementation specific to that user.

Starting with (Hussain and Malliaris 2000) several teams have evolved good hashing algorithms (Berarducci et al. 2004; Estebanez et al.; Karasek et al. 2011).

Paterson showed GP can create problem specific caching code (Paterson and Livesey 1997). O'Neill and Ryan (1999) used their Grammatical Evolution (O'Neill and Ryan 2001, 2003) approach also to create code. Whilst (Branke et al. 2006) looked at a slightly different problem: deciding which (variable length) documents to retain to avoid fetching them again across the Internet. (Following (Handley 1994) several authors have sped up genetic programming itself by caching partial fitness evaluations, including me (Langdon 1998). However here we are interested in improving software in general rather than just improving genetic programming.)

Many languages allow the programmer to allocate and free chunks of memory as their program runs, e.g. C, C++ and Java. Typically the language provides a dynamic memory manager, which frees the programmer of the tedium of deciding exactly which memory is used and provides some form of garbage collection whereby memory that is no longer in use can be freed for re-use. Even with modern huge memories, memory management can impose a significant overhead. Risco-Martin et al. (2010) showed the GP can generate an optimised garbage collector for the C language.

8.2.2 Mashups, Hyper-heuristics and Multiplicity Computing

The idea behind web services is that useful services should be easily constructed from services across the Internet. Such hacked together systems are known as web mashups. A classic example is a travel service which invokes web servers from a number of airlines and hotel booking and car hire services, and is thus able to provide a composite package without enormous coding effort in itself. Since web services must operate within a defined framework ideally with rigid interfaces, they would seem to be ideal building blocks with which genetic programming might construct high level programs. Starting with Rodriguez-Mier, several authors have reported progress with genetic programming evolving composite web services (Rodriguez-Mier et al. 2010; Fredericks and Cheng 2013; Xiao et al. 2012).

There are many difficult optimisation problems which in practise are efficiently solved using heuristic search techniques, such as genetic algorithms (Holland 1992; Goldberg 1989). However typically the GA needs to be tweaked to get the best for each problem. This has lead to the generation of hyper-heuristics (Burke et al. 2013), in which the GA or other basic solver is tweaked automatically. Typically genetic programming is used. Indeed some solvers have been evolved by GP combining a number of basic techniques as well as tuning parameters or even re-coding GA components, such as mutation operators (Pappa et al. 2014).

A nice software engineering example of heuristics is compiler code generation. Typically compilers are expected not only to create correct machine code but also that it should be in some sense be "good". Typically this means the code should be fast or small. Mahajan and Ali (2008) used GP to give better code generation heuristics in Harvard's MachineSUIF compiler.

Multiplicity computing (Cadar et al. 2010) seeks to over turn the current software mono-culture where one particular operating system, web browser, software company, etc., achieves total dominance of the software market. Not only are such monopolies dangerous from a commercial point of view but they have allowed widespread problems of malicious software (especially computer viruses) to prosper. Excluding specialist areas, such as mutation testing (DeMillo and Offutt 1991; Langdon et al. 2010), so far there has been only a little work in the evolution of massive numbers of software variants (Feldt 1998). Only software automation (perhaps by using genetic programming) appears a credible approach to N-version programming (with N much more than 3). N-version programming has also been proposed as a way of improving predictive performance by voting between three or more classifiers (Imamura and Foster 2001; Imamura et al. 2003) or using other non-linear combinations to yield a higher performing multi-classifier (Langdon and Buxton 2001; Buxton et al. 2001).

Other applications of GP include: creating optimisation benchmarks which demonstrate the relative strengths and weaknesses of optimisers (Langdon and Poli 2005) and first steps towards the use of GP on mobile telephones (Cotillon et al. 2012).

8.2.3 Genetic Programming and Non-Function Requirements

Andrea Arcuri was in at the start of inspirational work on GP showing it can create real code from scratch. Although the programs remain small, David White, he and John Clark (White et al. 2011) also evolved programs to accomplish real tasks such as creating pseudo random numbers for ultra tiny computers where they showed a trade off between "randomness" and energy consumption.

The Virginia University group (see next section) also showed GP evolving Pareto optimal trade offs between speed and fidelity for a graphics hardware display program (Sitthi-amorn et al. 2011). Evolution seems to be particularly suitable for exploring such trade-offs (Feldt 1999; Harman et al. 2012) but (except for the work described later in this chapter) there has been little research in this area.

Orlov and Sipper (2011) describe a very nice system, Finch, for evolving Java byte code. The initial program to be improved is typically a Java program, which is compiled into byte code. Effectively the GP population instead of starting randomly (Lukschandl et al. 1998) is seeded (Langdon and Nordin 2000) with byte code from the initial program. The Finch crossover operator acts on Java byte code to ensure the offspring program area also valid java byte code. Large benefits arise because there is no need to compile the new programs. Instead the byte code can be run immediately. As Java is a main stream language, the byte code can be efficiently executed using standard tools, such as Java virtual machines and just in time (JIT) compilers. Also after evolution, standard java tools can be used to attempt to reverse the evolved byte code into Java source code.

Archanjo and Von Zuben (2012) present a GP system for evolving small business systems. They present an example of a database system for supporting a library of books.

Ryan (1999) and Katz and Peled (2013) provide interesting alternative visions. In genetic improvement the performance, particularly the quality of the mutated program's output, is assessed by running the program. Instead they suggest each mutation be provably correct and thus the new program is functionally the same as the original but in some way it is improved, e.g. by running in parallel. Katz and Peled (2013) suggests combining GP with model checking to ensure correctness.

Zhu and Kulkarni (2013) suggest using GP to evolve fault tolerant programs. Schulte et al. (2014a) describes a nice system which can further optimise the low level Intel X86 code generated by optimising compilers. They show evolution can reduce energy consumption of non-trivial programs. (Their largest application contains 141,012 lines of code.)

8.2.4 Automatic Bug Fixing

As described in the previous two sections, recently genetic programming has been applied to the production of programs itself, however so far relatively small programs have been evolved. Nonetheless GP has had some great successes when applied to existing programs. Perhaps the best known work is that on automatic bug fixing (Arcuri and Yao 2008). Particularly the Humie award winning[2] work of Westley Weimer (Virginia University) and Stephanie Forrest (New Mexico) (Forrest et al. 2009). This has received multiple awards and best paper prizes (Weimer et al. 2009, 2010). GP has been used repeatedly to automatically fix most (but not all) real bugs in real programs (Le Goues et al. 2012a). Weimer and Le Goues have now shown GP bug fixing to be effective on several millions of lines of C++ programs. Once GP had been used to *do the impossible* others tried (Wilkerson and Tauritz 2010; Bradbury and Jalbert 2010; Ackling et al. 2011) and it was improved (Kessentini et al. 2011) and also people felt brave enough to try other techniques, e.g. Nguyen et al. (2013); Kim et al. (2013). Indeed their colleague, Eric Schulte, has shown GP can even work at abstraction levels other than source code. In Schulte et al. (2010) he showed bugs can be fixed at the level of the assembler code generated by the compiler or even machine code (Schulte et al. 2013). After Weimer and co-workers showed that automatic bugfixing was not impossible, people studied the problem more openly. It turns out, for certain real bugs, with modern software engineering support tools, such as bug localisation (e.g. Yoo 2012), the problem may not even be hard (Weimer 2013).

[2]Human-competitive results presented at the annual GECCO conference http://www.genetic-programming.org/combined.php.

Formal theoretical analysis (Cody-Kenny and Barrett 2013) of evolving sizable software is still thin on the ground. Much of the work presented here is based on GP re-arranging lines of human written code. In a very large study of open source software (Gabel and Su 2010) showed that excluding white space, comments and details of variable names, any human written line of code has probably been written before. In other words, given a sufficiently large feedstock of human written code, current programs could have been written by re-using and re-ordering existing lines of code. In many cases in this and the following sections, this is exactly what GP is doing. Schulte et al. (2014b) provides a solid empirical study which refutes the common assumption that software is fragile. (See also Fig. 8.2). While a single random change may totally break a program, mutation and crossover operations can be devised which yield populations of offspring programs in which some

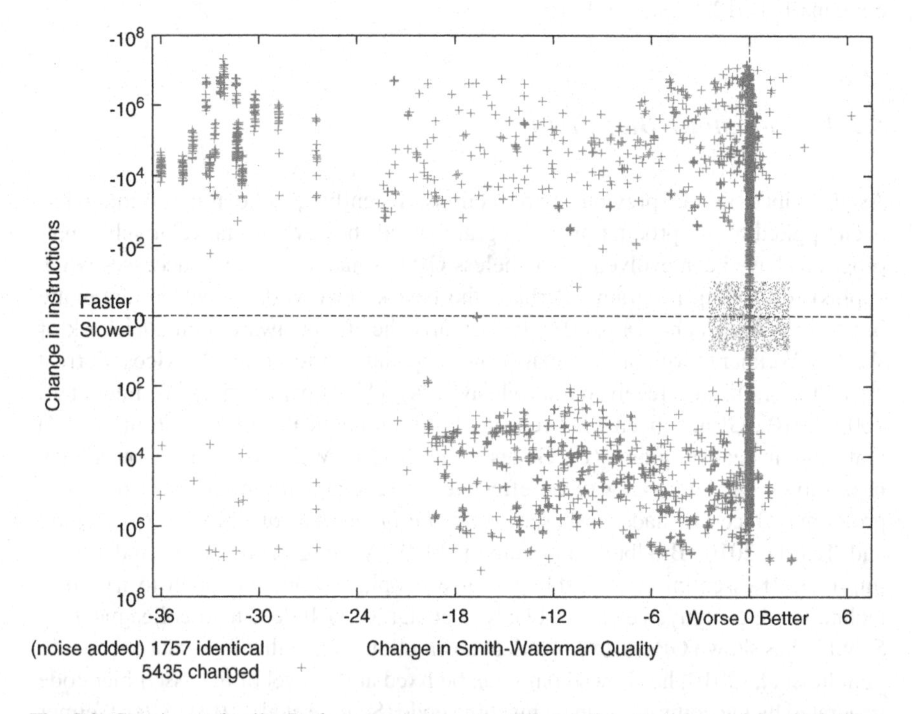

Fig. 8.2 C++ is not fragile. Performance versus speed for random mutations of Bowtie2. The *horizontal axis* shows the change in quality of Bowtie2 output, whilst the *vertical axis* (note non-linear scale) shows the change in the number of lines of code executed. As expected some mutations totally destroy the program, e.g. they fail to compile or abort (not plotted) or reduce the quality of the answer enormously (e.g. -36). Some are slower (*lower half*) and some are faster (*top*). However a large number have exactly the same quality as the original code (plotted above "0"). These may be either slower or faster. The *rectangle of dots* attempts to emphasis the 18 % that are identical (in terms of quality of answer and run time) to the original code. To the *right* of the "0", there are even a few random programs which produce slightly better answers than the original code. It is these Darwinian evolution selects and breeds the next generation from. Total 10,000 random program runs. Failed runs are not plotted

may be very bad but the population can also contains many reasonable programs and even a few slightly improved ones. Over time the Darwinian processes of fitness selection and inheritance (Darwin 1859) can amplify the good parts of the population, yielding greatly improved programs.

8.3 Auto Porting Functionality

The Unix compression utility gzip was written in C in the days of Digital Equipment Corp.'s mini-computers. It is largely unchanged. However there is one procedure (of about two pages of code) in it, which is so computationally intensive that it has been re-written in assembler for the Intel 86X architecture (i.e. Linux). The original C version is retained and is distributed as part of Software-artifact Infrastructure Repository sir.unl.edu (Hutchins et al. 1994). SIR also contains a test suite for gzip. In *Genetic Improvement*, as with Le Goues' bug-fixing work, we start with an existing program and a small number of test cases. In the case of the gzip function, we showed genetic programming could evolve a parallel implementation for an architecture not even dreamt of when the original program was written (Langdon and Harman 2010). Whereas Le Goues uses the original program's AST (abstract syntax tree) to ensure that many of the mutated programs produced by GP compile, we have used a BNF grammar. In the case of (Langdon and Harman 2010) the grammar was derived from generic code written by the manufacture of the parallel hardware. Note that it had nothing special to do with gzip. The original function in gzip was instrumented to record its inputs and its outputs each time it was called (see Fig. 8.3). When gzip was run on the SIR test suite, this generated more than a million test cases, however only a few thousand were used by the GP.[3] Essentially GP was told to create parallel code from the BNF grammar which when given a small number of example inputs returned the same answers. The resulting parallel code is functionally the same as the old gzip code.

8.4 Bowtie2^{GP} Improving 50,000 lines of C++

As Fig. 8.4 shows, genetic programming produces populations of programs which may have different abilities on different scales. While Fig. 8.4 shows speed versus quality, other tradeoffs have been investigated (Harman et al. 2012, see also Schulte et al. 2014a). For example it may be impossible to simultaneously minimise

[3]Later work used even fewer tests.

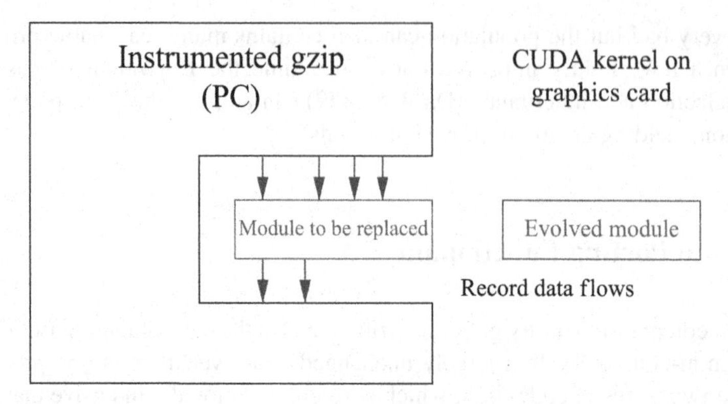

Fig. 8.3 Auto porting a program module to new hardware (a GPU). The original code is instrumented to record the inputs (*upper blue arrows*) to the target function (*red*) and the result (*lower blue arrows*) it calculates. Its inputs and outputs are logged every time every time it is called. These become the test suite and fitness function for the automatically evolved replacement module running on novel hardware. By inspecting the evolved CUDA code automatically generated by GP we can see that it is functionally identical to the C code inside gzip. Also it has been demonstrated by running back-to-back with the original code more than a million times (Langdon and Harman 2010) (Color figure online)

execution time, memory foot print and energy consumption. Yet, conventionally human written programs choose one trade-off between multiple objectives and it becomes infeasible to operate the program with another trade-off. For example, consider approximate string matching.

Finding the best match between (noisy) strings is the life blood of Bioinformatics. Huge amounts of people's time and computing resources are devoted every day to matching protein amino acid sequences against databases of known proteins from all forms of life. The acknowledge gold standard is the BLAST program (Altschul et al. 1997) which incorporate heuristics of known evolutionary rates of change. It is available via the web and can lookup a protein in every species which has been sequences in a few minutes. Even before the sequencing of the human genome, the volume of DNA sequences was exploding exponentially at a rate like Moore's Law (Moore 1965). With modern NextGen sequencing machines throwing out 100s of millions (even billions) of (albeit very noisy) DNA base-pair sequences, there is no way that BLAST can be used to process this volume of data. This has lead to human written look up tools for matching NextGen sequences against the human genome. Wikipedia list more than 140 programs (written by some of the brightest people on the planet) which do some form of Bioinformatics string matching.

The authors of all this software are in a quandary. For their code to be useful the authors have to chose a point in the space of tradeoffs between speed, machine resources, quality of solution and functionality, which will: (1) be important to the Bioinformatics community and (2) not be immediately dominated by other programs. In practise they have to choose a target point when they start, as once basic design choices (e.g. target data sources and computer resources) have been

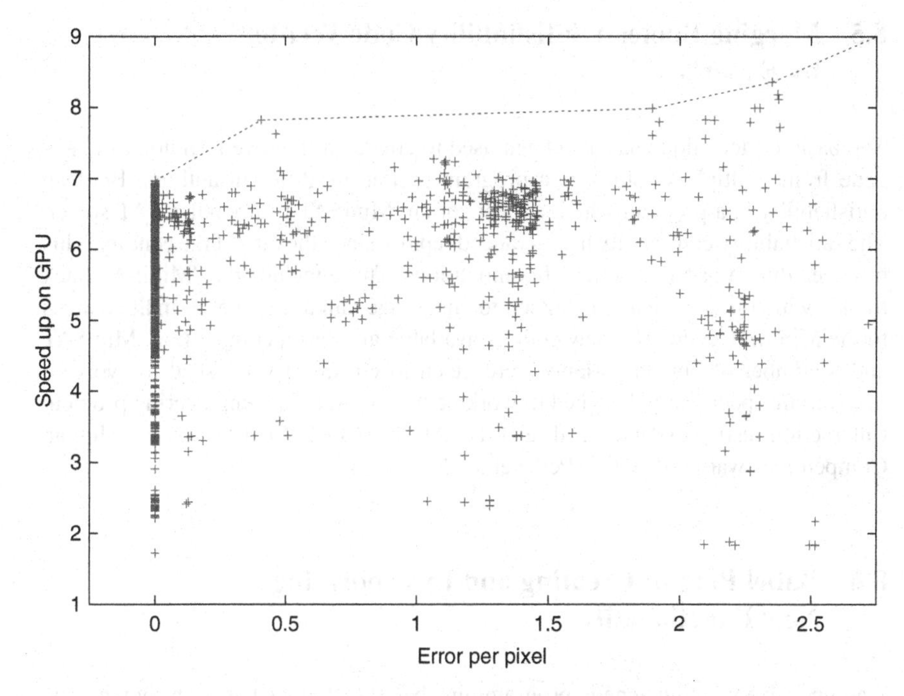

Fig. 8.4 Example of automatically generated Pareto tradeoff front (Harman et al. 2012). Genetic programming used to improve 2D Stereo Camera code (Stam 2008) for modern nVidia GPU (Langdon and Harman 2014b). *Left* (above 0) many programs are faster than the original code written by nVidia's image processing expert (human) and give exactly the same answers. Many other automatically generated programs are also faster but give different answers. Some (cf. *dotted blue line*) are faster than the best zero error program (Color figure online)

made, few people or even research teams have the resources to discard what they have written and start totally from scratch. Potentially genetic programming offers them a way of exploring this space of tradeoffs (Feldt 1999; Harman et al. 2012). GP can produce many programs across the trade-off space and so can potentially say "look here is a trade-off which you had not considered". This could be very useful to the human, even if they refuse to accept machine generated code and insist on coding the solution themselves.

We have made a start by showing GP can transform human written DNA sequence matching code, moving it from one tradeoff point to another. In our example, the new program is specialised to a particular data source and sequence problem for which it is on average more than 70 times faster. Indeed on this particular problem, we were fortunate that not only is the variant faster but indeed it gives a slight quality improvement on average (Langdon and Harman 2015).

8.5 Merging Boolean Satisfiability Code Written by Experts

The basic GI technique has also been used to create an improved version of C++ code from multiple versions of a program written by different authors. Boolean Satisfiability is a problem which appears often. MiniSAT is a popular SAT solver. The satisfiability community has advanced rapidly since the turn of the century. This has been due in part to a series of competitions. These include the "MiniSAT hack track", which is specifically designed to encourage humans to make small changes to the MiniSAT code. The new code is available after each competition. MiniSAT and a number of human variants were given to GI and it was asked to evolve a new variant specifically designed to work better on a software engineering problem (interaction testing) (Petke et al. 2014b). At GECCO 2014 it received a Human Competitive award (HUMIE) (Petke et al. 2014a).

8.6 Babel Pidgin: Creating and Incorporating New Functionality

Another prize winning genetic programming based technique has been shown to be able to extend the functionality of existing code (Harman et al. 2014). GP, including human hints, was able to evolved new functionality externally and then search based techniques (Harman 2011) were used to graft the new code into an existing program (pidgin) of more than 200,000 lines of C++.

8.7 Improving Parallel Processing Code Written by Experts

There is increasing use of parallelism both in conventional computing but also in mobile applications. At present the epitome of parallelism are dedicated multi-core machines based on gaming graphics cards (GPUs). Although originally devised for the consumer market, they are increasingly being used for general purpose computing on GPUs (GPGPU) (Owens et al. 2008) with several the world's fastest computers being based on GPUs. However, although support tools are improving, programming parallel computers continues to be a challenge (Langdon 2012) and simply leaving code generation to parallel compilers is often insufficient. Instead experts, e.g. Merrill et al. (2012), have advocated writing highly parameterised parallel code which can then be automatically tuned. Unfortunately this throws the load back on to the coder (Langdon 2011). In the rest of the chapter we explain how genetic programming (see Fig. 8.5) was able to automatically update for today's GPUs software written specifically by nVidia's image processing expert to show off the early generations of their graphics cards (Stam 2008). While originally

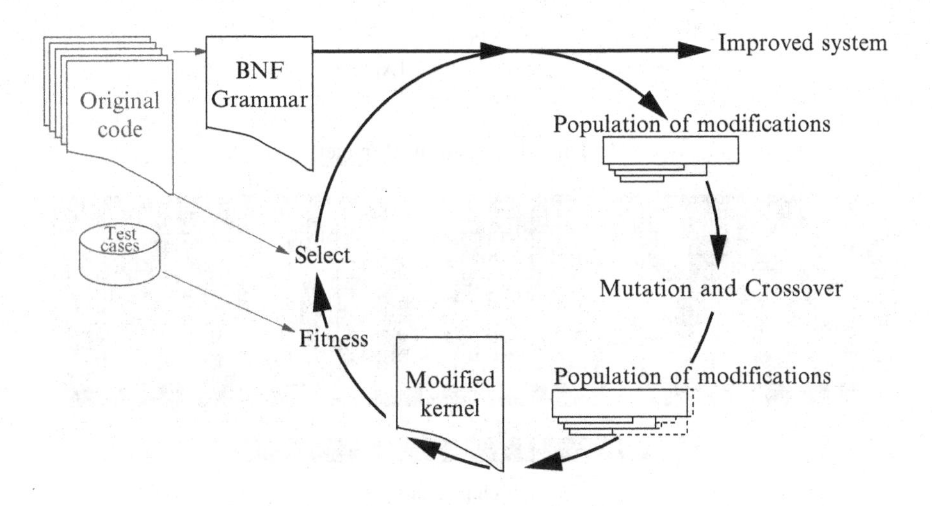

Fig. 8.5 Genetic Improvement of stereoKernel

(Langdon and Harman 2014b) we considered six types of hardware, in the interests of brevity we shall concentrate on the most powerful (Tesla K20c). Performance of the other five GPUs and more details can be found in Langdon and Harman (2014b) and technical report (Langdon and Harman 2014a). GP gave more than a six fold performance increase relative to the original code on the same hardware. (Each Tesla K20c contains 2496 processing elements, arranged in 13 blocks of 192 and running at 0.71GHz. Bandwidth to its on board memory is 140 Gbytes per second. See Fig. 8.6.)

In another example a combination of manual and automated changes to production 3D medical image processing code lead to the creation of a version of a performance critical kernel which (on a Tesla K20c) is more than 2000 times faster than the production code running on an 2.67 GHz CPU (Langdon et al. 2014).

The next sections briefly gives the StereoCamera CUDA code. This is followed by descriptions of the stereo images (page 194), and the code tuning process (pages 196–205). The changes made specifically for the K20c Tesla are described in Sect. 8.16 (page 206) whilst the Appendix (pages 211–214) holds the complete CUDA source code for the K20c Tesla. The code is also available in `StereoCamera_v1_1c.zip`.

8.8 Source Code: StereoCamera

The StereoCamera system was written by nVidia's stereo image processing expert Joe Stam (Stam 2008) to demonstrate their 2007 hardware and CUDA. StereoCamera was the first to show GPUs could give real time stereo image processing (> 30 frames per second). StereoCamera V1.0b was downloaded from SourceForge but,

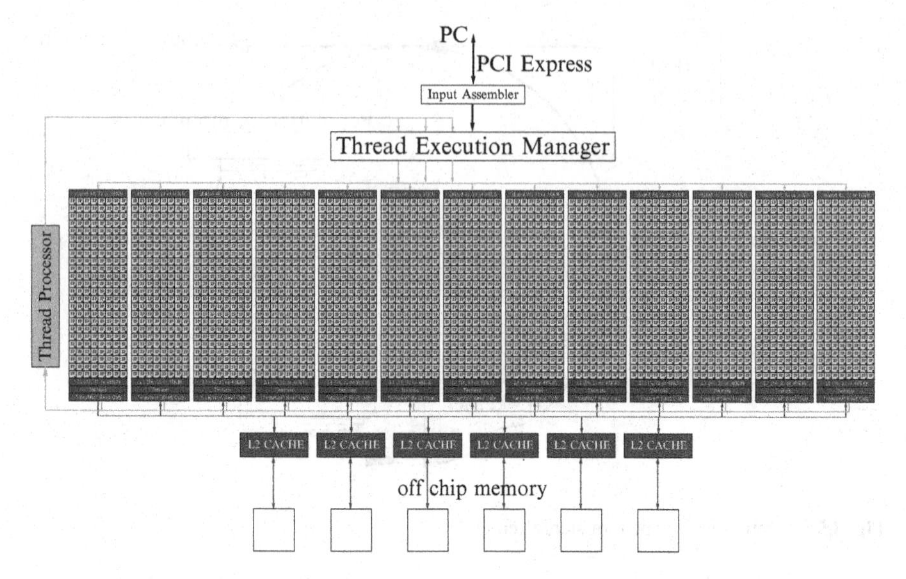

Fig. 8.6 Tesla K20c contains 13 SMX multiprocessors (each containing 192 stream processors), a PCI interface to the host PC, thread handling logic and 4800 MBytes of on board memory

despite the exponential increase in GPU performance, it had not been updated since 2008 (except for my bugfix). In the six years after it was written, nVidia GPUs went through three major hardware architectures whilst their CUDA software went through five major releases.

StereoCamera contains three GPU kernels plus associated host code. We shall concentrate upon one, stereoKernel which contains the main stereo image algorithm. For each pixel in the left image, GPU code stereoKernel reports the number of pixels the right image has to be shifted to get maximal local alignment (see Fig. 8.7). Stam (2008) notes that the parallel processing power of the GPU allows the local discrepancy between the left and right images to be calculated using the sum of *squares* of the difference (SSD) between corresponding pixels and this sum is taken over the relatively large 11×11 area. It does this by minimising the sum of squares of the difference (SSD) between the left and right images in a 11×11 area around each pixel. Once SSD has been calculated, the grid in the right hand image is displaced one pixel to the left and the calculation is repeated. Although the code is written to allow arbitrary displacements, in practice the right hand grid is move a pixel at a time. SSD is calculated for 0 to 50 displacements and the one with the smallest SSD is reported for each pixel in the left hand image. In principle each pixel's value can be calculated independently but each is surrounded by a "halo" of five others in each direction.

Even on a parallel computer, considerable savings can be made by reducing the total number of calculations by sharing intermediate calculations (Stam 2008, Fig. 3). Each SSD calculation (for a given discrepancy between left and right images) involves summing 11 columns (each of 11 squared discrepancy values).

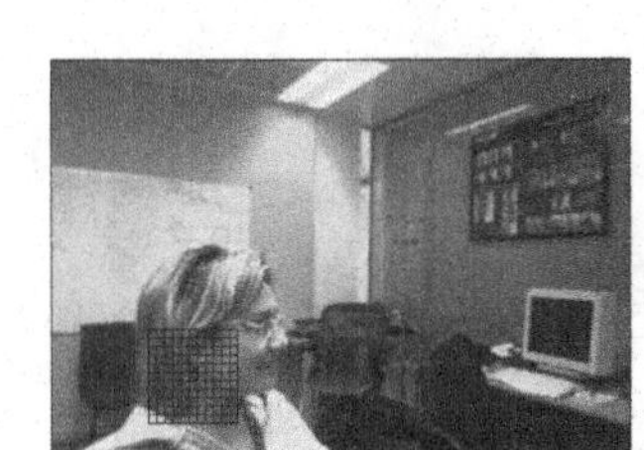

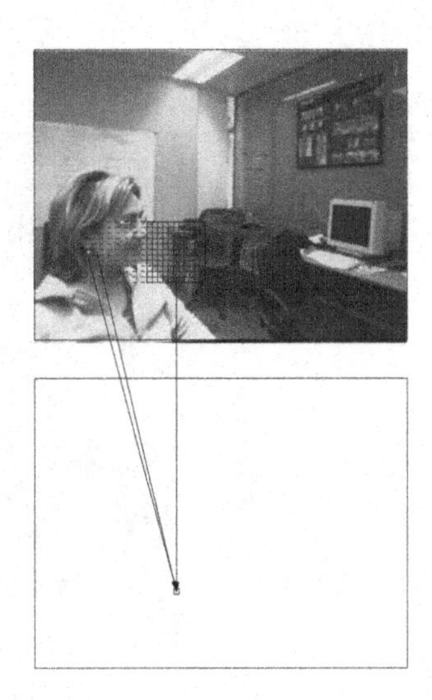

Fig. 8.7 Schematic of stereo disparity calculation. *Top*: *left* and *right* stereo images. *Bottom*: output. Not to scale. For each pixel stereoKernel calculates the sum of squared differences (SSD) between 11×11 regions centred on the pixel in the *left* image and the same pixel in the *right* hand image. This is the SSD for zero disparity. The *right* hand 11×11 region is moved one place to the *left* and new SSD is calculated (SSD for 1 pixel of disparity). This is repeated 50 times. Each time a smaller SSD is found, it is saved. Although the output pixel (*bottom*) may be updated many times, its final value is the distance moved by the 11×11 region which gives the smallest SSD. That is the distance between left and right images which gives the maximum similarity between them (across an 11×11 region). This all has to be done for every pixel. Real time performance is obtained by parallel processing and reducing repeated calculations

By saving the column sums in shared memory adjacent computational threads can calculate just their own column and then read the remaining ten column values calculated by their neighbouring threads.

After one row of pixel SSDs have been calculated, when calculating the SSD of the pixels immediately above, ten of the eleven rows of SSD values are identical. Given sufficient storage, the row values could be saved and then 10 of them could be reused requiring only one row of new square differences to be calculated. However fast storage was scare on GPUs and instead Stam compromised by saving the total SSD (rather than the per row totals). The SSD for the pixel above is then the total SSD plus the contribution for the new row *minus* the contribution from the lowest row (which is no longer included in the 11×11 area). Stam took care that the code avoids rounding errors. The more rows which share their partial results, the more efficient is the calculation but then there is less scope for performing calculations in parallel. To avoid re-reading data it is desirable that all the image data for both left and right images (including halos and discrepancy offsets) should fit within the GPU's texture caches. The macro `ROWSperTHREAD` (40) determines how many rows are calculated together in series. The macro `BLOCK_W` (64) determines how the image is partitioned horizontally (see Figs. 8.8 and 8.9). To fit the GPU architecture

Fig. 8.8 The *left* and *right* images (*solid rectangle*) are split into BLOCK_W×ROWSperTHREAD tiles. The *dashed lines* indicate the extra pixels outside the tile which must be read to calculate values for pixels in the tile. The *right* hand image is progressively offset by between zero and STEREO_MAXD pixels (50, *dotted lines*)

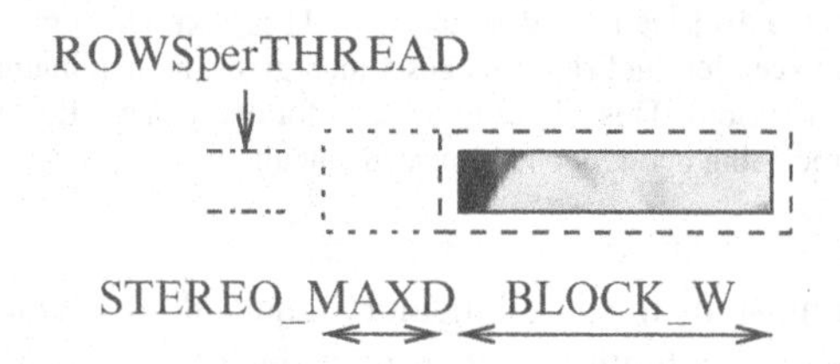

Fig. 8.9 An example of the part of the *right* hand of a stereo image pair which is processed by a block of CUDA threads. The area covered in the *right* image is eventually shifted STEREO_MAXD (50) pixels to the *left*. For most GPUs the original code did not use the optimal shape, see Fig. 8.10. Although the width (BLOCK_W, 64) was correct, the height (ROWSperTHREAD) should be reduced from 40 to 5

BLOCK_W will often be a multiple of 32. In practise all these factors interact in non-obvious (and sometimes undocumented) hardware dependent ways.

8.9 Example Stereo Pairs from Microsoft's I2I Database

Microsoft have made available for image processing research thousands of images. Microsoft's I2I database contains 3010 stereo images. Figure 8.7 (top) is a typical example. Many of these are in the form of movies taken in an office environment. Figure 8.1 shows the first pair from a typical example.

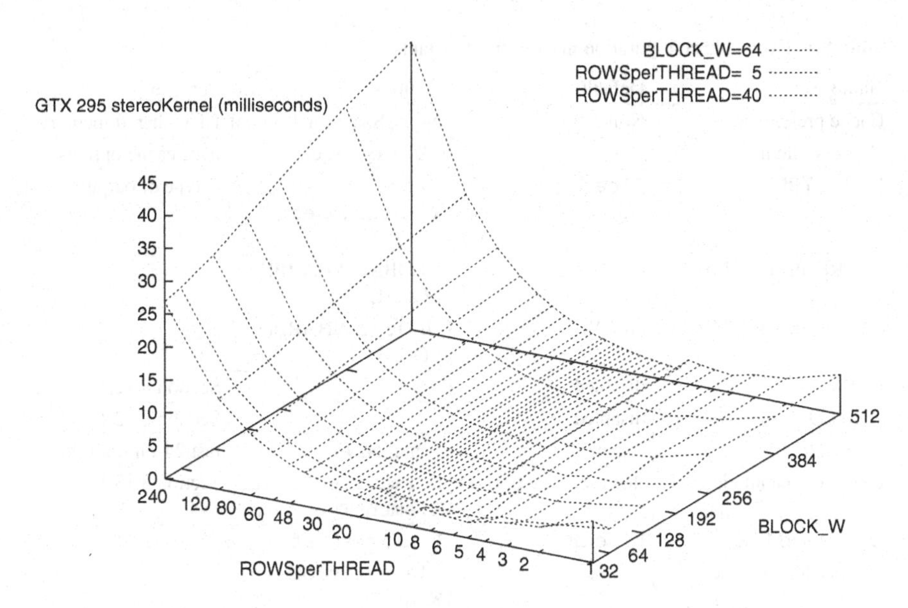

Fig. 8.10 The effect of changing the work done per thread (ROWSperTHREAD) and the block size (BLOCK_W) on CUDA kernel speed before it was optimised by GP. For all but one of the GPUs, stereoKernel is fastest at 5, 64 (the default is 40, 64)

We downloaded i2idatabase.zip[4] (1.3GB) and extracted all the stereo image pairs and converted them to grey scale. Almost images all are 320×240 pixels. We took (up to) the first 200 pairs for training leaving 2810 for validation. Notice we are asking the GP to create a new version of the CUDA stereoKernel GPU code which is tuned to pairs of images of this type. As we shall see (in Sect. 8.15) the improved GPU code is indeed tuned to 320×240 images but still works well on the other I2I stereo pairs.

8.10 Host Code and Baseline Kernel Code

The supplied C++ code is designed to read stereo images from either stereo webcams or pairs of files and using OpenGL, to display both the pair of input images and the calculated discrepancy between them on the user's monitor (see Fig. 8.1). This was adapted to both compare answers generated by the original code with those given by the tuned GP modified code and to time execution of the modified GPU kernel code. These data are logged to a file and the image display is disabled.

[4]http://research.microsoft.com/en-us/um/people/antcrim/data_i2i/i2idatabase.zip.

Table 8.1 Evolvable configuration macros and constants

Name	Default	Options	Purpose
Cache preference	None	None, Shared, L1, Equal	L1 v. shared memory
-Xptxas -dlcm		' ', ca, cg, cs, cv	nvcc cache options
OUT_TYPE	`float`	`float`, `int`, `short int`, `unsigned char`	C type of output
STORE_disparityPixel	GLOBAL	GLOBAL, SHARED, LOCAL	
STORE_disparityMinSSD	GLOBAL	GLOBAL, SHARED, LOCAL	
DPER	disabled		Section 8.12.2
XHALO	disabled		Section 8.12.1
__mul24(a,b)	__mul24	__mul24, *	fast 24-bit multiply
GPtexturereadmode	`NormalizedFloat`	`NormalizedFloat`, `ElementType`, none	Section 8.13.1.4
texturefilterMode	`Linear`	`Linear`, `Point`	
textureaddressMode		`Clamp`, `Mirror`, `Wrap`	
texturenormalized		0, 1	

The original kernel code is in a separately compiled file to ensure it is not affected by GP specified compiler options (particularly -Xptxas -dlcm, Table 8.1). For each pixel it generates a value in the range $0.0, 1.0, 2.0 \ldots 50.0$ being the minimum discrepancy between the left and right images. If a match between the left and right images cannot be found (i.e. SSD $\geq$ 500000) then it returns -1.0.

8.11 Pre- and Post- Evolution Tuning and Post Evolution Minimisation of Code Changes

In initial genetic programming runs, it became apparent that there are two parameters which have a large impact on run time but whose default settings are not suitable for the GPUs now available. Since there are few such parameters and they each have a small number of sensible values, it is feasible to run StereoCamera on all reasonable combinations and simply choose the best for each GPU. Hence the revised strategy is to tune `ROWSperTHREAD` and `BLOCK_W` before running the GP. (`DPER`, Sect. 8.12.2, is not initially enabled.) Figure 8.10 shows the effect of tuning `ROWSperTHREAD` and `BLOCK_W` for the GTX 295. As with (Le Goues et al. 2012b) and our GISMOE approach (Langdon and Harman 2015), after GP has run the best GP individual from the last generation is cleaned up by a simple one-at-a-time hill climbing algorithm. Langdon and Harman (2015) (Sect. 8.11) and finally `ROWSperTHREAD`, `BLOCK_W` and `DPER` are tuned again. (Often no further changes were needed.)

For each combination of parameters, the kernel is compiled and run. By recompiling rather than using run time argument passing, the nVidia nvcc C++ compiler is given the best chance of optimising the code (e.g. loop unrolling) for these parameters and the particular GPU.

`BLOCK_W` values were based on sizes of thread blocks used by nVidia in the examples supplied with CUDA 5.0. (They were 8, 32, 64, 128, 192, 256, 384 and 512.) All small `ROWSperTHREAD` values or values which divide into the image height (240) exactly were tested. (I.e., 1, ... 18, 20, 21, 24, 26, 30, 34, 40, 48, 60, 80, 120 and 240.) Autotuning reduced `ROWSperTHREAD` (see Fig. 8.9) from 40 to 5 before the GP was run. For the Tesla K20c, this gave a speed up of 2.373 ± 0.03 fold.

The best GP individual in the last generation is minimised by starting at its beginning and progressively removing each individual mutation and comparing the performance of the new kernel with the evolved one. For simplicity this is done on the last training stereo image pair. Unless the new kernel is worse the mutation is excluded permanently. To encourage removal of mutations with little impact, those that make less than 1 % difference to the kernel timing are also removed.

In the after evolution tuning, if GP had enabled `DPER` (Sect. 8.12.2) then as well as tuning `BLOCK_W` and `ROWSperTHREAD` the autotuner tried values 1, 2, 3 and 4 for `DPER`. (In the case of the Tesla K20c, GP enabled `DPER` but its default value, 2, turned out to be optimal.)

8.12 Alternative Implementations

8.12.1 Avoiding Reusing Threads: XHALO

As mentioned in Sect. 8.8 each row of pixels is extended by five pixels at both ends. The original code reused the first ten threads of each block to calculate these ten halo values. Much of the kernel code is duplicated to deal with the horizontal halo. GPUs have a special type of parallel architecture which means many identical operations can be run in parallel but if the code branches in different directions part of the hardware becomes idle. (This is known as thread divergence.) Thus diverting ten threads to deal with the halo causes all the remaining threads in the warp to become idle. (Each warp contains 32 threads.) Option XHALO allows GP to use ten additional threads which are dedicated to the halo. Thus each thread only deals with one pixel. In practise the net effect of XHALO is to disable the duplicated code so that instead of each block processing vertical stripes of 64 pixels, each block only writes stripes 54 pixels wide.

8.12.2 Parallel of Discrepancy offsets: DPER

The original code (Sect. 8.8) steps through sequentially 51 displacements of the right image with respect to the left. Modern GPUs allow many more threads and often it is best to use more threads as it allows greater parallelism and may improve throughput by increasing the overlap between computation and I/O. Instead of stepping sequentially one at a time through the 51 displacements, the DPER option allows 2, 3 or 4 displacement SSD values to be calculated in parallel. As well as increasing the number of threads, the amount of shared memory needed is also increased by the same factor. Nevertheless only one (the smallest) SSD value per pixel need be compared with the current smallest, so potentially saving some I/O. Although the volume of calculations is little changed, there are also potential saving since each DPER block uses almost the same data.

8.13 Parameters Accessible to Evolution

The GISMOE GP system (Langdon and Harman 2015) was extended to allow not only code changes but also changes to C macro `#defines`. The GP puts the evolved values in a C `#include .h` file, which is complied along with the GP modified kernel code and the associated (fixed) host source code.

Table 8.1 shows the twelve configuration parameters. Every GP individual chromosome starts with these 12, which are then followed by zero or more changes to the code. Figure 8.16 page 206 contains an example GP individual, whereas Figs. 8.12 page 202, 8.13 and 8.14 contain simplified schematics of GP individuals.

8.13.1 Fixed Configuration Parameters

8.13.1.1 OUT_TYPE

The return value should be in the range -1 to 50 (Sect. 8.10). Originally this is coded as a `float`. OUT_TYPE gives GP the option of trying other types. Notice, since the data will probably be used on the GPU, we do not use the fact that the smaller data types take less time to transfer between GPU and host. (I.e. all fitness times, Sect. 8.14.5.2, are on the GPU.)

8.13.1.2 STORE_disparityPixel and STORE_disparityMinSSD

disparityPixel and disparityMinSSD are major arrays in the kernel. Stam coded them to lie in the GPU's slow off chip global memory. These configuration options give evolution the possibility of trying to place them in either shared memory or in local memory. Where the compiler can resolve local array indexes, e.g. as a result of unrolling loops, it can use fast registers in place of local memory.

8.13.1.3 __mul24

For addressing purposes, older GPU's included a fast 24 bit multiply instruction, which is heavily used in the original code. It appears that in the newer GPUs __mul24 may actually be slower than ordinary (32 bit) integer multiply. Hence we give GP the option of replacing __mul24 with ordinary multiply.

8.13.1.4 Textures

CUDA textures are intimately linked with the GPU's hardware and provide a wide range of data manipulation facilities (normalisation, default values, control of boundary effects and interpolation) which the original code does not need but is obliged to use. The left and right image textures are principally used because they provide caching (which was not otherwise available on early generation GPUs.) We allowed the GP to investigate all texture options, including not using textures. Some combinations are illegal but the host code gives sensible defaults in these cases.

Unfortunately it is tricky to ensure access directly to the data and via a texture produce identical answers. Once cause of differences is there can be a $\frac{1}{2}$ pixel discrepancy between direct access (which treats the images as 2D arrays) and textures where reference point is the centre of the pixel. This leads to small differences between direct access and the original code. Whilst such slight differences make little difference to the outputs' appearance, even so such GP individuals are penalised by the fitness function (Sect. 8.14.5). This may have inhibited GP exploring all the data access options.

8.14 Evolvable Code

Following the standard GISMOE approach (Langdon and Harman 2015), the evolutionary cycle is amended so that we start by creating a BNF grammar from the supplied source code and the GP evolves linear patches to the code (applied via the grammar) rather than trees, cf. Fig. 8.5 page 191. The source code, including XHALO and DPER (Sects. 8.12.1 and 8.12.2), is automatically translated line by line into the grammar (see Fig. 8.11). Notice the grammar is not generic, it represents only one program, stereoKernel, and variants of it. The grammar contains 424 rules, 277 represent fixed lines of C++ source code. There are 55 variable lines, 27 IF and 10 of each of the three parts of C `for` loops. There are also five CUDA specific types:

```
<KStereo.cuh_52>    ::= "__attribute__((global)) " <launchbounds_KStereo.cuh_52>
                        " void KERNEL(\n"
#kernel
<launchbounds_KStereo.cuh_52> ::= ""
<launchbounds_K0>   ::= "\n" "#ifdef DPER\n" "__launch_bounds__(BLOCK_W*dperblock)\n"
                        "#else\n" "__launch_bounds__(BLOCK_W)\n"   "#endif /*DPER*/\n"
                    ...
<launchbounds_K5>   ::= "\n" "#ifdef DPER\n" "__launch_bounds__(BLOCK_W*dperblock,5)\n"
                        "#else\n" "__launch_bounds__(BLOCK_W,5)\n" "#endif /*DPER*/\n"
<optrestrict_KStereo.cuh_52> ::= " __restrict__ "
#kernelarg
<KStereo.cuh_53>    ::= "OUTYPE *" <optrestrict_KStereo.cuh_52> "disparityPixel,\n"
<KStereo.cuh_54>    ::= <optconst_KStereo.cuh_54> "size_t out_Pitch,\n"
<optconst_KStereo.cuh_54> ::= "const "
<KStereo.cuh_55>    ::= "#ifdef GLOBAL_disparityMinSSD\n"
<KStereo.cuh_56>    ::= "int *" <optrestrict_KStereo.cuh_52> "disparityMinSSD,\n"
<KStereo.cuh_57>    ::= "#if OUT_TYPE != float_ && OUT_TYPE != int_\n"
<KStereo.cuh_58>    ::= <optconst_KStereo.cuh_58> "size_t out_pitch,\n"
<optconst_KStereo.cuh_58> ::= "const "
<KStereo.cuh_59>    ::= "#endif\n"
<KStereo.cuh_60>    ::= "#endif /*GLOBAL_disparityMinSSD*/\n"
                    ...
<KStereo.cuh_72>    ::= ")\n"
                    ...
<KStereo.cuh_141>   ::= " if" <IF_KStereo.cuh_141>
                        "  extra_read_val = BLOCK_W+threadIdx.x;\n"
#"if
<IF_KStereo.cuh_141> ::= "(threadIdx.x < (2*RADIUS_H))"
                    ...
<KStereo.cuh_158>   ::= <pragma_KStereo.cuh_158> "for("
                        <for1_KStereo.cuh_158> ";" "OK()&&"
                        <for2_KStereo.cuh_158> ";"
                        <for3_KStereo.cuh_158> ") \n"
#for
<pragma_KStereo.cuh_158> ::= ""
#pragma
<pragma_K0>         ::= "#pragma unroll \n"
<pragma_K1>         ::= "#pragma unroll 1\n"
                    ...
<pragma_K11>        ::= "#pragma unroll 11\n"
<for1_KStereo.cuh_158> ::= "i = 0"
<for2_KStereo.cuh_158> ::= "i<ROWSperTHREAD && Y+i < height"
<for3_KStereo.cuh_158> ::= "i++"
<KStereo.cuh_159>   ::= "{\n"
<KStereo.cuh_160>   ::= "" <_KStereo.cuh_160> "\n"
#other
<_KStereo.cuh_160> ::= "init_disparityPixel(X,Y,i);"
<KStereo.cuh_161>   ::= "" <_KStereo.cuh_161> "\n"
<_KStereo.cuh_161> ::= "init_disparityMinSSD(X,Y,i);"
<KStereo.cuh_162>   ::= "}\n"
```

Fig. 8.11 Fragments of BNF grammar used by GP. Most rules are fixed but rules starting with `<_`, `<IF_`, `<for1_`, `<pragma_`, etc. can be manipulated using rules of the same type to produce variants of stereoKernel. Lines beginning with # are comments

1. `#pragma unroll` allows GP to control the nvcc compiler's loop unrolling. pragma rules are automatically inserted before each `for` loop but rely on GP to enable and set their values. Using the type constraints GP can either: remove it, set it to `#pragma unroll`, or set it to `#pragma unroll` *n* (where *n* is 1 to 11).
2. optvolatile CUDA allows shared data types to be marked as `volatile` which influences the compiler's optimisation. As required by the CUDA compiler,

the grammar automatically ensures all shared variables are either flagged as `volatile` or none are.

The remaining three CUDA types apply to the kernel's header.

3. optconst Each of kernel's scalar inputs can be separately marked as `const`.
4. optrestrict All of the kernel's array arguments can be marked with `__restrict__` This potentially helps the compiler to optimise the code. On the newest GPUs (SM 3.5) optrestrict allows the compiler to access read only arrays via a read only cache. Since both only apply if all arrays are marked `__restrict__`, the grammar ensures they all are or none are.
5. launchbounds is again a CUDA specific aid to code optimisation. By default the compiler must generate code that can be run with any numbers of threads. Since GP knows how many threads will be used, specifying it via `__launch_bounds__` gives the compiler the potential of optimising the code. `__launch_bounds__` takes an optional second argument which refers to the number of blocks that are active per streaming multiprocessor SMX. How it is used is again convoluted, but the grammar allows GP to omit it, or set it to 1, 2, 3, 4 or 5.

8.14.1 Initial Population

Each member of the initial population is unique. They are each created by selecting at random one of the 12 configuration constants (Table 8.1) and setting it at random to one of its non-default values. As the population is created it becomes harder to find unique mutations and so random code changes are included as well as the configuration change. Table 8.2 summarises the GP parameters.

Table 8.2 Genetic programming parameters for improving stereoKernel

Representation:	Fixed list of 12 parameter values (Table 8.1) followed by variable list of replacements, deletions and insertions into BNF grammar
Fitness:	Run on a randomly chosen 320×240 monochrome stereo image pair. Compare answer & run time with original code and time its execution. See Sects. 8.14.5 and 8.14.6.
Population:	Panmictic, non-elitist, generational. 100 members. New randomly chosen training sample each generation.
Parameters:	Initial population of random single mutants heavily weighted towards the kernel header and shared variables. 50 % truncation selection. 50 % crossover (uniform for fixed part, 2pt for variable). 50 % mutation 25 % mutation random change to fixed part. 25 % add code mutation (one of: delete, replace, insert, each equally likely). No size limit. Stop after 50 generations.

8.14.2 Weights

Normally each line of code is equally likely to be modified. However, only as part of creating a diverse initial population, the small number of rules in the kernel header (i.e. launchbounds, optrestrict, optconst and optvolatile) are 1000 times more likely to be changed than the other grammar rules. (Forcing each member of the GP population to be unique is also only done in the initial population.) In future, it might be worthwhile ensuring GP does not waste effort changing CUDA code which can have no effect by setting the weights of lines excluded by conditional compilation to zero.

8.14.3 Mutation

Half of mutations are made to the configuration parameters (Table 8.1). In which case one of the 12 configuration parameters is chosen uniformly at random and its current value is replaced by another of its possible values again chosen uniformly at random, see Fig. 8.12. The other half of the mutations are made to the code. In which case the mutation operator appends an additional code patch to the parent (see Fig. 8.13). There are three possible code mutations: delete a line of code, replace a line and insert a line. The replacement and inserted lines of code are copied from stereoKernel itself (via the grammar). Notice GP does not create code. It merely rearranges human written code.

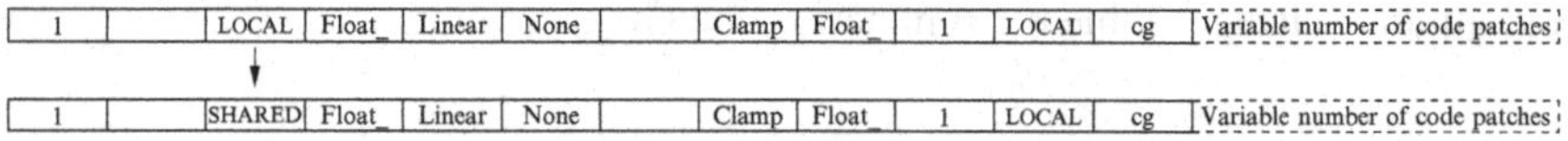

1		LOCAL	Float_	Linear	None		Clamp	Float_	1	LOCAL	cg	Variable number of code patches
↓												
1		SHARED	Float_	Linear	None		Clamp	Float_	1	LOCAL	cg	Variable number of code patches

Fig. 8.12 Example of mutation to the configuration part at the start of a GP individual. *Top*: parent *Bottom*: offspring. The 12 configuration parameters are given in Table 8.1

<284>+<194> volitile <247><186><180><231><358><154><174>+<176>

<284>+<194> volitile <247><186><180><231><358><154><174>+<176><288>+<161>

Fig. 8.13 Example of mutation to the variable length part of a GP individual. Patch <288>+<161> is appended to parent (*top*) causing in the child (*bottom*) a copy of source line 161 to be inserted before line 288 in the kernel source code. (For clarity the *left* hand part omitted and full grammar rule names simplified, e.g. to just the line numbers.)

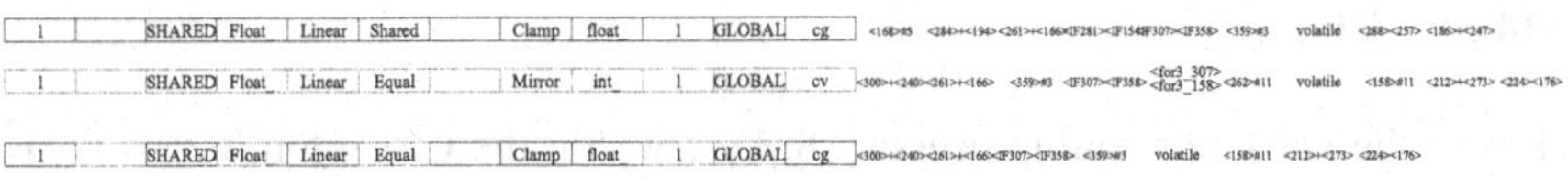

1		SHARED	Float	Linear	Shared		Clamp	float	1	GLOBAL	cg	<168>#5 <284>+<194><261>+<166><IF281><IF154><IF307><IF358> <359>#3 volatile <288><257> <186>+<247>
1		SHARED	Float_	Linear	Equal		Mirror	int_	1	GLOBAL	cv	<300>+<240><261>+<166> <359>#3 <IF307><IF358> <for3_307> <for3_158> <262>#11 volatile <158>#11 <212>+<273> <224><176>
1		SHARED	Float_	Linear	Equal		Clamp	float_	1	GLOBAL	cg	<300>+<240><261>+<166><IF307><IF358> <359>#3 volatile <158>#11 <212>+<273> <224><176>

Fig. 8.14 Example of crossover. Parts of two above median parents (*top and middle*) recombined to yield a child (*bottom*)

8.14.4 Crossover

Crossover creates a new GP individual from two different members of the better half (Sect. 8.14.6) of the current population. The child inherits each of the 12 fixed parameters (Table 8.1) at random from either parent (uniform crossover Syswerda 1989, see Fig. 8.14). Whereas in Langdon and Harman (2015) we used append crossover which deliberately increases the size of the offspring, here, on the variable length part of the genome, we use an analogue of Koza's tree GP crossover (Koza 1992). Two crossover points are chosen uniformly at random. The part between the two crossover points of the first parent is replaced by the patches between the two crossover points of the second parent to give a single child. On average, this gives no net change in length.

8.14.5 Fitness

To avoid over fitting and to keep run times manageable, each generation one of the two hundred training images pairs is chosen (Langdon 2010). Each GP modified kernel in the population is tested on that image pair.

8.14.5.1 CUDA memcheck and Loop Overruns

Normally each GP modified kernel is run twice. The first time it is run with CUDA memcheck and with loop over run checks enabled. If no problems are reported by CUDA memcheck and the kernel terminates normally (i.e. without exceeding the limit on loop iterations) it is run a second time without these debug aids. Both memcheck and counting loop iterations impose high overheads which make timing information unusable. Only in the second run are the timing and error information used as part of fitness. If the GP kernel fails in either run, it is given such a large penalty, that it will not be a parent for the next generation.

When loop timeouts are enabled, the GP grammar ensures that each time a C++ `for` loop iterates a per thread global counter is incremented. If the counter exceeds the limit, the loop is aborted and the kernel quickly terminates. If any thread reaches its limit, the whole kernel is treated as if it had timed out. The limit is set to 100× the maximum reasonable value for a correctly operating good kernel.

8.14.5.2 Timing

Each of the streaming multiprocessor (SMXs) within the GPU chip has its own independent clock. On some GPUs `cudaDeviceReset()` resets all the clocks, this is not the case with the C2050. To get a robust timing scheme, which applies to all GPUs, each kernel block records both its own start and end times and the SMX unit it is running on. After the kernel has finished, for each SMX, the end time of the last block to use it and the start time of the first block to use it are subtracted to give the accurate duration of usage for each SMX. (Note to take care of overflow `unsigned int` arithmetic is used.) Whilst we do not compare values taken from clocks on different SMXs, it turns out to be safe to assume that the total duration of the kernel is the longest time taken by any of the SMXs used. (As a sanity check this GPU kernel time is compared to the, less accurate, duration measured on the host CPU.) The total duration taken by the GP kernel (expressed as GPU clock tics divided by 1000) is the first component of its fitness.

8.14.5.3 Error

For each pixel in the left image the value returned by the GP modified kernel is compared with that given by the un-modified kernel. If they are different a per pixel penalty is added to the total error which becomes the second part of the GP individual's fitness.

If the unmodified kernel did not return a value (i.e. it was -1.0, cf. Sect. 8.10) the value returned by the GP kernel is also ignored. Otherwise, if the GP failed to set a value for a pixel, it gets a penalty of 200. If the GP value is infinite or otherwise outside the range of expected values (0..50) it attracts a penalty of 100. Otherwise the per pixel penalty is the absolute difference between the original value and the GP's value.

For efficiency, previously (Langdon and Harman 2010) we batched up many GP generated kernels into one file to be compiled in one go. For simplicity, since we are using a more advanced version of nVidia's nvcc compiler, and GP individuals in the same population may need different compiler options, we did not attempt this. Typically it takes about 3.3 s to compile each GP generated kernel. Whereas to run the resulting StereoCamera program (twice see Sect. 8.14.5.1) takes about 2.0 s,

8.14.6 Selection

At the end of each generation we compare each mutant with the original kernel's performance on the same test case and only allow it to be a parent if it does well. In detail, it must be both faster and be, on average, not more than 6.0 per pixel different from the original code's answer. However mostly the evolved code passes both tests. At the end of each generation the population is sorted first by their error and then by their speed. The top 50 % are selected to be parents of the next generation.

Each selected parent creates one child by mutation (Sect. 8.14.3) and another by crossover with another selected parent (Sect. 8.14.4). The complete GP parameters are summarised in Table 8.2.

8.15 Results

The best individual from the last generation (50) was minimised to remove unneeded mutations which contributed little to its overall performance and returned (Sect. 8.11). This reduced the length of the GP individual from 29 to 10. On the Tesla K20c, on average, across all 2516 I2I 320×240 stereo image pairs, GP sped up the original StereoCamera code almost seven fold. (The mean speed up is 6.837 ± 0.04.) By reducing `ROWSperTHREAD` from the original 40 to 5, pretuning (Sect. 8.11) itself gave a factor of 2.4 fold speed up. The original value of `BLOCK_W` (64) and the default value of DPER (2) were optimal for the Tesla K20c. I.e. the GP code changes gave another factor of almost three on top of the parameter tuning. The speedup of the improved K20c kernel on all of the I2I stereo images is given in Fig. 8.15. The speed up for the other five GPUs varied in a similar way

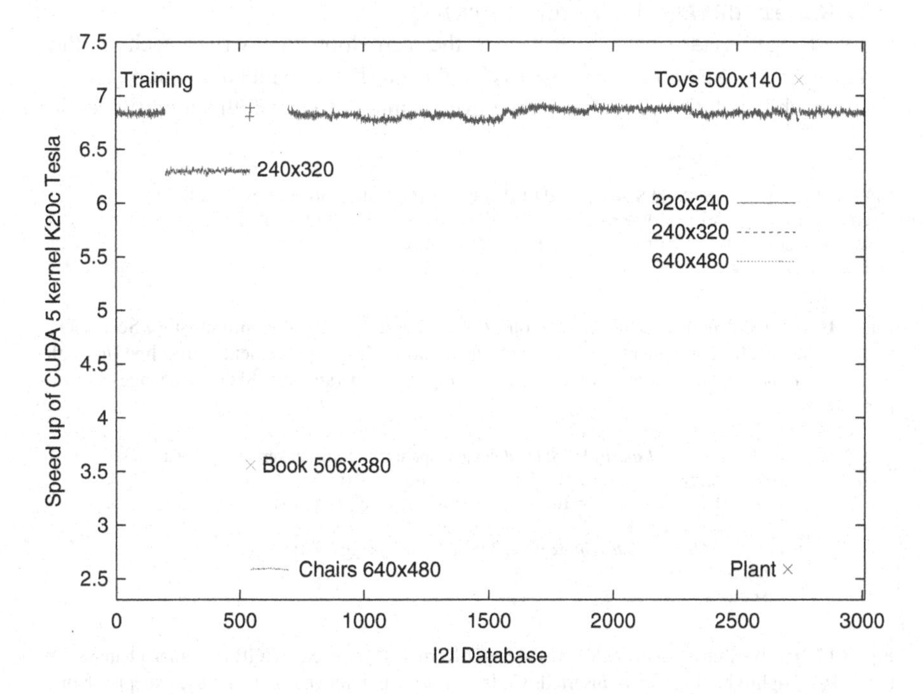

Fig. 8.15 Performance of GP improved K20c Tesla kernel on all 3010 stereo pairs in Microsoft's I2I database relative to original kernel on the same image pair on the same GPU. Fifty of first 200 pairs used in training. The evolved kernel is always much better, especially on images of the same size and shape as it was trained on

to the K20c. Finally, notice typically there is very little difference in performance across the images of the same size and shape as the training data

8.16 Evolved Tesla K20c CUDA Code

The best of generation 50 individual changes 6 of the 12 fixed configuration parameters (Table 8.1) and includes 23 grammar rule changes. After removing less useful components (Sect. 8.11) four configuration parameters were changed and there were six code changes. See Figs. 8.16 and 8.17. The complete code is given in the appendix (pages 211–214).

The evolved configuration parameters mean that DPER is enabled and the new kernel calculates two disparity values in parallel (Sect. 8.12.2), disparityPixel and disparityMinSSD are stored in shared memory (Sect. 8.13.1.2) and XHALO is enabled (Sect. 8.12.1).
The final code changes, Fig. 8.17, are:

- disable volatile, Sect. 8.14.
- insert `#pragma unroll 11` before the `for` loop that steps through the `ROWSperTHREAD` - 1 other rows (Sect. 8.8).
- insert `#pragma unroll 3` before the `for` loop that writes each of the `ROWSperTHREAD` rows of disparityPixel from shared to global memory. Its not clear why evolution chose to ask the nvcc compiler to unroll this loop (which is

```
DPER=1 STORE_disparityMinSSD=SHARED XHALO=1 STORE_disparityPixel=SHARED
<pragma_KStereo.cuh_359><pragma_K3> <_KStereo.cuh_161>+<_KStereo.cuh_224>
<_KStereo.cuh_348> <optvolatile_KStereo.cuh_86>
<pragma_KStereo.cuh_262><pragma_K11> <IF_KStereo.cuh_326><IF_KStereo.cuh_154>
```

Fig. 8.16 Best GP individual in generation 50 of K20c Tesla run after minimising, Sect. 8.11, removed less useful components. (Auto-tuning made no further improvements.) Top line (normal font) are four non-default values for the 12 fixed configuration parameters. Six code changes shown in `tt` font

```
int * __restrict__ disparityMinSSD, //Global disparityMinSSD not kernel argument
volatile extern __attribute__((shared)) int col_ssd[];
volatile int* const reduce_ssd = &col_ssd[(64 )*2 -64];
#pragma unroll 11
if(X < width && Y < height) replaced by if(dblockIdx==0)
__syncthreads();
#pragma unroll 3
```

Fig. 8.17 Evolved changes to K20c Tesla StereoKernel. (Produced by GP grammar changes in Fig. 8.16). Highlighted `code` is inserted. Code in *italics* is removed. For brevity, except for the kernel's arguments, disparityPixel and disparityMinSSD changes from global to shared memory are omitted. The appendix, pages 211–214, gives the complete source code

always executed 5 times) only 3 times. But then when nvcc decides to do loop unrolling is obscure anyway.

- Mutation <_KStereo.cuh_161>+<_KStereo.cuh_224> causes line 224 to be inserted before line 161. Line 224 potentially updates local variable ssd, however ssd is not used before the code which initialises it. It is possible that the compiler spots that the mutated code cannot affect anything outside the kernel and simply optimises it away. During minimisation removing this mutation gave a kernel whose run time was exactly on the removal threshold.
- Mutation <IF_KStereo.cuh_326><IF_KStereo.cuh_154> replaces X < width && Y < height by dblockIdx==0. This replace a complicated expression by a simpler (and so presumably faster) expression, which itself has no effect on the logic since both are always true. In fact, given the way if(dblockIdx==0) is nested inside another if, the compiler may optimise it away entirely. I.e. GP has found a way of improving the GPU kernel by removing a redundant expression.

  The original purposed of if(X < width && Y < height) was to guard against reading outside array bounds when calculating SSD. However the array index is also guarded by i < blockDim.x
- delete __syncthreads() on line 348. __syncthreads() forces all threads to stop and wait until all reach it. Line 348 is at the end of code which may update (with the smaller of two disparities values) shared variables disparityPixel and disparityMinSSD. In effect GP has discovered it is safe to let other threads proceed since they will not use the same shared variables before meeting other __syncthreads() elsewhere in the code. As well as reducing the number of instructions, removing synchronisation calls potentially allows greater overlapping of computation and I/O leading to an overall saving.

8.17 Discussion

Up to Intel's Pentium, Moore's Law (Moore 1965) had applied not only to the doubling of the number of transistors on a silicon chip but also to exponential rises in clock speeds. Since 2005 mainstream processor clock speeds have remained fairly much unchanged. However Moore's Law continues to apply to the exponential rise in the number of available logic circuits. This has driven the continuing rise of parallel multi-core computing. In mainstream computing, GPU computing continues to lead in terms of price v. performance. However GPGPU computing (Owens et al. 2008) (and parallel computing in general) is still held back by the difficulty of high-performance parallel programming (Langdon 2011; Merrill et al. 2012).

When programming the GPU, in addition to the usual programming tasks, there are other hardware specific choices, e.g. where to store data. Even for the expert it is difficult to find optimal choices for these while simultaneously programming. Merrill et al. (2012) propose heavy use of templates in kernel code in an effort to separate algorithm coding for data storage etc. However templates are in practise even harder to code and current versions of the compiler cannot optimally make choices for the programmer. As Sect. 8.11 shows, it can be feasible to remove the choice of key parameters (typically block size) from the programmer. Instead their code is run with all feasible values of the parameter and the best chosen. There are already tools to support this. Such enumerative approaches are only feasible with a small number of parameters. The GP approach is more scalable and allows mixing both parameter tuning and code changes. To be fair, we should say at present all the approaches are still at the research stage rather than being able to assist the average graphics card programmer.

Future new requirements of StereoCamera might be dealing with: colour, moving images (perhaps with time skew), larger images, greater frame rates and running on mobile robots, 3D telephones, virtual reality gamesets or other low energy portable devices. We can hope our GP system could be used to automatically create new versions tailored to new demands and new hardware.

In some cases modern hardware readily gives on line access to other important non-functional properties (such as power or current consumption, temperature and actual clock speeds). Potentially these might also be optimised by GP. White et al. (2008) showed it can be possible to use GP with a cycle-level power level simulator to optimise small programs for embedded systems. (Schulte et al. 2014a recently extended this to large open source every day programs.) Here we work with the real hardware, rather than simulators, however real power measurements are not readily available with all our GTX and Tesla cards.

Many computers, including GPUs, and especially in mobile devices, now have variable power consumption. Thus reducing execution time can lead to a proportionate reduction in energy consumption and hence increase in battery life, since as soon as the computation is done the computer can revert to its low power idle hibernating state. Yao et al. (1995); Han et al. (2010); Radulescu et al. (2014) consider other ways of tuning of the processor's clock speed (which might be combined with software improvements).

Another promising extension is the combined optimisation for multiple functional and non-functional properties (Colmenar et al. 2011). Initial experiments hinted that NSGA-II (Deb et al. 2002; Langdon et al. 2010) finds it hard to maintain a complete Pareto front when one objective is much easier than the others. Thus a population may evolve to contain many fast programs which have lost important functionality while slower functional program are lost from the population. Newer multi-objective GAs or alternative fitness function scalings may address this.

The newer versions of CUDA also include additional tools (e.g. CUDA race check) which might be included as part of fitness testing.

The supplied kernel code contains several hundred lines of code. It may be that this only just contains enough variation for GP's cut-and-past operations (Sect. 8.14.3). We had intended to allow GP to also use code taken from the copious examples supplied by nVidia with CUDA (see Sect. 8.5) but so far this has not been tried.

nVidia and other manufactures are continuing to increase the performance, economy and functionality of their parallel hardware. There are also other highly parallel and low power chips with diverse architectures (e.g. multicore CPUs, FPGAs, Intel Xeon Phi, mobile and RFID devices Andreopoulos 2013). These trends suggest the need for software to be ported (Langdon and Harman 2010) or to adapt to new parallel architectures will continue to increase.

One of the great success for modular system design has been the ability to keep software running whilst the underlying hardware platforms have gone through several generations of upgrades. In some cases this has been achieved by freezing the software, even to the extent of preserving binaries for years. In practise this is not sufficient and software that is in use is under continual and very expensive maintenance. There is a universal need for software to adapt.

8.18 Conclusions

We have reviewed published work on using genetic programming on software. Initially we showed examples where genetic programming was able to evolve real software from scratch. In some cases, e.g. by automatically creating bespoke applications tailored to particular tasks, the GP generated code improves on generic human written code.

Even now, code evolved from scratch tends to be small. The GGGP (grow and graft) system, described in Sect. 8.6, is a potential way around the problem. GGGP still evolves small new components but also uses GP to graft them into much bigger human written codes, thus create large hybrid software.

Similarly the CUDA gzip example (Sect. 8.3) showed small but valuable units of code can be effectively automatically ported by evolving new code to match the functionality of the existing code, even if it is written in a different language or executes on different hardware. Indeed auto bugfixing (Sect. 8.2.4), Bowtie2, NiftyReg and StereoCamera also use the existing code as the de facto specification of the functionality of the to be evolved software.

The work on miniSAT (Sect. 8.5) shows GP can potentially scavenge not just code from the program it is improving but code from multiple programs by multiple authors. This GP plastic surgery (Barr et al. 2014) created in a few hours an award winning version of miniSAT tailored to solving an import software engineering problem, for which it was better than generic versions of miniSAT which has been optimised by leading SAT solving experts for years.

Another promising area is evolving software to meet multiple conflicting requirements. Indeed GP's potential ability to present the software designer with a Pareto trade-off front of different measures of code performance (Harman et al. 2012), may be one avenue that leads most quickly to the wide spread adoption of genetic programming for software improvement. One can imagine a system which shows a range of programs with different speed versus memory requirements, which invites the software designer to choose a suitable trade-off *before* any manual coding starts. Few, if any projects, once the location of their implementation on the trade off space is known, i.e. coding is almost complete, can afford to reject their initial design choices and start again from scratch. Instead typically only relatively small performance changes can be made within the straight jacket of the original design. Further the GP system could consider not only conventional alternatives (e.g. speed v. memory) but also aspects required by mobile computing, such as network bandwidth, power, battery life, and even quality of solutions. It would be very useful to be able to see credible results of design decisions before implementation starts, even if the machine generated code is totally discarded and the designer insists on human coding.

GI is definitely a growth area with the first international event on Genetic Improvement, GI-2015, being held in Madrid along side the main evolutionary computation conference, GECCO, and GI papers being well represented in the SSBSE software engineering conference series as well as gaining best paper prizes in the top software engineering conference and human competitive awards.

Mostly we have described in detail an application of our BNF grammar based GP system, in which a population of code patches is automatically evolved to create new versions of parallel code to run on graphics hardware. The evolving versions are continuously compared with the original, which is treated as the de facto specification, by running regression tests on both and frequently changing the example test used. The fitness function penalises deviation from the original but rewards faster execution. The GP evolves code which exploits the abilities of the hardware the code will run on. The StereoCamera system was specifically written by nVidia's image processing expert to show off their hardware and yet GP is able to improve the code for hardware which had not even been designed when it was originally written yielding almost a seven fold speed up in the graphics kernel.

8.18.1 Sources and Datasets

Le Goues' bug fixing system (Sect. 8.2.4) is available on line: http:// genprog.cs.virginia.edu/ The grammar based genetic programming systems for gzip (Sect. 8.3), Bowtie2 (Sect. 8.4), StereoCamera (Sect. 8.7 onwards) and 3D Brain scan registration (NiftyReg Sect. 8.7 page 191) are available on line via `ftp.cs.ucl.ac.uk`. (For the MiniSAT genetic improvement code, Sect. 8.5, please contact Dr. Petke directly.) The StereoCamera code is in file genetic/gp-code/StereoCamera_1_1.tar.gz and training images are in StereoImages.tar.gz The new code is available in StereoCamera_v1_1c.zip.

Acknowledgements I am grateful for the assistance of njuffa, Istvan Reguly, vyas of nVidia, Ted Baker, and Allan MacKinnon.

GPUs were given by nVidia. Funded by EPSRC grant EP/I033688/1.

Appendix: StereoKernel Tuned for K20c Tesla

In addition to the complete Stereo Camera system `StereoCamera_v1_1c.zip` contains the following CUDA kernel. The modifications to the openVidia CUDA Stereo Camera code distributed by SourceForge are also described in Sect. 8.16 (pages 206 to 207).

```
/*******
stereoKernel
Now for the main stereo kernel:  There are four parameters:
disparityPixel points to memory containing the disparity value (d)
for each pixel.
width & height are the image width & height, and out_pitch
specifies the pitch of the output data in words (i.e. the number
of floats between the start of one row and the start of the next.).
disparityMinSSD removed by GP
*********/

__attribute__((global)) void stereoKernel(
  // pointer to the output memory for the disparity map
  float * __restrict__ disparityPixel,
  // the pitch (in pixels) of the output memory for the disparity
  map const size_t out_pitch,
  const int width,
  const int height,
  unsigned int * __restrict__ timer,    //For GP timing only
  int * __restrict__ sm_id              //For GP timing only
)
{
FIXED_init_timings(timer,sm_id);        //For GP timing only
extern __attribute__((shared)) float disparityPixel_S[];

int* const disparityMinSSD = (int*)&disparityPixel_S[ROWSper
     THREAD*BLOCK_W];
 // column squared difference functions
int* const col_ssd = &disparityMinSSD[ROWSperTHREAD*BLOCK_W];
float d; // disparity value
float d0,d1;
float dmin;

int diff;     // difference temporary value
int ssd;      // total SSD for a kernel
float x_tex; // texture coordinates for image lookup
float y_tex;
int row;      // the current row in the rolling window
int i;        // for index variable
const int dthreadIdx = threadIdx.x % BLOCK_W;
const int dblockIdx  = threadIdx.x / BLOCK_W;

//bugfix force subsequent calculations to be signed
const int X = (__mul24(blockIdx.x,(BLOCK_W-2*RADIUS_H)) +
  dthreadIdx);
const int ssdIdx = threadIdx.x;
```

```
int* const reduce_ssd = &col_ssd[(BLOCK_W )*dperblock-BLOCK_W];
const int Y = (__mul24(blockIdx.y,ROWSperTHREAD));

//int extra_read_val = 0; no longer used
//if(dthreadIdx < (2*RADIUS_H)) extra_read_val = BLOCK_W + ssdIdx;

// initialize the memory used for the disparity and the disparity
    difference
//Uses first group of threads to initialise shared memory
if(threadIdx.x<BLOCK_W-2*RADIUS_H)
if(dblockIdx==0)
if(X<width )
{
  for(i = 0;i<ROWSperTHREAD && Y+i < height;i++)
  {
    // initialize to -1 indicating no match
    disparityPixel_S[i*BLOCK_W +threadIdx.x] = -1.0f;
    //ssd += col_ssd[i+threadIdx.x];
    disparityMinSSD[i*BLOCK_W +threadIdx.x] = MIN_SSD;
  }
}
__syncthreads();

x_tex = X - RADIUS_H;
for(d0 = STEREO_MIND;d0 <= STEREO_MAXD;d0 += STEREO_DISP_STEP*
  dperblock)
{
  d = d0 + STEREO_DISP_STEP*dblockIdx;
  col_ssd[ssdIdx] = 0;

  // do the first row
  y_tex = Y - RADIUS_V;
  for(i = 0;i <= 2*RADIUS_V;i++)
  {
    diff = readLeft(x_tex,y_tex) - readRight(x_tex-d,y_tex);
    col_ssd[ssdIdx] += SQ(diff);
    y_tex += 1.0f;
  }
  __syncthreads();

  // now accumulate the total
  if(dthreadIdx<BLOCK_W-2*RADIUS_H)
  if(X < width && Y < height)
  {
    ssd = 0;
    for(i = 0;i<=(2*RADIUS_H);i++)
    {
      ssd += col_ssd[i+ssdIdx];
    }
  }
  if(dblockIdx!=0) reduce_ssd[threadIdx.x] = ssd;
  __syncthreads();

  //Use first group of threads to set ssd to smallest SSD for
```

```
    d1<d0+dperblock
  if(threadIdx.x<BLOCK_W-2*RADIUS_H)
  if(X < width && Y < height)
  {
    dmin = d;
    d1 = d + STEREO_DISP_STEP;
    for(i = threadIdx.x+BLOCK_W;i < blockDim.x;i += BLOCK_W) {
      if(d1 <= STEREO_MAXD && reduce_ssd[i] < ssd) {
        ssd = reduce_ssd[i];
        dmin = d1;
      }
      d1 += STEREO_DISP_STEP;
    }
    //if ssd is smaller update both shared data arrays
    if( ssd < disparityMinSSD[0*BLOCK_W +threadIdx.x])
    {
     disparityPixel_S[0*BLOCK_W +threadIdx.x] = dmin;
     disparityMinSSD[0*BLOCK_W +threadIdx.x] = ssd;
    }
  }
  __syncthreads();

  // now do the remaining rows
  y_tex = Y - RADIUS_V; // this is the row we will remove
  #pragma unroll 11
  for(row = 1;row < ROWSperTHREAD && (row+Y < (height+RADIUS_V));
    row++)
  {
    // subtract the value of the first row from column sums
    diff = readLeft(x_tex,y_tex) - readRight(x_tex-d,y_tex);
    col_ssd[ssdIdx] -= SQ(diff);

    // add in the value from the next row down
    diff =  readLeft(x_tex,  y_tex + (float)(2*RADIUS_V)+1.0f) -
           readRight(x_tex-d,y_tex + (float)(2*RADIUS_V)+1.0f);
    col_ssd[ssdIdx] += SQ(diff);
    y_tex += 1.0f;
    __syncthreads();

    if(dthreadIdx<BLOCK_W-2*RADIUS_H)
    if(X<width && (Y+row) < height)
    {
      ssd = 0;
      for(i = 0;i<=(2*RADIUS_H);i++)
      {
        ssd += col_ssd[i+ssdIdx];
      }
    }
    if(dblockIdx!=0) reduce_ssd[threadIdx.x] = ssd;
    __syncthreads();

    //Use 1st group threads to set ssd/dmin to smallest SSD for
      d1<d0+dperblock
    if(threadIdx.x<BLOCK_W-2*RADIUS_H)
```

```
    if(dblockIdx==0)
    {
      dmin = d;
      d1 = d + STEREO_DISP_STEP;
      for(i = threadIdx.x+BLOCK_W;i < blockDim.x;i += BLOCK_W) {
        if(d1 <= STEREO_MAXD && reduce_ssd[i] < ssd) {
          ssd = reduce_ssd[i];
          dmin = d1;
        }
        d1 += STEREO_DISP_STEP;
      }
      //if smaller SSD found update shared memory
      if(ssd < disparityMinSSD[row*BLOCK_W +threadIdx.x])
      {
        disparityPixel_S[row*BLOCK_W +threadIdx.x] = dmin;
        disparityMinSSD[row*BLOCK_W +threadIdx.x] = ssd;
      }
    }//endif first group of thread
  }// for row loop
}// for d0 loop

//Write answer in shared memory to global memory
if(threadIdx.x<BLOCK_W-2*RADIUS_H)
if(dblockIdx==0)
if(X < width) {
#pragma unroll 3
  for(row = 0;row < ROWSperTHREAD && (row+Y < height);row++)
  {
    disparityPixel[__mul24((Y+row),out_pitch)+X] =
                disparityPixel_S[row*BLOCK_W +threadIdx.x];
  }
}
FIXED_report_timings(timer,sm_id);      //For GP timing only
}
```

References

Thomas Ackling, Bradley Alexander, and Ian Grunert. Evolving patches for software repair. In Natalio Krasnogor et al., editors, *GECCO '11: Proceedings of the 13th annual conference on Genetic and evolutionary computation*, pages 1427–1434, Dublin, Ireland, 12-16 July 2011. ACM.

Stephen F. Altschul, Thomas L. Madden, Alejandro A. Schaffer, Jinghui Zhang, Zheng Zhang, Webb Miller, and David J. Lipman. Gapped BLAST and PSI-BLAST a new generation of protein database search programs. *Nucleic Acids Research*, 25(17):3389–3402, 1997.

Yiannis Andreopoulos. Error tolerant multimedia stream processing: There's plenty of room at the top (of the system stack). *IEEE Transactions on Multimedia*, 15(2):291–303, Feb 2013. Invited Paper.

Gabriel A. Archanjo and Fernando J. Von Zuben. Genetic programming for automating the development of data management algorithms in information technology systems. *Advances in Software Engineering*, 2012.

Andrea Arcuri and Xin Yao. A novel co-evolutionary approach to automatic software bug fixing. In Jun Wang, editor, *2008 IEEE World Congress on Computational Intelligence*, pages 162–168, Hong Kong, 1–6 June 2008. IEEE Computational Intelligence Society, IEEE Press.

Andrea Arcuri and Xin Yao. Co-evolutionary automatic programming for software development. *Information Sciences*, 259:412–432, 2014.

Wolfgang Banzhaf, Peter Nordin, Robert E. Keller, and Frank D. Francone. *Genetic Programming – An Introduction; On the Automatic Evolution of Computer Programs and its Applications*. Morgan Kaufmann, San Francisco, CA, USA, January 1998.

Earl T. Barr, Yuriy Brun, Premkumar Devanbu, Mark Harman, and Federica Sarro. The plastic surgery hypothesis. In Alessandro Orso, Margaret-Anne Storey, and Shing-Chi Cheung, editors, *22nd ACM SIGSOFT International Symposium on the Foundations of Software Engineering (FSE 2014)*, Hong Kong, 16–12 Nov 2014. ACM.

Patrick Berarducci, Demetrius Jordan, David Martin, and Jennifer Seitzer. GEVOSH: Using grammatical evolution to generate hashing functions. In R. Poli et al., editors, *GECCO 2004 Workshop Proceedings*, Seattle, Washington, USA, 26–30 June 2004.

Jeremy S. Bradbury and Kevin Jalbert. Automatic repair of concurrency bugs. In Massimiliano Di Penta et al., editors, *Proceedings of the 2nd International Symposium on Search Based Software Engineering (SSBSE '10)*, Benevento, Italy, 7–9 September 2010. Fast abstract.

Jurgen Branke, Pablo Funes, and Frederik Thiele. Evolutionary design of en-route caching strategies. *Applied Soft Computing*, 7(3):890–898, June 2006.

Edmund K Burke, Michel Gendreau, Matthew Hyde, Graham Kendall, Gabriela Ochoa, Ender Ozcan, and Rong Qu. Hyper-heuristics: a survey of the state of the art. *Journal of the Operational Research Society*, 64(12):1695–1724, December 2013.

B. F. Buxton, W. B. Langdon, and S. J. Barrett. Data fusion by intelligent classifier combination. *Measurement and Control*, 34(8):229–234, October 2001.

Cristian Cadar, Peter Pietzuch, and Alexander L. Wolf. Multiplicity computing: a vision of software engineering for next-generation computing platform applications. In Kevin Sullivan, editor, *Proceedings of the FSE/SDP workshop on Future of software engineering research*, FoSER '10, pages 81–86, Santa Fe, New Mexico, USA, 7–11 November 2010. ACM.

Brendan Cody-Kenny and Stephen Barrett. The emergence of useful bias in self-focusing genetic programming for software optimisation. In Guenther Ruhe and Yuanyuan Zhang, editors, *Symposium on Search-Based Software Engineering*, volume 8084 of *Lecture Notes in Computer Science*, pages 306–311, Leningrad, August 24–26 2013. Springer. Graduate Student Track.

J. Manuel Colmenar, Jose L. Risco-Martin, David Atienza, and J. Ignacio Hidalgo. Multi-objective optimization of dynamic memory managers using grammatical evolution. In Natalio Krasnogor et al., editors, *GECCO '11: Proceedings of the 13th annual conference on Genetic and evolutionary computation*, pages 1819–1826, Dublin, Ireland, 12–16 July 2011. ACM.

Alban Cotillon, Philip Valencia, and Raja Jurdak. Android genetic programming framework. In Alberto Moraglio et al., editors, *Proceedings of the 15th European Conference on Genetic Programming, EuroGP 2012*, volume 7244 of *LNCS*, pages 13–24, Malaga, Spain, 11–13 April 2012. Springer Verlag.

Charles Darwin. *The Origin of Species*. John Murray, penguin classics, 1985 edition, 1859.

K. Deb, A. Pratap, S. Agarwal, and T. Meyarivan. A fast and elitist multiobjective genetic algorithm: NSGA-II. *IEEE Transactions on Evolutionary Computation*, 6(2):182–197, Apr 2002.

Richard A. DeMillo and A. Jefferson Offutt. Constraint-based automatic test data generation. *IEEE Transactions on Software Engineering*, 17(9):900–910, 1991.

Cesar Estebanez, Yago Saez, Gustavo Recio, and Pedro Isasi. Automatic design of noncryptographic hash functions using genetic programming. *Computational Intelligence*. Early View (Online Version of Record published before inclusion in an issue).

Robert Feldt. Generating diverse software versions with genetic programming: an experimental study. *IEE Proceedings - Software Engineering*, 145(6):228–236, December 1998. Special issue on Dependable Computing Systems.

Robert Feldt. Genetic programming as an explorative tool in early software development phases. In Conor Ryan and Jim Buckley, editors, *Proceedings of the 1st International Workshop on Soft Computing Applied to Software Engineering*, pages 11–20, University of Limerick, Ireland, 12–14 April 1999. Limerick University Press.

Stephanie Forrest, ThanhVu Nguyen, Westley Weimer, and Claire Le Goues. A genetic programming approach to automated software repair. In Guenther Raidl et al., editors, *GECCO '09: Proceedings of the 11th Annual conference on Genetic and evolutionary computation*, pages 947–954, Montreal, 8–12 July 2009. ACM. Best paper.

Erik M. Fredericks and Betty H. C. Cheng. Exploring automated software composition with genetic programming. In Christian Blum et al., editors, *GECCO '13 Companion: Proceeding of the fifteenth annual conference companion on Genetic and evolutionary computation conference companion*, pages 1733–1734, Amsterdam, The Netherlands, 6–10 July 2013. ACM.

Alex A. Freitas. A genetic programming framework for two data mining tasks: Classification and generalized rule induction. In John R. Koza et al., editors, *Genetic Programming 1997: Proceedings of the Second Annual Conference*, pages 96–101, Stanford University, CA, USA, 13–16 July 1997. Morgan Kaufmann.

Mark Gabel and Zhendong Su. A study of the uniqueness of source code. In *Proceedings of the eighteenth ACM SIGSOFT international symposium on Foundations of software engineering*, FSE '10, pages 147–156, New York, NY, USA, 2010. ACM.

David E. Goldberg. *Genetic Algorithms in Search Optimization and Machine Learning*. Addison-Wesley, 1989.

Xin Han, Tak-Wah Lam, Lap-Kei Lee, Isaac K.K. To, and Prudence W.H. Wong. Deadline scheduling and power management for speed bounded processors. *Theoretical Computer Science*, 411(40-42):3587–3600, 2010.

S. Handley. On the use of a directed acyclic graph to represent a population of computer programs. In *Proceedings of the 1994 IEEE World Congress on Computational Intelligence*, volume 1, pages 154–159, Orlando, Florida, USA, 27–29 June 1994. IEEE Press.

Saemundur O. Haraldsson and John R. Woodward. Automated design of algorithms and genetic improvement: contrast and commonalities. In John Woodward et al., editors, *GECCO 2014 4th workshop on evolutionary computation for the automated design of algorithms*, pages 1373–1380, Vancouver, BC, Canada, 12–16 July 2014. ACM.

Mark Harman, William B. Langdon, Yue Jia, David R. White, Andrea Arcuri, and John A. Clark. The GISMOE challenge: Constructing the Pareto program surface using genetic programming to find better programs. In *The 27th IEEE/ACM International Conference on Automated Software Engineering (ASE 12)*, pages 1–14, Essen, Germany, September 3–7 2012. ACM.

Mark Harman, Yue Jia, and William B. Langdon. Babel pidgin: SBSE can grow and graft entirely new functionality into a real world system. In Claire Le Goues and Shin Yoo, editors, *Proceedings of the 6th International Symposium, on Search-Based Software Engineering, SSBSE 2014*, volume 8636 of *LNCS*, pages 247–252, Fortaleza, Brazil, 26–29 August 2014. Springer. Winner SSBSE 2014 Challange Track.

Mark Harman. Software engineering meets evolutionary computation. *Computer*, 44(10):31–39, October 2011. Cover feature.

John H. Holland. *Adaptation in Natural and Artificial Systems: An Introductory Analysis with Applications to Biology, Control and Artificial Intelligence*. MIT Press, 1992. First Published by University of Michigan Press 1975.

Daniar Hussain and Steven Malliaris. Evolutionary techniques applied to hashing: An efficient data retrieval method. In Darrell Whitley et al., editors, *Proceedings of the Genetic and Evolutionary Computation Conference (GECCO-2000)*, page 760, Las Vegas, Nevada, USA, 10–12 July 2000. Morgan Kaufmann.

M. Hutchins, H. Foster, T. Goradia, and T. Ostrand. Experiments on the effectiveness of dataflow- and control-flow-based test adequacy criteria. In *Proceedings of 16th International Conference on Software Engineering, ICSE-16*, pages 191–200, May 1994.

Kosuke Imamura and James A. Foster. Fault-tolerant computing with N-version genetic programming. In Lee Spector et al., editors, *Proceedings of the Genetic and Evolutionary Computation*

Conference (GECCO-2001), page 178, San Francisco, California, USA, 7–11 July 2001. Morgan Kaufmann.

Kosuke Imamura, Terence Soule, Robert B. Heckendorn, and James A. Foster. Behavioral diversity and a probabilistically optimal GP ensemble. *Genetic Programming and Evolvable Machines*, 4(3):235–253, September 2003.

Christian Jacob. *Illustrating Evolutionary Computation with Mathematica*. Morgan Kaufmann, 2001.

Jan Karasek, Radim Burget, and Ondrej Morsky. Towards an automatic design of non-cryptographic hash function. In *34th International Conference on Telecommunications and Signal Processing (TSP 2011)*, pages 19–23, Budapest, 18–20 August 2011.

Gal Katz and Doron Peled. Synthesizing, correcting and improving code, using model checking-based genetic programming. In Valeria Bertacco and Axel Legay, editors, *Proceedings of the 9th International Haifa Verification Conference (HVC 2013)*, volume 8244 of *Lecture Notes in Computer Science*, pages 246–261, Haifa, Israel, November 5–7 2013. Springer. Keynote Presentation.

Marouane Kessentini, Wael Kessentini, Houari Sahraoui, Mounir Boukadoum, and Ali Ouni. Design defects detection and correction by example. In *19th IEEE International Conference on Program Comprehension (ICPC 2011)*, pages 81–90, Kingston, Canada, 22–24 June 2011.

Dongsun Kim, Jaechang Nam, Jaewoo Song, and Sunghun Kim. Automatic patch generation learned from human-written patches. In *35th International Conference on Software Engineering (ICSE 2013)*, pages 802–811, San Francisco, USA, 18–26 May 2013.

Arthur K. Kordon. *Applying Computational Intelligence How to Create Value*. Springer, 2010.

Miha Kovacic and Bozidar Sarler. Genetic programming prediction of the natural gas consumption in a steel plant. *Energy*, 66(1):273–284, 1 March 2014.

John R. Koza. *Genetic Programming: On the Programming of Computers by Natural Selection*. MIT press, 1992.

W. B. Langdon and S. J. Barrett. Genetic programming in data mining for drug discovery. In Ashish Ghosh and Lakhmi C. Jain, editors, *Evolutionary Computing in Data Mining*, volume 163 of *Studies in Fuzziness and Soft Computing*, Chapter 10, pages 211–235. Springer, 2004.

W. B. Langdon and B. F. Buxton. Genetic programming for combining classifiers. In Lee Spector et al., editors, *Proceedings of the Genetic and Evolutionary Computation Conference (GECCO-2001)*, pages 66–73, San Francisco, California, USA, 7–11 July 2001. Morgan Kaufmann.

W. B. Langdon and M. Harman. Evolving a CUDA kernel from an nVidia template. In Pilar Sobrevilla, editor, *2010 IEEE World Congress on Computational Intelligence*, pages 2376–2383, Barcelona, 18–23 July 2010. IEEE.

W. B. Langdon and M. Harman. Genetically improved CUDA kernels for stereocamera. Research Note RN/14/02, Department of Computer Science, University College London, Gower Street, London WC1E 6BT, UK, 20 February 2014.

William B. Langdon and Mark Harman. Genetically improved CUDA C++ software. In Miguel Nicolau et al., editors, *17th European Conference on Genetic Programming*, volume 8599 of *LNCS*, pages 87–99, Granada, Spain, 23–25 April 2014. Springer.

William B. Langdon and Mark Harman. Optimising existing software with genetic programming. *IEEE Transactions on Evolutionary Computation*, 19(1):118–135, February 2015.

W. B. Langdon and J. P. Nordin. Seeding GP populations. In Riccardo Poli et al., editors, *Genetic Programming, Proceedings of EuroGP'2000*, volume 1802 of *LNCS*, pages 304–315, Edinburgh, 15–16 April 2000. Springer-Verlag.

William B. Langdon and Riccardo Poli. Evolving problems to learn about particle swarm and other optimisers. In David Corne et al., editors, *Proceedings of the 2005 IEEE Congress on Evolutionary Computation*, volume 1, pages 81–88, Edinburgh, UK, 2–5 September 2005. IEEE Press.

William B. Langdon, Mark Harman, and Yue Jia. Efficient multi-objective higher order mutation testing with genetic programming. *Journal of Systems and Software*, 83(12):2416–2430, December 2010.

William B. Langdon, Marc Modat, Justyna Petke, and Mark Harman. Improving 3D medical image registration CUDA software with genetic programming. In Christian Igel et al., editors,

GECCO '14: Proceeding of the sixteenth annual conference on genetic and evolutionary computation conference, pages 951–958, Vancouver, BC, Canada, 12–15 July 2014. ACM.

William B. Langdon. *Genetic Programming and Data Structures: Genetic Programming + Data Structures = Automatic Programming!*, volume 1 of *Genetic Programming*. Kluwer, Boston, 1998.

W. B. Langdon. Global distributed evolution of L-systems fractals. In Maarten Keijzer et al., editors, *Genetic Programming, Proceedings of EuroGP'2004*, volume 3003 of *LNCS*, pages 349–358, Coimbra, Portugal, 5–7 April 2004. Springer-Verlag.

W. B. Langdon. A many threaded CUDA interpreter for genetic programming. In Anna Isabel Esparcia-Alcazar et al., editors, *Proceedings of the 13th European Conference on Genetic Programming, EuroGP 2010*, volume 6021 of *LNCS*, pages 146–158, Istanbul, 7–9 April 2010. Springer.

W. B. Langdon. Graphics processing units and genetic programming: An overview. *Soft Computing*, 15:1657–1669, August 2011.

W.B. Langdon. Creating and debugging performance CUDA C. In Francisco Fernandez de Vega, Jose Ignacio Hidalgo Perez, and Juan Lanchares, editors, *Parallel Architectures and Bioinspired Algorithms*, volume 415 of *Studies in Computational Intelligence*, chapter 1, pages 7–50. Springer, 2012.

William B. Langdon. Genetic improvement of programs. In Franz Winkler et al., editors, *16th International Symposium on Symbolic and Numeric Algorithms for Scientific Computing (SYNASC 2014)*, pages 14–19, Timisoara, 22–25 September 2014. IEEE. Keynote.

Claire Le Goues, Michael Dewey-Vogt, Stephanie Forrest, and Westley Weimer. A systematic study of automated program repair: Fixing 55 out of 105 bugs for $8 each. In Martin Glinz, editor, *34th International Conference on Software Engineering (ICSE 2012)*, pages 3–13, Zurich, June 2–9 2012.

Claire Le Goues, ThanhVu Nguyen, Stephanie Forrest, and Westley Weimer. GenProg: A generic method for automatic software repair. *IEEE Transactions on Software Engineering*, 38(1):54–72, January-February 2012.

Jason D. Lohn and Gregory S. Hornby. Evolvable hardware using evolutionary computation to design and optimize hardware systems. *IEEE Computational Intelligence Magazine*, 1(1):19–27, February 2006.

Eduard Lukschandl, Magus Holmlund, and Eirk Moden. Automatic evolution of Java bytecode: First experience with the Java virtual machine. In Riccardo Poli et al., editors, *Late Breaking Papers at EuroGP'98: the First European Workshop on Genetic Programming*, pages 14–16, Paris, France, 14–15 April 1998. CSRP-98-10, The University of Birmingham, UK.

Anjali Mahajan and M S Ali. Superblock scheduling using genetic programming for embedded systems. In *7th IEEE International Conference on Cognitive Informatics, ICCI 2008*, pages 261–266, August 2008.

Duane Merrill, Michael Garland, and Andrew Grimshaw. Policy-based tuning for performance portability and library co-optimization. In *Innovative Parallel Computing (InPar), 2012*. IEEE, May 2012.

Gordon E. Moore. Cramming more components onto integrated circuits. *Electronics*, 38(8):114–117, April 19 1965.

Hoang Duong Thien Nguyen, Dawei Qi, Abhik Roychoudhury, and Satish Chandra. SemFix: program repair via semantic analysis. In Betty H. C. Cheng and Klaus Pohl, editors, *35th International Conference on Software Engineering (ICSE 2013)*, pages 772–781, San Francisco, USA, May 18–26 2013. IEEE.

Michael O'Neill and Conor Ryan. Automatic generation of caching algorithms. In Kaisa Miettinen et al., editors, *Evolutionary Algorithms in Engineering and Computer Science*, pages 127–134, Jyväskylä, Finland, 30 May - 3 June 1999. John Wiley & Sons.

Michael O'Neill and Conor Ryan. Grammatical evolution. *IEEE Transactions on Evolutionary Computation*, 5(4):349–358, August 2001.

Michael O'Neill and Conor Ryan. *Grammatical Evolution: Evolutionary Automatic Programming in a Arbitrary Language*, volume 4 of *Genetic programming*. Kluwer Academic Publishers, 2003.

Michael Orlov and Moshe Sipper. Flight of the FINCH through the Java wilderness. *IEEE Transactions on Evolutionary Computation*, 15(2):166–182, April 2011.

John D. Owens, Mike Houston, David Luebke, Simon Green, John E. Stone, and James C. Phillips. GPU computing. *Proceedings of the IEEE*, 96(5):879–899, May 2008. Invited paper.

Gisele L. Pappa, Gabriela Ochoa, Matthew R. Hyde, Alex A. Freitas, John Woodward, and Jerry Swan. Contrasting meta-learning and hyper-heuristic research: the role of evolutionary algorithms. *Genetic Programming and Evolvable Machines*, 15(1):3–35, March 2014.

Norman Paterson and Mike Livesey. Evolving caching algorithms in C by genetic programming. In John R. Koza et al., editors, *Genetic Programming 1997: Proceedings of the Second Annual Conference*, pages 262–267, Stanford University, CA, USA, 13–16 July 1997. Morgan Kaufmann.

Justyna Petke, Mark Harman, William B. Langdon, and Westley Weimer. Using genetic improvement & code transplants to specialise a C++ program to a problem class. 11th Annual Humies Awards 2014, 14 July 2014. Winner Silver.

Justyna Petke, Mark Harman, William B. Langdon, and Westley Weimer. Using genetic improvement and code transplants to specialise a C++ program to a problem class. In Miguel Nicolau et al., editors, *17th European Conference on Genetic Programming*, volume 8599 of *LNCS*, pages 137–149, Granada, Spain, 23–25 April 2014. Springer.

Bojan Podgornik, Vojteh Leskovsek, Miha Kovacic, and Josef Vizintin. Analysis and prediction of residual stresses in nitrided tool steel. *Materials Science Forum*, 681, Residual Stresses VIII:352–357, March 2011.

Riccardo Poli, William B. Langdon, and Nicholas Freitag McPhee. *A field guide to genetic programming*. Published via `http://lulu.com` and freely available at `http://www.gp-field-guide.org.uk`, 2008. (With contributions by J. R. Koza).

Vlad Radulescu, Stefan Andrei, and Albert M. K. Cheng. A heuristic-based approach for reducing the power consumption of real-time embedded systems. In Franz Winkler, editor, *16th International Symposium on Symbolic and Numeric Algorithms for Scientific Computing (SYNASC 2014)*, Timisoara, 22–25 September 2014. Pre-proceedings.

Craig Reynolds. Interactive evolution of camouflage. *Artificial Life*, 17(2):123–136, Spring 2011.

Jose L. Risco-Martin, David Atienza, J. Manuel Colmenar, and Oscar Garnica. A parallel evolutionary algorithm to optimize dynamic memory managers in embedded systems. *Parallel Computing*, 36(10-11):572–590, 2010. Parallel Architectures and Bioinspired Algorithms.

Pablo Rodriguez-Mier, Manuel Mucientes, Manuel Lama, and Miguel I. Couto. Composition of web services through genetic programming. *Evolutionary Intelligence*, 3(3-4):171–186, 2010.

Juan Romero, Penousal Machado, and Adrian Carballal. Guest editorial: special issue on biologically inspired music, sound, art and design. *Genetic Programming and Evolvable Machines*, 14(3):281–286, September 2013. Special issue on biologically inspired music, sound, art and design.

Conor Ryan. *Automatic Re-engineering of Software Using Genetic Programming*, volume 2 of *Genetic Programming*. Kluwer Academic Publishers, 1 November 1999.

Eric Schulte, Stephanie Forrest, and Westley Weimer. Automated program repair through the evolution of assembly code. In *Proceedings of the IEEE/ACM international conference on Automated software engineering*, pages 313–316, Antwerp, 20–24 September 2010. ACM.

Eric Schulte, Jonathan DiLorenzo, Westley Weimer, and Stephanie Forrest. Automated repair of binary and assembly programs for cooperating embedded devices. In *Proceedings of the eighteenth international conference on Architectural support for programming languages and operating systems*, ASPLOS 2013, pages 317–328, Houston, Texas, USA, March 16–20 2013. ACM.

Eric Schulte, Jonathan Dorn, Stephen Harding, Stephanie Forrest, and Westley Weimer. Post-compiler software optimization for reducing energy. In *Proceedings of the 19th International*

Conference on Architectural Support for Programming Languages and Operating Systems, ASPLOS'14, pages 639–652, Salt Lake City, Utah, USA, 1–5 March 2014. ACM.

Eric Schulte, Zachary P. Fry, Ethan Fast, Westley Weimer, and Stephanie Forrest. Software mutational robustness. *Genetic Programming and Evolvable Machines*, 15(3):281–312, September 2014.

Pitchaya Sitthi-amorn, Nicholas Modly, Westley Weimer, and Jason Lawrence. Genetic programming for shader simplification. *ACM Transactions on Graphics*, 30(6):article:152, December 2011. Proceedings of ACM SIGGRAPH Asia 2011.

Joe Stam. Stereo imaging with CUDA. Technical report, nVidia, V 0.2 3 Jan 2008.

Gilbert Syswerda. Uniform crossover in genetic algorithms. In J. David Schaffer, editor, *Proceedings of the third international conference on Genetic Algorithms*, pages 2–9, George Mason University, 4–7 June 1989. Morgan Kaufmann.

Westley Weimer, ThanhVu Nguyen, Claire Le Goues, and Stephanie Forrest. Automatically finding patches using genetic programming. In Stephen Fickas, editor, *International Conference on Software Engineering (ICSE) 2009*, pages 364–374, Vancouver, May 16–24 2009.

Westley Weimer, Stephanie Forrest, Claire Le Goues, and ThanhVu Nguyen. Automatic program repair with evolutionary computation. *Communications of the ACM*, 53(5):109–116, June 2010.

Westley Weimer. Advances in automated program repair and a call to arms. In Guenther Ruhe and Yuanyuan Zhang, editors, *Symposium on Search-Based Software Engineering*, volume 8084 of *Lecture Notes in Computer Science*, pages 1–3, Leningrad, August 24–26 2013. Springer. Invited keynote.

David R. White, John Clark, Jeremy Jacob, and Simon M. Poulding. Searching for resource-efficient programs: low-power pseudorandom number generators. In Maarten Keijzer et al., editors, *GECCO '08: Proceedings of the 10th annual conference on Genetic and evolutionary computation*, pages 1775–1782, Atlanta, GA, USA, 12–16 July 2008. ACM.

David R. White, Andrea Arcuri, and John A. Clark. Evolutionary improvement of programs. *IEEE Transactions on Evolutionary Computation*, 15(4):515–538, August 2011.

Josh L. Wilkerson and Daniel Tauritz. Coevolutionary automated software correction. In Juergen Branke et al., editors, *GECCO '10: Proceedings of the 12th annual conference on Genetic and evolutionary computation*, pages 1391–1392, Portland, Oregon, USA, 7–11 July 2010. ACM.

Liyuan Xiao, Carl K. Chang, Hen-I Yang, Kai-Shin Lu, and Hsin yi Jiang. Automated web service composition using genetic programming. In *36th Annual IEEE Computer Software and Applications Conference Workshops (COMPSACW 2012)*, pages 7–12, Izmir, 16–20 July 2012.

Frances Yao, Alan Demers, and Scott Shenker. A scheduling model for reduced cpu energy. In *36th Annual Symposium on Foundations of Computer Science*, pages 374–382. IEEE, Oct 1995.

Shin Yoo. Evolving human competitive spectra-based fault localisation techniques. In Gordon Fraser et al., editors, *4th Symposium on Search Based Software Engineering*, volume 7515 of *Lecture Notes in Computer Science*, pages 244–258, Riva del Garda, Italy, September 28–30 2012. Springer.

Ling Zhu and Sandeep Kulkarni. Synthesizing round based fault-tolerant programs using genetic programming. In Teruo Higashino et al., editors, *Proceedings of the 15th International Symposium on Stabilization, Safety, and Security of Distributed Systems (SSS 2013)*, volume 8255 of *Lecture Notes in Computer Science*, pages 370–372, Osaka, Japan, November 13–16 2013. Springer.

Chapter 9
Design of Real-Time Computer-Based Systems Using Developmental Genetic Programming

Stanisław Deniziak, Leszek Ciopiński, and Grzegorz Pawiński

9.1 Introduction

Computer-based system (CBS) is a system which uses microprocessors or computers for executing tasks. We may find computers almost everywhere, from simple microcontrollers or more complex embedded systems used in automotive, telecommunication, medical, home and other appliances, to powerful computing centers running cloud applications. Certain system features like cost, power consumption, performance are critical in most applications. Thus, CBS should be optimized by developing the dedicated architecture that satisfies all user requirements and expectations.

In many applications a system response is expected during the specified time period. Violation of the time limit causes system fail or degrades the quality of service. This class of systems, called real-time systems, is used in many domains. Most of embedded systems work in real time. Any CBS that interacts with the environment e.g. by controlling processes or electromechanical devices, also is a subject of real-time constraints. Recently, requirements for real-time features appeared also for some classes of cloud services.

Design of the dedicated architecture of real time CBS consists of the following tasks: resource allocation, task assignment and task scheduling. Usually the goal of optimization is to minimize a cost or power consumption, while satisfying all real time constraints. In general, this process is defined as a resource constrained process scheduling problem (RCPSP).

RCPSP is an NP-complete problem, which is computationally very hard (Blazewicz et al. 1983), therefore optimal solutions for real-life systems may

S. Deniziak (✉) • L. Ciopiński • G. Pawiński
Kielce University of Technology, Kielce, Poland
e-mail: s.deniziak@computer.org

A.H. Gandomi et al. (eds.), *Handbook of Genetic Programming Applications*,
DOI 10.1007/978-3-319-20883-1_9

be found only by using efficient heuristics. It was shown that for the design of real-time computer-based systems, approaches based on the developmental genetic programming are very effective.

9.2 Developmental Genetic Programming

Developmental genetic programming (DGP) (Keller and Banzhaf 1999; Koza et al. 2003) evolves the development process, instead of computer programs. In classical genetic approaches, the search space (genotype) is the same as a solution space (phenotype). DGP distinguishes between genotypes and phenotypes and uses a genotype to phenotype mapping prior to fitness evaluation of the phenotype (Fig. 9.1).

Genotypes usually are represented by trees. Nodes of the genotype are genes specifying the system construction functions. The edges indicate the order of execution of these functions. Thus, the genotype specifies the procedure of construction of the final solution (phenotype). Genotype to phenotype mapping is performed by the execution of this procedure, starting from the root. During mapping all constraints are taken into consideration, therefore only valid phenotypes will be obtained.

Motivations for the DGP approach are hard-constrained optimization problems. Genetic algorithms handle these problems by constraining genetic operators in the manner, which makes them to produce only legal individuals. However, constrained operators create infeasible regions in the search space, also eliminating sequences of genes, which may lead to high quality solutions. In the DGP the problem does not exists anyway. Because of separating the search space from the solution space, legal as well as illegal genotypes are evolved, while each genotype is mapped onto a legal phenotype. It is worth to notice that the evolution of an illegal genotype may lead to the legal genotype constructing the optimal result. Thus, the whole search space is explored.

DGP is a quite new and it is not fully studied, yet. However, it has already been successfully applied in the design of electronic circuits, control algorithms, strategy algorithms in computer games (Koza et al. 2003) etc. Many of the human-competitive results that were produced using runs of genetic programming that employed a developmental process are described in Koza (2010).

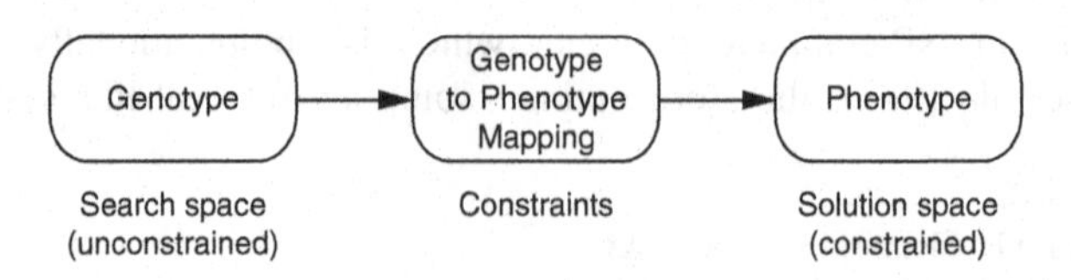

Fig. 9.1 Developmental approach

9.3 Resource Constrained Project Scheduling Problem

Researchers' attention has been focused on making the best use of scarce resources available since PERT (Program Evaluation and Review Technique) and CPM (Critical Path Method) developed in the late 1950s (Hendrickson and Tung 2008). Resource-constrained project scheduling problem (RCPSP) (Klein 2000) addresses the task of allocating limited resources over time, in order to perform a set of activities subject to constraints on the order, in which the activities may be executed.

9.3.1 Classical Approach

RCPSP attempts to schedule the project tasks, efficiently using limited renewable resources, minimizing the maximal completion time of all activities $V = \{v_1, \ldots, v_n\}$. Each activity $v_i \in V$ has a specific processing time p_i and it requires resources $R = \{r_1, \ldots, r_m\}$ to be processed. In general, activities may not be interrupted during their processing (non-preemption) and cannot be processed independently from each other, due to limited resource capacity and additional technological requirements. Technological requirements are represented by precedence relationships that specify a fixed processing order between pairs of activities. The finish–start relationship with zero time lags means that the activity can be started immediately after all its predecessors are completed. An example of a project plan with precedence constraints is shown in Fig. 9.2.

Resources are constrained due to limited number of available units. An activity v_i requires s_{ik} units of one or several resources $r_k \in R$. However, the resource capacity R_k is constant in each period. If an activity v_i is being executed by the resource r_k, then it consumes s_{ik} resource units, which cannot be used by another activity. Thus, a feasible solution only exists if, in each period, resource demands for all activities are not higher than resource capacities (Dorndorf et al. 2000). An example of the resource demands for activities is shown in Table 9.1.

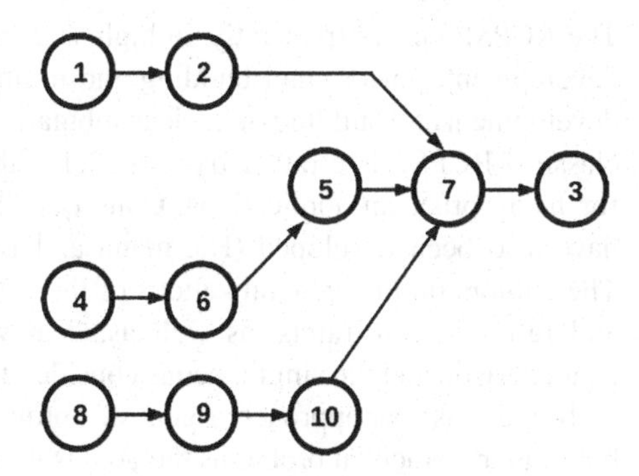

Fig. 9.2 Precedence constraints

Table 9.1 Resource demands for activities

v_i	p_i	s_i
1	5	2
2	4	4
3	2	3
4	3	3
5	3	3
6	3	1
7	2	1
8	2	3
9	4	1
10	1	2

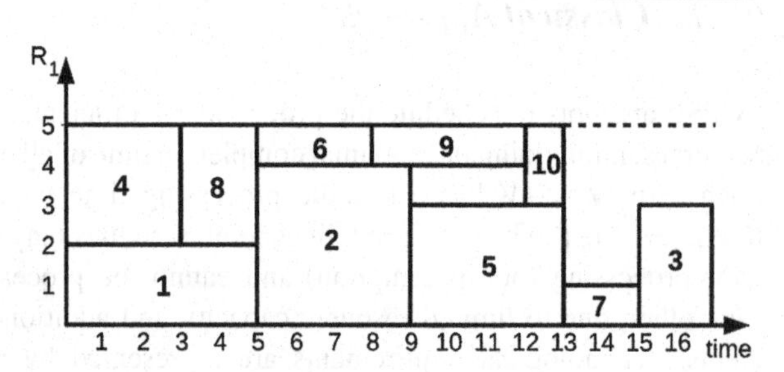

Fig. 9.3 Sample schedule

The objective of the RCPSP is to find feasible completion times for all activities such that the makespan of the project is minimized, while the precedence of activities and limits of resources are not violated (Kolisch and Hartmann 1999). Figure 9.3 presents a feasible schedule of a project comprising $n = 10$ activities (Fig. 9.2) which have to be scheduled, assuming that only one renewable resource with a capacity of five units is available.

9.3.2 RCPSP Extensions

The RCPSP occurs frequently, in high scale project management such as software development, power plant building and military industry projects such as design, development and building of nuclear submarines (Pinedo and Chao 1999). However, classical RCPSP is a rather basic model with assumptions that are too restrictive for many practical applications. Consequently, various extensions of the RCPSP have also been developed (Hartmann and Briskorn 2010; Węglarz et al. 2011). The authors outline generalizations of the activity concept, alternative precedence and resource constraints, as well as, deal with different objectives, task graph characteristic and the simultaneous consideration of multiple projects.

In the classical approach a goal of optimization is to minimize the makespan, but in many practical problems the goal is to minimize the cost i.e. to minimize the number of resources or the cost of using resources required for executing all tasks, while time constraints should be satisfied. Such problems are defined as resource investment problems (Drexl and Kimms 2001) or resource renting problems (Nubel 2001). Example of a practical application of this extension is the optimization of distributed embedded systems, especially implemented as a network on chip architectures or based on multi-core embedded processors. Moreover, existing RCPSP approaches do not take into consideration initial resource workload. Thus, resources are available in the whole time period. Such constraint better fits real-life project management problems. Dealing with more than one project is common in IT business, for example, where managers have to use a resource-sharing approach. An extension of the problem, where resources are only partially available, since they may be involved in many projects, was also investigated (Pawiński and Sapiecha 2014a, b). Finally, some architectures of future computing systems may also be modelled as the RCPSP. This concerns the so called "cloud computing". Results of the research will be crucial for optimization of real-time distributed applications for Internet of things and for designing of distributed systems implemented according to the IaaS (Infrastructure as a Service) model of the cloud computing (Bąk et al. 2013).

9.3.3 Solutions of the Problem

RCPSP has become a well-known standard of optimization, which has attracted numerous researchers who developed both, exact and heuristic scheduling algorithms (Brucker et al. 1998; Demeulemeester and Herroelen 1997, 2002). In most cases, branch-and-bound is the only exact method, which allows the generation of optimal solutions for scheduling rather small projects (usually containing less than 60 tasks and not highly constrained), within acceptable computational effort (Alcaraz and Maroto 2001; Demeulemeester and Herroelen 2002). Since the finding of the best solution is very complex, only efficient heuristics may be applied for real-life systems. In-depth study of the performance of the recent RCPSP heuristics can be found in Kolisch and Hartmann (2006). Heuristics described by the authors, include X-pass approach, also known as priority rule based heuristics, classical metaheuristics, such as Genetic algorithms, Tabu search, Simulated annealing (SA), and Ant systems (Dorigo and Stützle 2004). Results of the investigation showed that the best performing heuristics for solving the RCPSP were the Genetic algorithm (GA) of Hartmann (1998) and the Tabu search (TS) procedure of Bouleimen and Lecocq (1998).

Another metaheuristic algorithm, driven by a metric of the gain of optimization (MAO) (Deniziak 2004), was also applied to the RCPSP (Pawiński and Sapiecha 2012). The advantage of the algorithm is that it has a capacity of getting out of local minima. The authors adapted the algorithm to take into account specific features of human resources participating in a project schedule. The computational experiments showed significant efficiency of the approach in optimizing the RCPSP and an extension of the problem, where resources are only partially available, since they may be involved in many projects (Pawiński and Sapiecha 2014a).

Deiranlou and Jolai (2009) have published one of the latest review papers, which present exact methods and heuristics for solving the RCPSP. The authors paid particular attention to GAs. They introduced a new crossover operator and auto-tuning for adjusting the rates of crossover and mutation operators. Two approaches for solving the problem with GAs and Genetic Programming (GP) (Koza 1992) are given in Frankola et al. (2008). The authors achieved good quality results by the use of GAs. Yet, they state that GAs, as a technique, is inappropriate for dynamic environments and for projects with large number of activities, because of their uncertainty and amount of time required to obtain satisfactory results. The authors propose GP to find a solution of an acceptable quality within a reasonable time.

9.4 Application of the DGP to the Optimization of Computer-Based Real Time Systems

We assume that the behaviour of a system is described by a task graph $G = \{V,E\}$, which is an acyclic, directed graph. Each node $v_i \in V$ represents a task, describing a single thread of execution. An edge $e_{i,j} \in E$ describes a dependency between tasks v_i and v_j. Each edge is annotated with a number $d_{i,j}$ describing the amount of data that have to be transferred between the two connected tasks. With any node v_i a deadline c_i may be associated. The deadline is the time by which the given task must complete its execution. All deadlines create a set of constraints C, where each constraint has to be satisfied by the target system. A sample task graph with deadlines is presented in Fig. 9.4. Tasks *Start* and *Stop* are dummy tasks indicating the entry and exit points of the specified function. They may be omitted during the synthesis.

The function specified as a task graph with real time constraints may be implemented as various types of computer based systems: multicore and/or multiprocessor system, dedicated embedded system or real-time cloud. In all cases an

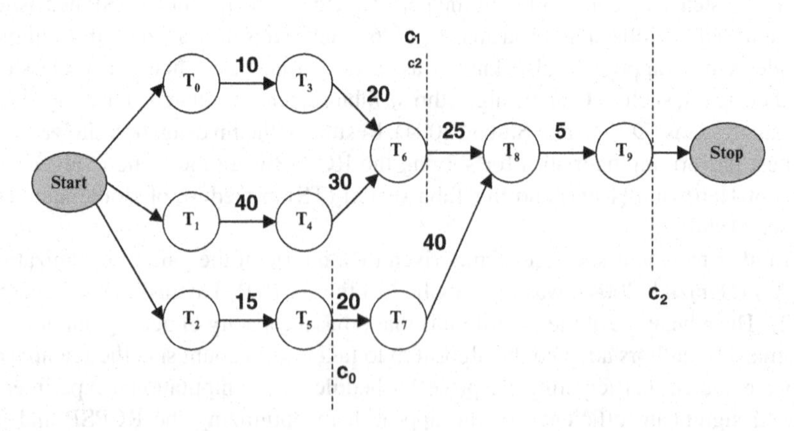

Fig. 9.4 Sample task graph

efficient optimization method should be applied, to minimize the cost of the system, while all real time constraints should be satisfied. In this chapter we will consider the following optimization problems of real time computer-based systems:

- scheduling of real-time tasks in multiprocessor systems,
- hardware/software co-design of distributed embedded systems,
- budget-aware real-time cloud computing.

We will show that the DGP may be efficiently applied for optimization purposes in all above problems.

9.4.1 DGP Approach to the RCPSP

The goal of the DGP is to find the optimal schedule of all tasks. Since the proper schedule has to fulfill all requirements given in the system specification, the system construction functions should be enough flexible to construct only feasible schedules.

9.4.1.1 Embryonic System

A root of the genotype tree specifies a construction of an embryonic system, while all other nodes correspond to functions that schedule tasks, according to the assigned strategies. The embryo may be a system executing one of the first tasks from the task graph (Deniziak and Górski 2008) or it may specify other design decisions e.g. it may partition the system into subsystems (Pawiński and Sapiecha 2014b; Sapiecha et al. 2014).

9.4.1.2 System Construction Functions

Functions that construct the target system consist of the following steps:

- resource allocation and task assignment, which selects an appropriate resource to execute a particular task,
- task scheduling (only when more than one task is assigned to the same resource).

A resource is allocated according to the strategy that is randomly selected with the given probability. We may define a strategy in terms of the resource types (processor, hardware core), resource parameters (the fastest, the cheapest, the smallest, etc.), resource usage (the longest idle time, the least frequently allocated), time constraints (start and finish times of activities). A set of strategies should be taken specific to the considered problem and the optimization goal. Strategies may also be combined, e.g. choose a resource which is the fastest and causes the smallest increase of the system cost.

9.4.1.3 Genotypes and Phenotypes

The genotype has a form of a tree corresponding to the procedure of synthesis of phenotypes (target solutions). In the case of real time systems, only schedules satisfying all time requirements are feasible. Two types of approaches are possible: hierarchical and sequential.

In the hierarchical approach (Pawiński and Sapiecha 2014b), two types of nodes are distinguished: internal nodes and leaves. The internal node defines a new level in the hierarchical design by dividing the part of the system into subsystems. The leave implements the corresponding subsystem. The edges represent the division of tasks into two subgroups, while nodes specify a location of the division d_i and the strategy of resource allocation s_i. With each node, a list of tasks is associated. A root node contains the list consisting of all tasks, ordered according to the level in a task graph. Next, the list is cut into two sublists, the first sublist will be associated with the left successor while the second one is passed to the right successor. The same operation is repeated for successor nodes. If a node is the internal node, d_i is used for dividing currently considered list of tasks into two sublists and strategies are assigned to them. The left child define a strategy for the first sublist and the right child for the other. Lists of tasks corresponding to leaves are scheduled on resources according to the strategy specified by the node. Strategies and cut positions for each node are randomly generated during the creation of the genotype.

Figure 9.5 gives a sample genotype and the corresponding sequence of strategies, each corresponding to the given task. For each node, the top number means the cut position (important only for the internal nodes) while the bottom number defines the

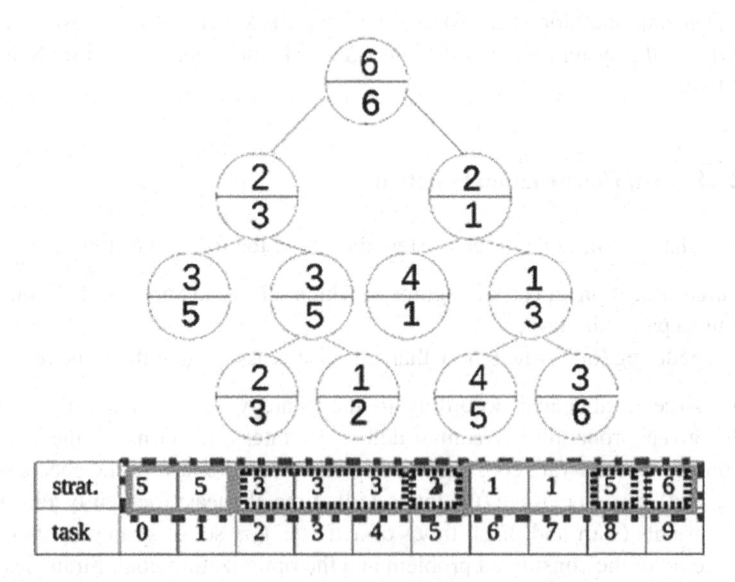

Fig. 9.5 Genotype using 6 decision strategies (0–5) and the corresponding sequence of strategies (strat.)

strategy (used only by leaves). The sequence of strategies is obtained by traversing the tree in the depth-first order starting from the top node (the root). It corresponds to the list of leaves starting from left to right. First, tasks are partitioned into groups: $\{T_0–T_5\}$ and $\{T_6–T_9\}$. Next, the first group is divided into $\{T_0–T_1\}$, $\{T_2–T_5\}$, and the second group is divided into $\{T_6–T_7\}$, $\{T_8–T_9\}$. The group $\{T_0–T_1\}$ is associated with the leaf, hence the strategy 5 is assigned to tasks T_0 and T_1. Group $\{T_2–T_5\}$ is partitioned into $\{T_2–T_4\}$ and $\{T_5\}$, then the strategies 3 and 2 are assigned to tasks T_2, T_3, T_4 and T_5 respectively. And so on, the result is given in Fig. 9.5.

A phenotype corresponds to the final system. It represents a resource allocation, task assignment and a task schedule. Phenotype is used for evaluation of the quality (fitness) of the corresponding genotype. The method of genotype to phenotype mapping guarantees that each phenotype will satisfy all requirements and constraints.

9.4.1.4 Genotype-to-Phenotype Mapping

The genotype to phenotype mapping constructs the target schedule, according to strategies specified by the genotype nodes. System is constructed by executing functions corresponding to nodes. Each function takes into consideration constraints. After assigning strategies to tasks, the following steps have to be carried out, to create a task schedule:

- search activities from the task graph, according to the precedence relationships, in order to find a list of ready-to-start tasks,
- assign the strategies with corresponding tasks and execute the strategy to calculate a resource to allocate,
- schedule tasks—calculate a start time for each task, based on the earliest precedence relationships and the feasible time of a resource,
- repeat the first step, until there are unassigned tasks.

Thus, each genotype specifies custom scheduling policy for the whole system. The goal of the evolution is to find the genotype giving the best result.

9.4.2 Scheduling of Real-Time Tasks in Multiprocessor Systems

A multiprocessor system consists of many multi-core processors. Each core is a resource and may have allocated tasks to execute. Like in classical RCPSP approach, tasks are precedence-related and have to be executed in a specific order. We consider the variant, where resources have already got their own schedule and are available only in particular time periods. Such tasks cannot be moved. The task graphs are created on working system and therefore, the current availability of resources has to be taken into consideration. The goal is to schedule real-time tasks and allocate

Table 9.2 Strategies for implementation of tasks

No.	Strategy
1	The fastest core
2	The cheapest processor
3	The earliest start of the task
4	The earliest finish of the task
5	The smallest local duration of schedule
6	The smallest local cost of the system

resources of multiprocessor systems, taking into consideration the availability of resources, in order to minimize the total cost (or power consumption) of the system and complete it before a deadline. We consider the architecture with shared memory i.e. transmissions between consecutive tasks will be neglected.

DGP approach that may be applied for scheduling tasks in multiprocessor systems is presented in Pawiński and Sapiecha (2014b). The method uses a list of possible strategies for resource assignment, chosen for the task, presented in Table 9.2. The first two strategies, search for a processor which is the fastest or the cheapest, its load is not taken into consideration. Strategies 3 and 4 refer to tasks execution time. Task may be assigned to a processor, which will start the task as fast as possible or execute it as soon as possible, respectively. We distinguish these two strategies, because execution time of task may differ for different processors. The last two strategies check, how the resource assignment and the task allocation affect current duration and current cost of the system that is being built.

The initial population consists of individuals generated randomly by recursively creating nodes until a pre-established maximum height of the genotype tree (H) is reached. Each node has one of the strategies, assigned with the same probability and a random cut point d_i, which is inversely proportional to H. However, it has to be verified whether nodes contain improper values of d_i. The location of the division cannot be greater than the number of tasks in the currently considered sublist. One of the repairing mechanisms could be a "deleting repair" that removes all children of the invalid node. The process is similar to withering of unused features in live organisms, like in the intron splicing (Watson et al. 1992). But we used a "replacing repair" that replaces the invalid node by any of its children, instead of removing the entire branch. Therefore, more genetic information will be kept in the genotype.

Let's assume that we have three processors with four cores each (resource capacity) and deadlines are the following: $c_0 = 5$, $c_1 = 10$, $c_2 = 15$. With each processor the following parameters are associated: processing speed, cost of task execution per time unit and unit cost (Table 9.3). For architectures with message passing also a communication cost and a throughput of communication channels should be given. Assume that the system is specified with the task graph given in Fig. 9.4 and resource demands are given in Table 9.1 (tasks are multithreaded). Then according to the genotype given in Fig. 9.5, the system is constructed as follows. First, a list of tasks without predecessors is created and a processor is assigned to

Table 9.3 Values of resource parameters

Resource	Processing speed	Execution cost per time	Unit cost
1	1000 MHz	0.95	20
2	1200 MHz	0.96	21
3	1600 MHz	1.07	25

them, according to the corresponding strategy. Afterwards, the list is updated and the process is continued. Thus, tasks are being assigned in the following order:

- T_0 (strategy 5)—assign a resource that causes the smallest increase of duration of the current schedule; all processors fits; the first one (R_1) is chosen,
- T_1 (strategy 5)—processors R_2 and R_3 are available, so R_2 is chosen,
- T_2 (strategy 3)—assign processor R_3, because it can start the task the earliest,
- T_3 (strategy 3)—processor R_3 will be available the earliest, but T_3 may not start before T_0 is completed, so all processors fits and R_1 is chosen,
- T_4 (strategy 3)—again, processor R_3 is available the earliest, but T_4 may not start before T_1 is completed, the second best processor is R_2,
- T_5 (strategy 2)—assign a resource that is the cheapest, that is processor T_2, but the deadline c_0 would be exceeded, so processor R_1 is assigned,
- T_6 (strategy 1)—assign a resource that is the fastest, that is R_3,
- T_7 (strategy 1)—assign processor R_3,
- T_8 (strategy 5)—all processors fits, because T_8 have to be started after T_6 and T_7 are completed, and T_6 finishes the latest, processor R_1 is chosen,
- T_9 (strategy 6)—assign a resource that causes the smallest increase of the current cost. Processor R_1 is chosen.

The result of the genotype-to-phenotype mapping is a feasible schedule illustrated in Fig. 9.6. The total cost of the system equals 224.46.

Efficiency of the DGP approach was tested on projects from PSPLIB (Kolish and Sprecher 1996). In our study we used project instances with 30 non-dummy activities because it is the hardest standard set of RCPSP instances, for which all optimal solutions are currently known (Demeulemeester and Herroelen 1997). The multiprocessor systems were randomly generated. A single group of 10 test instances was examined, in which 10 schedules were computed for each test case. Figure 9.7 presents the project cost averaged from 100 schedules. We used a tournament selection method with a tournament size equal to 3. At the beginning a population is the most various and its diversity lowers in further generations. Good quality results start to dominate in the population very quickly, along with the increasing number of generations and therefore the project cost decreases. The convergence of the method is fast. Only nine generations are enough to obtain good quality results. Further improvement is very slight, but it occurs till the last generation.

Usually, the project cost becomes lower along with increasing probabilities of mutation and crossover, because the operators produce more new genotypes and

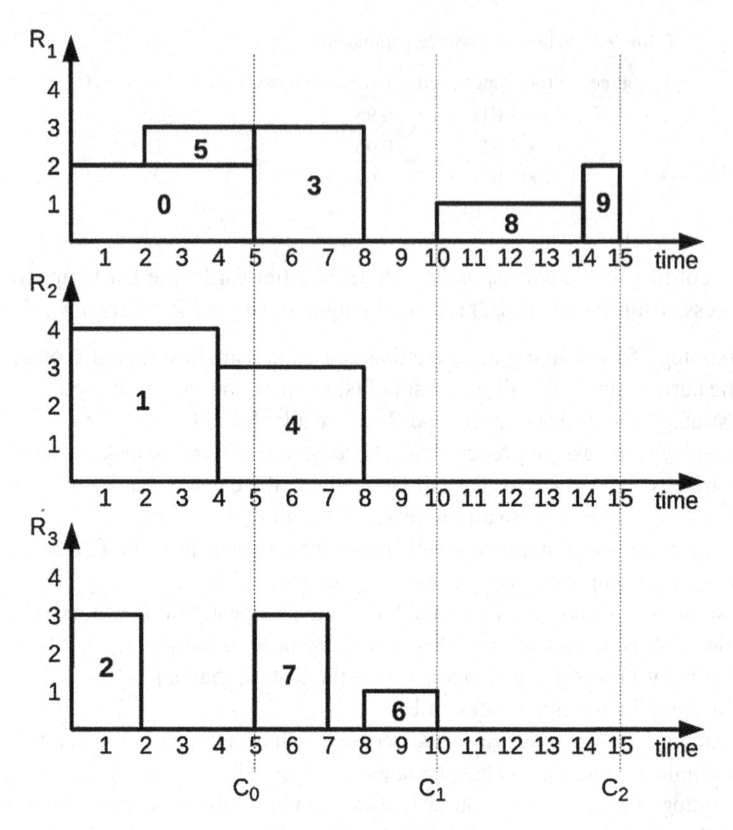

Fig. 9.6 A target solution obtained after genotype-to-phenotype mapping

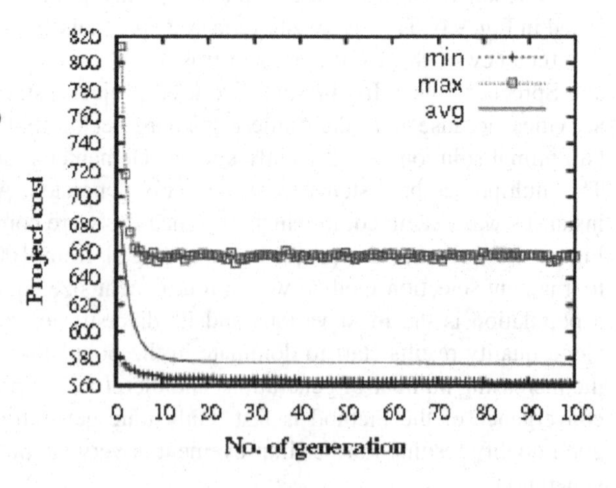

Fig. 9.7 Project cost in each generation for $P_{mut} = 0.8$ (mutations rate), $P_{cross} = 0.8$ (crossover rate), $POP_{size} = 30$ (population size), *min*—the lowest system cost, *max*—the highest system cost, *avg*—the average system cost from all individuals of a given generation

Table 9.4 Experimental results for different methods (DGP: $POP = 30$, $TS_{size} = 3$, $P_{mut} = 0.8$, $P_{cross} = 0.8$, $H = 0.8$)

Scheduling method	Cost	Duration	Computation time [ms]
$Greedy_{time}$	651.89	105.40	70
$Greedy_{cost}$	654.92	106.47	65
MAO	637.94	102.97	4,560
GA	619.89	98.57	10,062
DGP	599.90	92.69	13,650

Table 9.5 A comparison of the uncorrected sample standard deviation S_N

Scheduling method	S_N
GA	12.83
DGP	4.89

the population is more diverse. Thus, the chance of finding the optimal solution is greater. However, only the best genotypes will be selected to the next generation. The variety of individuals may also be increased by increasing their number in generations. Generally, if POP_{size} is bigger, then the results are better. Nevertheless, the slope of the cost reduction is similar.

Finally, we have performed efficiency test on all 480 instances, where 10 project schedules were computed for each test case. The results were averaged and compared with other methods (Table 9.4) (Pawiński and Sapiecha 2014b). Greedy procedures try to find an optimal resource for each task, according to the smallest increase of the project duration ($Greedy_{time}$) or according to the minimal growth of the total cost ($Greedy_{cost}$). MAO is given from Pawiński and Sapiecha (2014a). Genetic approaches have similar evolution process but they differ in a way of coding the genotype. GA is a classical genetic algorithm. In the GA, the genotype does not have a tree structure. Genetic operators are applied directly to a sequence of resources corresponding to the activities.

The comparison results showed that DGP is the slowest method, mainly because of the large number of generations. DGP in a comparison with other methods is 3-times slower than MAO and only 36 % slower than GA. On the other hand, it outperforms MAO in the project cost reduction by 5.5 %, greedy methods by 8 % and it outperforms MAO in the project time reduction by 6 % and greedy methods by 12 %. Furthermore, the uncorrected sample standard deviation of DGP is 3-times lower than the deviation of GA (Table 9.5).

9.4.3 Hardware/Software Co-design of Embedded Systems

Hardware/software co-synthesis (Yen and Wolf 1997) automatically generates architecture for an embedded system specified on the system level. The goal of the co-synthesis is to optimize certain system properties like a cost, performance or average power consumption. Since, modern embedded systems are implemented as multiprocessor systems usually realized as a single chip (System on Chip—SoC), most co-synthesis methods consider distributed target architectures composed of processing elements (PEs) and communication links (CLs).

We assume that a database of available PEs and CLs is given. For each $pe_i \in PE$ a worst-case execution time $t_{j,i}$ of each task v_j is given, as well as an area s_j occupied by this task. There are two basic kinds of pe_i: programmable processors (PP) and hardware cores (HC). Each $pp_i \in$ PP may execute all tasks which are compatible with it. Task areas specified for pp_i mean the size of a memory required to execute these tasks, while S_i is an area of pp_i itself. Hardware core $hc_i \in$ HC executes only task v_i, but more than one core may be available, each corresponding to another hardware implementation of this task. The area of a task implemented in hardware is the size of the corresponding hc_i. Communication links $cl_i \in$ CL are defined by the following parameters: a bandwidth b_i and an area $s_{i,j}$ occupied by this link connected to pe_j. Table 9.6 presents a sample resource database for the system described by the task graph from Fig. 9.4. Task T_7 is not compatible with PP_2, and task T_5 has only one hardware implementation. All other tasks have four alternative implementations. There are two communication links available.

Since the area occupied by the system corresponds to the cost of the system implemented as SOC, the goal of the co-synthesis is to find the architecture with the smallest area that satisfies all real time constraints.

The DGP approach for the co-synthesis of embedded systems is presented in Deniziak and Górski (2008). The embryo is a system implementing the first task from the given task graph. Each node of a genotype tree represents a function implementing one task from the task graph. Hence, all genotypes have the same structure, which is the spanning tree of the task graph. This corresponds to the sequential approach.

Table 9.6 Sample resource database

	PP_1 S = 100		PP_2 S = 200		HC_1		HC_2	
	t	s	t	s	t	S	t	S
T_0	30	3	10	2	3	50	4	10
T_1	50	5	20	4	6	80	5	20
T_2	40	4	10	3	3	60	5	20
T_3	10	3	8	1	1	20	2	5
T_4	30	3	15	2	4	70	10	30
T_5	30	5	30	3	5	110	–	–
T_6	40	3	15	2	10	70	12	15
T_7	30	3	–	–	5	50	8	18
T_8	8	3	5	1	2	30	3	10
T_9	10	3	5	1	3	40	4	12
CL_1 B = 8	s = 2		s = 1		s = 10			
CL_2 B = 16	s = 3		s = 4		s = 15			

Each system-construction function consists of the following steps:

1. PE allocation: allocates a new PE. This step is optional.
2. Task assignment: chooses a PE to execute the given task. This step must always be performed. If the first step is performed then the task will be assigned to the newly allocated PE, otherwise task will be assigned to any previously allocated PE.
3. CL allocation: allocates a new CL for the transmission. This step is optional.
4. Transmission assignment: chooses CL for the transmission. Steps 3 and 4 are repeated for each transmission, associated with incoming edges of the node corresponding to the task being implemented.
5. Task scheduling: this step is performed only when more than one task are assigned to one PP.

For each system-construction function, all steps are chosen randomly, according to the options presented in Table 9.7.

Figure 9.8 presents a sample genotype for the task graph from Fig. 9.3. Node numbers indicate the order of execution of the corresponding functions. The function may be executed only if all its predecessors were processed. Assume that deadlines are the following: $c_0 = 80$, $c_1 = 120$, $c_2 = 150$. Then the system is constructed as follows:

1. The embryo allocates processor pe_0 (of type PP_1) as a processor with the smallest cost, then task *Start* is assigned to it. None communication channel is allocated. Steps 4 and 5 are not applicable.

Table 9.7 System-construction options used for genotype to phenotype mapping, for each step one option is randomly selected according to a given probability P

Step	Option	P
1	a. None	0.6
	b. Smallest area	0.1
	c. Fastest	0.1
	d. Lowest t * S	0.1
	e. Least used	0.1
2	a. Smallest area	0.2
	b. Fastest	0.2
	c. Lowest utilization	0.2
	d. Idle for the longest time	0.2
	e. The same as a predecessor	0.2
3	a. None	0.5
	b. Smallest area	0.2
	c. Highest B	0.2
	d. Least used	0.1
4	a. Smallest area	0.3
	b. Fastest	0.3
	c. Lowest utilization	0.2
	d. Idle for the longest time	0.2
5	List scheduling	1.0

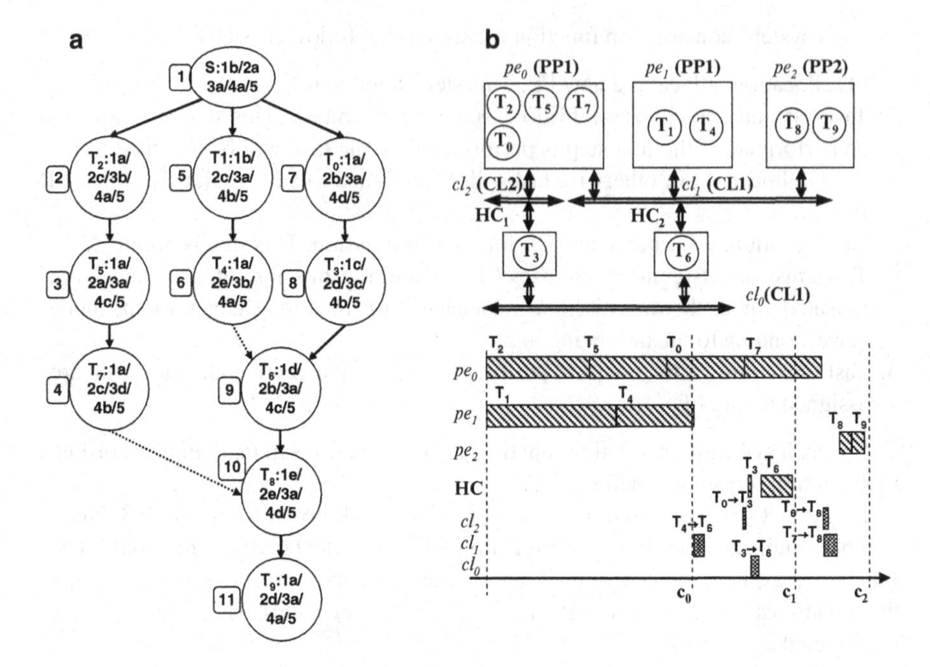

Fig. 9.8 Sample genotype (**a**) and the corresponding phenotype (**b**)

2. Task T_2 is assigned to pe_0 (only this processor is available) and CL_1 is allocated (as the cheapest channel), step 5 is not required, therefore it is omitted.
3. Task T_5 is also assigned to pe_0 (only this processor is available). Since T_2 and T_5 are assigned to the same processor, hence the transmission between these tasks is omitted. There is only one valid schedule, therefore task T_5 will be executed after finishing task T_2. Task T_5 will finish its execution at time 70, hence c_0 constraint will be satisfied.
4. T_7 will be also assigned to pe_0. Transmission is neglected and the only valid schedule is T_2, T_5, T_7.
5. The second PP_1 (pe_1) is allocated (smallest area) and T_1 is assigned to it, as a processor with the lowest utilization. Steps 3, 4 and 5 do not change the system.
6. T_4 will be also assigned the same processor as task T_1. The second CL_1 is allocated (as the cheapest channel) but none transmission will be assigned to it. The only valid schedule is T_1, T_4.
7. T_0 cannot be assigned to pe_1, because this will violate c_1 (T_4 or T_0 will finish its execution at 110 and it will not possible to implement T_6 with incoming transmissions to finish this task at 110 or earlier, even using the fastest resources). Thus, T_0 will be assigned to pe_0. The only valid schedule is T_2, T_5, T_0, T_7, T_0. All tasks will be finished at 100.
8. For task T_3 the fastest resource (HC_1) is allocated. The fastest communication channel (cl_2) is allocated for transmission between T_0 and T_3. Time of transmission is equal 1 and the T_3 will finish at 102.

Table 9.8 Experimental results (N—number of tasks, T_{max}—global time constraint, time—execution time of all tasks, area—the cost of system)

		Yen-Wolf		MAO		DGP-average		DGP-best	
N	T_{max} [ms]	Time	Area	Time	Area	Time	Area	Time	Area
10	400	315	1573	287	1517	395	1552	395	1545
20	450	196	3046	196	3046	396	2665	396	2649
30	500	499	4213	488	4361	473	3988	484	3823
40	800	794	5204	779	5188	782	4959	795	4595
50	1100	1099	6017	1092	5967	1079	5289	1090	5046
60	1400	1360	8218	1386	7316	1337	8086	1376	7784
70	1600	1548	6859	1590	6657	1554	5832	1599	5607
80	1900	1893	11,692	1878	8662	1745	10,243	1854	9918
90	2000	1917	13,184	1995	8257	1943	7012	1986	6599
100	2150	2115	10,800	2140	7240	2098	8524	2133	7941
110	2200	2167	12,171	2199	9030	2142	9098	2193	8499
Total			82,977		67,241		67,248		64,006

9. Task T_6 is implemented using the resource giving the lowest t*S factor. It will be obtained by HC_2. Incoming transmissions will be assigned to the lowest utilized channels. Task T_6 will finish its execution at 117, hence the c_1 will also be satisfied.
10. For T_8 the least used resource should be allocated. Since none instance exists only for the PP_2, therefore it is allocated. For incoming transmissions cl_1 and cl_2 are assigned. T_8 will finish at 140.
11. Finally, T_9 should be assigned to pe_1 but it will violate c_2, the same will be caused by assigning this task to pe_1. Thus, the only feasible solution is to assign T_9 to pe_2. Then it will finish its execution at 145.

Table 9.8 presents the experimental results obtained for some randomly generated task graphs. First, systems were synthesized using DGP, Yen-Wolf (Yen and Wolf 1995) and MAO (Deniziak 2014) methods. Yen-Wolf starts from the fastest architecture, where for each task the fastest PE is allocated. Then, the algorithm evaluates different solutions created by moving one task from one PE to another. The best one is selected for the next step. Iteration stops when there is no such improvement, which reduces the total system cost and which does not violate any constraint. MAO works in similar way, but it uses more sophisticated refinement methods, instead of moving only one task it allocates and/or removes one PE to/from the system architecture. In this way in one step many tasks can be moved to other PEs. Usually MAO produces significantly better results than Yen-Wolf method (Deniziak 2014). Moreover, this method found better or comparable solutions than ones, found with the help of genetic algorithm (Dick and Jha 1998), for the same task graphs. Table 9.8 presents average results (30 trials) obtained using DGP, as well as the best solution found in all experiments. For all systems the DGP method found comparable or better solutions in comparison with other heuristics.

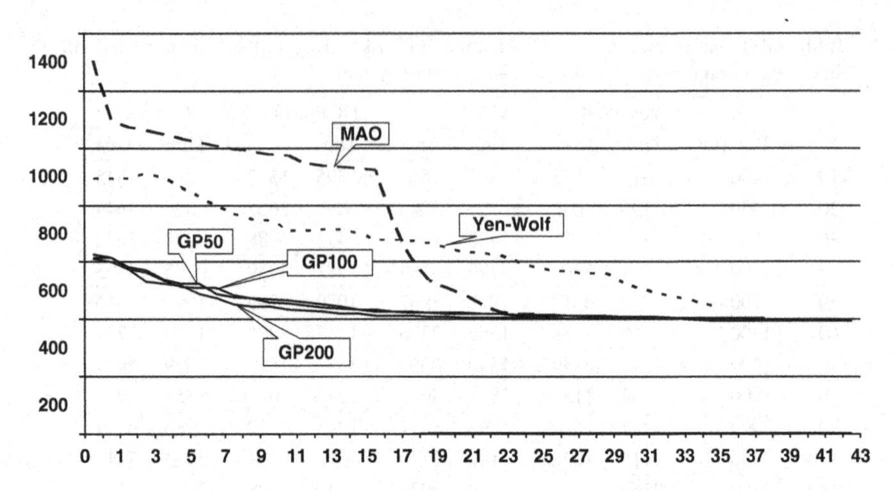

Fig. 9.9 Optimization flows for different co-synthesis methods. GP50, GP100 and GP200 represent the DGP method with population size equal 50*N, 100*N and 200*N, respectively

Figure 9.9 presents the comparison of the optimization flow for all methods mentioned above. Graph with 50 tasks was synthesized with constraint $T_{max} = 2200$. The Y axis represents the cost of a solution, found in the following optimizations steps (generations). For DGP approach the X axis represents the population number, while for MAO and Yen-Wolf it represents the following refinement steps. In the DGP methods solutions with costs: 4062, 4044, 3939 were found, while using iterative improvement methods the best solution found had a cost equal to 4191. It may be observed that DGP converges faster than heuristics based on iterative improvements.

9.4.4 Budget-Aware Scheduling for Real-Time Cloud Computing

Distributed Internet applications require expensive network platforms, consisting of servers, routers, communication links etc., to operate. The cost of such systems may be reduced by sharing the network resources between different applications. This is possible by using the Infrastructure as a Service (IaaS) model of the cloud computing services (Buyya et al. 2011). IaaS together with a real-time cloud environment seems the most suitable platform for many real-time cloud applications. But to guarantee the quality of service and minimize the cost of the system, efficient methods of mapping real-time applications onto cloud resources should be developed.

In Deniziak et al. (2014), the IaaS model of the real-time cloud computing is considered, where the user pays the cost of using the resources supported by the service provider. The authors present the methodology for the mapping real-time cloud applications, specified as a set of distributed echo algorithms, onto the IaaS cloud. The goal of the methodology is to find the mapping giving the minimal cost of IaaS services required for running the real-time applications in the cloud environment, while the level of QoS will be as high as possible. For this purpose an efficient algorithm based on developmental genetic programming was developed.

The methodology starts from the formal specification of the system. Next, the specification is converted into a set of task graphs. Then the optimal set of cloud resources is assigned to tasks. In this way the cost of outsourcing the network infrastructure to the IaaS cloud provider is minimized. Finally, all tasks are scheduled, taking into consideration real time constraints. Allocation of resources, task assignment and scheduling are optimized using developmental genetic programming.

Genotypes represent the hierarchical approach as in the classical RCPSP approach described above. Lists of tasks corresponding to leaves are scheduled on cloud resources according to the strategy specified by the node. All possible strategies are presented in Table 9.9. The last column indicates the probability of selection of the corresponding strategy. The strategy 4 selects an alternative node, i.e. a node that cannot be chosen by strategies 1–3.

In Bąk et al. (2013) an adaptive navigation system, as an example of real-time distributed application, was presented. The system was specified as a set of different distributed algorithms. First, each specification was converted into task graphs, than the system was mapped onto cloud resources. Figure 9.10 presents part of this system represented by 3 task graphs consisting of 12, 12 and 10 tasks, respectively.

Table 9.9 Strategies for implementation of tasks

No.	Strategy	P
1	As fast as possible	0.1
2	As cheap as possible	0.1
3	The lowest cost*execution time	0.2
4	Alternative node	0.2
5	First available node	0.2
6	The fastest finishing node	0.2

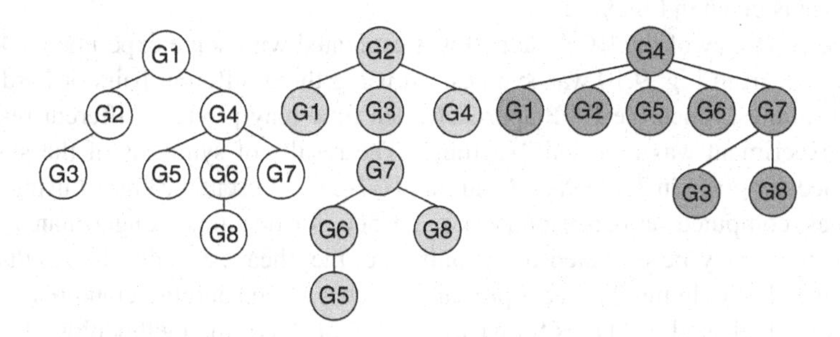

Fig. 9.10 Specification of the distributed system

Table 9.10 Cost of cloud resources

Server	Processors	Per hour	Link	Bandwidth [Mb/s]	Per hour
S1	1.7 GHz	0.004 $	L1	1	0.0001 $
S2	2.4 GHz	0.008 $	L2	5	0.0010 $
S3	2 × 1.7 GHz	0.007 $	L3	10	0.0028 $
S4	2 × 2.4 GHz	0.014 $	L4	20	0.0069 $
S5	4 × 1.7 GHz	0.013 $			
S6	4 × 2.4 GHz	0.025 $			

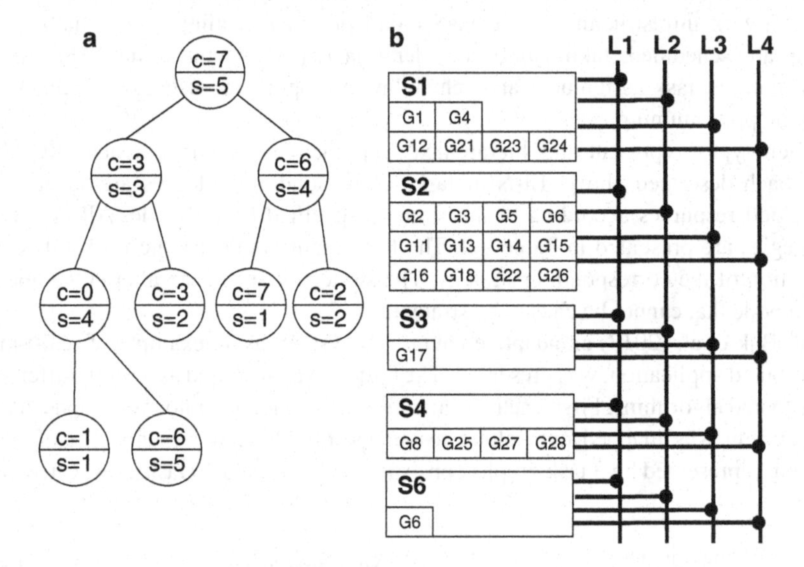

Fig. 9.11 Sample genotype (**a**) and the corresponding phenotype (**b**)

Assume that cloud consists of six different servers connected using four communication links. Detailed specification of cloud resources is given in Table 9.10. According to the DGP methodology presented in Deniziak et al. (2014) the sample genotype and the corresponding phenotype are presented in Fig. 9.11. The task schedule is given in Fig. 9.12.

The efficiency of the DGP method was estimated with some experiments. Each task graph from Fig. 9.10 was synthesized using three different pairs of hard and soft deadlines. Since the DGP approach each time may produce different results, each experiment was repeated 2–3 times. The results of synthesis of the system specified are given in Table 9.11. Columns *TG1–TG3* present the cost of using IaaS services, computed on per-resource basis. Since, for deadlines longer than 7 s all applications may be executed using only one, the cheapest node, the results are the same. The column *Total cost* presents the cost of outsourcing cloud resources, assuming dedicated resources for each application. Next, the methodology to map all applications, taking into consideration the resource sharing, was applied. The column *TG1* + *TG2* + *TG3* presents the cost of the optimized system. The last column presents the reduction of costs obtained using the DGP methodology.

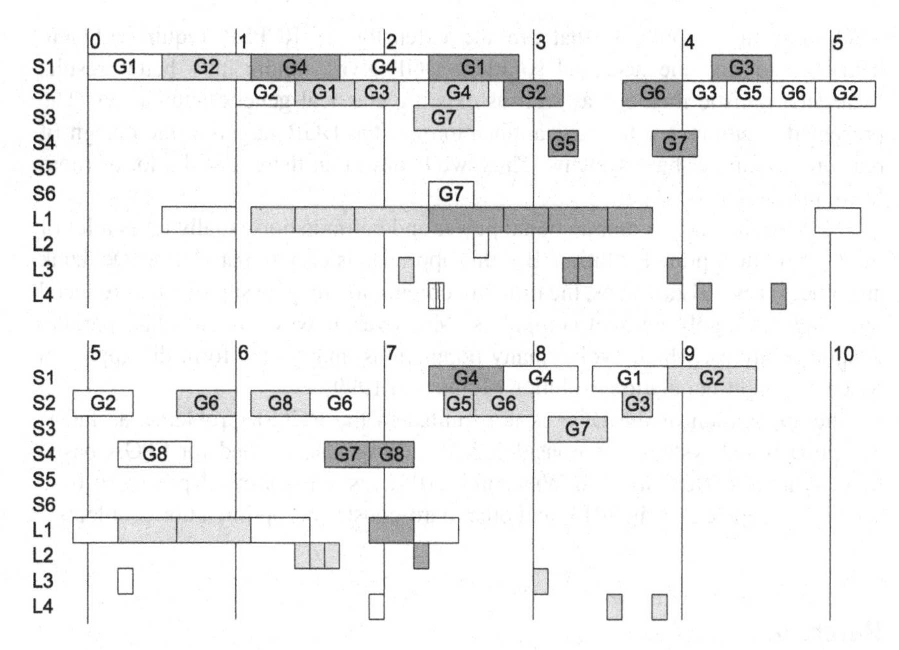

Fig. 9.12 Task schedule for shared cloud resources

Table 9.11 Results of optimization

Deadline	TG1	TG2	TG3	Total cost	TG1+TG2+TG3	Cost reduction [%]
6 s	17.07	17.07	12.79	46.93	27.21	42.02
6 s	17.07	17.07	17.07	51.21	25.53	50.15
6 s	17.07	17.07	12.07	46.21	25.52	55.23
10 s	4	4	4	12	10.43	13.08
10 s	4	4	4	12	10.45	12.92
10 s	4	4	4	12	10.24	15.50
12 s	4	4	4	12	8.01	33.25
12 s	4	4	4	12	7.44	38.00

9.5 Conclusions and Outlook

Developmental genetic programming proved to be very efficient in many domains (Koza 2010). We showed that it may also be successfully applied for optimization of real-time computer-based systems. Since it the DGP genotypes are evolved, while phenotype is used for fitness evaluation, this approach is very suitable for hard constrained problems.

Optimization problems, that are the extension of RCPSP, require efficient heuristics to find the accepted solution. DGP gives significantly better results than deterministic methods as well as existing classical genetic approaches. The presented solutions are the first applications of the DGP, targeting the design of real-time computer-base systems. Thus, we believe that there is still a lot of room for improvements.

DGP requires large computational power and optimization usually takes a lot of time to find the optimal solution. But this approach is easy to parallelize (Deniziak and Wieczorek 2012a). Thus, the time of computation may be significantly reduced by using the highly parallel computers. Moreover, it was observed that parallel genetic solutions, which evolve many populations, may outperform the approach based on single-population evolution (Tomassini 1999).

The application of the DGP is not limited to the RCPSP problems, as far as computer-based systems are considered. It may be also applied for FPGA-based logic synthesis (Deniziak and Wieczorek 2012b), synthesis of adaptive real time scheduler (Sapiecha et al. 2014) and other hard-constraints optimization problems.

References

Alcaraz J, Maroto C (2001) A robust genetic algorithm for resource allocation in project scheduling. Annals of Operations Research, 102, pp. 83-109.

Bąk S, Czarnecki R, Deniziak S (2013) Synthesis of real-time applications for internet of things. In: Pervasive Computing and the Networked World. Lecture Notes in Computer Science, Springer Berlin Heidelberg, p. 35-49.

Blazewicz J, Lenstra JK, Rinnooy Kan (1983) Scheduling subject to resource constraints: Classification and complexity, Discrete Applied Mathematics, No. 5, pp. 11–24.

Bouleimen K, Lecocq H (1998). A new efficient simulated annealing algorithm for the resource-constrained project scheduling problem, Technical Report, Service de Robotique et Automatisation, Universite de Liege.

Brucker P, Knust S, Schoo A, Thiele O (1998) A branch-and-bound algorithm for the resource-constrained project scheduling problem. European Journal of Operational Research, 107: 272–288.

Buyya R, Broberg J, Goscinski A (2011) Cloud Computing: Principles and Paradigms. Wiley Press, New York, USA

Deiranlou M, Jolai F (2009) A New Efficient Genetic Algorithm for Project Scheduling under Resource Constrains. World Applied Sciences Journal, 7 (8): pp. 987-997.

Demeulemeester EL, Herroelen WS (1997) New benchmark results for the resource-constrained project scheduling problem. Management Science, 43: 1485–1492

Demeulemeester EL, Herroelen WS (2002) Project Scheduling. A Research Handbook, Springer

Deniziak S (2004) Cost-efficient synthesis of multiprocessor heterogeneous systems. Control and Cybernetics 33: 341–355

Deniziak S, Górski A (2008) Hardware/Software Co-Synthesis of Distributed Embedded Systems Using Genetic Programming. Lecture Notes in Computer Science, Springer-Verlag, pp. 83-93.

Deniziak S, Wieczorek S (2012a) Parallel Approach to the Functional Decomposition of Logical Functions Using Developmental Genetic Programming. Lecture Notes in Computer Science 7203:406-415.

Deniziak S, Wieczorek S (2012b) Evolutionary Optimization of Decomposition Strategies for Logical Functions. Lecture Notes in Computer Science 7269, pp. 182-189

Deniziak S, Ciopiński L, Pawiński G et al (2014) Cost Optimization of Real-Time Cloud Applications Using Developmental Genetic Programming, IEEE/ACM 7th International Conference on Utility and Cloud Computing

Dick RP, Jha NK (1998) MOGAC: A Multiobjective Genetic Algorithm for the Co-Synthesis of Hardware-Software Embedded Systems. IEEE Trans. on Computer Aided Design of Integrated Circuits and Systems 17(10):920–935

Drexl A, Kimms A (2001) Optimization guided lower and upper bounds for the resource investment problem, Journal of the Operational Research Society 52 pp. 340–351

Dorndorf U, Pesch E and Toàn Phan-Huy (2000) Constraint propagation techniques for the disjunctive scheduling problem. Artificial intelligence 122.1 (2000): 189-240.

Dorigo M, Stützle T (2004) Ant Colony Optimization. Massachusetts Institute of Technology, USA

Frankola T, Golub M and Jakobovic D (2008) Evolutionary algorithms for the resource constrained scheduling problem. In Proceedings of 30th International Conference on Information Technology Interfaces 7269:715-722

Hartmann S (1998) A Competitive Genetic Algorithm for Resource-Constrained Project Scheduling. Naval Research Logistics, 45:733-750

Hartmann S, Briskorn D (2010) A survey of variants and extensions of the resource-constrained project scheduling problem. European journal of operational research : EJOR. - Amsterdam : Elsevier 207, 1 (16.11.), pp. 1-15

Hendrickson C, Tung A (2008) Advanced Scheduling Techniques. In: Project Management for Construction, cmu.edu (2.2 ed.), Prentice Hall

Keller R, Banzhaf W (1999) The Evolution of Genetic Code in Genetic Programming. In: Proc. of the Genetic and Evolutionary Computation Conference, pp. 1077–1082

Klein R, (2000) Scheduling of Resource-Constrained Projects. Springer Science & Business Media

Kolish R, Sprecher A (1996) Psplib - a project scheduling library. European journal of operational research, 96:205-216.

Kolisch R, Hartmann S (1999) Heuristic algorithms for the resource-constrained project scheduling problem: Classification and computational analysis. Springer US

Kolisch R, Hartmann S (2006) Experimental investigation of heuristics for resource-constrained project scheduling: An update. European journal of operational research, 174:23-37

Koza JR (1992) Genetic Programming: On the Programming of Computers by Means of Natural Selection, MIT Press, Cambridge, MA, USA

Koza J, Keane MA, Streeter MJ et al. (2003) Genetic Programming IV: Routine Human-Competitive Machine Intelligence. Kluwer Academic Publisher, Norwell

Koza JR (2010) Human-competitive results produced by genetic programming. In Genetic Programming and Evolvable Machines, pp. 251-284

Nubel H (2001) The resource renting problem subject to temporal constraints. OR Spektrum 23: 359–381

Pawiński G. Sapiecha K (2012) Resource allocation optimization in Critical Chain Method. Annales Universitatis Mariae Curie-Sklodowska sectio Informaticales, 12 (1), p 17–29

Pawiński G, Sapiecha K (2014a) Cost-efficient project management based on critical chain method with partial availability of resources. CONTROL AND CYBERNETICS, 43(1)

Pawiński G, Sapiecha K (2014b) A Developmental Genetic Approach to the cost/time trade-off in Resource Constrained Project Scheduling. IEEE Federated Conference on Computer Science and Information Systems

Pinedo M, Chao X (1999) Operations Scheduling with applications in Manufacturing. Irwin/McGraw-Hill, Boston, New York, NY, USA, 2nd edition.

Sapiecha K, Ciopiński L, Deniziak S (2014) An Application of Developmental Genetic Programming for Automatic Creation of Supervisors of Multitask Real-Time Object-Oriented Systems. IEEE Federated Conference on Computer Science and Information Systems, 2014.

Tomassini M (1999) Parallel and distributed evolutionary algorithms: A review. In P. Neittaanmki K. Miettinen, M. Mkel and J. Periaux, editors, Evolutionary Algorithms in Engineering and Computer Science, J. Wiley and Sons, Chichester

Watson JD, Hopkins NH, Roberts JW et al. (1992). Molecular Biology of the Gene. Benjamin Cummings. Menlo Park, CA.
Węglarz J et al. (2011) Project scheduling with finite or infinite number of activity processing modes–A survey. European Journal of Operational Research 208.3: 177-205.
Yen, TY, Wolf WH (1995) Sensitivity-Driven Co-Synthesis of Distributed Embedded Systems. In: Proc. of the Int. Symposium on System Synthesis, pp. 4–9
Yen, TY, Wolf WH (1997) Yen, T.-Y., Wolf, W.: Hardware-Software Co-synthesis of Distributed Embedded Systems. Springer, Heidelberg

Chapter 10
Image Classification with Genetic Programming: Building a Stage 1 Computer Aided Detector for Breast Cancer

Conor Ryan, Jeannie Fitzgerald, Krzysztof Krawiec, and David Medernach

10.1 Introduction

Image Classification (IC) is concerned with automatically classifying images based on their *features*, which are typically some sort of measurable/quantifiable property, such as brightness, interest points, etc. The term "feature" can have several meanings in Pattern Recognition (PR) and Machine Learning (ML) where it may be defined as either a location in an image that is relevant with respect to some classification/detection/analysis task or simply as a scalar value extracted from an image. For this work we adopt the latter meaning.

IC has been applied in fields as diverse as medicine (Petrick et al. 2013), military (Howard et al. 2006), security (Xie and Shang 2014), astronomy (Riess et al. 1998) and food science (Tan et al. 2000). Part of the success of IC stems from the fact that the same key steps are applied regardless of the application domain. This chapter describes IC in detail using mammography as a test problem.

Statistics produced by the Organisation for Economic Co-operation and Development (OECD) highlight the importance of the early detection of breast cancer, both in terms of extending the longevity of women and in reducing financial costs. Routine mammographic screening, particularly at a national level, is by far the most

---

Electronic supplementary material: The online version of this chapter (doi: 10.1007/978-3-319-20883-1_10) contains supplementary material, which is available to authorized users.

C. Ryan (✉) • J. Fitzgerald • D. Medernach
University of Limerick, Limerick, Ireland
e-mail: Conor.Ryan@ul.ie; Jeannie.Fitzgerald@ul.ie; David.Medernach@ul.ie

K. Krawiec
University of Poznań, Poznań, Poland
e-mail: krzysztof.krawiec@cs.put.poznan.pl

A.H. Gandomi et al. (eds.), *Handbook of Genetic Programming Applications*,
DOI 10.1007/978-3-319-20883-1_10

effective tool for the early detection and subsequent successful treatment of breast cancer (Tot et al. 2000; Tabar et al. 2000; Smith et al. 2012). It is essential to discover signs of cancer early, as survival is directly correlated with early detection (Tabar et al. 2000).

Screening is usually performed on asymptomatic women over a certain age (e.g. over 50 in many European countries) at regular periods, typically every 2 or 3 years. In national mammography screening, radiologists examine the mammograms of thousands of women (typically having only a few minutes to examine each image) to determine if there are early signs of a cancerous growth or a lesion that may require further examination.

The introduction of screening programs has contributed to a higher demand for radiologists and a world wide shortage of qualified radiologists who choose mammography as their area of specialisation (Bhargavan et al. 2002), particularly in the USA, has led to many radiologists being dangerously overworked (Berlin 2000). This is likely to lead to (i) there being insufficient time for radiologists to read and interpret mammograms (mammograms are notoriously difficult to read); (ii) an inability to provide redundant readings (more than one radiologist checking each mammogram); and (iii) radiologists being overly conservative, which in turn is likely to increase the number of patient call backs, thus resulting in unnecessary biopsies. This can lead to anxiety and mistrust of the system such that patients become disillusioned with the process and less inclined to participate. This work aims to improve the early detection of true positives by evolving detectors which, although accurate, are not overly conservative.

If breast cancer is diagnosed, further tests are usually carried out to determine the extent of the cancer. The disease is then assigned a "stage" depending on characteristics such size of the tumour, whether the cancer is invasive or non-invasive, whether lymph nodes are involved, and whether the cancer has spread to other areas of the body. These stages are numbered 0, 1, 2, 3 and 4, and there are various sub-stages in between. At stage 0 the cancer is localised and there is no evidence of cancerous cells outside the original site, while at stage 4 the cancer has spread to other organs of the body. According to the OECD, 75 % of patients diagnosed with breast cancer at Stage 0 are said to have close to 100 % survival rate, while at Stage 4 the survival rates drop between 20 and 40 %. Treatment cost is six times more when a diagnosis is made at Stage 4 than at Stage 0 (Hughes and Jacobzone 2003). There is a large body of scientific evidence supporting the view that mammography is currently the strongest tool available in the fight against breast cancer (Kopans 2003; Tabar et al. 2000).

A stage 1 detector examines mammograms and highlights *suspicious* areas that require further investigation. A too conservative approach degenerates to marking every mammogram (or *segment* of) as suspicious, while missing a cancerous area can be disastrous.

Various studies (Anttinen et al. 1993; Ciatto et al. 2005) have shown that second (redundant) reader functionality has a valuable role to play in breast cancer detection, offering increases in detection rates of between 4.5 and 15 % together with the possibility of discovering cancers at an earlier stage (Thurfjell et al. 1994). However, due to shortages of qualified personnel in several countries and the extra costs involved, the use of independent second readers in large scale screening programs is not always possible. Thus, the availability of a reliable *automated* stage 1 detector would be a very useful and cost effective resource.

We describe a fully automated work-flow for performing stage 1 breast cancer detection with GP (Koza 1990) as its cornerstone. Mammograms are by far the most widely used method for detecting breast cancer in women, and its use in national screening can have a dramatic impact on early detection and survival rates. With the increased availability of digital mammography, it is becoming increasingly more feasible to use automated methods to help with detection.

Our work-flow positions us right at the data collection phase such that we generate textural features ourselves. These are fed through our system, which performs feature analysis on them before passing the ones that are determined to be most salient on to GP for classifier generation. The best of these evolved classifiers produces results of 100 % sensitivity and a false positive per image rating of just 0.33, which is better than prior work. Our system can use GP as part of a feedback loop, to both select existing features and to help extract further features. We show that virtually identical work-flows (just with different feature extraction methods) can be applied to other IC tasks.

The following section provides a background to the work and outlines some of the important existing research, while Sect. 10.3 demonstrates how our proposed work-flow moves from raw mammograms to GP classifiers. The specifics of the GP experiments are detailed in Sect. 10.4 and the results are in Sect. 10.5. We finish with the conclusions and future work in Sect. 10.6.

10.2 Background

Image analysis with classification is a broad research area and there is a plethora of GP literature on the topic, from early work such as Koza (1993), Tackett (1993), Andre (1994) to more recent studies such as Bozorgtabar and Ali Rezai Rad (2011), Fu et al. (2014), and Langdon et al. (2014). Describing the full breath of the research is far beyond the scope of this article. Thus, we direct the interested reader to Krawiec et al. (2007) for a review of GP for general image analysis, and we choose to focus here on the most relevant aspects in the current context: classification, object detection and feature **extraction, detection** and **selection**.

In early work on image analysis, Poli (1996) presented an approach based on the idea of using GP to evolve effective image filters. They applied their method to the problem of segmentation of the brain in pairs of Magnetic Resonance images and reported that their GP system outperformed Neural Networks (NNs) on the same problem.

Agnelli et al. (2002) demonstrated the usefulness of GP in the area of document image understanding and emphasised the benefits of the understandability of GP solutions compared with those produced by NNs or statistical approaches. In other work Zhang et al. (2003) proposed a *domain independent* GP method for tackling object detection problems in which the locations of small objects of multiple classes in large images must be found. Their system applied a "moving window" and used pixel statistics to construct a *detection map*. They reported competitive results with a low false alarm rate.

Several novel fitness functions were investigated in Zhang and Lett (2006) together with a filtering technique applied to training data, for improving object localisation with GP. They reported fewer false alarms and faster training times for a weighted localisation method when this was compared with a clustering approach. The suitability of different search drivers was also studied by Krawiec (2015) who examined the effectiveness of several different fitness functions applied the problem of detection of blood vessels in ophthalmology imaging. Another example where preprocessing of training samples proved effective can be seen in Ando and Nagao (2009) where training images were divided into sub-populations based on predefined image characteristics.

An alternative approach, which leveraged the natural ability of evolutionary computation to perform feature selection, was suggested by Komosiński and Krawiec (2000) who developed a novel GA system which *weighted* selected features. They reported superior results when their approach was used for detection of central nervous system neuroepithelial tumours. In related work Krawiec (2002) investigated the effectiveness of a *feature construction* approach with GP, where features were constructed based on a measure of utility determined by their perceived effectiveness according to decision tree induction. The reported feature construction approach significantly outperformed standard GP on several classification benchmarks.

A grammar guided GP approach was used to locate the common carotid artery in ultrasound images in Benes et al. (2013), and this approach resulted in a significant improvement on the state of the art for that task.

Details of other object detection research of note may be found in for example Robinson and McIlroy (1995), Benson (2000), Howard et al. (2002), Isaka (1997), Zhang and Lett (2006), and Trujillo and Olague (2006).

A thorough review of feature detection approaches in the general literature can be found in Tuytelaars and Mikolajczyk (2008). In the field of GP, a wide variety of different types of features have been used to guide classification. "Standard" approaches include first, second and higher order statistical features which may be local (pixel based) (Howard et al. 2006) or global (area based) (Lam and Ciesielski 2004). *Wavelets* have been employed in various work including Chen and Lu (2007) and Padole and Athaide (2013). Texture features constructed from pixel grey levels were used in Song et al. (2002) to discriminate simple texture images. Cartesian GP was used to implement *Transform based Evolvable Features* (TEFs) in Kowaliw et al. (2009), which were used to evolve image transformations: an approach which improved classification of Muscular Dystrophy in cell nuclei by 38 % over previous methods (Zhang et al. 2013).

Recently, *local binary patterns* (LBP) were successfully used with GP for anomaly detection in crowded scenes (Xie and Shang 2014). LBPs were also previously used in, for example Al-Sahaf et al. (2013) and Oliver et al. (2007).

With regard to feature extraction and classification, Atkins et al. (2011) also suggested a *domain independent* approach to image feature extraction and classification where each individual was constructed using a three tier architecture, where each tier was responsible for a specific function: classification, aggregation, and filtering. This work was later developed to use a two tier architecture in Al-Sahaf et al. (2012). These researchers demonstrated that their automated system performed as well as a baseline GP-based classifier system that used human-extracted features. A review of pattern recognition approaches to cancer diagnosis is presented in Abarghouei et al. (2009), where the researchers reported competitive performance of GP on feature extraction when compared with other machine learning (ML) algorithms and feature extraction algorithms.

A multi-objective GP (MOGP) approach to feature extraction and classification was recently adopted in Shao et al. (2014) which constructed feature descriptors from low-level pixel primitives and evaluated individual performance based on classification accuracy and tree complexity. They reported superior performance of their method when compared with both a selection of hand-crafted approaches to feature extraction and several automated ML systems.

Of special note are the Hybrid Evolutionary Learning for Pattern Recognition (HELPR) (Rizki et al. 2002) and CellNet (Kharma et al. 2004) systems, both of which aspire to being fully autonomous pattern recognisers. HELPR combines aspects of evolutionary programming, genetic programming, and genetic algorithms (GAs) whereas CellNet employs a co-evolutionary approach using GAs.

10.2.1 Performance Metrics

In classification the *true positive rate* (TPR) is the proportion of positive instances which the radiologist or learning system classifies as positive, and the *false positive rate* (FPR) is the proportion of instances actually belonging to the negative class that are misclassified as positive. In the classification task of discriminating cancerous from non-cancerous instances, the objective is to maximize the TPR while at the same time minimizing the FPR. The TPR is of primary importance as the cost of missing a cancerous case is potentially catastrophic for the individual concerned. However, it is also very important to reduce the FPR as much as possible due to the various issues associated with false alarms, as outlined in Sect. 10.1. In the literature, when a classification task involves image processing, the number of false positives per image (FPPI) is usually reported. This is the number of false positives divided by the number of images.

In classification literature the TPR is often referred to as *sensitivity* or *recall*, whereas *specificity* is a term used to describe the true negative rate (TNR) and $FPR = 1 - specificity$.

The *Receiver Operating Characteristic* (ROC) is a tool which originates from World War II where it was used to evaluate the performance of radio personnel at accurately reading radar images. These days it is sometimes used to measure the performance of medical tests, radiologists and classifiers. It can also be used to examine the balance between the TPR and FPR as the decision threshold is varied.

Fig. 10.1 Comparison of ROC curves

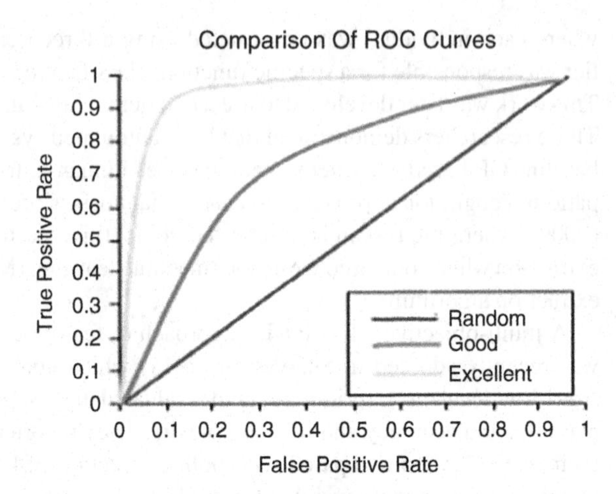

In a ROC curve, the TPR is plotted against the FPR for different cut-off points. Each point on the resulting plot represents a sensitivity/specificity pair corresponding to a particular decision threshold. A "perfect" classifier will have a ROC curve which passes through the upper left corner of the plot which represents 100 % sensitivity and 100 % specificity. Therefore the closer the ROC curve is to the upper left corner, the higher the overall accuracy of the classifier (Zweig and Campbell 1993), whereas a curve that splits the plot exactly across the diagonal is equivalent to random guessing. This is illustrated in Fig. 10.1.

The area under the ROC curve, known as the *AUC* is a scalar value which captures the accuracy of a classifier. The AUC is a non-parametric measure representing ROC performance independent of any threshold (Brown and Davis 2006). A perfect ROC will have an AUC of 1, whereas the ROC plot of a random classifier will result in an AUC of approximately 0.5.

In this work, we report the TPR, FPR, FPPI and AUC for each of the various configurations of mammographic image data included in our work-flow.

10.2.2 Mammography

A mammogram is a low-energy X-ray projection of a breast which is performed by compressing the breast between two plates which are attached to a mammogram machine: an adjustable plate on top with a fixed x-ray plate underneath. An image is recorded using either X-ray film or a digital detector located on the bottom plate. The breast is compressed to prevent it from moving, and to make the layer of breast tissue thinner.

Two views of each breast are recorded: the craniocaudal (CC) view, which is a top down view, and the mediolateral oblique (MLO) view, which is a side view taken at an angle. See Fig. 10.2 for examples of each view. Functional breast tissue is termed parenchyma and this appears as white areas on a mammogram, while the black areas are composed of adipose (non-functioning fatty) tissue which is transparent under X-rays.

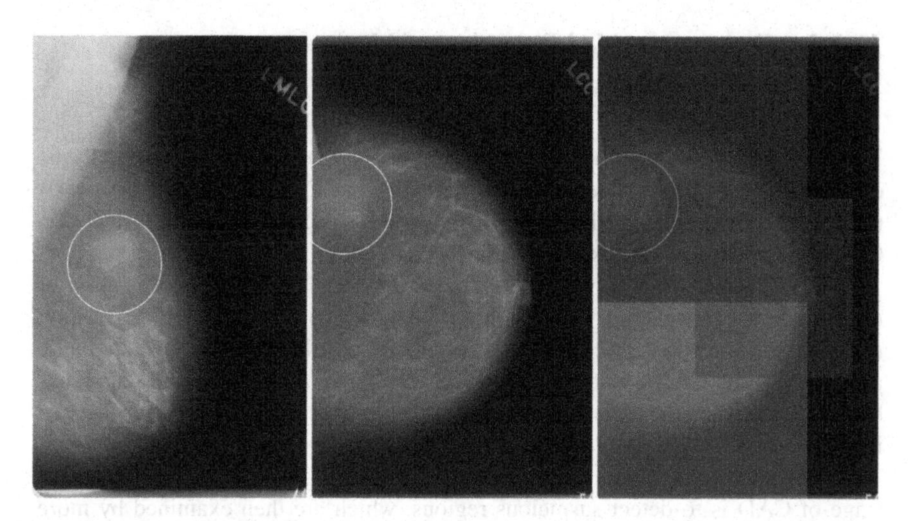

Fig. 10.2 Mammograms. On the *left* is the *MLO* view, with benign micro-calcifications magnified, while in the middle is the *CC* view, with a cancerous mass magnified. Notice the extra information in the background of the image, such as view labels. On the *right* is the same CC view divided into segments; each segment is examined separately for suspicious areas by the method proposed in this chapter

Mammographic images are examined by radiologists who search for *masses* or *architectural distortions*. A mass is defined in American College of Radiology (2003) as a space-occupying lesion that can be seen in at least two views. Architectural distortion is defined as an alteration in the direction of a normal area of the breast, such that it appears straight, pulled in, wavy or bumpy (Lattanzio et al. 2010). Mammograms often also contain micro-calcifications, which are tiny deposits of calcium that show up as bright spots in the images. With the exception of very dense breasts, where calcifications can be obscured, it is generally accepted that compared with other abnormalities, micro-calcifications are easier to detect both visually and by machine due to their bright and distinctive appearance and the fact that they are intrinsically very different from the surrounding tissue. Also, micro-calcifications are usually, but not always, benign.

Depending on the machine used for the mammogram, the resulting image is stored either as a plastic sheet of film or as an electronic image. Many machines in use today produce digital mammograms. With digital mammograms, the original images can be magnified and manipulated in different ways on a computer screen. Several studies have also found that digital mammograms are more accurate in finding cancers in women under the age of fifty, in peri-menopausal women, and in women with dense breast tissue (Pisano et al. 2005). Most importantly, the advent of digital mammography opens up huge opportunities for the development of computer aided analysis of mammograms.

10.2.3 Computer-Aided Detection of Mammographic Abnormalities

Various levels of automation exist in mammography, and these can generally be divided into Computer-Aided *Detection* (CAD) and Computer-Aided *Diagnosis* (CADx) (Sampat and Bovik 2010). In this work we concentrate exclusively on CAD, in particular, what is known as *Stage 1* detection.

In 1967 Winsberg et al. (1967) developed a system for automated analysis of mammograms. However, it was not until the late 1980s that improved digitisation methods and increases in computer power made the development of potentially useful CAD and CADx systems feasible. Since then, a large body of research has been undertaken on the topic, with many research groups currently active in the area internationally.

A typical work-flow for a computer-aided system is shown in Fig. 10.3. The first stage of CAD is to detect suspicious regions, which are then examined by more specialised routines in the second stage. The output of this stage is a set of *Regions of Interest* (ROIs) which are passed either to a radiologist or to a CADx system which outputs the likelihood of malignancy. The involvement of radiologists and/or later stages obviates the need for a perfectly understandable system, as any diagnostic action is ultimately determined by them.

As with many medical applications, mammography demands near-perfection, particularly in the identification of True Positives (TPs), where the true positive rate (TPR) is measured as the percentage of test cases containing cancerous areas identified. In general, Stage 1 detectors are quite conservative (Sampat and Bovik 2010) and often return a relatively high False Positives per Image (FPPI) rate, that is, the number of areas from an image that are incorrectly identified as having cancerous masses.

While an important function of stage 2 detectors is to reduce the FPPI in the output produced by the Stage 1 detector, the rate of FPPI can have an impact on the speed and quality of stage 2 detectors, as a too-conservative approach will degenerate to returning virtually every image. Although this would return a perfect TPR, the FPPI rate would render the system virtually useless.

The potential for CAD to improve screening mammography outcomes by increasing the cancer detection rate has been shown in several retrospective studies Vyborny (1994), Brake et al. (1998), Nishikawa et al. (1995), and

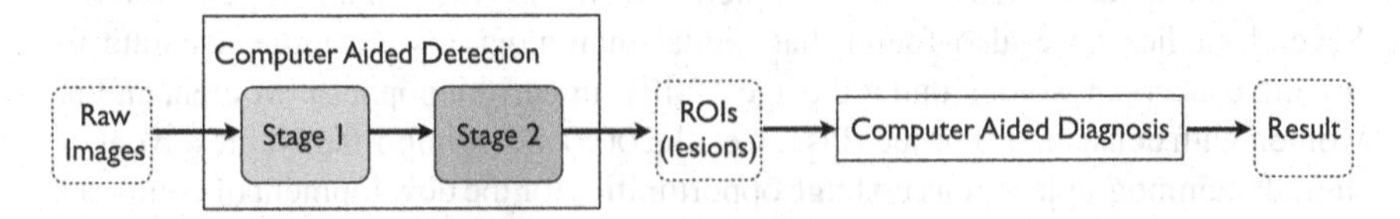

Fig. 10.3 A typical flowchart for computer aided detection and diagnosis. Stage 1 of the process aims to detect suspicious areas with high sensitivity, while Stage 2 tries to reduce the number of suspicious lesions without compromising sensitivity

Warren Burhenne and D'Orsi (2002). A more recent study (Cupples et al. 2005) reported an overall increase of 16 % in the cancer detection rates using CAD together with traditional detection methods. In this study, CAD increased the detection rate of small invasive cancers (1 cm or less) by 164 %. The study concluded that "increased detection rate, younger age at diagnosis, and significantly earlier stage of invasive cancer detection are consistent with a positive screening impact of CAD".

In general, most automated approaches to mammography divide the images into segments (Sampat and Bovik 2010) on which further analysis is undertaken. Each segment is examined for signs indicative of suspicious growths. This work takes a radically different approach by considering textural asymmetry *across* the breasts and between segments of the same breast as a potential indicator for suspicious areas. This is a reasonable approach because, although breasts are generally physically (in terms of size) *asymmetrical*, their *parenchymal* patterns (i.e., their mammographic appearance) and, importantly, the texture of their mammograms, are typically relatively uniform (Tot et al. 2000).

Density of breast tissue is an important attribute of the parenchyma and it has been established that mammograms of dense breasts are more challenging for human experts. At the same time, repeated studies have demonstrated that women with dense tissue in greater than 75 % of the breast are 4–6 times more likely to develop breast cancer compared with women with little to no breast density (Boyd et al. 1995, 1998; Byrne et al. 2001; McCormack and Santos Silva 2006). Douglas et al. (2008) highlighted a correlation between genetic breast tissue density and other known genetic risk factors for breast cancer, and concluded that the "shared architecture" of these features should be studied further. Given the importance of parenchymal density as a risk factor and the difficulty for human experts in identifying suspicious areas in this challenging environment, we believe that a stage 1 detector which focuses on textural asymmetry may have a strong decision support role to play in the identification of suspicious mammograms.

10.2.4 Feature Detection, Selection and Extraction

Feature detection, feature selection and feature extraction are crucial aspects of any image analysis or classification task. This importance is reflected in the volume of research that has been undertaken on the subject. Feature *detection* involves the extraction of possibly interesting features from image data, with a view to using them as a starting point to guide some detection or classification task. The objectives of feature *selection* are the extraction from a potentially large set of detected features those features that are most useful, in terms of discrimination, for the particular purpose and also for determining which combinations of features may work best. Finally, *feature extraction* is the process of extracting from detected features the non-redundant meaningful information that will inform a higher level task such as classification. This may involve reducing the number of features or combining features (or aspects thereof) to form new, more compact or useful features.

Mammograms are large (the images in this work are of the order 3600 × 5600 pixels) grey-scale images, but only minute parts of them contain diagnostically relevant information. Therefore, a detection process typically relies on the existence of *features*, which describe various properties of the image. Features are typically extracted using either area- or pixel-based measures. In this work we focus exclusively on area-based features as they are better suited to the identification of ROIs (because the images are so large) than their pixel-based counterparts, which are best suited for highly localized search. Section 10.3 below describes the features extracted.

10.2.5 Related Work

Although several CAD systems already exist, most are Stage 2 detectors (Sampat and Bovik 2010) and focus on particular kinds of masses, e.g. spiculated lesions. Of the more general systems, the best reported appears to be that of Ryan et al. (2014) which reports a best TPR of 100 % with an FPPI of just 1.5. Other good results were produced by Li et al. (2001) with 97.3 % TPR and 14.81 FPPI. Similar work by Polakowski et al. (1997) had a lower TPR (92 %) but with a much lower FPPI rate (8.39).

There has been a great deal of research undertaken in the area of detection and classification of micro-calcifications. Various approaches to feature detection have been proposed including texture features, gray level features (Dhawan et al. 1996), wavelet transforms (Strickland and Hahn 1996), identification of linear structures (Wu et al. 2008) and various statistical methods. In 2004 Soltanian-Zadeh et al. (2004) undertook a comparison of the most popular features used for micro-calcification detection including texture, shape and wavelet features. They concluded that the multi-wavelet approach was superior for the particular purpose. In more recent work using Cartesian GP, micro-calcifications were targeted by Volk et al. (2009), in a CADx application, where they took 128 × 128 pixel segments, each of which contained at least one micro-calcification and predicted the probability of it being malignant.

For the objectives of mass segmentation and detection, image features which capture aspects of shape (Rangayyan et al. 1997), edge-sharpness (Mudigonda et al. 2000) and texture (Bovis and Singh 2000) are frequently used. Nandi et al. (2006) reported a classification accuracy of 98 % on test data when using a combination of all three of these feature types. In that work, the researchers examined a database of 57 images, each of which already had 22 features detected, and used GP in combination with various feature selection methods to reduce the dimensionality of the problem.

Varying numbers of Haralick texture features were used in Woods (2008) to train a NN classifier to detect cancerous lesions in contrast enhanced magnetic resonance imaging (DCE-MRI) for both breast and prostate cancer. The results of that study showed that the proposed approach produced classifiers which were competitive to a human radiologist.

10.2.5.1 Learning Paradigms

Various research paradigms such as neural networks (Papadopoulos et al. 2005), fuzzy logic (Cheng et al. 2004) and Support Vector Machines (SVM) Dehghan et al. (2008), Cho et al. (2008) have been applied to the problem. In a review of various ML approaches for detecting micro-calcifications, Sakka et al. (2006) concluded that neural networks showed the most promise of the methods studied. In a contemporary review, Alanís-Reyes et al. (2012) employed feature selection using a GA and then compared the classification performance of various ML algorithms in classifying both micro-calcifications and other suspicious masses, using these features. Their results showed that SVM produced the best overall performance.

Given the success of GP in finding solutions to a wide range of problems, it is not surprising that the approach has been applied to problems relating to mammography. Quite a lot of the GP research effort has successfully demonstrated feature selection and classification of micro-calcifications and masses as either benign or malignant (Zheng et al. 1999; Nandi et al. 2006; Verma and Zhang 2007; Sánchez-Ferrero and Arribas 2007; Hernández-Cisneros et al. 2007). In this work the feature detection task is generally not handled by the genetic programs.

A genetic algorithm was used for feature selection in Sahiner et al. (1996), where a very large number of initial features were reduced to a smaller set of discriminative ones and then passed to either a NN or a linear classifier.

Other notable research using GP is that Ahmad et al. (2012) who designed a Stage 2 cancer detector for the well known Wisconsin Breast Cancer dataset, in which they used the features extracted from a series of fine needle aspirations (FNAs) and an evolved neural network. Ludwig and Roos (2010) used GP to estimate the prognosis of breast cancer patients from the same data set, initially using GP to reduce the number of features, before evolving predictors. Langdon and Harrison (2008) took a different approach, using biopsy gene chip data, but their system approached a similar level of automation.

Current work in mammography has been concerned with a combination of feature selection and classification (Ganesan et al. 2013). One such approach suggested by Ryan et al. (2014) reports a best TPR of 100 % with an FPPI of only 1.5. In other work, the best reported appears to be that of Li et al. (2001) which delivers a 97.3 % TPR with 14.81 FPPI. Similar work by Polakowski et al. (1997) reported a lower TPR (92 %) but with a much lower FPPI rate (8.39). The standard method of reporting results is the TP/FPPI breakdown, which is what we will also present here.

See Petrick et al. (2013) for an evaluation of the current state-of-the-art of computer-aided detection and diagnosis systems. In other work, Worzel et al. (2009) reported favourably on the application of GP in cancer research generally.

Most systems operate only at the *Classification* stage, although more recent work also considers *Feature Selection*. As we generate our own features, we can modify and parameterize them based on the analysis of our classifiers. While the focus of this chapter is on the classification system, because we extract the features from the images ourselves, GP will eventually form part of a feedback loop, instructing the system about what sorts of features are required. See Sect. 10.6 for more details on this.

Most previous work relies upon previously extracted features, and all the previous work mentioned above deals with a single breast in isolation (although using segmentation and multiple views). Our work leverages the research by Tot et al. (2000) which indicates that, in general, both breasts from the same patient have the same textural characteristics. Our hypothesis is that breasts of the same patient that differ texturally may contain suspicious areas.

In summary, the unique features of our approach are that we do not focus only on a single breast but address the problem by considering textural asymmetry *across* the breasts as well as between segments of the same breast and we do not confine our efforts simply to the classification step—rather we adopt an end-to-end strategy which focuses on area-based features and incorporates feature extraction, detection and selection.

10.3 Workflow

Part of the challenge in a project like this is to choose how to represent the data. A typical mammogram used in this study is 3575 × 5532 pixels and 16 bit gray-scale, which is a challenging volume of data to process. The following work-flow was created. Steps 1–5 are concerned with the raw images, while steps 6 and 7 use GP to build and test classifiers. The described work-flow provides a template for similar tasks, where steps 1 and 2 can be replaced with domain specific counterparts.

1. Background suppression
2. Image segmentation
3. Feature detection
4. Feature selection
5. Dataset construction
6. Model development
7. Model testing

10.3.1 Background Suppression

Figure 10.2 shows that much of the images consist of background, and clearly, this must first be removed before calculating segments and extracting features.

Removing the background is a non-trivial task, partly because the non-uniformity of breast size across patients, but also because of the difficulty in taking consistent mammograms. Due to the pliable nature of the breasts and the way in which the mammograms are photographed (by squeezing the breast between two plates), the same breast photographed more than once on the same machine (after a reset) may look different.

The background of the mammographic image is never perfectly homogeneous, and it includes at least one tag letter indicating if the image is either a right or left breast. This is sometimes augmented by a string of characters indicating which view (CC or MLO) is depicted. It is necessary to remove this background detail and replace it with a homogeneous one so that the image can be properly processed at a later stage.

Our first attempt was based on the Canny Edge Detector, but, although this method is efficient on raw imagery, the mammograms we dealt with had been processed to increase the contrast within the breast (to make them easier to read, but which has the side effect of reducing the contrast between the edge of the breast and the background). Canny Edge Detection revealed itself to be less efficient on these images.

Our most efficient technique was to use a threshold (average of the median pixel value and the average pixel value) such that any pixel (p_x, p_y) above the threshold level of was kept, i.e. $20^2 = (x-p_x)^2 + (y-p_y)^2, x < p_x$. We used local thresholding with the threshold defined as an average of mean and median, calculated from each pixel's circular neighbourhood of radius 20. We scan each horizontal line right to left. Once three consecutive pixels are brighter than the threshold calculated in the above way, those pixels and the pixels to the left of it are considered as belonging to the breast.

10.3.2 Image Segmentation

Our approach is to divide each image into three segments, and to examine each segment separately. As there can be more than one suspicious area in an image, we return *true* for as many segments as the systems finds suspicious, meaning that a single mammogram can have several positives returned. With Stage 1 detectors such as ours, this is described by the FPPI of an image, as discussed in Sect. 10.2.5.

Of course, the maximum FPPI is capped by the number of segments that the breast is divided into. Using fewer segments means that the FPPI will be lower, but the cost of the detection of the TPs is substantially more difficult because the area is larger.

Using the same algorithm outlined in Ryan et al. (2014), we segmented the breast images into three overlapping sub-images of roughly similar size, as shown in Fig. 10.2. The first of these captures the nipple area and the other two cover the top and bottom sections of the rest of the breast. The three segments intersect, to help reduce the possibility of a mass going unnoticed.

In summary, each patient has two breasts, and mammograms are taken for two views (CC and MLO) of each breast—giving a total of four mammograms per patient. From these four mammograms we obtain sixteen images: four images of the full breast (left CC, left MLO, right CC, right MLO) and three sub images (top, bottom, nipple) for each of these four images. We construct our training and test data with features obtained from these sixteen images/sub-images.

10.3.3 Textural Features

As with most image classification systems, before attempting classify mammograms as suspicious or not we must first extract *features* for GP. In this study, we use Haralick's Texture Features (Haralick et al. 1973) as we believe that textural features are appropriate in this case because we are examining parenchymal patterns, and our hypothesis is that suspicious areas are likely to be texturally dissimilar to normal areas. However, different features may be suited for other problem domains, in which case the detection and selection of these problem specific features can simply slot into the work-flow at this juncture.

The seminal work of Haralick et al. (1973) described a method of generating 14 measures which can be used to form 28 textural features from a set of co-occurrence matrices or "grey tone spatial dependency matrices". When applied to pixel grey levels, the Grey Level Co-occurrence Matrix (GLCM) is defined to be the distribution of co-occurring values at a given offset. In other words, GLCM is a joint distribution (histogram) of brightness of two pixels bound by a given spatial relationship. That relationship is typically specified by assuming that the second pixel is at a specific offset with respect to the first one.

Given a neighbourhood relationship r, an element $c(i,j)$ of a GLCM of image m is the probability that a pixel p and its neighbour pixel q have brightness values i and j respectively, i.e., $\Pr(r(p,q) \wedge m(p) = i \wedge m(q) = j)$.

Using the co-occurrence matrix, different properties of the pixel distribution can be generating by applying various calculations to the matrix values.

Given the image matrix in Table 10.1 which handles three grey levels, the co-occurrence matrix below is obtained by moving over the image matrix and calculating $f(i,j)$ where $f(i,j)$ is the frequency that grey levels i and j occur with at a given distance and direction.

Table 10.1 Pixel grey levels

0	0	0	1	2
1	1	0	1	1
2	2	1	0	0
1	1	0	2	0
1	0	1	0	0

Table 10.2 Co-occurrence matrix

	0	**1**	**2**
0	8	8	2
1	8	6	2
2	2	2	2

For example, $f(0, 0) = 8$ is obtained by scanning the image matrix, and for each pixel with a grey value of zero incrementing $f(0, 0)$ every time one of its neighbours on the horizontal direction at a distance of 1, also has a value of zero. Co-occurrence matrices can also be generated in other directions: 90, 135 and 45 degrees (vertical and diagonal), and for distances other than one. In this work we examine a neighbourhood of one and average the feature values for the four orientations (Table 10.2).

Haralick et al. (1973) showed that GLCMs conveniently lend themselves to efficient calculation of various informative measures, including:

1. Angular Second Moment
2. Contrast
3. Correlation
4. Sum of squares
5. Inverse Difference Moment
6. Sum Average
7. Sum Variance
8. Sum Entropy
9. Entropy
10. Difference Variance
11. Difference Entropy
12. Information Measure of Correlation 1
13. Information Measure of Correlation 2
14. Maximal Correlation Coefficient

For a chosen distance there are four spatial dependency matrices corresponding to the four directions 0°, 45°, 90° and 135°, giving four values for each of the 14 Haralick texture measures listed. There are some issues with this approach. The amount of data in the co-occurrence matrices varies with the range and number of values chosen for neighbourhood and direction and will be significantly higher that the amount of data in the original image. Simple examples of the method found in the literature typically use few gray levels for ease of explanation. However, in real-life applications the number of grey levels is likely to be significant. This obviously greatly increases the volume of matrix data: there will be an $n \times n$ matrix for each direction and each distance chosen, where n is the number of gray levels. Also, the resulting matrices are often very sparse as certain combinations of brightness may never occur in an image. In spite of these obvious downsides, Haralick features are widely used in the research.

To quantitatively describe the textural characteristics of breast tissue, we calculate a GLCM for each segment and for each breast. To keep the GLCM size manageable, we first reduce the number of gray levels to 256 (from 65,535 in the original images) via linear scaling. Because textures in mammograms are often anisotropic (directionally dependent), we independently calculate GLCMs for four orientations corresponding to two adjacent and two diagonal neighbours. Next, we calculate 13 Haralick features (Haralick et al. 1973) (we exclude the 14th feature: Maximal Correlation Coefficient as it can be computationally unstable Woods 2008). By doing this for each orientation, we obtain 52 features per segment, which may subsequently be passed to a ML system for classification. This down-sampling of gray levels, construction of GLCMs and extraction of Haralick features is achieved using MATLAB (2013).

Segments are rectangular and often extend beyond the breast, which means that they may contain some background. A GLCM calculated from such a segment in the normal way would register very high values for black pixels ($m(p) = 0$ or $m(q) = 0$) which may distort the values of Haralick features. As many mammographic images contain useful information captured in black pixels, such as sections of adipose tissue (fat), which appears black in mammograms, it would not be correct to simply ignore black pixels. Therefore, before calculating the GLCM, we increase by one the intensity of every pixel within the breast, using the information resulting from the segmentation stage (see previous subsection). The pixels that already had the maximal value retain it (this causes certain information loss, albeit negligible one, as there are typically very few such pixels). Then, once the GLCM has been calculated, we simply "hoist" the GLCM up and to the left to remove the impact of the *unmodified* background pixels.

Feature Selection As previously mentioned, the *neighbourhood relation* of the GLCM can be varied, such that the calculation is conducted on pixels further away from each other, but, each extra neighbourhood examined produces another 52 features per segment. In this work we examine the neighbourhoods composed of direct neighbours only (i.e., at a distance of 1 from the reference pixel) and averaged the feature values for the four orientations.

We conducted a preliminary analysis of the 13 computed Haralick features where we initially examined variance across and between both classes and then carried out a more formal analysis using several ranker methods (Hall et al. 2009) which ranked the attributes according to the concept of *information gain*. In this context information gain can be thought of as a measure of the value of an attribute which describes how well that attribute separates the training examples according to their target class labels. Information gain is also known as Kullback Leibler divergence (Kullback and Leibler 1951), information divergence or relative entropy. Information gain employs the idea of *entropy* as used in information theory. These feature selection steps suggested that the most promising features in terms of discrimination were *contrast* and *difference entropy*. Accordingly, we discarded the other features and let GP focus on those two.

10.4 Experimental Setup

In this section, we describe the construction and distributions of the datasets used, together with details of configurations of that data for specific experiments. We also provide details of the GP parameters and classification algorithm employed.

10.4.1 Dataset Construction

This work employs University of South Florida Digital Database for Screening Mammography (DDSM) (Heath et al. 2001) which is a collection of 43 "volumes" of mammogram cases. A volume is a collection of mammogram cases (typically about 50–100 different patients) and can be classified as either *normal*, *positive*, *benign* or *benign without callback*. All patients in a particular volume have the same classification. We use cases from the **cancer02** and three of the **normal** volumes (volumes 1–3). For this study we do not use images from either the **benign** or **benign without callback** volumes.

The incidence of positives within mammograms is roughly 5 in 1000,[1] giving a massively imbalanced data set. To ensure that our training data maintains a more realistic balance, we deliberately select only a single volume of positive cases.

Several images were discarded either because of image processing errors or because we were unable to confidently identify which segment/s were cancerous for a particular positive case. In the current work, this latter task was performed manually. We will automate this step in the next iteration. This initial processing resulted in a total of 294 usable cases, 75 of which contain cancerous growths (which we call *positive* in the remainder of this document). Each *case* initially consists of images for the left and right breasts and for the MLO and CC views of each breast. Once the segmentation step has been completed images are added for each of the three segments (nipple/top/bottom) for each view of each breast. Thus, there are a total of four images per view for each breast: one for the entire breast (A), and one for each of the three segments (A_t, A_b, A_n).

If we count the numbers of positives and negatives in terms of breasts rather than cases, which is reasonable, given that each is examined independently (i.e. most, but not all, patients with cancerous growths do not have them in both breasts), then the number of non-cancerous images increases significantly: giving two for each non-cancerous case and one for most cancerous growths. For the volumes studied, of the 75 usable positive cases, 3 have cancer in both breasts. Thus, considering full breast CC images only, we have 78 positive images and 510 (219 * 2 + 72) negative ones

Turning our attention to segments (A_t, A_b, A_n) (excluding full breast images), and again considering only CC segments for the moment, for each non-cancerous case

[1]The actual incidence over a patient's lifetime is closer to 1 in 7 (Kerlikowske et al. 1993).

we have 3 segments for each breast (left and right) together with 2 non-cancerous segments for each cancerous breast which gives a total of 1686 non-cancerous segments and 78 cancerous segments. Similarly, for the MLO view there are 1686 non-cancerous segments and 78 cancerous ones.

Thus, we obtain three different distributions, one for the non-segmented single views (CC or MLO) *full breast* images (78 positives (P), and 510 negatives, (N)), one for the *segmented* single views (78 Ps and 1686 Ns) and one for segmented combined CC MLO views (156 Ps and 3372 Ns). Each of these three distributions exhibit very significant class imbalance which, in and of itself, increases the level of difficulty of the classification problem. The imbalance in the data was handled in all cases by using Proportional Individualised Random Sampling (Fitzgerald and Ryan 2013), as described in Sect. 10.4.2

Based on this master dataset, we consider several setups representing different configurations of breasts, segments and views (see Table 10.3). The following terminology is used to describe the composition of *instances* for a given setup, where an instance is a single training or test example in a dataset: B*X*S*Y*V*Z*, where *X* is the number of breasts, *Y* the number of segments and *Z* the number of views *for a given instance*. In the cases where there is just one view (B1S1V1, B2S2V1,

Table 10.3 Experimental configurations

Name	Ps	Ns	Description
B1S0V1	78	510	1 breast, unsegmented image,1 view, uses CC view only
B1S1V1	78	1686	1 breast, 1 seg., 1 view; uses CC view only
B1S2V2	156	3372	1 breast, 2 segs., 2 views; uses both CC and MLO views
B1S3V1	78	1686	1 breast, 3 segs., 1 view; CC view only
B2S0V1	78	510	2 breasts, unsegmented image, 1 view, uses CC view only
B2S2V1	78	1686	2 breasts, 2 segs., 1 view; both CC views, one segment from each
B2S4+0V1	78	1686	2 breasts, 4 segs.,1 view, CC views, 1 segment + unsegmented from each
B2S3+0V1	78	1686	2 breasts, 3 segs., 1 view, CC views, 1 segment from each + unsegmented from first
B2S4V1	78	1686	2 breasts, 4 segs., 1 view; CC views, three segments + one segment
B2S6V1	78	1686	2 breasts, 6 segs., 1 view; CC views, three segments + one segment + 2 unsegmented

Each was generated from the *same* master data set

B1S3V1, B2S4V1) we use the CC views, while in the cases where the breast has been segmented, the system attempts to classify whether or not the *segment* has a suspicious area or not. In particular, the two breast (B2S*Y*V1) special setups which investigate the use of asymmetry. These rely solely on the CC view: each instance is comprised of selected features from one breast CC segment/s together with the same features taken from the corresponding other breast CC segment/s for the same patient.

When it comes to processing the data, we want to exploit any differences between a segment and the rest of the breast (i.e. between A and A_x) but also between a segment and the corresponding segment from the opposite breast, (say B and B_x), with the objective of evolving a classifier capable of *pinpointing a specific cancerous segment*. To facilitate this process, where we use more than one segment for a particular setup, features from the segment of interest are the first occurring data items in each instance for the dataset for that setup, where the segment of interest is the segment for which we want to obtain a prediction. Details of he specific setups used in the current study are as follows:

10.4.1.1 B1S0V1

This dataset configuration has an instance for the selected features of each full breast image. It uses the CC view only and has an instance for each breast for each patient. It has 78 Ps and 501 Ns.

10.4.1.2 B1S1V1

The BIS1V1 configuration also uses only the CC view, but this setup uses each of the three segments (A_t, A_b, A_n) separately, i.e each instance is comprised of the feature values for a single segment. Again there is an instance for each breast for each segment. This results in 78 Ps and 1686 Ns.

10.4.1.3 B1S2V2

Both views are used in the B1S2V2 setup. For each segment, excluding the full breast image, for each breast, each instance contains feature values for that segment and the corresponding segment for the other view (CC or MLO), i.e each instance has information for both views of a single breast. So for a given segment, say A_t, there are instances for the following:

A_t LEFT_CC, A_t LEFT_MLO
A_t LEFT_MLO, A_t LEFT_CC

A_t RIGHT_CC, A_t RIGHT_MLO
A_t RIGHT_MLO, A_t RIGHT_CC

In this setup the *segments of interest* are A_t *LEFT_CC*, A_t *LEFT_MLO*, A_t *RIGHT_CC* and A_t *RIGHT_MLO* respectively, i.e the segment whose features occur *first*. This principle applies to all of the remaining setups, where more than one segment is used.

10.4.1.4 B1S3V1

This configuration uses three CC segments (A_t, A_b, A_n) for a single breast, where the first segment is alternated in successive instances For example, for a given single breast there are three training instances. Similar to:

A_t LEFT_CC, A_b LEFT_CC, A_n LEFT_CC
A_b LEFT_CC, A_n LEFT_CC, A_t LEFT_CC
A_n LEFT_CC A_t LEFT_CC, A_b LEFT_CC

Where the order of the remaining two segments does not matter.

10.4.1.5 B2S2V1

In this configuration we investigate the simplest case of symmetry: each entry consists of the feature values for a single CC segment from one breast combined with those of the corresponding CC segment from the other breast, for the same patient. In this case there are two entries for each segment: (A_x LEFT_CC, A_x RIGHT_CC) and (A_x RIGHT_CC, A_x LEFT_CC), where x represents a particular segment (A_t, A_b, A_n).

10.4.1.6 B2S3+0V1

There are two set-ups which deviate slightly from the naming scheme above, namely, B2S3+0V1 and B2S4+0V1. Here, +0 indicates that features for a non-segmented image have been included. Each instance in this setup is comprised of feature data from segmented and unsegmented images. It consists of information for a segment, the unsegmented image and the corresponding segment from the other breast. For example:

A_t LEFT_CC, A LEFT_CC, B_t LEFT_CC

10.4.1.7 B2S4+0V1

Similar to B2S3+0V1, each instance in this setup is again comprised of feature data from segmented and unsegmented images. It consists of information for a segment, the unsegmented image and the corresponding segment from the other breast together with the unsegmented image data for the other breast. For example:

A_t LEFT_CC, A LEFT_CC,B_t LEFT_CC,B LEFT_CC

10.4.1.8 B2S4V1

The B2S4V1 experimental setup is a combination of B1S3V1 and B2S2V1 where each training instance is comprised of the feature values for the three segments for a single breast (A) combined with the corresponding segment from the other breast (B) for the leftmost, first occurring segment of A. For example:

A_t LEFT_CC, A_b LEFT_CC, A_n LEFT_CC, B_t LEFT_CC

Where in this instance A_t LEFT_CC is the segment of interest.

10.4.1.9 B2S6V1

The final experimental setup is an extension of B2S4V1 where feature values for the full breast segment for the right and left breasts are added. For example:

A_t LEFT_CC, A_b LEFT_CC, A_n LEFT_CC,A LEFT_CC, B_t LEFT_CC, B LEFT_CC

Where in this instance A_t LEFT_CC is the segment of interest.

It is important to note here is that where more than one segment is used the segment of interest is the first occurring leftmost one, for example, A_t LEFT_CC in the B2S6V1 setup example above. If that segment is diagnosed as cancerous then the training/test instance in which it occurs is marked as positive, and if it is not diagnosed as cancerous then the entire instance is marked as negative *regardless of the cancer status of any other segments used in that particular instance*. Thus, excluding the B1S0V1 setup, the objective is not simply to determine if a given breast is positive for cancer, but rather to pinpoint which segments are positive. If successful, this capability could pave the way for further diagnosis.

10.4.2 Proportional Individualised Random Sampling

In each of our experimental configurations there is significant disparity in the number of positive to negative instances. Greater disparity makes classification problems much more challenging, as there is an inherent bias towards the class which has greater representation in the dataset—in this case, the negative class. When a ML algorithm, designed for general classification tasks and scored according to classification accuracy, is faced with significant imbalance, the "intelligent" thing for it to do is to predict that all instances belong to the majority class. Ironically, it is frequently the case that the minority class is the one which contains the most important or interesting information. In datasets from the medical domain, such as our mammographic data it is often the case that instances which represent malignancy or disease are far fewer than those which do not.

Various approaches to mitigating class imbalance problems have been proposed in the literature. In general, methods can be divided into those which tackle the imbalance at the data level, and those which propose an algorithmic solution. In addition, several hybrid approaches have been advanced which combine aspects of the other two.

Methods which operate on the data level attempt to repair the imbalance by rebalancing training data. This is usually achieved by either under-sampling the majority class or over-sampling the minority class, where the former involves removing some examples of the majority class and the latter is achieved by adding duplicate copies of minority instances until such time as some predefined measure of balance is achieved. Over- or under-sampling may be random in nature (Batista et al. 2004) or "informed" (Kubat et al. 1997), where in the latter, various criteria are used to determine which instances from the majority class should be discarded. An interesting approach called SMOTE (Synthetic Minority Oversampling Technique) was suggested by Chawla et al. (2002) in which rather than over sampling the minority class with replacement they generated new synthetic examples.

At the algorithmic level Joshi et al. (2001) modified the well known AdaBoost (Freund and Schapire 1996) algorithm so that different weights were applied for boosting instances of each class. Akbani et al. (2004) modified the kernel function in a Support Vector Machine implementation to use an adjusted decision threshold. Class imbalance tasks are closely related to cost based learning problems, where misclassification costs are not the same for both classes. Adacost (Fan et al. 1999) and MetaCost (Domingos 1999) are examples of this approach. See Kotsiantis et al. (2006), He and Garcia (2009) for a thorough overview of these and various other methods described in the literature.

There several disadvantages to the application of over or under sampling strategies. The obvious downside to under-sampling is that it discards potentially useful information. The main drawback with standard over sampling is that exact copies of minority instances are introduced into the learning system, which may increase the potential for over-fitting. Also, the use of over-sampling generally results in increased computational cost because of the increased size of the dataset. In this study, we have employed a proportional sampling approach (Fitzgerald and Ryan 2013) which eliminates or mitigates these disadvantages.

Using this approach the size of the dataset remains unchanged so there is no extra computational cost, as is generally the case with random over sampling. Instead, the number of instances of each class is varied. At each generation and for *each individual* in the population the percentage of majority instances is randomly selected in the range between the percentages of minority (positive) and majority (negative) instances in the original distribution. Then, *that particular individual* is evaluated on that percentage of majority instances with instances of the minority class making up the remainder of the data. In both cases, each instance is randomly selected *with replacement*. In this way, individuals within the population are evaluated with different distributions of the data within the *range of the original distribution*.

The benefit of the method from the under sampling perspective is that while the majority class may not be fully represented at the level of the individual, all of the data for that class is available to *the population* as a whole. Because all of the available knowledge is spread across the population the system is less likely to suffer from the loss of useful data that is normally associated with under sampling techniques. From the under sampling viewpoint, over-fitting may be less likely as the distribution of instances of each class is varied for each individual at every generation. Also, as all sampling is done with replacement, there may be duplicates of negative as well as positive instances.

Previous work (Liu and Khoshgoftaar 2004) has shown that aside from the consideration of balance in the distribution of instances, the use of random sampling techniques may have a beneficial effect in reducing over-fitting.

10.4.3 GP Methodology and Parameters

All experiments used a population 200 individuals, running for 60 generations, with a crossover rate of 0.8 and mutation rate of 0.2. The minimum initial depth was four, while the maximum depth was 17. The instruction set was small, consisting of just $+, -, *, /$. The tree terminals (leaves) fetch the Haralick features as defined in Sect. 10.3.3, with two available per segment.

To transform a continuous output of a GP tree into a nominal decision (Positive, Negative), we binarize it using the method described in Fitzgerald and Ryan (2012), which optimizes the binarization threshold individually for each GP classifier.

We employed an NSGA-II (Deb et al. 2002) algorithm as updated in Fortin and Parizeau (2013) as the selection and replacement strategy. Using a population of 200, at each generation, 200 new offspring are generated, then parents and offspring are merged into one population pool before running pareto-based selection to select the best 200). During evolution, we aim to minimize **three** fitness objectives, where AUC is a the area under ROC, calculated using the Mann-Whitney (Stober and Yeh 2007) test, where the false positive rate (FPR) and TPR are calculated with the output threshold set using the binarization technique mentioned above:

- Objective 1: FPR;
- Objective 2: 1−TPR;
- Objective 3: 1−AUC.

The chosen multi-objective fitness function is specifically tailored to suit the mammography task. However, it would be quite straightforward to modify the work-flow and GP system to accommodate a different set of objectives or a single valued or composite single objective fitness function to suit a problem from an alternative domain.

We performed stratified fivefold cross-validation (CV, Geisser 1993; Hastie et al. 2009) for all setups. However, we also retained 10 % of the data as a "hold out" (HO) test set, where for each meta run (each consisting of 5 CV runs) this HO test set data was separated from the CV data prior to the latter's allocation to folds for CV. The data partitioning was carried out using the sci-kit learn ML toolkit (Pedregosa et al. 2011). We conducted 50 cross-validated runs (each consisting of 5 runs) with identical random seeds for each configuration outlined in Table 10.3.

10.5 Results

In this section we present our experimental results firstly with regard to AUC measure on the training and test partitions of the CV phase. Secondly we examine the TPR and FPR for this data. Finally we explore the results for each performance metric, adopting various approaches to model selection, this time taking the performance on hold-out data into consideration.

10.5.1 AUC

Figure 10.4, left plot shows the change in average AUC over generations on the CV training partitions averaged over all cross validated runs, whereas Fig. 10.4 right plot shows the development of the best training AUC also averaged over all cross validated runs. Similarly, Fig. 10.5 left plot shows the change in average AUC over generations on the CV test partition averaged over all cross validated runs, and Fig. 10.5 right plot shows the development of the best population test AUC also averaged over all cross validated runs.

It appears that the best performing setups from the perspectives of both training and test partitions are those which leverage information from both breasts, the single breast configuration which uses all three segments or single breast setup which uses features from the unsegmented image. The B2S2V1 configuration delivers "middle of the road" AUC figures: better results than the two worst performing setups but worse that the better ones.

Clearly some of the worst AUC results are achieved with the configurations which use segments from a single breast, particularly that which uses two views (CC and MLO) of the same area (top, bottom or nipple). The latter is not very surprising as the features contain essentially the same information and having features which are strongly correlated with each other is known to be detrimental to accurate classification.

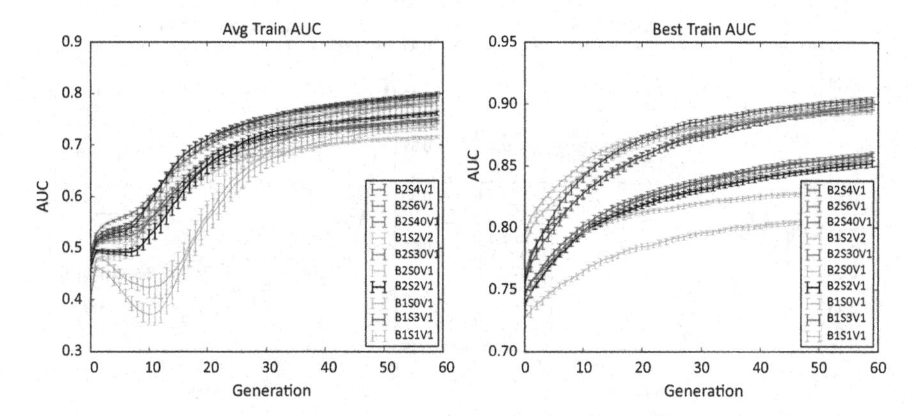

Fig. 10.4 (*Left*) Average population training AUC over five validation folds, averaged over all cross validated runs. (*Right*) Best training AUC averaged over all cross validated runs. Whiskers represent the standard error of the mean

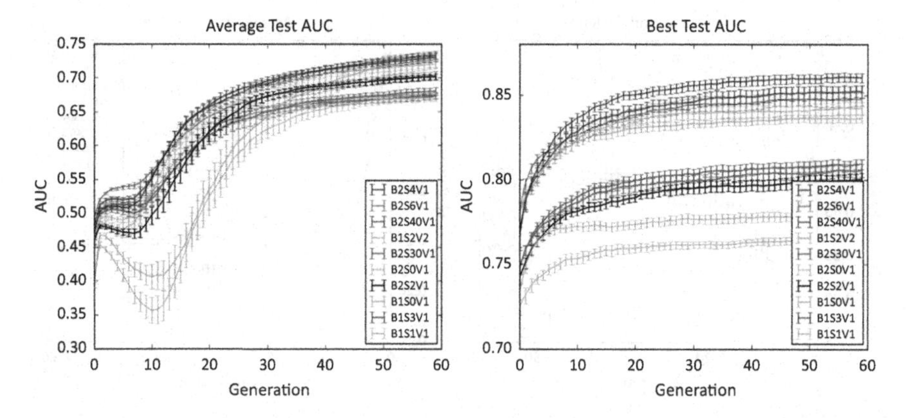

Fig. 10.5 (*Left*) Average population training AUC over five validation folds, averaged over all cross validated runs. (*Right*) Best training AUC averaged over all cross validated runs. Whiskers represent the standard error of the mean

Overall, the results suggest that simply increasing the number of segments gives a significant boost to performance in terms of training fitness but that the strategy does not necessarily improve results on test data.

10.5.2 TP/FPRs

Population average TPR and FPRs for training data are shown in Fig. 10.6 and the corresponding rates on test data can be seen in Fig. 10.7. The plots exemplify the tension which exists in the population between the two competing objectives of

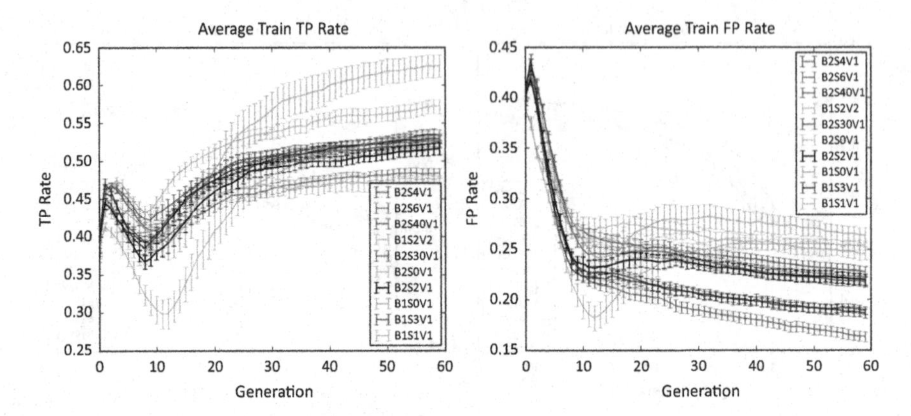

Fig. 10.6 (*Left*) Average CV training TPR averaged over all cross validated runs. (*Right*) Average CV training FPR averaged over all 50 cross validated runs. Whiskers represent the standard error of the mean

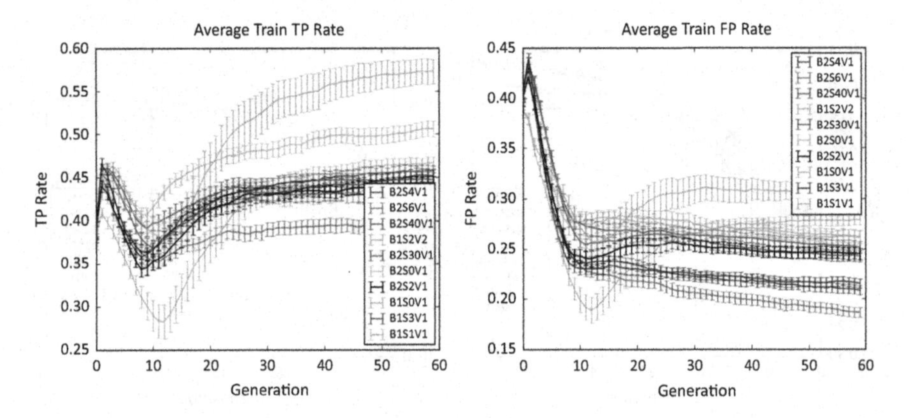

Fig. 10.7 (*Left*) Average CV TPR on test partitions averaged over all cross validated runs. (*Right*) Average CV FPR on test partitions averaged over all 50 cross validated runs. Whiskers represent the standard error of the mean

maximizing the TPR while simultaneously trying to minimize the FPR. In general, a configuration which produces a higher than average TPR will also produce a correspondingly higher FPR. For any configuration, there will always be individuals within the population which classify all instances as either negative or positive. In order to accurately distinguish which configurations are likely to deliver a usable classification model it is more useful to examine the results of the best performing individuals in the population on the various metrics: TPR, FPR and AUC. We explore this aspect in Sect. 10.5.4.

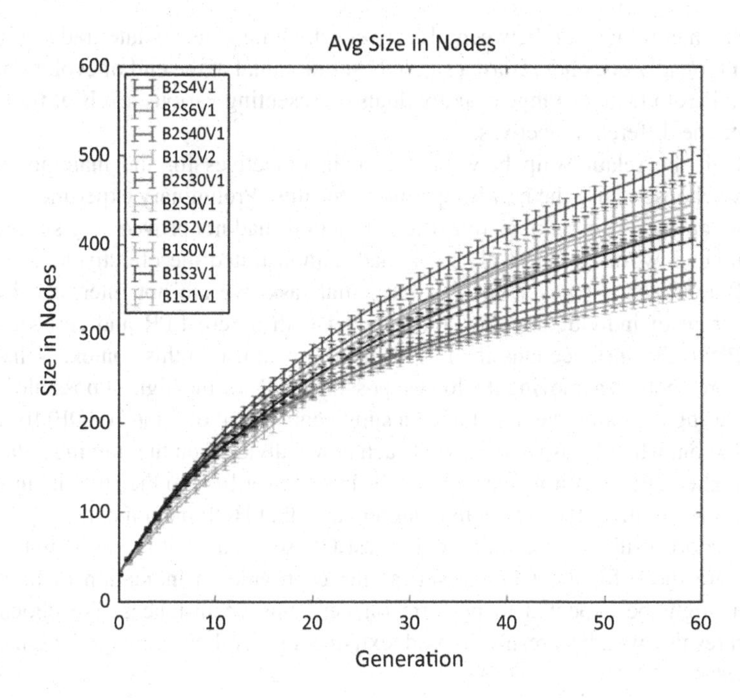

Fig. 10.8 Average population size for each configuration. Whiskers represent the standard error of the mean

10.5.3 Program Size

Turning our attention to the average size of the individuals produced by each configuration as shown in Fig. 10.8, we can see that there is a substantial difference in average size between the smallest and the largest, and the difference appears to increase as evolution progresses. The smallest individuals are produced by the non-segmented configurations, and the next smallest by the most feature rich B2S6V1 setup. The largest programs result from the B1S3V1 configuration which, interestingly, has half as many feature values for each instance as B2S6V1 does. We can hypothesise that this may be because with fewer feature values, the system needs to synthesize them itself, which would be fairly typical evolutionary behaviour.

10.5.4 Model Selection

As described earlier in Sect. 10.4.3 the NSGA-II multi-objective GP (MOGP) algorithm (Deb et al. 2002; Fortin and Parizeau 2013) was used to drive selection according to performance against our three objectives of maximizing AUC and TPR while also minimizing FPR. When using this type of algorithm for problems where there is a natural tension between objectives which may necessitate trade-offs, the system typically does not return a single best individual at the end of evolution, but rather a Pareto front or range of individuals representing various levels of trade-off between the different objectives.

Due to the relationship between the main objectives for the mammography task, we choose to use the multi-objective algorithm. Preliminary experiments with various composite single objective fitness functions had not proved very successful and previous work (Ryan et al. 2014) had demonstrated the effectiveness of the MOGP approach. However, for this particular task, we are not interested in the pareto front of individuals, after all, a model with a zero FPR and zero or very low TPR (every instance classified as N) is not much use in this context. What we really care about is achieving the lowest possible FPR for the highest possible TPR. Thus, during evolution we maintained a single entry "hall of fame" (HOF) for each CV iteration, whereby as we evaluated each new individual on the training data, if it had a higher TPR or if it had an equal TPR but a lower FPR to the HOF incumbent for that CV iteration, the new individual replaced that HOF incumbent.

We report results on the training and test CV segments but the most important results are those for the HO test set, as these provide an indication of how the system might be expected to perform on new, unseen instances. We choose to present results, with best results in bold text, under several different model selection schemes:

- Mean average best trained individual: results for each HOF are firstly averaged for each CV run and then averaged across the 50 cross validated runs. See Table 10.4.
- Average best trained individual: the best trained individual is chosen from each CV run and the results for these 50 best individuals are averaged. See Table 10.5.
- Average best test individual: the best performing individual on the test dataset is chosen from amongst the 5 HOF members for each CV run and the results of this 50 individuals are averaged. See Table 10.6.
- Best overall trained individual: the single best trained individual. See Table 10.7.
- Best overall test individual: the single best individual on the CV test data chosen from amongst the 250 best *trained solutions*. See Table 10.8.

The results on training data shown in Tables 10.4, 10.5, 10.6, 10.7 and 10.8 show that simply adding features gives a boost to performance. The configuration with the greatest number of features (B2S6V1) consistently produces the lowest FPR and the best AUC score on the training data. However, this setup appears to suffer from over-fitting, as the excellent training results do not translate into good test results, as evidenced by the low TPR on the hold out test data. This configuration has the largest number of segments, and, as each added segment contributes two extra features—it also has more features than the others.

Regardless of which model selection approach we choose to adopt for evaluating performance on the hold out test data, the best evolved model is produced by the two breast non-segmented configuration (B2S0V1) which has a best result TPR of 1 with a very low FPR of 0.19. With the exception of B1S3V1, the single breast segmented setups perform worst overall.

Table 10.4 Mean average training, test and hold out TPR, FPR AUC of best trained individuals

Method	Train TPR	Train FPR	Train AUC	Test TPR	Test FPR	Test AUC	HO TPR	HO FPR	HO AUC
B1S0V1	1	0.60	0.78	0.92	0.63	0.73	1	0.66	0.76
B1S1V1	1	0.62	0.74	0.94	0.65	0.69	1	0.65	0.80
B1S2V2	1	0.72	0.71	0.97	0.74	0.68	0.97	0.72	0.70
B1S3V1	1	0.48	0.82	0.93	0.50	0.77	0.96	0.51	0.76
B2S0V1	1	0.49	0.81	0.92	0.54	0.75	1	**0.51**	**0.83**
B2S2V1	1	0.55	0.77	0.96	0.58	0.74	1	0.57	0.82
B2S3+0V1	1	0.57	0.76	0.93	0.59	0.73	0.96	0.55	0.77
B2S4+0V1	1	0.54	0.76	0.92	0.57	0.71	0.96	0.52	0.78
B2S4V1	1	0.48	0.82	0.92	0.52	0.76	0.97	0.52	0.78
B2S6V1	1	0.40	0.84	0.92	0.45	0.77	0.86	0.46	0.73

Table 10.5 Average best training, test and hold out TPR, FP AUC according to best *trained* individuals for each run

Method	Train TPR	Train FPR	Train AUC	Test TPR	Test FPR	Test AUC	HO TPR	HO FPR	HO AUC
B1S0V1	1	0.47	0.83	0.86	0.51	0.75	0.99	0.51	0.79
B1S1V1	1	0.53	0.77	0.90	0.56	0.69	1	0.54	0.85
B1S2V2	1	0.67	0.74	0.95	0.68	0.72	0.96	0.66	0.74
B1S3V1	1	0.37	0.86	0.90	0.39	0.81	0.95	0.41	0.80
B2S0V1	1	0.39	0.84	0.85	0.41	0.77	1	**0.38**	**0.86**
B2S2V1	1	0.48	0.79	0.93	0.51	0.75	0.99	0.48	**0.86**
B2S3+0V1	1	0.47	0.80	0.90	0.51	0.74	0.98	0.41	0.80
B2S4+0V1	1	0.45	0.80	0.88	0.48	0.72	0.96	0.43	0.82
B2S4V1	1	0.37	0.85	0.88	0.39	0.79	0.96	0.39	0.83
B2S6V1	1	0.32	0.87	0.86	0.38	0.76	0.82	0.36	0.76

Table 10.6 Average training, test and hold out TPR, FP AUC according to best *test* individuals for each run

Method	Train TPR	Train FPR	Train AUC	Test TPR	Test FPR	Test AUC	HO TP	HO FPR	HO AUC
B1S0V1	1	0.61	0.77	0.98	0.64	0.76	1	0.67	0.76
B1S1V1	1	0.62	0.73	1	0.65	0.71	1	0.66	0.79
B1S2V2	1	0.72	0.73	1	0.73	0.71	0.96	0.72	0.71
B1S3V1	1	0.49	0.82	1	0.48	0.84	0.96	0.50	0.77
B2S0V1	1	0.51	0.80	1	0.55	0.78	1	**0.51**	0.82
B2S2V1	1	0.55	0.78	1	0.57	0.77	1	0.55	**0.84**
B2S3+0V1	1	0.57	0.76	0.99	0.57	0.76	0.97	0.55	0.79
B2S4+0V1	1	0.56	0.76	0.99	0.56	0.77	0.98	0.52	0.80
B2S4V1	1	0.48	0.81	0.99	0.51	0.78	0.96	0.51	0.78
B2S6V1	1	0.42	0.83	1	0.44	0.82	0.89	0.45	0.75

Table 10.7 Training, test and hold out TPR, FPR AUC of single best trained individual

Method	Train TPR	Train FPR	Train AUC	Test TPR	Test FPR	Test AUC	HO TPR	HO FPR	HO AUC	Size
B1S0V1	1	0.30	0.84	0.80	0.38	0.77	0.75	0.29	0.80	687
B1S1V1	1	0.43	0.79	0.86	0.47	0.66	1	0.41	0.88	473
B1S2V2	1	0.61	0.78	0.89	0.62	0.73	1	0.61	0.77	537
B1S3V1	1	0.26	0.90	0.80	0.26	0.82	1	0.28	0.88	271
B2S0V1	1	0.31	0.89	0.79	0.27	0.87	1	**0.19**	**0.91**	979
B2S2V1	1	0.36	0.88	0.94	0.49	0.79	1	0.48	0.88	441
B2S3+0V1	1	0.37	0.85	0.92	0.63	0.70	1	0.58	0.78	211
B2S4+0V1	1	0.35	0.81	0.85	0.35	0.80	1	0.34	0.84	805
B2S4V1	1	0.27	0.88	0.80	0.29	0.82	1	0.30	0.89	341
B2S6V1	1	0.20	0.93	0.69	0.38	0.68	0.92	0.38	0.83	613

Table 10.8 Training, test and hold out TP, FP AUC of single best test individual

Method	Train TPR	Train FPR	Train AUC	Test TPR	Test FPR	Test AUC	HO TPR	HO FPR	HO AUC	Size
B1S0V1	1	0.56	0.78	1	0.37	0.88	1	0.44	**0.88**	193
B1S1V1	1	0.59	0.77	1	0.50	0.82	1	0.58	0.87	391
B1S2V2	1	0.67	0.73	1	0.66	0.76	0.95	0.66	0.74	401
B1S3V1	1	0.35	0.81	1	0.34	0.89	0.80	0.37	0.73	21
B2S0V1	1	0.43	0.87	1	0.38	0.83	1	**0.37**	0.83	341
B2S2V1	1	0.48	0.79	1	0.50	0.81	1	0.46	0.87	171
B2S3+0V1	1	0.44	0.75	1	0.39	0.81	1	0.39	0.82	509
B2S4+0V1	1	0.40	0.82	1	0.37	0.84	1	0.38	0.86	599
B2S4V1	1	0.36	0.85	1	0.36	0.85	1	0.38	0.79	419
B2S6V1	1	0.30	0.87	1	0.29	0.92	0.77	0.31	0.79	349

Both B2S3+0V1 and B2S4+0V1 are two breast configurations which combine features of segmented and unsegmented images. They are essentially combinations of B2S2V1 and B2S0V1—both of which deliver good results on the hold out test data. The augmented methods do not appear to contribute a huge improvement, although we see that for several views of the data, they produce a very competitive low FPR.

When we compare the figures for average program size between the single best individuals selected based on training or CV test partitions it is interesting to note that those selected on test performance are almost universally smaller than the ones selected based on training results. However, the best overall individual is the largest, at 979 nodes.

To compare with results from the literature we convert the FPRs into FPPI which we report in Table 10.9. Here, the average results reported result from the data in Table 10.4 which represents the mean average results for *all* of the best trained individuals. The best results use the TP and FP data of the single best individuals selected based on performance on CV test partitions. Results refer to performance on the crucial hold out test data.

Clearly the best results are produced by the two breast non-segmented approach B2S2V1 with a TPR of 1 and an FPPI of 0.33. This is closely followed by its single breast counterpart B1S0V1 which again delivered a perfect TPR and an FPP1 of 0.41.

Of the segmented setups the two augmented configurations of B2S3+0V1 and B2S4+0V1 also produced good results with perfect TPRs combined with good FPPIs of 1.11 and 1.08 respectively. Also the B2S4V1 method did very well with a TPR of 1 and FPP1 of 1.11. Contrast these figures with the results reported in Sect. 10.2.5 with scores of 97 % TP and FPPIs of 4–15.

Overall, several of our configurations proved capable of correctly classifying 100 % of the cancerous cases while at the same time having a low FPPI, and the best results were delivered by individuals trained to view breast asymmetry.

Table 10.9 Mean average TPR and FPPI of best trained individual, TPR and FPPI of single best trained individual, both on HO data

Method	Avg TPR	Avg FPPI	Best TPR	Best FPP1
B1S0V1	1	0.61	1	0.41
B1S1V1	1	1.88	1	1.68
B1S2V2	0.97	2.03	0.95	1.86
B1S3V1	0.96	1.49	0.80	1.08
B2S0V1	1	0.45	1	0.33
B2S2V1	1	1.67	1	1.34
B2S3+0V1	0.96	1.57	1	1.11
B2S4+0V1	0.96	1.48	1	1.08
B2S4V1	0.97	1.52	1	1.11
B2S6V1	0.86	1.28	0.77	1.06

10.6 Conclusions and Future Work

We have presented an entire work-flow for automated mammogram analysis with GP as its cornerstone. Our system operates with raw images, extracts the features and presents them to GP, which then evolves classifiers. The result is a Stage 1 cancer detector that achieves 100 % accuracy on unseen test data from the USF mammogram library, with a lowest reported FPPI of 0.33.

This work-flow can be applied to virtually any image classification task with GP, simply by using task-specific features. In this chapter we use textural features which can be directly employed by any problem with similar images, which means that the only modification required to the system is the manner in which the images are segmented.

The experimental set up that had the lowest FPPI was the one that compared both entire breasts, showing that we successfully leveraged textural breast asymmetry as a potential indicator for cancerous growths. Additionally, several of the segmented configurations also produced very good results, indicating that the system is capable of not only identifying with high accuracy which breasts are likely to have suspicious lesions but also which segments contain suspicious areas. The first of these capabilities could prove useful in providing second reader functionality to busy radiologists, whereas the second supply inputs to an automated diagnostic system where further analysis can be undertaken.

One minor limitation of this work is that all of the positive cases examined came from the same volume. However, it is reasonable to assume that for any automated system, a classifier will be generated for a specific type of X-ray machine used by a screening agency. Digital mammograms come in the DICOM (Whitcher et al. 2011) format which contains much meta-data, including the specific machine and location where the mammogram was taken. This means that it is feasible to produce machine-specific classifiers which are trained to deal with the particular idiosyncrasies of various machines. However, our next step will be to train the system across multiple volumes to test the impact on the TPR/FPPI scores.

Most prior work examines just a single step in the typical work-flow, i.e. the *classification* step, assuming the existence of previously selected features and concentrating on extracting the best possible results from those features. We are, however, positioned to leverage the ability of GP to produce solutions that are in some way human-readable, and treat GP as *part* of the work-flow, rather than the entire focus of the work. This means that, as the work progresses, we can create a feedback loop which examines the GP individuals to ascertain which terminals (features) are most useful, and extract more information related to those from the data. This is possible because data acquisition is also part of our work-flow; this is a system that accepts raw mammograms and outputs adjudicated segments.

GP is the essential element of the work-flow, as it is responsible for synthesizing the classifier by processing the previously selected and transformed image features. One of the recognised drawbacks of GP, compared with several other ML algorithms, is that the approach requires significantly longer training times. However, as the roles envisaged for the evolved GP classifiers involve providing second reader functionality and/or supplying input to a computer aided diagnosis module, both of which are off-line functions—the requirement for long training times is a non-issue.

For this study, feature selection was conducted through analysis of the initial Haralick features which dramatically reduced the number of features, and indicated that the contrast and sum of squares variance features were promising in terms of their potential to discriminate. Current work is examining the ways in which GP is combining these features and initial results are very positive; these results will then be compared with other ML systems, specifically SVMs and C4.5, to investigate the specific impact that GP has, particularly as the number of features increase.

Our segments are relatively large. While we were still able to maintain a 100 % TPR with them, there is a case to be made for examining smaller segments, as the smaller these are, the better it is for the later analysis stage.

Finally, although the Haralick textural measures are powerful, they are not the only features that have been used in image analysis. Our system also extracts Hu invariants (Hu 1962) and Local Binary Patterns (Ojala et al. 1994); we will use GP to combine these with the Haralick features to further decrease the FPPI.

Acknowledgements Krzysztof Krawiec acknowledges support from the Ministry of Science and Higher Education (Poland) grant 09/91/DSPB/0572. The remaining authors gratefully acknowledge the support of Science Foundation Ireland, grand number 10/IN.1/I3031.

References

Amir Atapour Abarghouei, Afshin Ghanizadeh, Saman Sinaie, and Siti Mariyam Shamsuddin. A survey of pattern recognition applications in cancer diagnosis. In *International Conference of Soft Computing and Pattern Recognition, SOCPAR '09*, pages 448–453, December 2009.

Davide Agnelli, Alessandro Bollini, and Luca Lombardi. Image classification: an evolutionary approach. *Pattern Recognition Letters*, 23(1–3):303–309, 2002.

Arbab Masood Ahmad, Gul Muhammad Khan, Sahibzada Ali Mahmud, and Julian Francis Miller. Breast cancer detection using cartesian genetic programming evolved artificial neural networks. In Terry Soule et al, editor, *GECCO '12: Proceedings of the fourteenth international conference on Genetic and evolutionary computation conference*, pages 1031–1038, Philadelphia, Pennsylvania, USA, 7–11 July 2012. ACM.

Rehan Akbani, Stephen Kwek, and Nathalie Japkowicz. Applying support vector machines to imbalanced datasets. In *Machine Learning: ECML 2004*, pages 39–50. Springer, 2004.

Harith Al-Sahaf, Andy Song, Kourosh Neshatian, and Mengjie Zhang. Extracting image features for classification by two-tier genetic programming. In Xiaodong Li, editor, *Proceedings of the 2012 IEEE Congress on Evolutionary Computation*, pages 1630–1637, Brisbane, Australia, 10–15 June 2012.

Harith Al-Sahaf, Mengjie Zhang, and Mark Johnston. Binary image classification using genetic programming based on local binary patterns. In *28th International Conference of Image and Vision Computing New Zealand (IVCNZ 2013)*, pages 220–225, Wellington, November 2013. IEEE Press.

Edén A Alanís-Reyes, José L Hernández-Cruz, Jesús S Cepeda, Camila Castro, Hugo Terashima-Marín, and Santiago E Conant-Pablos. Analysis of machine learning techniques applied to the classification of masses and microcalcification clusters in breast cancer computer-aided detection. *Journal of Cancer Therapy*, 3:1020, 2012.

Jun Ando and Tomoharu Nagao. Image classification and processing using modified parallel-ACTIT. In *IEEE International Conference on Systems, Man and Cybernetics, SMC 2009*, pages 1787–1791, October 2009.

David Andre. Automatically defined features: The simultaneous evolution of 2-dimensional feature detectors and an algorithm for using them. In Kenneth E. Kinnear, Jr., editor, *Advances in Genetic Programming*, chapter 23, pages 477–494. MIT Press, 1994.

I Anttinen, M Pamilo, M Soiva, and M Roiha. Double reading of mammography screening films-one radiologist or two? *Clinical Radiology*, 48(6):414–421, 1993.

Daniel Atkins, Kourosh Neshatian, and Mengjie Zhang. A domain independent genetic programming approach to automatic feature extraction for image classification. In Alice E. Smith, editor, *Proceedings of the 2011 IEEE Congress on Evolutionary Computation*, pages 238–245, New Orleans, USA, 5–8 June 2011. IEEE Computational Intelligence Society, IEEE Press.

Gustavo EAPA Batista, Ronaldo C Prati, and Maria Carolina Monard. A study of the behavior of several methods for balancing machine learning training data. *ACM Sigkdd Explorations Newsletter*, 6(1):20–29, 2004.

Radek Benes, Jan Karasek, Radim Burget, and Kamil Riha. Automatically designed machine vision system for the localization of CCA transverse section in ultrasound images. *Computer Methods and Programs in Biomedicine*, 109(1):92–103, 2013.

Karl A Benson. Evolving finite state machines with embedded genetic programming for automatic target detection within SAR imagery. In *Proceedings of the 2000 Congress on Evolutionary Computation CEC00*, pages 1543–1549, La Jolla Marriott Hotel La Jolla, California, USA, 6–9 July 2000. IEEE Press.

Leonard Berlin. Liability of interpreting too many radiographs. *American Journal of Roentgenology*, 175:17–22, 2000.

Mythreyi Bhargavan, Jonathan H. Sunshine, and Barbara Schepps. Too few radiologists? *American Journal of Roentgenology*, 178:1075–1082, 2002.

Keir Bovis and Sameer Singh. Detection of masses in mammograms using texture features. In *Pattern Recognition, 2000. Proceedings. 15th International Conference on*, volume 2, pages 267–270. IEEE, 2000.

NF Boyd, JW Byng, RA Jong, EK Fishell, LE Little, AB Miller, GA Lockwood, DL Tritchler, and Martin J Yaffe. Quantitative classification of mammographic densities and breast cancer risk: results from the canadian national breast screening study. *Journal of the National Cancer Institute*, 87(9):670–675, 1995.

Norman F Boyd, Gina A Lockwood, Jeff W Byng, David L Tritchler, and Martin J Yaffe. Mammographic densities and breast cancer risk. *Cancer Epidemiology Biomarkers & Prevention*, 7(12):1133–1144, 1998.

Behzad Bozorgtabar and Gholam Ali Rezai Rad. A genetic programming-PCA hybrid face recognition algorithm. *Journal of Signal and Information Processing*, 2(3):170–174, 2011.

G. M. Brake, N. Karssemeijer, and J. H. Hendricks. Automated detection of breast carcinomas not detected in a screening program. *Radiology*, 207:465–471, 1998.

Christopher D Brown and Herbert T Davis. Receiver operating characteristics curves and related decision measures: A tutorial. *Chemometrics and Intelligent Laboratory Systems*, 80(1):24–38, 2006.

Celia Byrne, Catherine Schairer, Louise A Brinton, John Wolfe, Navin Parekh, Martine Salane, Christine Carter, and Robert Hoover. Effects of mammographic density and benign breast disease on breast cancer risk (united states). *Cancer Causes & Control*, 12(2):103–110, 2001.

Nitesh V Chawla, Kevin W Bowyer, Lawrence O Hall, and W Philip Kegelmeyer. Smote: Synthetic minority over-sampling technique. *Journal of Artificial Intelligence Research*, 16:321–357, 2002.

Zheng Chen and Siwei Lu. A genetic programming approach for classification of textures based on wavelet analysis. In *IEEE International Symposium on Intelligent Signal Processing, WISP 2007*, pages 1–6, October 2007.

Heng-Da Cheng, Jingli Wang, and Xiangjun Shi. Microcalcification detection using fuzzy logic and scale space approaches. *Pattern Recognition*, 37(2):363–375, 2004.

Sunil Cho, Sung Ho Jin, Ju Won Kwon, Yong Man Ro, and Sung Min Kim. Microcalcification detection system in digital mammogram using two-layer svm. In *Electronic Imaging 2008*, pages 68121I–68121I. International Society for Optics and Photonics, 2008.

S Ciatto, D Ambrogetti, R Bonardi, S Catarzi, G Risso, M Rosselli Del Turco, and P Mantellini. Second reading of screening mammograms increases cancer detection and recall rates. results in the florence screening programme. *Journal of medical screening*, 12(2):103–106, 2005.

Tommy E. Cupples, Joan E. Cunningham, and James C. Reynolds. Impact of computer-aided detection in a regional screening mammography program. *American Journal of Roentgenology*, 186:944–950, 2005.

Kalyanmoy Deb, Amrit Pratap, Sameer Agarwal, and TAMT Meyarivan. A fast and elitist multiobjective genetic algorithm: Nsga-ii. *Evolutionary Computation, IEEE Transactions on*, 6(2):182–197, 2002.

F Dehghan, H Abrishami-Moghaddam, and M Giti. Automatic detection of clustered microcalcifications in digital mammograms: Study on applying adaboost with svm-based component classifiers. In *Engineering in Medicine and Biology Society, 2008. EMBS 2008. 30th Annual International Conference of the IEEE*, pages 4789–4792. IEEE, 2008.

Atam P Dhawan, Yateen Chitre, and Christine Kaiser-Bonasso. Analysis of mammographic microcalcifications using gray-level image structure features. *Medical Imaging, IEEE Transactions on*, 15(3):246–259, 1996.

Pedro Domingos. Metacost: A general method for making classifiers cost-sensitive. In *Proceedings of the fifth ACM SIGKDD international conference on Knowledge discovery and data mining*, pages 155–164. ACM, 1999.

Julie A Douglas, Marie-Hélène Roy-Gagnon, Chuan Zhou, Braxton D Mitchell, Alan R Shuldiner, Heang-Ping Chan, and Mark A Helvie. Mammographic breast density evidence for genetic correlations with established breast cancer risk factors. *Cancer Epidemiology Biomarkers & Prevention*, 17(12):3509–3516, 2008.

Wei Fan, Salvatore J Stolfo, Junxin Zhang, and Philip K Chan. Adacost: misclassification cost-sensitive boosting. In *ICML*, pages 97–105. Citeseer, 1999.

Jeannie Fitzgerald and Conor Ryan. Exploring boundaries: optimising individual class boundaries for binary classification problem. In *Proceedings of the fourteenth international conference on Genetic and evolutionary computation conference*, GECCO '12, pages 743–750, New York, NY, USA, 2012. ACM.

Jeannie Fitzgerald and Conor Ryan. A hybrid approach to the problem of class imbalance. In *International Conference on Soft Computing*, Brno, Czech Republic, June 2013.

Félix-Antoine Fortin and Marc Parizeau. Revisiting the nsga-ii crowding-distance computation. In *Proceedings of the 15th Annual Conference on Genetic and Evolutionary Computation*, GECCO '13, pages 623–630, New York, NY, USA, 2013. ACM.

Yoav Freund and Robert E. Schapire. Experiments with a new boosting algorithm. In *Thirteenth International Conference on Machine Learning*, pages 148–156, San Francisco, 1996. Morgan Kaufmann.

Wenlong Fu, Mark Johnston, and Mengjie Zhang. Low-level feature extraction for edge detection using genetic programming. *IEEE Transactions on Cybernetics*, 44(8):1459–1472, 2014.

Ganesan, K. et al. Decision support system for breast cancer detection using mammograms. *Proceedings of the Institution of Mechanical Engineers, Part H: Journal of Engineering in Medicine*, 227(7):721–732, January 2013.

Seymour Geisser. *Predictive Inference*. Chapman and Hall, New York, NY, 1993.

Mark Hall, Eibe Frank, Geoffrey Holmes, Bernhard Pfahringer, Peter Reutemann, and Ian H. Witten. The weka data mining software: an update. *SIGKDD Explor. Newsl.*, 11(1):10–18, November 2009.

R. et al Haralick. Texture features for image classification. *IEEE Transactions on Systems, Man, and Cybernetics*, 3(6), 1973.

Trevor Hastie, Robert Tibshirani, Jerome Friedman, T Hastie, J Friedman, and R Tibshirani. *The elements of statistical learning*, volume 2. Springer, 2009.

Haibo He and Edwardo A Garcia. Learning from imbalanced data. *Knowledge and Data Engineering, IEEE Transactions on*, 21(9):1263–1284, 2009.

Michael Heath, Kevin Bowyer, Daniel Kopans, Richard Moore, and W. Philip Kegelmeyer. The digital database for screening mammography. In M.J. Yaffe, editor, *Proceedings of the Fifth International Workshop on Digital Mammography*, pages 212–218. Medical Physics Publishing, 2001.

Rolando R Hernández-Cisneros, Hugo Terashima-Marín, and Santiago E Conant-Pablos. Comparison of class separability, forward sequential search and genetic algorithms for feature selection in the classification of individual and clustered microcalcifications in digital mammograms. In *Image Analysis and Recognition*, pages 911–922. Springer, 2007.

Daniel Howard, Simon C. Roberts, and Conor Ryan. The boru data crawler for object detection tasks in machine vision. In Stefano Cagnoni, Jens Gottlieb, Emma Hart, Martin Middendorf, and G"unther Raidl, editors, *Applications of Evolutionary Computing, Proceedings of EvoWorkshops2002: EvoCOP, EvoIASP, EvoSTim/EvoPLAN*, volume 2279 of *LNCS*, pages 222–232, Kinsale, Ireland, 3–4 April 2002. Springer-Verlag.

Daniel Howard, Simon C. Roberts, and Conor Ryan. Pragmatic genetic programming strategy for the problem of vehicle detection in airborne reconnaissance. *Pattern Recognition Letters*, 27(11):1275–1288, August 2006. Evolutionary Computer Vision and Image Understanding.

MK Hu. Visual pattern recognition by moment invariants. *Trans. Info. Theory,*, IT-8:179–187, 1962.

M. Hughes and S. Jacobzone. Ageing-related diseases project: Comparing treatments, costs and outcomes for breast cancer in oecd countries. *OECD Health Working Papers*, 2003.

Satoru Isaka. An empirical study of facial image feature extraction by genetic programming. In John R. Koza, editor, *Late Breaking Papers at the 1997 Genetic Programming Conference*, pages 93–99, Stanford University, CA, USA, 13–16 July 1997. Stanford Bookstore.

Mahesh V Joshi, Vipin Kumar, and Ramesh C Agarwal. Evaluating boosting algorithms to classify rare classes: Comparison and improvements. In *Data Mining, 2001. ICDM 2001, Proceedings IEEE International Conference on*, pages 257–264. IEEE, 2001.

Karla Kerlikowske, Deborah Grady, John Barclay, Edward A. Sickles, Abigail Eaton, and Virginia Ernster. Positive predictive value of screening mammography by age and family history of breast cancer. *Journal of the American Medical Association*, 270:2444–2450, 1993.

N Kharma, Taras Kowaliw, E Clement, Chris Jensen, A Youssef, and Jie Yao. Project cellnet: Evolving an autonomous pattern recognizer. *International Journal of Pattern Recognition and Artificial Intelligence*, 18(06):1039–1056, 2004.

Maciej Komosiński and Krzysztof Krawiec. Evolutionary weighting of image features for diagnosing of cns tumors. *Artificial Intelligence in Medicine*, 19(1):25–38, 2000.

Daniel B. Kopans. The most recent breast cancer screening controversy about whether mammographic screening benefits women at any age:nonsense and nonscience. *American Journal of Roentgenology*, 180:21–26, 2003.

Sotiris Kotsiantis, Dimitris Kanellopoulos, Panayiotis Pintelas, et al. Handling imbalanced datasets: A review. *GESTS International Transactions on Computer Science and Engineering*, 30(1):25–36, 2006.

Taras Kowaliw, Wolfgang Banzhaf, Nawwaf Kharma, and Simon Harding. Evolving novel image features using genetic programming-based image transforms. In *Evolutionary Computation, 2009. CEC'09. IEEE Congress on*, pages 2502–2507. IEEE, 2009.

J. Koza. Genetic programming: A paradigm for genetically breeding populations of computer programs to solve problems. Technical Report STAN-CS-90-1314, Dept. of Computer Science, Stanford University, June 1990.

John R. Koza. Simultaneous discovery of detectors and a way of using the detectors via genetic programming. In *1993 IEEE International Conference on Neural Networks*, volume III, pages 1794–1801, San Francisco, USA, 1993. IEEE.

Krzysztof Krawiec. Genetic programming-based construction of features for machine learning and knowledge discovery tasks. *Genetic Programming and Evolvable Machines*, 3(4):329–343, 2002.

Krzysztof Krawiec. Genetic programming with alternative search drivers for detection of retinal blood vessels. In *Applications of Evolutionary Computation*, Lecture Notes in Computer Science, Copenhagen, Denmark, 2015. Springer.

Krzysztof Krawiec, Daniel Howard, and Mengjie Zhang. Overview of object detection and image analysis by means of genetic programming techniques. In *Frontiers in the Convergence of Bioscience and Information Technologies, 2007. FBIT 2007*, pages 779–784. IEEE, 2007.

Miroslav Kubat, Stan Matwin, et al. Addressing the curse of imbalanced training sets: one-sided selection. In *ICML*, volume 97, pages 179–186. Nashville, USA, 1997.

Solomon Kullback and Richard A Leibler. On information and sufficiency. *The Annals of Mathematical Statistics*, pages 79–86, 1951.

S. A. Feig L. J. Warren Burhenne, C. J. D'Orsi. The potential contribution to computer-aided detection to the sensitivity of screening mammography. *Radiology*, 215:554–562, 2002.

Brian Lam and Vic Ciesielski. Discovery of human-competitive image texture feature extraction using genetic programming. In K. Debs et al., editor, *LNCS*, volume 3103, pages 1114–1125. GECCO, Springer-Verlag Berlin Heidelberg, 2004.

W.B. Langdon and A.P. Harrison. Gp on spmd parallel graphics hardware for mega bioinformatics data mining. *Soft Computing*, 12(12):1169–1183, 2008.

William B Langdon, Marc Modat, Justyna Petke, and Mark Harman. Improving 3d medical image registration cuda software with genetic programming. In *Proceedings of the 2014 conference on Genetic and evolutionary computation*, pages 951–958. ACM, 2014.

Vincenzo Lattanzio, C Di Maggio, and G Simonetti. *Mammography: Guide to Interpreting, Reporting and Auditing Mammographic Images-Re. Co. RM (From Italian Reporting and Codifying the Results of Mammography)*. Springer Science & Business Media, 2010.

Huai Li, Yue Wang, KJ Ray Liu, S-CB Lo, and Matthew T Freedman. Computerized radiographic mass detection. i. lesion site selection by morphological enhancement and contextual segmentation. *Medical Imaging, IEEE Transactions on*, 20(4):289–301, 2001.

Yi Liu and Taghi Khoshgoftaar. Reducing overfitting in genetic programming models for software quality classification. In *Proceedings of the Eighth IEEE Symposium on International High Assurance Systems Engineering*, pages 56–65, Tampa, Florida, USA, 25–26 March 2004.

Simone A. Ludwig and Stefanie Roos. Prognosis of breast cancer using genetic programming. In Rossitza Setchi, Ivan Jordanov, Robert J. Howlett, and Lakhmi C. Jain, editors, *14th International Conference on Knowledge-Based and Intelligent Information and Engineering Systems (KES 2010), Part IV*, volume 6279 of *LNCS*, pages 536–545, Cardiff, UK, September 8–10 2010. Springer.

M Markey M. Sampat and A Bovik. Computer-aided detection and diagnosis in mammography. In Alan C. Bovik, editor, *Handbook of Image and Video Processing*. Elsevier Academic Press, 2010.

MATLAB. *version 8.2 (R2012a)*. MathWorks Inc., Natick, MA, 2013.

Valerie A McCormack and Isabel dos Santos Silva. Breast density and parenchymal patterns as markers of breast cancer risk: a meta-analysis. *Cancer Epidemiology Biomarkers & Prevention*, 15(6):1159–1169, 2006.

Naga R Mudigonda, Rangaraj M Rangayyan, and JE Leo Desautels. Gradient and texture analysis for the classification of mammographic masses. *Medical Imaging, IEEE Transactions on*, 19(10):1032–1043, 2000.

R. J. Nandi, A. K. Nandi, R. Rangayyan, and D. Scutt. Genetic programming and feature selection for classification of breast masses in mammograms. In *28th Annual International Conference of the IEEE Engineering in Medicine and Biology Society, EMBS '06*, pages 3021–3024, New York, USA, August 2006. IEEE.

R. W. Nishikawa, K. Doi, and M. L. Geiger. Computerised detection of clustered microcalcifications: evaluation of performance on mammograms from multiple centers. *Radiographics*, 15:445–452, 1995.

American College of Radiology. *ACR BIRADS Mammography, Ultrasound & MRI, 4th ed.* American College of Radiology, Reston VA, 2003.

T. Ojala, M. Pietikäinen, and D. Harwood. Performance evaluation of texture measures with classification based on kullback discrimination of distributions. In *Proceedings of the 12th IAPR International Conference on Pattern Recognition (ICPR 1994)*, pages 582–585. IEEE, 1994.

Arnau Oliver, Xavier Lladó, Jordi Freixenet, and Joan Martí. False positive reduction in mammographic mass detection using local binary patterns. In *Medical Image Computing and Computer-Assisted Intervention–MICCAI 2007*, pages 286–293. Springer, 2007.

Chandrashekhar Padole and Joanne Athaide. Automatic eye detection in face images for unconstrained biometrics using genetic programming. In Bijaya Ketan Panigrahi, Ponnuthurai Nagaratnam Suganthan, Swagatam Das, and Subhransu Sekhar Dash, editors, *Proceedings of the 4th International Conference on Swarm, Evolutionary, and Memetic Computing (SEMCCO 2013), Part II*, volume 8298 of *Lecture Notes in Computer Science*, pages 364–375, Chennai, India, December 19–21 2013. Springer.

Athanasios Papadopoulos, Dimitrios I Fotiadis, and Aristidis Likas. Characterization of clustered microcalcifications in digitized mammograms using neural networks and support vector machines. *Artificial Intelligence in Medicine*, 34(2):141–150, 2005.

F. Pedregosa, G. Varoquaux, A. Gramfort, V. Michel, B. Thirion, O. Grisel, M. Blondel, P. Prettenhofer, R. Weiss, V. Dubourg, J. Vanderplas, A. Passos, D. Cournapeau, M. Brucher, M. Perrot, and E. Duchesnay. Scikit-learn: Machine learning in Python. *Journal of Machine Learning Research*, 12:2825–2830, 2011.

Nicholas Petrick, Berkman Sahiner, Samuel G Armato III, Alberto Bert, Loredana Correale, Silvia Delsanto, Matthew T Freedman, David Fryd, David Gur, Lubomir Hadjiiski, et al. Evaluation of computer-aided detection and diagnosis systemsa). *Medical physics*, 40(8):087001, 2013.

Etta D Pisano, Constantine Gatsonis, Edward Hendrick, Martin Yaffe, Janet K Baum, Suddhasatta Acharyya, Emily F Conant, Laurie L Fajardo, Lawrence Bassett, Carl D'Orsi, et al. Diagnostic performance of digital versus film mammography for breast-cancer screening. *New England Journal of Medicine*, 353(17):1773–1783, 2005.

W. E. Polakowski, D. A. Cournoyer, and S. K. Rogers. Computer-aided breast cancer detection and diagnosis of masses using difference of gaussians and derivative-based feature saliency,. *IEEE Trans. Med. Imag.*, 16:811–819, 1997.

Riccardo Poli. Genetic programming for image analysis. In *Proceedings of the First Annual Conference on Genetic Programming*, pages 363–368. MIT Press, 1996.

Rangaraj M Rangayyan, Nema M El-Faramawy, JE Leo Desautels, and Onsy Abdel Alim. Measures of acutance and shape for classification of breast tumors. *Medical Imaging, IEEE Transactions on*, 16(6):799–810, 1997.

Adam G Riess, Alexei V Filippenko, Peter Challis, Alejandro Clocchiatti, Alan Diercks, Peter M Garnavich, Ron L Gilliland, Craig J Hogan, Saurabh Jha, Robert P Kirshner, et al. Observational evidence from supernovae for an accelerating universe and a cosmological constant. *The Astronomical Journal*, 116(3):1009, 1998.

Mateen M. Rizki, Michael A. Zmuda, and Louis A. Tamburino. Evolving pattern recognition systems. *Evolutionary Computation, IEEE Transactions on*, 6(6):594–609, 2002.

Gerald Robinson and Paul McIlroy. Exploring some commercial applications of genetic programming. Project 4487, British Telecom, Systems Research Division, Martelsham, Ipswitch, UK, 9/3/95 1995.

Conor Ryan, Krzysztof Krawiec, Una-May O'Reilly, Jeannie Fitzgerald, and David Medernach. Building a stage 1 computer aided detector for breast cancer using genetic programming. In M. Nicolau, K. Krawiec, M. I. Heywood, M. Castelli, P. Garci-Sanchez, J. J. Merelo, V. M. R. Santos, and K. Sim, editors, *17th European Conference on Genetic Programming*, volume 8599 of *LNCS*, pages 162–173, Granada, Spain, 23–25 April 2014. Springer.

Conor Ryan, Krzysztof Krawiec, Una-May O'Reilly, Jeannie Fitzgerald, and David Medernach. Building a stage 1 computer aided detector for breast cancer using genetic programming. In *Genetic Programming*, pages 162–173. Springer, 2014.

Berkman Sahiner, Heang-Ping Chan, Datong Wei, Nicholas Petrick, Mark A Helvie, Dorit D Adler, and Mitchell M Goodsitt. Image feature selection by a genetic algorithm: Application to classification of mass and normal breast tissue. *Medical Physics*, 23(10):1671–1684, 1996.

E Sakka, A Prentza, and D Koutsouris. Classification algorithms for microcalcifications in mammograms (review). *Oncology reports*, 15(4):1049–1055, 2006.

Gonzalo V Sánchez-Ferrero and Juan Ignacio Arribas. A statistical-genetic algorithm to select the most significant features in mammograms. In *Computer Analysis of Images and Patterns*, pages 189–196. Springer, 2007.

Ling Shao, Li Liu, and Xuelong Li. Feature learning for image classification via multiobjective genetic programming. *IEEE Transactions on Neural Networks and Learning Systems*, 25(7):1359–1371, July 2014.

Robert A Smith, Stephen W Duffy, and László Tabár. Breast cancer screening: the evolving evidence. *Oncology*, 26(5):471–475, 2012.

Hamid Soltanian-Zadeh, Farshid Rafiee-Rad, and Siamak Pourabdollah-Nejad D. Comparison of multiwavelet, wavelet, haralick, and shape features for microcalcification classification in mammograms. *Pattern Recognition*, 37(10):1973–1986, 2004.

Andy Song, Vic Ciesielski, and Hugh E Williams. Texture classifiers generated by genetic programming. In *Computational Intelligence, Proceedings of the World on Congress on*, volume 1, pages 243–248. IEEE, 2002.

Paul Stober and Shi-Tao Yeh. An explicit functional form specification approach to estimate the area under a receiver operating characteristic (roc) curve. *Avaialable at, http://www2.sas.com/proceedings/sugi27/p226--227.pdf, Accessed March*, 7, 2007.

R Strickland and Hee Il Hahn. Wavelet transforms for detecting microcalcifications in mammograms. *Medical Imaging, IEEE Transactions on*, 15(2):218–229, 1996.

Tabar, L. et al. A new era in the diagnosis of breast cancer. *Surgical oncology clinics of North America*, 9(2):233–77, April 2000.

Walter Alden Tackett. Genetic programming for feature discovery and image discrimination. In Stephanie Forrest, editor, *Proceedings of the 5th International Conference on Genetic Algorithms, ICGA-93*, pages 303–309, University of Illinois at Urbana-Champaign, 17–21 July 1993. Morgan Kaufmann.

FJ Tan, MT Morgan, LI Ludas, JC Forrest, and DE Gerrard. Assessment of fresh pork color with color machine vision. *Journal of animal science*, 78(12):3078–3085, 2000.

Erik L Thurfjell, K Anders Lernevall, and AA Taube. Benefit of independent double reading in a population-based mammography screening program. *Radiology*, 191(1):241–244, 1994.

T. Tot, L. Tabar, and P. B. Dean. The pressing need for better histologic-mammographic correlation of the many variations in normal breast anatomy. *Virchows Archiv*, 437(4):338–344, October 2000.

Leonardo Trujillo and Gustavo Olague. Synthesis of interest point detectors through genetic programming. In *Proceedings of the 8th annual conference on Genetic and evolutionary computation*, pages 887–894. ACM, 2006.

Tinne Tuytelaars and Krystian Mikolajczyk. Local invariant feature detectors: a survey. *Foundations and Trends® in Computer Graphics and Vision*, 3(3):177–280, 2008.

Brijesh Verma and Ping Zhang. A novel neural-genetic algorithm to find the most significant combination of features in digital mammograms. *Applied Soft Computing*, 7(2):612–625, 2007.

Katharina Volk, Julian Miller, and Stephen Smith. Multiple network CGP for the classification of mammograms. In Mario Giacobini et al, editor, *Applications of Evolutionary Computing, EvoWorkshops2009.*, volume 5484 of *LNCS*, pages 405–413, Tubingen, Germany, 15–17 April 2009. Springer Verlag.

C. J. Vyborny. Can computers help radiologists read mammograms? *Radiology*, 191:315–317, 1994.

Brandon Whitcher, Volker J. Schmid, and Andrew Thornton. Working with the DICOM and NIfTI data standards in R. *Journal of Statistical Software*, 44(6):1–28, 2011.

F. Winsberg, M. Elkin, J. Macy, V. Bordaz, and W. Weymout. Detection of radiographic abnormalities in mammograms by means of optical scanning and computer analysis. *Radiology*, 89:211–215, 1967.

Brent J Woods. *Computer-aided detection of malignant lesions in dynamic contrast enhanced MRI breast and prostate cancer datasets*. PhD thesis, The Ohio State University, 2008.

William P Worzel, Jianjun Yu, Arpit A Almal, and Arul M Chinnaiyan. Applications of genetic programming in cancer research. *The international journal of biochemistry & cell biology*, 41(2):405–413, 2009.

Zhi Qing Wu, Jianmin Jiang, and YH Peng. Effective features based on normal linear structures for detecting microcalcifications in mammograms. In *Pattern Recognition, 2008. ICPR 2008. 19th International Conference on*, pages 1–4. IEEE, 2008.

Cheng Xie and Lin Shang. Anomaly detection in crowded scenes using genetic programming. In Carlos A. Coello Coello, editor, *Proceedings of the 2014 IEEE Congress on Evolutionary Computation*, pages 1832–1839, Beijing, China, 6–11 July 2014.

Mengjie Zhang, Victor B Ciesielski, and Peter Andreae. A domain-independent window approach to multiclass object detection using genetic programming. *Genetic and Evolutionary Computation for Signal Processing and Image Analysis*, 8:841–859, 2003.

Mengjie Zhang, Mario Koeppen, and Sergio Damas. Special issue on computational intelligence in computer vision and image processing. *IEEE Computational Intelligence Magazine*, 8(1): 14–15, February 2013. Guest Editorial.

Mengjie Zhang and Malcolm Lett. Genetic programming for object detection: Improving fitness functions and optimising training data. *The IEEE Intelligent Informatics Bulletin*, 7(1):12–21, December 2006.

Bin Zheng, Yuan-Hsiang Chang, Xiao-Hui Wang, Walter F Good, and David Gur. Feature selection for computerized mass detection in digitized mammograms by using a genetic algorithm. *Academic radiology*, 6(6):327–332, 1999.

Mark H Zweig and Gregory Campbell. Receiver-operating characteristic (roc) plots: a fundamental evaluation tool in clinical medicine. *Clinical chemistry*, 39(4):561–577, 1993.

Chapter 11
On the Application of Genetic Programming for New Generation of Ground Motion Prediction Equations

Mehdi Mousavi, Alireza Azarbakht, Sahar Rahpeyma, and Ali Farhadi

11.1 An Introduction to Ground Motion Prediction Equation

Seismic hazard analyses are traditionally used to estimate ground motion intensities in a specific site for design purposes. These can be done through two different approaches i.e., Deterministic Seismic Hazard Analysis (DSHA) and Probabilistic Seismic Hazard Analysis (PSHA). In the first method uncertainties are neglected to identify a "worst-case" ground motion, whereas the second approach considers all possible earthquake events to estimate ground shaking levels associated with its probability of exceedance (McGuire 1995). However, it should be emphasized that estimating strong ground motion intensities, based on the seismological parameters such as: earthquake moment magnitude and distance from earthquake rupture zone to the site of interest, is definitely required for using both of the abovementioned methodologies. Generally, Ground Motion Prediction Equations (GMPEs), known also as attenuation relationships, are used in order to approximately calculate the intensity of strong ground motions. It has been proved that the attenuation of earthquake ground motions is a key element in estimating ground motion parameters for assessment and design purposes. GMPEs are analytical expressions describing how energy of seismic waves attenuates during propagation. It is noteworthy that GMPEs permit the estimation of both the ground shaking level at a site

Electronic supplementary material The online version of this chapter (doi: 10.1007/978-3-319-20883-1_11) contains supplementary material, which is available to authorized users.

M. Mousavi (✉) • A. Azarbakht • A. Farhadi
Arak University, Arak, Iran
e-mail: m-mousavi@araku.ac.ir

S. Rahpeyma
Earthquake Engineering Research Center, University of Iceland, Reykjavík, Iceland

A.H. Gandomi et al. (eds.), *Handbook of Genetic Programming Applications*,
DOI 10.1007/978-3-319-20883-1_11

from a specific earthquake and the uncertainty associated with the prediction. This predicted value is one of the most potent elements in any probabilistic and deterministic Seismic Hazard Analysis (SHA).

During recent decades, multitude numbers of GMPEs have been developed by different researchers. The most common intensity measures (IMs) used in GMPEs are PGA, PGV, PGD, and SA. GMPEs for such intensity measures have been incorporated into PSHA correspondences (McGuire 2004; OpenSHA 2009) are used to employ in PSHA or other structural earthquake engineering (Petersen et al. 2008) for specific sites.

During recent decades, the number of strong motion recording equipment has dramatically proliferate which provided more complete and reliable data sets for different seismic regions. In general, the functional form of ground motion attenuation relationships is as follows:

$$\log(Y) = \log(b_1) + \log[f_1(M)] + \log[f_2(R)] + \log[f_3(M, R)] + \log[f_4(E_i)] + \log(\varepsilon) \quad (11.1)$$

where Y is the ground motion intensity measure, and b_1 is a scaling factor. The terms f_1 to f_4 are functions of earthquake magnitude M, distance to the site R, and other influential parameters. The term E denotes uncertainties associated with predicted values. Equation (11.1) is an additive function based on the model for ground motion regression equations defined by Campbell (1985). It is suggested that the fault type and the focal depth should be taken into account for the source characteristics (McGarr 1984; Youngs et al. 1997). In addition, for the path effects, the appropriate definition of the distance in the near-source area is in controversy (Campbell 1985). Fourth, for the site effects, it is also suggested that the qualitative evaluation using soil types is not sufficient (Fukushima and Tanaka 1990; Midorikawa et al. 1994). The abovementioned equation provides logarithmic value of the ground motion intensity which has a log normal distribution.

Worldwide, there are a vast number of reviews of attenuation studies which provide beneficial information regarding the methods for obtaining GMPEs and their results (e.g. Trifunac and Brady 1976; Idriss 1978; Boore and Joyner 1982; Campbell 1985; Joyner and Boore 1988, 1996; Ambraseys and Bommer 1995; Power et al. 2008). A comprehensive global summary of GMPEs was given by J. Douglas in 2011, which reviews all empirical models for the prediction of PGA and elastic-response spectral ordinates between 1964 and 2010 (Douglas 2011). The growing quantity and quality of ground-motion information on recordings, in different databases, have resulted in numerous regional and worldwide GMPEs and continue reviewed and updated attenuation relationships through recent decades. Furthermore, it is worth to mention that a few studies concerned with relating ground-motion parameters to an intensity scale compared with magnitude based GMPEs. These studies were conducted in the latter part of the twentieth century (Elnashai and Di Sarno 2008); however, nowadays, some of researches put a great effort in order to use as much as available seismic parameters into the relationships (Abrahamson et al. 2013).

11.2 Different Approaches for Driving GMPEs

As it is aforementioned, GMPEs are the key element in any seismic hazard analysis. Two basic different methodologies have been developed by different researchers, i.e. empirical and physical relationships, for attaining prediction equation models according to the site geology and distribution of events. The first method, known as empirical models, describes the observations by means of regression analysis on a specific site with abundant data set. It should be noted that this type of approach requires a vast number of data set in order to achieve the best predictive trend between Ground Motion Records (GMRs); however, these types of models simply describe observed information and do not necessarily reflect the realistic process of an earthquake. Obviously, obtaining a valid empirical model requires availability of a vast collection of data. Consequently, empirical GMPEs have been derived and effectively applied in specific regions where abundant GMRs were available, such as California and Japan. The main drawback of this method is its inextricable sensitivity to, and dependency on, the data sets. Also, the reliability and robustness of these empirical predictions depend largely on how data are classified and what regression methods were applied. Even in regions with abundant data, enrichment of databases could lead to large changes in the model predictions. In addition, physical models, which describe seismic wave's generation and propagation, are used in a specific site with lack of observations. In physical relationships sparse data are used to calibrate a model which describes realistic process of an earthquake. Physical models have been usually derived through stochastic manner and based on random vibration theory, as a result, further information can be obtained from studies concerned with the applied methodology and basic theory (see e.g. Papageorgiou and Aki 1983 and references therein). More advanced physical models, which try to model the realistic process of faulting through numerical modeling of fault rupture model and wave propagation, have also been developed (e.g. Komatitsch et al. 2004; Krishnan et al. 2006). Apparently, physical GMPEs can only be applied in specific regions where comprehensive information about soil structure and faulting mechanisms are available, such as Los Angeles Basin.

The objective of having computers automatically solve problems is central to artificial intelligence, machine learning, and the broad area encompassed by what called 'machine intelligence'. Arthur Samuel as the pioneer in the area of machine learning in his 1983 talk entitled 'AI: Where It Has Been and Where It Is Going' (Samuel 1983), stated that the main goal of the fields of machine learning and artificial intelligence is:

> To get machines to exhibit behavior, which if done by humans, would be assumed to involve the use of intelligence.

There are various parameters both known and unknown affecting induced hazard of a specific event. As a result, developing a valid physical model capable to describe such a complex behavior may not be possible with traditional engineering methodologies. On the other hand, Artificial Intelligence (AI) is becoming more popular and particularly amenable to model the complex behavior of most geotechnical engineering materials because of its superiority over conventional methods in predicting more accurate outputs (Fister et al. 2014; Yang et al. 2013; Gandomi and Alavi 2012). Among the available AI techniques are artificial neural networks (ANNs), Genetic programming (GP), Evolutionary polynomial regression (EPR), support vector machines, M5 model trees, and K-nearest neighbors (Elshorbagy et al. 2010). Utilization of ANNs in deriving GMPEs is shown in different studies (e.g. Alavi and Gandomi 2011a, b; Gandomi et al. 2015).

Genetic programming is an Evolutionary Computation (EC) technique that automatically solves problems without having to tell the computer explicitly how to do it. At the most abstract level, GP is a systematic, dominion dependent method for getting computers to automatically solve problems starting from a high-level statement of what needs to be done (Langdon et al. 2008).

During recent decades, beside the two mentioned traditional approaches, methods of information processing known as soft computing techniques, such as Evolutionary Algorithms (EA), have been used in order to obtain attenuation relationships as a modern and beneficial approach. Nowadays, one of the most practical methods in this area is GP which incredibly affects the methodologies for obtaining the new GMPEs. EA algorithms, specifically genetic programming and genetic algorithm (GA), are optimization techniques based on the rules of natural selection (Koza 1992; Goldberg 1989). It has been proved that although using GP methods does not reduce the uncertainties; however, there is more complicated interaction among the observation and prediction values (Cabalar and Cevik 2009). It has been proved by different researchers (e.g. Rahpeyma et al. 2013; Mohammadnejad et al. 2012; Alavi et al. 2011a, b; Gandomi et al. 2011; Cabalar and Cevik 2009) that GP approach can be successfully applied to derive new GMPEs comparable with the previous attenuation relationships. In this chapter, an introduction to the genetic programming assumptions is presented and some of the mentioned studies beside mathematical tools for applying the GP programs are discussed comprehensively.

11.3 A Brief Introduction to the Genetic Programming

In artificial intelligence, Genetic Programming (GP) is an Evolutionary Algorithm-based (EA) methodology inspired by biological evolution and principles of Darwinian natural selection. It tries to find computer programs that perform a user-defined task which was introduced by John Koza (1992). It should be recalled that GP firstly was used by Friedberg (1958) to put a program in progress in stepwise manner. Much later, Cramer (1985) applied Gas and tree-like structures to evolve programs. In the latter part of 1980s experiments of Koza (1992) resulted in extension of GA as the major breakthrough in the area of GP (Elshorbagy et al. 2010).

Basically, GP is a collection of instructions and a fitness function for evaluating performance of a computer in doing a specific task. It should be emphasized that GP is a specialization of GA where each individual is a computer program. GP develops computer programs, traditionally represented in memory as tree structures (parse tree) (Koza 1992; Alavi et al. 2011a, b). Trees can be easily evaluated in a recursive manner. Every tree node has an operator function and every terminal node has an operand, making mathematical expressions easy to evolve and evaluate. GP and GA are different when it comes to the evolving programs (individuals), individuals in GP are parse trees instead of fixed length binary strings. GA as a traditional optimization technique often used to obtain best values for a given set of input parameters of a model. However, GP is a machine learning technique used to optimize a population of computer programs regarding a fitness landscape determined by capability of a program to carry out specific computational task. On the other hand, it gives the basic structure of the approximation model together with the values of its parameters. In GP structure, a random population of individuals (parse trees) is created to achieve high diversity and include as much as possible multiplicity of individuals. The symbolic optimization algorithms present the potential solutions by structural ordering of several symbols (Torres et al. 2009; Gandomi et al. 2010).

The computer programs provided by standard GP are represented as tree structures (Koza 1992; Gandomi et al. 2011). This conventional scheme is also referred to the tree-based genetic programming (TGP). Linear genetic programming (LGP) is a particular subset of TGP (Brameier and Banzhaf 2007). LGP evolves programs of an imperative language or machine language instead of the standard TGP expressions of a functional programming language (Brameier and Banzhaf 2001, 2007). LGP has shown to be an efficient alternative to the traditional TGP (Oltean and Grossan 2003). Moreover, Graph-based GP (parallel) programs are another type of traditional TGP (Alavi et al. 2011a, b) (see Fig. 11.1).

As it is abovementioned, a population member in GP is a hierarchically structured tree comprising functions and terminals. The functions and terminals are selected from a set of functions and a set of terminals. For instance, the function set can include the basic arithmetic operations (e.g. +, _, _, /, etc.), Boolean logic functions (e.g. AND, OR, NOT, etc.), or any other mathematical functions. The terminal set contains the arguments for the functions and can consist of numerical constants, logical constants, variables, etc. The functions and terminals are chosen

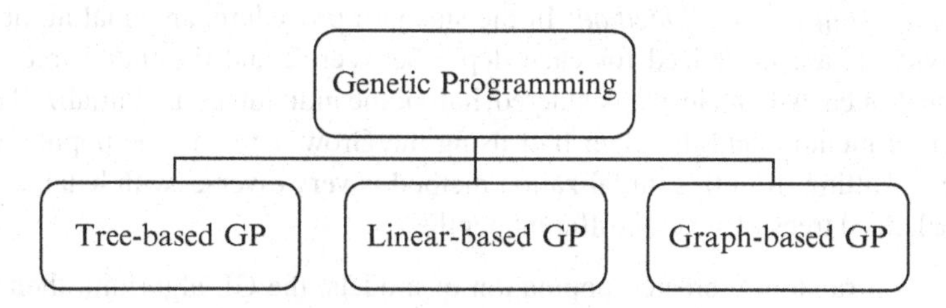

Fig. 11.1 Different types of GP (Gandomi et al. 2013; Alavi and Gandomi 2011a, b)

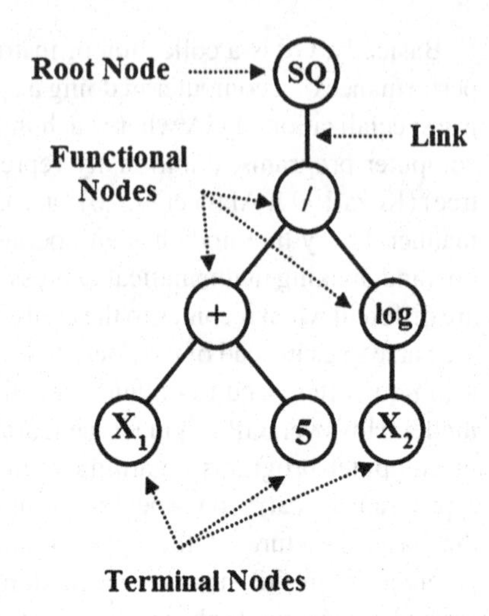

Fig. 11.2 Tree representation of a GP model $((X_1 + 5) / \log (X_2)^2)$

at random and constructed together to form a computer model in a tree-like structure with a root point with branches extending from each function and ending in a terminal. An example of a simple tree representation of a GP model is illustrated in Fig. 11.2. The creation of the initial population is a blind random search for solutions in the large space of possible solutions. The importance of initialization in GP leads to define a multitude numbers of approaches to create the initial population. There are three different prime methods in GP which each of them uses either the standard procedure based on depth (Koza 1992), or the new variation based on size, i.e., number of nodes (Silva 2007), depending on the parameter depth of nodes (Silva 2007):

- *Full method*: In the standard procedure, the new tree receives non terminal (internal) nodes until the initial tree depth is reached—the last depth level is limited to terminal nodes. As a result, trees initialized with this method will be perfectly balanced with all the branches of the same length.
- *Grow method*: in the standard procedure, each new node is randomly chosen between terminals and non-terminals, except nodes at the initial tree depth level, which must be terminals. Trees created with this method may be very unbalanced, with some branches much longer than others.
- *Ramped Half-and-Half method*: In the standard procedure, an equal number of individuals are initialized for each depth between 2 and the initial tree depth value. For each depth level considered, half of the individuals are initialized using the Full method, and the other half using the Grow method. The population of trees resulting from this initialization method is very diverse, with balanced and unbalanced trees of several different depths.

Having a randomly created population of models, the GP algorithm then evaluates the individuals, selects them for reproduction, generates new individuals by mutation, crossover, and direct reproduction, and finally creates the new generation

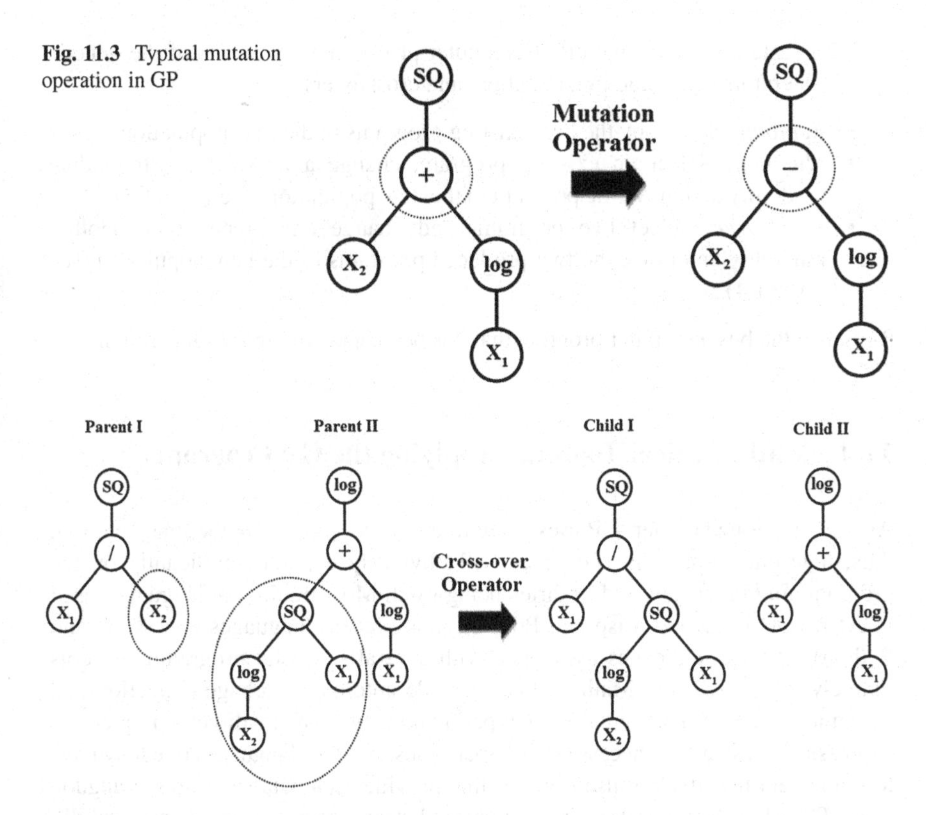

Fig. 11.3 Typical mutation operation in GP

Fig. 11.4 Typical cross-over operation in GP

in all iterations (Koza 1992; Gandomi et al. 2010). During the mutation procedure, a point on a branch of each solution (program) selected at random and the set of terminals and/or functions from each program are then swapped to create two new programs as can be seen in Fig. 11.3. The evolutionary process continues by evaluating the fitness of the new population and starting a new round of reproduction and crossover. During this process, the GP algorithm occasionally selects a function or terminal from a model at random and mutates it. This operator is illustrated in Fig. 11.4. Consequently, the best program that appeared in any generation, the best-so-far solution, defines the output of the GP algorithm (Koza 1992). The following three main steps are summarized form for the abovementioned introduction for operating within any GP procedure (Koza 1992):

1- Generate an initial population of random compositions of functions and terminals.
2- Repeat (below) steps 2.1 and 2.2 until the establishment of the program's suitable and final condition:

   2.1 Executes each program and assigns a fitness value to it according to the fitness function.

2.2 Create a new population of computer pro- grams by means of the genetic operators (reproduction, mutation, and cross-over).

- *Reproduction*: Copy the best existing programs in the new population.
- *Mutation*: Select an existing program, change a node of the individual randomly and move the program to the new population (see Fig. 11.3)
- *Cross-over*: Select two programs and change one branch with another randomly and move the two produced programs to the new population (see Fig. 11.4).

3- Select the best computer program that has been appeared in any generation.

11.4 Mathematical Tools for Applying the GP Programs

As it is mentioned earlier, GP uses parse tree in order to operate the final solution; thus, how one implements GP trees will obviously depend significantly on the programming languages and libraries being used. Most traditional languages used in AI research (such as Lisp and Prolog), many recent languages (say Ruby and Python), and the languages associated with several scientific programming tools (namely, MATLAB and Mathematica) provide automatic garbage collection and dynamic lists as fundamental data types making it easy to directly implement expression trees and the necessary GP operations. In other languages one may have to implement lists/trees or use libraries that provide such data structures (Langdon et al. 2008); however, non-tree representations have been suggested and successfully implemented, such as linear genetic programming which suits the more traditional imperative languages (see, Banzhaf et al. 1998). While GP often evolves computer programs, the solutions can be executed without post-processing. Conversely, the coded binary strings typically evolved by GA require post-processing.

Additionally, it should be noted that there is multitude numbers of different GP implementations in order to resolve different types of problems (see http://www.cosc.brocku.ca/Offerings/5P71/). Among all the practical computational software methods, for instance, the "*Discipulus*" uses automatic induction of binary machine code "AIM" to achieve better performance and generates code in most high level languages (Peter Nordin 1997; Banzhaf et al. 1998, Sections 11.6.2–11.6.3). The "*MicroGP* (μ*GP*)" uses directed multi-graphs to generate programs that fully exploit the syntax of a given assembly language which is known as a general purpose tool, mostly exploited for assembly language generation. The "*GeneXproTools*" is an extremely flexible modeling tool designed for Function Finding, Classification, Time Series Prediction, and Logic Synthesis. Moreover, "*GPLAB*" *and* "*GPTIPS*" are genetic programming toolboxes for MATLAB. These toolboxes are very flexible that different functions can be adapted easily even by non professional users. It was tested on different MATLAB versions and computer platforms, and it does not require any additional toolboxes (Silva 2007).

11.5 Genetic Programming; A Capable Tool for Driving GMPEs

11.5.1 Genetic Programming for Linear-in-Parameters Models

In 2011, Gandomi et al. employed a novel hybrid method coupling genetic programming and orthogonal least squares, called GP/OLS, to develop a new GMPE in order to predict the ground-motion parameters of the Iranian plateau earthquakes. The principal ground-motion parameters formulated were peak ground acceleration (PGA), peak ground velocity (PGV) and peak ground displacement (PGD). The developed GMPE relate abovementioned intensity measures to seismological parameters including earthquake magnitude, earthquake source to site distance, average shear-wave velocity (Vs30), and faulting mechanisms. The equations were established based on an abundant collection of GMRs released by Pacific Earthquake Engineering Research Center (PEER). Furthermore, contribution of influential parameters which affect predicted values of intensity measures was determined by sensitivity analysis. Also, parametric analysis was performed to assess the sensitivity of the models to the variations of the influential parameters. The most appropriate GMPEs were determined according to a multi-objective strategy as below:

(i) The simplicity of the model, although this was not a predominant factor.
(ii) Providing the best fitness value on a learning set of data.
(iii) Providing the best fitness value on a validation set of data.

The first objective can be controlled by the user through the parameter settings (e.g., maximum program depth). For the other objectives, the following objective function (OBJ) was constructed as a measure of how well the model predicted output agrees with the measured output. The selections of the best models were deduced by the minimization of the Eq. (11.2):

$$OBJ = \left(\frac{NO_{Learning} - NO_{Validation}}{NO_{Training}}\right)\frac{RMSE_{Learning} + MAE_{Learning}}{R^2_{Learning}} + \frac{2NOValidation}{NO_{Training}}\frac{RMSE_{Validation} + MAE_{Validation}}{R^2_{Validation}} \tag{11.2}$$

where $\mathrm{No}_{\mathrm{Learning}}$, $\mathrm{No}_{\mathrm{Validation}}$ and $\mathrm{No}_{\mathrm{Training}}$ are the number of learning, validation and training data in turn; R, RMSE and MAE are, respectively, correlation coefficient, root mean squared error and mean absolute. The peak ground acceleration (PGA), velocity (PGV) and displacement (PGD) prediction equations, for the best results by the GP/OLS algorithm, are as following equations:

Table 11.1 The coefficients derived for different fault types

	F		
Fault type	PGA	PGV	PGD
Reverse	0.046	0.001	−0.037
Normal	−0.059	−0.371	−0.372
Strike-slip	−0.101	0.080	0.236

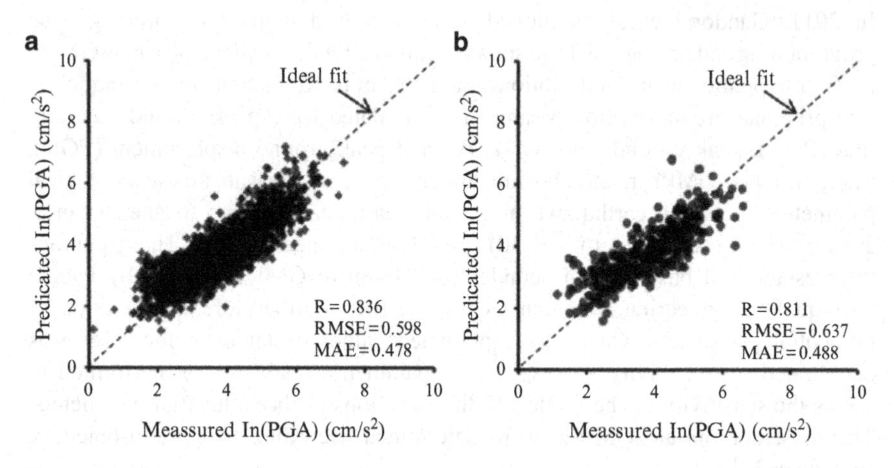

Fig. 11.5 Measured versus predicted PGA values using the GP/OLS model: (**a**) training data and (**b**) testing data (Gandomi et al. 2011)

$$\ln(PGA)\left({}^{cm}/{}_{s^2}\right) = 7.673 - 1.28\left(\ln\left(V_{s30}\right)^{\left(\ln\left(R_{jb}\right)\right)/M_w}\right) + F_{PGA} \tag{11.3}$$

$$\ln(PGV)\left({}^{cm}/{}_{s}\right) = -3.37 + 1.059M_w - 0.02\Big(\ln\left(R_{jb}\right) + \ln\left(R_{jb}\right)^2 \ln(Vs30)\Big) + F_{PGV} \tag{11.4}$$

$$\ln(PGD)(cm) = -1.973 + 1.917M_w + 7.82\frac{\ln\left(R_{jb}\right)}{\ln\left(V_{s30}\right)} + F_{PGD} \tag{11.5}$$

where M_w, R_{jb}, and V_{s30}, respectively, denote the earthquake magnitude, earthquake source to site distance, and average shear-wave velocity. F_{PGA}, F_{PGV}, and F_{PGD} are the empirical coefficients derived for different fault types. These coefficients are presented in Table 11.1. PGD Comparisons of the measured and predicted PGA, PGV and PGD values are, respectively, shown in Figs. 11.5, 11.6, and 11.7.

The results indicate that the obtained GMPEs are effectively capable of estimating the site ground-motion parameters. In comparison with the previous GMPEs, the obtained attenuation relationships predict values with better similarity to the

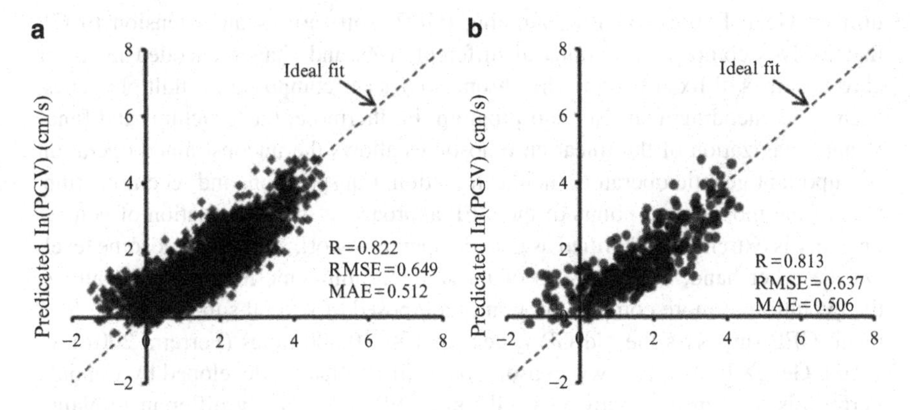

Fig. 11.6 Measured versus predicted PGV values using the GP/OLS model: (**a**) training data and (**b**) testing data (Gandomi et al. 2011)

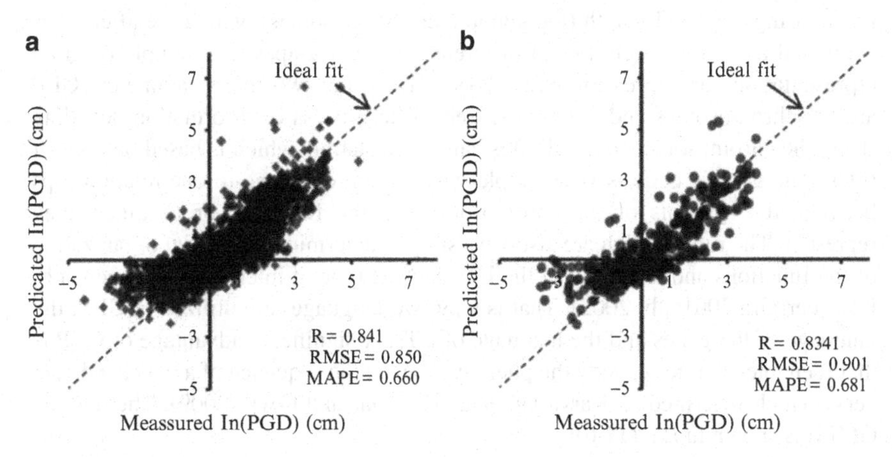

Fig. 11.7 Measured versus predicted PGD values using the GP/OLS model: (**a**) training data and (**b**) testing data (Gandomi et al. 2011)

observed data than those found in literature. Developed GMPEs can be reliably implemented in design procedure due to the fact of having very simple functional form (Gandomi et al. 2011).

11.5.2 Gene Expression Programming-Based GMPE

In 2009, Cabalar and Cevik introduced a genetic programming-based GMPE for predicting PGA for Turkey seismic zone based on earthquakes in Turkey. The proposed GP based GMPE obtained from the most reliable set of data in Turkey and predicts PGA values as output. It should be emphasized that the researchers utilized Gene Expression Programming (GEP) software as an extension to GP that evolves computer programs of different sizes and shapes encoded in linear chromosomes of fixed length. The chromosomes are composed of multiple genes, each gene encoding a smaller sub-program. Furthermore, the structural and functional organization of the linear chromosomes allows the unconstrained operation of important genetic operators such as mutation, transposition, and recombination. One of the most potent points of the GEP approach is that the creation of genetic diversity is extremely simplified as genetic operators work at the chromosome level. On the other hand, GEP consists of its unique, multigenic nature, which allows the evolution of more complex programs composed of several sub-programs. As a result GEP surpasses the old GP system in 100–10,000 times (Ferreira 2001a, b, 2002). GeneXTools (see www.gepsoft.com); GEP software developed by Candida Ferreira is used in this study. GA; GP, and GEP are basically different in nature of individuals, in GA the individuals are fixed length binary strings, in GP the evolving programs are parse trees,; and in GEP the individuals are encoded as linear strings of fixed length (the genome or chromosomes), which are afterwards expressed as nonlinear entities of different sizes and shapes (i.e., simple diagram representations or expression trees (ETs)). Thus, the two main parameters GEP are the chromosomes and expression trees. The process of information decoding (from the chromosomes to the ETs) is called translation, which is based on a set of rules. The genetic code is very simple where there exist one-to-one relationships between the symbols of the chromosome and the functions or terminals they represent. The rules, which are also very simple, determine the spatial organization of the functions and terminals in the ETs and the type of interaction between sub-ETs (Ferreira 2001a, b, 2002). That is why two languages are utilized in GEP: the language of the genes and the language of ETs. A significant advantage of GEP is that it enables to infer exactly the phenotype given the sequence of a gene, and vice versa which is termed as Karva language (Cabalar and Cevik 2009). The derived GMPE is shown in Eq. (11.6):

$$PGA = \left(\frac{5.7}{A}\right)^5 + \frac{(B)\,(M_w)\,(\log\, M_w)}{\sqrt[3]{R}}$$

where

$$A = \sqrt[4]{V_{S30}} + \left(R - \sqrt[9]{V_{S30}^4 + 238^5}\right)^2$$

$$B = \sqrt[3]{\sqrt[4]{\frac{651}{V_{S30}}} - \frac{\log\, R}{3}} \quad (11.6)$$

The results indicate that proposed GP based GMPE provides relatively high consistency among predicted output and observed data. Additionally, this model yields more accurate PGA in comparison with an existing attenuation relationship.

11.5.3 Capability of Adjusting Novel and Modern Fitness Function

In 2013, Rahpeyma et al. represent a new attenuation relationship for predicting PGA parameter based on Iranian plateau database by means of GPLAB, a genetic programming toolbox (Silva 2007). This toolbox is an operational and practical application for different types of users. Recently, some researchers have used this toolbox for obtaining predictive equations (Kermani et al. 2009; Johari et al. 2006). For using GPLAB, the database is divided into the training set (80 % of the data set) and the testing set (20 % of the data set), chosen randomly (uniformly distributed). The programs in GPLAB (tree structures), are initialized with one of the three accessible initializing methods "Full, Grow, and Ramped Half-and-Half" (Koza 1992). In this study, initial population is produced based on Ramped Half-and-Half method. In the standard procedure, an equal number of individuals is initialized for each depth between two and the initial tree depth value (Silva 2007). The population of trees resulting from this initialization method is very diverse, with balanced and unbalanced trees of several different depths (Silva 2007). One of the important features of GPLAB is some appropriate restrictions on tree's depth or size to avoid bloat that is a phenomenon consisting of an excessive code growth without any corresponding Improvement in the fitness (Koza 1992; Silva 2007). In GPLAB, parents are selected for reproduction according to four usual sampling methods (Koza 1992; Silva 2007). In this paper, Lexictour sampling approach was used for selecting parents. In this approach, a random number of individuals are chosen from the population and the best of them is chosen (Silva 2007). In this study, the GP fitness function is defined based on information theoretic method, the average sample log likelihood (LLH) (Scherbaum et al. 2009) is proposed in order to quantitatively assess the predictive models. After obtaining the initial predictive model, by GP, in order to reduce the bias toward different earthquake parameters and likewise for reducing the sensitivity of the initial attenuation model to the considered database, the GA fitness function is defined according to Eq. (11.7) as a combination of LLH criterion and re-sampling analysis, known as RSA method (Azarbakht et al. 2014).

$$\begin{aligned} \textit{Fitness Function} &= \omega_1 \times LLH_{\textit{Training Data}} \\ &+ \omega_2 \times \sum_{n}^{90,\,110,\,130,\,145} \frac{\sum_{i=1}^{Nd} \sum^{M,R,Vs30} \left| 1 - PV_j^{S_i^n} \right|}{3 \times Nd} \end{aligned} \tag{11.7}$$

where Nd is the uniformly distributed random databases (in this study, $Nd = 100$), S_i^n is the ith sample with n records, S_i^n is the residuals' P-value of S_i^n versus jth parameter, M is the earthquake moment magnitude, R is the distance measure, $Vs30$ is the shear wave velocity, $w_1 = 0.25$ and $w_2 = 0.125$ are the weighting constants based on authors judgment. The final form of LnPGA is shown in Eq. (11.8) and

Table 11.2 The constant coefficients obtained by GA from final optimized model (Rahpeyma et al. 2013)

a_1	1.00	a_8	1.00
a_2	3.44	a_9	1.00
a_3	0.72	a_{10}	1.00
a_4	1.00	a_{11}	0.056
a_5	0.11	a_{12}	1.00
a_6	1.00	a_{13}	1.00
a_7	2.33	$\sigma_{\ln Y}$	0.9276

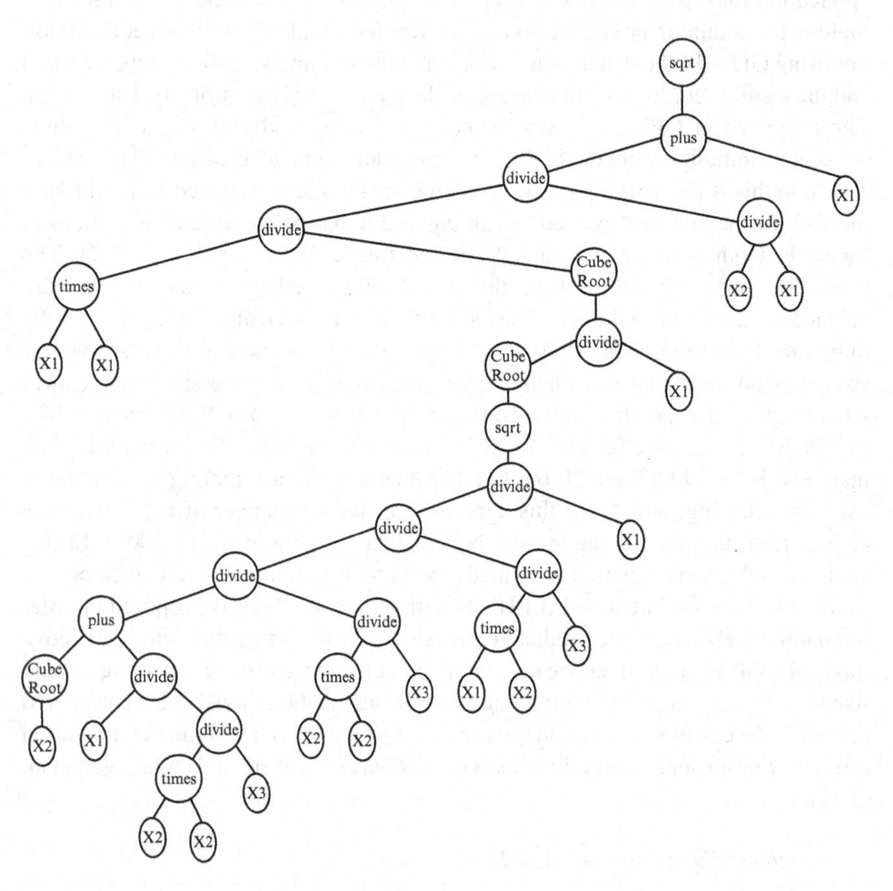

Fig. 11.8 Tree-expression for the proposed GP model (X1, X2, and X3 are moment magnitude, distance measure, and shear wave velocity, respectively)

Table 11.2 shows the result of the coefficients achieved by GA. Figure 11.8 shows tree-expression of the derived model.

$$\ln\left(PGA\right) = Sqrt\left(a_1 \frac{M_w^{a_2}}{R^{a_3}} \times \frac{a_4}{V_{S30}^{a_5}\left(a_6\, R^{a_7} + a_8\, M_w^{a_9} \times V_{S30}^{a_{10}}\right)^{a_{11}}} + a_{12} M_w^{a_{13}}\right) + \sigma \tag{11.8}$$

Furthermore, the proposed GP model is compared with a set of existing attenuation relationships via several traditional and modern mathematical and statistical methods. Explicitly, the obtained model shows clearly more consistency with the local data in comparison with the other selected models. As a plain evidence, Table 11.3 indicates the progress of obtained GP. It should be noted that R will not change significantly by shifting the output values of a model equally, and error functions (e.g. RMSE and MAE) only shows the error not correlation. As a result, for illustrating suitable comparison among all models a new criteria ($\rho = \mathrm{RRMSE}/(\mathrm{R} + 1)$) is also computed within Table 11.3.

11.6 Conclusion

Seismic hazard studies estimate ground motion intensities in a specific site by means of GMPEs. These equations are functions of seismological parameters such as earthquake magnitude and source to site distance. Attenuation relationships as one of the most potent elements in any seismic hazard analysis can be developed through two basic approaches. The first method, known as empirical models or mathematical relationships which have been fit to a database of recorded ground motions using regression methods. Empirical ground motion prediction equations have been developed and successfully implemented for regions where there are adequate and sufficient strong ground-motion data. However, obtaining the best predictive equation requires a large set of data and depends largely on the performance of regression and classification methods. Additionally, this methodology simply results in a mathematical relationship that only describes observed information and does not necessarily reflect the realistic process of an earthquake. The second approach, however, is concerned with physical modeling of the problem. In this methodology, limited observed data in stochastic manner are used for calibration of physical model. As a result, these GMPEs are only applicable in specific regions where comprehensive information about the soil structure mechanical characteristics and the active dynamic fault systems are available. During recent decades, beside the two abovementioned methodologies soft computing techniques, such as Evolutionary Algorithms (EA), have been used for obtaining attenuation relationships with more complicated interaction among the observation and prediction values, whereas, physically based engineering approaches usually fail to capture such complex behavior of predicting relationships. EA algorithms, specifically genetic programming (GP) and genetic algorithm (GA), are optimization techniques based on the rules of natural selection. GA is generally used in parameter optimization to evolve the best values for a given set of model parameters. However, in GP as a machine learning technique, random population of individuals is created and then selected for reproduction of different generations to obtain the best program that appeared in any generation as the output. GP is one of the most capable approaches which incredibly affected the methodologies for deriving the new GMPEs. GMPEs obtained by means of GP are remarkably simple and straightforward and provide a

Table 11.3 Result of error terms (RMSE, MAE), LLH, E, and R^2

					RMSE			MAE		
Model	LLH	R^2	P	E	rij (total)	ri (inter)	rij (intra)	rij (total)	ri (inter)	rij (intra)
GP model	1.9368	0.9454	0.4567	33.4252	0.9264	0.5643	0.7858	0.7470	0.4269	0.6272
Setal12	2.2913	0.9399	0.4725	22.9985	0.9598	0.6383	0.7704	0.7787	0.4891	0.6085
Zetal12	1.9712	0.9460	0.4542	34.1801	0.9212	0.5750	0.7651	0.7385	0.4135	0.6049
Getal07	2.9947	0.9393	0.4795	24.4972	0.9742	0.6774	0.7771	0.7855	0.5424	0.6310
AB10	2.5466	0.9341	0.5072	22.9090	1.0320	0.6920	0.7949	0.8524	0.5133	0.6388
AC10	3.2676	0.8651	0.7014	−64.362	1.4556	1.2127	0.7918	1.2117	1.1202	0.6251
Aetal05	2.4133	0.9364	0.4974	25.7504	1.0115	0.6697	0.7907	0.8234	0.4865	0.6359
Ozetal05	3.6427	0.9032	0.6005	−17.829	1.2324	0.9556	0.7841	0.9839	0.8423	0.61152
KG04	2.2981	0.9444	0.4604	32.3028	0.9342	0.6179	0.7716	0.7568	0.4760	0.6127
Bindi10	2.2288	0.9310	0.5167	19.1787	1.0522	0.6815	0.8675	0.8593	0.4886	0.6787
CB08	2.7452	0.9433	0.4648	30.9763	0.9433	0.6211	0.7710	0.7588	0.4607	0.6167
BA08	2.6496	0.9396	0.4793	28.8792	0.9737	0.7443	0.6953	0.7660	0.5649	0.5382
CY08	2.5322	0.9378	0.4865	24.5607	0.9889	0.6403	0.7816	0.8164	0.4568	0.6183
AS08	2.2605	0.9425	0.4680	29.9721	0.9501	0.6120	0.7771	0.7708	0.4301	0.6199

prediction performance better than or comparable with the attenuation relationships derived through two traditional methodologies. In this chapter, further information on variety of GP based GMPEs including those developed by Gene Expression Programming (GEP) software, a hybrid method coupling genetic programming and orthogonal least squares (GP/OLS) and GPLAB are discussed in more details.

References

Ambraseys, N. N., and J. J. Bommer. "Attenuation relations for use in Europe: an overview." *Proceedings of Fifth SECED Conference on European Seismic Design Practice*. of, 1995.

Abrahamson, N. A., W. J. Silva, and R. Kamai. (2013). "Update of the AS08 Ground-Motion Prediction equations based on the NGA-west2 data set." *Pacific Engineering Research Center Report* 4.

Alavi, A.H., Ameri, M., Gandomi, A.H., Mirzahosseini, M.R., 2011. Formulation of flow number of asphalt mixes using a hybrid computational method. Construction and Building Materials 25 (3), 1338–1355.

Alavi A.H., Gandomi A.H., "Prediction of Principal Ground-Motion Parameters Using a Hybrid Method Coupling Artificial Neural Networks and Simulated Annealing." Computers & Structures, 89 (23–24): 2176–2194, 2011.

Alavi A.H., Gandomi A.H., "A Robust Data Mining Approach for Formulation of Geotechnical Engineering Systems." International Journal for Computer-Aided Engineering and Software-Engineering Computations, 28(3): 242–274, 2011.

Alavi A.H., Gandomi A.H., Modaresnezhad M., Mousavi M., "New Ground-Motion Prediction Equations Using Multi Expression Programming." Journal of Earthquake Engineering, 15(4): 511–536, 2011.

Azarbakht, Alireza, Sahar Rahpeyma, and Mehdi Mousavi. (2014). "A New Methodology for Assessment of the Stability of round-Motion prediction Equations." *Bulletin of the Seismological Society of America*. 104(3), 1447–1457.

Banzhaf, Wolfgang, et al. *Genetic programming: an introduction*. Vol. 1. San Francisco: Morgan Kaufmann, 1998.

Brameier M. and Banzhaf W. (2007) Linear Genetic Programming. Number XVI in Genetic and Evolutionary Computation. Springer.

Brameier M. and Banzhaf W. (2001) A comparison of linear genetic programming and neural networks in medical data mining, IEEE Trans. Evol. Comput., 5(1), 17–26.

Boore, David M., and William B. Joyner. "The empirical prediction of ground motion." *Bulletin of the Seismological Society of America* 72.6B (1982): S43-S60.

Campbell, K.W (1985), "Strong motion attenuation relations: a ten-year perspective", *Earthquake Spectra, Vol. 1,* pp. 759-804_1985.

Cabalar, A.F. and Cevik, A. (2009). Genetic Programming-Based Attenuation Relationship: An Application of Recent Earthquakes in Turkey, *Computers and Geosciences*, **35**(9), 1884-1896.

Cramer, N. L. (1985, July). A representation for the adaptive generation of simple sequential programs. In *Proceedings of the First International Conference on Genetic Algorithms* (pp. 183-187).

Douglas, J. (2011). Ground Motion Estimation Equations 1964–2010, Pacific Earthquake Engineering Research Center College of Engineering, University of California, Berkeley.

Elnashai, Amr S., and Luigi Di Sarno. *Fundamentals of earthquake engineering*. Chichester, UK: Wiley, 2008.

Elshorbagy, A., Corzo, G., Srinivasulu, S., & Solomatine, D. P. (2010). Experimental investigation of the predictive capabilities of data driven modeling techniques in hydrology-Part 2: Application. *Hydrology and Earth System Sciences*, *14*(10), 1943-1961.

Fukushima, Y. and T. Tanaka (1990), "A new attenuation relation for peak horizontal acceleration of strong earthquake ground motion in Japan", *Bull. Seism. Soc. Am., Vol. 80,* pp. 757-783.

Friedberg, R. M. (1958). A learning machine: Part I. *IBM Journal of Research and Development, 2*(1), 2-13.

Ferreira, C., 2001b. Gene expression programming: a new adaptive algorithm for solving problems. Complex Systems 13 (2), 87–129.

Ferreira, C., 2001a. Gene expression programming in problem solving. In: Proceedings of the Sixth Online World Conference on Soft Computing in Industrial Applications, 10–24. /http://www.gene-expression-programming.com/author.asp.

Ferreira, C., 2002. Gene Expression Programming: Mathematical Modeling by an Artificial Intelligence. Angra do Heroisma, Portugal. /http://www.gene-expression-programming.com/author.asp.

Fister, I., Gandomi, A.H., Fister, I.J., Mousavi, M., Farhadi, A., (2014), Soft Computing in Earthquake engineering: A short overview. International Journal of Earthquake Engineering and Hazard Mitigation 2 (2), 42-48.

Gandomi, A. H., A. H. Alavi, M. Mousavi and S.M. Tabatabaei, (2011), A hybrid computational approach to derive new ground-motion prediction equations, Engineering Applications of Artificial Intelligence 24 (4), 717-732.

Gandomi, A. H., Yang, X. S., Talatahari, S., & Alavi, A. H. (Eds.). (2013). *Metaheuristic applications in structures and infrastructures.*

Gandomi M, Soltanpour M, Zolfaghari MR, Gandomi AH. (2015). Prediction of peak ground acceleration of Iran's tectonic regions using a hybrid soft computing technique. Geoscience Frontiers.

Gandomi A.H., Alavi A.H. (2012). A New Multi-Gene Genetic Programming Approach to Nonlinear System Modeling. Part II: Geotechnical and Earthquake Engineering Problems. Neural Computing and Applications 21(1), 189–201.

Gepsoft- Modeling made easy-data mining software www.gepsoft.com (accessed 15.07.08).

Goldberg, D.E. (1989). Genetic Algorithms in Search, Optimization, and Machine Learning, First Edition, Addison-Wesley publication, Inc. Reading, MA, 432.

Gandomi, A.H., Alavi, A.H., Arjmandi, P., Aghaeifar, A., Seyednour, R., 2010. Modeling of compressive strength of hpc mixes using a combined algorithm of genetic programming and orthogonal least squares. Journal of Mechanics of Materials and Structures 5 (5), 735–753.

Idriss, I. M. (1978). Characteristics of earthquake ground motions, state-of-the-arts report, earthquake engineering and soil dynamics'. In *Proceeding of the ASCE Geotechnical Engineering Division specialty conference: Earthquake Engineering and Soil Dynamic, Vol.III. pp:1151–1265*

Joyner, William B., and David M. Boore. "Measurement, characterization, and prediction of strong ground motion." *Earthquake Engineering and Soil Dynamics II, Proc. Am. Soc. Civil Eng. Geotech. Eng. Div. Specialty Conf.* 1988.

Joyner, W. B., and D. M. Boore. "Recent developments in strong-motion attenuation relationships." *NIST SPECIAL PUBLICATION SP* (1996): 101-116.

Johari, A., Habibagahi, G., and Ghahramani, A. (2006). Prediction of Soil-Water by Characteristic Curve Using Genetic Programming, *Journal of Geotechnical and Geoenvironmental Engineering*, **132**(5), 661-665.

Krishnan, S., Chen, J., Komatitsch, D., Tromp, J., 2006. Case studies of damage to tall steel moment-frame buildings in Southern California during large San Andreas earthquakes. Bulletin of the Seismological Society of America 96, 1523–1537.

Komatitsch D, Liu Q, Tromp J, Su ss P, Stidham C, Shaw JH. Simulations of ground motion in the Los Angeles basin based upon the spectral-element method. Bull Seismol Soc Am 2004;94:187–206.

Koza, J.R. (1992). Genetic Programming: On the Programming of Computers by Means of Natural Selection, MIT Press, Cambridge, Massachusetts, 840.

Kermani, E., Jafarian, Y., and Baziar, M.H. (2009). New Predictive Models for the Ratio *v* of Strong Ground Motions using Genetic Programming, *International Journal of Civil Engineering*, **7**(4), 246-239.

Langdon, W.B., et al. (2008). "Genetic programming: An introduction and tutorial, with a survey of techniques and applications." *Computational Intelligence: A Compendium*. Springer Berlin Heidelberg, pp. 927–1028.

McGuire, R.K. (1995). Probabilistic Seismic Hazard Analysis and Design Earthquakes: Closing the Loop, *Bulletin of the Seismological Society of America*, **85**(5), 1275-1284.

McGuire, R. K., 2004. *Seismic Hazard and Risk Analysis*, Monograph MNO-10, Earthquake Engineering Research Institute, Oakland, CA.

McGarr, A. (1984), "Scaling of ground motion parameters, state of stress, and focal depth", *J. Geophys. Res., Vol. 89,* pp. 6969-6979.

Midorikawa, S., M. Matsuoka and K. Sakugawa (1994), "Site effects on strong-motion records during the 1987 Chibaken-toho-oki_Japan earthquake", *The 9thJapan Earthquake Engineering Symposium, Vol. 3,* pp. 85-90.

Mohammadnejad A.K., Mousavi S.M., Torabi M., Mousavi M., Alavi A.H. (2012). Robust Attenuation Relations for Peak Time-Domain Parameters of Strong Ground-Motions. Environmental Earth Sciences, 67(1):53-70.

Nordin, P. (1997). *Evolutionary program induction of binary machine code and its applications*. Munster: Krehl.

Oltean, M. and, Gross, C., 2003. A comparison of several linear genetic programming techniques. Advances in Complex Systems 14(4), 1–29.

OpenSHA (2009). *Open Seismic Hazard Analysis Computer Platform*, http://www.opensha.org/

Petersen, M. D., Frankel, A. D., Harmsen, S. C., Mueller, C. S., Haller, K. M., Wheeler, R. L., Wesson, R. L., Zeng, Y., Boyd, O. S., Perkins, D. M., Luco, N., Field, E. H., Wills, C. J., and Rukstales, K. S., 2008. *Documentation for the 2008 update of the United States national seismic hazard maps, USGS Open-File Report 2008-1128.*

Power, Maurice, et al. "An overview of the NGA project." *Earthquake Spectra* 24.1 (2008): 3-21.

Papageorgiou AS, Aki K. A specific barrier model for the quantitative description of inhomogeneous faulting and the prediction of strong ground motion, Part I: description of the model. Bull Seismol Soc Am 1983;73:693–722.

Rahpeyma, S., A. Azarbakht and M. Mousavi. (2013) "A new Peak-Ground-Acceleration prediction model by using genetic optimization techniques for Iran's plateau database". Journal of Seismology and Earthquake Engineering, 15(3) p153-170

Samuel AL (1983) AI, where it has been and where it is going. In: IJCAI, pp 1152–1157

Silva, S. (2007). GPLAB_A Genetic Programming Toolbox for MATLAB, Version3, http://gplab.sourceforg.net

Scherbaum, F., Delavaud, E., and Riggelsen, C. (2009). Model Selection in Seismic Hazard Analysis: An Information-Theoretic Perspective, *Bulletin of the Seismological Society of America*, **99**(6), 3234-3247.

Trifunac, M. D., and A. G. Brady. "Correlations of peak acceleration, velocity and displacement with earthquake magnitude, distance and site conditions." *Earthquake Engineering & Structural Dynamics* 4.5 (1976): 455-471.

Torres, R. D. S., Falcão, A. X., Gonçalves, M. A., Papa, J. P., Zhang, B., Fan, W., & Fox, E. A. (2009). A genetic programming framework for content-based image retrieval. *Pattern Recognition*, *42*(2), 283-292.

Yang, X. S., Gandomi, A. H., Talatahari, S., & Alavi, A. H. (Eds.). (2013). Metaheuristics in Water, Geotechnical and Transportation Engineering. Elsevier, Waltham, MA, 2013.

Youngs, R.R., S.J. Chiou, W.J. Silva, and J.R. Humphrey (1997), "Strong ground motion attenuation relationships for subduction zone earthquakes", *Seism. Res. Lett., Vol. 68,* pp. 58-73.

Chapter 12
Evaluation of Liquefaction Potential of Soil Based on Shear Wave Velocity Using Multi-Gene Genetic Programming

Pradyut Kumar Muduli and Sarat Kumar Das

12.1 Introduction

Seismic hazards can be categorized as ground shaking, structural hazards, liquefaction, landslides, retaining structure failures, lifeline hazards, tsunamis. Out of the above, seismically induced liquefaction of soil is a major cause of both loss of life and damage to infrastructures and lifeline systems. The soil liquefaction phenomenon was known in early stage of development of soil mechanics by Terzhagi and Peck (1948) to explain the phenomenon of sudden loss of strength in loose sand deposit. It was recognized as the main cause of slope failure in saturated sandy deposit. Though, soil liquefaction phenomena have been recognized since long, it was more comprehensively brought to the attention of engineers, seismologists and scientific community of the world by several devastating earthquakes around the world; Niigata and Alaska (1964), Loma Prieta (1989), Kobe (1995), Kocaeli (1999), and Chi-Chi (1999) earthquakes (Baziar and Jafarian 2007). Since then, a numerous investigations on field and laboratory revealed that soil liquefaction may be better described as a disastrous failure phenomenon in which saturated soil loses strength due to increase in pore water pressure and reduction in effective stress under rapid loading and the failed soil acquires a degree of mobility sufficient to permit movement from meters to kilometers. Soil liquefaction can cause ground failure in the way of sand boils,

Electronic supplementary material The online version of this chapter (doi: 10.1007/978-3-319-20883-1_12) contains supplementary material, which is available to authorized users.

P.K. Muduli • S.K. Das (✉)
Civil Engineering Department, National Institute of Technology Rourkela, Rourkela 769008, India
e-mail: pradyut.muduli@gmail.com; saratdas@rediffmail.com; sarat@nitrkl.ac.in

A.H. Gandomi et al. (eds.), *Handbook of Genetic Programming Applications*,
DOI 10.1007/978-3-319-20883-1_12

major landslides, surface settlement, lateral spreading, lateral movement of bridge supports, settling and tilting of buildings, failure of waterfront structure and severe damage to the lifeline systems etc.

The liquefaction hazard evaluation involves liquefaction susceptibility analysis, liquefaction potential evaluation, assessment of effect of liquefaction (i.e., the extent of ground failure caused by liquefaction) and study of response of various foundations in liquefied soil. These are the major concern of geotechnical engineers. In the present study, the focus is on liquefaction potential evaluation, which determines the likelihood of liquefaction triggering in a particular soil in a given earthquake. Evaluation of the liquefaction potential of a soil subjected to a given seismic loading is an important first step towards mitigating liquefaction-induced damage. Though, different approaches like cyclic strain-based, energy- based and cyclic stress-based approaches are in use, the stress based approach is the most widely used method for evaluation of liquefaction potential of soil (Kramer 1996). Thus, the focus of present study is on the evaluation of liquefaction potential on the basis of the cyclic stress-based approach.

There are two types of cyclic stress based-approach available for assessing liquefaction potential. One is by means of laboratory testing (e.g., cyclic tri-axial test and cyclic simple shear test) of undisturbed samples, and the other involves the use of empirical relationships that relate observed field behavior with in-situ tests such as standard penetration test (*SPT*), cone penetration test (*CPT*), shear wave velocity (V_s) measurement and the Becker penetration test (*BPT*).

The methods like finite element, finite difference, statistically-derived empirical methods based on back-analyses of field earthquake case histories are used for liquefaction analysis. Finite element and finite difference analyses are the most complex and accurate of the above methods. However, liquefied sediments are highly variable over short distances, developing a sufficiently accurate site model for a detailed numerical model requires extensive site characterization effort. Desired constitutive modeling of liquefiable soil is very difficult, even with considerable laboratory testing. Hence, in-situ tests along with the post liquefaction case histories-calibrated empirical relationships have been used widely around the world. The cyclic stress-based simplified methods based on in-situ test such as *SPT*, *CPT*, V_{S} measurements and BPT are commonly preferred by the geotechnical engineer to evaluate the liquefaction potential of soils throughout most part of world.

The stress-based simplified procedure is pioneered by Seed and Idriss (1971). The SPT-based simplified method, developed by Seed and Idriss (1971), has been modified and improved through several revisions (Seed and Idriss 1982; Seed et al. 1983, 1985; Youd et al. 2001) and remains the most widely used methods around the world. However, SPT method cannot detect thin layers and the result depends upon the efficiency of the machine and operator. Robertson and Campanella (1985) first developed a CPT based method for evaluation of liquefaction potential, which is a conversion from the SPT based method using empirical correlation of SPT-CPT and follows the same stress-based approach of Seed and Idriss (1971). Thereafter, various CPT-based methods of soil liquefaction potential evaluation using statistical and regression analysis techniques have been developed (Seed and de Alba 1986; Olsen 1997; Robertson and Wride 1998; Youd et al. 2001). Though CPT is most effective in most of the cases but, difficult to penetrate through gravelly soil.

A potential alternative to the above penetration based methods is the in-situ measurements of small-strain shear-wave velocity V_s. The use of V_s as an index of liquefaction resistance is firmly based on the fact that both V_s and liquefaction resistance are similarly controlled by many of the same factors (e.g., void ratio, state of stress, stress history, and geological age). Some advantages of V_s method can be summarized as: (1) the V_s measurements are possible in soils, which are difficult to sample, such as gravelly soils where penetration tests are mostly unreliable; (2) V_s measurements can also be conducted on small laboratory specimens, allowing direct comparisons between laboratory and field behavior; (3) V_s is a basic mechanical property of soil materials, directly related to small-strain shear modulus, G_{max} as given below:

$$G_{max} = \rho V_s^2 \tag{12.1}$$

where ρ is the mass density of soil (4) G_{max}, or V_s, is in general a required property in earthquake site response and soil-structure interaction analyses; and (5) V_s can be measured by the spectral analysis of surface waves (SASW) technique at sites where borings may not be permitted, such as, sites that extend for great distances where rapid evaluation is required, sites composed of gravels, cobbles, even boulders etc., where sampling is difficult. There are certain difficulties also arise when using V_s to evaluate liquefaction resistance as because (1) seismic testing does not have the provision of collection of samples for classification of soils and determination of non-liquefiable soft clay- rich soils (2) thin, low V_s strata may not be detected if the measurement interval is too large and (3) measurements are made at small strains, whereas pore-water pressure buildup and liquefaction are medium- to high-strain phenomena. The last concern as mentioned above may be significant for cemented soils, because small-strain measurements are highly sensitive to weak inter-particle bonding that is eliminated at medium and high strains. It also can be significant in silty soils above the water table where negative pore-water pressures can increase V_s (Andrus and Stokoe 2000; Youd et al. 2001).

Over the past three decades, a number of investigations have been performed to study the relationship between V_s and liquefaction resistance. These studies include field performance observations (Stokoe II and Nazarian 1985; Kayen et al. 1992; Andrus and Stokoe 1997, 2000; Andrus et al. 2004), penetration-V_s correlations (Seed et al. 1983), analytical investigations (Bierschwale and Stokoe 1984; Stokoe et al. 1988), and laboratory tests (Dobry et al. 1982; de Alba et al. 1984; Tokimatsu and Uchida 1990). Most of the above liquefaction evaluation procedures have been developed with limited field performance data and following the general simplified procedure of the Seed-Idriss, where V_s is corrected to a reference overburden stress and correlated with the cyclic stress ratio.

For a given soil resistance index, such as the V_s, the boundary curve yields liquefaction resistance of a soil, which is usually expressed as the cyclic resistance ratio (*CRR*). Under a given seismic loading, usually expressed as the cyclic stress ratio (*CSR*), the liquefaction potential of a soil is evaluated in terms of a factor of safety (F_s), which is defined as the ratio of *CRR* to *CSR*. The approach of expressing liquefaction potential of soil in terms of F_s is referred to as a deterministic method and is very much preferred by geotechnical professionals due its simplicity for use. National Center for Earthquake Engineering Research (NCEER) workshop, 1998, published the reviews of the above in-situ test based deterministic methods for evaluation of liquefaction potential of soil (Youd et al. 2001).

However, due to parameter and model uncertainties, $Fs \geq 1$ does not always indicate non-liquefaction and also does not necessarily guarantee zero chance of soil being liquefied. Similarly $Fs \leq 1$ may not always correspond to liquefaction and may not guarantee 100 % chance of being liquefied. The boundary surface that separates liquefaction and non-liquefaction cases in the deterministic methods is considered as a performance function or "limit state function" and is generally biased towards the conservative side by encompassing most of the liquefied cases. But, the degree of conservatism is not quantified (Juang et al. 2000). Thus, attempts have been made by several researchers (Haldar and Tang 1979; Liao et al. 1988; Youd and Nobble 1997; Toprak et al. 1999; Juang et al. 2001, 2005, 2006; Muduli and Das 2013a, b; Muduli et al. 2014) to assess liquefaction potential in terms of probability of liquefaction (P_L) using statistical or probabilistic approaches.

Most common statistical techniques like; logistic regression (Liao et al. 1988; Juang et al. 2001; Gandomi and Alavi 2013), decision tree (Gandomi and Alavi 2013) and well known soft computing techniques such as; artificial neural network (ANN) (Goh 1994, 2002; Juang and Chen 2000; Hanna et al. 2007), support vector machine (SVM) (Pal 2006; Goh and Goh 2007; Samui and Sitharam 2011) and relevance vector machine (RVM) (Samui 2007) have been used to develop liquefaction prediction models based on an in-situ test database, which are found to be very efficient. However, the ANN has poor generalization, attributed to attainment of local minima during training and needs iterative learning steps to obtain better learning performances. The SVM has better generalization compared to ANN, but the parameters 'C' and insensitive loss function (ε) needs to be fine tuned by the user. Moreover, these techniques will not produce a comprehensive relationship between the inputs and output and are also called as 'black box' system.

In the recent past, genetic programming (GP) based on Darwinian theory of natural selection is being used as an alternate soft computing technique. The GP is defined as the next generation soft computing technique and also called as a 'grey box' model (Giustolisi et al. 2007) in which the mathematical structure of the model can be derived, allowing further information of the system behaviour. The GP models have been applied to some difficult geotechnical engineering problems (Yang et al. 2004; Javadi et al. 2006; Rezania and Javadi 2007; Alavi et al. 2011; Gandomi and Alavi 2012a; Muduli et al. 2013) with success. However, use of GP and its variant in liquefaction susceptibility assessment are limited. Alavi and Gandomi (2012) have used promising variants of GP; linear genetic programming (LGP) and multi expression programming (MEP) to develop strain energy-based models for evaluation of the liquefaction resistance of sandy soils. Gandomi and Alavi (2012b) developed a classification model based on CPT database using multi-gene genetic programming (MGGP), a variant of GP. But, the performance of the developed model is not compared with that of the existing models based on other soft computing techniques. A first order reliability-based model for evaluation of liquefaction potential in terms of probability of liquefaction is developed using the MGGP by Muduli and Das (2013a) based on CPT database. Muduli and Das (2013b) described the development of a probabilistic model using the MGGP based on SPT database and Bayesian theory of conditional probability. Muduli and Das (2014a) developed a deterministic model, using the MGGP based on post earthquake SPT database, which out performs the available ANN-based model. Two deterministic models have been developed by Muduli and Das (2014b) using the MGGP based on CPT database, which are found to be more efficient than the available ANN-based model and at par with the available SVM-based model respectively. Muduli et al. (2014) presented a Bayesian mapping function using the MGGP based on post liquefaction CPT database for evaluation of liquefaction potential within probabilistic framework. The main advantage of GP and its variant multi-gene genetic programming (MGGP) over traditional statistical methods and other soft computing techniques is its ability to develop a compact and explicit prediction equation in terms of different model variables.

The objective of the present study is to develop deterministic and probabilistic models to evaluate the liquefaction potential of soil using multi-gene genetic programming (MGGP) based on available post liquefaction V_s database. Here, the liquefaction potential is evaluated and expressed in terms of liquefaction field performance indicator, referred as a liquefaction index (*LI*) and factor of safety against the occurrence of liquefaction (F_s). Further, the developed LI_p models have been used to develop both V_s-based *CRR* model. These developed *CRR* models in conjunction with the widely used $CSR_{7.5}$ model, form the proposed MGGP-based deterministic method. The efficiency of the developed V_s-based deterministic model has been compared with that of available statistical and ANN-based model on the basis of independent database. And also the probabilistic evaluation of liquefaction potential has been performed where liquefaction potential is expressed in terms of probability of liquefaction (P_L) and the degree of conservatism associated with developed deterministic model is quantified in terms of P_L. Using Bayesian theory of conditional probability the F_s is related with the P_L through the developed mapping function. The development of compact and comprehensive model equation using deterministic method based on V_s data will enable geotechnical professional to use it with confidence and ease. The presentation of probabilistic methods in conjunction with deterministic factor of safety *(F_s)* value gives the measure of probability of liquefaction corresponding to particular F_s. The developed V_s -based deterministic as well as probabilistic model has been compared with that of the available ANN-based models through two examples one from liquefied case and the other from non-liquefied case to show the robustness of the developed models.

12.2 Deterministic Approach

In deterministic approach, the F_s, which is defined as the ratio of *CRR* to *CSR*, is calculated on the basis of prediction of single values of load (*CSR*) and resistance (*CRR*) as shown in Fig. 12.1 without considering the uncertainty associated in

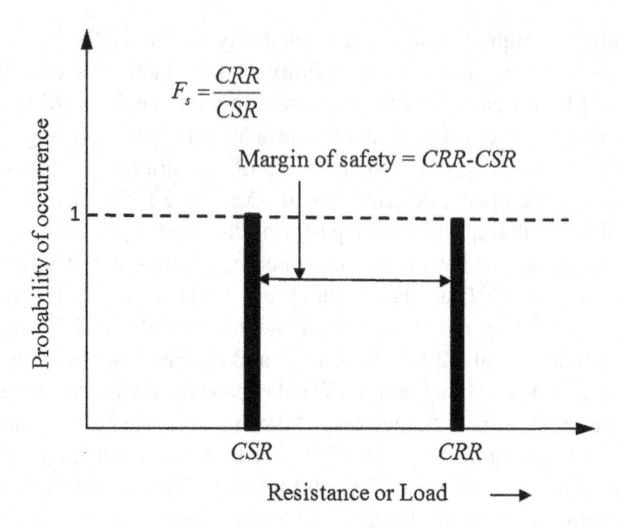

Fig. 12.1 Deterministic approach used in liquefaction potential evaluation (modified from Becker 1996)

prediction of loading and resistance. It is assumed that there is 100 % probability of occurrence of calculated *CRR* and *CSR*. In deterministic approach, $F_S > 1$ corresponds to non-liquefaction and $F_S \leq 1$ corresponds to liquefaction. Here in this approach, only single F_S based on past experience is used to account for all the uncertainties associated with the load and resistance parameters. Though, this method of analysis does not provide adequate information about the behaviour of variables causing liquefaction, is still very much preferred by the geotechnical professionals due to its simple mathematical approach with minimum requirement of data, time and effort.

In the present study, multi-gene genetic programming (MGGP), the variant of GP is used to develop prediction model for evaluation of liquefaction potential of soil within the framework of deterministic approach.

12.2.1 Genetic Programming

Genetic Programming is a pattern recognition technique where the model is developed on the basis of adaptive learning over a number of cases of provided data, developed by Koza (1992). It mimics biological evolution of living organisms and makes use of the principles of genetic algorithms (GA). In traditional regression analysis the user has to specify the structure of the model, whereas in GP, both structure and the parameters of the mathematical model are evolved automatically.

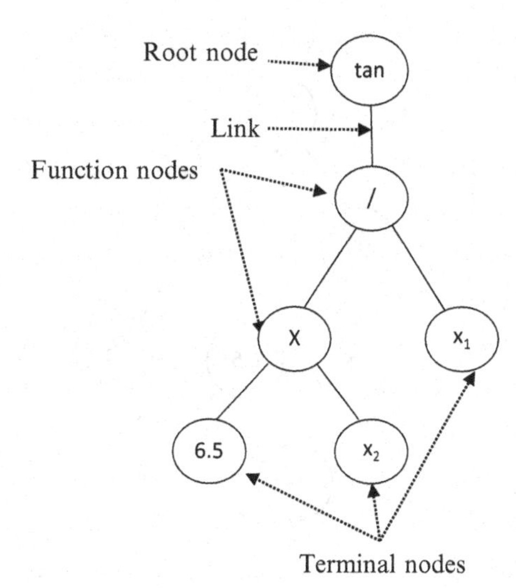

Fig. 12.2 Typical GP tree representing a mathematical expression: tan $(6.5x_2/x_1)$

It provides a solution in the form of a tree structure or in the form of a compact equation using the given dataset. A brief description about GP is presented for the completeness, but the details can be found in Koza (1992).

GP model is composed of nodes, which resembles a tree structure and thus, it is also known as GP tree. Nodes are the elements either from a functional set or terminal set. A functional set may include arithmetic operators ($+$, $\times$, $\div$, or $-$), mathematical functions (sin(.), cos(.), tanh(.) or ln(.)), Boolean operators (AND, OR, NOT, etc.), logical expressions (IF, or THEN) or any other suitable functions defined by the user. The terminal set includes variables (like x_1, x_2, x_3, etc.) or constants (like 3, 5, 6, 9, etc.) or both. The functions and terminals are randomly chosen to form a GP tree with a root node and the branches extending from each function nodes to end in terminal nodes as shown in Fig. 12.2.

Initially a set of GP trees, as per user defined population size, is randomly generated using various functions and terminals assigned by the user. The fitness criterion is calculated by the objective function and it determines the quality of each individual in the population competing with the rest. At each generation a new population is created by selecting individuals as per the merit of their fitness from the initial population and then, implementing various evolutionary mechanisms like reproduction, crossover and mutation to the functions and terminals of the selected GP trees. The new population then replaces the existing population. This process is iterated until the termination criterion, which can be either a threshold fitness value or maximum number of generations, is satisfied. The best GP model, based on its fitness value that appeared in any generation, is selected as the result of genetic programming. The description of various evolutionary mechanisms (i.e., reproduction, crossover and mutation) in GP are presented in Muduli and Das (2013b).

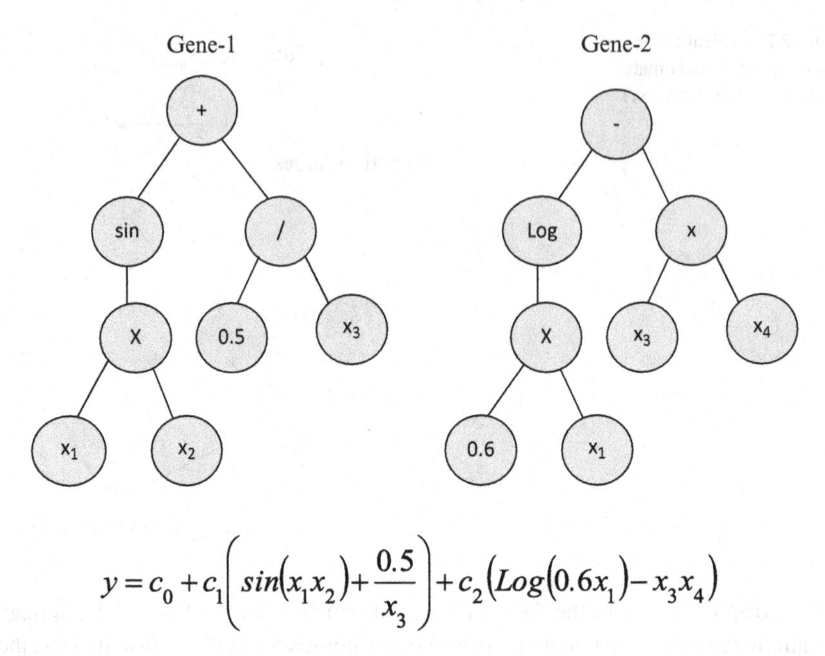

Fig. 12.3 An example of typical multi-gene GP model

12.2.2 Multi-Gene Genetic Programming

MGGP is a variant of GP and is designed to develop an empirical mathematical model, which is a weighted linear combination of a number of GP trees. It is also referred to as symbolic regression. Each tree represents lower order non-linear transformations of input variables and is called a 'gene'. "Multi-gene" refers to the linear combination of these genes.

Figure 12.3 shows an example of MGGP model where the output is represented as a linear combination of two genes (Gene-1 and Gene-2) that are developed using four input variables (x_1, x_2, x_3, x_4). Each gene is a nonlinear model as it contains nonlinear terms (sin(.) / log(.)). The linear coefficients (weights) of Gene-1 and Gene-2 (c_1 and c_2) and the bias (c_0) of the model are obtained from the training data using statistical regression analysis (ordinary least square method).

In MGGP procedure, initial population is generated by creating individuals that contain randomly evolved genes from the user defined functions and variables. In addition to the standard GP evolution mechanisms there are some special MGGP crossover mechanisms (Searson et al. 2010), which allow the exchange of genes between individuals. Similarly, MGGP also provides six methods of mutation for genes (Gandomi and Alavi 2012a): (1) sub-tree mutation, (2) mutation of constants using additive Gaussian perturbation, (3) substitution of a randomly selected input node with another randomly selected input node, (4) substitute a randomly selected constant with another randomly generated constant (5) setting of randomly selected constant to zero, (6) setting a randomly selected constant one. The probabilities of the each of the re-combinative processes (evolutionary mechanisms) can be set by the users for achieving the best MGGP model. These processes are grouped into categories referred to as events. Therefore, the probability of crossover, mutation and the direct reproduction event are to be specified by the user in such a way that the sum of these probabilities is 1.0. The probabilities of the event subtypes can also be specified by the user. For example, once the probability of crossover event is selected, it is possible to define the probabilities of a two point high-level crossover and low-level crossover keeping in mind that the sum of these event subtype probabilities must be equal to one.

Various controlling parameters such as function set, population size, number of generations, maximum number of genes allowed in an individual (G_{max}), maximum tree depth (d_{max}), tournament size, probabilities of crossover event, high level crossover, low level crossover, mutation events, sub-tree mutation, replacing input terminal with another random terminal, Gaussian perturbation of randomly selected constant, reproduction, and ephemeral random constants are involved in MGGP predictive algorithm. The generalization capability of the model to be developed by MGGP is affected by selection of these controlling parameters. These parameters are selected based on some previously suggested values (Searson et al. 2010) and after following a trial and error approach for the problem under consideration. The function set (arithmetic operators, mathematical functions etc.) is selected by the user on the basis of physical knowledge of the system to be analysed. The number of programs or individuals in the population is fixed by the population size. The number of generation is the number of times the algorithm is used before the run terminates. The proper population size and number of generations often depend on the complexity of the problems. A fairly large number of population and generations are tested to find the best model. The increase in G_{max} and d_{max} value increases the fitness value of training data whereas the fitness value of testing data decreases, which is due to the over-fitting to the training data. The generalisation capability of the developed model decreases. Thus, in the MGGP-model development it is important to make a tradeoff between accuracy and complexity in terms G_{max} and d_{max}. There are optimum values of G_{max} and d_{max}, which produce a relatively compact model (Searson et al. 2010). The success of MGGP algorithm usually increases by using optimal values above of controlling parameters.

In the MGGP procedure a number of potential models are evolved at random and each model is trained and tested using the training and testing data respectively. The fitness of each model is determined by minimizing the root mean square error (*RMSE*) between the predicted and actual value of the output variable (*LI*) as the objective function (*f*),

$$RMSE = f = \sqrt{\frac{\sum_{i=1}^{n} \left(LI - LI_p\right)^2}{n}} \tag{12.2}$$

where n = number of cases in the fitness group, LI_p = predicted value of liquefaction field performance indicator (LI) and $LI = 1$ for liquefaction and $LI = 0$ for non-liquefaction field manifestations. If the errors calculated by using Eq. (12.2) for all the models in the existing population do not satisfy the termination criteria, the evolution of a new generation of the population continues till the best model is developed.

12.2.3 Formulation of the V_s-Based Method

The general form of MGGP-based model for LI_p based on V_s database can be presented here as:

$$LI_p = \sum_{i=1}^{n} F\left[X, f(X), c_i\right] + c_0 \tag{12.3}$$

F = the function created by the MGGP process referred herein as liquefaction index function, X = vector of input variables = {Vs, σ_v', FCI, $CSR_{7.5}$} where, Vs = corrected blow count, σ'_v = vertical effective stress of soil at the depth under consideration, FCI = fines content index ($FCI = 1$, for fines content of soil, $FC \leq 5$ %; $FCI = 2$, for 5 % $\leq FC \leq 35$ %; $FCI = 3$, for $FC \geq 35$ %) (Juang et al. 2001). Here, in the present study, the general formulation of CSR as presented by Seed and Idriss (1971) and Youd et al. (2001) is adopted with minor modification, i.e., CSR is adjusted to the benchmark earthquake (moment magnitude, M_w, of 7.5) by using the parameter, magnitude scaling factor (MSF).

$$CSR_{7.5} = 0.65\left(\frac{\sigma_v}{\sigma_v'}\right)\left(\frac{a_{\max}}{g}\right)(r_d)/MSF \tag{12.4}$$

where a_{max} = peak horizontal ground surface acceleration, g = acceleration due to gravity, r_d = shear stress reduction factor which is determined as per Youd et al. (2001):

$$\begin{aligned} r_d &= 1.0 - 0.00765\,z, \text{ for } z \leq 9.15\text{m} \\ &= 1.174 - 0.0267z, \text{ for } 9.15 \leq z \leq 23\text{m} \end{aligned} \tag{12.5}$$

where z is depth under consideration.

The adopted MSF equation is presented below according to Youd et al. (2001).

$$MSF = \left(\frac{M_w}{7.5}\right)^{-2.56} \tag{12.6}$$

c_i is a constant, $f(X)$ = function defined by the user from the functional set of MGGP, n is the number of terms of model equation and c_0 is the bias. The MGGP as per Searson et al. (2010) is used and the present model is developed and implemented using Matlab (MathWorks Inc. 2005).

12.2.4 Database and Preprocessing

In the present study, V_s -based dataset of post liquefaction case histories from various earthquakes is used (Juang and Chen 2000). It contains information about soil and seismic parameters: depth (d), measured V_s, soil type, σ_v, σ'_v, a_{max}, M_w and $CSR_{7.5}$, with field performance observations (LI). The soil in these cases ranges from grave and gravelly sand with soils having 5–10 % fines content, clean sand with less than 5 % fines, sand mixtures to sand with fines content between 5 and 15 %, sandy silt to silty sand with FC between 15 and 35 % and silty sand to sandy and clayey silt with $FC > 35$ %. The depths at which V_s measurements are reported in the database range from 2 to 14.8 m. The V_s values range from 28.7 to 1230 m/s. The FCI values are in the range of 1–3. The a_{max}, M_w and $CSR_{7.5}$ values are in the range of [0.02, 0.51 g], [5.9, 8.3] and [0.01, 0.41] respectively. The database consists of total 186 cases, 88 out of them are liquefied cases and other 98 are non-liquefied cases. Out of the above data 130 cases are randomly selected for training and remaining 56 data are used for testing the developed model. Juang and Chen (2000) also used the above databases with the above number of training and testing data while developing ANN-based liquefaction model. Here, in the MGGP approach normalization or scaling of the data is not required which is an advantage over ANN approach.

12.3 Results and Discussion

The results of the deterministic and probabilistic models are presented separately as follows. First the deterministic approach is presented in terms of determination of factor of safety against liquefaction and then the probabilistic approach is developed based on the results of deterministic approach.

12.3.1 Deterministic Approach

In this section, the result of deterministic model based on post liquefaction V_s database is presented. A limit state function that separates liquefied cases from the non-liquefied cases and also represents cyclic resistance ratio (CRR) of soil is also

Table 12.1 Controlling parameter settings for MGGP-based LI_p model development

Parameters	Ranges	Resolution	Selected optimum values
Population size	1000–4000	200	3000
Number of generations	100–300	50	200
Maximum number of genes (G_{max})	2–4	1	3
Maximum tree depth (d_{max})	2–5	1	3
Tournament size	2–8	1	7
Reproduction probability	0.01–0.07	0.02	0.05
Crossover probability	0.75–0.9	0.05	0.85
Mutation probability	0.05–0.15	0.05	0.1
High level cross over probability	0.1–0.4	0.1	0.2
Low level cross over probability	0.5–0.9	0.1	0.8
Sub-tree mutation	0.6–0.9	0.05	0.85
Substituting input terminal with another random terminal	0.05–0.2	0.05	0.05
Gaussian perturbation of randomly selected constant	0.05–0.2	0.05	0.1
Ephemeral random constant	[−10, 10]	–	–

developed by using MGGP. The developed *CRR* model in conjunction with widely used $CSR_{7.5}$ is used to evaluate liquefaction potential in terms of F_s and the results are presented in following sequence.

The selection of controlling parameters affects the efficacy of the model generated by the MGGP. Thus, optimum values of the parameters are selected for the development of LI_p model based on some previously suggested values (Searson et al. 2010) and after following a trial and error approach and are presented in Table 12.1.

Using the optimum values of controlling parameters as given in Table 12.1 different LI_p models were developed running the MGGP code several times. These models are analyzed with respect to physical interpretation of LI_p as well as their rate of successful prediction capability and the “best” LI_p model was selected. The developed model is presented below as Eq. (12.7).

$$LI_p = 1.779\,\tanh(8.249CSR_{7.5}) - 0.0067V_s + \frac{0.0069}{CSR_{7.5}} - \frac{7.694CSR_{7.5}\exp(FCI)}{\sigma_v^{'}} + 0.221 \quad (12.7)$$

The developed LI_p model has been characterized by Figs. 12.4, 12.5, and 12.6. Figure 12.4 shows the variation of the best fitness (log values) and mean fitness with number of generations. It can be seen from this figure, the fitness values decrease with increasing the number of generations and its decrements. The best fitness was found at the 149th generation (fitness = 0.349). The statistical significance of each of the three genes and bias of the developed model is shown in Fig. 12.5. As shown in Fig. 12.5 the weight (coefficient) of the gene-2 is higher than the other genes

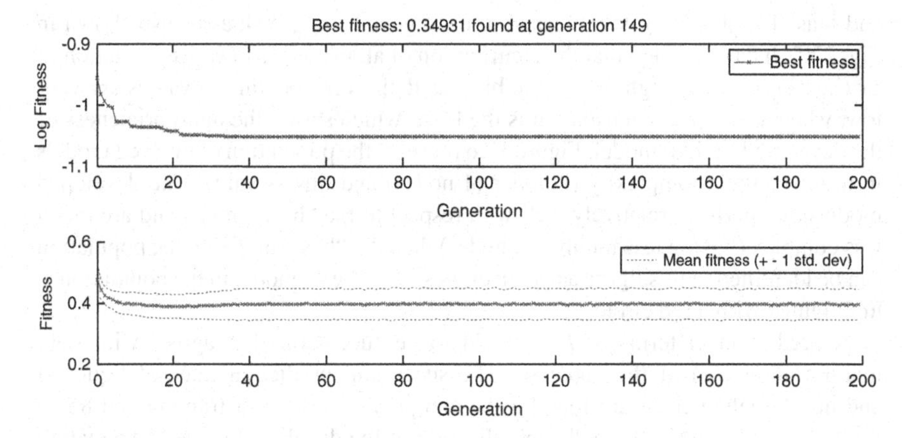

Fig. 12.4 Variation of the best and mean fitness with the number of generation

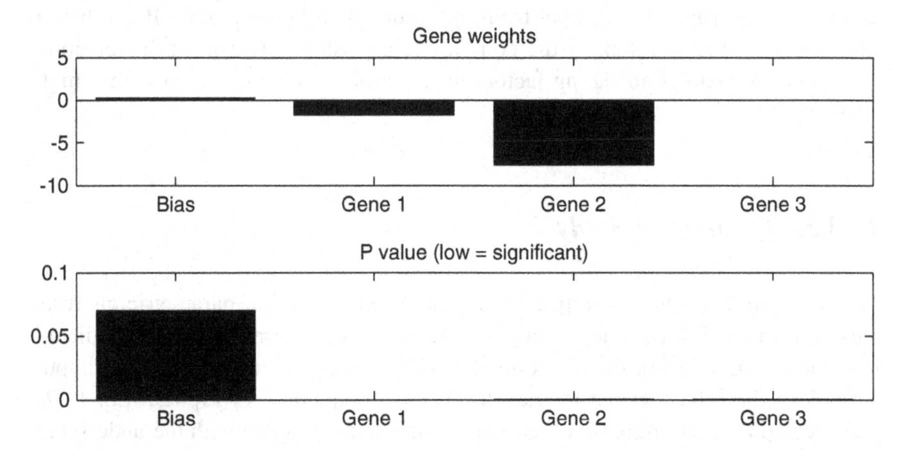

Fig. 12.5 Statistical properties of the evolved MGGP-based LI_p model (on training data)

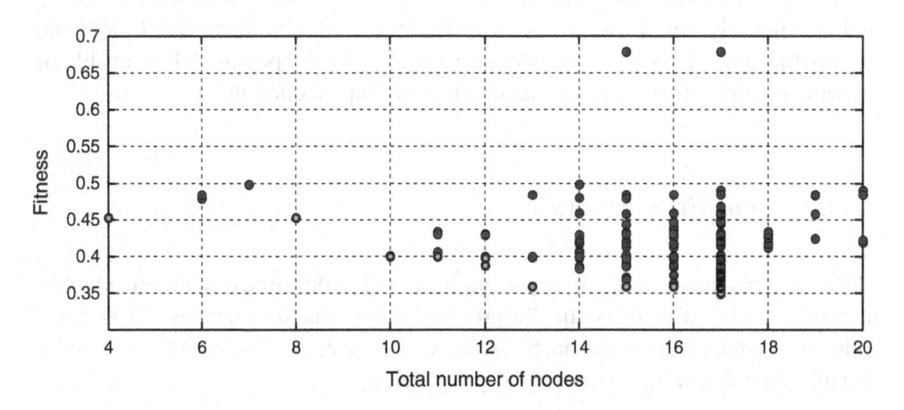

Fig. 12.6 Population of evolved models in terms of their complexity and fitness

and bias. The degree of significance of each gene using p values is also shown in Fig. 12.5. It can be noted that the contribution of all the genes towards prediction of LI (i.e., LI_p) is very high except the bias, as their corresponding p values are very low, whereas the bias contribution is the least, which shows the appropriateness of the developed MGGP model. Figure 12.6 presents the population of evolved models in terms of their complexity (number of nodes) and fitness value. The developed models that perform relatively well with respect to the "best" model and are much less complex (having less number of nodes) than the "best" model in the population can be identified in this figure as green circles. The "best" model in the population is highlighted with a red circle.

A prediction in terms of LI_p is said to be successful if it agrees with field manifestation (LI) of the database. The successful prediction rates of liquefied and non-liquefied cases are found to be comparable, 88 % for training and 86 % for testing data, showing good generalization of the developed model. The overall success rate of the trained model in predicting liquefaction and non-liquefaction cases is 87 %. Thus, it is evident from the results that the proposed MGGP based LI_p model is able to establish the complex relationship between the liquefaction index and its main contributing factors in terms of a model equation with a high accuracy.

12.3.2 Parametric Study

For verification of the developed MGGP-based LI_p model, a parametric analysis was performed. The parametric analysis investigates the response of the predicted liquefaction index from the above model with respect to the corresponding input variables. The robustness of the developed model equation for LI_p (i.e. Eq. 12.7) is evaluated by examining how well the predicted values agree with the underlying physical behavior of occurrence of liquefaction.

As it can be observed from Fig. 12.7, that LI_p decreased with increase in V_s, and FCI linearly. But, it can be seen that LI_p increased with increasing $CSR_{7.5}$ and σ'_v nonlinearly. The above results confirm that the developed model is capable of showing the important physical characteristics of liquefaction index.

12.3.3 Sensitivity Analysis

The sensitivity analysis is an important aspect of a developed model to identify important input parameters. In the present study, sensitivity analysis was made following Gandomi et al. (2013a, b). As per Gandomi et al. (2013a, b) the sensitivity (S_i) of each parameter, is expressed as given below:

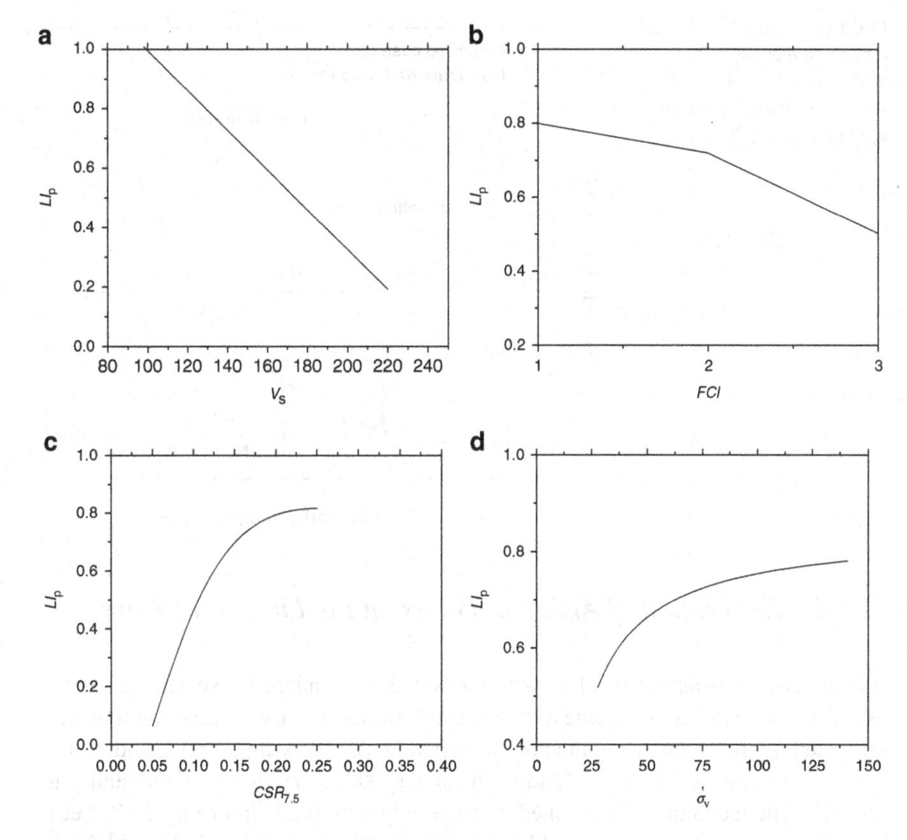

Fig. 12.7 Parametric analysis of LI_p for developed MGGP-based model

Table 12.2 Sensitivity analysis of inputs for the developed MGGP-based LI_p model

Parameters	V_s	FCI	σ'_v	$CSR_{7.5}$
Sensitivity (%)	46.2	10.5	11.3	32.0
Rank	1	4	3	2

$$N_i = f_{\max}(x_i) - f_{\min}(x_i) \tag{12.8a}$$

$$S_i = \frac{N_i}{\sum_{j=1}^{n} N_j} \times 100 \tag{12.8b}$$

where $f_{\max}$ (x_i) and $f_{\min}$ (x_i) are respectively the maximum and minimum of the predicted output (i.e. LI_p) over the ith input domain, where other variables are equal to their average values. Table 12.2 presents the results of above analysis for the proposed MGGP model. Thus, V_s is the most important parameter. The other important inputs are $CSR_{7.5}$, and σ'_v with FCI is the least important parameter.

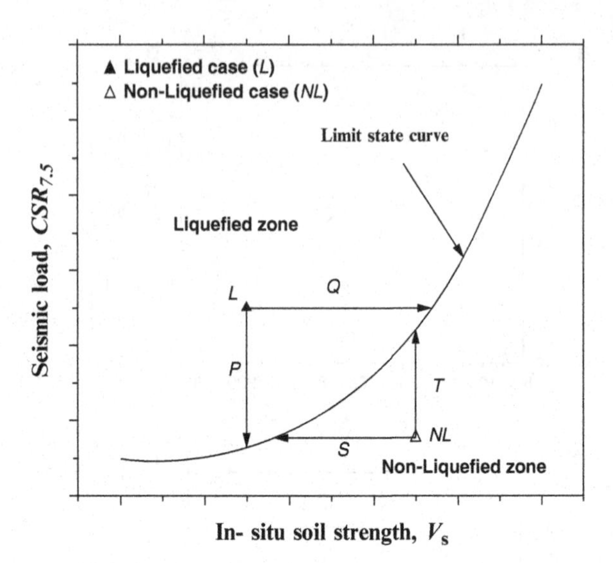

Fig. 12.8 Conceptual model for search technique for artificial data points on limit state curve (modified from Juang and Chen 2000)

12.3.4 Generation of Artificial Points on the Limit State Curve

As discussed earlier artificial data points on the boundary curve are generated using Eq. (12.7) to approximate a function, referred as limit state function that will separate liquefied cases from the non–liquefied ones, following a simple and robust search technique developed by Juang and Chen (2000) and used by Muduli and Das (2013b). The technique is explained conceptually with the help of Fig. 12.8. Let a liquefied case, '*L*' (target output $LI=1$) of the database as shown in Fig. 12.8 can be brought to the boundary or limit state curve [i.e. when the case becomes just non-liquefied as per the evaluation by Eq. (12.7)] if $CSR_{7.5}$ is allowed to decrease (path *P*) or V_s is allowed to increase (path *Q*). Further, for a non-liquefied case, '*NL*' (target output $LI=0$) of the database, the search for a point on the boundary curve involves an increase in $CSR_{7.5}$ (path *T*) or a decrease in V_s(path *S*) and the desired point is obtained when the case just becomes liquefied as adjudged by Eq. (12.7). Figure 12.9 shows the detailed flowchart of this search technique for path '*P*' and '*T*'. A multi-dimensional (Vs, $\sigma_v{}'$, *FCI*, $CSR_{7.5}$) data point on the unknown boundary curve is obtained from each successful search. In this study, the limit state is defined as the 'limiting' $CSR_{7.5}$, which a soil can resist without occurrence of liquefaction and beyond which the soil will liquefy. Thus, for a particular soil at it's in-situ conditions, this limit state specifies its *CRR* value. A total of 210 multi-dimensional artificial data points (Vs, $\sigma_v{}'$, *FCI*, *CRR*), which are located on the boundary curve are generated using the developed model (Eq. 12.7) and the technique explained in Figs. 12.8 and 12.9. These data points are used to approximate the limit state function in the form of $CRR=f(Vs, \sigma_v{}', FCI)$ as per MGGP and is presented below.

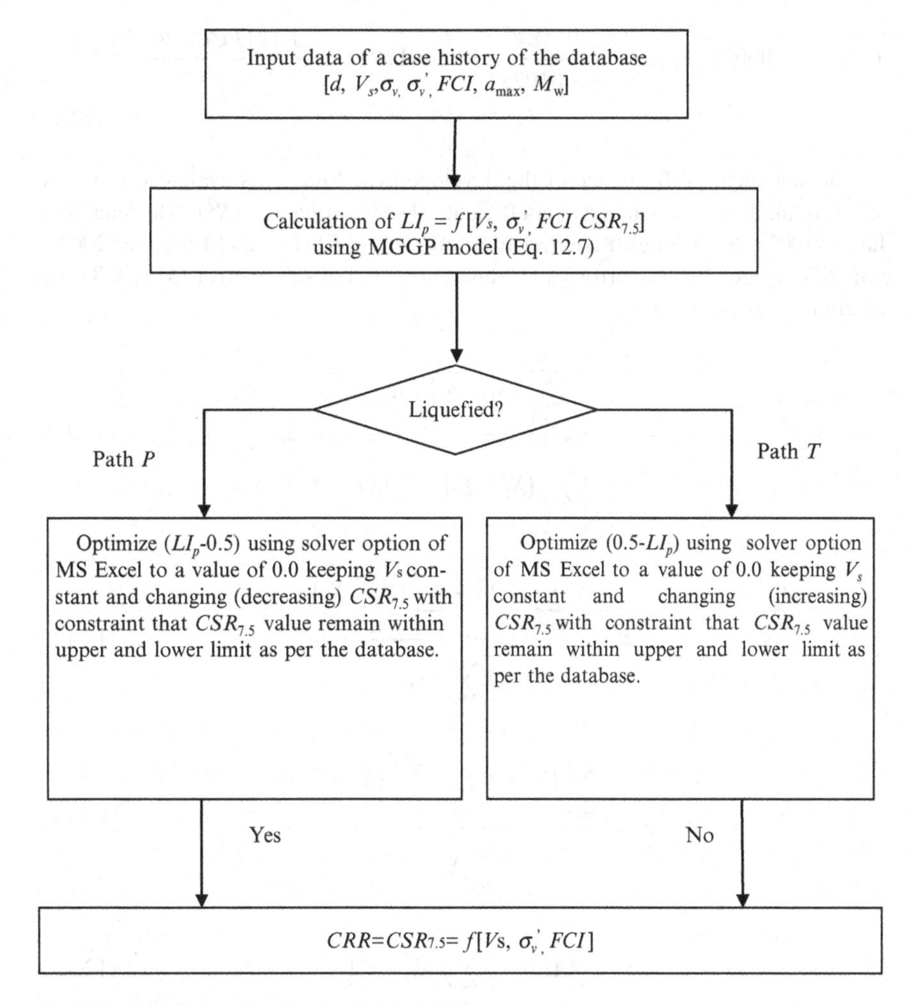

Fig. 12.9 Search algorithm for data point on limit state curve

12.3.5 MGGP Model for CRR

The multi-gene GP is also used for development of *CRR* model using 210 artificially generated data points, out of which 140 data points are selected randomly for training and rest 70 numbers for testing the developed model. The optimum values of the controlling parameters are obtained as explained above using the range of values given in Table 12.1. Several *CRR* models were obtained with the optimum values of controlling parameters by running the MGGP program several times. Then, the developed models were analyzed with respect to physical interpretation of *CRR* of soil and after careful consideration of various alternatives the following expression (Eq. 12.9) was found to be most suitable prediction model for *CRR*.

$$CRR = 0.0000144V_s^2 - \frac{0.000158V_s^2}{\sigma_v^{'} - 8.010754} - 0.001704V_s + \frac{1.171FCI \times V_s}{\sigma_v^{'2}} + 0.0458 \tag{12.9}$$

The statistical performance of the developed *CRR* model is evaluated in terms of correlation coefficient (*R*), coefficient of determination (R^2) (Rezania and Javadi 2007), Nash-Sutcliff coefficient of efficiency (*E*) (Das and Basudhar 2008), *RMSE*, average absolute error (*AAE*) and maximum absolute error (*MAE*). These coefficients are defined as:

$$R = \frac{\sum_{i=1}^{n} (X_t - \overline{X_t})(X_P - \overline{X_P})}{\sqrt{\sum_{i=1}^{n} (X_t - \overline{X_t})^2 \sum_{i=1}^{n} (X_P - \overline{X_P})^2}} \tag{12.10}$$

$$R^2 = \frac{\sum_{i=1}^{n} (X_t)^2 - \sum_{i=1}^{n} (X_t - X_p)^2}{\sum_{i=1}^{n} (X_t)^2} \tag{12.11}$$

$$E = \frac{\sum_{i=1}^{n} (X_t - \overline{X_t})^2 - \sum_{i=1}^{n} (X_t - X_p)^2}{\sum_{i=1}^{n} (X_t - \overline{X_t})^2} \tag{12.12}$$

$$AAE = \frac{1}{n}\sum_{i=1}^{n} |X_t - X_p| \tag{12.13}$$

$$MAE = \max |X_t - X_p| \tag{12.14}$$

$$RMSE = \sqrt{\frac{\sum_{i=1}^{n} (X_t - X_p)^2}{n}} \tag{12.15}$$

where n is the number of case histories and X_t and X_p are the measured (i.e., target) and predicted values (of *CRR* in this case), respectively. $\overline{X_t}$ is the average of measured values. In addition, another criterion the performance index (ρ), which is a combination of *R* and *RMSE* as proposed by Gandomi et al. (2014), is also used to evaluate the performance of the developed model because *R* will not change

Table 12.3 Statistical performances of developed MGGP based *CRR* model

Data	R^2	E	AAE	MAE	$RMSE$	ρ
Training (140)	0.94	0.83	0.02	0.21	0.04	0.14
Testing (170)	0.95	0.85	0.02	0.13	0.03	0.13

significantly by shifting the output values of a model equally, and error functions (e.g. *RMSE* and *AAE*) only shows the error not the correlation and is presented below.

$$\rho = \frac{RMSE}{\overline{X_t}} \times \frac{1}{R+1} \tag{12.16}$$

Thus, statistical performances: R^2, E, $RMSE$, AAE, MAE and ρ of the developed *CRR* model as presented in Table 12.3 for training and testing data are comparable showing good generalization capability of the *CRR* model, which also ensures that there is no over-fitting. The developed *CRR* model is found to be very compact and comprehensive for use by the geotechnical professionals. The performance of the proposed *CRR* model is also evaluated by calculating the F_s for each case of the present database. In deterministic approach $F_s \leq 1$ predicts occurrence of liquefaction and $F_s > 1$ refers to non-liquefaction. A prediction (liquefaction or non-liquefaction) is considered to be successful if it agrees with the field manifestation. The deterministic approach is preferred by the geotechnical professionals and various design decisions for further works to be taken up at the site under consideration are taken on the basis of F_s. In the present study, Eq. (12.9) in conjunction with the model for $CSR_{7.5}$ (Eq. 12.4) forms the proposed V_S-based deterministic method for evaluation of liquefaction potential. The performance of the proposed MGGP-based deterministic model is compared with that of the ANN-based model (Juang and Chen 2000) and statistical model (Andrus and Stokoe 1997) and the results are presented in Table 12.4. It is noted from Table 12.4 that the success rate in prediction of liquefied cases is 91 % and that for non-liquefied cases is 85 % and the overall success rate is found to be 88 % by the present MGGP model, whereas the accuracies in prediction of liquefied cases, non-liquefied cases and for overall cases are 88, 88 and 88 %, respectively by the available ANN-based deterministic model (Juang and Chen 2000); 99, 40 and 68 % respectively by the available statistical model (Andrus and Stokoe 1997). This clearly indicates the robustness of the proposed deterministic model as it is at par with the available ANN-based model and better than available statistical model. However, it may be mentioned here that the ANN is a 'black box' system and the expression for F_s is not comprehensive for future use.

Table 12.4 Comparison of performance of the developed MGGP-based deterministic model with ANN-based model of Juang and Chen (2000) and Statistical model of Andrus and Stokoe (1997)

Performance in terms of successful prediction (%)								
Liquefied cases(88)			Non-liquefied cases(98)			Overall (186)		
MGGP	ANN	Statistical	MGGP	ANN	Statistical	MGGP	ANN	Statistical
91	88	99	85	88	40	88	88	68

12.3.6 Parametric Study

For verification of the developed MGGP-based *CRR* model, a parametric analysis was performed. The robustness of the developed model equation for *CRR* (Eq. 12.9) is evaluated by examining how well the predicted values agree with the underlying physical behavior of cyclic resistance ratio of soil.

It is observed from Fig. 12.10, that *CRR* increased with increasing $V_{s,}$ nonlinearly whereas it increased linearly with *FCI*. But, it can be noted that *CRR* decreased with increasing σ'_v nonlinearly. The above results confirm that the developed model is capable of showing the underlying physical behavior of *CRR*.

12.3.7 Sensitivity Analysis

The sensitivity analysis of the developed MGGP-based *CRR* model was done following Gandomi et al. (2013a, b) as explained in the previous section. Table 12.5 presents the results of above analysis. It is found that V_s is the most important parameter. The other important input parameter is σ'_v with *FCI* is the least important parameter.

The developed methodology is presented with examples for non-liquefied and liquefied cases separately for easier in using the results of the present study.

Example No. 1: Deterministic Evaluation of a Non-liquefied Case

This example is of a non-liquefied case. Field observation of the site, which is designated as Salinas River, north of *1989 Loma Prieta California* earthquake (as cited in Juang and Chen 2000), indicated no occurrence of liquefaction. The mean values of seismic and soil parameters at the critical depth (9.85 m) are given as follows: $Vs = 177$ m/s, $\sigma_v = 178.2$ kPa, $\sigma_v{}' = 140.8$ kPa, soil type index/ soil class number = 1.5, $a_{max} = 0.15$ g and $M_w = 7.1$.

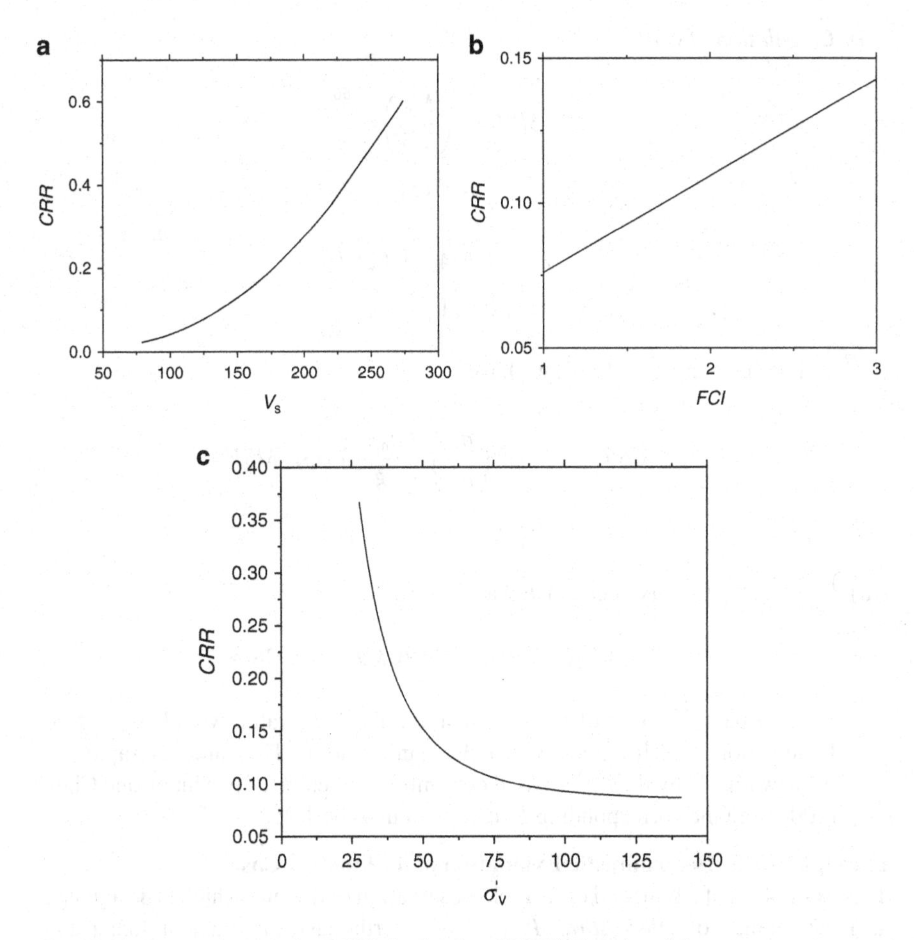

Fig. 12.10 Parametric analysis of *CRR* for developed MGGP-based model

Table 12.5 Sensitivity analysis of inputs for the developed MGGP-based *CRR* model

Parameters	V_s	*FCI*	σ'_v
Sensitivity (%)	51.4	10.0	38.6
Rank	1	3	2

(i) *Calculation of CRR*
Soil class number 1.5 corresponds to silt to sand mixture with fines content more than 35 %, which corresponds to *FCI* value of 3. *CRR* is calculated using Eq. (12.9)

$$CRR = 0.0000144V_s^2 - \frac{0.000158V_s^2}{\sigma'_v - 8.010754} - 0.001704V_s + \frac{1.171FCI \times V_s}{\sigma'^2_v} + 0.0458 = 0.189$$

(ii) *Calculation of CSR*

$$MSF = \left(\frac{M_w}{7.5}\right)^{-2.56}$$
$$= 1.151$$

$$r_d = 1.174 - 0.0267z,$$
$$= 0.911$$

Finally using Eq. (12.4), we have

$$CSR_{7.5} = 0.65\left(\frac{\sigma_v}{\sigma_v^{'}}\right)\left(\frac{a_{\max}}{g}\right)(r_d)/MSF$$
$$= 0.098$$

(iii) The factor of safety is calculated as follows:

$$F_s = CRR/CSR = 0.189/0.098 = 1.928 > 1$$

As a back analysis of the case history, above F_s value would suggest no liquefaction, which agrees with field manifestation. The same example has been evaluated by the ANN-based deterministic method as per Juang and Chen (2000) and the corresponding F_s is found out to be 1.262.

Example No. 2: Deterministic Evaluation of a Liquefied Case

This example is of a liquefied case. Field observation of the site, which is designated as Pence Ranch of *1983 Borah Peak, Idaho* earthquake (as cited in Juang and Chen 2000), indicated occurrence of liquefaction. The mean values of seismic and soil parameters at the critical depth (2.35 m) are given as follows: $Vs = 131$ m/s, $\sigma_v = 39.4$ kPa, $\sigma_v^{'} = 33.8$ kPa, soil type index/ soil class number = 4, $a_{\max} = 0.36$ g and $M_w = 6.9$.

(i) *Calculation of CRR*

Soil class number 4 corresponds to gravel and gravelly sand with fines content equal to 5 %, which corresponds to *FCI* value of 1. *CRR* is calculated using Eq. (12.9)

$$CRR = 0.0000144V_s^2 - \frac{0.000158V_s^2}{\sigma_v^{'} - 8.010754} - 0.001704V_s$$
$$+ \frac{1.171FCI \times V_s}{\sigma_v^{'2}} + 0.0458 = 0.099$$

(ii) *Calculation of CSR*

$$MSF = \left(\frac{M_w}{7.5}\right)^{-2.56}$$
$$= 1.238$$
$$r_d = 1.0 - 0.00765\,z,$$
$$= 0.982$$

Finally using Eq. (12.4), we have

$$CSR_{7.5} = 0.65\left(\frac{\sigma_v}{\sigma_v'}\right)\left(\frac{a_{\max}}{g}\right)(r_d)/MSF$$
$$= 0.216$$

(iii) The factor of safety is calculated as follows:

$$F_s = CRR/CSR = 0.099/0.216 = 0.458 < 1$$

As a back analysis of the case history, this F_s value would suggest liquefaction, which agrees with field manifestation. The above example has also been evaluated by the ANN-based deterministic method as per Juang and Chen (2000) and the corresponding F_s is found out to be 0.600, which confirms the finding of MGGP-based deterministic method.

12.3.8 Probabilistic Approach

Because of the parameter and model uncertainties, in liquefaction potential evaluation, $F_s > 1$ does not always correspond to non-liquefaction that it cannot guarantee a zero chance of occurrence of liquefaction and similarly, $F_s \leq 1$ does not always correspond to liquefaction. This can be explained considering the variability of *CRR* and *CSR* as shown in Fig. 12.11. If F_s is evaluated considering the mean values of *CRR* and *CSR* then, F_s is greater than 1.0. But, as per the distributions of *CSR* and *CRR* shown in Fig. 12.11, there is some probability that the *CRR* will be less than *CSR* as indicated by the shaded region of the figure, which will yield $F_s < 1$, proving the previous prediction wrong and a non-liquefied case may turn out to be a liquefied case. Thus, in recent years efforts have been made to assess the liquefaction potential in terms of probability of liquefaction (P_L) as discussed earlier.

Here, in the present study the V_s-based deterministic method as proposed in the previous section is calibrated with the liquefaction field performance observations

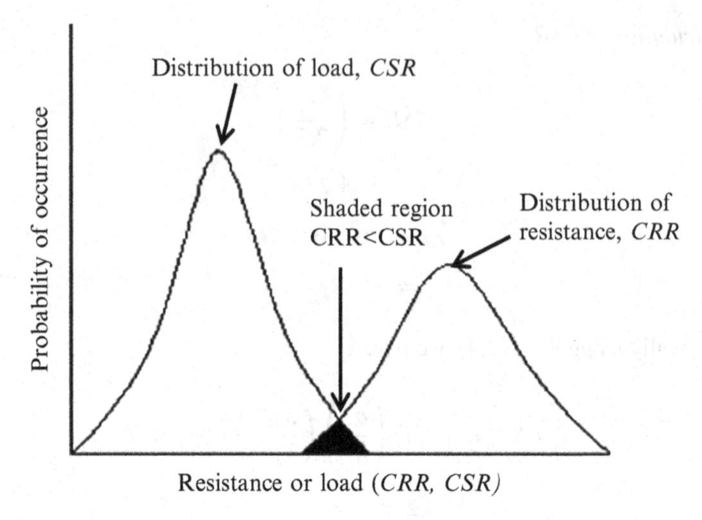

Fig. 12.11 Shows the possible distribution of *CRR* and *CSR* in liquefaction potential evaluation

using Bayesian theory of conditional probability and case histories of post liquefaction V_s database to develop a probabilistic model, referred herein as Bayesian mapping function, which is used to correlate F_s with P_L.

12.3.9 Development of Bayesian Mapping Function

According to Juang et al. (1999) the probability of liquefaction occurrence of a case in the database, for which the F_s has been calculated, can be found out using Bayes' theorem of conditional probability as given below.

$$P(L/F_s) = \frac{P(F_s/L)\,P(L)}{P(F_s/L)\,P(L) + P(F_s/NL)\,P(NL)} \tag{12.17}$$

where $P(L/F_s)$ = probability of liquefaction for a given F_s; $P(F_s/L)$ = probability of F_s, assumed that liquefaction did occur; $P(F_s/\mathrm{NL})$ = probability of F_s, assumed that liquefaction did not occur; $P(L)$ = prior probability of liquefaction; and $P(NL)$ = prior probability of non-liquefaction. $P(F_s/L)$ and $P(F_s/NL)$ can be obtained by using Eqs. (12.8a) and (12.8b), respectively.

$$P(F_s/L) = \int_{F_s}^{F_s+\Delta F_s} f_L(x)dx \tag{12.18a}$$

$$P(F_s/NL) = \int_{F_s}^{F_s+\Delta F_s} f_{NL}(x)dx \tag{12.18b}$$

where $f_L(x)$ and $f_{NL}(x)$ are the probability density functions of F_s for liquefied cases and non-liquefied cases of the database respectively. As $\Delta F_s \rightarrow 0$ Eq. (12.17) can be expressed as Eq. (12.19).

$$P(L/F_s) = \frac{f_L(F_s)\,P(L)}{f_L(F_s)\,P(L) + f_{NL}(F_s)\,P(NL)} \tag{12.19}$$

If the information of prior probabilities $P(L)$ and $P(NL)$ is available, Eq. (12.19) can be used to determine the probability of liquefaction for a given F_s. In absence of $P(L)$ and $P(NL)$ values it can be assumed that $P(L) = P(NL)$ on the basis of the maximum entropy principle (Juang et al. 1999). Thus, under the assumption that $P(L) = P(NL)$, Eq. (12.19) can be presented as Eq. (12.20).

$$P_L = \frac{f_L(F_s)}{f_L(F_s) + f_{NL}(F_s)} \tag{12.20}$$

where $f_L(F_s)$ and $f_{NL}(F_s)$ are the probability density functions (PDFs) of F_s for liquefied cases and non-liquefied cases respectively.

In the present investigation, the calculated F_s values, using the V_S-based deterministic method as presented in the previous section, for different cases of the database (Juang and Chen 2000) are grouped according to the field performance observation of liquefaction (*L*) and non-liquefaction (*NL*). Several different probability density functions are considered and out of them the three best fitting curves (Lognormal, Weibull and Rayleigh) to the histogram for both *L* and *NL* groups are shown in Fig. 12.12a–f. It is found on the basis of chi-square test for goodness-of-fit that histograms of the factor of safety of both *L* and *NL* group are best fitted by lognormal probability density function with mean (μ) and standard deviation (σ): $\mu = -0.622$, $\sigma = 0.535$ and $\mu = 0.732$, $\sigma = 1.018$ respectively as shown in Fig. 12.12a, d, respectively.

For the present V_S database, $f_L(F_s)$ and $f_{NL}(F_s)$ are the lognormal probability density functions of F_s for liquefied cases and non-liquefied cases, respectively. Based on the obtained probability density functions, P_L is calculated using Eq. (12.20) for each case in the database. The F_s and the corresponding P_L of the total 186 cases of database are plotted and the mapping function is approximated through curve (logistic) fitting as shown in Fig. 12.13. The mapping function is presented as Eq. (12.21) with a high value of R^2 (0.99).

$$P_L = \frac{1}{1 + \left(\frac{F_s}{a}\right)^b} \tag{12.21}$$

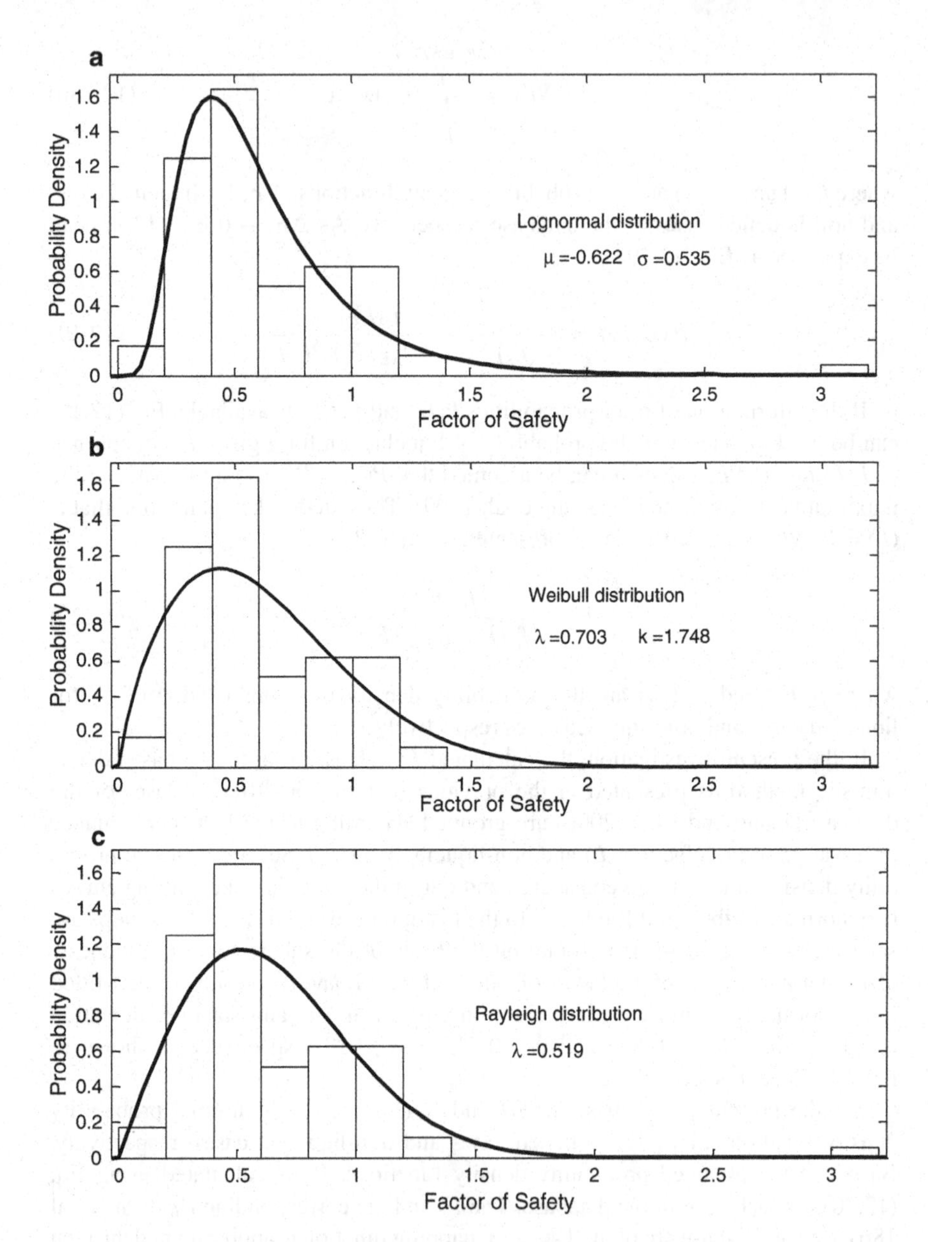

Fig. 12.12 Histogram showing the PDFs of calculated factor of safeties: (**a**),(**b**), (**c**) Liquefied (*L*) cases; (**d**),(**e**),(**f**) Non-liquefied (*NL*) cases

where a (1.04) and b (3.8) are the parameters of the fitted logistic curve. The F_s is calculated using the proposed MGGP-based deterministic method and the corresponding P_L can be found out using the developed mapping function.

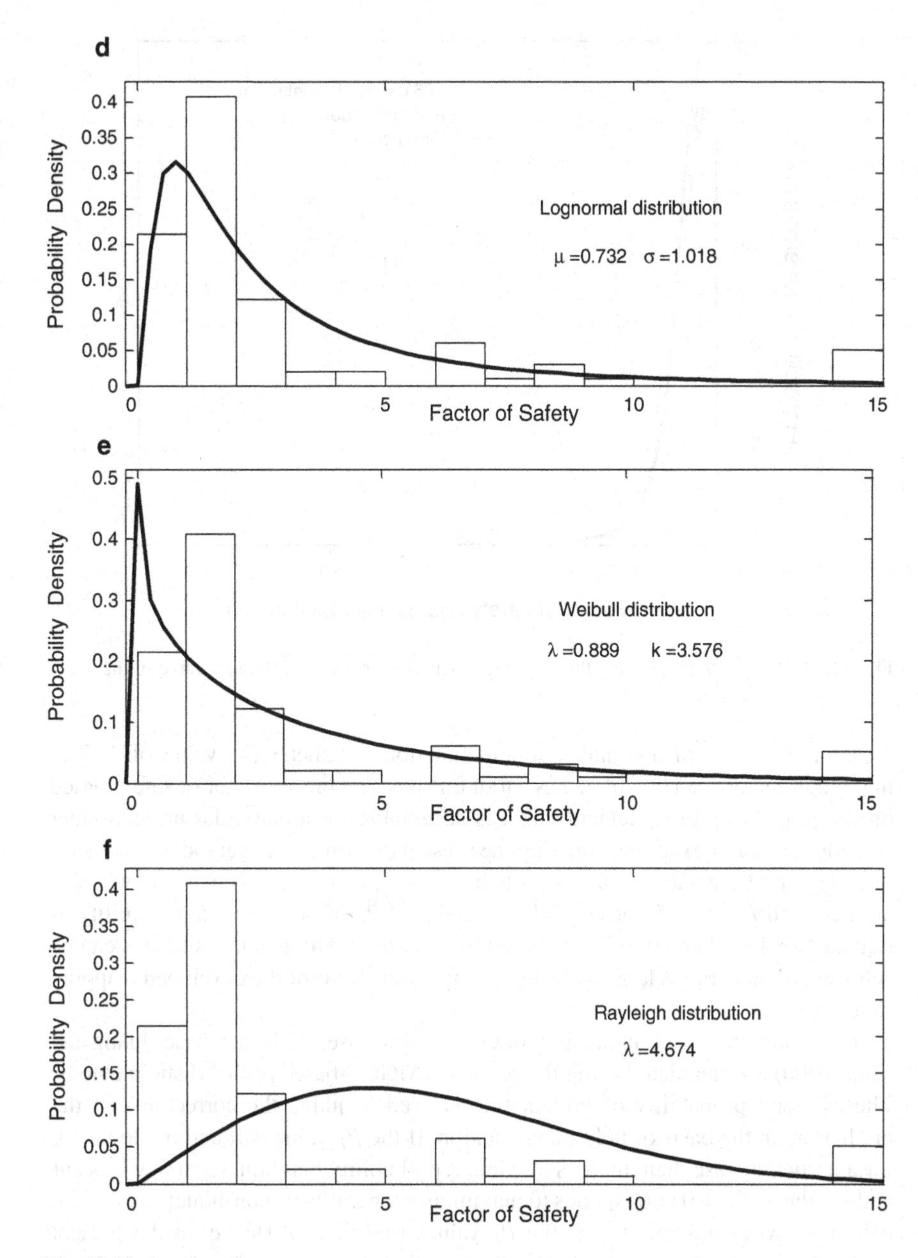

Fig. 12.12 (continued)

The proposed deterministic limit state surface ($F_s = 1.0$) corresponds to a probability of 53.7 % according to Eq. (12.21). This is close to the 50 % probability that is associated with a completely unbiased limit state curve corresponding to

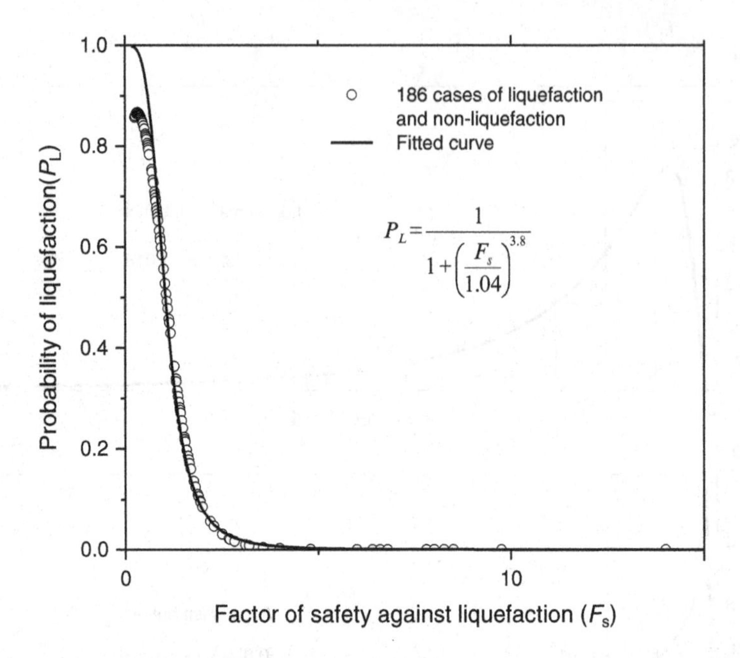

Fig. 12.13 Plot of P_L-F_s showing the mapping function approximated through curve fitting

a factor of safety of 1.0 and mapping function parameter (a) value of 1. The mapping function can be utilized as a tool for selecting proper factor of safety based on the probability of liquefaction that is acceptable for a particular project under consideration. For example, applying the present deterministic method with a factor of safety of 1.04 would result in a probability of liquefaction (P_L) of 50 %, whereas an increased F_s of 1.15 corresponds to a reduced P_L of 40.5 %. If a probability of liquefaction less than 40.5 % is required for a particular project in a site, this can be achieved by selecting a larger factor of safety on the basis of the developed mapping function.

The probabilities of liquefaction for the total 186 cases of the database (Juang and Chen 2000) are calculated using the proposed MGGP-based probabilistic methods. The assessed probability of liquefaction is used to judge the correctness of the prediction on the basis of field manifestation. If the P_L value is found out to be 1.0 for a particular case then, there is maximum probability that liquefaction will occur and similarly, $P_L = 0$ corresponds to maximum probability of non-liquefaction. But, it is not always possible to get the P_L values as 1.0 or 0. Hence, in the present study, the success rate of prediction of liquefied cases is measured on the basis of three different limits of P_L values and are as follows: [0.85–1.0], [0.65–1] and [0.5–1.0]. Similarly, for non-liquefied cases the three P_L limits considered are in the range [0, 0.15], [0, 0.35] and [0, 0.5] (Muduli et al. 2014). As per results presented in Table 12.6, the rate of successful prediction by the proposed MGGP-based probabilistic model for liquefied cases on the basis of above three different P_L limits are (38 %), (73 %) and (91 %), for non-liquefied cases the rate of successful

Table 12.6 Results of proposed MGGP-based probabilistic model based on the present database (Juang and Chen 2000)

Criterion (P_L range)	No. of successful prediction	Rate of success full prediction (%)
Based on 88 liquefied cases		
0.85–1.00	33	38
0.65–1.00	64	73
0.5–1.00	80	91
Based on 98 non-liquefied cases		
0–0.15	47	48
0–0.35	66	67
0–0.5	79	81
Based on all 186 cases		
0.85–1.00 and 0–0.15	80	43
0.65–1.00 and 0–0.35	130	70
0.5–1.00 and 0–0.5	159	85

prediction are (48 %), (67 %) and (81 %) and for overall cases of the database are (43 %), (70 %) and (85 %). The above probabilistic analysis of the present database compliments the results as obtained by the deterministic approach presented in the previous section.

Similar to deterministic approach, the above probabilistic liquefaction analysis is also presented with examples.

Example No. 3: Probabilistic Evaluation of a Non-liquefied case

This example is of a non-liquefied case that was analysed previously using the deterministic approach (see Example No. 1). As describe previously, field observation of the site, indicated no occurrence of liquefaction during *1989 Loma Prieta, California* earthquake. The mean values of seismic and soil parameters at the critical depth (9.85 m) are given as follows: $Vs = 177$ m/s, $\sigma_v = 178.2$ kPa, $\sigma_v{}' = 140.8$ kPa, soil type index/soil class number = 1.5, $a_{\max} = 0.15$ g and $M_w = 7.1$ (as cited in Juang and Chen 2000).

The F_s of the above example is found out to be 1.928.

Using the developed mapping function (Eq. 12.21) the probability of liquefaction is calculated:

$$
\begin{aligned}
P_L &= \frac{1}{1 + \left(\frac{F_s}{1.04}\right)^{3.8}} \\
&= \frac{1}{1 + \left(\frac{1.928}{1.04}\right)^{3.8}} \\
&= 0.087
\end{aligned}
$$

As the obtained P_L is within the range [0, 0.15] this is a case of non-liquefaction, which agrees with the field observation as well as the result of the deterministic approach presented in Example No. 1. The above example has also been evaluated by the ANN-based probabilistic method as per Juang et al. (2001) and the corresponding P_L is found out to be 0.15, which confirms the finding of MGGP-based probabilistic method. As the P_L (0.087) according to MGGP model is found be less than the P_L (0.15) as per the ANN-based model and also more close to 0, the prediction capability of MGGP is considered to better than that of ANN-based method.

Example No. 4: Probabilistic Evaluation of a Liquefied Case
This example is of a liquefied case that was analyzed previously using the deterministic approach (see Example No. 2). Field observation of the site, indicated occurrence of liquefaction during *1983 Borah Peak, Idaho* earthquake. The mean values of seismic and soil parameters at the critical depth (2.35 m) are given as follows: $Vs = 131$ m/s, $\sigma_v = 39.4$ kPa, $\sigma_v' = 33.8$ kPa, soil type index/soil class number = 4, $a_{max} = 0.36$ g and $M_w = 6.9$ (as cited in Juang and Chen 2000).

The F_s of the above example is found out to be 0.458.

Using the developed mapping function (Eq. 12.21) the probability of liquefaction is calculated as given below:

$$P_L = \frac{1}{1 + \left(\frac{F_s}{1.04}\right)^{3.8}} = \frac{1}{1 + \left(\frac{0.458}{1.04}\right)^{3.8}} = 0.957$$

As the obtained P_L is within the range [0.85, 1] this is a case of liquefaction, which agrees with the field observation as well as the result of the deterministic approach presented in Example No. 2. The above example has also been evaluated by the ANN-based probabilistic method as per Juang et al. (2001) and the corresponding P_L is found out to be 0.637, which is in the range [0.5–1] of a liquefied case. As the P_L (0.957) as per MGGP model is found be greater than the P_L (0.637) of ANN-based model and more close to 1, the prediction capability of MGGP is considered to better than that of ANN-based method.

The present findings from the above two examples show that the MGGP-based probabilistic method is more accurate than ANN-based method considering the non-liquefied [P_L of MGGP method (0.087) < P_L of ANN method (0.15)] as well as liquefied case [P_L of MGGP method (0.957) > P_L of ANN method (0.637)].

12.4 Conclusion

12.4.1 Conclusions Based on V_s-Based Deterministic Approach

The following conclusions are drawn from the results and discussion of the V_s-based liquefaction potential evaluation studies by the proposed deterministic approach:

1. V_S-based post liquefaction database available in literature is analyzed using multi-gene genetic programming to predict the liquefaction potential of soil in terms of liquefaction field performance indicator, *LI*.
2. The efficacy of the developed MGGP-based LI_p model in terms of rate of the successful prediction of liquefied and non-liquefied cases are comparable, 88 % for training and 86 % for testing data, showing good generalization of the developed model.
3. From the parametric study it is found that LI_p decreased with increasing V_S, and *FCI* linearly. But, LI_p increased with increasing $CSR_{7.5}$ and σ'_v nonlinearly. The above results confirm that the developed model is capable of showing the important physical characteristics of liquefaction index.
4. Based on sensitivity analyses the measured V_S are found to be "most" important parameter contributing to the prediction of liquefaction index (LI_p) as well as *CRR*.
5. For the proposed deterministic method based on developed *CRR* model and widely used $CSR_{7.5}$ model, the rates of successful prediction of liquefaction and non-liquefaction cases are 91, and 85 % respectively. The overall success rate of the proposed method for all 186 cases in the present database is found to be 88 %. The performance of the present deterministic method is at par with that of the ANN-based method (Juang and Chen 2000) and better than that of statistical method (Andrus and Stokoe 1997). But, unlike ANN based method the present methodology presents a compact expression for easier in prediction.
6. From the parametric study it is found that *CRR* increased with increasing V_S, nonlinearly whereas it increased with increasing *FCI* linearly. But, *CRR* decreased with increasing σ'_v nonlinearly.

12.4.2 Conclusions Based on V_s-Based Probabilistic Analysis

The following conclusions are drawn based on the results and discussion as presented above for V_S-based probabilistic evaluation of liquefaction potential.

1. The proposed V_S-based deterministic method is characterized with a probability of 53.7 % by means of the developed Bayesian mapping function relating F_s to P_L. The developed Bayesian mapping function can be utilized as a tool for selecting proper factor of safety in deterministic approach based on the

probability of liquefaction that is acceptable for a particular project under consideration. For example, applying the present deterministic method with a factor of safety of 1.04 would result in a probability of liquefaction (P_L) of 50 %, whereas an increased F_s of 1.15 corresponds to a reduced P_L of 40.5 %. If a probability of liquefaction of less than 40.5 % is required, it can be achieved by selecting a larger F_s based on the proposed mapping function.

2. The present findings from the two examples show that the MGGP-based probabilistic method is more accurate than that of available ANN-based method for both the non-liquefied [P_L of MGGP method (0.087) < P_L of ANN method (0.15)] as well as liquefied [P_L of MGGP method (0.957) > P_L of ANN method (0.637)] cases.

References

Alavi, A. H., Aminian, P., Gandomi, A. H., and Esmaeili, M. A. (2011). "Genetic-based modeling of uplift capacity of suction caissons." *Expert Systems with Applications*, 38, 12608–12618.

Alavi, A. H., and Gandomi, A. H. (2012). "Energy-based numerical models for assessment of soil liquefaction." *Geoscience Frontiers*, 3(4), 541–555.

Andrus, R. D., and Stokoe, K. H. (1997). "Liquefaction resistance based on shear wave velocity." *Proc., NCEER Workshop on Evaluation of Liquefaction Resistance of Soils, Tech. Rep. NCEER-97-0022,* T. L. Youd and I. M. Idriss, eds., Nat. Ctr. for Earthquake Engrg. Res., State University of New York at Buffalo, Buffalo, 89–128.

Andrus, R. D., and Stokoe, K. H. (2000). "Liquefaction resistance of soils from shear-wave velocity." *Journal of Geotechnical and Geoenvironmental Engineering,* 126 (11), 1165–1177.

Andrus, D.A., P. Piratheepan, B.S. Ellis, J. Zhang, and C.H. Juang (2004). "Comparing Liquefaction Evaluation Methods Using Penetration-VS Relationships." *Soil Dynamics and Earthquake Engineering*, 24(9–10), 713–721.

Baziar, M. H., and Jafarian, Y. (2007). "Assessment of liquefaction triggering using strain energy concept and ANN model: Capacity Energy." *Soil Dynamics and Earthquake Engineering,* 27, 1056–1072.

Becker, D.E. (1996). "Eighteenth Canadian Geotechnical Colloquium: Limit states Design for foundations, Part I. An overview of the foundation design process." *Canadian Geotechnical Journal*, 33, 956–983.

Bierschwale, J. G., and Stokoe, K. H. (1984). "Analytical evaluation of liquefaction potential of sands subjected to the 1981 Westmorland earthquake." *Geotech. Engrg. Report 95-663*, University of Texas, Austin, Texas.

Das, S. K. and Basudhar, P. K. (2008). "Prediction of residual friction angle of clays using artificial neural network." *Engineering Geology*, 100 (3–4), 142–145.

de Alba, P., Baldwin, K., Janoo, V., Roe, G., and Celikkol, B. (1984). "Elastic-wave velocities and liquefaction potential." *Geotech. Testing J.*, 7(2), 77–87.

Dobry, R., Ladd, R. S., Yokel, F. Y., Chung, R. M., and Powell, D. (1982).*Prediction of Pore Water Pressure Buildup and Liquefaction of Sands during Earthquakes by the Cyclic Strain Method.* National Bureau of Standards, Publication No. NBS-138, Gaithersburg, MD.

Gandomi, A. H., and Alavi, A. H., (2012a). "A new multi-gene genetic programming approach to nonlinear system modeling, Part I: materials and structural Engineering Problems. " *Neural Computing and Application,* 21 (1), 171–187.

Gandomi, A. H., and Alavi, A. H. (2012b). "A new multi-gene genetic programming approach to nonlinear system modeling, Part II: Geotechnical and Earthquake Engineering Problems." *Neural Computing and Application,* 21 (1), 189–201.

Gandomi, A. H., Yun, G. J., and Alavi, A. H. (2013a). An evolutionary approach for modeling of shear strength of RC deep beams. *Materials and Structures*, 46(12), 2109–2119

Gandomi, A. H, Fridline, M. M., and Roke, D. A. (2013b). "Decision Tree Approach for Soil Liquefaction Assessment." *The Scientific World Journal*, 2013, 1–8.

Gandomi, M., Soltanpour, M., Zolfaghari, M. R., and Gandomi, A. H. (2014). "Prediction of peak ground acceleration of Iran's tectonic regions using a hybrid soft computing technique." *Geoscience Frontiers*, 1–8.

Giustolisi, O., Doglioni, A., Savic, D. A., and Webb, B.W. (2007). "A multi-model approach to analysis of environmental phenomena." *Environmental Modelling and Software*, 5, 674–682.

Goh, A. T. C. (1994). "Seismic liquefaction potential assessed by neural networks." *Journal of Geotechnical Engineering,* 120 (9), 1467–1480.

Goh, A. T. C. (2002). "Probabilistic neural network for evaluating seismic liquefaction potential." *Canadian Geotechnical Journal,* 39, 219–232.

Goh, T. C., and Goh, S. H. (2007). "Support vector machines: Their use in geotechnical engineering as illustrated using seismic liquefaction data." *Journal of Computers and Geomechanics.,* 34, 410–421.

Haldar, A., and Tang, W. H. (1979). "Probabilistic evaluation of liquefaction potential." *Journal of Geotechnical Engineering Division,* ASCE, 105(GT2), 145–163.

Hanna, A. M., Ural, D., and Saygili, G. (2007). "Neural network model for liquefaction potential in soil deposits using Turkey and Taiwan earthquake data." *Soil Dynamics and Earthquake engineering*, 27, 521–540.

Javadi, A. A., Rezania, M., and Nezhad, M. M. (2006). "Evaluation of liquefaction induced lateral displacements using genetic programming." *Journal of Computers and Geotechnics,* 33, 222–233.

Juang, C. H., Rosowsky D. V., and Tang, W. H. (1999). "Reliability based method for assessing liquefaction potential of soils." *Journal of Geotechnical and Geoenvironmental Engineering,* ASCE, 125 (8), 684–689.

Juang, C. H., and Chen, C. J., (2000) "A Rational Method for development of limit state for liquefaction evaluation based on shear wave velocity measurements." *International Journal for Numerical and Analytical methods in Geomechanics,* 24, 1–27.

Juang, C. H., Chen, C. J., Jiang, T., and Andrus, R. D. (2000). "Risk-based liquefaction potential evaluation using standard penetration tests." *Canadian Geotechnical Journal,* 37, 1195–1208.

Juang, C. H., Chen, C. J., and Jiang, T. (2001). "Probabilistic framework for liquefaction potential by shear wave velocity." *Journal of Geotechechnical and Geoenvironmental Engineering,* ASCE, 127 (8), 670–678.

Juang, C. H., Yang, S. H., and Yuan, H. (2005).Model uncertainty of shear wave velocity-based method for liquefaction potential evaluation." *Journal of Geotechnical and Geoenvironmental Engineering,* ASCE, 131 (10), 1274–1282.

Juang, C. H., Fang, S. Y., and Khor, E. H. (2006). "First order reliability method for probabilistic liquefaction triggering analysis using CPT." *Journal of Geotechechnical and Geoenvironmental Engineering* ASCE, 132 (3), 337–350.

Kayen, R. E., Mitchell, J. K., Seed, R. B., Lodge, A., Nishio, S., and Coutinho, R. (1992). "Evaluation of SPT-, CPT-, and shear wave-based methods for liquefaction potential assessment using Loma Prieta data." *Proc., 4th Japan-U.S. Workshop on Earthquake Resistant Des. of Lifeline Fac. and Countermeasures for Soil Liquefaction, Tech. Rep. NCEER-92-0019, M. Hamada and T. D. O'Rourke, eds., National Center for Earthquake Engineering Research, Buffalo, Vol. 1,* 177–204.

Koza, J. R. (1992). Genetic programming: on the programming of computers by natural selection, The MIT Press, Cambridge, Mass.

Kramer, S. L. (1996). *Geotechnical earthquake engineering,* Pearson Education (Singapore) Pte. Ltd., New Delhi, India.

Liao, S. S. C., Veneziano, D., and Whitman, R. V. (1988). "Regression models for evaluating liquefaction probability." *Journal of Geotechnical Engineering Division,* ASCE, 114(4), 389–411.

MathWorks Inc. (2005), *MatLab User's Manual, Version 6.5*, The MathWorks Inc., Natick.
Muduli, P. K., and Das, S. K. (2013a). "First order reliability method for probabilistic evaluation of liquefaction potential of soil using genetic programming". *International Journal of Geomechanics*, doi:10.1061/(ASCE)GM.1943-5622.0000377.
Muduli, P. K., and Das, P. K. (2013b). "SPT-Based Probabilistic Method for Evaluation of Liquefaction Potential of Soil Using Multi- Gene Genetic Programming". *International Journal of Geotechnical Earthquake Engineering,* 4(1), 42–60.
Muduli, P. K., Das, M. R., Samui, P., and Das, S. K. (2013). "Uplift capacity of suction caisson in clay using artificial intelligence techniques". *Marine Georesources and Geotechnology*, 31(4), 375–390.
Muduli, P. K., and Das, P. K. (2014a). "Evaluation of liquefaction potential of soil based on standard penetration test using multi-gene genetic programming model." *Acta Geophysica*, 62 (3), 529–543.
Muduli, P. K., and Das, P. K. (2014b). "CPT-based Seismic Liquefaction Potential Evaluation Using Multi-gene Genetic Programming Approach" *Indian Geotech Journal,* 44(1), 86–93.
Muduli, P. K., Das, P. K., and Bhattacharya, S. (2014). "CPT-based probabilistic evaluation of seismic soil liquefaction potential using multi-gene genetic programming". *Georisk: Assessment and Management of Risk for Engineered Systems and Geohazards,* 8(1), 14–28.
Olsen, R. S. (1997). "Cyclic liquefaction based on the cone penetration test." *Proceedings of the NCEER Workshop of Evaluation of liquefaction resistance of soils. Technical report No. NCEER-97-0022,* T. L. Youd and I. M. Idriss, eds., Buffalo. NY: National center for Earthquake Engineering Research. State University of New York at Buffalo. 225–276.
Pal, M. (2006). "Support vector machines-based modeling of seismic liquefaction potential." *Journal of Numerical and Analytical Methods in Geomechanics*, 30, 983–996.
Rezania, M., and Javadi, A. A. (2007). "A new genetic programming model for predicting settlement of shallow foundations." *Canadian Geotechnical Journal*, 44, 1462–1473
Robertson, P. K., and Campanella, R. G. (1985). "Liquefaction potential of sands using the CPT." *Journal of Geotechnical Engineering,* ASCE, 111(3), 384–403.
Robertson, P. K., and Wride, C. E. (1998). "Evaluating cyclic liquefaction potential using cone penetration test." *Canadian Geotechnical journal,* 35(3), 442–459
Samui, P. (2007). "Seismic liquefaction potential assessment by using Relevance Vector Machine. " *Earthquake Engineering and Engineering Vibration*, 6 (4), 331–336.
Samui.P., and Sitharam, T. G. (2011). "Machine learning modelling for predicting soil liquefaction susceptibility." *Natural Hazards and Earth Sciences,* 11, 1–9.
Searson, D. P., Leahy, D. E., and Willis, M. J. (2010). "GPTIPS: an open source genetic programming toolbox from multi-gene symbolic regression." *In: Proceedings of the International multi conference of engineers and computer scientists,* Hong Kong.
Seed H. B., and Idriss, I. M. (1982).*Ground Motions and Soil Liquefaction During Earthquakes*, Earthquake Engineering Research Institute, Oakland, CA, 134.
Seed, H. B., and de Alba, P. (1986). "Use of SPT and CPT tests for evaluating liquefaction resistance of sands." *Proc., Specialty Conf. on Use of In Situ Testing in Geotechnical Engineering, Geotechnical Special Publ. No. 6*, ASCE, New York, 281–302.
Seed, H. B., and Idriss, I. M. (1971). "Simplified procedure for evaluating soil liquefaction potential." *Journal of the Soil Mechanics and Foundations Division,* ASCE, 97(SM9), 1249–1273.
Seed, H. B., Idriss, I. M., and Arango, I. (1983). "Evaluation of liquefaction potential using field performance data." *Journal of Geotechnical Engineering Division,* ASCE, 109 (3), 458–482.
Seed, H. B., Tokimatsu, K., Harder, L. F., and Chung, R. (1985). "Influence of SPT procedures in soil liquefaction resistance evaluations." *Journal of Geotechnical Engineering,* ASCE, 111(12), 1425–1445.
Stokoe, K. H., II, and Nazarian, S. (1985). "Use of Rayleigh waves in liquefaction studies." *Measurement and use of shear wave velocity for evaluating dynamic soil properties*, R. D. Woods, ed., ASCE, New York, 1–17.

Stokoe, K. H., II, Roesset, J. M., Bierschwale, J. G., and Aouad, M. (1988). "Liquefaction potential of sands from shear wave velocity." *Proc., 9th World Conf. on Earthquake Engrg.*, Vol. III, 213–218.

Terzaghi, K., and Peck, R.B. (1948). *Soil mechanics in engineering practice,* 1st Edition, John Wiley & Sons, New York.

Tokimatsu, K., and Uchida, A. (1990). "Correlation between liquefaction resistance and shear wave velocity." *Soils and Foundations*, Tokyo, 30(2), 33–42.

Toprak, S., Holzer, T. L., Bennett, M. J. and Tinsley, J. C. III, (1999). "CPT- and SPT-based probabilistic assessment of liquefaction." *Proceedings 7th U.S.–Japan Workshop on Earthquake Resistant Design of Lifeline Facilities and Counter measures against Liquefaction*, Seattle, Multidisciplinary Center for Earthquake Engineering Research, Buffalo, N.Y., 69–86.

Yang, C.X., L.G. Tham, X.T. Feng, Y.J. Wang, and P.K.K. Lee (2004), Two stepped evolutionary algorithm and its application to stability analysis of slopes, J. Comput. Civil. Eng. 18, 145–153.

Youd, T. L., and Nobble, S. K. (1997). "Liquefaction criteria based statistical and probabilistic analysis." *In: Proceedings of NCEER workshop on Evaluation of Liquefaction Resistance of Soils, technical Report No. NCEER-97-0022*, State University of New York at Buffalo, Buffalo, New York, 201–216.

Youd, T. L., Idriss I. M., Andrus R. D., Arango, I., Castro, G., Christian, J. T., Dobry, R., Liam Finn, W. D., Harder Jr, L. F., Hynes, M. E., Ishihara, K., Koester, J. P., Liao, S. S. C., Marcuson III W. F., Martin, G. R., Mitchell, J. K., Moriwaki, Y., Power, M. S., Robertson, P. K., Seed, R. B., and Stokoe II, K. H. (2001). "Liquefaction resistance of soils: summary report from the 1996 NCEER and 1998 NCEER/NSF workshops on evaluation of liquefaction resistance of soils." *Journal of Geotechnical and Geoenvironmental Engineering,* ASCE, 127 (10), 817–833.

Chapter 13
Site Characterization Using GP, MARS and GPR

Pijush Samui, Yıldırım Dalkiliç, and J Jagan

13.1 Introduction

Geotechnical site characterization is an important task for any civil engineering project. Civil engineers use different in-situ tests {Standard Penetration Test (SPT), Cone Penetration Test (CPT) and shear wave velocity technique} for site characterization purpose. The main aim of site characterization is the prediction of soil properties at any point in site based on experimental data. Researchers use random field method and geostatic for geotechnical site characterization (Yaglom 1962; Lumb 1975; Vanmarcke 1977; Tang 1979; Wu and Wong 1981; Asaoka and Grivas 1982; Vanmarcke 1983; Baecher 1984; Kulatilake and Miller 1987; Kulatilake 1989; Fenton 1998; Phoon and Kulhawy 1999; Uzielli et al. 2005; Kulatilake and Ghosh 1988; Kulatilake 1989; Soulie et al. 1990; Chiasson et al. 1995; DeGroot 1996). However, the act of arbitrary field method and geostatic are not promising (Juang et al. 2001). Artificial Neural Network (ANN) has been adopted for site characterization (Juang et al. 2001; Samui and Sitharam 2010). However, ANN undergoes various restraints such as black box approach, arriving at local minima, low generalization capability, etc (Park and Rilett 1999; Kecman 2001). The problem of ANN was solved by using Support Vector Machine (SVM) (Samui and Das 2011). However, SVM has the various limitations (Tipping 2001).

P. Samui (✉) • J. Jagan
Centre for Disaster Mitigation and Management, VIT University,
Vellore 632014, Tamilnadu, India
e-mail: pijush.phd@gmail.com; janyfriends57@gmail.com

Y. Dalkiliç
Faculty of Engineering, Civil Engineering Department, Erzincan University,
Erzincan, Turkey
e-mail: yildirim.dalkilic@gmail.com

A.H. Gandomi et al. (eds.), *Handbook of Genetic Programming Applications*,
DOI 10.1007/978-3-319-20883-1_13

This article adopts Genetic Programming (GP), Multivariate Adaptive Regression Spline (MARS) and Gaussian Process Regression (GPR) for developing site characterization model of Bangalore (India) based on corrected Standard Penetration Test (SPT) value (N_c). This article uses the database collected from the work of Samui and Sitharam (2010). GP is constructed based on genetic algorithm (Koza 1992). There are lots of applications of GP in the literatures (Londhe and Charhate 2010; Guven and Kişi 2011; Alavi and Gandomi 2012; Gandomi and Alavi 2011, 2012; Danandeh et al. 2013; Zahiri and Azamathulla 2014; Yang et al. 2013; Fister et al. 2014; Langdon 2013; Alavi and Gandomi 2011, 2012). MARS is developed by Friedman (1991). It is a non-parametric regression technique. It has been successfully applied for solving different problems (Harb et al. 2010; Mao et al. 2011; Zhan et al. 2012; Kumar and Singh 2013; Garcia and Alvarez 2014). The formulation of GPR is Bayesian. GPR assumes covariance function for final prediction. It has been previously applied in different problems (Shen and Sun 2010; Xu et al. 2011; Kongkaew and Pichitlamken 2012; Fairchild et al. 2013; Holman et al. 2014). The results of GP, MARS and GPR have compared with each other. The developed GP, MARS and GPR give the spatial variability of N_c.

13.2 Methods

13.2.1 Details of GP

GP is developed based on the concept of 'survival of the fittest'. In 1st step, a random population of equations is created. The fitness value of each equation is determined in 2nd step. 'Parents' are selected in this step. In 3rd step, 'offspring's are created through the procedure of reproduction, mutation and crossover. In 4th step, the best equation is obtained.

The above flowchart exposes the simplified working procedure of the genetic programming. Crossover is one of the genetic operators used to fuse the population. This operator gently selects preferable parent's chromosome and couple them to produce an offspring. The crossover point was elected in an aimless manner. Figure 13.1 explains the process of crossover.

Mutation is another genetic operator which is identical to the crossover only in selecting the chromosomes of the parent's (i.e.) the mutation point is randomly selected. It changes the building blocks helping to breakout from the local minimal traps. The size of the mutation is also arbitrary, which helps to get the best individual.

From Fig. 13.2 we can get the exposure on the technique of mutation. These are the general details of the GP.

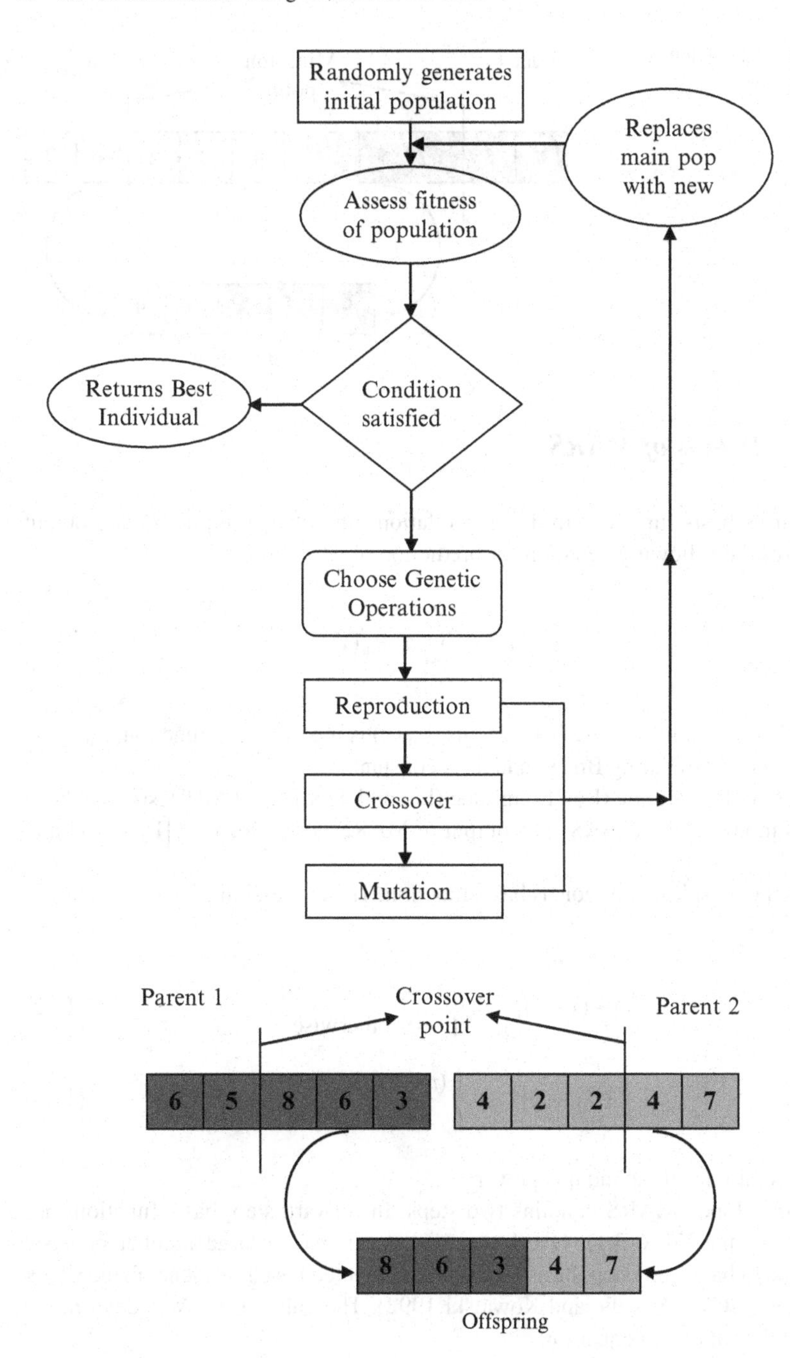

Fig. 13.1 Process of crossover

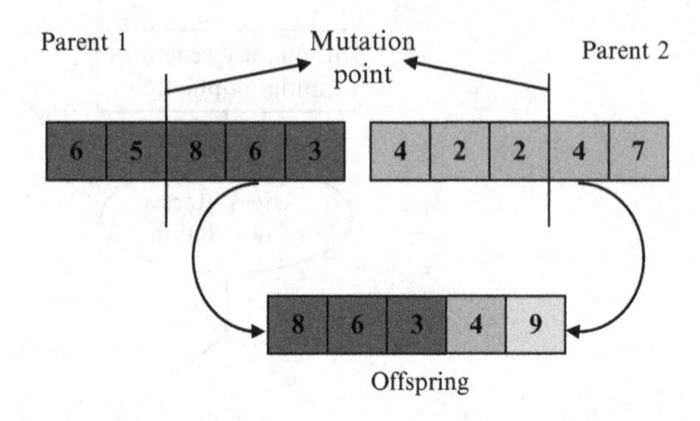

Fig. 13.2 Methodology of mutation

13.2.2 *Details of MARS*

MARS uses basis functions to develop relationship between input (x) and output (y). It uses the following equation for prediction of y.

$$y = a_0 + \sum_{m=1}^{M} a_m B_m(x) \quad (13.1)$$

where M is the number of basis functions, $B_m(x)$ is the mth basis function, a_m is the coefficient corresponding Bm(x) and a_0 is constant.

In this study, Latitude (L_x), Longitude (L_y) and depth (z) of SPT tests have been taken as inputs of the MARS. The output of MARS is N_c. So, $x = [L_x, L_y, z]$ and $y = [N_c]$.

MARS uses spline function as basis function. The expression of basis function is given below.

$$[-(x-t)]_+^q = \begin{cases} (t-x)^q, \text{if } x < t \\ 0, \text{ otherwise} \end{cases} \quad (13.2)$$

$$[+(x-t)]_+^q = \begin{cases} (t-x)^q, \text{if } x \geq t \\ 0, \text{ otherwise} \end{cases} \quad (13.3)$$

where t is knot location and q is power.

The developed MARS contains two steps. In forward step, basis functions are added to define Eq. (13.1). Overfitting can occur due to large number of basis functions. In backward step, basis functions are deleted based on Generalized Cross Validation (GCV) (Sekulic and Kowalski 1992). The value of GCV is determined by using the following equation.

$$GCV = \frac{\frac{1}{N}\sum_{i=1}^{N}[y_i - f(x_i)]^2}{\left[1 - \frac{C(B)}{N}\right]^2} \tag{13.4}$$

where N is the number of data and C(B) is a complexity penalty. The value of C(B) is given by the following expression.

$$C(B) = (B + 1) + dB \tag{13.5}$$

where B is number of basis function and d is the penalty for each basis function. These depicted details gives the basic information of MARS.

13.2.3 Details of GPR

This section will serve a detail methodology of GPR. Let us consider the following dataset (D).

$$D = \{(x_i, y_i)\}_{i=1}^{N} \tag{13.6}$$

where N is number of dataset, x is input and y is output.

In this article, L_x, L_y and z of SPT tests have been taken as inputs of the GPR. The output of GPR is N_c. So, $x = [L_x, L_y, z]$ and $y = [N_c]$.

In GPR, the relation between x and y is given below.

$$y_i = f(x_i) + \varepsilon_i \tag{13.7}$$

where f is latent real-valued function and ε is observational error.

GPR uses the following relation for prediction of new output (y_{N+1})

$$\begin{pmatrix} y \\ y_{N+1} \end{pmatrix} \sim N(0, K_{N+1}) \tag{13.8}$$

where K_{N+1} is covariance matrix and the expression of K_{N+1} is given below.

$$K_{N+1} = \begin{bmatrix} [K] & [k(x_{N+1})] \\ [k(x_{N+1})^T] & k1(x_{N+1}) \end{bmatrix} \tag{13.9}$$

where $k(x_{N+1})$ denotes covariances between training inputs and the test input and $k1(x_{N+1})$ denotes the autocovariance of the test input.

The distribution of y_{N+1} is Gaussian with the following mean (μ) and variance (σ) respectively.

Table 13.1 Statistical parameters of the dataset

Variable	Mean	Standard deviation	Skewness	Kurtosis
L_x(degree)	77.59	0.03	0.39	2.51
L_y(degree)	12.97	0.03	0.39	2.57
z(m)	5.29	3.64	5.44	1.41
N	31.49	19.23	1.07	4.08

$$\mu = k(x_{N+1})^T K^{-1} y \quad (13.10)$$

$$\sigma = k\,(x_{N+1}) - k\Big((x_{N+1})^T K^{-1} k\,(x_{N+1})\quad (13.11)$$

Radial basis function has been used as covariance function. The above mentioned information's will give the basic knowledge of GPR.

13.3 Experimental Data

In Bangalore, during the period 1995–2005 major projects were carried out. The collaborative work on Standard Penetration Test (SPT) was administered by Torsteel Research Foundation (India) and the Indian Institute of Science (Bangalore) and the data was collected from their vault. The data gathered are on average up to a depth of 40 m below the ground level. The borelogs enclose knowledge about depth, density of the soil, N values, fines content and depth of ground water table. These obtained data have been used for this study. In this study, we use three input parameters L_x, L_y, and z which represents the latitude, longitude and depth of a point.

For the construction of GP, MARS and GPR the dataset has been branched into two sets.

Training Dataset This is used to develop the model. It adopts 90 % boreholes as training dataset.

Testing Dataset It is used to evaluate the developed model. The well left 10 % boreholes are considered as testing dataset.

The dataset is normalized against their maximum values. The statistical parameters of the dataset have been tabulated in Table 13.1. The programs of GP, MARS and GPR were constructed by using MATLAB.

13.4 Model Development

In order to construct GP, the number of population is kept to 800. The number of generation is set to 200. The developed GP gives the best performance for mutation frequency = 60 and crossover frequency = 40. The final equation of the GP is given below.

Table 13.2 Details of basis functions

$B_m(x)$	Equation	a_m
$B_1(x)$	$\max(0, z - 0.387)$	0.239
$B_2(x)$	$\max(0, 0.387 - z)$	−0.891
$B_3(x)$	$\max(0, L_x - 0.387)$	0.579
$B_4(x)$	$B_2(x)\max(0, L_y - 0.971)$	60.658
$B_5(x)$	$\max(0, 0.678 - L_x)\max(0, L_y - 0.599)$	1.008
$B_6(x)$	$\max(0, 0.678 - L_x)\max(0, 0.599 - L_y)$	1.423
$B_7(x)$	$\max(0, L_y - 0.884)$	−2.0353
$B_8(x)$	$\max(0, 0.884 - L_y)$	−0.230

$$N_c = 1.24\sin\left(\sin\left(L_y + z\right)\right) + 0.9874L_y(\exp(L_x))^2 - 0.203\exp(L_x)\left(8.22L_y + z\right) + 0.272\exp(L_x)\sin\left(8.22L_y\right) - 0.897\sin\left(8.059L_y\right)\tanh(L_x) - 1.641L_xL_y(2L_x - \tanh(z)) + 0.0216 \quad (13.12)$$

The value of N_c has been predicted by using the above Eq. (13.12) for training and testing datasets.

In the event of developing MARS, 12 basis functions have been recommended in over heading step. 4 basis functions have been excluded in backward step. Finally, MARS model encloses eight basis functions. The expression of MARS (by putting $y = N_c$, $M = 8$ and $a_0 = 0.510$ in Eq. (13.1) is given below.

$$N_c = 0.510 + \sum_{m=1}^{8} a_m B_m(x) \quad (13.13)$$

Table 13.2 shows the details of a_m and $B_m(x)$.

The performance of GPR depends on ε and width (σ) of radial basis function. The design values of ε and σ have been determined by trial and error approach. The developed GPR gives better reaction at $\varepsilon = 0.04$ and $\sigma = 0.03$.

13.5 Results and Discussion

This section covers the results of the developed GP, MARS and GPR models. The following figure establishes the achievement of the GP.

The Coefficient of Correlation (R) value has been used for determining the achievement of GP. When the value of R is near to one, then we can say the developed model is good. As shown in Fig. 13.3, the performance of GP is not so good. Figure 13.4 shows the variation of N_c on the plane of $z = 1.5$ m by using the GP model. This figure helps in predicting the N_c values at different places without performing several tests.

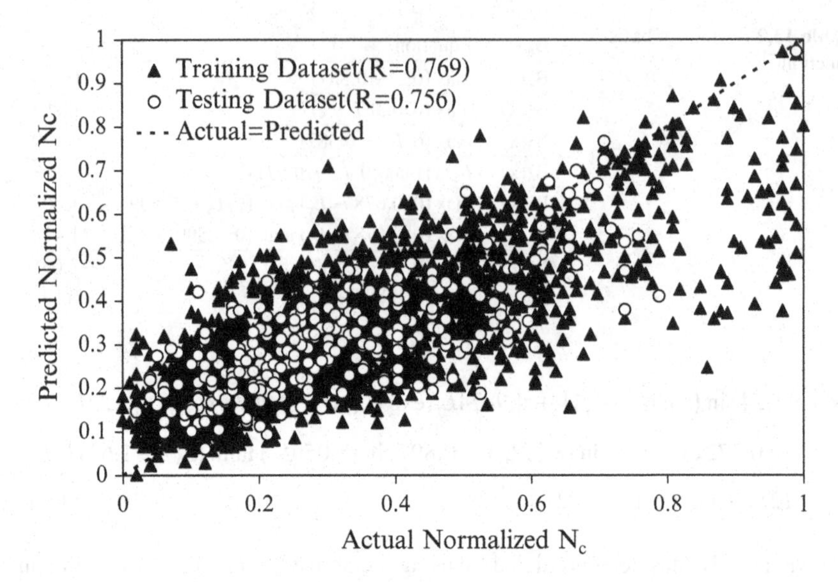

Fig. 13.3 Performance of GP

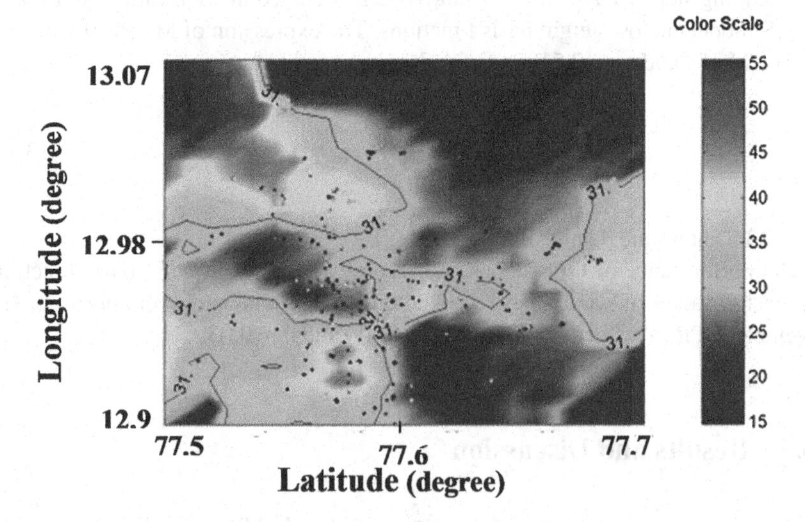

Fig. 13.4 Variation of N_c on the plane of $z = 1.5$ m by using the GP model

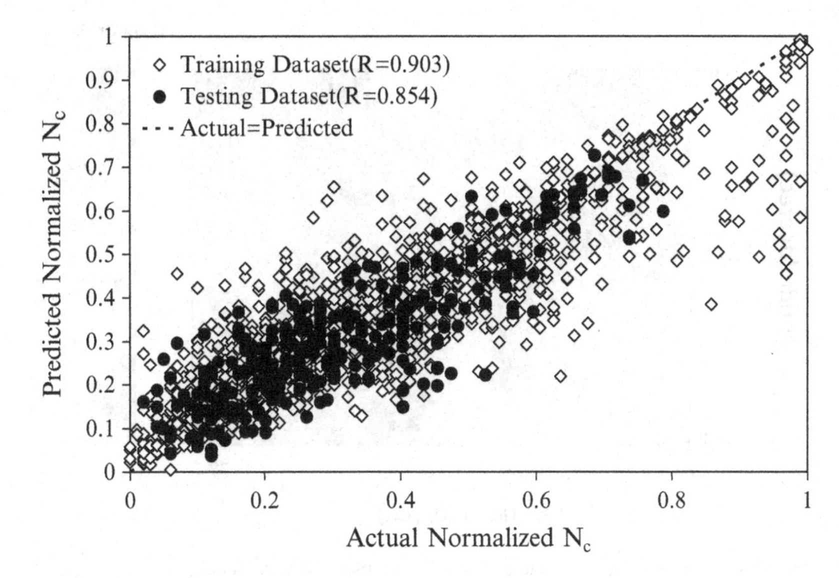

Fig. 13.5 Performance of MARS

The developed model can be evaluated by its performance. Figure 13.5 reveals the performance of MARS.

The Coefficient of Correlation (R) value is used to conclude the achievement of the developed MARS model. When the R value is close to one, we can say the developed model is a good model. Figure 13.5 shows the capability of MARS in determining the N_c value. Figure 13.6 depicts the spatial variability of N_c on the plane of $z = 1.5$ m by using MARS model. This figure helps in finding the N_c values at different places without conducting any other tests.

The GPR model also performed to its capacity it is illustrated in Fig. 13.7. When the Coefficient of Correlation (R) value is close to one, then the built model is a good model.

Figure 13.8 shows spatial variability of N_c on the plane of $z = 1.5$ m using GPR model. This figure provides the guidance for forecasting the N_c values at different places without performing several other tests.

The developed MARS and GPR uses two tuning parameter, whereas the GP use four tuning parameters for predicting N_c. The GP model gives the direct equation when compared with the other models, which could be a benefit of this model. The developed MARS gives better performance than the GPR and GP models. In GPR, it is assumed that the dataset should follow Gaussian distribution. GP and MARS do to assume any data distribution.

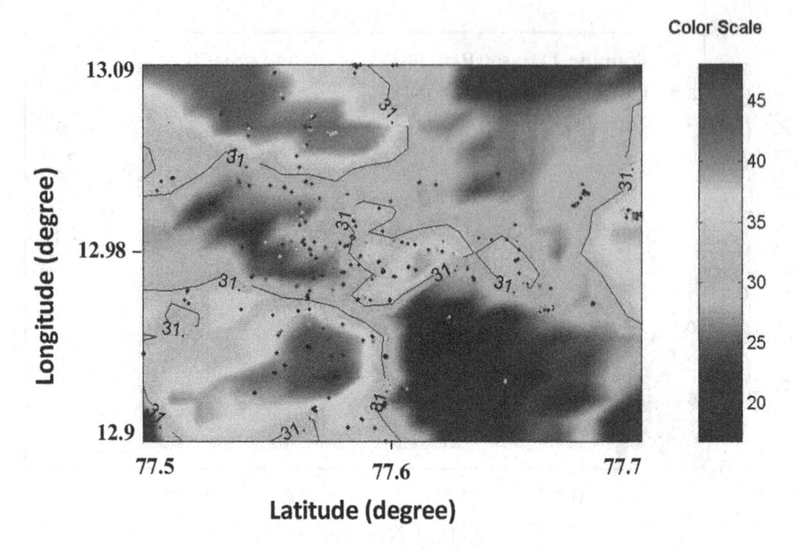

Fig. 13.6 Variation of N_c on the plane of z = 1.5 m by using MARS model

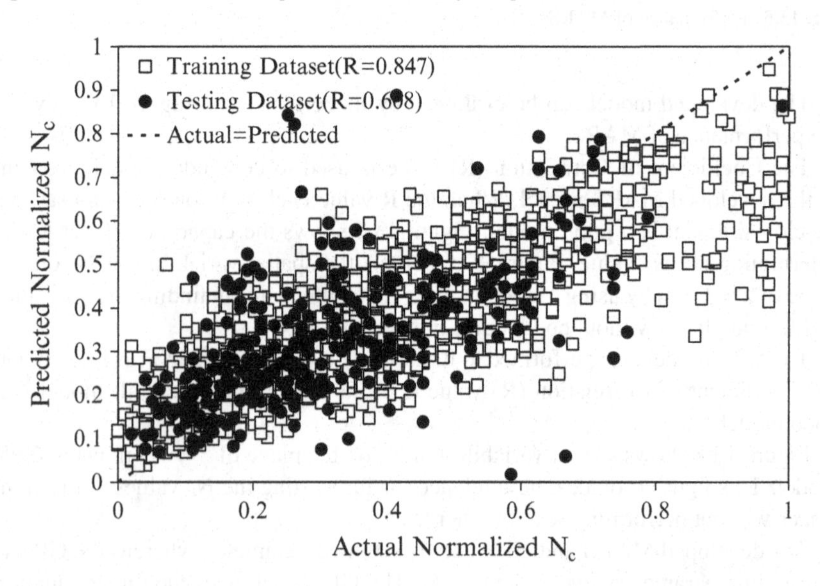

Fig. 13.7 Performance of the GPR

13.6 Sensitivity Analysis

Sensitivity analysis is an analysis used to determine the sensitivity of an output when there is a change in the input while keeping the other input constant. It assists the reviewer to resolve which parameter is the influential parameter of a models result. The sensitivity analysis (S) of each input parameter is calculated and tabulated in Table 13.3 of the input parameters

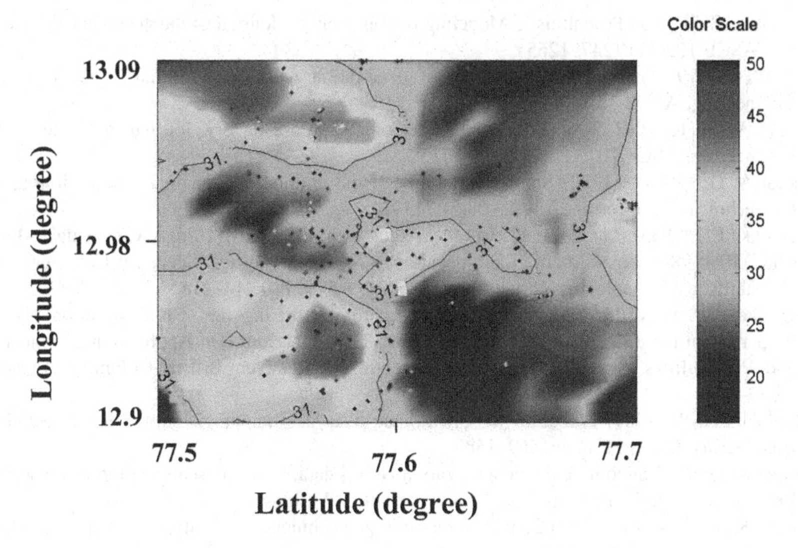

Fig. 13.8 Variation of N_c on the plane of $z = 1.5$ m using GPR model

Table 13.3 Sensitivity analysis

Input	L_x	L_y	z
GP, S(%)	37.40	35.04	27.49
MARS, S(%)	40.29	34.69	25.1
GPR, S(%)	39.58	36.27	24.15

13.7 Conclusion

This article examines the capability of GPR, MARS and GP for developing site characterization model of Bangalore based on N_c. The performance of MARS is best. Spatial variability of N_c in Bangalore has been obtained from the developed models. MARS uses basis function for final prediction. GPR adopts kernel function for developing the model. GP employs any function for final prediction. User can use the developed equations for prediction of N_c at any point in Bangalore. The developed models give N_c values without performing any SPT tests.

Acknowledgement Authors thank to T.G. Sitharam for providing the dataset.

References

Yaglom AM (1962) Theory of stationary random functions. Prentice-Hall, Inc., Englewood Cliffs, N.J.

Lumb P (1975) Spatial variability of soil properties. In: Proc. Second Int. Conf. on Application of Statistics and Probability in Soil and struct. Engrg, Aachen, Germany, p 397–421.

Vanmarcke EH (1977) Probabilistic Modeling of soil profiles. Journal of Geotechnical Engineering, ASCE 102(11):1247–1265
Tang WH (1979) Probabilistic evaluation of penetration resistance. Journal of Geotechnical Engineering, ASCE 105(GT10):1173–1191
Wu TH, Wong K (1981) Probabilistic soil exploration: a case history. Journal of Geotechnical Engineering, ASCE 107(GT12):1693–1711
Asaoka A, Grivas DA (1982) Spatial variability of the undrained strength of clays. Journal of Geotechnical Engineering, ASCE 108(5):743–745
Vanmarcke EH (1983) Random fields: Analysis and synthesis. The MIT Press, Cambridge, Mass
Baecher GB (1984) On estimating auto-covariance of soil properties. Specialty Conference on Probabilistic Mechanics and Structural Reliability, ASCE 110:214–218
Kulatilake PHSW, Miller KM (1987) A scheme for estimating the spatial variation of soil properties in three dimensions. In: Proc. Of the Fifth Int. Conf. On Application of Statistics and Probabilities in Soil and Struct Engineering, Vancouver, British Columbia, Canada, p 669–677
Kulatilake PHSW (1989) Probabilistic Potentiometric surface mapping. Journal of Geotechnical Engineering ASCE 115(11):1569–1587
Fenton GA (1998) Random field characterization NGES data. Paper presented at the workshop on Probabilistic Site Characterization at NGES, Seattle, Washington
Phoon KK, Kulhawy FH (1999) Characterization of geotechnical variability. Canadian Geotechnical Journal 36(4):612–624
Uzielli M, Vannucchi G, Phoon KK (2005) Random filed characterization of stress-normalised cone penetration testing parameters. Géotechnique 55(1):3–20
Kulatilake PHSW, Ghosh A (1988) An investigation into accuracy of spatial variation estimation using static cone penetrometer data. In: Proc. Of the First Int. Symp. On Penetration Testing, Orlando, Fla., p 815–821
Soulie M, Montes P, Sivestri V (1990) Modelling spatial variability of soil parameters. Canadian Geotechnical Journal 27:617–630
Chiasson P, Lafleur J, Soulie M et al (1995). Characterizing spatial variability of clay by geostatistics. Canadian Geotechnical Journal 32:1–10
Degroot DJ (1996) Analyzing spatial variability of in situ soil properties. In: ASCE proceedings of uncertainty '96, uncertainty in the geologic environment: from theory to practice, vol 58, ASCE, geotechnical special publications, p 210–238
Juang CH, Jiang T, Christopher RA (2001) Three-dimensional site characterisation: neural network approach. Géotechnique 51(9):799–809
Samui P, Sitharam TG (2010) Site characterization model using artificial neural network and kriging. Int. J. Geomech 10(5):171–180
Park D, Rilett LR (1999) Forecasting freeway link travel times with a multi-layer feed forward neural network. Computer Aided Civil and Infra Structure Engineering 14:358–367
Kecman V (2001) Learning and Soft Computing: Support Vector Machines, Neural Networks, and Fuzzy Logic Models. MIT press, Cambridge, Massachusetts, London, England
Samui P, Das S (2011) Site Characterization Model Using Support Vector Machine and Ordinary Kriging. Journal of Intelligent Systems 20(3):261–278
Tipping ME (2001) Sparse Bayesian learning and the relevance vector machine. J. Mach. Learn 1:211–244
Koza JR (1992) Genetic programming: on the programming of computers by means of natural selection. MIT Press, Cambridge.
Londhe S, Charhate S (2010) Comparison of data-driven modelling techniques for river flow forecasting. Hydrological Sciences Journal 55(7):1163–1174
Guven A, Kişi O (2011) Estimation of Suspended Sediment Yield in Natural Rivers Using Machine-coded Linear Genetic Programming. Water Resources Management 25(2):691–704
Alavi AH, Gandomi AH (2011) A Robust Data Mining Approach for Formulation of Geotechnical Engineering Systems." International Journal for Computer-Aided Engineering-Engineering Computations, Emerald, 28(3): 242–274

Alavi AH, Gandomi AH (2012) Energy-based numerical models for assessment of soil liquefaction. Geoscience Frontiers 3(4):541–555
Danandeh Mehr A, Kahya E, Olyaie E (2013) Streamflow prediction using linear genetic programming in comparison with a neuro-wavelet technique. Journal of Hydrology 505: 240–249
Zahiri A, Azamathulla HM (2014) Comparison between linear genetic programming and M5 tree models to predict flow discharge in compound channels. Neural Computing and Applications 24(2):413–420
Friedman JH (1991) Multivariate adaptive regression splines. The Annals of Statistics 19(1):1–67
Harb R, Su X, Radwan E (2010) Empirical analysis of toll-lane processing times using proportional odds augmented MARS. Journal of Transportation Engineering 136(11):1039–1048
Mao HP, Wu YZ, Chen LP (2011) Multivariate adaptive regression splines based simulation optimization using move-limit strategy. Journal of Shanghai University 15(6):542–547
Zhan X, Yao D, Zhan X (2012) Exploring a novel method among data mining methods and statistical methods to analyse risk factors of peripheral arterial disease in type 2 diabetes mellitus. International Journal of Digital Content Technology and its Applications 6(23): 243–253
Kumar P, Singh Y (2013) Comparative analysis of software reliability predictions using statistical and machine learning methods. International Journal of Intelligent Systems Technologies and Applications 12(3-4):230–253
García Nieto PJ, Álvarez Antón JC (2014) Nonlinear air quality modeling using multivariate adaptive regression splines in Gijón urban area (Northern Spain) at local scale. Applied Mathematics and Computation 235(25):50–65
Shen Q, Sun Z (2010) Online learning algorithm of gaussian process based on adaptive natural gradient for regression. Advanced Materials Research 139-141:1847–1851
Xu Y, Choi J, Oh S (2011) Mobile sensor network navigation using Gaussian processes with truncated observations. IEEE Transactions on Robotics 27(6):1118–1131
Kongkaew W, Pichitlamken J (2012) A Gaussian process regression model for the traveling salesman problem. Journal of Computer Science 8(10):1749–1758
Fairchild G, Hickmann KS, Mniszewski SM et al (2013) Optimizing human activity patterns using global sensitivity analysis. Computational and Mathematical Organization Theory 1–23
Holman D, Sridharan M, Gowda P et al (2014) Gaussian process models for reference ET estimation from alternative meteorological data sources. Journal of Hydrology 517:28–35
Sekulic S, Kowalski BR (1992) MARS: A tutorial. Journal of Chemometrics 6(4):199–216
Yang XS, Gandomi AH, Talatahari S et al (2013) Metaheuristis in Water, Geotechnical and Transportation Engineering. Elsevier, Waltham, MA
Fister I, Gandomi AH, Fister IJ et al (2014) Soft Computing in Earthquake engineering: A short overview. International Journal of Earthquake Engineering and Hazard Mitigation 2(2):42–48
Gandomi AH, Alavi AH (2012) A New Multi-Gene Genetic Programming Approach to Nonlinear System Modeling. Part II: Geotechnical and Earthquake Engineering Problems. Neural Computing and Applications 21(1):189–201
Gandomi AH, Alavi AH (2011) A robust data mining approach for formulation of geotechnical engineering systems. Engineering Computations 28(3):242–274
Langdon W (2013) Hybridizing Genetic Programming with Orthogonal Least Squares for Modeling of Soil Liquefaction. International Journal of Earthquake Engineering and Hazard Mitigation 1(1):2–8

Chapter 14
Use of Genetic Programming Based Surrogate Models to Simulate Complex Geochemical Transport Processes in Contaminated Mine Sites

Hamed Koohpayehzadeh Esfahani and Bithin Datta

14.1 Introduction

Reactive transport of chemical species, in contaminated groundwater system, especially with multiple species, is a complex and highly non-linear process. Simulation of such complex geochemical processes using efficient numerical models is generally computationally intensive. In order to increase the model reliability for real field data, uncertainties in hydrogeological parameters and boundary conditions are needed to be considered as well. Also, often the development of an optimal contaminated aquifer management and remediation strategy requires repeated solutions of complex and nonlinear numerical flow and contamination process simulation models. To address these combination of issues, trained ensemble Genetic Programming (GP) surrogate models can be utilized as approximate simulators of these complex physical processes in the contaminated aquifer. For example, use of trained GP surrogate models can reduce the computational burden in solving linked simulation based groundwater aquifer management models

Electronic supplementary material The online version of this chapter (doi: 10.1007/978-3-319-20883-1_14) contains supplementary material, which is available to authorized users.

H.K. Esfahani (✉)
Discipline of Civil Engineering, College of Science Technology and Engineering, James Cook University, Townsville, QLD 4811, Australia
e-mail: hamed.koohpayehzadehesfahani@my.jcu.edu.au

B. Datta (✉)
Discipline of Civil Engineering, College of Science Technology and Engineering, James Cook University, Townsville, QLD 4811, Australia

CRC-CARE, Mawson Lakes, SA 5095, Australia
e-mail: bthin.datta@jcu.edu.au

A.H. Gandomi et al. (eds.), *Handbook of Genetic Programming Applications*,
DOI 10.1007/978-3-319-20883-1_14

(Sreekanth and Datta 2011a, b) by orders of magnitude. Ensemble GP models trained as surrogate models can also incorporate various uncertainties in modelling the flow and transport processes. The development and performance evaluation of ensemble GP models to serve as computationally efficient approximate simulators of complex groundwater contaminant transport process with reactive chemical species under aquifer parameters uncertainties are presented. Performance evaluation of the ensemble GP models as surrogate models for the reactive species transport in groundwater demonstrates the feasibility of its use and the associated computational advantages. In order to evolve any strategy for management and control of contamination in a groundwater aquifer system, a simulation model needs to be utilized to accurately describe the aquifer properties in terms of hydrogeochemical parameters and boundary conditions. However, the simulation of the transport processes becomes complex and extremely non-linear when the pollutants are chemically reactive. In many contaminated groundwater aquifer management scenarios, an efficient strategy is necessary for effective and reliable remediation and control of the contaminated aquifer. Also, in a hydrogeologically complex aquifer site e.g., mining site, acid mine drainage (AMD) and the reactive chemical species together with very complex geology complicates the characterization of contamination source location and pathways.

In such contamination scenarios, it becomes necessary to develop optimal source characterization models, and strategies for future remediation. Solution of optimization models either for source characterization, or optimal management strategy development requires the incorporation of the complex physical processes in the aquifer. Also, most of the developed optimization models for source characterization or remediation strategy development require repeated solution of the numerical simulation models within the optimization algorithm. This process is enormously time consuming and often restricts the computational feasibility of such optimization approaches.

In order to overcome these computational restrictions, and to ensure computational feasibility of characterizing sources and pathways of contamination it is computationally advantageous to develop surrogate models which can be trained using solutions obtained from rigorous numerical simulation models. A number of attempts have been reported by researchers to develop surrogate models for approximately simulating the physical processes. Especially the use of trained Artificial Neural Network (ANN) models has been reported by a number of researchers (Ranjithan et al. 1993). However, the architecture of an ANN model needs to be determined by extensive trial and error solutions, and may not be suitable to deal with the simulation of very complex geochemical processes in contaminated aquifer site such as mine sites. Genetic Programming (GP) based surrogate models may overcome some of the limitations of earlier reported surrogate models. Therefore, this study develops GP model to approximately simulate three-dimensional, reactive, multiple chemical species transport in contaminated aquifers.

Trained and tested GP models based surrogate models are developed using the simulated response of a complex contaminated aquifer to randomly generated source fluxes. An ensemble GP model is an extension of the GP modelling technique capable of incorporating various uncertainties in a contaminated aquifer system data.

These ensemble GP models are trained and tested utilizing transient, three dimensional groundwater flow and transport simulation models for an illustrative study area hydrogeologically representing an abandoned mine site in Australia. Performance of the developed surrogate models is also evaluated by comparing GP model solutions with solution results obtained by using a rigorous numerical simulation of the aquifer processes. The three dimensional finite element based transient flow and contaminant transport process simulator, HYDROGEOCHEM 5.0 (Sun 2004) is used for this purpose. Reactive transport processes incorporating acid mine drainage in a typical mine site is simulated. Comparison of the solutions obtained with the surrogate models and the numerical simulation model solution results show that the ensemble GP surrogate models can provide acceptable approximations of the complex transport process in contaminated groundwater aquifers, with a complex geochemical scenario.

The performance of the developed surrogate models is evaluated for an illustrative study area to establish the suitability of GP models as surrogate models for such complex geological processes. These surrogate models if suitable will ensure the computational feasibility of developing optimization based models for source characterization, and help in the development of optimum strategies for remediation of large contaminated aquifer study areas. This study will demonstrate the utility and feasibility of using trained and tested ensemble GP models as a tool for approximate simulation of the complex geochemical processes in contaminated mine sites.

Aquifer contamination by reactive chemical species is widespread especially in mining sites. Numerical simulation models incorporating both chemical and physical behaviours are essential to describe reactive chemical transport process accurately. The numerical simulation model using the chemical reactive transport processes in aquifer contamination was addressed by (Parkhurst et al. 1982) and also implemented by (Herzer and Kinzelbach 1989; Tebes-Stevensa et al. 1998; Prommer et al. 2002). Coupled physical–chemical transport processes was developed using non-reactive transport model like MT3DMS (Zheng and Wang 1999) incorporating with various reactive transport numerical models (Prommer 2002; Parkhurst and Appelo 1999; Parkhurst et al. 2004; Waddill and Widdowson 1998; Mao et al. 2006) to simulate more realistic chemical reactive transport processes.

HYDROGEOCHEM (Yeh and Tripathi 1991) as a comprehensive numerical simulation model of flow and geochemically reactive transport in saturated–unsaturated media incorporates wide range of aquatic chemical equations as well as complex physical processes effectively. Heat, reactive geochemical and biochemical transport processes along with flow equations for the subsurface (saturated and unsaturated zones) are solved by three-dimensional model, HYDROGEOCHEM 5.0 (Sun 2004). In the proposed study, HYDROGEOCHEM 5.0 (HGCH) is used to simulate groundwater flow and transport processes with chemically reactive pollutants for an illustrative subsurface study area utilizing actual hydrogeologic data and synthetic hydro-geochemical data. Trained and tested ensemble GP based surrogate models are then utilized to approximately model complex geological and geochemical processes to improve the computational efficiency as well as reasonably accurate solutions.

One of the most hazardous contaminants for water resources is acid mine drainage (AMD) and its related compounds spatially distributed which are the products of mining activities (Kalin et al. 2006). Generally AMD or acid rock drainage (ARD) is produced by various sulphide rocks' surface chemical weathering in presence of water, oxygen and microorganisms. Mining activities accelerate AMD production by increasing the rocks' surface as well as distributing wastewater and waste deposit of sulphide minerals such as pyrite (FeS2), pyrrhotite (Fe1-xS), chalcopyrite (CuFeS2), arsenopyrite (FeAsS), etc. in mine sites (Nordstrom and Alpers 1999). These contaminants pollute water resources widely as well as decrease the water pH which leads to increase in the concentration of other hazardous metals and heavy metals in water (Kalin et al. 2006). In this study, the transport process of sulphate, iron and copper, hazardous AMD's compounds, along with their chemical reactions through the contaminated aquifer is considered.

Recently surrogate models have been proposed as approximate replacement for numerical simulation model for developing linked simulation optimization models (Bhattacharjya and Datta 2005) for groundwater quality management. Replacing aquifer responses simulation by linear surrogate models developed using response matrix approach was initially reported (Zhou et al. 2003; Abarca 2006). Recently, Artificial Neural Network (ANN) (Ranjithan et al. 1993) and Genetic Programming (GP) based surrogate models have been proposed as efficient non-linear surrogate models (Koza 1994).

Artificial Neural Networks (ANN) has been widely used as approximate surrogate models for groundwater simulation (Aly and Peralta 1999). Rogers et al. (1995) presented one of the earliest attempts using ANN as a surrogate for a coastal groundwater flow model. They demonstrated the substantial saving in terms of computation time by using ANN and Genetic Algorithmic (GA) based meta-model (surrogate model) within a linked simulation-optimization model for evolving optimal groundwater management strategies. Replacing groundwater simulation models with ANN-base surrogate models were developed by Bhattacharjya and Datta (2005, 2009) and Bhattacharjya et al. (2007) and Dhar and Datta (2009). McPhee and Yeh (2006) used ordinary differential equation surrogates to approximating simulate of groundwater flow and transport processes. Optimizing the surrogate model parameters related on fixed initial surrogate model structure is the main concept of most of these surrogate modelling approaches to obtain the best between the explanatory and response variables. Even the most popularly used trained ANN-based surrogate modelling approach obtains the optimal model formulation by trial and error (Bhattacharjya and Datta 2005).

Bhattacharjya et al. (2007) used ANN as an approximate simulation for substitutes the three dimensional flow and transport simulation model to simulate the complex flow and transport process in a coastal aquifer. Bhattacharjya and Datta (2009) used the trained ANN-based surrogate models for approximating density depended saltwater intrusion process in coastal aquifer to predict the complex flow and transport processes. Dhar and Datta (2009) used ANN as a surrogate model for simulation of flow and transport in the multiple objective non-dominated front search process resulting in saving a huge amount of computational time.

Genetic Programming (GP), proposed by Koza (1994) is an evolutionary algorithm which is capable approximate simulation of complex models effectively using stochastic search methods. Compared to other regression techniques, the most important advantage of GP is its ability to optimize both the variables and constants of the candidate models without initial model structure definition. This approach makes GP a strong surrogate model to characterize the model structure uncertainty. Recently genetic programming has been utilized in hydrological applications in several researches (Dorado et al. 2002; Makkeasorn et al. 2008; Wang et al. 2009). Trained GP-based surrogate models has been used to substitutes the simulation models for runoff prediction, river stage and real-time wave forecasting (Whigham and Crapper 2001; Savic et al. 1999; Khu et al. 2001; Babovic and Keijzer 2002; Sheta and Mahmoud 2001; Gaur and Deo 2008). In addition, GP has been applied to approximate modelling of different geophysical processes including flow over a flexible bed (Babovic and Abbott 1997); urban fractured-rock aquifer dynamics (Hong and Rosen 2002); temperature downscaling (Coulibaly 2004); rainfall-recharge process (Hong et al. 2005); soil moisture (Makkeasorn et al. 2006); evapotranspiration (Parasuraman et al. 2007b); saturated hydraulic conductivity (Parasuraman et al. 2007a); and for modelling chemical entropy (Bagheri et al. 2012, 2013, 2014). Zechman et al. (2005) developed a trained GP-based surrogate models as an approximate simulation of groundwater flow and transport processes in a groundwater pollutant source identification problem.

Sreekanth and Datta (2010) implemented GP as meta-model to replace the flow and transport simulation of density dependent saltwater intrusion in coastal aquifers for ultimate development of optimal saltwater intrusion management strategies. Sreekanth and Datta (2011b, 2012) compared two non-linear surrogate models based on GP and ANN models, respectively and showed that the GP based models perform better in some aspects. These advantages include: simpler surrogate models, optimizing the model structure more efficiently, and parsimony of parameters. Datta et al. (2013) described the utilization of trained GP surrogate models for groundwater contamination management, and development of a monitoring network design methodology to develop optimal source characterization models. Replacing simulation groundwater model by GP-based ensemble surrogate models in linked simulation-optimization developed methodology was addressed by Datta et al. (2014) and Sreekanth and Datta (2011a) which improve the computational efficiency and obtains reasonably accurate results under aquifer hydrogeologic uncertainties.

In this study our main objectives is to develop ensemble genetic programming based surrogate models to approximately simulate the complex transport process in a complex hydrogeologic system with reactive chemical species, and to illustrate its efficiency and reliability in a contaminated aquifer resembling an abandoned mine site. The numerical model's formulations as well as using ensemble genetic programming based surrogate models are described in Sect. 14.2 and the results are presented and discussed in Sect. 14.3.

14.2 Methodology

The methodology developed includes two main components. In the first step, the simulation model for the flow and transport processes is described, and complex chemical reactive transport process is simulated by the HGCH, a three-dimensional coupled physical and chemical transport simulator, to realize the reactive contaminants behaviours is contaminated aquifers. The hydrogeochemical data and boundary conditions at the illustrative study site are similar to an abandoned mine site in Queensland, Australia. Trained ensemble GP based surrogate models are then developed to approximately obtain concentrations of the chemical contaminants at different times in specified locations while incorporating uncertainties in hydrogeological aquifer parameters like hydraulic conductivity. Comparison of the spatio-temporal concentrations obtained as solution by solving the implemented numerical three dimensional reactive contaminant transport simulation model (HGCH) and those obtained using ensemble GP models are then presented to show the potential applicability and the efficiency of using GP ensemble surrogate models under aquifer uncertainties.

14.2.1 Simulation Model of Groundwater Flow and Geochemical Transport

HYDROGEOCHEM 5.0 (HGCH), consisting of the numerical flow simulator and physio-chemical transport simulator HGCH is a computer program that numerically solves the three-dimensional groundwater flow and transport equations for a porous medium. The finite-element method is used in this simulation model.

The general equations for flow through saturated–unsaturated media are obtained based on following components: (1) fluid continuity, (2) solid continuity, (3) Fluid movement (Darcy's law), (4) stabilization of media, and (5) water compressibility (Yeh et al. 1994). Following governing equation is used:

$$\frac{\rho}{\rho_0}F\frac{\partial h}{\partial t} = \nabla.\left[K.\left(\nabla h + \frac{\rho}{\rho_0}\nabla z\right)\right] + \frac{\rho^*}{\rho_0}q \tag{14.1}$$

F is the generalized storage coefficient (1/L) defined as:

$$F = \alpha'\frac{\theta}{n_e} + \beta'\theta + n_e\frac{ds}{dh} \tag{14.2}$$

K is the hydraulic conductivity tensor (L/T) is:

$$K = \frac{\rho g}{\mu}k = \frac{(\rho/\rho_0)}{(\mu/\mu_0)}\frac{\rho_0 g}{\mu_0}k_s\,k_r = \frac{(\rho/\rho_0)}{(\mu/\mu_0)}K_{so}\,k_r \tag{14.3}$$

V is the Darcy's velocity (L/T) described as:

$$V = -K\left[\frac{\rho}{\rho_0}\nabla h + \nabla z\right] \tag{14.4}$$

Where:
θ: effective moisture content (L3/L3);
h: pressure head (L);
t: time (T);
z: potential head (L);
q: source or sink of fluid [(L3/L3)/T];
ρ_0: fluid density without biochemical concentration (M/L3);
ρ: fluid density with dissolved biochemical concentration (M/L3);
ρ^*: fluid density of either injection ($=\rho^*$) or withdraw ($=\rho$) (M/L3);
μ_0: fluid dynamic viscosity at zero biogeochemical concentration [(M/L)/T];
μ: the fluid dynamic viscosity with dissolved biogeochemical concentrations [(M/L)/T];
α': modified compressibility of the soil matrix (1/L);
ß: modified compressibility of the liquid (1/L);
ne: effective porosity (L3/L3);
S: degree of effective saturation of water;
G: is the gravity (L/T2);
k: permeability tensor (L2);
k_s: saturated permeability tensor (L2);
K_{so}: referenced saturated hydraulic conductivity tensor (L/T);
k_r: relative permeability or relative hydraulic conductivity (dimensionless)

When combined with appropriate boundary and initial conditions, the above equations are used to simulate the temporal-spatial distributions of the hydrological variables, including pressure head, total head, effective moisture content, and Darcy's velocity in a specified study area.

The contaminant transport equations used in the HG model can be derived based on mass balance and biogeochemical reactions (Yeh 2000). The general transport equation using advection, dispersion/diffusion, source/sink, and biogeochemical reaction as the major transport processes can be written as follows:

$$\frac{D}{Dt}\int_{v}\theta C_i dv = -\int_{\Gamma} n.\left(\theta C_i\right)V_i d\Gamma - \int_{\Gamma} n.J_i d\Gamma + \int_{v}\theta r_i dv + \int_{v} M_i dv,\ i \in M \tag{14.5}$$

Where
C_i: the concentration of the ith species in mole per unit fluid volume (M/L^3);
ν: the material volume containing constant amount of media (L3);
Γ: the surface enclosing the material volume ν (L2);
n: the outward unit vector normal to the surface Γ;

Ji: the surface flux of the ith species due to dispersion and diffusion with respect to relative fluid velocity [(M/T)/L2];

θri: the production rate of the ith species per unit medium volume due to all biogeochemical reactions [(M/L3)/T];

Mi: the external source/sink rate of the ith species per unit medium volume [(M/L3)/T];

M: the number of biogeochemical species;

Vi: the transporting velocity relative to the solid of the ith biogeochemical species (L/T).

14.2.2 Genetic Programming Based Ensembles Surrogate Model

GP models are used in this study to evolve surrogate models for approximately simulating flow and transport processes in a contaminated mine site. Trained GP models are developed using the simulated response of the aquifer to randomly generated source fluxes. GP, a branch of genetic algorithms (Koza 1994), is an evolutionary algorithm-based methodology inspired by biological evolution to find computer programs that perform a user-defined task (Sreekanth and Datta 2011b). Essentially, GP is a set of instructions and a fitness function to measure how well a computer model has performed a task. The main difference between GP and genetic algorithms is the representation of the solution. GP creates computer programs in the lisp or scheme computer languages as the solution. Genetic algorithms create a string of numbers that represent the solution.

The main operators applied in genetic programming as in evolutionary algorithms are crossover and mutation. Crossover is applied on an individual by simply replacing one of the nodes with another node from another individual in the population. With a tree-based representation, replacing a node means replacing the whole branch (Fig. 14.1). This adds greater effectiveness to the crossover operator. The expressions resulting from crossover are very different from their initial parents. Mutation affects an individual in the population. It can replace a whole node in the selected individual, or it can replace just the node's information. To maintain integrity, operations must be fail-safe or the type of information the node holds must be taken into account. For example, mutation must be aware of binary operation nodes, or the operator must be able to handle missing values.

GP utilizes a set of input–output data which are generated randomly by using the flow and contaminant transport simulation models. The numerical Simulation model creates M number of out-put sets from M number of input sets, which is generated by using random Latin Hypercube sampling in defined ranges. The performance of each GP program is an evaluated formulation in terms of training, testing the validation using the set of input–output patterns. The testing data evaluates the

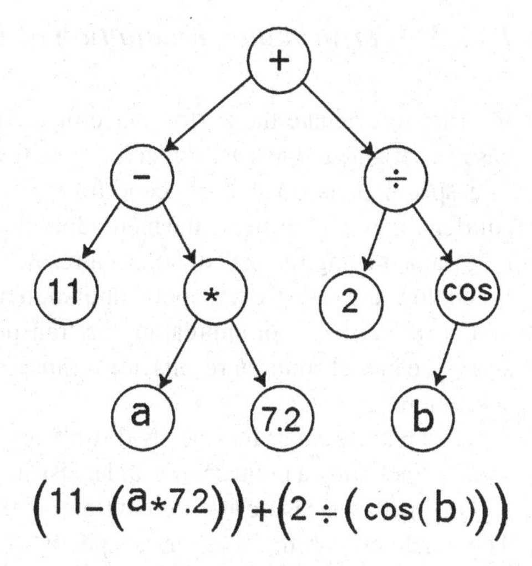

Fig. 14.1 Function represented as a tree structure

model performance for new data using the fitness function obtained in the training phase. Non-tree representations have been proposed and successfully implemented, such as linear genetic programming which suits the more traditional imperative languages (Banzhaf et al. 1998). The commercial GP software Discipulus (Francone 1998) performs better by using automatic induction of binary machine code. In the proposed methodology, Discipulus GP software is used to solve and generate GP models. Discipulus uses Linear Genetic Programming (LGP) which utilizes input variables in line-by-line approach. This objective of this program is minimizing difference in value between the output estimated by GP program on each pattern and the actual outcome. The fitness objective functions are often absolute error or minimum squared error. Almost two-thirds of the input–output data sets obtained from the numerical simulation model are utilized for training and testing the GP model. The remaining data sets are used to validate the GP models. The r-square value shows the fitness efficiency to the GP models (Sreekanth and Datta 2010).

14.2.2.1 Performance Evaluation

The trained ensemble GP surrogate models are evaluated to verify the performance of the surrogate models approximating flow and transport processes simulation with reactive chemical species, under hydrogeological uncertainties. Input data sets are generated randomly by Latin Hypercube sampling in defined ranges. The aquifer hydrogeological uncertainties include uncertainties in estimating hydraulic conductivity, water content and constant groundwater label in boundary conditions.

14.2.3 Performance Evaluation of Developed Methodology

In order to evaluate the performance of the proposed methodology, ensemble GP based surrogate models are utilized for an illustrative study area shown in Fig. 14.2. The specified hydrogeologic conditions resemble a homogeneous and isotropic aquifer. In order to evaluate the methodology, the ensemble GP surrogate models are first trained using the sets of solution results obtained using the 3-D finite element based flow and reactive transport simulation model. Once trained and tested, the GP models are utilized for simulating the transport process in the study site. Then the surrogate model solution results are compared with the actual numerical simulation solution results.

The areal extent of the specified study area is 10,000 m^2 with complex pollutant sources including a point source and a distributed source. The spatial concentrations are assumed to measure at different times at ten arbitrary observation well locations. The thickness of the aquifer is specified as 50 m with anisotropic hydraulic conductivity in the three directions. The boundaries of the study area are no-flow for top and bottom sides while left and right sides of the aquifer have constant head boundaries with specified hydraulic head values. The total head decreases from top to bottom and left to right gradually. The aquifer system is shown in Fig. 14.2.

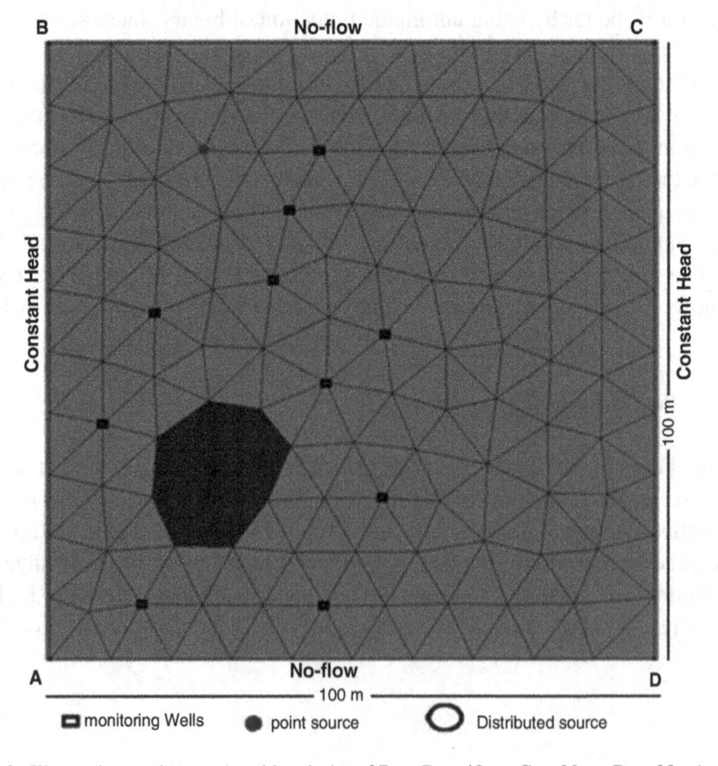

Fig. 14.2 Illustrative study area (total head: A = 37 m, B = 40 m, C = 33 m, D = 30 m)

As shown in Fig. 14.2, the dark blue area represent the contaminant sources S(i) which include distributed and point sources. Concentration data from monitoring well locations, shown as black rectangular points, are used to train, test and validate the GP model formulations.

Table 14.1 shows dimensions, hydrogeological properties, and boundary conditions of the study area which are utilized for numerical models to simulate groundwater flow and chemical reactive transport processes. The synthetic concentration measurement data used for the specified polluted aquifer facilitates evaluation of the developed methodology. These synthetic concentration measurement data at specified observation locations are obtained by solving the numerical simulation model with known pollution sources, boundary conditions, initial conditions, and hydrogeologic as well as geochemical parameter values. In the incorporated scenario, copper (Cu^{2+}), Iron (Fe^{2+}) and sulphate ($SO4^{2-}$) are specified as the chemical species in the pollutant sources. The associated chemical reactions are listed in Table 14.2.

Nine different scenarios are defined based on different hydraulic conductivity and boundary conditions with maximum 10 % differences between maximum

Table 14.1 Aquifer's properties

Aquifer parameter	Unit	Value
Dimensions (length * width * thickness) study area	m * m * m	100 * 100 * 50
Number of nodes		387
Number of elements		1432
Hydraulic conductivity, Kx, Ky, Kz	m/d	10.0, 5.0, 3.0
Effective porosity, Θ		0.3
Longitudinal dispersivity, αL	m/d	10.0
Transverse dispersivity, αT	m/d	6.0
Horizontal anisotropy		1
Initial contaminant concentration	Mole/lit	0–5
Diffusion coefficient		0

Table 14.2 Typical chemical reactions during the contaminant transport process

Chemical reaction equations	Constant rate (Log k)[a]
Equilibrium reactions	
(1) $Cu^{2+} + H_2O \leftrightarrow Cu(OH)^+ + H^+$	−9.19
(4) $Cu^{2+} + SO_4{}^{2-} \leftrightarrow CuSO_4$	2.36
(7) $Fe^{2+} + SO_4{}^{2-} \leftrightarrow FeSO_4$	2.39
(9) $4Fe^{2+} + 4H^+ \leftrightarrow 4Fe^{3+} + 2H_2O$	8.5
(14) $Fe^{3+} + SO_4{}^{2-} \leftrightarrow FeSO_4{}^+$	4.05
(15) $Fe^{3+} + SO_4{}^{2-} + H^+ \leftrightarrow FeHSO_4{}^{2+}$	2.77
Kinetic reactions	
(17) $FeOOH_{(s)} + 3H^+ \leftrightarrow Fe^{3+} + 3H_2O$	$K_f = 0.07$

[a]Constant rates are taken from Ball and Nordstrom (1992)

and minimum values, and with the mean value assumed as the actual value for simulating the synthetic concentration observation (N11, N12,..., N21, ...N33). First digit indicates an index for hydraulic conductivity values and second one represents an index for the hydraulic head as boundary condition. 1 illustrates parameters with 5 % less than the actual definition as well as 2 and 3 shows the exact data and 5 % more than actual parameters in illustrative aquifer respectively.

14.2.3.1 Generation of Training and Testing Patterns for the Ensemble GP Models

The total time of source activities is specified as 800 days, subdivided into eight similar time intervals of 100 days each. The actual pollutant concentration from each of the sources is presumed to be constant over each stress period. The pollutant concentration of copper, iron as well as sulphate in the pit is represented as Cpit(i), Fepit(i) and Spit(i) respectively, where i indicates the stress period number, and also C(i), Fe(i) and S(i) represent copper, iron and sulphate concentrations in the point sources, respectively at different time steps.

An overall of sixteen concentration values for each contaminant are considered as explicit variables in the simulation model. The concentration measurements are simulated for a time horizon of 800 days since the start of the simulation. The pollutant concentration are assumed to be the resulting concentrations at the observation wells at every 100 days interval and this process is continued at all the observation locations till $t = 800$ days. Only for this methodology evaluation purpose, these concentration measurements are not obtained from field data, but are synthetically obtained by solving the numerical simulation model for specified initial conditions, boundary conditions and parameter values. In actual application these measurement data need to be simulated using a calibrated flow and contaminant simulation model. However, using field observations for calibration, and then for evaluation of a proposed methodology results in uncertain evaluation results as the quality of the available measurement data cannot be quantified most of the time. Therefore as often practiced, synthetic aquifer data is used for this evaluation of the methodology proposed.

The comprehensive three-dimensional numerical simulation model was used to simulate the aquifer flow and chemical reactive transport processes due to complex pollutant sources in this study area. Different random contaminant source fluxes as well as different realization of boundary conditions and hydraulic conductivities were generated using Latin hypercube sampling. For random generation purpose, 10 % initial aquifer properties are considered as Maximum error for the uncertainties of aquifer parameters. HGCH was utilized to obtain the concentrations resulting from each of these concentration patterns. The simulated concentration measured data at monitoring network and the corresponding concentration of contaminants at sources form the input–output pattern. Totally, 8000 concentration patterns for all the ten concentration observation locations were used in this evaluation. Eight input–output patterns were defined based on different time steps.

Genetic programming models were obtained using each of these data sets to create ensemble GP based surrogate models. Each data set was split into halves for training and testing the genetic programming-based surrogate models.

Surrogate models were developed for simulating pollutant concentrations at the observation locations at different times resulting from the specified pollutant sources at different times under hydrogeological uncertainties. All the GP models used a population size of 1000, and mutation frequency of 95. The Discipulus, commercial Genetic Programming software, was used to develop the surrogate models. The model was developed using default parameters values of Discipulus. The GP fitness function was the squared deviation between GP model generated and actually simulated concentration values at the observation locations and times.

14.3 Evaluation Results and Discussion

The flow and concentration simulation results for the study area obtained using the numerical HGCH simulation model are shown in Figs. 14.3, 14.4, 14.5, and 14.6. The flow movement, total head contours in top layer and also velocity vectors are shown in Figs. 14.3, 14.4 and 14.5 respectively. Figures 14.3, 14.4 and 14.5 show the hydraulic heads for flow. The contours show a gentle slope from point B towards D. Figure 14.6 shows the copper concentration distribution in the study area which shows the complex transport processes with reactive chemical species.

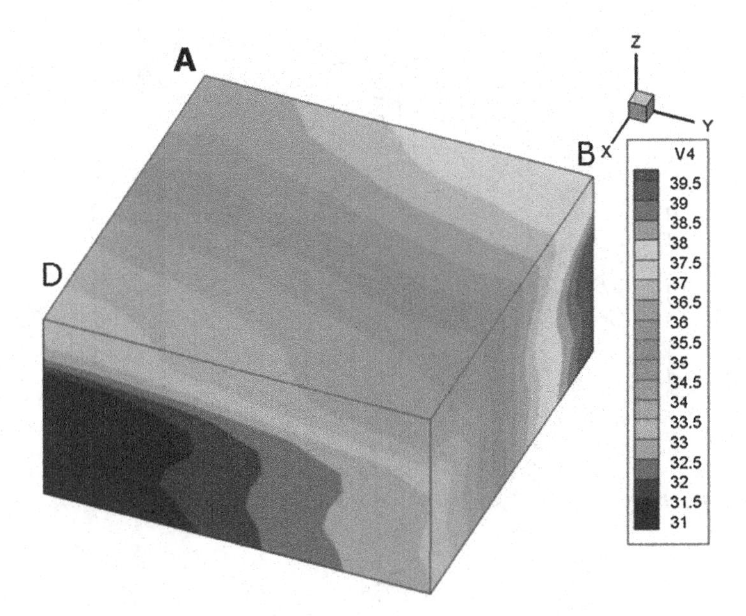

Fig. 14.3 3-D view of hydraulic head distribution

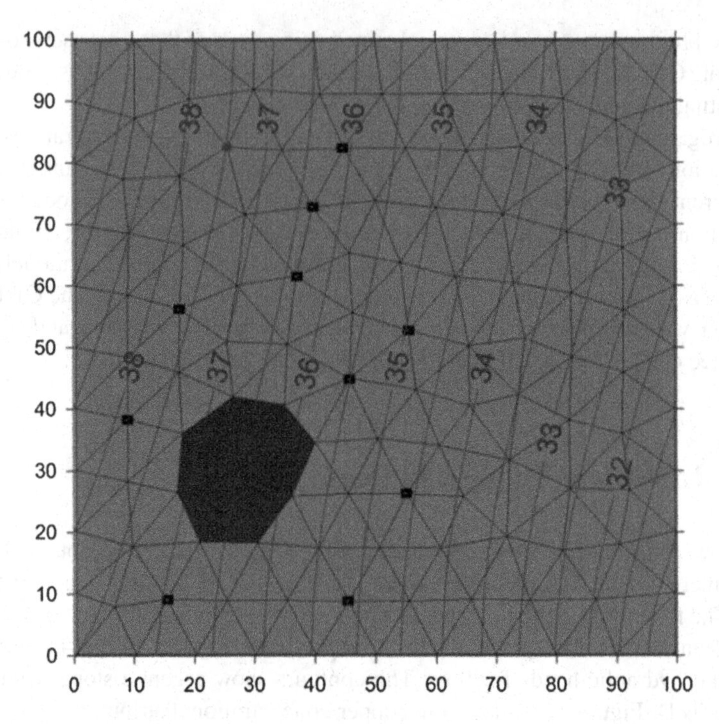

Fig. 14.4 Hydraulic head contours (m)

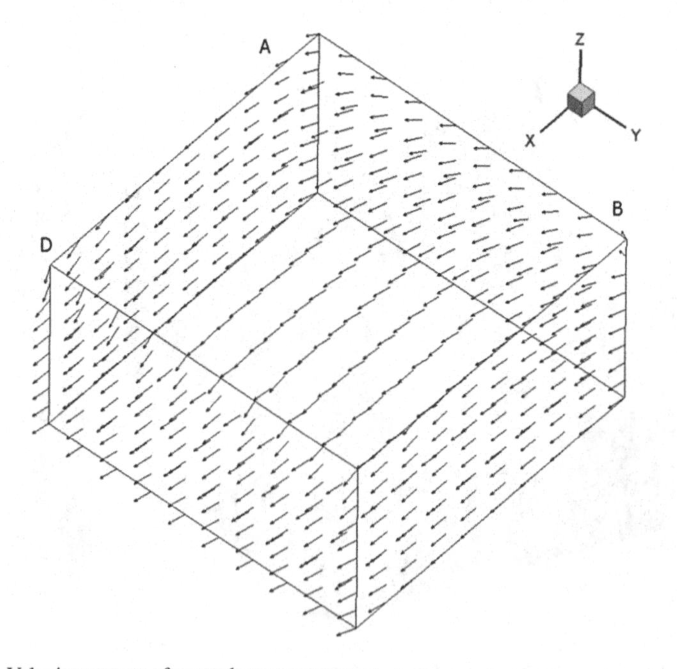

Fig. 14.5 Velocity vectors of groundwater movement

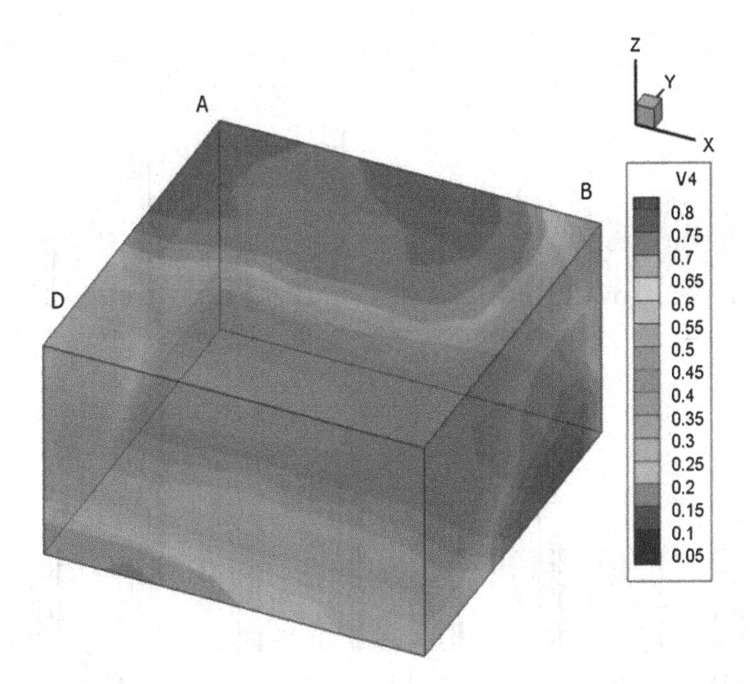

Fig. 14.6 Copper concentration (mole/lit) distribution in the study area

The concentration of sulphate remains almost the same while iron concentration is lower in groundwater. Based on pH changes the iron can react and cease to be in solute phase, thus removed form groundwater.

The results obtained using the developed ensemble genetic programming based surrogate models for approximate simulation of pollutant concentrations are compared with the numerical simulation results obtained using the HGCH. Nine different scenarios are considered. These nine scenarios are characterized by different hydraulic conductivity value realizations and hydraulic head boundary conditions. Each randomized within $10\,(\pm 5)\%$ errors in the mean values (assumed same as the actual values) for hydrogeological parameters and boundary conditions. Incorporation of these scenarios together with the Latin Hypercube based randomization to achieve the efficiency of ensemble GP based surrogate models. The uncertainties in the parameter values of the scenarios are within the range of input data which are used to create the ensemble GP models. Figure 14.7a–c illustrate these comparison results in which one scenario for one particular hydraulic conductivity is selected for obtaining simulated output data from HGCH model at each monitoring networks. Each time step is marked on the x-axis. Each of the bars corresponds to contaminant concentration in each well, obtained by HGCH and ensemble GP models.

Figure 14.7 shows that the results obtained from the ensemble GP based surrogate models are very close to the simulated results obtained using the numerical

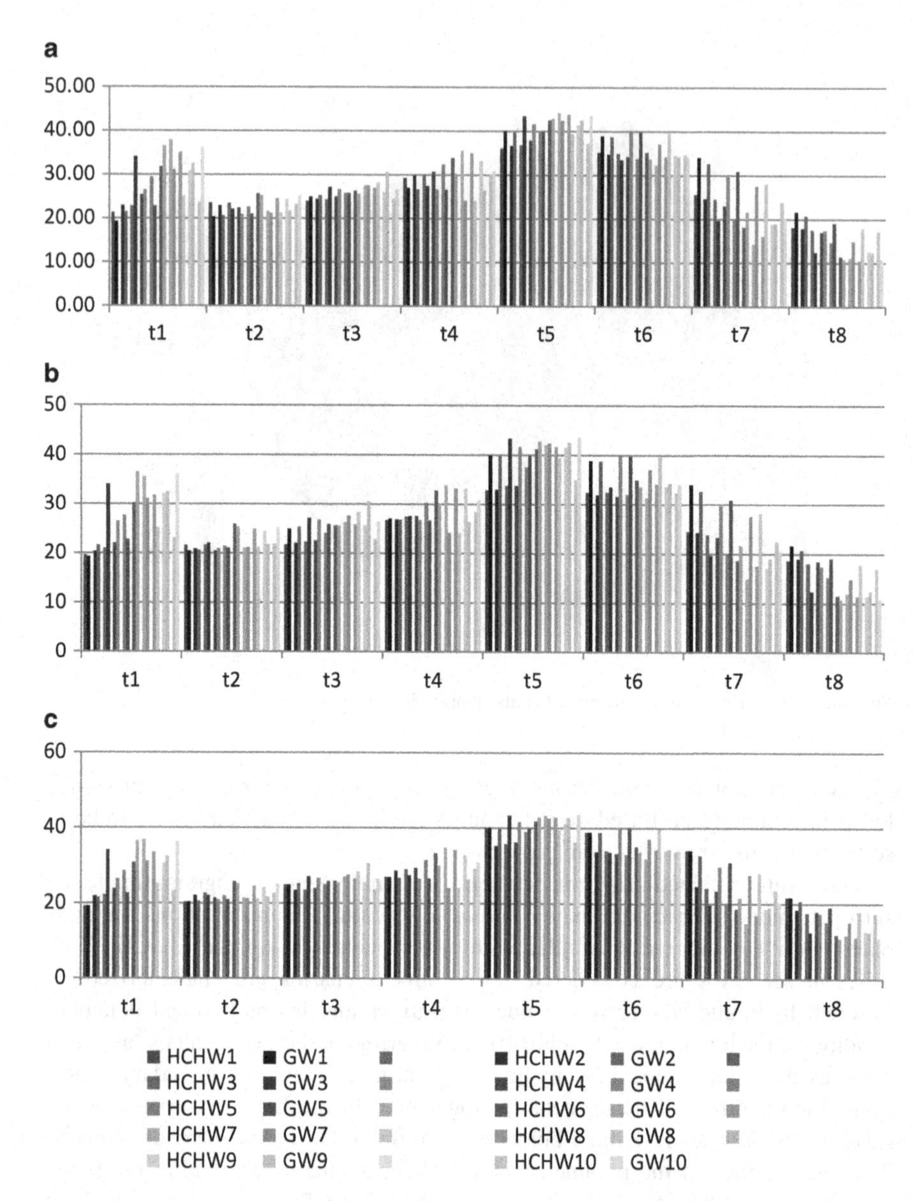

Fig. 14.7 Comparison of ensemble GP model solutions with HGCH simulation results for specified parameter values defined by (**a**) lower bound on uncertain aquifer parameter values, (**b**) actual or mean parameters values and (**c**) upper bound on aquifer parameter values (GW1: concentration data at well number 1 based on GP formulation, HCHW1: concentration data at well number 1 based on HGCH simulation)

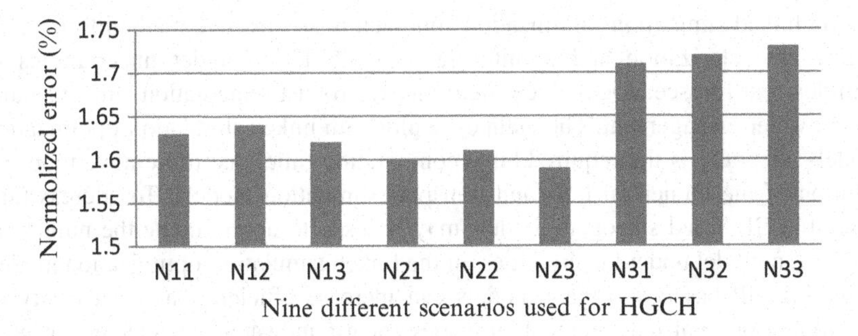

Fig. 14.8 Normalized error for all scenarios under uncertainties

simulation model, and also incorporates der uncertainties. Figure 14.8 shows the summation of normalized error at each of the observation locations for each monitoring network averaged over the 8 time periods. It is noted that, the ensemble GP models provide relatively accurate results for concentrations at observation locations. Although the boundary conditions are different, the normalized errors for all the three scenarios with same hydraulic conductivity are almost the same. The most important advantage of using the developed GP models is that the numerical simulation model requires long computational time usually several hours for a typical study area, while ensemble genetic programming surrogate models deliver the solution results in typically fraction of a second. Also the ensemble GP models directly incorporate hydrogeologic uncertainties in the modelled system. Therefore the computational advantage of using the ensemble GP for approximate simulation of complex reactive transport processes in aquifers is enormous if the errors in simulation are within acceptable range. Especially, this computational time saving could be critical in development and solution of linked simulation-optimization models (Datta et al. 2014) for management of contaminated aquifers.

14.4 Conclusion

Although surrogate models are widely used in solving groundwater management problems replacing the actual complex numerical models, often the main issue is the accuracy and reliability of surrogate model predictions under input data uncertainties. This study developed a methodology based on ensemble GP surrogate models to substitute numerical simulation for approximate simulation of the chemically reactive multiple species transport process in a contaminated aquifer resembling the geochemical characteristics of an abandoned mine site. The evaluation results show the applicability of this methodology to approximating the complex reactive transport process in an aquifer. The developed ensemble GP models result in increasing the computation efficiency and computational feasibility, while providing acceptable results.

The linked simulation-optimization approach is an effective method to identify source characterization and monitoring network design under uncertainties in complex real life scenarios which important for robust remediation strategies and groundwater management. The main difficulty with linked simulation-optimization models generally is the required huge computation time, due to iterative repeated solution of the numerical flow and transport simulation models. To address this, ensemble GP based surrogate models may be used to approximate the numerical simulation model under uncertainties, in the linked simulation-optimization model. Ensemble GP based surrogate models can increase efficiency and feasibility of developing optimal management strategies for groundwater management in geochemically complex contaminated aquifers such as mine sites, while at the same time incorporating uncertainties in defining the hydrogeologic system. The evaluations results show that it is feasible to use ensemble GP models as approximate simulators of complex hydrogeologic and geochemical processes in a contaminated groundwater aquifer incorporating uncertainties in describing the physical system.

References

Abarca, E., Vazquez-Sune, E., Carrera, J., Capino, B., Gamez, D., Battle, F., (2006) *Optimal design of measures to correct seawater intrusion.* Water Resource Research., 42.

Aly AH, and Peralta, R. C. (1999) Optimal design of aquifer cleanup systems under uncertainty using a neural network and a genetic algorithm. Water Resour Res 35 (8):10

Babovic V and Abbott M B (1997) Evolution of equation from hydraulic data. part I: Theory. J Hydraul Res 35:14

Babovic V, and Keijzer, M., (2002) Rainfall runoff modelling based on genetic programming. Nord Hydrol 33 (5):15

Bagheri M, Borhani T.N.G., Gandomi A.H., and Manan Z.A., (2014) A simple modelling approach for prediction of standard state real gas entropy of pure materials. SAR and QSAR in Environmental Research, 25 (9): 695–710

Bagheri M., Gandomi A.H., Bagheri M., and Shahbaznezhad M., (2013) Multi-expression programming based model for prediction of formation enthalpies of nitro-energetic materials. Expert Systems, 30 (1): 66

Bagheri M., Bagheri M., Gandomi A.H., Golbraikhc A., (2012) Simple yet accurate prediction method for sublimation enthalpies of organic contaminants using their molecular structure. Thermochimica Acta 543: 96–106.

Ball JW, Nordstrom, D.K., (1992) User's manual for WATEQ4F, with revised thermodynamic database and test cases for calculating speciation of major, trace and redox elements in natural waters. US Geol Surv Open-File Rep 91-183 (Revised and reprinted August 1992)

Banzhaf W, Nordin, P., Keller, R. E., and Francone, F. D. (1998) Genetic Programming: An Introduction. Morgan Kaufmann, Inc, San Francisco, USA

Bhattacharjya R, and Datta, B., (2005) Optimal management of coastal aquifers using linked simulation optimization approach. Water Resour Manage 19 (3):25

Bhattacharjya R, K., Datta, B., and Satish, M., (2007) Artificial neural networks approximation of density dependent saltwater intrusion process in coastal aquifers. Journal of Hydrologic Engineering 12 (3):10

Bhattacharjya RK, and Datta, B., (2009) ANN-GA-based model for multiple objective management of coastal aquifers. Water Resour Planning Manage 135 (5):8

Coulibaly P (2004) Downscaling daily extreme temperatures with genetic programming, Geophys. Res Lett 31 (L16203)

Datta B, Prakash, O., Campbell, S., Escalada, G., (2013) Efficient Identification of Unknown Groundwater Pollution Sources Using Linked Simulation-Optimization Incorporating Monitoring Location Impact Factor and Frequency Factor. Water Resour Manage 27:18

Datta B, Prakash, O., Sreekanth, J (2014) Application of genetic programming models incorporated in optimization models for contaminated groundwater systems management. EVOLVE - A Bridge between Probability, Set Oriented Numerics, and Evolutionary Computation V Advances in Intelligent Systems and Computing 288:16

Dhar A, Datta, B. (2009) Saltwater Intrusion Management of Coastal Aquifers. I: Linked Simulation-Optimization. Journal Hydrology Engineering 14:9

Dorado J, Rabunal, J. R., Puertas, J., Santos, A., and Rivero, D., (2002) Prediction and modelling of the flow of a typical urban basin through genetic programming. Appl Artifici Intel 17 (4):14

Whigham PA, Craper, P. (2001) Modelling rainfall-runoff using genetic programming. Math Comput Modell 33:14

Francone FD (1998) Discipulus Software Owner's Manual, version 3.0 draft. Machine Learning Technologies Inc, Littleton, CO, USA

G MJaYW (2006) Experimental design for groundwater modeling and management. Water Resources Research 42 (2):1-13

Gaur S, and Deo, M. C., (2008) Real-time wave forecasting using genetic programming. Ocean Eng 35 (11-12):7

Herzer j, Kinzelbach, W., (1989) Coupling of Transport and Chemical Processes in Numerical Transport Models. Geoderma 44:13

Hong Y S aRMR (2002) Identification of an urban fractured rock aquifer dynamics using an evolutionary self-organizing modelling. J Hydrol 259:15

Hong, Y.S., White, P. A., Scott, D. M. (2005) *Automatic rainfall recharge model induction by evolutionary computational intelligence.* Water Resour. Res., 41.

Kalin M, Fyson, A., Wheeler, W. N., (2006) Review The chemistry of conventional and alternative treatment systems for the neutralization of acid mine drainage Science of the Total Environment 366:14

Khu S T, Liongs S Y, Babovic V, Madsen H, and Muttil N, (2001) Genetic programming and its application in real-time runoff forecasting. J Am Water Resour Assoc 8:20

Koza JR (1994) Genetic programming as a means for programming computers by natural-selection. Statistics and computing 4:26

Makkeasorn A, Chang, N. B., and Zhou, X. (2008) Short-term streamflow forecasting with global climate change implications—A comparative study between genetic programming and neural network models. J Hydrol 352 (3-4):19

Makkeasorn A CNB, Beaman M, Wyatt C and Slater C (2006) Soil moisture estimation in semiarid watershed using RADARSAT-1 satellite imagery and genetic programming. Water Resour Res 42 (W09401)

Mao X, Prommer, H., Barry, D.A., Langevin, C.D., Panteleit, B., Li, L. (2006) Three-dimensional model for multi-component reactive transport with variable density groundwater flow. Environmental Modelling & Software 21 ((5)):14

McPhee, J., Yeh, W. G. (2006) Experimental design for groundwater modeling and management. WATER RESOURCES RESEARCH. 42. 2. P:1–13

Nordstrom D.K., Alpers C.N., (1999) Geochemistry of acid mine waters, in Plumlee, G.S., and Logsdon, M.J., eds., The environmental geochemistry of mineral deposits, Part A: Processes, techniques, and health issues, Reviews Economic Geology, 6A:28

Parasuraman K, Elshorbagy, A, and Carey S. K., (2007b) Modelling the dynamics of evapotranspiration process using genetic programming. Hydrol Sci J 52:15

Parasuraman K., Elshorbagy A., and Si, B. C., (2007a) Estimating saturated hydraulic conductivity using genetic programming. Soil Sci Soc Am J 71:9

Parkhurst DL, Appelo, C.A.J. (1999) User's guide to PHREEQC—A computer program for speciation, reaction-path, 1D-transport, and inverse geochemical calculations. Technical Report 99-4259, US Geol Survey Water-Resources Investigations Report
Parkhurst DL, Kipp, K.L., Engesgaard, P., Charlton, S.R. (2004) PHAST e A Program for Simulating Ground-water Flow, Solute transport, and Multicomponent Geochemical Reactions. US Geological Survey, Denver, Colorado
Parkhurst DL, Thorstenson, D.C. and Plummet, L.N., (1982) PHREEQE - A computer program for geochemical calculations. Water Resour Invest 210:16
Prommer H (2002) PHT3D—A reactive multi-component transport model for saturated porous media. Version 10 User's Manual, Technical report, Contaminated Land Assessment and Remediation Research Centre, The University of Edinburgh
Prommer H, Barry, D.A., Davis, G.B. (2002) Modelling of physical and reactive processes during biodegradation of a hydrocarbon plume under transient groundwater flow conditions. Journal of Contaminant Hydrology 59:19
Ranjithan S, Eheart, J. W., and Garrett, J. H. (1993) Neural network based screening for groundwater reclamation under uncertainty. Water Resour Res 29 (3):12
Rogers LL, Dowla, F. U. and Johnson, V. M. (1995) Optimal field-scale groundwater remediation using neural networks and the genetic algorithm. Environmental Science & Technology 29:11
Savic D A WGAaDJW (1999) Genetic programming approach to rainfall-runoff modelling. Water Resour Manage 13:12
Sheta AF, and Mahmoud, A., (2001) Forecasting using genetic programming. 33rd Southeastern Symposium on System Theory: 5
Sreekanth J, Datta, B., (2010) Multi-objective management models for optimal and sustainable use of coastal aquifers. Journal of Hydrology 393:11
Sreekanth J, Datta, B., (2011a) Comparative Evaluation of Genetic Programming and Neural Network as Potential Surrogate Models for Coastal Aquifer Management Water Resour Manage (25):18
Sreekanth J, Datta, B., (2011b) Coupled simulation-optimization model for coastal aquifer management using genetic programming-based ensemble surrogate models and multiple-realization optimization. Water resource management 47 (W04516):17
Sreekanth J, Datta, B., (2012) Genetic programming: efficient modelling tool in hydrology and groundwater management. Genetic Programming - New Approaches and Successful Applications, Dr Sebastian Ventura Soto (Ed), InTech 01/2012
Sun J (2004) A Three-Dimensional Model of Fluid Flow, Thermal Transport, and Hydrogeochemical Transport through Variably Saturated Conditions. M S Thesis Department of Civil and Environmental Engineering, University of Central Florida, Orlando, FL 32816
Tebes-Stevensa C, Valocchia, A. J., VanBriesenb J. M., Rittmannb, B. E. (1998) Multicomponent transport with coupled geochemical and microbiological reactions: model description and example simulations. Journal of Hydrology 209:16
Waddill DW, Widdowson, M.A. (1998) SEAM3D: A numerical model for three-dimensional solute transport and sequential electron acceptor-based bioremediation in groundwater. Technical report, Virginia Tech, Blacksburg, Virginia
Wang WC, K. W. Chau, Cheng, C. T., and Qiu, L., (2009) A comparison of performance of several artificial intelligence methods for forecasting monthly discharge time series. J Hydrol 374 (3–4):12
Yeh GT, Cheng, J.R. and Lin, H.C. (1994) 3DFEMFAT: User's Manual of a 3-Dimensional Finite Element Model of Density Dependent Flow and Transport through Variably saturated Media. Technical Report submitted to WES, US Corps of Engineers, Vicksburg, Mississippi Department of Civil and Environmental Engineering, Penn State University, University Park, PA 16802
Yeh GT, Tripathi, V. S (1991) HYDROGEOCHEM: A coupled model of hydrologic transport and geochemical multicomponent equilibria in reactive systems. Environmental Science Division Publication No. 3170, Oak Ridge National Laboratory, Oak Ridge, TN

Yeh, G.T., Cheng, J.R., Lin, H.C.,(2000) *Computational Subsurface Hydrology Reactions, Transport, and Fate of Chemicals and icrobes.* Kluwer Academic Publishers.
Zechman E, Baha, M., Mahinthakumar, G., and Ranjithan, S. R., (2005) A genetic programming based surrogate model development and its application to a groundwater source identification problem. ASCE Conf 173 (341)
Zheng C, Wang, P.P. (1999) MT3DMS: A modular three-dimensional multispecies model for simulation of advection, dispersion and chemical reactions of contaminants in groundwater systems; Documentation and User's Guide. Contract Report SERDP-99-1, US Army Engineer Research and Development Center, Vicksburg, MS
Zhou, X., Chen, M., Liang, C., (2003) *Optimal schemes of groundwater exploitation for prevention of seawater intrusion in the Leizhou Peninsula in southern China.* Environmental Geology, 43: p. 8.

Chapter 15
Potential of Genetic Programming in Hydroclimatic Prediction of Droughts: An Indian Perspective

Rajib Maity and Kironmala Chanda

15.1 Introduction

15.1.1 Backgound

The presence of hydroclimatic teleconnection between large-scale atmospheric-oceanic circulation patterns and hydrologic variables has been established through previous research. Particularly, the impacts of El Niño-Southern Oscillation (ENSO) on the rainfall anomalies across the world (Nicholls 1983; Kousky et al. 1984) have been a subject of research for quite some time.

Other large scale oceanic-atmospheric phenomena such as ENSO, Equatorial Indian Ocean Oscillation (EQUINOO), Pacific Decadal Oscillation (PDO), Atlantic Multi-decadal Oscillation (AMO), Indian Ocean Dipole (IOD) etc. have also been found to play significant roles in influencing hydrological variables worldwide (Chiew and McMahon 2002; Terray et al. 2003; Gadgil et al. 2004; Goswami et al. 2006; Maity and Nagesh Kumar 2006, 2008; Feng and Hu 2008; Li et al. 2008; Mo and Schemm 2008; Ting et al. 2011; Singhrattna et al. 2012; Oubeidillah et al. 2012; Jiang et al. 2013; Rogers 2013). Apart from the influence of these large-scale circulation patterns, local meteorological variables also add to the complexity of hydroclimatic association at smaller spatio-temporal scales (Maity and Kashid 2010, 2011). Studies have indicated that changes in air temperature, relative humidity, evaporation and land use pattern often act as triggers of drought events (Giannini

Electronic supplementary material The online version of this chapter (doi: 10.1007/978-3-319-20883-1_15) contains supplementary material, which is available to authorized users.

R. Maity (✉) • K. Chanda
Indian Institute of Technology Kharagpur, Kharagpur, West Bengal, India
e-mail: rajib@civil.iitkgp.ernet.in; kironmala.iitkgp@gmail.com

A.H. Gandomi et al. (eds.), *Handbook of Genetic Programming Applications*,
DOI 10.1007/978-3-319-20883-1_15

et al. 2008; Wong et al. 2011; Maity et al. 2013). The multi-year droughts that swept over the Northern hemisphere in 1998–2002 are known to have been aggravayed by the persistent above normal surface temperature and the accompanying increase in moisture demand (Hoerling and Kumar 2003). Thus, the contribution of local and global climatic factors in shaping regional hydrological extremes is intricate and dynamic, thus making the prediction of such extremes a very challenging task. In the Indian context, such predictions are even more tricky since rainfall extremes in India are one of the most complex, continental-scale hydrologic phenomena and very much unpredictable (Yuan and Wood 2013).

15.1.2 Genetic Programming

Genetic Programming (GP) is one of the many artificial intelligence techniques available presently. It is a problem solving approach similar to Genetic Algorithm (GA). However, it operates on computer programs, while GA operates on (coded) strings of numbers. GP does not assume any functional form of the solution; it can optimize both the structure of the model and its parameters. The computer programs that are created in GP are generally represented as tree structures (Gandomi et al. 2008; Alavi et al. 2012 and the references therein). Hence it is often referred as tree-based GP (TGP). LGP is a subset of TGP where the evolved programs are in imperative language like C/C++ or in machine language rather than in functional programming language like List Processing (LISP) (Alavi et al. 2010a, 2012). Figure 15.1 depicts the representation of a function in GP. It shows a distinct *tree* structure rather than a flat one dimensional string. The structure consists of simple

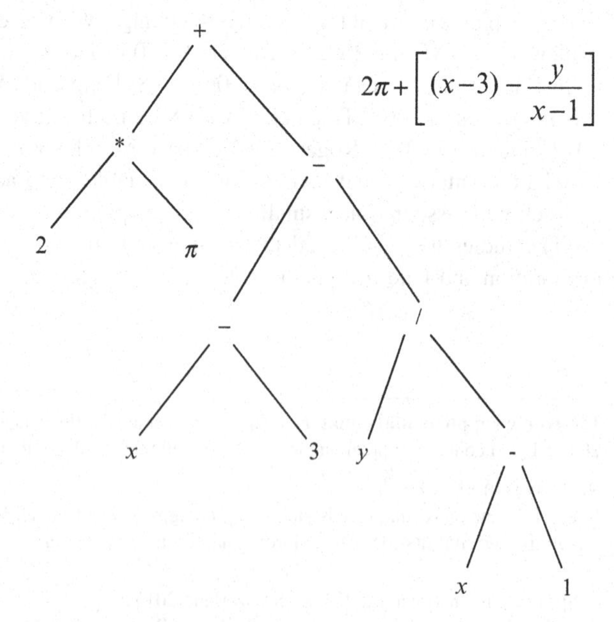

Fig. 15.1 Representation of a function in GP (*Source*: Kashid and Maity 2012)

functions that can be encoded using a high level computer programming language that supports routines for *tree manipulation*. GP provides a methodology where a set of computer programs evolve automatically to represent the model relating input and output variables. GP automatically selects input variables that contribute favourably to the model and disregards those that do not (Jayawardena et al. 2005). GP is based on the Darwinian principles of *reproduction* and *survival of the fittest*. Reproduction is the process of transferring a certain number of parent programs into the population of the next generation. The actual number of programs transferred depends on the reproduction rate. In GP, evolution of better programs occurs via the genetic operators *Mutation* and *Crossover*. Mutation is the process that causes random changes in programs, while crossover is the process of exchange of some sequences of instructions (a 'branch' of a tree structure as shown in Fig. 15.1) between two programs. The exchange results in two offspring programs that are then inserted into the new population replacing the worse programs of the lot. The basic steps involved in GP are depicted in the form of a flow chart in Fig. 15.2.

GP has been successfully utilized by many researchers to solve a multitude of classification problems. In engineering, a number of variants of GP has been used for solving problems ranging from soil classification (Alavi et al. 2010b) to simulation of pH neutralization process (Gandomi and Alavi 2011) to behavioural modelling of structural engineering systems (Gandomi and Alavi 2012; Pérez et al. 2012). In water engineering, GP has been extensively used in flood routing and flood control (Orouji et al. 2012; Yang et al. 2013; Gandomi et al. 2013) as well as optimisation of water supply systems (Xu et al. 2013; Ahn and Kang 2014). GP has also found use in hydrological applications such as rainfall-runoff modelling (Drecourt 1999; Babovic and Keijzer 2002; Jayawardena et al. 2005), flow discharge computations (Sivapragasam et al. 2008; Azamathulla and Zahiri 2012), ground water level simulation (Fallah-Mehdipour et al. 2013), rainfall prediction (Kashid and Maity 2012) and streamflow forecasting in the annual scale (Ni et al. 2010), monthly scale (Mehr et al. 2014) as well as weekly scale (Maity and Kashid 2010). GP was found to have one of the best predictive capabilities among various data driven modeling techniques (Elshorbagy et al. 2010).

In this chapter, the potential of LGP in modelling and prediction of extreme hydrological events is explored. While the hydrological extremes could refer to extremes of many variables such as precipitation, soil moisture, runoff etc., in this study, meteorological droughs (i.e., extremes in precipitation) are in focus. The goal of the study is to develop an LGP based model that uses global climatic information for the detection of ensuing extreme (dry and wet) precipitation events at a target region. The immense volume of global climatic information is first arranged into a reduced dataset by establishing a Global Climate Pattern (GCP) (explained in Sect. 15.2.3) that is influential in triggering precipitation extremes in the target area i.e., India. Using the observed GCP as input, and observed standardized precipitation anamalies (in terms of an index discussed in Sect. 15.2.2) as output, the LGP tool is trained to predict a time series of index values representing standardized precipitation anaomalies. The predicted index values are further classified to obtain the categories of extreme events. The performance of the raw prediction as well as the final categorization is evaluated in terms of standard statistical measures.

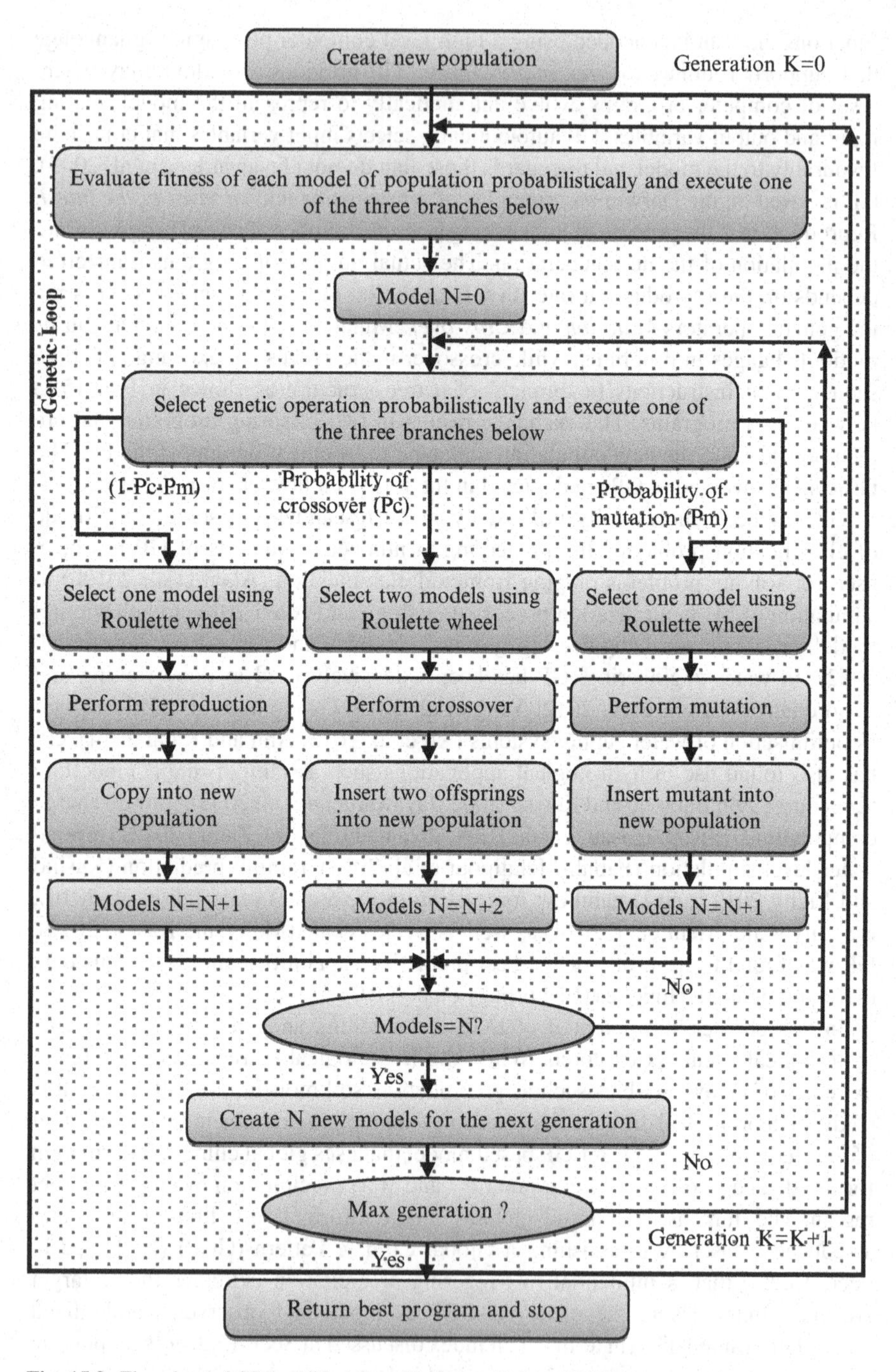

Fig. 15.2 Flowchart of GP (modified from Kashid and Maity 2012)

15.2 Materials and Methods

15.2.1 Data

Monthly precipitation data of India for the period 1959–2010 is obtained from India Meteorological Department (IMD). It is retrieved from the website of Indian Institute of Tropical Meteorology (IITM) (www.tropmet.res.in). Monthly gridded (2.5° lat × 2.5° lon) climate data (Reanalysis I) for four global climate variables namely—Surface Pressure (SP), Air Temperature (AT), Wind Speed (WS) and Total Precipitable Water (TPW) for the period 1958–2010 are obtained from *National Oceanic and Atmospheric* Administration (NOAA) (http://www.esrl.noaa.gov/psd/data/gridded/data.ncep.reanalysis.surface.html). Kaplan Sea Surface Temperature (SST) anomalies having spatial resolution of 5° lat × 5° lon are obtained from NOAA (http://www.esrl.noaa.gov/psd/data/gridded/data.kaplan_sst.html) for the same period.

15.2.2 Standardization of Precipitation Anomaly for Quantification of Extreme Events

To identify extreme events, some method of standardization of precipitation of the target area is necessary. Standardized Precipitation Index (SPI) is one of the popular standardization techniques. However, for monsoon dominated regions with strongly seasonal precipitation pattern, the recently developed Standardized Precipitation Anomaly Index (SPAI) fares better (Chanda and Maity 2015a, b). In case of the popular SPI, the range of index values would be similar for each of the 12 months and equally extreme index values may arise in both monsoon months (high rainfall months) as well as non-monsoon months (very low rainfall months). For instance, let us consider an extreme SPI value of −2. From the statistical point of view, this particular value of SPI may occur in any month of the year and correspond to similarly rare events across the months. However, an SPI of −2 in the month of July (peak monsoon month) is socio-economically very different from the same SPI occurring in the month of January (traditionally driest month). This is because, heavy rainfall during the monsoon period is customary in a monsoon dominated region such as India. Rainfall in monsoon drives agriculture and this economic growth. On the other hand, in traditionally dry months, small rainfall deficits are harmless since the community is used to meeting agricultural and other water requirements during these periods from other sources. Again, it is worthwhile to note that an index value of −2 does not always correspond to identical hydroclimatic conditions. The global climatic conditions that lead to an extreme SPI in monsoon in the target region is generally very different from that causing similar SPI in non-monsoon period. Trying to find a pattern in global climate fields considering all extreme events manifested by SPI together would dilute the signal pool and hinder the detection of the GCP that triggers extremes in high rainfall months. To address the issues mentioned above, two methodological differences are adopted in SPAI with respect to SPI. Firstly, instead of absolute precipitation series, SPAI uses anomaly series of precipitation and secondly, instead of 12 different probability distributions used in SPI, a single distribution is fitted in case of SPAI.

For the present study, the target area is India which experiences strongly seasonal precipitation pattern; hence, the SPAI is used here for standardization of precipitation. The range of SPAI is less for non-monsoon months and high for monsoon months. This helps in putting more prominence to monsoon months which are much critical for drought related vagaries for a region with strongly periodic rainfall.

The SPAI can be computed for any temporal scale; in this study, it is computed at a temporal scale of 3 months. Hence, SPAI-3 is used throughout this study and it is referred as simply SPAI. If t represents the current month, the sum of the precipitation of the tth, $(t-1)^{th}$ and $(t-2)^{th}$ month is standardized with respect to the long-term precipitation totals of those three consecutive months of the year. Similarly, the total precipitation for each three consecutive overlapping months in the study period is standardized with respect to the long-term climatology of the respective 3 months. For example, the first entry in the dataset, i.e, January 1959, which consists of precipitation totals of November 1958, December 1958 and January 1959 is standardized with respect to the long-term climatology of November-December-January. The base period is considered to be 1961–1990, which represents the long-term climatology of the study area. The anomalies are calculated based on the long term means of the aforementioned period. The anomalies of precipitation are given by:

$$y_{i,j} = \left(x_{i,j} - \overline{x}_j\right) / s_j \tag{15.1}$$

where, $y_{i,j}$ is the precipitation anomaly for the ith year and j^{th} time step of the year, $x_{i,j}$ is the precipitation value for the ith year and j^{th} time step of the year, $\overline{x}_j$ and s_j are the long-term mean and standard deviation of precipitation for the jth time step of the year. Hence, if $y_{i,j}$ denotes the precipitation anomaly of July–August–September (JAS) 1990, $x_{i,j}$ denotes the total precipitation of JAS, 1990 and $\overline{x}_j$ and s_j are the long-term mean and standard deviation of precipitation for the months JAS.

Next, a t Location-Scale probability distribution is fitted to the entire anomaly series (y).The pdf of the t Location-Scale distribution is given by:

$$f\left(x|\,\mu,\sigma,\nu\right) = \frac{\Gamma\left(\frac{\nu+1}{2}\right)}{\sigma\sqrt{\nu\pi}\,\Gamma\left(\frac{\nu}{2}\right)}\left[\frac{\nu+\left(\frac{x-\mu}{\sigma}\right)^2}{\nu}\right]^{-\left(\frac{\nu+1}{2}\right)} \quad ; \quad -\infty < x < \infty \tag{15.2}$$

where μ, σ and ν are the location, scale ($\sigma > 0$) and shape ($\upsilon > 0$) parameters respectively. From the fitted Cumulative Distribution Function (CDF) of *t* Location-Scale distribution, the quantiles [$q_i \in (0, 1)$] of each of the observed anomaly values are computed. These quantiles (q) are then transformed to standard normal variates (Z) to obtain the SPAI values.

Thus,

$$Z_i = \Phi^{-1}(q_i) \tag{15.3}$$

where Φ^{-1} is the cumulative inverse standard normal distribution.

As in case of SPI, SPAI values also range from $-\infty$ to ∞. The values less than zero indicate dryness while those greater than zero indicate wetness. In the present study, *dry* events are defined as the events with SPAI ≤ -0.8 and *wet* events are defined as those with SPAI ≥ 0.8.

15.2.3 Identification and Characterization of Input Dataset

For the target region, the dry and wet events (as defined before) are identified first. For each dry event, standardized (with long-term standard deviation) anomalies of the climate variables (except SST, where Kaplan SST anomalies are used directly without standardization) are obtained at 3 months lag at all grid locations across the globe. The lag period is selected as 3 months to reflect the climate conditions immediately before the extreme event. Any temporal overlap is avoided so that the climate indicators may be used for prediction of extremes. Thus, the anomalies of the climate variables are obtained from January–February–March for a *dry* event of the months April–May–June. Similarly, for each wet event, anomalies of the climate variables are obtained at 3 months lag. Now, the global anomaly fields of a given climate variable are averaged across all *dry* events to get a mean global anomaly field corresponding to dry events at the target area. Similarly, a mean global anomaly field corresponding to *wet* events is also obtained. For each of the climate variables considered, the maps of the mean global anomaly fields during dry and wet events are inspected and one or more zones showing significantly contrasting anomaly values during opposite extremes (dry and wet) are identified. These zones may serve as possible predictors in a model concerned with detection and prediction of the extreme events. Thus, together they form the input dataset where each input refers to a climate anomaly from a specific location on the globe. The identified inputs are characterized as the Global Cimate Pattern (GCP). Further details on GCP can be found elsewhere (Chanda and Maity 2015b). The focus of this chapter is to demonstrate the efficacy of LGP in predicting droughts using hydroclimatic information.

15.2.4 Hydroclimatic Prediction of Droughts Using LGP

The potential of LGP in hydroclimatic prediction of extreme events is explored and the climate anomaly series at the selected zones constituting the GCP is used as input. The period 1959–1990 is designated as the training period while 1991–2010 is designated as the validation period.

15.2.4.1 Data Processing

The target output is the SPAI series which follows standard normal distribution. For a standard normal distribution, the values theoretically lie between $(-\infty, \infty)$ and 99.73 % of them lie between $(-3, 3)$. The target series is rescaled from $(-3, 3)$ to $(-30, 30)$ by multiplying each value with 10. The analysis presented in this study was also performed without this rescaling. However, the performance was found to improve with this rescaling. This might be because an improved evolution of programs through LGP was possible due to the wider numerical gap (between dry and wet events) after rescaling. The input data set consists of anomalies of SST and standardized anomalies of AT, WS and TPW. The later also range from $(-\infty, \infty)$ theoretically. However, they are not rescaled since it is not necessary before determination of the most impactful inputs through random program generation in GP.

15.2.4.2 Prediction Using LGP

Generally, for a LGP based prediction, the input dataset is selected based on prior knowledge about the nature of the problem. In the present study, prediction of hydrological extremes is attempted using hydroclimatic information. Hence, an input dataset (characterized as the GCP) consisting of global climate anomaly fields has been developed based on statistical evidence from historical extreme events (explained in Sect. 15.2.3).

In a LGP, the evolved function that relates the input and output variables may be expressed as

$$Y^m = f(X^n) \tag{15.4}$$

where X^n is an n-dimensional input vector and Y^m is a m-dimensional output vector. In this study, the selected climate anomaly zones, characterized by the GCP, represent the input vector and the SPAI series represents the output vector. Freely downloadable software named *Discipulus Lite* (Frankone 1998) is used as the LGP tool in this study. Before running this software, the 'Problem Category' must be set. 'Function Fitting' and 'Classification' are the two options available. In the first one, a regression or curve fitting model is built, while in the second one, a binary classification model is built which can distinguish category X from category Y. There is an option to change the 'Error Measurement' for 'Fitness Calculation'. Either 'Linear' or 'Square' may be used to determine how *fit* an evolving program is. 'Linear' indicates that Mean Absolute Error (MAE) would be used as the fitness function while 'Square' indicates that Mean Squared Error (MSE) would be used for the same. Some of the important control parameters in Discipulus are Population Size, Maximum Tournaments, Mutation Frequency and Crossover Frequency. The 'Population Size' indicates the number of programs in the population that Discipulus will evolve. There is no upper limit as such for this parameter; rather it is only limited by the RAM available on the computer being used and the maximum length of the programs in the population. The larger the population, the higher will be the time required to complete the run, however, Discipulus is fast enough to evolve very large populations within acceptable time frames. In LGP, a number (k) of individual programs are chosen at random and the best programs (based on fitness criteria) among them are selected for crossover. Here k is called the Tournament Size and the maximum number of times such random program selections occur is called the 'Maximum Tournaments'. The parameter 'Mutation Frequency' is used to set the probability of mutation of the programs that have been selected as winners in a tournament by Discipulus. The allowable range for 'Mutation Frequency' is 0–100 %. The 'Crossover Frequency' parameter is used to set the probability that crossover will occur between the two winners in a tournament by Discipulus. The allowable range for this parameter is 0–100 %. Though many LGP applications use a very low mutation rate, it has been established (Frankone 1998) that Discipulus benefits from a high (in excess of 90 %) mutation rate. If the values of these control parameters are not set specifically, Discipulus will run with the default values. The values used in this study are reported later in Sect. 15.3.2.1.

It may be noted here that the term 'linear' in Linear Genetic Programming refers to the structure of the imperative programme representation and does not refer to functional genetic programs restricted to a linear list of nodes only. In fact, highly nonlinear solutions are generally represented through LGP (Brameier and Banzhaf 2004).

15.2.4.3 Evaluation of Prediction Performance

The predicted SPAI series must be rescaled to $(-3, 3)$ before the prediction performance is evaluated. The predicted and observed SPAI is visually compared to evaluate the correspondence between them. A very good match is not expected since prediction of anomaly series is always challenging. However, rather than one-to-one comparison of SPAI values, it may be worthwhile to investigate whether extreme events are identified correctly by the GP model. Hence, a categorical classification of dry, wet and normal events is attempted. Since multi-category classification is not directly supported in GP, the 'Function Fitting' problem type in GP is availed. After obtaining a predicted SPAI series, they may be classified based on the criteria that

SPAI ≤ -0.8 correspond to dry events, SPAI ≥ 0.8 correspond to wet events, and the SPAI values in between correspond to normal events. Thus, the predicted SPAI series may be evaluated directly as also the classification performance.

In order to evaluate the prediction performance, 3-way contingency tables are constructed for both the training and the validation periods. The cell X_{ij}, in the contingency table, refers to the number of events falling in i^{th}observed and j^{th} predicted category. The contingency table is visually inspected and the model performance is quantitatively assessed in terms of the Contingency Coefficient (C) (Pearson 1904). This coefficient measures the degree of association in a contingency table for N samples (Gibbons and Chakraborti 2011) and may be expressed as

$$C = \sqrt{\frac{Q}{Q+N}} \quad (15.5)$$

where Q is a statistic that tests the null hypothesis that there is no association between observed and predicted categories. The statistic Q may be expressed as

$$Q = \sum_{i=1}^{m}\sum_{j=1}^{n} \frac{\left(NX_{ij} - X_{i\otimes}Y_{\otimes j}\right)^2}{NX_{i\otimes}Y_{\otimes j}} \quad (15.6)$$

where m and n are the number of categories, X_{ij} is the number of cases falling in i^{th} observed and j^{th} predicted category, and $X_{i\otimes} = \sum_{j=1}^{n} X_{ij}$ and $Y_{\otimes j} = \sum_{i=1}^{m} Y_{ij}$.

The statistic Q approximately follows chi-square distribution with ν degrees of freedom, where $\nu = (m-1)(n-1)$. The null hypothesis (no association between observed and predicted categories) can be rejected if the p-value is very low. The higher the value of C, the better the association between the observed and predicted categories. Theoretically, the maximum value that C can attain is 1. The upper bound of C is given by

$$C_{\max} = \sqrt{\frac{t-1}{t}} \quad (15.7)$$

where $t = \min(m, n)$(Gibbons and Chakraborti 2011). The ratio $C/C_{\max}$ is often used as a measure of the degree of association. The higher the value of $C/C_{\max}$, the better the prediction performance.

15.3 Results and Discussions

15.3.1 Determination of the Input Dataset

As stated earlier, in this demonstration, the SPAI is computed on a trimonthly scale for the entire study period. Using the criteria SPAI ≤ -0.8, a total of 98 dry events (66 during 1959–1990 and 32 during 1991–2010) are identified. Similarly, using the criteria SPAI ≥ 0.8, a total of 135 wet events (86 during 1959–1990 and 49 during 1991–2010) are identified. The trimonthly anomaly series of the five global climate variables are obtained for the entire study period. For all the dry events, the global anomaly fields of each climate variable at 3 months lag (as explained in Sect. 15.2.3) is averaged event-wise to identify the mean global field of that variable corresponding to dry events. Similarly, the mean global anomaly field corresponding to *wet* events is also obtained for each of the climate variables. As mentioned earlier, the maps of the mean global anomaly fields during dry and wet events are inspected and the grid-wise difference in anomalies are computed for each climate variable. The zones where the differences in mean anomalies (during dry and wet events) pass the test for statistical significance at 95 % confidence level are identified. When the identified zone is contiguous and statistically significant, the core area is selected. However, if the spatial extent of an identified zone is very small with respect to the global scale, then it is ignored even though it may be statically significant. Finally, 14 zones from five climate variables are selected for characterization of the Global Climate Pattern (GCP). A map depicting the SST anomaly zones selected along with the corresponding p-values is shown in Fig. 15.3. Since the number of zones of significance for the other climate variables is much less, they are not shown separately in maps, rather they are summarized along with SST zones in Table 15.1.

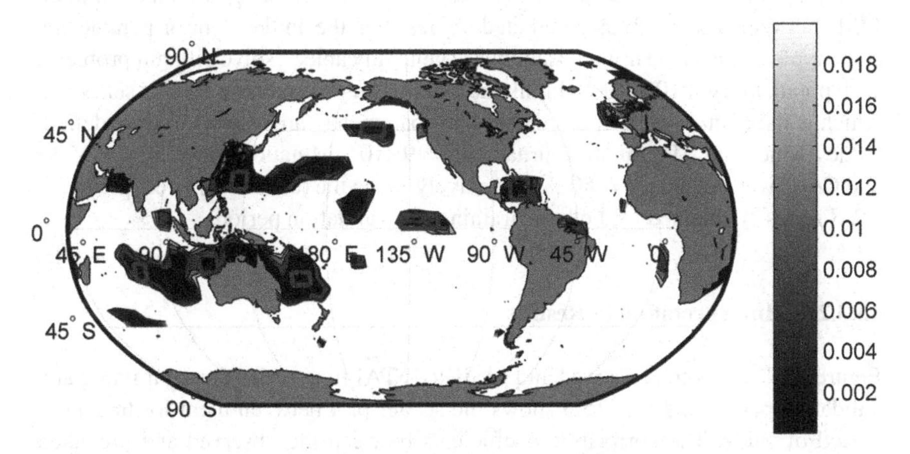

Fig. 15.3 p-values of the differences in mean Kaplan SST anomalies during dry events (SPAI − 3 ≤ -0.8) and wet events (SPAI − 3 ≥ 0.8). Only those regions are shown where p-value is less than 0.02. Selected input zones are demarcated with rectangles

Table 15.1 Descriptions of zones of climate anomalies used to characterize the global pattern

Sl. no.	Physical variable	Symbol	Latitudinal extent	Longitudinal extent
1	Air temperature	AT	20–30°N	85–95°E
2	Wind speed	WS	0°–5°N	70–85°E
3	Total precipitable water	TPW	5–15°N	55–65°E
4	Surface pressure	SP1	15–30°N	145–160°W
5	Surface pressure	SP2	30–40°S	0°–10°W
6	Surface pressure	SP3	55–65°N	145–160°W
7	Surface pressure	SP4	10–20°N	55–65°E
8	Sea surface temperature	SST1	42–48°N	150–164°W
9	Sea surface temperature	SST2	13–23°N	70–80°W
10	Sea surface temperature	SST3	20–26°S	160–170°E
11	Sea surface temperature	SST4	0°–6°N	120–140°W
12	Sea surface temperature	SST5	16–24°S	74–80°E
13	Sea surface temperature	SST6	12–20°S	110–120°E
14	Sea surface temperature	SST7	18–26°N	127–135°E

15.3.2 Potential of LGP in Hydroclimatic Prediction of Droughts

15.3.2.1 Running the LGP Tool

The 14 variable input dataset and the rescaled target series for the period 1959–1990 is used for training the LGP and that for the period 1991–2010 is used for validation. The 'Problem Category' in *Discipulus* is set as 'Function Fitting' to obtain a predicted SPAI series. The criterion for function fitting is set as minimum MSE between observed and predicted values. For the initial control parameters, some typical values are used. Discipulus is generally able to solve difficult problems with populations of 100–1000 (Frankone 1998). Thus 'Population Size' is set as 500 which is the default value. The other control parameters are also set to their default values which are Maximum Tournaments = 9×10^6, Mutation Frequency = 95 % and Crossover Frequency = 50 % respectively. Once the run is completed, predicted SPAI series is obtained for both the training and validation period.

15.3.2.2 Interpretation of Results

Figure 15.4 shows the observed and predicted SPAI time series during training and validation period and Fig. 15.5 shows the scatter plot between them for the entire period of study. The correlation coefficients between the observed and predicted series are found to be 0.49 and 0.35 during the training and validation period respectively. It may be noted that the present study is focussed on the investigation

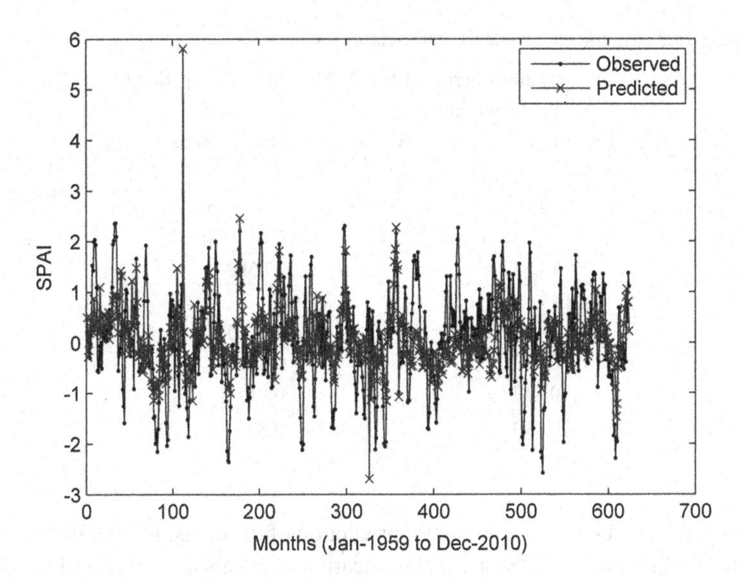

Fig. 15.4 Timeseries of observed and predicted SPAI

Fig. 15.5 Scatter plot of observed and predicted SPAI

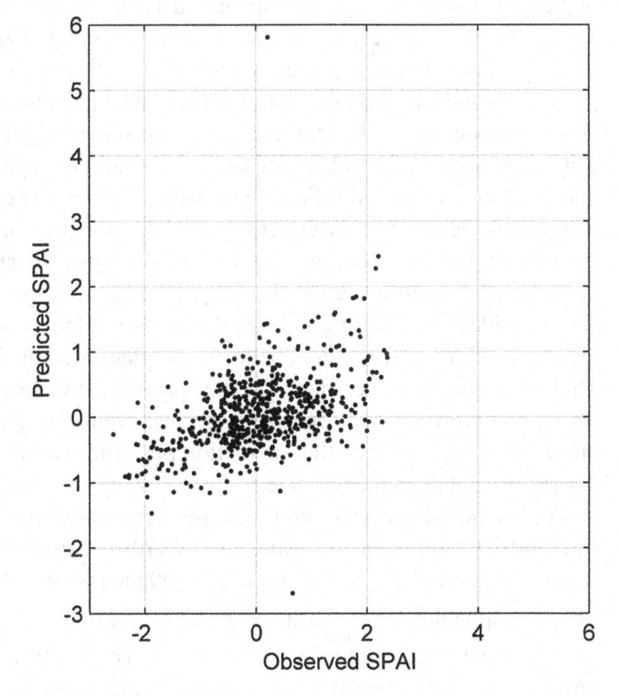

Table 15.2 Contingency table for multi-class classification

	Training period (1959–2000)			Validation period (2001–2010)		
	Predicted category					
Observed category	Dry	Normal	Wet	Dry	Normal	Wet
Dry	14	52	0	5	27	0
Normal	7	212	13	1	150	8
Wet	0	61	25	0	46	3
Q	83.387			27.582		
ν	4			4		
p-value	$\sim <10^{-4}$			$\sim <10^{-4}$		
C	0.422			0.321		
C_{max}	0.817			0.817		
C/C_{max}	0.517			0.393		

of the potential of GP in hydroclimatic prediction of droughts. Hence, based on the predicted SPAI values, a categorical classification of events into dry, wet and normal is attempted as explained earlier. Table 15.2 shows the contingency table for multi-class classification for both training and validation period. It is observed that 14 out of the 66 dry events are correctly identified during the training period. The remaining 52 dry events are falsely classified as normal events. In the validation period, 5 out of the 32 dry events are correctly identified and the remaining 27 dry events are falsely classified as normal events. On the other side of the extreme, 25 out of the 86 wet events are correctly identified during the training period while 3 out of the 49 wet events are correctly identified during the validation period. It may be observed that for both training and validation periods, all the corner elements in the secondary diagonal are zero. Thus, the extremes which are not identified correctly are classified as 'normal'. For normal events, the rate of correct identification is quite high for both training and validation periods. During the former, 212 out of 232 normal events are correctly identified; during the latter, 150 out of 159 normal events are correctly identified. Thus, in brief, the extremes are either identified correctly or as normal. They are generally not predicted as the opposite extremes (i.e, dry as wet, or wet as dry). The performance of predicting normal events are quite satisfactory. Another interesting observation is that prediction performance of extremes is better on the dry side than that on the wet side.

The quantitative evaluation of the prediction performance is done through the Contingency Coefficient C, which is described earlier. During the training and validation periods, the values of C are obtained as 0.422 and 0.321 respectively. The corresponding C/C_{max} values are 0.517 and 0.393 respectively.

It may be noted (Fig. 15.5) that the range of SPAI values in the predicted series lies mostly between −1.5 and 2 while that of the observed series lies mostly between −2.5 and 2.5. This may be because of the fact that the fitness criterion is 'minimization of MSE'. Since, the number of normal events is much more than that of extreme events; it is possible that minimization of the MSE for all the normal events makes the evolved programs less accurate towards the extremes. This is supported by a superior prediction performance for normal events. Hence, careful selection of the fitness function might be necessary depending on the focus of the problem at hand. To put higher prominence to hydrologic extremes, as needed for the problem discussed in the chapter, identification/development of more suitable fitness criteria apart from just MSE/MAE (as available in *Discipulus Lite*), may yield even better results. This can be considered as future research scope.

15.4 Summary and Conclusions

The potential of LGP in the prediction of dry and wet events in India using global climatic information is explored in this study. GP is known to possess good predictive capabilities due to its flexible functional nature. While GP has been successfully used in a number of hydrological applications previously, this study presents a GP-based approach for hydroclimatic prediction of extreme precipitation events in the Indian context. Unlike most earlier studies which use the well-known large scale circulation patterns (such as ENSO, EQUINOO) that are known to affect rainfall anomalies worldwide, an extensive global climate field is considered in this study. Zones of importance are identified and characterized as Global Climate Pattern (GCP). The GCP, consisting of 14 variables, is used as the input dataset and values of SPAI are predicted using LGP. The predicted output series obtained through LGP is further processed to obtain a categorical classification of rainfall events as *dry*, *wet* and *normal*. Considering the complexity and challenge involved in drought prediction, the application of LGP is found to hold promise in capturing the inherent dependence structure between GCP and hydrological extremes. Many of the extreme events are identified and predicted correctly. It is noted that there may be scope for more practically useful results if some problem-specific customizations in fitness functions can be incorporated in GP based modelling.

References

Ahn, J. and Kang, D. 2014. Optimal planning of water supply system for long-term sustainability. *Journal of Hydro-Environment Research*, 8(4), doi: 10.1016/j.jher.2014.08.001.

Alavi A.H., Gandomi A.H. and Gandomi M. 2010a. Comment on Genetic Programming Approach for Flood Routing in Natural Channels. *Hydrological Processes*, Wiley-VCH, 24(6).

Alavi, A.H., Gandomi, A.H., Sahab, M. G. and Gandomi, M. 2010b. Multi expression programming: a new approach to formulation of soil classification. *Engineering with Computers*, 26 (2), 10.1007/s00366-009-0140-7.

Alavi A.H., Gandomi A.H., Bolury J. and Mollahasani A. 2012. Linear and Tree-Based Genetic Programming for Solving Geotechnical Engineering Problems. Chapter 12 in *Metaheuristics in Water, Geotechnical and Transportation Engineering*, XS Yang et al. (Eds.), Elsevier, 2012.

Azamathulla, H.M. and Zahiri, A. 2012. Flow discharge prediction in compound channels using linear genetic programming. *Journal of Hydrology*, 454–455, doi:10.1016/j.jhydrol.2012.05.065.

Babovic, V., Keijzer, M., 2002. Rainfall runoff modeling based on genetic programming. *Nord. Hydrol.* 33, 331–343.

Brameier, M., Banzhaf, W., 2004. Evolving teams of predictors with linear genetic programming. *Genetic Programming and Evolvable Machines.* 2 (4), 381–407.

Chanda, K. and Maity, R. 2015a. Meteorological Drought Quantification with Standardized Precipitation Anomaly Index (SPAI) for the Regions with Strongly Seasonal and Periodic Precipitation. *Journal of Hydrologic Engineering*, ASCE.

Chanda, K. and Maity, R. 2015b. Uncovering Global Climate Fields Causing Local Precipitation Extremes. *Hydrological Sciences Journal*, doi: 10.1080/02626667.2015.1006232.

Chiew, F.H.S., and McMahon, T.A. 2002. Global ENSO-streamflow teleconnection, streamflow forecasting and interannual variability. *Hydrological Sciences Journal*. 47(3), 505–522.

Drecourt, J.P. 1999. Application of neural networks and genetic programming to rainfall runoff modeling. Danish Hydraulic Institute (Hydro-Informatics Technologies HIT). 3rd DHI Software Conference & DHI Software Courses.

Elshorbagy, A. Corzo, G., Srinivasulu, S., Solomatine, D. P. 2010. Experimental investigation of the predictive capabilities of data driven modeling techniques in hydrology - Part 2: Application. *Hydrol. Earth Syst. Sci.*, 14, 1943–1961.

Fallah-Mehdipour, E., Bozorg Haddad, O. and Mariño, M.A. 2013. Prediction and simulation of monthly groundwater levels by genetic programming. Journal of Hydro-Environment Research, 7 (4), doi:10.1016/j.jher.2013.03.005.

Feng, S., and Hu, Q. 2008. How the North Atlantic Multidecadal Oscillation may have influenced the Indian summer monsoon during the past two millennia. *Geophys. Res. Lett.*, 35, L01707, doi:10.1029/2007GL032484.

Frankone, F. D., 1998: Discipulus owner's manual, fast genetic programming based on AIML technology. RML Rep., 196 pp. [Available online at http://www.rmltech.com/.]

Gadgil, S., Vinayachandran, P.N., Francis, P.A., and Gadgil, S. 2004. Extremes of the Indian summer monsoon rainfall, ENSO and equatorial Indian Ocean oscillation. *Geophys. Res. Lett.*, 31, L12213, doi:10.1029/2004GL019733.

Gandomi A.H., Alavi A.H. and Sadat Hosseini S.S. 2008. Discussion on Genetic Programming for Retrieving Missing Information in Wave Records Along the West Coast of India. Applied Ocean Research, Elsevier, 30(4).

Gandomi A.H., Yang X.S., Talatahari S., Alavi A.H. 2013. Metaheuristic Applications in Structures and Infrastructures. Elsevier, Waltham, MA.

Gandomi, A.H. and Alavi, A.H. 2011. Multi-stage genetic programming: A new strategy to nonlinear system modelling. *Information Sciences*, 181 (23), doi:10.1016/j.ins.2011.07.026.

Gandomi, A.H. and Alavi, A.H. 2012. A new multi-gene genetic programming approach to nonlinear system modeling. Part I: materials and structural engineering problems, *Neural Computing and Applications*, 21 (1), doi: 10.1007/s00521-011-0734-z.

Giannini, A., Biasutti, M., and Verstraete, M.M., 2008. A climate model-based review of drought in the Sahel: desertification, the re-greening and climate change. *Glob Planetary Change*, 64, 119–128.

Gibbons, J.D., and Chakraborti, S. 2011. Nonparametric Statistical Inference, 5th ed., Chapman and Hall, Boca Raton, Fla.

Goswami, B.N., Madhusoodanan, M.S., Neema, C.P., and Sengupta, D. 2006. A physical mechanism for North Atlantic SST influence on the Indian summer monsoon. *Geophys. Res. Lett.*, 33, L02706, doi:10.1029/2005GL024803.

Hoerling, M., and Kumar, A. 2003. The Perfect Ocean for Drought. *Science*, 299 (691), doi: 10.1126/science.1079053.

Jayawardena, A. W., N. Muttil, and T. M. K. G. Fernando, 2005: Rainfall-runoff modelling using genetic programming. Proceedings of International Congress on Modelling and Simulation, A. Zerger and R. M. Argent, Eds., Modelling and Simulation Society of Australia and New Zealand (MODSIM 2005), 1841–1847.

Jiang, P., Gautam, M., Zhu, J., and Yu, Z. 2013. How well do the GCMs/RCMs capture the multi-scale temporal variability of precipitation in the Southwestern United States? *Journal of Hydrology*, 479, 75-85, doi: 10.1016/j.jhydrol.2012.11.041.

Kashid, S.S. and Maity, R. 2012. Prediction of monthly rainfall on homogeneous monsoon regions of India based on large scale circulation patterns using Genetic Programming, *Journal of Hydrology*, 454–455, 26–41, doi:10.1016/j.jhydrol.2012.05.033

Kousky, V.E., Kagano, M.T., and Cavalcanti, I.F.A. 1984. A review of the Southern Oscillation: oceanic-atmospheric circulation changes and related rainfall anomalies. 36A (5), 490–504, doi: 10.1111/j.1600-0870.1984.tb00264.x.

Li, S., Perlwitz, J., Quan, X., and Hoerling, M.P. 2008. Modelling the influence of North Atlantic multidecadal warmth on the Indian summer rainfall. *Geophys. Res. Lett.*, 35, L05804, doi:10.1029/2007GL032901.

Maity, R., and Nagesh Kumar, D. 2006. Hydroclimatic association of the monthly summer monsoon rainfall over India with large-scale atmospheric circulations from tropical Pacific Ocean and the Indian Ocean region. *Atmos. Sci. Lett.*, 7, 101– 107, doi:10.1002/asl.141(RMetS).

Maity, R., and Nagesh Kumar, D. 2008. Basin-scale stream-flow forecasting using the information of large-scale atmospheric circulation phenomena. *Hydrol. Processes*, 22(5), 643–650, doi:10.1002/ hyp.6630.

Maity, R., and Kashid, S.S. 2010. Short-term basin-scale streamflow forecasting using large-scale coupled atmospheric oceanic circulation and local outgoing longwave radiation. *Journal of Hydrometeorology*, American Meteorological Society (AMetSoc), 11(2), 370–387, doi: 10.1175/2009JHM1171.1.

Maity, R., and Kashid, S.S. 2011. Importance Analysis of Local and Global Climate Inputs for Basin-Scale Streamflow Prediction. *Water Resources Research*, American Geophysical Union, 47, W11504, doi:10.1029/2010WR009742.

Maity, R., Ramadas, M., and Rao G.S. 2013. Identification of hydrologic drought triggers from hydroclimatic predictor variables. *Water Resources Research*, American Geophysical Union, 49, doi:10.1002/wrcr.20346.

Mehr, A.D., Kahya, E., Yerdelen, C. 2014. Linear genetic programming application for successive-station monthly streamflow prediction. *Computers & Geosciences*, 70, 63–72.

Mo, K.C., and Schemm, J.E., 2008. Relationships between ENSO and drought over the southeastern United States. *Geophysical Research Letters*, Vol. 35, L15701, doi:10.1029/2008GL034656,

Nicholls, N. 1983. Predicting Indian monsoon rainfall from sea-surface temperature in the Indonesia–north Australia area. *Nature*, 306, 576 – 577, doi:10.1038/306576a0.

Ni, Q. Wang, L. Ye, R. Yang, F., Sivakumar, M. 2010. *Environmental Engineering Science*. 27(5): 377–385. doi:10.1089/ees.2009.0082.

Orouji, H., Haddad, O.B., Fallah-Mehdipour, E. Mariño, M.A. 2012. *Proceedings of the ICE - Water Management*, 167(2), 115 –123.

Oubeidillah, A.A., Tootle, G., and Anderson, S.-R., 2012. Atlantic Ocean sea-surface temperatures and regional streamflow variability in the Adour-Garonne basin, France. *Hydrological Sciences Journal*, 57(3), 496–506.

Pearson, K. 1904. On the Theory of Contingency and its Relation to Association and Normal Correlation, Draper's Comp. Res. Mem. Biometric Ser. I, Dulau and Co., London, U.K.

Pérez, J.L., Cladera, A., Rabuñal, J.R. and Martínez-Abella, F. 2012. Optimization of existing equations using a new Genetic Programming algorithm: Application to the shear strength of reinforced concrete beams, Advances in Engineering Software, 50 (1), doi:10.1016/j.advengsoft.2012.02.008.

Rogers, J.C. 2013. The 20th century cooling trend over the southeastern United States. *Climate Dynamics*, 40(1-2), 341–352.

Singhrattna, N., Babel, M.S. and Perret, S.R. 2012. Hydroclimate variability and long-lead forecasting of rainfall over Thailand by large-scale atmospheric variables. *Hydrological Sciences Journal*, 57 (1), 26–41.

Sivapragasam, C., Maheswaran, R. and Venkatesh, V. 2008. Genetic programming approach for flood routing in natural channels. *Hydrol. Process.*, 22: 623–628. doi: 10.1002/hyp.6628.

Terray, P., Delecluse, P., Labattu, S., and Terray, L. 2003. Sea surface temperature associations with the late Indian summer monsoon. *Climate Dynamics*, doi: 10.1007/s00382-003-0354-0.

Ting, M., Kushnir, Y., Seager, R., and Li C. 2011. Robust features of Atlantic multi-decadal variability and its climate impacts. *Geophys. Res. Lett.*, 38, L17705, doi:10.1029/2011GL048712.

Wong, W.K., Beldring, S., Engen-Skaugen, T., Haddeland, I., and Hisdal, H. 2011. Climate change effects on spatiotemporal patterns of hydroclimatological summer droughts in Norway. *Journal of Hydrometeorology*, 12(6), doi: 10.1175/2011JHM1357.1.

Xu, Q., Chen, Q., Ma, J. and Blanckaert, K. 2013. Optimal pipe replacement strategy based on break rate prediction through genetic programming for water distribution *Journal of Hydro-Environment Research*, 7 (2), doi:10.1016/j.jher.2013.03.003.

Yang, X. S., Gandomi, A. H., Talatahari, S. and Alavi, A. H. 2013. Metaheuristis in Water, Geotechnical and Transportation Engineering. Elsevier, Waltham, MA.

Yuan, X., and Wood E.F. 2013. Multimodel seasonal forecasting of global drought onset. *Geophysical Research Letters*, 40, doi:10.1002/grl.50949.

Chapter 16
Application of Genetic Programming for Uniaxial and Multiaxial Modeling of Concrete

Saeed K. Babanajad

16.1 Introduction

New generations of concrete, normal strength concrete (NSC), high strength concrete (HSC), high performance fiber reinforced concrete (HPFRC), and slurry infiltrated fiber concrete (SIFCON) are the most applied types of concrete in civil projects. This would force the investigations to be concentrated in achieving comprehensive understanding of their behaviors under different loading conditions.

To discover the stress–strain behavior of a solid material such as concrete, different types of stress paths should be considered to address the comprehensive understanding of that material. There are four general stress state paths previously introduced and investigated: (1) Uniaxial, (2) Biaxial, (3) Triaxial, and (4) True-triaxial. Each of these paths is also categorized in separate compressive and tensile conditions. Generally, multiaxial stress states are created in construction applications, such as in anchorage zones, shell or dams structures (Hampel et al. 2009). For simplicity those types are replaced by uniaxial cases. Therefore, a decrease due to combined compressive-tensile load or an increase in ultimate strength due to multiaxial compressive may be ignored (Hampel et al. 2009). In order to overcome this problem, each stress path must be studied separately to discover the behavior of concrete under multiaxial loading conditions.

Electronic supplementary material The online version of this chapter (doi: 10.1007/978-3-319-20883-1_16) contains supplementary material, which is available to authorized users.

S.K. Babanajad (✉)
Department of Civil and Materials Engineering, College of Engineering, University of Illinois at Chicago, Chicago, IL, USA
e-mail: skarim8@uic.edu

A.H. Gandomi et al. (eds.), *Handbook of Genetic Programming Applications*, DOI 10.1007/978-3-319-20883-1_16

Most of the constitutive models achieved by analytical/numerical calculations (such as Elasticity and Plasticity theories), are verified using different stress states. These test points could be obtained using different test set ups and loading paths. The Uniaxial path is used as the most important characteristic of hardened concrete in design and construction stages. Uniaxial Compressive Strength and Uniaxial Tensile Strength are two components of this categorization. In these paths, the axial stress is applied to the concrete specimen from only one axis while the other axes are not carrying any stresses ($\sigma_1 \neq 0,\ \sigma_2 = \sigma_3 = 0$). The Biaxial stress path is used to scrupulously extend the understanding of concrete under biaxial loading, in which the concrete cubic specimen is loaded in two surfaces while the third face is free of stress. The test is performed in different stress ratios ($^{\sigma_2}/_{\sigma_1},\ \sigma_1 > \sigma_2,\ \sigma_3 = 0$) to complete the stress–strain curve. Moreover, Triaxial test results have been used in several investigations and national building codes to consider the effect of confinement pressures. This stress state is the most applied stress conditions in the world since it occurs in many actual concrete elements used in structures and inside soil. Triaxial Compressive Strength ($\sigma_1 > \sigma_3 = \sigma_2$) and Triaxial Tensile Strength ($\sigma_1 < \sigma_3 = \sigma_2$) are two components of this categorization. Most of known triaxial data in the literature were determined by using Triaxial cells. As explained, only experiments with combinations of stress ratios of equal stresses in two directions ($\sigma_3 = \sigma_2$) are possible with these cylindrical cells (Hoek and Brown 1980). As mentioned, the Triaxial test is the most conventional and useful test to derive the general behavior of concrete under different loading paths; however, it still has some shortcomings in defining all the experimental points at a failure envelope. In contrast, True-triaxial test is able to provide all the stress components required for establishing failure envelopes. However, due to high-tech instruments needed for testing concrete under True-triaxial loading and since it is expensive, there are few technical references performed the experimental tests with this technique. It must be highlighted that the aforementioned stress state is general and can account all types of Uniaxial, Biaxial and Triaxial tests either in compression or tension. For design and industrial applications using this system will not be economical and the comprehensive information provided by this test will not be required for typical design purposes. However, to develop new material models it is necessary to include the True-triaxial test results.

To solve engineering problems, the currently used empirical, analytical, and numerical methods are simplified with different assumptions and in most cases accompanied with approximations. These simplifications have resulted in high rates of errors which made a large gap between the actual and calculated results (Boukhatem et al. 2011; Khan 2012; Chou and Tsai 2012; Sobhani et al. 2010). However, the machine learning based methods use the natural rules existing in the nature and since many of the features of natures are encountered within the calculations they can release more precise and reliable results (Boukhatem et al. 2011). The machine learning techniques directly learn from raw experimental (or field) data inserted and release the functional relationships among the data, even if the underlying relationships are unknown or the physical meaning is difficult to be explained (Gandomi and Alavi 2011; Gandomi et al. 2011a, 2013a, b). Contrary to these types of methods, many of conventional empirical/statistical/numerical/-analytical methods require prior knowledge about the nature of the relationships among the data as the start point (Alavi and Gandomi 2011). Therefore, the machine learning techniques will be more proper to be applied for modeling the complex behavior of many engineering problems with extreme variability in their natures (Shahin et al. 2009).

To explore the relationships between the performance characteristics of concrete and the affecting parameters, Pattern Recognition (PR) systems made a great opportunity to investigate new models. As known, the basis of these systems' performances relies on the training from experience and develops various discriminators. As a subcategorize of PR systems, Artificial Neural Networks (ANNs) has been included in solving great portions of engineering problems (e.g., Alavi et al. 2010a). Unlike these systems result in high efficiencies, they are usually unable to be ended by a certain input–output function (Gandomi and Alavi 2011). The ANNs have been utilized to discover different features of cement based materials such as concrete (Gupta et al. 2006; Khan 2012; Sobhani et al. 2010; Cheng et al. 2012; Kewalramani and Gupta 2006; Ilc et al. 2009). Tang (2010) recently used radial basis function neural network to predict the peak stress and strain in plain concrete while confinement pressures have been applied. As much as ANNs are successful in prediction, they are not successful in producing practical prediction equations and also the structure of a neural network should be predefined by researcher (Alavi and Gandomi 2011).

Genetic Programming (GP) as the advanced subcategorize of machine learning methods owns the ability to model the mechanical behavior of concrete without any prior assumptions whatsoever regarding material behavior. In this manner, Koza (1992) mentioned that GP is rather a new developed method for the modeling of structural engineering issues. As highlighted by Banzhaf et al. (1998), GP generates the solutions which are computer programs rather than binary strings. It could be accounted as a supervised machine learning technique while searching a program space instead of a data space (Gandomi et al. 2010a; Alavi et al. 2010a).

This chapter presents the feasibility of using PR method with an emphasize on GP for modeling the behavior of concrete in uniaxial condition and also under multi confinement pressures. Different configurations of confinement paths are considered. The chapter is organized as follows: Sect. 16.2 represents the main aspects and the features of the employed GP algorithms. In following Sect. 16.3, the modeling processes are described for the mentioned methods. Later on, numerical examples and the experimental results were also compared. Finally, some concluding remarks are provided in Sect. 16.4.

16.2 Genetic Programming

In a simple comparison, Genetic Algorithm (GA) creates a string of numbers representing the solution, however, the GP solutions are computer programs demonstrated as tree structures and expressed in a functional programming language (like

LISP) (Koza 1992; Gandomi et al. 2010a; Javadi and Rezania 2009; Alavi and Gandomi 2011). In GP, the evolving programs (individuals) are parse trees. These components are not fixed length and are able to be changed throughout the runs.

In parameter optimization, many of the traditional GA optimization techniques have been usually utilizing to evolve the suitable results for a given set of model parameters. In contrast, the values of model's parameters accompanied by the basic structure of the approximation model are generated by the GP. In GP, to properly optimize a population of computer programs a fitness seeking path determined by a program is used. Using a fitness function the fitness of each program in the population is evaluated. In this manner, the fitness function is the objective function GP aims to optimize (Alavi and Gandomi 2011). Also, GP proposed simplified prediction equations without any assumption regarding the form of the existing relationships. This specification will rise up the GP over the conventional statistical and ANN techniques. Furthermore, to discover complex relationships among experimental data the GP-developed equations could be very strong candidates to be easily verified in practical circumstances (e.g., Gandomi et al. 2010a).

Many researchers have employed GP and its variants to discover complex relationships between experimental data (Alavi and Gandomi 2011). In this chapter, Traditional Genetic Programming (TGP), Gene Expression Programming (GEP), Linear Genetic Programming (LGP), Hybrid GP-Orthogonal Least Squares algorithm (GP/OLS), Macroevolutionary Algorithm Genetic Programming (MAGP), Improved Grammatical Evolution Combined with Macrogenetic Algorithm (GEGA), Multi Expression Programming (MEP), and Genetic Operation Tree (GOT) approaches are discussed since they recently have been used in the area of concrete modeling (Gandomi et al. 2012; Babanajad et al. 2013). Each method is briefly introduced to give schematics of their performances in modeling concrete behavior.

16.2.1 Traditional Genetic Programming

TGP is introduced by Koza (1992) as the classical GP. Once a population of individuals (models) has been created at random, the TGP algorithm starts the evaluation of the individuals' fitnesses. Then, it selects individuals for reproduction, crossover and mutation. The reproduction operation generates new individuals to create a higher probability of selection toward more successful individuals. These individuals, without any change, are directly copied into the next generation. Then, the crossover operation is applied to exchange the genetic materials between the evolved programs. Within this procedure, a point is randomly chosen on a branch of each solution (program). Later on, from each program the set of terminals and/or functions are swapped in order to create two new solutions (programs). During the mutation process, occasionally a function or terminal from a model is randomly chosen to be mutated. If the randomly selected node is a terminal, it is replaced by another terminal. If the randomly selected node is a function, depending on the type of mutation (point or tree type), a new function is assigned. More details are listed in Gandomi and Alavi (2011).

16.2.2 Gene Expression Programming

As a recent extension and natural development of GP, GEP first invented by Ferreira (2001), has been developed to evolve computer programs with different sizes and shapes. The GEP approach can be utilized as an efficient alternative to the TGP for the numerical experiments (Ferreira 2001, 2006). Gandomi et al. (2012, 2013c) directed valuable efforts at applying GEP to civil engineering tasks. Many of the GAs' genetic operators could be used as GEP operators with few changes. Function set, terminal set, fitness function, control parameters, and termination condition are the five main components of the GEP structure.

A fixed length of character strings, called chromosomes, have been used by GEP to develop solutions to the problems as the parse trees of different sizes and shapes which are called GEP expression trees (ETs). However, the TGP methods use parse-tree representation. It is also noted by Gandomi et al. (2011a) that the generation of genetic diversity is extremely simplified because of the capability of genetic operators in working at the chromosome level. GEP is unique and has multigenic nature which allows evolutions of more complex programs composed of several subprograms (Gandomi et al. 2012). Further information regarding the detailed technical processes used in GEP method are available in Ferreira (2001, 2006), Alavi and Gandomi (2011) and Gandomi et al. (2012, 2014a).

16.2.3 Linear Genetic Programming

LGP is placed as a subcategorize of GP with a linear representation of individuals. It is explained that the expressions of a functional programming language (like LISP) are replaced by programs of an imperative language (like C/C++) (Brameier and Banzhaf 2001; Gandomi et al. 2010a; Alavi et al. 2013), which made LGP to differ from TGP. Figure 16.1 demonstrates the above mentioned difference between LGP and TGP.

In TGP, the data flow is more rigidly determined by the tree structure of the program while inside linear genetic program, the data flow graph generated by multiple usage of register content (Brameier and Banzhaf 2001, 2007; Gandomi et al. 2010a; Gandomi and Alavi 2011; Alavi and Gandomi 2011; Alavi et al. 2010b). That is, on the functional level the evolved imperative structure indicates a special directed graph. Comprehensive descriptions of the basic parameters used to direct a search for a linear genetic program can be found in Brameier and Banzhaf (2007), Gandomi et al. (2010a, 2011b, 2014a, b), Oltean and Grossan (2003a), Arıoglu et al. (2006) and Babanajad et al. (2013).

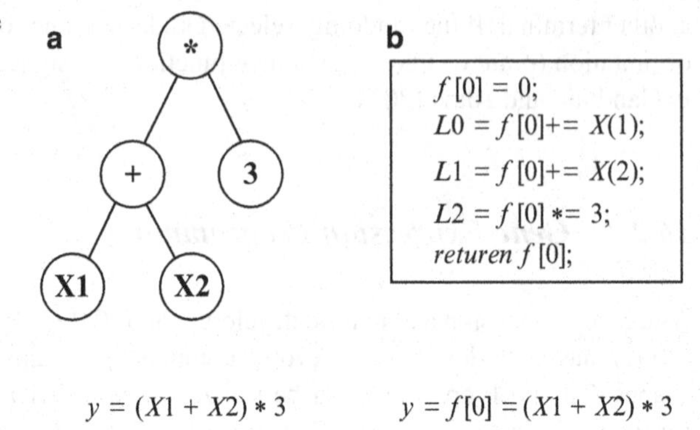

Fig. 16.1 Comparison of the GP program structures, (**a**) TGP; (**b**) LGP

16.2.4 Hybrid GP-Orthogonal Least Squares Algorithm

It is well known that the combination of Orthogonal Least Squares (OLS) and GP algorithms creates significant performance improvements of the GP (Mousavi et al. 2010; Billings et al. 1988; Chen et al. 1989; Madár et al. 2005; Gandomi and Alavi 2013; Gandomi et al. 2010b, 2010c). OLS algorithm could be effectively applied to determine the significant terms of a linear-in-parameters model (Billings et al. 1988; Chen et al. 1989). The OLS algorithm uses the error reduction ratio to monitor the variance of output by a given term. Madár et al. (2005) created a hybrid algorithm with the combination of GP and OLS techniques to achieve higher efficiency. It was proven that applying OLS into GP results in more robust and interpretable models (Madár et al. 2005). To determine the structure and parameters of the model the GP/OLS is relied on the data alone. Different references stated that the discussed technique has rarely been applied to the structural engineering issues and they have concluded that GP/OLS approach can be helpful in deriving empirical models by directly extracting the knowledge contained in the experimental data, such as concrete compressive strength (Mousavi et al. 2010; Gandomi and Alavi 2013, Gandomi et al. 2010b). The method transforms the trees to simpler trees with more transparencies and close accuracies to the original trees, which is the main point of the proposed method. First GP generates several solutions in the form of a tree structure and then OLS estimates the contribution of the branches of the tree to the accuracy of the model. The terms (sub-trees) with smallest error reduction ratio are selected to be eliminated from the main tree and this approach is conducted in every fitness evaluation before the calculation of the fitness values of the trees (Pearson 2003). This action is named as Pruning proposed by Pearson (2003). As the next step, the mean square error values and error reduction ratios are calculated again. After pruning the improved model obviously releases a higher mean square error including a more adequate structure.

16.2.5 *Macroevolutionary Algorithm Genetic Programming*

GP basics are relied on a transparent and structured way to obtain a fittest function type of experimental results in an automatic process. On the other hand, the GP has still issues regarding premature convergence to the local optimum. Therefore, based on the concept and also performance of Macroevolutionary Algorithm (MA) in solving similar issues, it was considered as a proper candidate (Chen 2003). MA has been introduced as a new concept of species evolution at the higher level which was presented to replace the conventional selection scheme of GP (Chen 2003; Marin and Sole 1999). They have concluded that MA has superior capability over the rest of techniques. Therefore, it was claimed that the combination of MA and GP improves the capability of searching global optima.

16.2.6 *Improved Grammatical Evolution Combined with Macrogenetic Algorithm*

Grammatical Evolution (GE) is classified as one of the subcategorize of evolutionary based techniques. GE proposes a novel technique of using grammars in the process of automatic programming. Backus–Naur form (BNF) is utilized by GE to express computer programs (Oltean and Grosan 2003a; Ryan et al. 1998; Ryan and O'Neill 1998). BNF refers to a notation that permits a computer program to be expressed as a grammar. Terminal and non-terminal symbols form the BNF grammar and the grammar symbols could be re-explained in other terminal and non-terminal symbols. The GE individual has variable length types of binary string and contains the vital information to select a production rule from a BNF grammar in its codons (groups of eight bits). This model was proposed by Chen and Wang (2010) in order to estimate the compressive strength of HPC. They have concluded that GEGA has shown less error in comparison with TGP and two popular types of traditional multiple regression analysis methods.

16.2.7 *Multi Expression Programming*

MEP is a subcategorize of GP that was developed by Oltean and Dumitrescu (2002). Linear chromosomes are used for solution encoding. MEP is capable to encode multiple solutions (computer programs) inside a single chromosome (Mohammadi Bayazidi et al. 2014). The best of the encoded solutions is selected based on the fitness values of the individuals, which represent the chromosomes. Other GP variants (such as GEP, GE, and LGP) store a single solution in a chromosome while MEP encodes multiple solutions in a chromosome (Oltean and Groşan 2003a, b). MEP translates mathematical expressions into machine code in a very similar way used in C and Pascal compilers (Aho et al. 1986). The number of genes within a single chromosome is constant and a function symbol (an element in the function set) or a terminal (an element in the terminal set) is encoded by each gene (Heshmati et al. 2010).

16.2.8 Genetic Operation Tree

GOT is a variant of GP that consists GA and operation trees (OT). Operation tree is designed in a tree shaped in order to represent a mathematical formula (Yeh et al. 2010). The designed tree could be optimized to create a self-organized regression formula. The discrete type of optimization mandates using GA instead of mathematical programming. Due to the inherit features of the GA based techniques global optimization, flexible, nonlinear, and parallelism problems could be easily addressed. Furthermore, some investigations have attempted to improve the performance of the GOT by combining with other techniques. For example, Cheng et al. (2014) integrated the Weighted Operation Structure (WOS) and Pyramid Operation Tree (POT) in order to extend and improve the efficiency of the prediction. They introduced Genetic Weighted Pyramid Operation Tree (GWPOT) as the improved version of GOT that consists of the Weighted Operation Structure, Pyramid Operation Tree, and GA methods.

The advantages and disadvantages of all of the variants of GP technique which were explained in previous paragraphs are completely addressed in Oltean and Grossan (2003a).

16.3 Strength Modeling of Concrete Using GP Techniques

As known, empirical models have been developed and verified based on the data points gathered from different experimental tests. As the numbers of various tests and their repeatability increase, the proposed model is more reliable and comprehensive to be used in different conditions. However, there are some major shortcomings in developing empirical models. The models are initializing based on very simple assumptions and observations. In many cases the proposed models can only handle specific and few parameters involved in the behavior of concrete. In the case of increasing the parameters, achieving a proper model would be more difficult. In addition, the proposed models are driven for specific range of concrete strengths (range of f'_c). In contrast, the analytical and numerical models release more reliable results. They are established based on theoretical and analytical rules (such as Elasticity, Plasticity theories, etc.) which could consider complex behavior of concrete under different loading paths. However, to validate these models different loading paths with different configurations must be available. This will result in time and cost consuming experiments. Like empirical tests, analytical models can only account few parameters in the modeling of concrete behavior (Farnam et al. 2010; Babanajad et al. 2012). Both empirical and analytical models only use principal stresses and strains, except some models considered also temperature effects in their modeling (He and Song 2010). While the PR models, specifically GP models, enable the researchers to consider other effective parameters in concrete behavior. Besides, these models do not need to be evaluated by specific experimental tests. They can use the available databases in the literature. However, their accuracies depend on the size and variety of the database. In the case of concrete modeling since there are different and very large databases, using PR models is very efficient. Later on, these models could be used as a proper tool for validation and verification of the existing models used by the national building codes.

As the first step in using GP techniques for modeling the strength of concrete, the main parameters of concrete previously used in different models are explained. The influenced variables include as follows:

$$\sigma_1 = f\left(\sigma_2, \sigma_3, f'_c, K, Z, W, C, P\left(B, F\right), S, MSA, Fib, Ln(A)\right) \tag{16.1}$$

where,

σ_1 (MPa): principal stress
σ_2 (MPa): principal stress
σ_3 (MPa): principal stress
$\sigma_1 \geq \sigma_2 \geq \sigma_3$
$f' c(MPa)$: Uniaxial Compressive Strength
K: Ratio of water and superplasticizer summation to binder $((W + S) / Binder)$
$Binder$: Binder content $(C + P)$
$W(kg/m^3)$: Water content
$C(kg/m^3)$: Cement content
$P(kg/m^3)$: Pozzolan content (B: Blast furnace slag, F: Fly Ash)
$S(kg/m^3)$: Superplasticizer content
Z: Ratio of coarse aggregate (CA) to fine aggregate (FA) content
MSA (mm): Maximum Size Aggregate
Fib (%): Fiber index ($Fib = V_f \frac{l_f}{d_f}$)
$V_f(\%)$: Steel fiber volume fraction
$\frac{l_f}{d_f}$: Fiber aspect ratio
l_f (mm): Fiber length
d_f (mm): Fiber diameter
A (day): Age of specimens at testing

The mix design variables were chosen as the input variables on the basis of literature review (Mousavi et al. 2010; Yeh and Lien 2009; Chen and Wang 2010; Babanajad et al. 2013) and after a trial study. In the following, the aforementioned models are listed, including:

- **TGP** (Gandomi and Alavi 2011): Uniaxial Compressive Strength

  $$\sigma_1 = f\left(W, C, B, F, Z, A\right); \ \sigma_1(compressive) = f'_c \neq 0; \sigma_2 = \sigma_3 = 0$$

- **GEP** (Mousavi et al. 2012): Uniaxial Compressive Strength

  $$\sigma_1 = f\left(W, C, B, F, S, CA, FA, A\right); \ \sigma_1(compressive) = f'_c \neq 0; \sigma_2 = \sigma_3 = 0$$

- **GP/OLS_I** (Mousavi et al. 2010): Uniaxial Compressive Strength

  $\sigma_1 = f\,(K, Z, A)\,;\ \sigma_1(compressive) = f'_c \neq 0; \sigma_2 = \sigma_3 = 0$

- **GP/OLS_II** (Mousavi et al. 2010): Uniaxial Compressive Strength

  $\sigma_1 = f\,(K, A)\,;\ \sigma_1(compressive) = f'_c \neq 0; \sigma_2 = \sigma_3 = 0$

- **GEGA** (Chen and Wang 2010): Uniaxial Compressive Strength

  $\sigma_1 = f\,(K, A,\ CA, FA)\,;\ \sigma_1(compressive) = f'_c \neq 0; \sigma_2 = \sigma_3 = 0$

- **MEP_I** (Heshmati et al. 2010): Uniaxial Compressive Strength

  $\sigma_1 = f\,(W, C, S,\ CA, FA)\,;\ \sigma_1(compressive) = f'_c \neq 0; \sigma_2 = \sigma_3 = 0$

- **MEP_II** (Heshmati et al. 2010): Uniaxial Compressive Strength

  $\sigma_1 = f\,(W, S, C,\ CA, FA)\,;\ \sigma_1(compressive) = f'_c \neq 0; \sigma_2 = \sigma_3 = 0$

- **MAGP_I** (Chen 2003): Uniaxial Compressive Strength

  $\sigma_1 = f\,(W, Binder)\,;\ \sigma_1(compressive) = f'_c \neq 0; \sigma_2 = \sigma_3 = 0$

- **MAGP_II** (Chen 2003): Uniaxial Compressive Strength

  $\sigma_1 = f\,(W, Binder, A,\ FA)\,;\ \sigma_1(compressive) = f'_c \neq 0; \sigma_2 = \sigma_3 = 0$

- **TGP** (Chen and Wang 2010**; Chen** 2003): Uniaxial Compressive Strength

  $\sigma_1 = f\,(K, A,\ CA, FA)\,;\ \sigma_1(compressive) = f'_c \neq 0; \sigma_2 = \sigma_3 = 0$

- **LGP** (Gandomi and Alavi 2011): Uniaxial Compressive Strength

  $\sigma_1 = f\,(W, C, B, F, S, Z, A)\,;\ \sigma_1(compressive) = f'_c \neq 0; \sigma_2 = \sigma_3 = 0$

- **GOT** (Yeh et al. 2010): Uniaxial Compressive Strength

  $\sigma_1 = f\,(W, K, Binder, S, CA, FA, A)\,;\ \sigma_1(compressive) = f'_c \neq 0; \sigma_2 = \sigma_3 = 0$

- **GOT** (Yeh and Lien 2009): Uniaxial Compressive Strength

  $\sigma_1 = f\,(W, Binder, S, A)\,;\ \sigma_1(compressive) = f'_c \neq 0; \sigma_2 = \sigma_3 = 0$

- **GOT** (Peng et al. 2010) **:** Uniaxial Compressive Strength

  $\sigma_1 = f\,(W, C, Binder, S, CA, A)\,;\ \sigma_1(compressive) = f'_c \neq 0; \sigma_2 = \sigma_3 = 0$

- **GEP** (Gandomi et al. 2012): Triaxial Compressive Strength

  $\sigma_1 = f\,(K, \sigma_3\,(= \sigma_2)\,, Fib, Z, A)\,;\ \sigma_1 \neq 0; \sigma_2 = \sigma_3 \neq 0$

- **LGP** (Babanajad et al. 2013): Triaxial Compressive Strength

  $\sigma_1 = f\left(\sigma_3\,(= \sigma_2)\,, f'_c\right); \sigma_1 \neq 0; \sigma_2 = \sigma_3 \neq 0$

- **GEP_I** (Babanajad et al. submitted): True-triaxial Strength

  $$\sigma_1 = f\,(K, \sigma_2, \sigma_3, Z, MSA)\,;\ \sigma_1 > \sigma_2 > \sigma_3 \neq 0$$

- **GEP_II** (Babanajad et al. submitted): True-triaxial Strength

  $$\sigma_1 = f\left(f'_c, \sigma_2, \sigma_3\right);\ \sigma_1 > \sigma_2 > \sigma_3 \neq 0$$

- **GEP_III** (Babanajad et al. submitted): True-triaxial Strength

$$\sigma_1 = f\left(f'_c, \sigma_2, \sigma_3\right); \; \sigma_1 > \sigma_2 > \sigma_3 \neq 0$$

16.3.1 Experimental Database

In this part, as an example, seven of previously established GP based models are briefly discussed. These models were established using GEP, TGP and LGP concepts to come up with the solution. Twice of the models were applied to predict the Uniaxial Compressive Strength ($f'_c = \sigma_1 \neq 0,\ \sigma_3 = \sigma_2 = 0$). Gandomi and Alavi (2011) developed a TGP and a LGP models to predict the Uniaxial Compressive Strength of concrete. The used database contained 1133 compressive strength of concrete test results presented by Yeh (2006a, b). Two more models were also generated to calculate the Triaxial Strength of concrete ($\sigma_1 \neq 0,\ \sigma_3 = \sigma_2 \neq 0$). Gandomi et al. (2012) developed a GEP model and later Babanajad et al. (2013) proposed a LGP model. They used the same database contained 330 Triaxial experimental results of concrete tests under equal confined pressures. Recently, Babanajad et al. (submitted) established three GEP models to predict the True-triaxial strength of concrete under multiaxial stresses ($\sigma_1 \neq \sigma_2 \neq \sigma_3 \neq 0$; General form). They have used 300 True-triaxial test results to verify the models. In the literature, there are few models explaining the behavior of concrete under biaxial loading conditions and there is no GP based model to predict the biaxial behavior of concrete. This is mainly because of the reason that Biaxial test is not common and widely applied stress state. Hence, in this chapter the Biaxial modeling of concrete is not covered.

To visualize the distribution of the samples, the input and output variables are presented by frequency histograms presented in Fig. 16.2 for models TGP and LGP (Gandomi and Alavi 2011), as an example. As can be observed from Fig. 16.2, the distributions of the predictor variables are not uniform. For the cases where the frequencies of the variables are higher the derived models are expected to provide better predictions.

It is well known that the GP based models have a capability to properly predict the target value within the data range used for their development. Thus, the size of training database for these algorithms is a critical issue, and guarantees the accuracy and reliability of the extracted models. Comprehensive data sets have to be used for training the algorithms. Hence, a reliable database consisting of tests on mixtures with a wide range of water/cement rations, coarse and fine aggregates gradation and properties has to be obtained from literature to develop the generalized models (Gandomi and Alavi 2011).

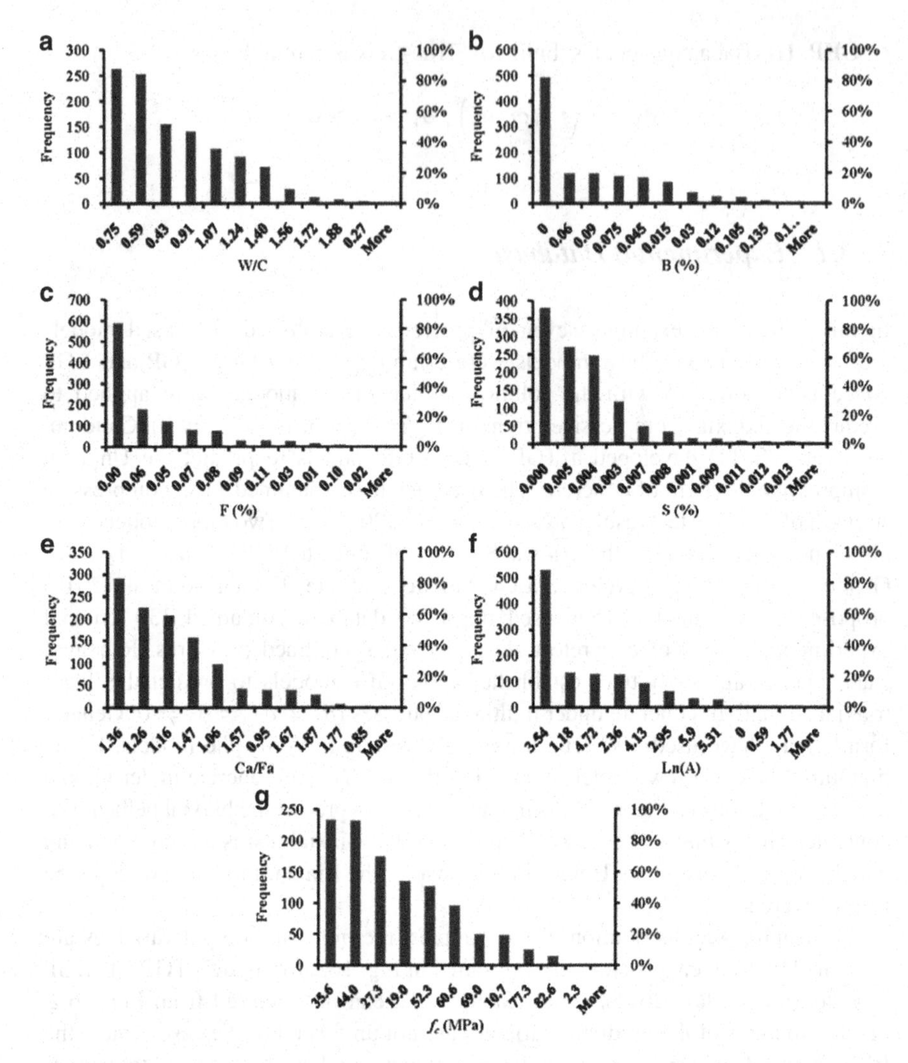

Fig. 16.2 Variables histogram used in the TGP and LGP development (Gandomi and Alavi 2011)

16.3.2 GP-Based Model Development

In the area of GP, the final selection of the best model is established on the basis of a multi-objective strategy as below (Gandomi et al. 2012; Babanajad et al. 2013):

1. The simplicity of the model,
2. Providing the best fitness value on the learning set of data,
3. And providing the best fitness value on the validation set of data,

The first objective is conducted by the user with selection of the parameter settings (e.g., number of genes or head size). However, this is not considered as a predominant factor. In the case of the second and third objectives, the objective function (OBJ) has to be constructed in a manner to monitor the closeness of the model predicted outputs with the experimental results. Then, the best model is deduced by minimizing the OBJ function.

In order to develop the GP based concrete model, different input parameters $(\sigma_2, \sigma_3, f'_c, K, Z, W, C, P\,(B, F)\,, S, MSA, Fib, and\ Ln\,(A))$ were previously used to create the GP models for the strength modeling of concrete. Various parameters involved in the GP based predictive algorithm were selected. The parameter selection affects the model generalization capability of the GP models. Therefore, several runs were conducted to come up with a parameterization of GP models that provided enough robustness and generalization to solve the problem. Different parameter settings such as number of Chromosomes, Genes, Head Size, Tail Size, Dc Size, Gene Size, etc. have been examined until there was no longer significant improvement in the performance of the proposed models.

In the following, the GP-based formulations existing in the entire technical literature are listed and the ultimate strength (σ_1) of concrete are given as below:

TGP (Gandomi and Alavi 2011): Uniaxial Compressive Strength; $\sigma_1\,(compressive) = f'_c \neq 0; \sigma_2 = \sigma_3 = 0$

$$\sigma_1 = \frac{C}{W}\Bigg((F+4).\bigg(Z + Ln(A) + \bigg(F + B.\frac{W}{C}\bigg) .\bigg(Ln(A).\,(Ln(A)+5) + 13B\Big(\frac{W}{C} + F + Ln(A)\Big)\bigg)\bigg)\Bigg) + 1$$

GEP (Mousavi et al. 2012): Uniaxial Compressive Strength; $\sigma_1(compressive) = f'_c \neq 0; \sigma_2 = \sigma_3 = 0$

$$\sigma_1 = \Bigg(-5.69\bigg(2.89 + \frac{2.64B - 3.19A + F}{CA}\bigg).\,(8.72S + 1.93A - 33.6 + C) .\,(16.08S + A - B - 196.26 - FA - F)\Bigg) \Bigg/ \Bigg(FA.\bigg(W - \frac{C - CA + A}{A}\bigg)\Bigg)$$

GP/OLS_I (Mousavi et al. 2010): Uniaxial Compressive Strength; $\sigma_1\,(compressive) = f'_c \neq 0; \sigma_2 = \sigma_3 = 0$

$$\sigma_1 = -39.36K - 3.94\left(Z - \frac{Ln(A)}{K}\right) + 30.88$$

GP/OLS_II (Mousavi et al. 2010): Uniaxial Compressive Strength; $\sigma_1\,(compressive) = f'_c \neq 0; \sigma_2 = \sigma_3 = 0$

$$\sigma_1 = 6.46\left(\frac{Ln(A)}{K} - K\right) + 11.37$$

MAGP_I (Chen 2003): Uniaxial Compressive Strength; $\sigma_1(compressive){=}f'_c \neq 0$; $\sigma_2 = \sigma_3 = 0$

$$\sigma_1 = \frac{860.914}{\left({}^{w}\!/_{Binder}\right)^2} + 1508.4$$

MAGP_II (Chen 2003): Uniaxial Compressive Strength; $\sigma_1(compressive) = f'_c \neq 0; \sigma_2 = \sigma_3 = 0$

$$\sigma_1 = 514.63\left(\frac{w}{Binder}\right)^{-1.27}(Ln(A) + 0.0312) + FA^{\cos(0.216FA)}$$

TGP (Chen and Wang 2010; **Chen** 2003): Uniaxial Compressive Strength; $\sigma_1(compressive) = f'_c \neq 0; \sigma_2 = \sigma_3 = 0$

$$\sigma_1 = \left(\frac{0.425}{K} + 0.345 * Ln(A) + 0.015\right) * \left(Ln\left(\frac{CA}{18.989}\right) + \frac{Ln(FA)}{K}\right)$$

LGP (Gandomi and Alavi 2011): Uniaxial Compressive Strength; σ_1 $(compressive) = f'_c \neq 0; \sigma_2 = \sigma_3 = 0$

$$\sigma_1 = \frac{C}{W}.Ln(A)\left(36F + 36B - \frac{2Ln(A)}{3}\left(1 - \frac{2W}{3C}\right) + Z + 5\right) - \frac{W}{C} - Ln(A) - S + 7$$

GEGA (Chen and Wang 2010): Uniaxial Compressive Strength; σ_1 $(compressive) = f'_c \neq 0; \sigma_2 = \sigma_3 = 0$

$$\sigma_1 = \frac{0.397}{K}\left(\frac{1}{K} + Ln(A)\right)(Ln\,(Ln(A) \times CA) + FA)$$

MEP_I (Heshmati et al. 2010): Uniaxial Compressive Strength; σ_1 $(compressive) = f'_c \neq 0; \sigma_2 = \sigma_3 = 0$

$$\sigma_1 = \frac{320C}{CA}\cos\left(cos\left(\frac{FA}{4C} - \frac{S}{5} - 2\frac{w}{c}\right)^2\right)$$

MEP_II (Heshmati et al. 2010): Uniaxial Compressive Strength; σ_1 $(compressive) = f'_c \neq 0; \sigma_2 = \sigma_3 = 0$

$$\sigma_1 = 80\left(0.1S\left(\frac{FA - CA}{C} + 2\right) - \frac{FA}{C} - \frac{CA}{2C} + 4\frac{w}{c} + 2\right)$$

GOT (Yeh et al. 2010): Uniaxial Compressive Strength; $\sigma_1(compressive) = f'_c \neq 0; \sigma_2 = \sigma_3 = 0$

$$\sigma_1 = -43.24 + 4.21 \times \frac{Ln(A) + 6.9}{K + \frac{W+S}{Binder+CA+FA}} \quad (K < 0.5)$$

$$\sigma_1 = 329.3 - 4.20 \times \frac{0.51 \times Ln\left(\frac{W}{C}\right) + 39 - Ln(A)}{K} \quad (0.35 < K < 0.65)$$

$$\sigma_1 = 2.02 + 363.5 \times \frac{Ln(A)}{0.85K} \quad (K > 0.5)$$

GOT (Yeh and Lien 2009): Uniaxial Compressive Strength; $\sigma_1(compressive) = f'_c \neq 0; \sigma_2 = \sigma_3 = 0$

$$\sigma_1 = -2366.78 + 400.82 \times \left(\frac{Binder}{S + W - 50.9}\right) \times [Ln(A) + Ln(35.63)]$$

GOT (Peng et al. 2010): Uniaxial Compressive Strength; $\sigma_1(compressive) = f'_c \neq 0; \sigma_2 = \sigma_3 = 0$

$$\sigma_1 = -2.98 - 56.95 \times \left(\frac{C \times (Ln(A) + W)}{Ln(A) - Ln(W)}\right)$$

$$early\ age\ \ (Age < 14\ days) - model\ I$$

$$\sigma_1 = -14.12 - 1194.54 \times \left(\frac{\frac{C}{-72.05} - Ln(A)}{W}\right)$$

$$early\ age\ \ (Age < 14\ days) - model\ II$$

$$\sigma_1 = -6.04 - 12.74 \times \left(\frac{Ln(W) \times W^2}{Ln\left(\frac{Ln(A)}{W}\right)}\right)$$

$$early\ age\ \ (Age < 14\ days) - model\ III$$

$$\sigma_1 = -15.42 + 3.86 \times \left(\frac{Binder}{2S + W - 50.69}\right) \times [Ln(A) + 1.57]$$

$$medium\ age\ \ (14 < Age < 56\ days) - model\ I$$

$$\sigma_1 = -12.74 + 5.41 \times \left(\frac{Binder}{2S + W - 50.69}\right) \times Ln(A)$$

medium age (14 < *Age* < 56 *days*) – *model II*

$$\sigma_1 = -15.31 + 17.52 \times \left(\frac{Binder}{S + Ln(A) + W - 50.73}\right)$$

medium age (14 < *Age* < 56 *days*) – *model III*

$$\sigma_1 = -4.61 + 1.74 \times \left(\frac{Binder}{2S + W - 52.15}\right) \times [Ln(CA) + Ln(47.8)]$$

Late age (56 < *Age* < 365 *days*) – *model I*

$$\sigma_1 = -5.33 + 18.98 \times \left(\frac{Binder}{2S + W - 52.15}\right)$$

Late age (56 < *Age* < 365 *days*) – *model II*

$$\sigma_1 = -9.07 + 2.38 \times \left(\frac{Binder}{2S + W - 52.15}\right) \times [Ln(A) + Ln(47.8)]$$

Late age (56 < *Age* < 365 *day*) – *model III*

GEP (Gandomi et al. 2012): Triaxial Compressive Strength;$\sigma_1 \neq 0$; $\sigma_2 = \sigma_3 \neq 0$

$$\sigma_1 = \left(\frac{Ln(A)}{K} + Fib\right).\left(\sqrt[4]{\sigma_3} + 9\right) + 7\left(\sqrt{Z + \sigma_3} - Ln(A) - 2\right) + 1.19\left(K^2 - 2\right)\left(Fib - Z - \sigma_3\right) + Z$$

LGP (Babanajad et al. 2013): Triaxial Compressive Strength; $\sigma_1 \neq 0$; $\sigma_2 = \sigma_3 \neq 0$

$$\sigma_1 = \sqrt{\sigma_3} + \sqrt{2\sigma_3.f'_c} + 2\sigma_3 - 4f'_c$$

GEP_I (Babanajad et al. submitted): True-triaxial Strength; $\sigma_1 > \sigma_2 > \sigma_3 \neq 0$

$$\sigma_1 = \sigma_3 + (Z + K)^{1/3} - \left(MSA - \frac{10}{K} - 9\right) + MSA.\left[(Z + \sigma_2)^{\frac{1}{3}}.\left(\frac{2}{K} - Z\right)\right]^{0.5} + \sigma_3.\left(4 - K.Z^2\right)$$

GEP_II (Babanajad et al. submitted): True-triaxial Strength;$\sigma_1 > \sigma_2 > \sigma_3 \neq 0$

$$\sigma_1 = f'_c + 2\sigma_3 + (3\sigma_2)^{1/3} + \sigma_2^{1/3} + (\sigma_3 + 4\sigma_3 . f'_c)^{1/2}$$

GEP_III (Babanajad et al. submitted): True-triaxial Strength;$\sigma_1 > \sigma_2 > \sigma_3 \neq 0$

$$\sigma_1 = (f'_c) . \left[\left(18\frac{\sigma_3}{f'_c} + \left(\frac{\sigma_2}{f'_c}\right)^{1/3} + 1 \right)^{0.5} + 0.25\left(\frac{\sigma_3}{f'_c} . \left(\left(\frac{\sigma_2}{f'_c}\right)^{0.5} - \frac{\sigma_3}{f'_c} + 4\right)\right) \right]$$

As an example, a comparison of the experimental against predicted triaxial strength values for GEP (Gandomi et al. 2012) is shown in Fig. 16.3 and reported in Table 16.1, and also the expression tree of the derived equation is given in Fig. 16.4. The formula creates a complex arrangement of operators, variables, and constants to predict σ_1. Based on the Fig. 16.4, the proposed equation is a combination of three independent components (subprograms or genes) which are connected by an additional function. Any individual aspect of the problem is highlighted by each of these three subprograms such that a meaningful overall solution is developed (Ferreira 2001). In other words, important information about the physiology of the final model could be achieved through each of the evolved subprograms. Ferreira (2001) and Gandomi et al. (2012) stated that each gene of the final equation is

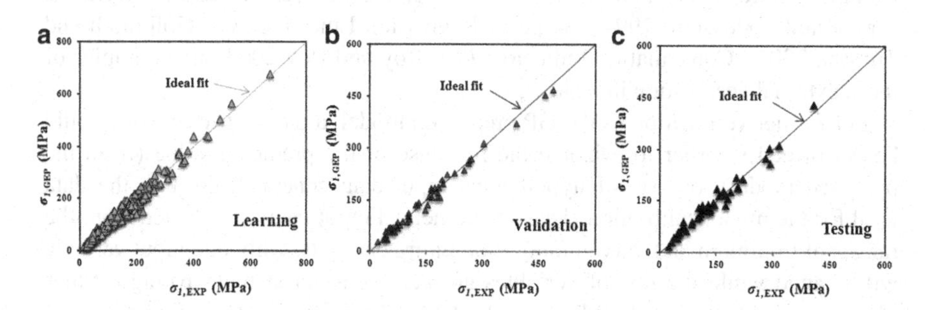

Fig. 16.3 Experimental versus predicted triaxial strength values using the GEP model (Gandomi et al. 2012)

Table 16.1 Statistical values of the GEP prediction model (Gandomi et al. 2012)

	MAE	*RMSE*	*R*
Learning	13.28	17.10	0.9850
Validation	10.46	12.60	0.9947
Testing	10.86	14.08	0.9849

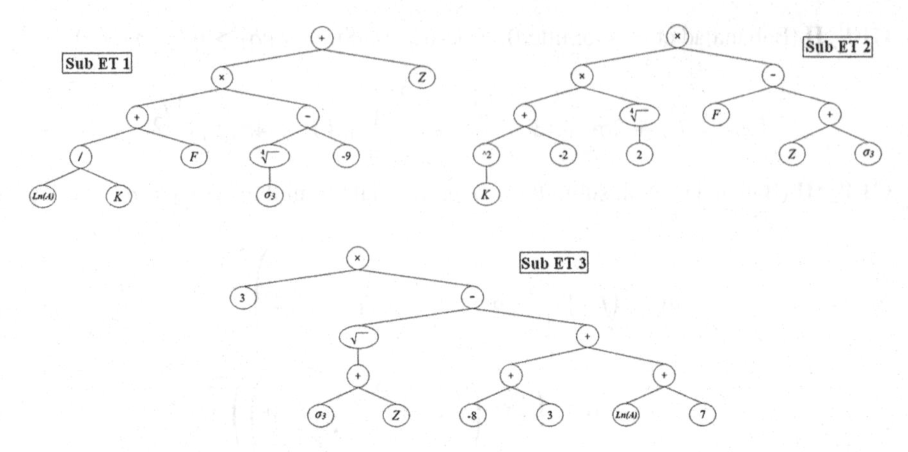

Fig. 16.4 Expression tree for strength modeling of concrete under triaxial compression (ETs = $\sum$ETi); GEP model (Gandomi et al. 2012)

committed to resolve a particular facet of the problem. They have concluded that the provided critical information will make the further scientific discussion at genetic and chromosomal level to be continued.

Based on a logical hypothesis (Smith 1986), it is concluded that for a model with $R > 0.8$ and minimum *MAE* value strong correlation between the predicted and measured values exists. The model can therefore be judged as very good (Gandomi et al. 2011c). The results shown in Table 16.1 prove the reliability and accuracy of the proposed GP model in predicting the Triaxial strength of concrete. In order to validate the proposed models different indices have been introduced in the literature. Minimum ratio of the number of objects over the number of selected variables (Frank and Todeschini 1994); Slope of Regression Line (k or k') (Golbraikh and Tropsha 2002); Confirmation Indicator (R_m) (Roy and Roy 2008) are examples of those external verification indicators.

For further verification of the GP prediction model, a parametric analysis could be performed in order to evaluate the response of the predicted value (from the proposed model) to a set of hypothetical input data generated through the data used for the model calibration. These hypothetical input data are selected from the ranges of minimum and maximum of their entire data sets. Only one input variable will change while the rest of variables are kept constant at their average values of their entire data sets (Gandomi et al. 2012). By increasing the value of varied parameter in increments, a set of synthetic data for that parameter will be generated. In this way, the inputs are entered to the model and the output is calculated. The entire above mentioned procedure is repeated for all input variables of the proposed model. Therefore, with concentrating on the proximity of the predicted model output values and the underlying properties of the simulated case, robustness of the design equations will be more highlighted (Kuo et al. 2009).

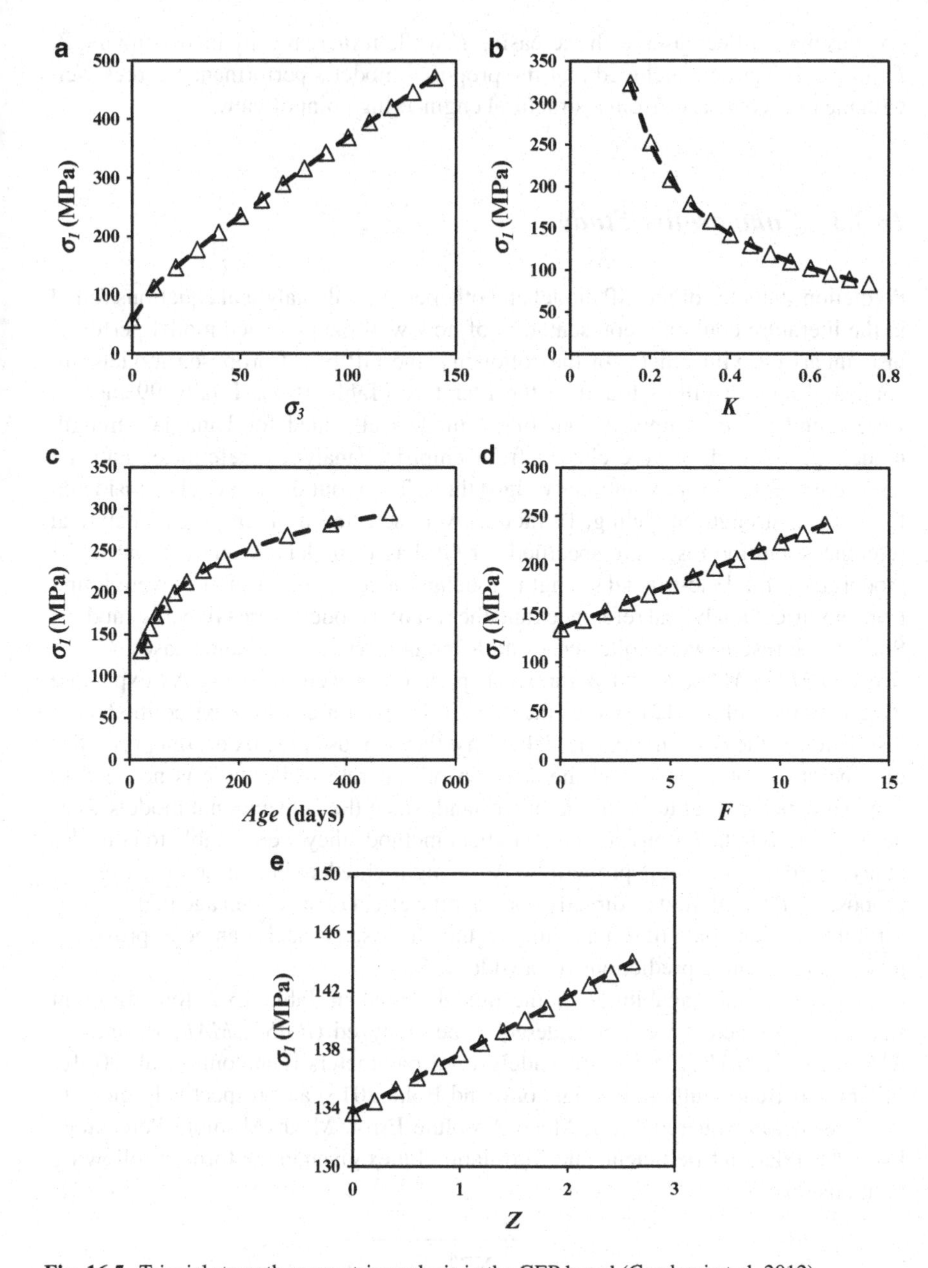

Fig. 16.5 Triaxial strength parametric analysis in the GEP based (Gandomi et al. 2012)

In order to clarify the parametric analysis applied for verification of the GP based models, Fig. 16.5 represents the ultimate Triaxial strength parametric study in the GEP model proposed by Gandomi et al. (2012). As presented in Fig. 16.5, it is obvious that σ_1 decreases with increasing K while it increases by increasing σ_3, Z, F, and A. It can be concluded that the proposed model's performance agrees well with the expected cases from a structural engineering point of view.

16.3.3 Comparative Study

Prediction statistics of the GP model and other empirical/analytical equations found in the literature could be representation of how well the proposed model performs in comparison with others. In the following, the GP based proposed models are compared against others found in the literature (Table 16.2). Totally 99 models were found to be compared. Out of 64 models allocated for Uniaxial strength modeling, 16 models were chosen from empirical/analytical references and the rest were specified for evolutionary algorithms. Then, out of 21 models considered for Triaxial strength modeling, 19 models were selected from empirical/analytical references and the rest were specified for GP-based models. Moreover, out of 14 models used for True-triaxial strength modeling of concrete, 11 models were found from empirical/analytical references and the rest of 3 models were GP based models. Since these results were collected from different references, in some cases, one or two of *RMSE*, *MAE*, R and ρ statistical parameters were missing. As explained by Gandomi et al. (2012), one of the powerful capabilities of GP based models in distinction to the conventional models is that they can use the mix design properties of concrete in order to model the strength of concrete while there is no need to conduct experimental tests. In the other hand, since the conventional models were derived through the empirical and analytical methods they were unable to consider many of the mix design properties. As a highlighted point, in some cases the proposed GP based models directly consider the effect of specimen age in their final formula. As more data become available, the GP based models can be improved to make more accurate predictions for a wider range.

To evaluate the capabilities of the models listed in Table 16.2, four different statistical parameters were considered to be analyzed (*RMSE*, *MAE*, R, and ρ). *RMSE*, *MAE*, *MAPE*, and ρ are widely used parameters (Gandomi et al. 2011c; Milani and Benasciutti 2010; Gandomi and Roke 2013) and respectively indicate the Root Mean Squared Error, Mean Absolute Error, Mean Absolute Percentage Error, Correlation Coefficient, and Correlation Index given in the form of following relationships:

$$RMSE = \sqrt{\frac{\sum_{i=1}^{n} (h_i - t_i)^2}{n}} \tag{16.2}$$

$$MAE = \frac{\sum_{i=1}^{n} |h_i - t_i|}{n} \tag{16.3}$$

Table 16.2 Comparison of proposed model with other models

Model ID	Researcher	Load path	*RMSE*	*MAE*	*R*	*ρ*
1*	LLSR[a] (Ryan 1997; Mousavi et al. 2012)	Uniaxial	–	7.25	0.840	–
2*	NLSR[b] (Ryan 1997; Mousavi et al. 2012)	Uniaxial	–	5.34	0.907	–
3*	RA I[c] (Chen and Wang 2010;Yeh 1998)	Uniaxial	15.28	–	–	–
4*	RA II[c] (Chen and Wang 2010;Yeh 1998)	Uniaxial	22.86	–	–	–
5*	NR[d] (Yeh and Lien 2009)	Uniaxial	12.80	–	0.966	6.51
6*	LSR I[e] (Ryan 1997; Mousavi et al. 2010)	Uniaxial	–	6.03	0.877	–
7*	LSR II[e] (Ryan 1997; Mousavi et al. 2010)	Uniaxial	–	6.05	0.875	–
8	RBF[f] (Gandomi and Alavi 2011)	Uniaxial	–	4.37	0.930	–
9	SLNN[g] (Rajasekaran and Amalraj 2002; Heshmati et al. 2010)	Uniaxial	7.82	6.90	0.898	4.12
10	MEP_I (Heshmati et al. 2010)	Uniaxial	8.63	7.04	0.924	4.49
11	MEP_II (Heshmati et al. 2010)	Uniaxial	8.32	6.86	0.908	4.36
12	BPN[h] (Chen and Wang 2010)	Uniaxial	10.36	–	–	–
13	GOT (Yeh and Lien 2009)	Uniaxial	9.30	–	0.931	4.82
14	MLP[i] (Gandomi and Alavi 2011)	Uniaxial	–	4.31	0.935	–
15*	NR[d]—$K < 0.5$ (Yeh et al. 2010)	Uniaxial	11.34	–	–	–
16*	NR[d]—$0.35 < K < 0.65$ (Yeh et al. 2010)	Uniaxial	7.56	–	–	–
17*	NR[d]—$K > 0.5$ (Yeh et al. 2010)	Uniaxial	5.38	–	–	–
18	ANN—$K < 0.5$ (Yeh et al. 2010)	Uniaxial	10.98	–	–	–
19	ANN—$0.35 < K < 0.65$ (Yeh et al. 2010)	Uniaxial	7.32	–	–	–
20	ANN—$K > 0.5$ (Yeh et al. 2010)	Uniaxial	5.27	–	–	–
21	GOT-medium age (Yeh et al. 2010)	Uniaxial	11.05	–	–	–
22	GOT-medium age (Yeh et al. 2010)	Uniaxial	7.49	–	–	–
23	GOT-medium age (Yeh et al. 2010)	Uniaxial	5.48	–	–	–
24	SVM[j] (Cheng et al. 2014)	Uniaxial	7.17	5.30	–	–
25	ANN (Cheng et al. 2014)	Uniaxial	7.00	5.42	–	–
26	ESIM[k] (Cheng et al. 2014)	Uniaxial	6.57	4.16	–	–
27	GWPOT[l] (Cheng et al. 2014)	Uniaxial	6.38	4.79	–	–
28	GOT (Cheng et al. 2014)	Uniaxial	7.12	5.51	–	–
29	WOS[m] (Cheng et al. 2014)	Uniaxial	6.89	5.23	–	–
30	BPN[h] (Yeh and Lien 2009; Yeh 2006c)	Uniaxial	5.7	–	0.966	2.90
31	ANN (Erdal et al. 2013)	Uniaxial	5.57	4.18	0.953	2.85
32	BANN[n] (Erdal et al. 2013)	Uniaxial	4.87	3.60	0.963	2.48
33	GBANN[o] (Erdal et al. 2013)	Uniaxial	5.24	4.09	0.963	2.67
34	WBANN[p] (Erdal et al. 2013)	Uniaxial	4.54	3.30	0.970	2.30
35	WGBANN[q] (Erdal et al. 2013)	Uniaxial	5.75	4.83	0.976	2.91
36	ANN (Chou et al. 2011)	Uniaxial	5.03	–	0.953	2.58
37*	MR[r] (Chou et al. 2011)	Uniaxial	10.43	–	0.782	5.85
38	SVM[j] (Chou et al. 2011)	Uniaxial	5.62	–	0.941	2.90
39*	MART[s] (Chou et al. 20115)	Uniaxial	4.95	–	0.954	2.53
40*	BRT[t] (Chou et al. 2011)	Uniaxial	5.57	–	0.943	2.87

(continued)

Table 16.2 (continued)

Model ID	Researcher	Load path	*RMSE*	*MAE*	*R*	*ρ*
41	GOT-early age-average (Peng et al. 2010)	Uniaxial	8.28	–	0.862	4.45
42	GOT-medium age-average (Peng et al. 2010)	Uniaxial	9.74	–	0.895	5.14
43	GOT-late age-average (Peng et al. 2010)	Uniaxial	8.13	–	0.925	4.22
44*	NR[d]-early age (Peng et al. 2010)	Uniaxial	7.4	–	–	–
45*	NR[d]-medium age (Peng et al. 2010)	Uniaxial	10.53	–	–	–
46*	NR[d]-late age (Peng et al. 2010)	Uniaxial	9.41	–	–	–
47	BPN[h]-early age (Peng et al. 2010)	Uniaxial	5.74	–	0.938	2.96
48	BPN[h]-medium age (Peng et al. 2010)	Uniaxial	9.56	–	0.899	5.03
49	BPN[h]-late age (Peng et al. 2010)	Uniaxial	7.12	–	0.943	3.66
50	ANN (Topcu and Sarıdemir 2008)	Uniaxial	2.81	–	0.998	1.41
51	FL[u] (Topcu and Sarıdemir 2008)	Uniaxial	2.02	–	0.999	1.01
52	ANFIS[v] (Ramezanianpour et al. 2004; Fazel Zarandi et al. 2008)	Uniaxial	14.21	–	–	–
53	FPNN[w] (Fazel Zarandi et al. 2008)	Uniaxial	9.56	–	–	–
54	GEGA (Chen and Wang 2010)	Uniaxial	9.49	–	–	–
55	TGP (Gandomi and Alavi 2011)	Uniaxial	–	6.14	0.881	–
56	LGP (Gandomi and Alavi 2011)	Uniaxial	–	5.71	0.906	–
57	GP/LOS I (Mousavi et al. 2010)	Uniaxial	–	5.72	0.889	–
58	GP/LOS II (Mousavi et al. 2010)	Uniaxial	–	6.19	0.876	–
59	MAGP_I (Chen 2003)	Uniaxial	7.38	–	–	–
60	MAGP_II (Chen 2003)	Uniaxial	4.64	–	–	–
61	TGP_I (Chen 2003)	Uniaxial	10.6	–	–	–
62	TGP_II (Chen 2003)	Uniaxial	8.27	–	–	–
63	TGP (Chen and Wang 2010; Chen 2003)	Uniaxial	10.87	–	–	–
64	GEP (Mousavi et al. 2012)	Uniaxial	–	5.20	0.914	–
65*	Richart et al. (1929)	Triaxial	22.83	14.80	0.975	11.56
66*	Balmer (1949)	Triaxial	26.79	17.68	0.985	13.50
67*	Martinez et al. (1984)	Triaxial	22.27	14.63	0.977	11.26
68*	Saatcioglu and Razvi (1992) (I)	Triaxial	31.60	20.66	0.981	15.95
	Saatcioglu and Razvi (1992) (II)	Triaxial	22.94	14.30	0.987	11.55
	Setunge et al. (1993) (I)	Triaxial	22.31	14.37	0.980	11.27
	Setunge et al. (1993) (II)	Triaxial	31.60	20.33	0.977	15.98
69*	Setunge et al. (1993) (III)	Triaxial	27.57	18.56	0.988	13.87
70*	Xie et al. (1995)	Triaxial	19.07	13.18	0.986	9.60
71*	Légeron and Paultre (2003)	Triaxial	34.30	22.60	0.979	17.33
72*	Attard and Setunge (1996)	Triaxial	18.02	11.79	0.984	9.08
73*	Girgin et al. (2007)	Triaxial	18.86	12.78	0.989	9.48
74*	Johnston (1985)	Triaxial	28.65	18.94	0.983	14.45
75*	Lan and Guo (1997)	Triaxial	43.74	26.95	0.960	22.32
76*	Ansari and Li (1998)	Triaxial	32.63	21.39	0.981	16.47
77*	Li and Ansari (2000)	Triaxial	31.11	20.08	0.976	15.74
78*	Bohwan et al. (2007)	Triaxial	77.55	35.54	0.739	44.59

(continued)

Table 16.2 (continued)

Model ID	Researcher	Load path	*RMSE*	*MAE*	*R*	*ρ*
79*	Chinn and Zimmerman (1965)	Triaxial	40.85	26.50	0.966	20.78
80*	Mullar (1973)	Triaxial	49.51	23.98	0.814	27.29
81*	Avram et al. (1981)	Triaxial	15.36	10.36	0.988	7.73
82*	Tang (2010)	Triaxial	18.14	13.56	0.984	9.14
83*	Samaan et al. (1998)	Triaxial	58.59	36.77	0.966	29.80
84	GEP (Gandomi et al. 2012)	Triaxial	14.08	10.86	0.985	7.09
85	LGP (Babanajad et al. 2013)	Triaxial	13.72	10.35	0.992	6.89
86*	Ottosen (1977)	True-Triaxial	26.84	19.90	0.890	14.20
87*	Hsieh et al. (1982)	True-Triaxial	77.82	43.27	0.692	45.99
88*	Li and Ansari (1999)	True-Triaxial	67.03	54.03	0.606	41.74
89*	Hampel et al. (2009)	True-Triaxial	9.390	6.340	0.988	4.72
90*	Willam and Warnke (1975)	True-Triaxial	47.73	28.17	0.840	25.94
91*	He and Song (2010)	True-Triaxial	14.08	10.25	0.984	7.10
92*	Mills and Zimmerman (1970)	True-Triaxial	27.50	20.03	0.906	14.43
93*	Wang et al. (1987)	True-Triaxial	54.22	37.23	0.880	28.84
94*	Seow and Swaddiwudhipong (2005)	True-Triaxial	42.45	33.48	0.955	21.71
95*	Desai et al. (1986)-Hinchberger (2009)	True-Triaxial	98.98	80.58	0.583	62.53
96*	Bresler and Pister (1958)	True-Triaxial	137.5	100.8	0.559	88.20
97	GEP_I (Babanajad et al. submitted)	True-Triaxial	13.58	9.82	0.984	6.84
98	GEP_II (Babanajad et al. submitted)	True-Triaxial	7.87	5.91	0.990	3.95
99	GEP_III (Babanajad et al. submitted)	True-Triaxial	11.14	7.11	0.988	5.60

Note: (–) indicates that there is no available results

Note: All of the *RMSE*, *MAE*, and *R* values correspond to the testing evaluation of dataset.

Note: For the models starred in the Model ID column, the formulas were achieved by empirical and analytical methods.

[a] Linear Least Squares Regression (LLSR)

[b] Nonlinear Least Squares Regression (NLSR)

[c] Regression Analysis (RA)

[d] Nonlinear Regression (NR)

[e] Least Squares Regression (LSR)

[f] ANN-based Radial Basis Function (RBF)

[g] Sequential Learning Neural Network (SLNN)

[h] ANN with back-propagation algorithm, called Back Propagation Network (BPN)

[i] ANN-based Multi-Layer Perceptron (MLP)

[j] Support Vector Machine (SVM)

[k] Evolutionary Support Vector Machine Inference Model (ESIM)

[l] Genetic Weighted Pyramid Operation Tree (GWPOT)

[m] Weighted Operation Structure (WOS)

[n] Bagged ANN (BANN)

[o] Gradient Boosted ANN (GBANN)

[p] Wavelet Bagged ANN (WBANN)

[q] Wavelet Gradient Boosted ANN (WGBANN)

[r] Multiple Regression (MR)

[s] Multiple Additive Regression Tree (MART)

[t] Bagging Regression Tree (BRT)

[u] Fuzzy Logic (FL)

[v] Adaptive Network-based Fuzzy Inference System (ANFIS)

[w] Fuzzy Polynomial Neural Netwrok (FPNN)

$$R = \frac{\sum_{i=1}^{n} \left(h_i - \overline{h_i}\right) . \left(t_i - \overline{t_i}\right)}{\sqrt{\sum_{i=1}^{n} \left(h_i - \overline{h_i}\right)^2 . \sum_{i=1}^{n} \left(t_i - \overline{t_i}\right)^2}} \quad (16.4)$$

$$\rho = \frac{RMSE}{1 + R} \quad (16.5)$$

in which h_i and t_i are respectively the actual and calculated outputs for the ith output, $\overline{h_i}$ and $\overline{t_i}$ are the average of actual and predicted outputs, respectively, and n is the number of samples.

By equal shifting of the output values of a model, R value does not change. Therefore, R could not be as a good indicator for predicting the accuracy of a model. The constructed OBJ function takes into account the changes of *RMSE*, *MAE* and R simultaneously. The combination of lower *RMSE* and *MAE* and higher R values results in lowering OBJ; and hence indicates a more precise model. However, it was concluded that the models have to be compared using a suitable criterion not only using R or error functions. It is because R will not change significantly by shifting the output values of a model equally, and error functions (e.g. *RMSE* and *MAE*) only show the error not correlation. Therefore, the criteria should be the combination of R, *RMSE* and/or *MAE*. In this way, Gandomi and Roke (2013) have proposed a new criterion (ρ) in order to address the mentioned issues. The lower ρ indicates more precise model.

As an example, comparisons of predicted versus experimental Uniaxial Compressive Strength values using TGP, LGP, Multi Layer Perceptron (MLP), and Radial Basis Function (RBF) are illustrated in Fig. 16.6 (Gandomi and Alavi 2011). MLP and RBF are categorized as the variants of ANN technique (Gandomi and Alavi 2011). It is obvious that the proposed models could greatly encompass the influencing variables which are not yet considered by existing models (Gandomi and Alavi 2011). The results of Table 16.2 indicate that the GP based models are able to predict the Uniaxial Compressive Strength of concrete with high degrees of accuracy and reliability. Comparing the performance of the GP based methods, it can be observed from Fig. 16.6 and Table 16.2 that LGP has produced better outcomes than TGP. The ANN-based models (i.e. MLP and RBF) have shown better results while there are lacks in preparing any applicable input–output formula.

In the case of Triaxial strength of concrete, the accuracy levels of LGP (Babanajad et al. 2013) and GEP (Gandomi et al. 2012) highlight the performance of GP based models from the conventional ones. Among all the conventional proposed models, Xie et al. (1995), Attard and Setunge (1996), and Girgin et al. (2007) released good performance measurements, however, they are not as good as those of the GP models. Besides, the mechanical characteristics of concrete (e.g., cylindrical Uniaxial compressive strength) are required as the input of empirical/numerical models, therefore, at least one set of laboratory tests has to be performed. While the proposed GEP can be implemented only by using the mix design parameters and confining pressure and there is no need to conduct any experimental program to obtain Uniaxial compressive strength (f'_c) or Uniaxial tensile strength (f'_t). The

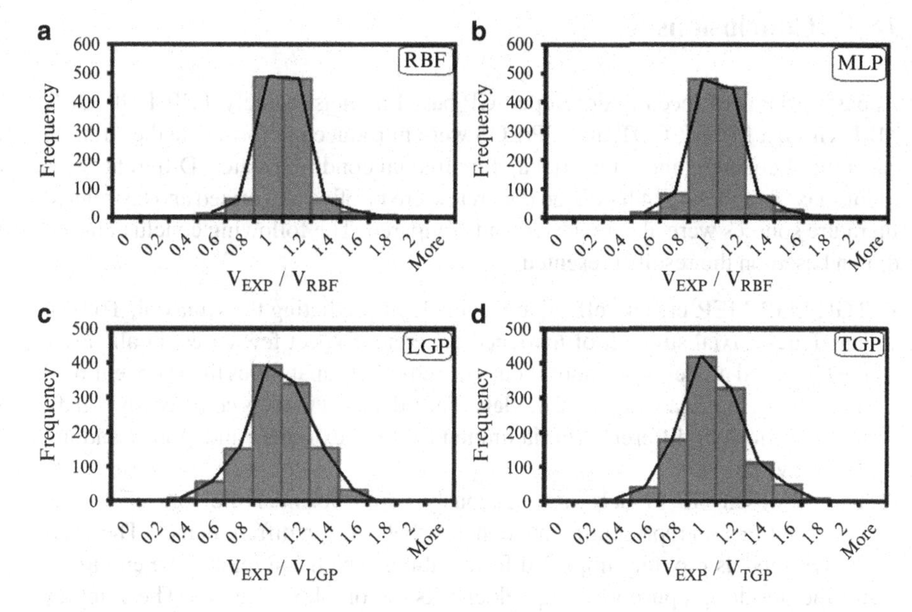

Fig. 16.6 Histograms of (**a**) RBF, (**b**) MLP, (**c**) LGP, and (**d**) TGP models (Gandomi and Alavi 2011)

model also can directly take into account the effect of specimen age (*A*) (Gandomi et al. 2012). Regarding the LGP model, it is concluded by Babanajad et al. (2013) that LGP based models require remarkably simple steps to come up with the solution. In addition to their reasonable accuracies they can also be implemented for design practice via hand calculations. The LGP based models are capable to derive explicit relationships for the strength of concrete without assuming prior forms of the existing relationships. This capacity highlights the superiority of LGP over the traditional analyses.

For True-triaxial modeling of concrete under different levels of confining pressures, the GP based models reported in Table 16.2 (GEP_I, GEP_II, GEP_III; proposed by Babanajad et al. submitted) outperform the rest of conventional models. Among all models (except GP based models) for True-triaxial modeling of concrete, the models proposed by Ottosen (1977), Hampel et al. (2009), He and Song (2010), and Mills and Zimmerman (1970) provide good *RMSE*, *R*, *MAE*, and ρ values. However, the performance measurements for these models are not as good as those of the GEP models. Furthermore, since the conventional empirical/analytical models were established based on the complex Elasticity and Plasticity concepts of materials behaviors, these models considered more input parameters in their formulations. This made the proposed formulas to have explicit relationships among the input parameters. For instance, using Hsieh et al. (1982) to calculate σ_1, the $f(\sigma_1, \sigma_2, \sigma_3, f'_c)$ has to be solved explicitly. While the GEP models release the formula in an implicit way, in which σ_1 is ordered separately from known values of σ_2, σ_3, f'_c.

16.4 Conclusions

Robust variants of recently developed GP based models, namely TGP, LGP, GEP, MEP, GEGA, MAGP, GOT, and GP/LOS were introduced to formulate the strength capacity of concrete under uniaxial and multiaxial condition states. Different types of multiaxial modeling of hardened concrete were briefly categorized and the related literature sources were also reviewed and compared. The following conclusions are drawn based on the results presented:

- TGP, LGP, MEP, etc. are effectively capable of predicting the Uniaxial, Triaxial and True-triaxial strength of hardened concrete. Expect few cases, in all cases, the GP based models give more accurate prediction capabilities than conventional numerical/empirical/analytical models. The GP based models could easily satisfy the conditions of different criteria limited for their external validation (Gandomi and Alavi 2011).
- The proposed GP models simultaneously could consider the role of several important factors representing the concrete behavior in different ages. The studied GP models provide simplified formulations which can reliably be employed for the pre-design purposes or quick checks of complex solutions. The complex solutions could be developed by more time consuming and in-depth deterministic analyses or national building codes.
- GP models provide the concrete compressive strength formula based on the mix design basic properties. In this manner, it is not necessary to conduct sophisticated and time consuming laboratory tests. Also, some of the reviewed GP models could consider the effect of specimen age and can be used for predicting of the uniaxial and multixial strength of concrete in different time intervals.
- The powerful abilities of GP models in tracking the complex behavior of the concrete without any need to predefined equations or simplifications makes it to be a substantial distinction model to the conventional empirical/statistical/numerical/analytical techniques (Gandomi and Alavi 2011).
- Based on the first principles (e.g., Elasticity, Plasticity, or Finite Element theories) have used in deriving the conventional constitutive models, the GP based constitutive models are basically formed based on different concepts. Artificial Intelligence and Machine Learning methods (e.g. GP, ANN, etc.) rely on the experimental data rather than on the assumptions which are made in developing the conventional models (Alavi and Gandomi 2011). This feature makes these techniques to be distinctive from the conventional ones. Consequently, these material models will be more accurate and reliable by re-training as more data becomes available.

Acknowledgment The author would like to appreciate Dr. Amir H. Gandomi, The University of Akron, for his great assistance throughout the preparation of the current book chapter.

Notation

f'_c = Compressive Strength
f'_t = Tensile Strength
$\sigma_1, \sigma_2, \sigma_3$ = Compressive Strength

References

Aho A, Sethi R, Ullman JD (1986) Compilers: Principles, Techniques and Tools. Reading (MA): Addison Wesley.

Alavi AH, Gandomi AH (2011) A robust data mining approach for formulation of geotechnical engineering systems. Eng Comput 28(3): 242–274

Alavi AH, Gandomi AH, Heshmati AAR (2010b) Discussion on soft computing approach for real-time estimation of missing wave heights. Ocean Eng 37(13):1239–1240

Alavi AH, Gandomi AH, Mollahasani A, Bolouri J (2013) Linear and Tree-Based Genetic Programming for Solving Geotechnical Engineering Problems. In: Yang XS et al. (eds) Metaheuristics in Water Resources, Geotechnical and Transportation Engineering, Elsevier, Chapter 12, p 289–310

Alavi AH, Gandomi AH, Mollahasani A, Heshmati AAR, Rashed A (2010a) Modeling of Maximum Dry Density and Optimum Moisture Content of Stabilized Soil Using Artificial Neural Networks. J Plant Nutr Soil Sc 173(3): 368–379

Ansari F, Li Q (1998) High-Strength Concrete Subjected to Triaxial Compression. ACI Mater J 95(6): 747–55

Arıoglu N, Girgin ZC, Arıoglu E (2006) Evaluation of Ratio between Splitting Tensile Strength and Compressive Strength for Concretes up to 120 MPa and its Application in Strength Criterion. ACI Mater J 103(1):18–24

Attard MM, Setunge S (1996) Stress–strain Relationship of Confined and Unconfined Concrete. ACI Mater J 93(5): 1–11

Avram C, Facadaru RE, Filimon I, Mîrşu O, Tertea I (1981) Concrete strength and strains. Elsevier Scientific, Amsterdam, The Netherland:156–178

Babanajad SK, Gandomim AH, Mohammadzadeh DS, Alavi AH (submitted) Comprehensive Strength Modeling of Concrete under True-Triaxial Stress States.

Babanajad SK, Farnam Y, Shekarchi M (2012) Failure Criteria and Triaxial Behaviour of HPFRC Containing High Reactivity Metakaolin and Silica Fume. Cnstr Bld Mater 29: 215–229

Babanajad SK, Gandomi AH, Mohammadzadeh DS, Alavi AH (2013) Numerical Modeling of Concrete Strength under Multiaxial Confinement Pressures Using LGP. Autom Cnstr 36: 136–144

Balmer GG (1949) Shearing Strength of Concrete under High Triaxial Stress—Computation of Mohr's Envelope as a Curve. Structural Research Laboratory, Report SP-23, U.S. Bureau of Reclamation, Denver, USA 1949

Banzhaf W, Nordin P, Keller R, Francone F (1998) Genetic programming - an introduction. In: on the automatic evolution of computer programs and its application, Heidelberg/San Francisco: dpunkt/Morgan Kaufmann 1998

Billings S, Korenberg M, Chen S (1988) Identification of nonlinear outputaffine systems using an orthogonal least-squares algorithm. Int J Syst Sci 19(8): 1559–1568

Bohwan O, Myung-Ho L, Sang-John P (2007) Experimental Study of 60 MPa Concrete Under Triaxial Stress. Structural Engineers World Congress, Bangalore, India 2007

Boukhatem B, Kenai S, Tagnit-Hamou A, Ghrici M (2011) Application of new information technology on concrete: an overview. J Civ Eng Manage 17 (2): 248–258

Brameier M, Banzhaf W (2001) A comparison of linear genetic programming and neural networks in medical data mining. IEEE Tran Evol Comput 5(1): 17–26

Brameier M, Banzhaf W (2007) Linear Genetic Programming. Springer Science + Business Media, NY, USA, 2007
Bresler B, Pister KS (1958) Strength of concrete under combined stresses. ACI J 55: 321–46
Candappa DC, Setunge S, Sanjayan JG (1999) Stress versus strain relationship of high strength concrete under high lateral confinement. Cem Concr Res 29: 1977–82
Candappa DC, Sanjayan JG, Setunge S (2001) Complete Triaxial Stress Strain Curves of High-Strength Concrete. J Mater Civ Eng 13(3): 209–15
Chen L (2003) A study of applying macroevolutionary genetic programming to concrete strength estimation. J Comput Civ Eng 17(4): 290–294
Chen S, Billings S, Luo W (1989) Orthogonal least squares methods and their application to non-linear system identification. Int J Ctrl 50(5): 1873–1896
Chen L, Wang TS (2010) Modeling Strength of High-Performance Concrete Using an Improved Grammatical Evolution Combined with Macrogenetic Algorithm. J Comput Civ Eng 24(3): 281–8
Cheng MY, Chou JS, Roy AFV, Wu YW (2012) High performance concrete compressive strength prediction using time-weighted evolutionary fuzzy support vector machines inference model. Autom Cnstr 28: 106–115
Cheng MY, Firdausi PM, Prayogo D (2014) High performance concrete compressive strength prediction Genetic Weighted Pyramid Operation Tree (GWPOT). Eng Appl Art Intel 29: 104–113
Chern JC, Yang HJ, Chen HW (1992) Behavior of Steel Fiber Reinforced Concrete in Multiaxial Loading. ACI Mater J 89(1): 32–40
Chinn J, Zimmerman RM (1965) Behavior of plain concrete under various high triaxial compression loading conditions. Technical Rep. No. WL TR 64–163, University of Colorado, Denver
Chou JS, Chiu CK, Farfoura M; Al-Taharwa I (2011) Optimizing the Prediction Accuracy of Concrete Compressive Strength Based on a Comparison of Data-Mining Techniques. J Comput Civ Eng 25(3); 242–253
Chou JS, Tsai CF (2012) Concrete compressive strength analysis using a combined classification and regression technique. Autom Cnstr 24: 52–60
Cordon WA, Gillespie HA (1963) Variables in concrete aggregates and Portland cement paste which influence the strength of concrete. ACI Struct J 60(8): 1029–1050
Desai CS, Somasundaram S, Frantziskonis G (1986) A hierarchical approach for constitutive modelling of geologic materials. Int J Numer Analyt Meth Geomech 10(3): 225–257
Erdal HI, Karakurt O, Namli E (2013) High performance concrete compressive strength forecasting using ensemble models based on discrete wavelet transform. Eng Appl Art Intel 26: 1246–1254
Farnam Y, Moosavi M, Shekarchi M, Babanajad SK, Bagherzadeh A (2010) Behaviour of Slurry In filtrated Fibre Concrete (SIFCON) under triaxial compression. Cem Concr Res 40(11): 1571–1581
Fazel Zarandi MH, Turksen IB, Sobhani J, Ramezanianpour AA (2008) Fuzzy polynomial neural networks for approximation of the compressive strength of concrete. Appl Soft Comput 8: 488–498
Ferreira C (2001) Gene expression programming: a new adaptive algorithm for solving problems. Cmplx Sys 13(2): 87–129
Ferreira C (2006) Gene Expression Programming: Mathematical Modeling by an Artificial Intelligence. Springer-Verlag Publication, 2nd Edition, Germany
Frank IE, Todeschini R (1994) The data analysis handbook. Elsevier, Amsterdam, the Netherland
Gandomi AH, Alavi AH (2013) Hybridizing Genetic Programming with Orthogonal Least Squares for Modeling of Soil Liquefaction. In: Computational Collective Intelligence and Hybrid Systems Concepts and Applications. IGI Global Publishing, in press
Gandomi AH, Alavi AH (2011) Applications of Computational Intelligence in Behavior Simulation of Concrete Materials. In: Yang XS, Koziel S (eds) Computational Optimization & Applications, Springer-Verlag Berlin Heidelberg, SCI 359, p 221–243
Gandomi AH, Alavi AH, Arjmandi P, Aghaeifar A, Seyednoor M (2010c) Genetic Programming and Orthogonal Least Squares: A Hybrid Approach to Modeling the Compressive Strength of

CFRP-Confined Concrete Cylinders. J Mech Mater Struct 5(5), Mathematical Sciences, UC Berkeley: 735–753
Gandomi AH, Alavi AH, Kazemi S, Gandomi M (2014a) Formulation of shear strength of slender RC beams using gene expression programming, part I:Without shear reinforcement. Automa Cnstr 42:112–121
Gandomi AH, Alavi AH, Mirzahosseini MR, Nejad FM (2011c) Nonlinear genetic-based models for prediction of flow number of asphalt mixtures. J Mater Civil Eng 23(3): 248–263
Gandomi AH, Alavi AH, Sahab MG (2010a) New formulation for compressive strength of CFRP confined concrete cylinders using linear genetic programming. Mater Struct 43(7): 963–983
Gandomi AH, Alavi AH, Sahab MG, Arjmandi P (2010b) Formulation of Elastic Modulus of Concrete Using Linear Genetic Programming. J Mech Sci Tech 24(6): 1273–8
Gandomi AH, Alavi AH, Ting TO, Yang XS (2013c) Intelligent Modeling and Prediction of Elastic Modulus of Concrete Strength via Gene Expression Programming. Advances in Swarm Intelligence, Lecture Notes in Computer Science 7928: 564–571
Gandomi AH, Alavi AH, Yun GJ (2011b) Nonlinear Modeling of Shear Strength of SFRC Beams Using Linear Genetic Programming. Struct Eng Mech, Techno Press 38(1): 1–25
Gandomi AH, Babanajad SK, Alavi AH, Farnam Y (2012) A Novel Approach to Strength Modeling of Concrete under Triaxial Compression. J Mater Civ Eng 24(9): 1132–43
Gandomi AH, Mohammadzadeh SD, Pérez-Ordóñez JL, Alavi AH (2014b) Linear genetic programming for shear strength prediction of reinforced concrete beams without stirrups. Appl Soft Comput 19: 112–120
Gandomi AH, Roke DA (2013) Intelligent formulation of structural engineering systems. Seventh M.I.T. Conference on Computational Fluid and Solid Mechanics — Focus: Multiphysics & Multiscale, Massachusetts Institute of Technology, Cambridge, MA
Gandomi AH, Roke DA, Sett K (2013b) Genetic programming for moment capacity modeling of ferrocement members. Eng Struct 57: 169–176
Gandomi AH, Tabatabaei SM, Moradian MH, Radfar A, Alavi AH (2011a) A new prediction model for the load capacity of castellated steel beams. J Cnstr Steel Res 67(7): 1096–1105
Gandomi AH, Yun GJ, Alavi AH (2013a) An evolutionary approach for modeling of shear strength of RC deep beams. Mater Struct 46: 2109–2119
Girgin ZC, Anoglu N, Anoglu E (2007) Evaluation of Strength Criteria for Very-High-Strength Concretes under Triaxial Compression. ACI Mater J 104(3): 278–84
Golbraikh A, Tropsha A (2002) Beware of q2. J Molecular Graphics M 20: 269–276
Gupta R, Kewalramani MA, Goel A (2006) Prediction of concrete strength using neural-expert system. J Mater Civil Eng 18(3): 462–466
Guven A, Aytek A (2009) New approach for stage–discharge relationship: Gene-expression programming. J Hydrol Eng 14(8): 812–820
Hampel T, Speck K, Scheerer S, Ritter R, Curbach M (2009) High-Performance concrete under biaxial and triaxial loads. J Eng Mech 135(11): 1274–1280
He Z, Song Y (2010) Triaxial strength and failure criterion of plain high-strength and high-performance concrete before and after high temperatures. Cem Concr Res 40: 171–178
Heshmati AAR, Salehzade H, Alavi AH, Gandomi AH, Mohammad Abadi M (2010) A Multi Expression Programming Application to High Performance Concrete. World Appl Sci J 11(11): 1458–66
Hinchberger SD (2009) Simple Single-Surface Failure Criterion for Concrete. J Eng Mech 135(7): 729–32
Hoek E, Brown ET (1980) Underground excavations in rock. Institution of Mining and Metallurgy Press, London, 1980
Hsieh SS, Ting EC, Chen WF (1982) Plasticity-fracture model for concrete. Int J Solids Struct 18(3): 577–93
Ilc A, Turk G, Kavčič F, Trtnik G (2009) New numerical procedure for the prediction of temperature development in early age concrete structures. Autom Cnstr 18(6): 849–855
Imran I, Pantazopoulou SJ (1996) Experimental Study of Plain Concrete under Triaxial Stress. ACI Mater J 93(6): 589–601

Javadi AA, Rezania M (2009) Applications of artificial intelligence and data mining techniques in soil modeling. Geomech Eng 1(1): 53–74

Johnston IW (1985) Strength of Intact Geomechanical Materials. J Geotech Eng 111(6): 730–748

Kewalramani MA, Gupta R (2006) Concrete compressive strength prediction using ultrasonic pulse velocity through artificial neural networks. Autom Cnstr 15(3): 374–379

Khan IM (2012) Predicting properties of high performance concrete containing composite cementitious materials using artificial neural networks. Autom Cnstr 22: 516–524

Koza JR (1992) Genetic programming: On the programming of computers by means of natural selection. Cambridge (MA), MIT Press.

Kraslawski A, Pedrycz W, Nyström L (1999) Fuzzy Neural Network as Instance Generator for Case-Based Reasoning System An Example of Selection of Heat Exchange Equipment in Mixing. Neural Comput Appl 8: 106–113

Kuo YL, Jaksa MB, Lyamin AV, Kaggwa WS (2009) ANN-based model for predicting the bearing capacity of strip footing on multi-layered cohesive soil. Comput Geotech 36: 503–516

Lahlou K, Aitcin PC, Chaallal O (1992) Behavior of high strength concrete under confined stresses. Cem Concr Compos 14: 185–193

Lan S, Guo Z (1997) Experimental investigation of multiaxial compressive strength of concrete under different stress paths. ACI Mater J 94(5): 427–433

Légeron F, Paultre P (2003) Uniaxial Confinement for Normal- and High-Strength Concrete Columns. J Struct Eng 129(2): 241–252

Li Q, Ansari F (1999) Mechanics of Damage and Constitutive Relationships for High-Strength Concrete in Triaxial Compression. J Eng Mech 125(1): 1–10

Li Q, Ansari F (2000) High-Strength Concrete in Triaxial Compression by Different Sizes of Specimens. ACI Mater J 97(6): 684–9

Liu HY, Song YP (2010) Experimental study of lightweight aggregate concrete under multiaxial stresses. J Zhejiang Uni-Science A (Applied Physics & Engineering) 11(8): 545–554

Lu X, Hsu CTT (2006) Behavior of high strength concrete with and without steel fiber reinforcement in triaxial compression. Cem Concr Res 36: 1679–85

Lu X, Hsu CTT (2007) Stress–strain relations of high strength concrete under triaxial compression. J Mater Civ Eng 19(3): 261–268

Madár J, Abonyi J, Szeifert F (2005) Genetic Programming for the Identification of Nonlinear Input–output Models. Ind Eng Chem Res 44(9): 3178–3186

Marin J, Sole RV (1999) Macroevolutionary algorithms A new optimization method on fitness landscapes. IEEE Trans Evol Comput, Piscataway, NJ, 3(4), pp 272–285

Martinez S, Nilson AH, Slate FO (1984) Spirally Reinforced High-Strength Concrete Columns. ACI J 81(5): 431–442

Mei H, Kiousis PD, Ehsani MR, Saadatmanesh H (2001) Confinement effects on high strength concrete. ACI Struct J 98(4): 548–553

Milani G, Benasciutti D (2010) Homogenized limit analysis of masonry structures with random input properties Polynomial response surface approximation and Monte Carlo simulations. Struct Eng Mech 34(4): 417–445

Mills LL, Zimmerman RM (1970) Compressive strength of plain concrete under multiaxial loading conditions. ACI J 67(10):802–807

Mohammadi Bayazidi A, Wang GG, Bolandi H, Alavi AH, Gandomi AH (2014) Multigene Genetic Programming for Estimation of Elastic Modulus of Concrete. Mathematical Problems in Engineering. Hindawi Publishing Corporation, NY, USA

Mousavi SM, Aminian P, Gandomi AH, Alavi AH, Bolandi H (2012) A new predictive model for compressive strength of HPC using gene expression programming. Adv Eng SW 45: 105–114

Mousavi SM, Gandomi AH, Alavi AH, Vesalimahmood M (2010) Modeling of compressive strength of HPC mixes using a combined algorithm of genetic programming and orthogonal least squares. Struc Eng Mech 36(2): 225–241

Mullar KF (1973) Dissertation Technische Universitat Munchen Germany

Nielsen CV (1998) Triaxial behavior of high-strength concrete and mortar. ACI Mater J 95(2): 144–51

Oltean M, Dumitrescu D (2002) Multi expression programming. Technical report UBB-01-2002, Babes-Bolyai University
Oltean M, Groşan CA (2003a) Comparison of several linear genetic programming techniques. Adv Cmplx Sys 14(4): 1–29
Oltean M, Grossan CA (2003b) Evolving evolutionary algorithms using multi expression programming. In: Banzhaf W et al. (eds). 7th European conference on artificial life. Dortmund, LNAI, pp 651–658
Ottosen NS (1977) A failure criterion for concrete. J Eng Mech Div 103(4): 527–535
Pearson (2003) Selecting nonlinear model structures for computer control. J Process Contr 13(1): 1–26
Peng CH, Yeh IC, Lien LC (2010) Building strength models for high-performance concrete at different ages using genetic operation trees, nonlinear regression, and neural networks. Eng Comput 26: 61–73
Rajasekaran S, Amalraj R (2002) Predictions of design parameters in civil engineering problems using SLNN with a single hidden RBF neuron. Comput Struct 80 (31): 2495–2505
Ramezanianpour AA, Sobhani M, Sobhani J (2004) Application of network based neuro-fuzzy system for prediction of the strength of high strength concrete. AKU J Sci Technol 15(59-C): 78–93
Richart E, Brandtzaeg A, Brown RL (1929) Failure of Plain and Spirally Reinforced Concrete in Compression. Bulletin 190, University of Illinois Engineering Experimental Station, Champaign, Illinois
Roy PP, Roy K (2008) On Some Aspects of Variable Selection for Partial Least Squares Regression Models. QSAR Comb Sci 27(3): 302–313
Ryan TP (1997) Modern regression methods. Wiley, New York
Ryan C, Collins JJ, O'Neill M (1998) Grammatical Evolution: Evolving Programs for an Arbitrary Language. In: Banzhaf W, Poli R., Schoenauer M, Fogarty TC (eds), First European Workshop on Genetic Programming, Springer-Verlag, Berlin
Ryan C, O'Neill M (1998) Grammatical Evolution: A Steady State Approach. In: Koza JR (ed) Late Breaking Papers Genetic Programming, University of Wisconsin, Madison, Wisconsin
Saatcioglu M, Razvi SR (1992) Strength and Ductility of Confined Concrete. J Struct Eng 118(6): 1590–1607
Samaan M, Mirmiran A, Shahawy M (1998) Model of Concrete Confined by Fiber Composites. J Struct Eng 124(9): 1025–1031
Seow PEC, Swaddiwudhipong S (2005) Failure Surface for Concrete under Multiaxial Load—a Unified Approach. J Mater Civ Eng 17(2): 219–228
Setunge S, Attard MM, Darvall PL (1993) Ultimate Strength of Confined Very High-Strength Concretes. ACI Struct J 90(6): 632–41
Sfer D, Carol I, Gettu R, Etse G (2002) Study of the Behavior of Concrete under Triaxial Compression. J Eng Mech 128(2): 156–63
Shahin MA, Jaksa MB, Maier HR (2009) Recent advances and future challenges for artificial neural systems in geotechnical engineering applications. Adv Artif Neur Syst, Hindawi, Article ID 308239
Smith GN (1986) Probability and statistics in civil engineering. Collins, London
Sobhani J, Najimi M, Pourkhorshidi AR, Parhizkar T (2010) Prediction of the compressive strength of no-slump concrete a comparative study of regression, neural network and ANFIS models. Cnstr Bld Mater 24 (5): 709–718
Tang CW (2010) Radial Basis Function Neural Network Models for Peak Stress and Strain in Plain Concrete under Triaxial Stress Technical Notes. J Mater Civ Eng 22(9): 923–934
Topcu IB, Sarıdemir M (2008) Prediction of compressive strength of concrete containing fly ash using artificial neural networks and fuzzy logic. Comput Mater Sci 41: 305–311
Wang CZ, Guo ZH, Zhang XW (1987) Experimental Investigation of Biaxial and Triaxial Compressive Concrete Strength. ACI Mater J: 92–100

Willam K, Warnke E (1975) Constitutive model for triaxial behavior of concrete. In: Seminar on Concrete Structure Subjected to Triaxial Stresses, International Association for Bridge and Structural Engineering, Bergamo, Italy, 17–19 May 1974, pp 1–30

Xie J, Elwi AE, Mac Gregor JG (1995) Mechanical properties of three high-strength concretes containing silica fume. ACI Mater J 92(2): 1–11

Yeh IC (1998) Modeling concrete strength with augment-neuron networks. J Mater Civ Eng 10(4): 263–268

Yeh IC (2006a) Analysis of strength of concrete using design of experiments and neural networks. ASCE J Mater Civil Eng 18(4): 597–604

Yeh IC (2006b) Exploring concrete slump model using artificial neural networks. J Comput Civ Eng 20(3): 217–21

Yeh IC (2006c) Generalization of strength versus water-cementations ratio relationship to age. Cem Concr Res 36(10): 1865–1873

Yeh IC, Lien LC (2009) Knowledge discovery of concrete material using Genetic Operation Trees. Expert Sys Appl 36: 5807–5812

Yeh IC, Lien CH, Peng CH, Lien LC (2010) Modeling Concrete Strength Using Genetic Operation Tree. Proceedings of the Ninth International Conference on Machine Learning and Cybernetics, Qingdao, July 2010

Chapter 17
Genetic Programming for Mining Association Rules in Relational Database Environments

J.M. Luna, A. Cano, and S. Ventura

17.1 Introduction

Association rule mining (ARM) was first introduced by Agrawal and Srikant (1994). This data mining task seeks frequent and strong relations among items in a single relation—in database terminology, 'relation' refers to a table in the database. Let $\mathscr{I} = \{i_1, i_2, i_3, \ldots, i_n\}$ be the set of items or patterns, and let $\mathscr{D} = \{t_1, t_2, t_3, \ldots, t_n\}$ be the set of all transactions in the relation. An association rule is an implication of the form $X \Rightarrow Y$ where $X \subset \mathscr{I}$, $Y \subset \mathscr{I}$, and $X \cap Y = \emptyset$. An association rule means that if the antecedent X is satisfied, then it is highly probable that the consequent Y will be satisfied.

Most existing ARM proposals discover association rules in a single relation by using methodologies based on support–confidence frameworks (Agrawal et al. 1996; Han et al. 2004). The support of an item-set X is defined as the probability of all the items in X appearing in $\mathscr{D}$. Similarly, the support of an association rule $X \Rightarrow Y$ is the proportion of transactions in $\mathscr{D}$ satisfying all the items in both X and Y. On the other hand, the confidence of $X \Rightarrow Y$ is defined as the support of this rule divided by the support of Y, measuring the reliability of the rule.

With the growing interest in the storage of information, databases have become essential (Konan et al. 2010). Whereas the extraction of association rules in a single relation is well-studied, only a few proposals have been made for mining

J.M. Luna • A. Cano • S. Ventura (✉)
Department of Computer Science and Numerical Analysis,
University of Crdoba, Rabanales Campus, 14071, Cordoba, Spain
e-mail: jmluna@uco.es; acano@uco.es; sventura@uco.es

A.H. Gandomi et al. (eds.), *Handbook of Genetic Programming Applications*,
DOI 10.1007/978-3-319-20883-1_17

id	name	city
1	Ben	Nottingham
2	Erwin	Leeds
3	Erwin	Manchester
4	Ben	Leeds

Customer relation

id	vegetable	fruit	household	basket price
1	beans	apple	broom	200
2	onion	pear	broom	125
2	carrot	apple	mop	212
2	beans	banana	dustpan	321
3	onion	orange	mop	251
4	garlic	banana	broom	132

Product relation

id	name	city	vegetable	fruit	household	basket price
1	Ben	Nottingham	beans	apple	broom	200
2	Erwin	Leeds	onion	pear	broom	125
2	Erwin	Leeds	carrot	apple	mop	212
2	Erwin	Leeds	beans	banana	dustpan	321
3	Erwin	Manchester	onion	orange	mop	251
4	Ben	Leeds	garlic	banana	broom	132

Customer ⋈ Product

Fig. 17.1 Sample market basket comprising two relations and the result of joining both relations. (**a**) Customer relation. (**b**) Product relation. (**c**) Customer ⋈ Product

association rules in relational databases due to its difficulty to be extracted by using exhaustive search models (Alashqur 2010; Goethals et al. 2010; Jiménez et al. 2011). Relational databases have a more complex structure and store more information than raw datasets. Therefore, existing ARM proposals for mining rules in single relations cannot directly be applied. Instead, data has to be transformed by joining all the relations into a single relation Ng et al. (2002) so existing algorithms can be applied to this relation. However, this transformation technique requires an in depth analysis: (1) a high computational time could be required, especially with the increment of the size of both $\mathscr{I}$ and $\mathscr{D}$, and (2) a detailed attention should be paid to avoid support deviation—each transaction in a relation could be duplicated as the result of a join, so the same item could be read and stored multiple times.

In existing relational approaches Crestana-Jensen and Soporkar (2000), Goethals et al. (2010), the support quality measure is defined as the number of transactions in the result of a join of the relations in the database. In this definition, it is crucial to clarify that the support of an item-set strongly depends on how well its items are connected. For instance, having the relations *Customer* and *Product* in a market basket relational database (see Fig. 17.1a and b), the city *Leeds* appears in the relation *Customer* with a probability of 0.50, i.e. satisfies 50 % of the transactions. On the other hand, the same city appears in 66 % of the transactions in the result of a join (see Fig. 17.1c). Therefore, joining different relations into a single relation could introduce distortions in their support values so counting rows to calculate the support measure is not correct. One row is not identified with one customer.

The aim of this paper is to propose an approach that maintains the database structure—not requiring a joining of the relations—and solve the problem of preserving the support deviation in the discovered associations. To this end, we introduce an interesting grammar-guided genetic programming (G3P)

(Espejo et al. 2010; McKay et al. 2010; Ratle and Sebag 2000) algorithm that represents the solutions in a tree-shape conformant to a context free grammar (CFG). This CFG enables syntax constraints to be considered (Gruau 1996). Additionally, solutions could be defined on any domain, so it is of great interest to apply G3P (Freitas 2002) to the field of relational databases in order to mine both quantitative and negative rules. It represents a great advantage with respect to existing algorithms in this field, which usually mine only positive and discrete items. G3P was previously applied to the ARM field (Luna et al. 2012), achieving promising results and reducing both the computational cost and the memory requirements. The use of an evolutionary methodology and searching for solutions by means of a good heuristic strategy (Mata et al. 2002; Papè et al. 2009) enables the computational and memory requirements to be solved. All these benefits are especially important in ARM over relational databases.

To study the effectiveness of our proposal, we exemplify the utility of the proposed approach by illustrating how it enables discovery of interesting relations and how it performs with an artificial generated database. Additionally, we apply the approach to a real case study, discovering interesting students' behaviors from a moodle database. Experiments show the usefulness and efficiency of the proposed algorithm, which discovers associations that comprise any kind of items, i.e. positive and negative, and also discrete and quantitative. This type of items could be hardly discovered by exhaustive search algorithms, which can not deal with quantitative and negative items due to the search space size. Finally, the computational complexity is analysed, stating that it is linear with regard to the instances and number of attributes.

This paper is structured as follows: Sect. 17.2 presents some related work; Sect. 17.3 describes the model and its main characteristics; Sect. 17.4 describes the experimental analysis and the relational databases used to this end; finally, some concluding remarks are outlined in Sect. 17.5.

17.2 Related Work

Association rules and measures for evaluating them were first explored by means of logical and statistical foundations provided by the GUHA method (Hájek et al. 1966). However, most existing studies in ARM were based on the *Apriori* algorithm (Agrawal and Srikant 1994), which is based on an exhaustive search methodology and only mines discrete items (also known as patterns in the ARM field). The main motivation of the *Apriori*-like algorithms is the reduction of the computational time and the memory requirements. Han et al. (2004) proposed the FP-Growth algorithm, which stores the frequent patterns mined into a frequent pattern tree (FP-tree) structure. Then, the algorithm works on the FP-tree instead of the whole dataset, employing a divide-and-conquer method to reduce the computational time. However, all these algorithms follow an exhaustive search methodology, being hard to maintain with an increase in the number of attributes and transactions.

The current increasing need for the storage of information has prompted the use of different ways of storing this information, relational databases being one of the most widely used. More and more organizations are using this type of storage to save their information. Under these circumstances, the mining of association rules using exhaustive search algorithms becomes a hard task. Despite the fact that the mining of patterns and association rules in relational databases has not been explored in depth yet, a few approaches already exist for this task (Goethals and Van den Bussche 2002; Goethals et al. 2012; Spyropoulou and De Bie 2011). For instance, Wang et al. (Ng et al. 2002) examined the problem of mining association rules from a set of relational tables. In particular, they focused on the case where the tables form a star structure. This structure consists of a fact table in the centre and multiple dimension tables. The mining over the joined result from a star structure is not an easy task, and some problems could appear. Using large dimensional tables, the join of all related tables into a single table would be hard to compute. Even if the join could be computed, the resulting relation extremely increases in the number of both attributes and transactions. As studied in depth by many other researchers (Han et al. 2004; Luna et al. 2012), the mining of association rules is very sensitive to the number of attributes and transactions.

In a different way, there are proposals that mine relational patterns directly from relational databases. Goethals et al. (2010) proposed the SMuRFIG (Simple Multi-Relational Frequent Item-set Generator) algorithm, an efficient algorithm in both time and memory requirements, since it does not work over a single join table. They determined that the absolute support of a relational item-set in a relation R is the number of distinct values of the key K, so the relative support of a relational item-set is the number of values in the KeyID list divided by the number of element in R. This approach was conceived for mining a set of frequent patterns in relational databases, so a subsequent step is required for mining association rules from these patterns. Hence, this second step hampers the association rule mining process.

RDB-MINER (Alashqur 2010) is another algorithm for mining frequent item-sets directly from relational databases. The main feature of this algorithm is the use of SQL statements to compute the support count of the item-sets discovered. Therefore it uses the GROUP BY clause along with the COUNT aggregate function. In order to discover item-sets whose support is greater than or equal to a minimum value, the algorithm uses the HAVING clause. The main advantage of this algorithm is its portability, i.e., it can be applied to any relational database, because of the specific use of SQL. However, similarly to *Apriori*, this algorithm mines only categorical attributes and works in two steps, discovering frequent patterns first and then mining association rules.

Actually, the proposals focused on the discovery of association rules from the original database have been not enough yet. An important research in this field was described in Jiménez et al. (2011), who based their approach on the representation of relational databases as sets of trees, for which they proposed two different representation schemes: key-based and object-based tree representations.

Each of these representations starts from a particular relation, namely the target relation or the target table, which is selected by the end user according to the user's specific goals. The key-based tree representation is based on the fact that each transaction is identified by its primary key. Therefore, a root node is represented by its primary key and the children of this primary key will be the remaining attribute values from the transaction in the relation. On the contrary, in the object-based tree, the root represents a transaction, and the children represent all the attribute values, including the primary key.

Based on the previous idea of representing relational databases as sets of trees, and bearing in mind the problems of using exhaustive search methodologies in the ARM field, the mining of associations in relational environments could be considered from the genetic programming (GP) (Koza 1992, 2008) point of view. In this evolutionary technique, solutions are represented in a tree-shape structure, where the shape, size and structural complexity are not constrained a priori. A major feature of GP, especially alluring in the ARM field, is its ability to constrain the GP process by searching for solutions with different syntax forms. Methods to implement such restrictions include using some form of constrained syntax (Freitas 2002) or using a grammar (McKay et al. 2010) to enforce syntactic and semantic constraints on the GP trees. This is known as grammar-guided genetic programming (G3P) (Ratle and Sebag 2000), an extension of GP where each individual is a derivation tree that generates and represents a solution using the language defined by the grammar. The tree is built up by applying productions to leaf nodes and a maximum tree depth is usually specified to avoid bloat, a well-known problem in GP. G3P has been successfully used in educational data mining for providing feedback to instructors (Romero et al. 2004) by mining association rules in a single relation (Luna et al. 2012).

17.3 Genetic Programming in Relational Databases

In short, this proposal represents association rules in the form of a tree through the use of G3P, not requiring a previous mining of frequent patterns or even any joining of the relations into a join table. Instead, the mining process is carried out on the original relational database regardless of its structure. The strength of representing individuals conforming to a CFG is twofold. Firstly, individuals are represented using a tree structure, having the goal of representing solutions on the original relational database, and preserving the original support value (Jiménez et al. 2011). Secondly, the CFG provides expressiveness, flexibility, and the ability to restrict the search space. Finally, the use of an evolutionary methodology provides a powerful ability to global search and optimization, performing well in terms of scalability, computational time, and memory requirements.

17.3.1 Encoding Criterion

The algorithm proposed in this paper represents the individuals by a genotype and a phenotype. The genotype is defined by using a tree structure, having different shape and size, whereas the phenotype represents the association rule previously defined by the genotype. The tree structure is obtained by the definition of a CFG (see Definition 1), which is depicted in Fig. 17.2.

Definition 1 (Context-Free Grammar). A context-free grammar (CFG) is defined as a four-tuple $(\Sigma_N, \Sigma_T, P, S)$. Here, Σ_N is the non-terminal symbol alphabet, Σ_T is the terminal symbol alphabet, P is the set of production rules, S is the start symbol, and Σ_N and Σ_T are disjoint sets (i.e., $\Sigma_N \cap \Sigma_T = \emptyset$). Any production rule follows the format $\alpha \to \beta$ where $\alpha \in \Sigma_N$, and $\beta \in \{\Sigma_T \cup \Sigma_N\}^*$.

To obtain new individuals, a sequence of steps is carried out by applying the set P of production rules and starting from the start symbol S. This process begins from the start symbol, which always has a child node representing the target relation and also the antecedent and consequent. Once the tree is defined through the application of production rules, the general structure of the individual is obtained, and the next step is to assign values to the leaf nodes. By using trees (Jiménez et al. 2011), each node represents a relation and the edges represent relationships between keys in two relations.

Once the tree structure is established according to the CFG, the next step is to assign values to the leaf nodes. These values depend on the specific relational database used. First, the target relation or target table is randomly selected from the set of tables in the relational database. Second, for each leaf node 'Condition' related to this target table, a condition of the form *attribute operator value* is assigned to this table. Finally, a table name is randomly assigned to the leaf node 'Table' selected from all tables related to this target relation. The process continues iteratively until

$G = (\Sigma_N, \Sigma_T, P, S)$ with:

S = AssociationRule
Σ_N = {AssociationRule, Antecedent, Consequent }
Σ_T = {'AND', 'Condition', 'TargetTable', 'Table' }
P = {
AssociationRule = 'TargetTable', Antecedent, Consequent ;
Antecedent = 'Condition' | 'AND', 'Condition', Antecedent |
'AND', 'Condition', 'Table', Antecedent ;
Consequent = 'Condition' | 'AND', 'Condition', Consequent |
'AND', 'Condition', 'Table', Consequent ;
}

Fig. 17.2 Context-free grammar expressed in extended BNF notation

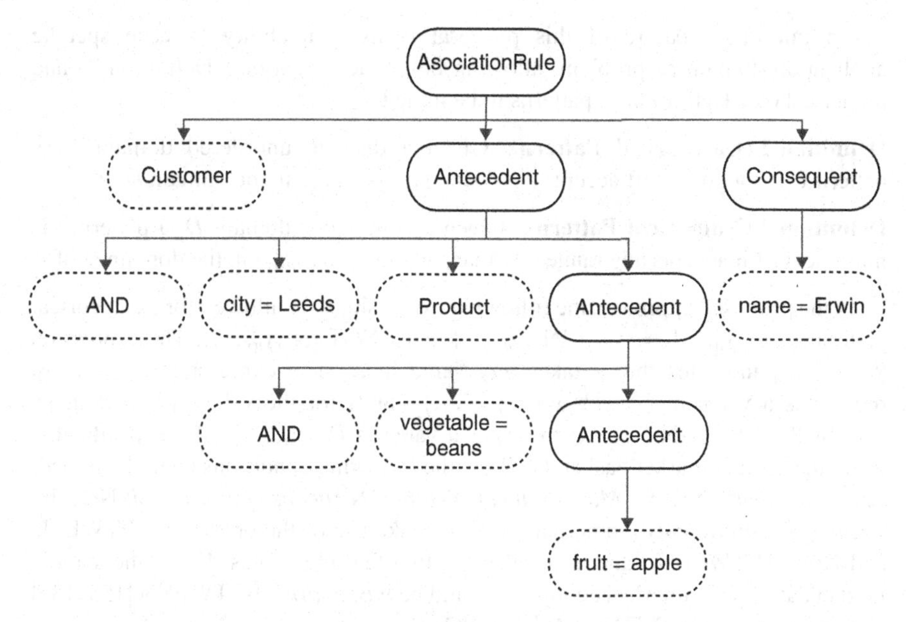

Fig. 17.3 Genotype of a sample individual obtained by using both the CFG defined in Fig. 17.2 and the relational database described in Fig. 17.1

a value is assigned to all leaf nodes. To understand the tree structure, a depth-first traversal of the tree is carried out. Thus, the 'Target Table' node is the root for any antecedent and consequent. At the same time, each 'Condition' belongs to the last 'Table' read.

For a better understanding, consider again the sample database depicted in Fig. 17.1, where there is a $1 : N$ relationship between tables *Customer* and *Product*, i.e. each customer could buy n products, and the individual depicted in Fig. 17.3. As mentioned above, the target table establishes the meaning of the association rule represented by the tree since this table is used as the fact table of the relations. Therefore, the meaning of the resulting association rule is *from all the customers in the database, if their city is Leeds and they have a market basket comprising beans and apples, then they should be named Erwin*, which is obtained from the following genotype:

$$\begin{aligned}
&\textbf{IF } \{city = Leeds\}_{customer} \\
&\text{AND } \{vegetable = beans\}_{customer.Product} \\
&\text{AND } \{fruit = apple\}_{customer.Product} \\
&\textbf{THEN } \{name = Erwin\}_{customer}
\end{aligned}$$

An important feature of this proposal is its adaptability to each specific application domain or problem, allowing both categorical (see Definition 2) and numerical (see Definition 3) patterns to be mined.

Definition 2 (Categorical Pattern). Given a discrete unordered domain D, a pattern x is categorical iff accepts any value $y_1, y_2, \ldots, y_n$ in the domain D of x.

Definition 3 (Numerical Pattern). Given a continuous domain D, a pattern x is numerical iff it accepts any range of values within (y_{min}, y_{max}) in the domain D of x.

In the proposed approach, the following expressions are allowed for a categorical pattern: $x = y_n$, $x\ != y_n$, x IN $\{y_n, \ldots\}$ and x NOT IN $\{y_n, \ldots\}$. The expression $x = y_n$ indicates that x takes any value in D. The expression $x\ != y_n$ represents a value in $D \backslash \{y_n\}$, x IN $\{y_n, \ldots\}$ stands for the values $\{y_n, \ldots\}$ in D, and finally x NOT IN $\{y_n, \ldots\}$ indicates a value in $D \backslash \{y_n, \ldots\}$. More specifically, using the sample market basket database, the following conditions could be mined: *name* = *Ben*, *city* ! = *Manchester*, *vegetable* IN {*beans*, *onion*}, *fruit* NOT IN {*banana*}. Additionally, numerical patterns make use of the operators BETWEEN and NOT BETWEEN, randomly selecting two feasible values. Using the sample market database, numerical conditions could be *basket price* BETWEEN [155, 135] or *basket price* NOT BETWEEN [210, 282].

17.3.2 Evaluation Process

Generally, the support of an association rule is defined as the proportion of the number of transactions including both the antecedent and consequent in a dataset. When dealing with relational databases, this definition requires a redefinition to avoid the support deviation. Therefore, the number of transactions is calculated based on the distinct keys instead of on the complete set of transactions. In relational databases, the support of an association rule is formally described in Eq. (17.1) as the proportion of the number of distinct keys K including the antecedent X and consequent Y in a relational database *RDB*. On the other hand, the confidence is formally defined in Eq. (17.2) as the proportion of the distinct keys which include X and Y among all the different keys that include X. Notice that the interest of an association rule can not be measured by a support–confidence framework, as described by Berzal et al. (2002). Even when the support and confidence thresholds are satisfied by a rule, this rule could be misleading if it acquires a confidence value less than its consequent support. Therefore, the lift measure is receiving increasing interest since it measures how many times more often the antecedent and consequent are related in a dataset than expected if they were statistically independent. Lift is formally defined in Eq. (17.3) as the difference between the support of the rule and the expected support under independence. In this regard, it is calculated by using both sides of the rule. Greater lift values ($\gg 1$) indicate stronger relations.

$$support(X \rightarrow Y) = \frac{|\{X \cup Y \subseteq K, K \in RDB\}|}{|K|} \quad (17.1)$$

$$confidence(X \rightarrow Y) = \frac{|\{X \cup Y \subseteq K, K \in RDB\}|}{|\{X \subseteq K, K \in RDB\}|} \quad (17.2)$$

$$lift(X \rightarrow Y) = \frac{confidence(X \rightarrow Y)}{support(Y)} \quad (17.3)$$

The proposed GP approach uses support, confidence and lift to evaluate each new individual obtained. In evolutionary algorithms each individual is evaluated by means of a fitness function, which determines the individual's ability to solve the computational problem. In the algorithm here presented, the evaluation process comprises a series of different steps. The first one verifies that the individual—given by the association rule $X \Rightarrow Y$—does not have the same pattern in the antecedent and consequent, i.e. $X \cap Y = \emptyset$. If the antecedent and consequent of an individual are not disjoint item-sets, then a 0 fitness function value is assigned. The second step calculates the lift value to determine the individual's interest value. In such a way, lift values equal to or less than unity imply a 0 fitness function value. Finally, the support value is calculated for the rule defined by the individual, and the final fitness function F is defined by:

$$F(rule) = \begin{cases} 0 & \text{if } X \cap Y \neq \emptyset \\ 0 & \text{if } lift(rule) \leq 1 \\ support(rule) & \text{otherwise} \end{cases} \quad (17.4)$$

Focusing on the support and confidence measures, notice that the confidence of a rule is always greater than or equal to its support, and its value is maximal if the transactions covered by the antecedent of an association rule are a subset of the transactions covered by the consequent of the rule. In addition, the confidence measure is symmetric iff $sup(A) = sup(C)$. Therefore, using the fitness function described above, the higher the support value, the higher the confidence value, so maximizing the fitness function value it is possible to increase both support and confidence values.

For the sake of calculating these three quality measures, the relational database is manipulated by running specific SQL expressions—automatically generated by the proposed algorithm—based on the relations of the database and the individual representation. For instance, using the individual depicted in Fig. 17.3, its support value is automatically calculated by the following two SQL statements:

```
SELECT COUNT(DISTINCT customer.id)
FROM customer, product
WHERE customer.city = 'Leeds' AND
   customer.name = 'Erwin' AND
   product.vegetable = 'beans' AND
   product.fruit = 'apple' AND
   customer.id = product.customer;
```

```
SELECT COUNT(*) FROM customer;
```

The first SQL statement calculates $|\{X \cup Y \subseteq K, K \in RDB\}|$, and the second one obtains $|K|$, see Eq. (17.1).

Using the same individual, the confidence value is calculated by using both $|\{X \cup Y \subseteq K, K \in RDB\}|$ and $|\{X \subseteq K, K \in RDB\}|$. The first clause was already obtained when calculating the support measure. The second one is calculated by the following SQL statement. Notice that this statement could be computed with the first SQL statement, so we have considered the optimization as a major issue in our proposal.

```
SELECT COUNT(DISTINCT customer.id)
FROM customer, product
WHERE customer.city = 'Leeds' AND
  product.vegetable = 'beans' AND
  product.fruit = 'apple' AND
  customer.id = product.customer;
```

Finally, the lift measure requires calculating both the confidence value and the consequent support value. Whereas the confidence was previously obtained, the consequent support value needs to be calculated by means of $|\{Y \subseteq K, K \in RDB\}|$ and $|K|$. $|K|$ was already calculated, while the former clause should be still obtained from the following SQL statement:

```
SELECT COUNT(DISTINCT customer.id)
FROM customer
WHERE customer.name = 'Erwin';
```

17.3.3 Genetic Operator

As any evolutionary algorithm, the proposal here presented has the aim of solving the computational problem by improving the fitness function values along the evolutionary process. Therefore, in each generation of the evolutionary process, new individuals having better fitness values are desired. For this reason, a genetic operator is presented (see the pseudo-code in Listing 1), with the aim of maintaining genetic diversity. This operator has the goal of obtaining conditions with higher support values than the originals. The support value of an association rule depends on the frequency of its attributes, so the greater their frequency of occurrence, the greater the probability of increasing the support value of the entire rule.

In this process, a substitution of the lowest support condition within every individual from the set of parents *parents* is carried out. For each of these individuals, a random value from 0 to 1 is obtained, carrying out the genetic operator if this value is lower than a predefined value $probability - threshold$. The operator provides two possibilities of changing the condition selected: (*a*) replacing the whole condition with a new condition, *newCondition*, randomly obtaining a new attribute, a new operator, and a new value; (*b*) allowing of replacing the logic operator used with its opposite operator: = and ! =; IN and NOT IN; BETWEEN and NOT BETWEEN. The use of opposite operators allows of obtaining a higher

Listing 1 Genetic operator

```
Require: parents
Ensure: newIndividualSet
 1: newIndividualSet ← Ø
 2: for all individual ∈ parents do
 3:    if random() < probability-threshold then
 4:       condition ← getLowerSupportCondition(individual)
 5:       if random() < 0.5 then
 6:          newCondition ← getNewCondition(condition)
 7:       else
 8:          switch (operator ← getOperator(condition))
 9:          case '=':
10:             newOperator = '!='
11:          case '!=':
12:             newOperator = '='
13:          case 'IN':
14:             newOperator = 'OUT'
15:          case 'OUT':
16:             newOperator = 'IN'
17:          case 'BETWEEN':
18:             newOperator = 'NOT BETWEEN'
19:          case 'NOT BETWEEN':
20:             newOperator = 'BETWEEN'
21:          end switch
22:          newCondition ← changeOperator(condition,
             operator,newOperator)
23:       end if
24:       newIndividual ← changeCondition(individual,
          condition,newCondition)
25:       newIndividualSet ← newIndividualSet ∪ newIndividual
26:    end if
27: end for
28: return newIndividualSet
```

support value. For example, for a categorical attribute A in a domain $D = \{a, b, c, d\}$, the support of A IN $\{a, c\}$ is 0.31, whereas the support of A NOT IN $\{a, c\}$ will be 0.69.

17.3.4 *Proposed Algorithm*

The algorithm here proposed follows a general evolutionary schema, as depicted in flowchart of Fig. 17.4. It starts by randomly generating individuals conforming to the CFG previously defined in Fig. 17.2. In this initial step, the algorithm guarantees that each individual covers at least one transaction. Therefore, if an individual generated conformant to the CFG does not satisfy any transaction, then a new random individual is generated. The process continues until the population is reached.

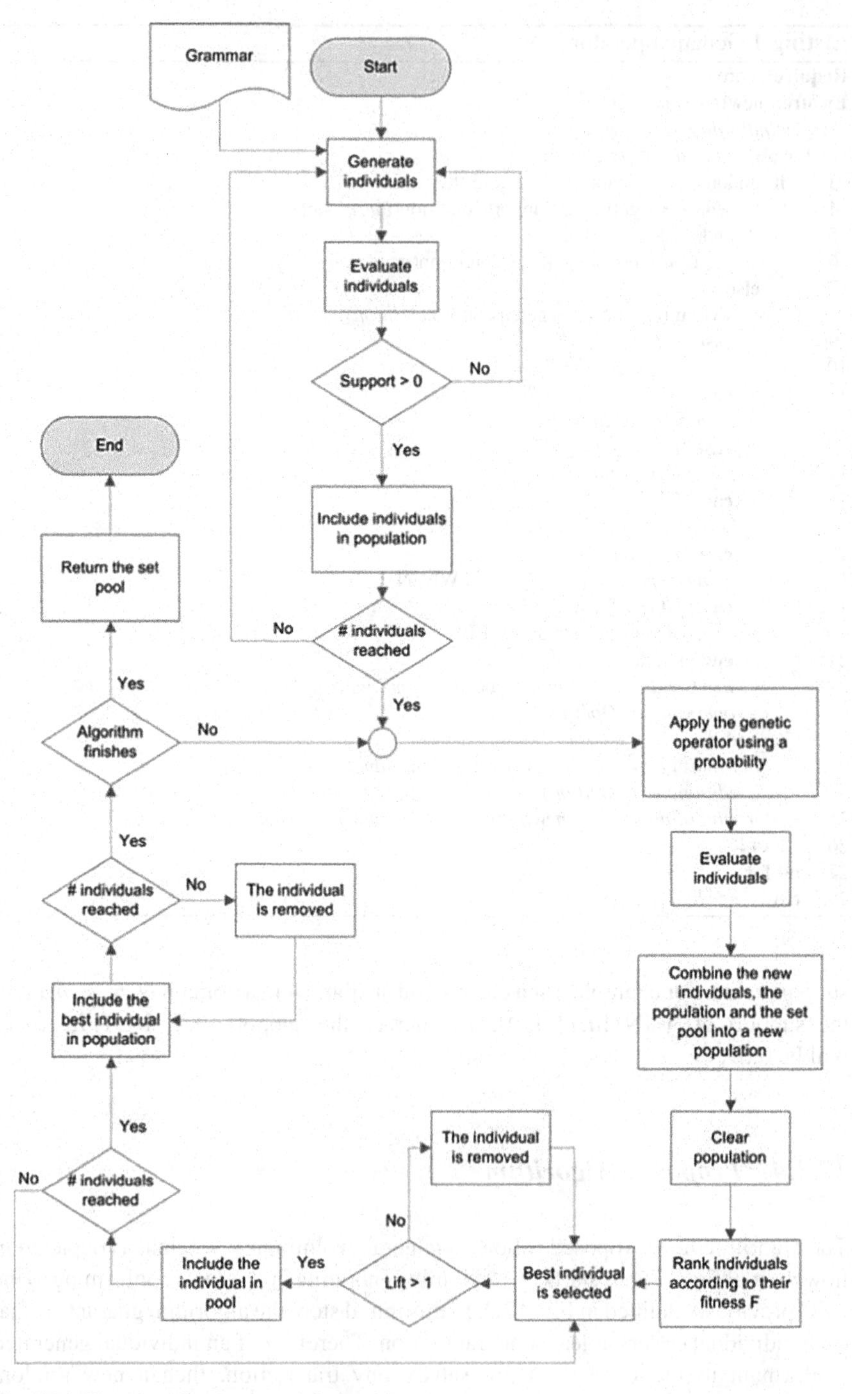

Fig. 17.4 Flowchart for the proposed algorithm

Once the population is completed, the evolutionary process works over this population along a number of generations. In each of these generations, and in order to obtain new individuals, the genetic operator is applied to the individuals by using a certain probability threshold. The new set of individuals is evaluated, calculating the support, the confidence, and the lift values for each individual.

The population set and the new set of individuals obtained and evaluated are merged into a new set, which allows of updating both a pool of individuals (it works as an elite population) and the population set. The first one comprises those n individuals having the best support values and lift values greater than unity, n being the pool size. Furthermore, this updating process guarantees that the pool does not include individuals having the same genotype. Despite the fact that n is the pool size, it is possible to not reach this number of individuals if there are not n distinct individuals having lift values greater than unity. On the other hand, the pool updating process keeps the m best individuals in the population: those having the best support values, regardless of their lift values. Notice that the proposed algorithm does not require any support and confidence thresholds to discover association rules. It instead discovers those rules having the best support values.

After these updating processes, the algorithm checks the number of generations, returning the pool when the maximum number of generations is reached. In such a way, the end user obtains at most n individuals, these individuals being the best ones discovered in the evolutionary process.

17.4 Experimental Study

In this paper, a number of experiments were carried out to check the effectiveness and performance of the algorithm, using both an artificial relational database and a real-world database. First, the artificial relational database used and the experiments performed will be given in detail in order to exhibit the behaviour of the proposed GP algorithm. Finally, a real-world relational database obtained from moodle database at the University of Córdoba (Spain) is analysed as a real case study to discover interesting students' behaviors. It is our understanding that any comparison to existing algorithms in this field is unfair, since no existing algorithms deal with quantitative and negative patterns, and none of them follows a G3P methodology.

The proposed approach was written by using JCLEC (Ventura et al. 2007), a Java library for evolutionary computation. All the experiments used in this study were performed on an Intel Core i7 machine with 12 GB main memory, running CentOS 5.4.

17.4.1 Artificial Relational Database

In this section, a synthetic relational database[1] and the experimental stage are described. By using this artificial database, the goal is to show the effectiveness of the proposed algorithm for mining association rules over different types of relationships, i.e. 1:1, 1:N, or even N:M. This relational database comprises information about different department stores and their customers. A department store is defined as an establishment which satisfies a wide range of consumers' personal product needs. Each department store was founded in a country, whereas each country can only found one department store. Focusing on the customers and their purchases, a number of customers buy in each department store. However, each customer can only buy in one department store. Each of these customers buys different items, which could be either food or clothes, grouped in a market basket, storing the price and the day of these purchases. The relational model (see Table 17.1) comprises eight tables, one per each relation and an additional one to represent the $B - I$ relationship. Each of these tables comprises a different number of attributes and transactions, giving rise to a heterogeneous and large database, the sum of all transactions being close to 220,000.

17.4.1.1 Parameter behaviour study

The evolutionary proposal here presented only requires four parameters to be determined by the expert: the maximum number of generations, the population size, the genetic operator probability, and the pool size. In this section, a sensitivity analysis is carried out, which aims to identify the influence of the parameters in the behaviour of the approach. Notice that this behaviour depends on the own characteristics of the problem.

Table 17.1 Entities for the synthetic relational database

Relation	Attributes	# Transactions
Country	countryName, inhabitants, capital	14
Department store	store, name, employees, **countryName**	14
Customer	passport, city, age, **store**	20,000
Basket market	id, day, price, **passport**	50,000
Item	code, price	100
B-I (Basket market-Item)	**code**, **id**	149,739
Clothes	**code**, name, season	50
Food	**code**, name, weight	50

[1] This synthetic relational database is available for download at http://www.uco.es/grupos/kdis/kdiswiki/index.php/Mining_Association_Rules_in_Relational_Databases.

The proposed algorithm keeps in a pool the best individuals obtained along the evolutionary process, i.e. those having the best fitness function values—determined by the support measure—until this pool is completed. Therefore, the average fitness value improves with the elapse of generations until the optimal individuals are discovered. Figure 17.5a depicts how the average value of both support and confidence improve in a linear way until the generation number 75 is reached. On the other hand, the lift value does not significantly improve, obtaining very similar results. Additionally, the execution time (see Fig. 17.5) grows in a linear way with the growth of generations. Therefore, it is necessary to reach a trade-off between the average quality measure values and the execution time required, and a value of 75 seems to be a good option.

A second parameter that must be previously fixed by the expert is the population size (see Fig. 17.6). Analysing Fig. 17.6a, it is depicted how the confidence value slightly improves by using a population size greater than 50 individuals. Focusing on the quality measures, their average values obtained slightly improve, so an analysis of the execution time (Fig. 17.6b) is required. The analysis shows that the execution time exponentially increases when more than 50 individuals are used, so this value could be fixed as optimal.

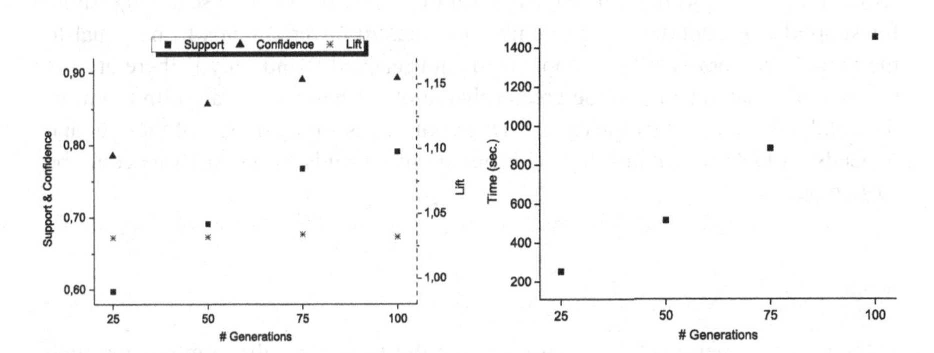

Fig. 17.5 Average values obtained by setting different number of generations

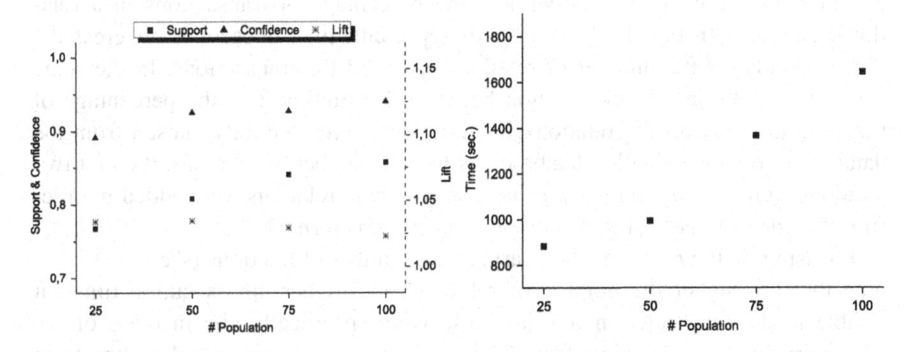

Fig. 17.6 Average values obtained by setting different population size

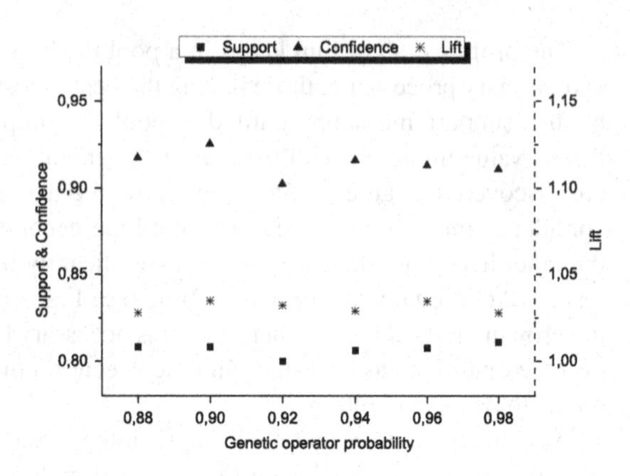

Fig. 17.7 Average measure values obtained by setting different genetic operator probabilities. Fitness values are represented by support

Another parameter to be set is the genetic operator probability, which is analysed in depth in Fig. 17.7. Regarding the support, confidence and lift measures, they slightly improve by using different probabilities, obtaining the best results with a probability of 0.90.

Finally, the pool size parameter establishes the final number of individuals to be returned by the algorithm. Since the algorithm proposed does not use any threshold for support and confidence, the number of rules discovered tends to be equal to the predefined pool size. This pool size is not reached if and only if there are not enough rules having a lift value greater than unity: otherwise the algorithm returns the number of rules set by the expert. This pool size is a prerequisite of the user and depends on his/her own aim for the problem under study, so no analysis could be carried out.

Scalability

A number of experiments were also carried out to analyse the computation time of the proposed algorithm. Figure 17.8 shows the relation between the runtime and both the number of relations and the percentage of transactions in a relational database. In Fig. 17.8a, the Y-axis represents time in seconds, whereas the X-axis stands for the number of relations using all the transactions. In the same way, Fig. 17.8b depicts the relation between the runtime and the percentage of transactions using all the relations. Transactions were randomly chosen from the database to obtain a smaller database. As for the number of relations, the first two relations were *Country* and *Department store*. Then, relations were added in order from *Customer* to *Food*, till the complete database is formed.

Focusing on the runtime when varying the number of relations (see Fig. 17.8a), note that the higher the number of relations, the higher the execution time—it is able to discover rules in few minutes. More specifically, the increase of the execution time when varying from 2 to 4 relations seems to be higher than from

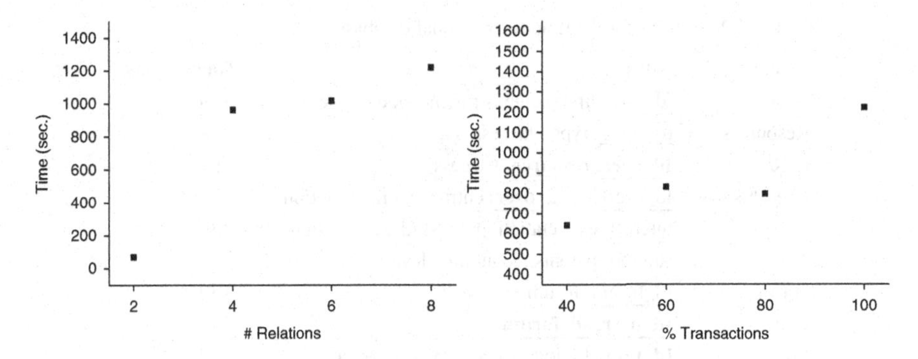

Fig. 17.8 Graphics for the scalability of the runtime

4 to 6 relations, where this increase is more linear. Something similar occurs when varying the number of transactions (see Fig. 17.8b). There is a small increase when increasing the number of transactions, up to 80 %. From this point on, the execution time increases in an exponential way. However, it should be noted that the whole increase, i.e., from 40 % to 100 %, tends to increase in a linear way.

17.4.2 Real-World Relational Database

Data used in this experimental stage was gathered from moodle at the University of Córdoba, Spain. Data correspond to 139 different students from a specific course of the degree in Computer Science. The relational model (see Table 17.2) comprises thirteen tables, one per each relation. Each of these tables comprises a different number of attributes and transactions, giving rise to a heterogeneous relational database.

Once the algorithm is run by using the real database, a series of interesting association rules are discovered (see Table 17.3), describing the behavior of students in this specific course. The first rule discovered describes the level of dedication of students based on their expectation at the beginning of the course. Students had to fill in a questionnaire at the beginning of the course, describing their interest in the course, their expectation to pass the course, etc., and this information was stored in the *user_history* relation. The first rule, ***IF*** *User_history.expectation_pass* $\neq$ *High* ***THEN*** *User_history.dedication* $\neq$ *Low*, describes that 98 % of the students that did not have high expectations that they will pass the course, they will dedicate enough time to the course. This description is quite interesting since determines that the expectations present a high relation to the time spent in the course.

The second and third rules describe behaviors in connection with specific questions and resources. For instance, rule #2 describes that questions that are not related to computational data tend to obtain a grade lower than 0.5 (in per unit basis) or higher than 0.88, i.e. ***IF*** *Questions.question* $\neq$ *computational data*

Table 17.2 Entities for the moodle relational database

Relation	Attributes	# Transactions
User	Id_user, first_name, surname, age, grade	139
Resources	resource, type, #accesses	8
U-R	**Id_user**, **resource**, #access	367
User_history	Id_userHis, **Id_user**, entrance_grade, option,	
	interest, expectation_pass, study, dedication,	139
	support_parents, economic_level	
Forum	Id_forum, forum	139
U-F	**Id_user**, **Id_forum**	82
Log	**Id_user**, Id_log, #access<30seg, #access	139
Log_action	Id_action, action, module	26
L-A	Id_action, **Id_log**	1575
Quizzes	name_quiz, grade	6
U-Q	**Id_user**, **name_quiz**, grade	576
Questions	Id_question, grade, question	253
U-Ques	**Id_user**, **Id_question**, grade	11501

Table 17.3 Some of the rules obtained by running the algorithm on the real database

# Rule	Support	Confidence	Lift
1 **IF** User_history.expectation_pass $\neq$ High **THEN** User_history.dedication $\neq$ Low	0.87	0.98	1.12
2 **IF** Questions.question $\neq$ computational data **THEN** Questions.grade NOT BETWEEN [0.5, 0.88]	0.59	0.79	1.03
3 **IF** Resources.type = file data **THEN** Resources.#accesses BETWEEN [123, 152]	0.50	0.80	1.07
4 **IF** User.age BETWEEN [20, 23] **THEN** User.grade NOT IN (Absent, Distinction),	0.39	0.81	1.17

THEN *Questions.grade NOT BETWEEN [0.5, 0.88]*, and this descriptive rule has an accuracy of 79 %. As for the rule #3, it describes that resources of the type '*file*' are highly downloaded by students, and this assertion is right with an accuracy of 80 %. In fact, since there are 139 students enrolled in the course, almost all the students tend to download this type of resources.

Finally, a very interesting association rule is obtained, i.e. rule #4, denoting that 81 % of the students with an age in the range [20, 23] tend to pass the course, and they rarely are absent or pass the course with distinction. In other words, since an age in the range [20, 23] is the common age for this course, only students that did not pass the course in previous years will pass the course with distinction.

17.5 Concluding Remarks

This paper presents the first G3P approach to discover association rules over relational databases. The advantage of representing individuals conforming to a CFG is twofold. Firstly, the individuals are represented in a tree structure, with the goal of maintaining the original relational database, and not requiring any joining of the relations into a unique table. Secondly, the CFG allows of defining syntax constraints in the trees and obtaining solutions in different attribute domains (positive, negative, discrete and continuous). More specifically, the use of a grammar provides expressiveness, flexibility, and the ability to restrict the search space. Additionally, the CFG allows the rules to be adapted to the expertise of the data miner.

This methodology was apply to the ARM field (Luna et al. 2012), achieving promising results and reducing both the computational cost and the memory requirements. The main advantage of using heuristic stochastic search is its ability to restrict the search space and to obtain negative and quantitative solutions, which could be a hard process by means of exhaustive search.

In the experimental stage, a detailed analysis of the proposed approach was carried out by using an artificial dataset, considering both the mined rules and the execution time. The results have shown the capability of the proposed algorithm for extracting highly reliable and frequent association rules. Finally, a real-world database has been used, extracting interesting descriptions about students' behaviors in a moodle course from the University of Córdoba, Spain.

Acknowledgements This research was supported by the Spanish Ministry of Science and Technology, project TIN-2011-22408, and by FEDER funds. This research was also supported by the Spanish Ministry of Education under FPU grants AP2010-0041 and AP2010-0042.

References

R. Agrawal, H. Mannila, R. Srikant, H. Toivonen, and A. Verkamo. Fast Discovery of Association Rules. In *Advances in Knowledge Discovery and Data Mining*, pages 307–328. American Association for Artificial Intelligence, Menlo Park, CA, USA, 1996.

R. Agrawal and R. Srikant. Fast Algorithms for Mining Association Rules in Large Databases. In J. B. Bocca, M. Jarke, and C. Zaniolo, editors, *VLDB'94, Proceedings of 20th International Conference on Very Large Data Bases, Santiago de Chile, Chile*, pages 487–499. San Francisco: Morgan Kaufmann, September 1994.

A. Alashqur. RDB-MINER: A SQL-Based Algorithm for Mining Rrue Relational Databases. *Journal of Software*, 5(9):998–1005, 2010.

V. Crestana-Jensen and N. Soporkar. Frequent Itemset Counting Across Multiple Tables. In *Proceedings of the 4th Pacific-Asia Conference on Knowledge Discovery and Data Mining (PADKK '00), Kyoto, Japan*, pages 49–61, April 2000.

P.G. Espejo, S. Ventura, and F. Herrera. A Survey on the Application of Genetic Programming to Classification. *IEEE Transactions on Systems, Man and Cybernetics: Part C*, 40(2):121–144, 2010.

F. Berzal and I. Blanco and D. Sánchez and M.A. Vila. Measuring the Accuracy and Interest of Association Rules: A new Framework. *Intelligent Data Analysis*, 6(3):221–235, 2002.

A. A. Freitas. *Data Mining and Knowledge Discovery with Evolutionary Algorithms*. Springer-Verlag Berlin Heidelberg, 2002.

B. Goethals and J. Van den Bussche. Relational Association Rules: Getting WARMeR. In *Proceedings of 2002 Pattern Detection and Discovery, ESF Exploratory Workshop, London, UK*, pages 125–139, September 2002.

B. Goethals, D. Laurent, W. Le Page, and C. T. Dieng. Mining frequent conjunctive queries in relational databases through dependency discovery. *Knowledge and Information Systems*, 33(3):655–684, 2012.

B. Goethals, W. Le Page, and M. Mampaey. Mining Interesting Sets and Rules in Relational Databases. In *Proceedings of the ACM Symposium on Applied Computing, Sierre, Switzerland*, pages 997–1001, March 2010.

F. Gruau. On using Syntactic Constraints with Genetic Programming. *Advances in genetic programming*, 2:377–394, 1996.

P. Hájek, I. Havel, and M. Chytil. The GUHA Method of Automatic Hypotheses Determination. *Computing*, 1(4):293–308, 1966.

J. Han, J. Pei, Y. Yin, and R. Mao. Mining Frequent Patterns without Candidate Generation: A Frequent-Pattern Tree Approach. *Data Mining and Knowledge Discovery*, 8:53–87, 2004.

A. Jiménez, F. Berzal, and J.C. Cubero. Using Trees to Mine Multirelational Databases. *Data Mining and Knowledge Discovery*, pages 1–39, 2011.

A.R. Konan, T.I. GÜndem, and M.E. Kaya. Assignment query and its implementation in moving object databases. *International Journal of Information Technology and Decision Making*, 9(3):349–372, 2010.

J. R. Koza. *Genetic Programming: On the Programming of Computers by Means of Natural Selection (Complex Adaptive Systems)*. The MIT Press, December 1992.

J. R. Koza. Introduction to Genetic Programming: Tutorial. In *GECCO'08, Proceedings of the 10th Annual Conference on Genetic and Evolutionary Computation, Atlanta, Georgia, USA*, pages 2299–2338. ACM, July 2008.

J. M. Luna, J. R. Romero, and S. Ventura. Design and Behavior Study of a Grammar-guided Genetic Programming Algorithm for Mining Association Rules. *Knowledge and Information Systems*, 32(1):53–76, 2012.

J. Mata, J. L. Alvarez, and J. C. Riquelme. Discovering Numeric Association Rules via Evolutionary Algorithm. *Advances in Knowledge Discovery and Data Mining*, 2336/2002:40–51, 2002.

R. McKay, N. Hoai, P. Whigham, Y. Shan, and M. ONeill. Grammar-based Genetic Programming: a Survey. *Genetic Programming and Evolvable Machines*, 11:365–396, 2010.

E. Ng, A. Fu, and K. Wang. Mining Association Rules from Stars. In *Proceedings of the 2002 IEEE International Conference on Data Mining (ICDM 2002), Maebashi City, Japan*, December 2002.

N.F. Papè, J. Alcalá-Fdez, A. Bonarini, and F. Herrera. *Evolutionary Extraction of Association Rules: A Preliminary Study on Their Effectiveness*, volume 5572/2009 of *Lecture Notes in Computer Science*, pages 646–653. 2009.

A. Ratle and M. Sebag. Genetic Programming and Domain Knowledge: Beyond the Limitations of Grammar-Guided Machine Discovery. In *PPSN VI, Proceedings of the 6th International Conference on Parallel Problem Solving from Nature, Paris, France*, pages 211–220, September 2000.

C. Romero, S. Ventura, and P. De Bra. Knowledge Discovery with Genetic Programming for Providing Feedback to Courseware Authors. *User Modeling and User-Adapted Interaction*, 14:425–464, 2004.

E. Spyropoulou and T. De Bie. Interesting Multi-relational Patterns. In *ICDM 2011, Proceedings of 11th IEEE International Conference on Data Mining, Vancouver, Canada*, pages 675–684, December 2011.

S. Ventura, C. Romero, A. Zafra, J.A. Delgado, and C. Hervás. *JCLEC: A Framework for Evolutionary Computation*, volume 12 of *Soft Computing*, pages 381–392. Springer Berlin / Heidelberg, 2007.

Chapter 18
Evolving GP Classifiers for Streaming Data Tasks with Concept Change and Label Budgets: A Benchmarking Study

Ali Vahdat, Jillian Morgan, Andrew R. McIntyre, Malcolm I. Heywood, and Nur Zincir-Heywood

18.1 Introduction

A traditional view of learning from data is most often characterized by the supervised learning 'classification' task. However, as we are increasingly encountering data rich environments, the basis for such a characterization are becoming less relevant. Decision making from streaming data is one such application area (e.g., stock market data, utility utilization, behavioural modelling, sentiment analysis, process monitoring etc.). Under a 'streaming' scenario for constructing models of classification, data arrives on a continuous basis, thus there is no concept of a 'beginning' or an 'end'. It is not possible to see all the data at once and it therefore becomes impossible to guarantee that the data 'seen' at any point in time are representative of the 'whole' task. Indeed, the process generating the data are frequently non-stationary. Applications display properties such as concept drift (a gradual change in the process creating the data) or concept shift (sudden changes to the process creating the data). Concept change in general implies that a model that functions effectively at one point in the stream will not necessarily function effectively later on. Moreover, in the general case it is not possible to provide labels for all the stream. Instead decisions need to be made regarding what to label without calling upon an oracle. Indeed, the real-time nature of many streaming classification tasks implies that the number of label requests needs to be very much lower than the throughput of the stream itself. When combined with the issue of non-stationarity, this makes it much more difficult to recognize when models need to be rebuilt.

A. Vahdat • J. Morgan • A.R. McIntyre • M.I. Heywood (✉) • N. Zincir-Heywood
Faculty of Computer Science, Dalhousie University, 6050 University Av., Halifax, NS, Canada
e-mail: ali.vahdat@dal.ca; jillian.morgan@dal.ca; armcintyre@gmail.com; mheywood@cs.dal.ca; zincir@cs.dal.ca

A.H. Gandomi et al. (eds.), *Handbook of Genetic Programming Applications*,
DOI 10.1007/978-3-319-20883-1_18

Finally, we note that an 'anytime' nature to the task exists. Irrespective of the state of the model building process itself, a champion individual (classifier) must be available to suggest labels for the current content of the data stream at any given time.

A distinction is drawn between regression (function approximation) and classification under streaming data. Regression under a streaming data context is most synonymous with the task of forecasting. As such the true value for the dependent variable is known a short time after a prediction is made by the model. This means that the issue of label budgets is not as prevalent, and models can therefore be much more reactive. Conversely, having to explicitly address the issue of label budgets implies that independent processes need to be introduced to prompt for label information (e.g. change detection).

Various proposals have been made for what properties GP might need to assume under environments that are in some way 'dynamic'. Several researchers have made a case for adopting modular frameworks for model building under dynamic scenarios. For example, environments that change their objective dynamically over the course of evolution (e.g., Kashtan et al. 2007). Likewise, modularity might be deemed useful from the perspective of delimiting the scope of variation operators, thus making credit assignment more transparent and facilitating incremental modification (e.g., Wagner and Altenberg 1996). Diversity maintenance represents a reoccurring theme, and is frequently emphasized by research in (non-evolutionary) ensemble learning frameworks applied to streaming tasks (Brown and Kuncheva 2010; Minku et al. 2010; Stapenhurst and Brown 2011). Finally, we note that 'evolvability' is typically defined in terms of a combination of the ability to support (phenotypic) variability and the fitness of the resulting offspring (in a future environment) (e.g., Parter et al. 2008). This can be viewed through the perspective of Baldwin mechanisms for evolution. Thus, it is desirable to retain parents that are most likely to lead to fit offspring on a regular basis.

This work undertakes a benchmarking study of a framework previously proposed for evolving modular GP individuals under streaming data contexts with label budgets (Vahdat et al. 2014). The specific form of GP assumed takes the form of Symbiotic Bid-Based GP (SBB) and is hereafter denoted **StreamSBB**. The framework consists of three elements: a sampling policy, a data subset, and a data archiving policy. The combination of sampling policy and the data subset achieve a decoupling between the rate at which the stream passes and the rate at which evolution commences. This also provides the basis for changing the distribution of data from that present in the stream at any point in time. In changing the distribution of data, we are in a position to, for example, resist the impact of class imbalance. Finally, in order to address the issue of model building under a limited **label budget**, a stochastic querying scheme is assumed. Thus, for any given window location, a fixed number of label requests are permitted. The selection of an appropriate label budget being a reflection of the cost of making label requests.

Benchmarking will be performed with both artificially created datasets (which provide the ability to embed known forms of concept drift and shift) as well as real-world datasets (electricity utilization/demand and forest cover types). Such datasets display a wide range of real world properties, with cardinality measured close to the millions, dummy attributes, class imbalance, and changing relationships between attribute and label. Moreover, benchmarking practices for streaming data are explicitly identified. In particular rather than assuming a 'prequential' accuracy metric, a formulation of (average) multi-class detection rate is assumed and estimated incrementally. This enables us to avoid the caveats that appear with accuracy style metrics under class imbalance. Comparator algorithms are included from the MOA toolbox, representing current state of the art in non-evolutionary approaches to streaming data classification. The benchmarking study demonstrates the appropriateness of assuming the StreamSBB framework, with specific recommendations made regarding the utility of: modularity, pre-training, and generations per sample of labelled data.

Section 18.2 provides a summary of related streaming data research. The StreamSBB framework is detailed in Sect. 18.3 with the experimental methodology discussed in Sect. 18.4. Section 18.5 presents results of the benchmarking study where this is designed to illustrate the contribution from various components of StreamSBB. Section 18.6 discusses the resulting findings and concludes the work.

18.2 Related Work

A significant body of work has developed regarding the application of machine learning (ML) to various streaming classification tasks (Quinonero-Candela et al. 2009; Gama 2010; Bifet 2010; Gama 2012; Heywood 2015). For brevity we concentrate on the issue of change detection which lies at the centre of building ML frameworks capable of operating under label budgets. Indeed, classification under label budgets represents the most recent trend in streaming data classification. We identify three broad categories of interest, outlined as follows:

Properties specific to the model of classification imply that measurements specific to an ML framework are made and compared to a prior characterization. For example, changes to the frequency of leaf node utility in decision trees might signify change, thus trigger label requests (Fan et al. 2004; Huang and Dong 2007).

Properties of the input data imply that change detection now focuses on characterizing behaviour relative to sliding window content. The principle design decision is with regards to what statistic to adopt. For example, Chernoff bounds (Kifer et al. 2004), entropy (Dasu et al. 2006; Vorburger and Bernstein 2006), Kullback–Leibler divergence (Sebastio and Gama 2007), Hoeffding bounds (Bifet and Gavalda 2007), Fractal correlation dimension (Folino and Papuzzo 2010) or Hellinger divergence metric (Ditzler and Polikar 2011). Potential drawbacks of pursuing such an approach include: (1) it is often necessary to label the data (i.e., metrics are estimated class-wise); and (2) changes to the association between label and input are not detected (Žliobaitė et al. 2014).

Properties of the label space imply that the classifier output 'behaviour' is quantified. For example, statistical characterizations of class boundary information (cf. classifier confidence) have been proposed (Lindstrom et al. 2010; Žliobaitė et al. 2011). Thus, thresholds might be used to detect changes in classifier certainly (Lindstrom et al. 2013), or changes to the number of confident predictions (Lanquillon 1999).

However, none of the above approaches are able to detect when a previously encountered input, $P(x)$, is associated with a new or different class label. Thus, under this scenario a label space characterization would still associate $P(x)$ with the previous label. Likewise change detection based on an input data formulation would not register any change either, i.e. $P(x)$ has not changed. Under these scenarios generating label requests uniformly (i.e., independently of the measurable properties) has been shown to be surprisingly effective (Zhu et al. 2010); as have hybrid approaches combining label space and uniform sampling (Žliobaitė et al. 2014).

Under the guise of evolutionary computation (EC) in general, a body of research has been developed regarding dynamic optimization tasks (e.g., Blackwell and Branke 2006). However, such tasks are distinct from streaming data classification in that the emphasis is with regards to tracking and identifying multiple optima; thus, there is no concept of operating under label budgets. From a genetic programming (GP) perspective, most developments come with respect to the specifics of evolving trading agents for financial applications (e.g., Dempsey et al. 2009). Although change detection is certainly important to building effective trading agents, the rebuilding of models is either performed on a continuous basis (as in function approximation) (Dempsey et al. 2009) or incrementally based on task specific properties such as an unacceptable loss (Heywood 2015). Thus, the issue of label budgets does not appear in frameworks for evolving trading agents. Finally, we note that in the special case of learning classifier systems (LCS), an explicitly online variant has been proposed in which probabilistic heuristics are used to fill 'gaps' in the provision of label information (Behdad and French 2013).

18.3 Methodology

Figure 18.1 provides a summary of the general architecture assumed for applying GP to streaming data under finite labelling budgets (Vahdat et al. 2014; Heywood 2015). We assume a non-overlapping 'sliding window' as the generic interface to the stream. For a given window location a fixed number of samples are taken. Let $SW(i)$ denote the location of the window and 'gap' denote the data sampled from this location, or $Gap(i) \in SW(i)$; where $|Gap(i)| \leq |SW(i)|$. The sampling policy determines which exemplars are selected to appear in $Gap(i)$ for a given window location $SW(i)$. Note that it is only the $|Gap(i)|$ exemplars chosen that have their corresponding labels requested and are added to the Data Subset ($DS(i)$). An archiving policy determines which exemplars are replaced from DS at each update. Note also that the rate at which GP training epochs are performed, $gen(k)$, is a function of the rate at which DS is updated or $j = i \times k; k \in \{1, 2, \ldots\}$. This means that for each update in DS content (index i), at least a single GP training epoch is performed ($k = 1$). Naturally, increasing the number of training epochs potentially increases the capacity to react to changes to stream content, but may potentially result in over learning (w.r.t. current DS content). Hereafter we will refer to this as **DS oversampling**. Section 18.5 will explicitly investigate this property in more detail.

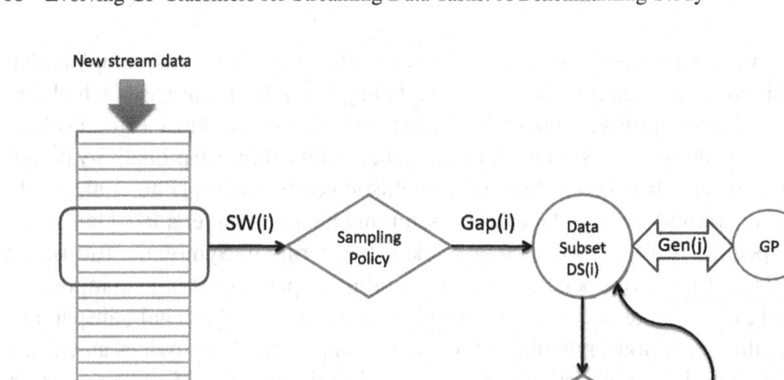

Fig. 18.1 Components of generic architecture for applying GP to streaming data under a label budget

The StreamSBB framework adopts symbolic bid-based GP (SBB) as the GP architecture (Doucette et al. 2012b). Specifically, SBB evolves teams of bid-based GP individuals to cooperatively decompose the classification task without having to specify team size. Supporting modularity in general has been deemed to be useful for dynamic task environments (Sect. 18.2), a property we explicitly verify in Sect. 18.5. Secondly, SBB assumes a Pareto archiving policy with diversity maintenance heuristics for enforcing a finite archive size. In effect, the concept of Pareto dominance is used to identify exemplars for retaining within DS. As such this gives them a 'lifetime' beyond the current location of the sliding window.

18.3.1 Sampling Policy

Rather than evaluating a GP classifier with respect to all data within $SW(i)$, a sampling policy is assumed to control entry into the Data Subset ($DS(i)$). This decouples the cost of any single training epoch and enforces the labelling budget, i.e. we control the cardinality of the data subset, but cannot control the throughput of the stream. Note, however, that the decision regarding the sampling of 'gap' exemplars from sliding window location $SW(i)$ to a data subset can only be performed *without* label information. It is only after identifying the exemplars included in $Gap(t)$ that labels are requested.

Two basic approaches for defining sampling policies have been identified in the wider literature (Sect. 18.2): stochastic sampling or classification confidence information. Classifier confidence information implies that as the certainty of the class label suggested by a classifier decreases (i.e. approaches ambiguity), then a label request is made (Lindstrom et al. 2013; Žliobaitė et al. 2014). In the case of stochastic label requests, this is performed uniformly relative to exemplars that are classified with certainty. The objective being to confirm that cases which are classified with certainty have not undergone some form of shift into a different class. Moreover, we also note that even requesting labels with a uniform probability (under a label budget) is often better than more sophisticated heuristics (Zhu et al. 2010). In this work we will assume the uniform sampling heuristic under a label budget.

The specific form of GP assumed takes the form of Symbiotic Bid-Based GP (SBB) and therefore benefits from the ability to perform task decomposition (construct a classifier as a team of programs). Aside from the additional transparency of the resulting solutions, pursuing a GP teaming approach also provides an elegant solution to multi-class classification. A short description of SBB is provided in Sect. 18.3.4, whereas readers are referred to the earlier papers for further details (Lichodzijewski and Heywood 2008; Doucette et al. 2012b).

18.3.2 Data Archiving Policy

The scheme assumed for prioritizing DS content for replacement is defined by a **data archiving policy**. Specifically, Pareto archiving is used to identify exemplars that 'distinguish' between the performance of GP classifiers. Such a set of exemplars are said to be non-dominated (de Jong 2007). One of the drawbacks of assuming a Pareto archiving policy, however, is that the archive of exemplars distinguishing between different GP classifiers increases to $P^2 - P$; where P is the size of the GP population. This would have implications for the overall efficiency of the algorithm. Hence, we limit the size of DS to a suitable finite value and employ a DS exemplar diversity/aging heuristic (Atwater and Heywood 2013). The process for choosing exemplars from DS for replacement switches between the following cases, depending on which condition is satisfied. Let the exemplars from DS forming 'distinctions' be d and those not supporting a distinction be $\bar{d}$:

Case 1 *The number of exemplars forming a distinction is less than or equal to* $|DS| - |Gap|$ (i.e. $|d| \leq |DS| - |Gap|$). This implies that the number of exemplars that *do not* support distinctions is greater than or equal to $|Gap|$ (i.e. $|\bar{d}| \geq |Gap|$). Hence, the DS exemplars replaced by $Gap(i)$ are selected from $\bar{d}$ alone.

Case 2 *The number of exemplars forming distinctions is more than* $|DS| - |Gap|$. Any exemplars not forming distinctions ($\bar{d}$) will be replaced. In addition $|Gap| - |\bar{d}|$ exemplars forming distinctions will also be replaced, potentially resulting in the loss of GP classifiers (i.e., no longer identified as being non-dominated). The exemplars forming distinctions are now ranked in accordance with how many other points form the same distinction and how long an exemplar has been in the archive (Atwater and Heywood 2013). In effect exemplars supporting: (1) unique distinctions see more priority than those forming more common distinctions i.e., a form of fitness sharing or diversity maintenance, and (2) older exemplars are more likely to removed in favour of those forming more recent distinctions.

Such preference schemes were previously shown to be useful under GP streaming classification, albeit without label budgeting (Atwater et al. 2012; Atwater and Heywood 2013). Further details of Pareto archiving as applied to GP are available in Doucette et al. (2012b).

18.3.3 Anytime Classifier Operation

In order to predict class labels for exemplars of the non-stationary stream a single GP individual must be present at any point in time to perform this task, or **anytime classifier operation**. To do so, we assume that the current content of the data subset $DS(i)$ is suitably representative of the classification task. That is to say, it is only the content of DS that is labelled, and the content is incrementally updated from each SW location with the data archiving policy enforcing a finite archive size (Sect. 18.3.2). A metric is now necessary for identifying the champion individual relative to the GP individuals identified as *non-dominated*. In limiting the available candidate GP classifiers to the non-dominated set, we reduce the likelihood of selecting degenerate classifiers. Given that class balance is not enforced on $SW(i)$ content, it is desirable to assume a metric that is robust to class imbalance (skew). With this in mind the following definition for average detection rate is assumed:

$$DR = \frac{1}{C} \sum_{c=[1,\ldots,C]} DR_c$$

$$DR_c = \frac{tp_c}{tp_c + fn_c} \tag{18.1}$$

where C is the number of classes observed in the dataset so far and tp_c and fn_c denote true positive and false negative counts w.r.t. class c respectively.

18.3.4 Symbiotic Bid-Based (SBB) GP

Symbiotic Bid-Based GP, or SBB for short, is a generic coevolutionary GP framework originally developed to facilitate task decomposition under discrete decision making tasks (Lichodzijewski and Heywood 2008). SBB has been applied to a wide range of problem categories such as reinforcement learning (e.g. Doucette et al. 2012a) and classification (e.g. Doucette et al. (2012b)).

SBB maintains two populations: symbiont and host. Symbionts (*sym*) take the form of bid-based GP individuals (Lichodzijewski and Heywood 2008). They specify a 'program' (p) and a scalar 'action' (c). The action in a classification context takes the form of a class label, assigned when each program is initialized. Individuals from the host population index some subset of the individuals from the symbiont population under a variable length representation.

Host (h) evaluation w.r.t. training exemplar (x) involves executing the program of each of the symbionts associated with that host, and identifying the bid-based GP with highest output, or:

$$sym^* = \arg\max_{sym \in h} \left(sym(p, x) \right) \quad (18.2)$$

This 'winning' bid-based GP individual (sym^*) suggests its corresponding action or class label $sym^*(c)$. The only constraint on host membership is that there must be at least two symbionts with different class labels per host, or:

$$\forall\ sym_{i,j} \in h;\ \ \exists\ sym_i(c) \neq sym_j(c) \quad (18.3)$$

where i, j are symbiont indexes and $i \neq j$. Hence, multiple symbionts might co-operate to represent a single class. Moreover, previous research with SBB under streaming tasks (without label budgets) indicated that class membership could be incrementally evolved over the course of a stream (Atwater et al. 2012; Atwater and Heywood 2013). This incremental evolution of class membership avoids the requirement for teams to solve all aspects of the class assignment task simultaneously.

Variation and selection operators remain unchanged from the original formulation of SBB (Lichodzijewski and Heywood 2008; Doucette et al. 2012b), and without loss of generality the form of GP assumed for symbiont programs is that of linear GP. The instruction set includes: $\{+, -, \times, \div, \cos(\cdot), \exp(\cdot), \log(\cdot)\}$, although others can be added. Readers are referred to the earlier SBB papers for further details of operators and instruction set of SBB (Lichodzijewski and Heywood 2008; Doucette et al. 2012b).

18.4 Experimental Methodology

This section begins by establishing the approach to benchmark dataset selection (Sect. 18.4.1). Section 18.4.2 discusses parameter setting and characterizes StreamSBB design decisions. Section 18.4.3.1 outlines the approach adopted to performance evaluation. Finally, the properties of an alternate streaming classifier (an adaptive form of Naive Bayes with label budgeting) is summarized in Sect. 18.4.3.2.

18.4.1 Streams/datasets

Four streams/datasets will be employed for the purposes of benchmarking: (1) two artificially created and therefore with known degrees of non-stationary behaviour[1]; "Gradual Concept Drift" and "Sudden Concept Shift" streams, and; (2) two well known real world datasets; "Electricity Demand" (Harries 1999), and "Forest Cover Types" (Bache and Lichman 2013) that have frequently been employed for streaming data benchmarking tasks. The basic properties of the datasets are summarized by Table 18.1.

Gradual Concept Drift stream (Fan et al. 2004): Hyperplanes are defined in a 10-dimensional space. Initial values of the hyperplane parameters are selected with uniform probability. This Dataset has 150,000 exemplars and every 1000 exemplars, half of the parameters are considered for modification with a 20 % chance of change, hence creating the gradual drift of class concepts. Class labels are allocated as a function of hyperplanes exceeding a class threshold.

Sudden Concept Shift stream (Zhu et al. 2010): The Dataset Generator tool[2] is used to construct decision trees that specify a partitioning of the attribute space into a 5-class classification task based on randomly generated thresholds. Data is generated with a uniform p.d.f. and then assigned a class using the decision tree. A total of two concept generator decision trees (C1, C2) are used to create two sources of data. A single stream of data is then constructed block-wise with data integrated from each of the two concept generator decision trees.

The process used to create sudden changes in the concept of classes of the stream has the following form. The stream is created 'block-wise' with 13 blocks and each block consists of 500,000 exemplars. Consider a concept generator tuple of the form: $\langle C1\,\%, C2\,\%\rangle$. We can now define the stream in terms of the transition of exemplars from 100 % $C1$ to 100 % $C2$ in 10 % increments: $\langle 100, 0\rangle$, $\langle 100, 0\rangle$, $\langle 100, 0\rangle$, $\langle 90, 10\rangle$, $\langle 80, 20\rangle$, ... $\langle 0, 100\rangle$. For example, $\langle 80, 20\rangle$ denotes a block consisting of exemplars in which 80 % are from $C1$ and 20 % are from $C2$. A uniform probability is used to determine the exemplar order in each block.

Table 18.1 Benchmarking dataset properties

Stream/Dataset	D	N	k	≈ Class Distribution (%)
Gradual concept drift (drift)	10	150,000	3	[16, 74, 10]
Sudden concept shift (shift)	6	6,500,000	5	[37, 25, 24, 9, 4]
Electricity demand (elec)	8	45,312	2	[58, 42]
Forest cover types (cover)	54	581,012	7	[36, 49, 6, 0.5, 1.5, 3, 4]

D denotes dimensionality, *N* refers to the total exemplar count, and *k* is the number of classes

[1]Publicly available at http://web.cs.dal.ca/~mheywood/Code/SBB/Stream/StreamData.html.

[2]Gabor Melli. The 'datgen' Dataset Generator. http://www.datsetgenerator.com/.

Electricity Demand characterizes the rise and fall of electricity demand in New South Wales, Australia, using consumption and price information for the target and neighbouring regions (Harries 1999). As such it is a two class dataset (demand will either increase or decrease relative to the previous period), moreover, unlike the other three datasets the distribution of classes is almost balanced.

Forest Cover Types defines forest cover type from cartographic variables (Bache and Lichman 2013). The actual forest cover type for a given observation (30×30 meter cell) was determined from US Forest Service (USFS) Region 2 Resource Information System (RIS) data. Forest cover type is a 7-class dataset with 54 attributes, 44 of which are binary. The distribution of classes are very imbalanced with the largest class covering almost 50 % of dataset and the smallest class covering a mere 0.5 %, almost $\frac{1}{100}$ of the majority class. In order to provide a temporal property, previous research sorted the dataset based on the elevation of the 30×30 meter cells to give it characteristics of streaming data (Žliobaitė et al. 2014). The same approach is adopted in this paper. What makes this dataset interesting from a streaming data perspective is that there are a comparatively large number of classes, and the seventh class does not appear until roughly half way through the stream. Thus, any classifier working under a label budget would need to discover the new class and react accordingly without disrupting its performance on the other six classes.

18.4.2 Parameterization of GP

Relative to the earlier work (Atwater et al. 2012; Atwater and Heywood 2013; Vahdat et al. 2014), the following represents a much more through experimental evaluation. Indeed, the StreamSBB framework of Fig. 18.1 was only proposed in Vahdat et al. (2014) and then benchmarked under a very restrictive scenario (i.e. limited to choices for the label budget). In this work, the objective is to identify what properties of StreamSBB contribute to specific capabilities under streaming data with non-stationary properties. With this in mind, the following three generic parameterizations will be assumed through out:

Model initialization is performed using the first S_{init} % of the stream during the first i_{init} % of generations. This represents the **pre-training budget**, with the remainder of the label budget being consumed across the remainder of the stream. Given that the interface to the stream assumed by StreamSBB is a non-overlapping window, then this assumption just defines the initial window length and assumes that i_{init} % of the generations are performed against this window location. Thereafter, the sliding window advances at a fixed rate through the stream.

A non-overlapping **sliding window** of length S_{max}/i_{max} exemplars is assumed after model initialization. The remainder of the stream passes through at a constant rate. The window content defines the pool from which the new $|Gap(i)|$ training exemplars are sampled and labels requested (Fig. 18.1). This results in a 'non-overlapping' sliding window. However, GP is evaluated w.r.t. the content of the

Data Subset, $|DS|$, (Fig. 18.1), and only $|Gap|$ exemplars are introduced per window location, hence, there is still a 'gentle' turnover in new to old exemplar content between consecutive generations. Parameters are set to $|Gap| = 20$ and $|DS| = 120$ in all experiments.

Label budget is the ratio of points whose labels are requested to the total stream length, or:

$$label\ budget(LB) = \frac{i_{max} \times |Gap|}{S_{max}} \tag{18.4}$$

In other words only $i_{max} \times |Gap|$ exemplars are requested for their label in a stream of length $S_{max}(\equiv N)$. Under the sudden concept shift stream with $S_{max} = 6,500,000$ exemplars and $i_{max} = 1000$ generations, the non-overlapping window defines 6500 exemplars between updates of the GP population. In the parameterization assumed here only $|Gap| = 20$ exemplars are added to $DS(i)$ (by the Sampling Policy) at each generation. Hence, the label budget in this example would be:

$$\frac{1000 \times 20}{6,500,000} \approx 0.3\,\%$$

Given the rather different stream lengths of the benchmarking datasets (Table 18.1), different parameterizations for i_{max} will be assumed per dataset as follows:

Case 1 For Concept Drift and Electricity Demand streams: $i_{max} \in \{500; 1000\}$
Case 2 For Concept Shift and Forest Cover Type streams: $i_{max} \in \{1000; 10,000\}$

This defines two label budgets (LB) per dataset, as summarized by Table 18.2. Note that i_{max} is taken to include the pre-training budget i_{init} %.

Table 18.3 summarizes the remaining generic SBB parameter settings assumed in this study e.g., population size, variation and selection operator frequencies. The following new questions are addressed to illustrate the role of design decisions made during StreamSBB and have not previously been considered:

Let the initial period of fixed sliding window over 10 % of stream (during which the model is constructed) be called 'pre-training' stage and the period of non-overlapping sliding window over S_{max}/i_{max} exemplars be called 'post-training' stage. Two **pre-training bias** experiments are considered in an effort to assess the impact of initial model construction on performance attained across the rest of the

Table 18.2 Stream/datasets with different generation count and label budgets (*LB*)

Stream/Dataset	S_{max}	i_{max}	*LB*	i_{max}	*LB*
Gradual concept drift (drift)	150, 000	500	6.7 %	1000	13.3 %
Sudden concept shift (shift)	6, 500, 000	1, 000	0.3 %	10, 000	3.1 %
Electricity demand (elec)	45, 312	500	22.1 %	1000	44.2 %
Forest cover types (cover)	581, 012	1, 000	3.4 %	10, 000	34.4 %

Table 18.3 Generic SBB parameters

Parameter	Value
DS size (*DS*)	120
Host population size (M_{size})	120
Probability of symbiont deletion (p_d)	0.7
Probability of symbiont addition (p_a)	0.7
Probability of action mutation (μ_a)	0.1
Maximum symbionts per host (ω)	20
DS gap size (*Gap*)	20
Host population gap size (M_{gap})	60

Symbiont population varies dynamically, hence no size parameter is defined. SBB assumes a 'breeder' model of evolution in which M_{gap} hosts are removed per generation (Doucette et al. 2012b)

stream. In effect we are asking whether spending more resource on the pre-training period has a negative impact on the ability to react to the content in the post-training period (remainder of the stream). Specifically, we are interested in the impact of: (1) longer model construction time with same percentage of exemplar labels, and (2) more label budget during model construction period with the same amount of time.

DS oversampling reflects the ability of StreamSBB to decouple the rate at which GP training epochs are performed from the rate at which the data subset content is updated. Specifically, updates to data subset (*DS*) and window location (*SW*) are synchronized and therefore assume the same index, i (Fig. 18.1). Conversely, for each *DS*(i) we consider the case of performing multiple training epochs, or $j = i \times k$ where $k = 1$ implies one GP generation per *DS*(i), whereas a parameterization of $k = 5$ implies five GP generations per *DS*(i). Decoupling GP training epochs in this way potentially provides SBB more time to learn form each data subset. Note also that this does not change the label budget.

Monolithic vs. modular models: The original StreamSBB is allowed to represent/model each class label using an evolved mix of symbionts (programs). This implies that multiple programs might coevolve to represent the same class label (task decomposition) (Lichodzijewski and Heywood 2010; Doucette et al. 2012b). We are interested in investigating the contribution of such open-ended modularity under the context of non-stationary streaming tasks. To do so, we introduce a constrained version of StreamSBB in which a host cannot have more than one symbiont (program) with the same action (class label). All other properties are unchanged; hereafter this will be referred to as the monolithic model.

18.4.3 Evaluation

Two performance metrics will be adopted for characterizing performance: a "prequential accuracy" and "incremental class-wise detection rate". Such metrics are applied relative to the champion individual assumed for labelling stream content. Likewise two comparator classifiers are assumed for the purpose of comparison: a "no-change" model (Bifet et al. 2013) and an Adaptive Naive Bayes classifier with fixed label budgeting (Žliobaitė et al. 2014). A summary of each follows:

18.4.3.1 Performance Metrics

Prequential accuracy (Dawid 1984) represents the most widely used performance metric for streaming data benchmarking. Specifically, the prequential accuracy at exemplar t in the stream is 'weighted' relative to all past $t - 1$ exemplars as well as exemplar t, or

$$preq_t = \frac{(t-1) \times preq_{t-1} + R_t}{t} \tag{18.5}$$

where $R_t = 1$ denotes a correct classification of exemplar t, and $R_t = 0$ denotes otherwise. The ratio of time indexes acts as a weighting factor, enforcing a decay for older updates (Gama et al. 2013). The resulting prequential accuracy takes the form of a curve, although current benchmarking practice also tends to emphasize the reporting of the final prequential accuracy estimate for $t = S_{max}$ as the indication of overall model quality.

A second performance metric, **incremental class-wise detection rate** will also be assumed. The basic motivation is to reduce the sensitivity of the performance metric to class imbalance. This is particularly important under streaming data situations as models are updated incrementally and therefore sensitive to the distribution of current window content (typically a skewed distribution of classes even when the overall class distribution is balanced). The incremental class-wise detection rate can be estimated directly from stream content as follows:

$$DR(t) = \frac{1}{C} \sum_{c=[1,\ldots,C]} DR_c(t)$$

$$DR_c(t) = \frac{tp_c(t)}{tp_c(t) + fn_c(t)} \tag{18.6}$$

where t is the exemplar index, $tp_c(t), fn_c(t)$ are the respective running totals for true positive and false negative rates up to this point in the stream.

18.4.3.2 Comparator Models

The **No-Change Classifier** requires complete label information, but represents a naive 'devils advocate' solution. The no-change 'classifier' is actually a 1-bit finite state machine in which the state is seeded by the class label, $l(t) = c$, for the present exemplar, $\mathbf{x}(t)$. The state machine 'predicts' this class for the next exemplar(s) until there is a change in the exemplar class label. A change in the label for exemplar $\mathbf{x}(t+n)$ to $l(t+n) \neq c$ results in a change in the 'prediction' for exemplar $\mathbf{x}(t+n+1)$ to that of the new class, say, c'. The process then repeats with each change in class label for the current exemplar being assumed as the prediction for the next exemplar. Such a predictor achieves a high accuracy when there are continuous sequences of exemplars in the stream with the same label. Naturally, such a no-change classifier provides a 'feel' for how much implicit class variation exists in a stream.

The second comparator classifier is documented in a recent study of streaming data classification under label budgets and drift detection (Žliobaitė et al. 2014), and has been made available in the Massive Online Analysis (MOA) toolbox.[3] Specifically, the **Naive Bayes** classifier with budgeted active learning and drift detection under the prequential evaluation task is employed. The drift detection mechanism selected was the DDM (Drift Detection Method) algorithm from Gama et al. (2004) with default values for threshold (1) and step parameters (0.01). The 'random' active learning strategy was selected as it provided the baseline in Žliobaitė et al. (2014) and is closest to the stochastic sampling policy adopted in this work. Finally the label budget parameter was selected according to the label budget settings of the StreamSBB method. Thus, a new classifier is built when the current classifier's performance begins to degrade, i.e. the current classifier is replaced when drift is explicitly detected by the DDM. Active learning with budgeting is managed under a random exemplar selection policy in which stream data is queried for labels with frequency set by the budget parameter. We also note that both the Naive Bayes and the StreamSBB classifier are configured to label exemplars based on each exemplar instance. Thus, no use is made of features designed to represent temporal properties such as tapped delay lines (see Vahdat et al. 2014 for StreamSBB configured under this scenario).

18.5 Results

Section 18.4.2 discussed configuration of StreamSBB in terms of the generic parameterization, and higher level design decisions. We start by adopting the basic parameterization decisions for the duration of pre-training, label budget, sliding window size, gap size and illustrate the utility of the stream performance metrics (Sect. 18.4.3.1). A common minimal label parameterization is then adopted to

[3]MOA prerelease 2014.03; http://moa.cms.waikato.ac.nz/overview/.

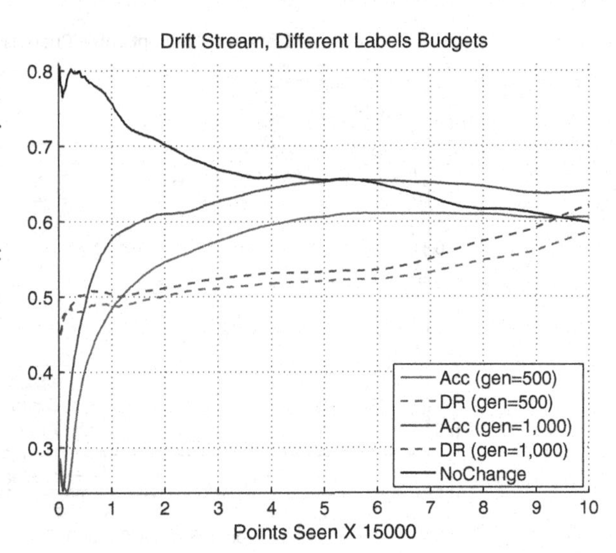

Fig. 18.2 StreamSBB on gradual concept **drift** stream. Curve of GP accuracy (*solid*) and DR (*dash*) during stream. First 10 % of stream (1.5×10^4 exemplars) are used to construct the initial model. *Red*: Label budget $\approx 6.7\,\%$ or $i_{max} = 500$; *Blue*: Label budget of $\approx 13.3\,\%$ or $i_{max} = 1000$; *Black*: No-Change model (Color figure online)

enable us to review the impact of making higher level design decisions regarding: model initialization, oversampling, and support for modularity in GP. Having established a preferred set of design decisions, we then compare against the Adaptive Naive Bayesian classifier from the MOA toolbox.

Model initialization and label budget: Figures 18.2, 18.3, 18.4, and 18.5 provide a behavioural summary of StreamSBB in terms of prequential accuracy (Eq. (18.5)) and incremental detection rate (Eq. (18.6)) w.r.t. different labelling budgets (Table 18.2) over the four datasets. In all cases each curve is the result of averaging the performance of GP over 50 runs for each configuration. A common parameterization will be assumed throughout for the generic GP parameters of StreamSBB (Table 18.3). Given that the sliding window follows a non-overlapping definition, then the size of the window is effectively parameterized by the label budget and corresponding frequency of gap sampling (Table 18.2). Likewise, the number of labels per sliding window location is fixed ($|Gap| = 20$) for a data subset of $|DS| = 120$. Moreover, at this point we will not consider the impact of more advanced features (oversampling etc), thus one training epoch is performed per window location.

As mentioned during Sect. 18.4.2, the first 10 % of the stream is made available for initial model construction.[4] The performance curves reflect the operation of the champion classifier (Sect. 18.3.3) as the stream data passes. Insight into the degree of mixing of class labels as the stream progresses is provided by the 'no-change' classifier curve (black solid curve). Thus, the artificial **drift** dataset begins with continuous sequences of the same class and then experiences a 20 % reduction

[4]Implying that 10 % of the label budget is consumed in pre-training.

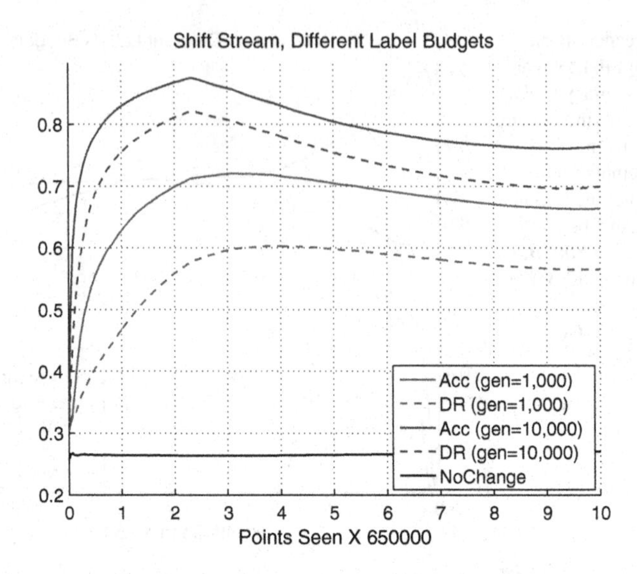

Fig. 18.3 StreamSBB on sudden concept **shift** stream. Curve of GP accuracy (*solid*) and DR (*dash*) during stream. First 10 % of stream (6.5 × 10^5 exemplars) are used to construct the initial model. *Red*: Label budget ≈ 0.3 % or i_{max} = 1000; *Blue*: Label budget of ≈ 3.1 % or i_{max} = 10, 000; *Black*: No-Change model (Color figure online)

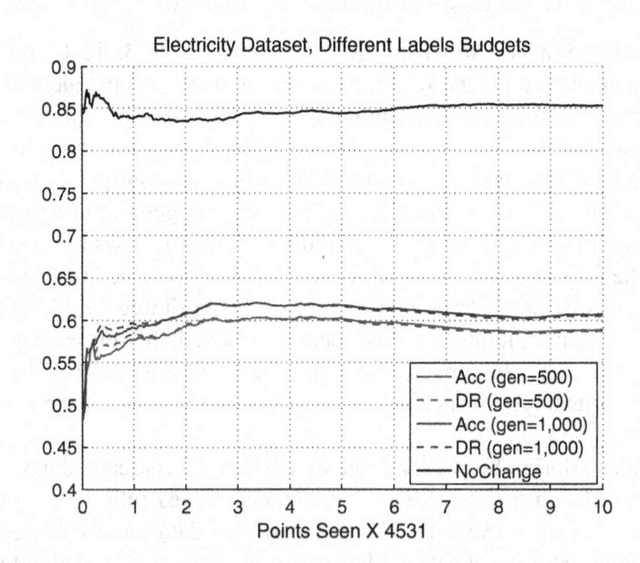

Fig. 18.4 StreamSBB on **electricity** dataset. Curve of GP accuracy (*solid*) and DR (*dash*) during stream. First 10 % of stream (4.5 × 10^3 exemplars) are used to construct the initial model. *Red*: Label budget ≈ 22.1 % or i_{max} = 500; *Blue*: Label budget of ≈ 44.2 % or i_{max} = 1000; *Black*: No-Change model (Color figure online)

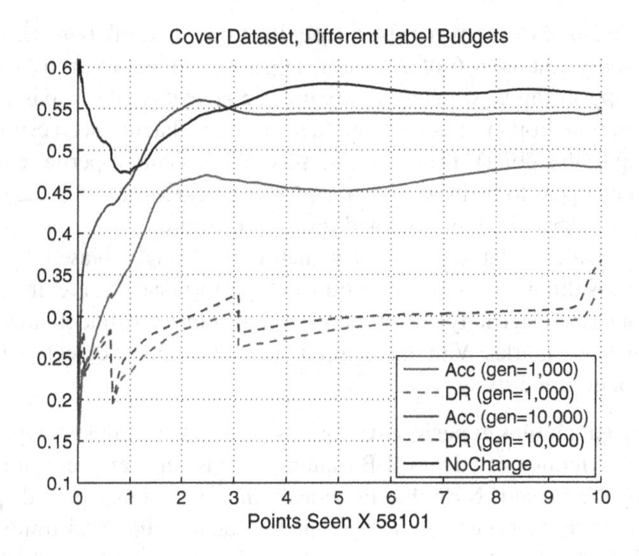

Fig. 18.5 StreamSBB on **cover type** dataset. Curve of GP accuracy (*solid*) and DR (*dash*) during stream. First 10 % of stream (5.8×10^4 exemplars) are used to construct the initial model. *Red*: Label budget ≈ 3.4 % or $i_{max} = 1000$; *Blue*: Label budget of ≈ 34.4 % or $i_{max} = 10,000$; *Black*: No-Change model (Color figure online)

as the stream progresses (Fig. 18.2). However, the artificial **shift** dataset experiences a high degree of mixing of class label throughout (Fig. 18.3). As previously been pointed out (Bifet et al. 2013), the **electricity** dataset has a low degree of mixing (Fig. 18.4) whereas the **cover type** dataset sees both periods of variation and continuity in the label during the stream (Fig. 18.5).

During the artificial concept **drift** dataset, steady gradual improvements are made throughout the course of the stream, resulting in performance eventually surpassing/reaching that of the no-change classifier under both metrics. Note that the no-change classifier performance is always described in terms of accuracy. The decaying trend of the no-change classifier implies that label mixing increases as the stream progresses.

Under the artificial **shift** dataset, there is insufficient continuity in labels for the no-change classifier to approach the performance of StreamSBB. Note also that under the shift dataset, the first 1,500,000 exemplars of the stream are drawn from concept $C1$. This lasts for nearly twice as long as the initial period of pre-training. Thus, as the concept generating the five classes shifts from $C1$ to $C2$ during the remaining course of the stream, a decay of ≈ 10 % in either metric appears irrespective of total label budget.

The accuracy and detection rate curves for GP are almost identical for the **electricity** demand dataset and quickly reach a plateau under this configuration of StreamSBB (Fig. 18.4). Performance of the no-change classifier benefits from continuous sequences of exemplars carrying the same class label. Note that the electricity dataset is a 2 class dataset with 58 %–42 % class distribution (almost balanced). Finally, the **forest cover type** dataset illustrates several dynamic properties. Pre-training only results in 6 of 7 classes appearing. Thus, when instances of the 7th class do appear (approximately where index 3 appears in the *x*-axis), then there is a corresponding drop in detection rate (detection rate having been estimated over six classes up to this point). There are also several transient properties earlier in the stream, which appear to be indicative of sudden context switches in the underlying process as they impact both metrics and forms of classifier.

It is also evident that the accuracy metric is strongly biased by the class distribution, resulting in an 'over optimistic' performance curve in all but the balanced data set (electricity); a property widely observed under non-streaming classification benchmarks. With this in mind, we will adopt the detection rate metric in the remainder of the study.

Pre-training epoch bias experiment: Figures 18.6 and 18.7 illustrate the impact of varying the distribution of StreamSBB training epochs between pre-training and the remainder of the stream. Note that previously the pre-training period (performed against 10 % of the data) consumed an equal amount of the total training epochs (10 %). In the case of these experiments, pre-training is still performed against 10 % of the data (thus still only utilizing a 10 % label budget), but consumes 20 % or 40 % of the training epochs. Naturally, the total number of training epochs per stream is unchanged, leading to the following three configurations:

Case 1 The default case, i.e. 10 % of evolution time is dedicated to model construction and 90 % after.
Case 2 20 % of evolution time is dedicated to model construction and 80 % after.
Case 3 40 % of evolution time is dedicated to model construction and 60 % after.

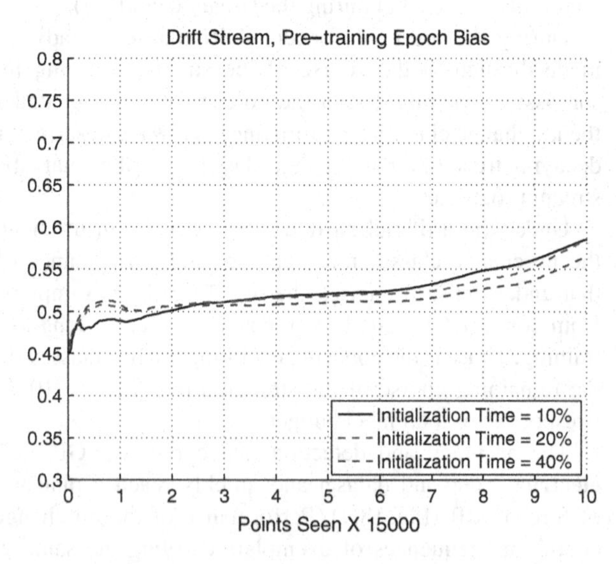

Fig. 18.6 Pre-training epoch bias experiment. *preqDR* on **drift** stream. *Black*: default 10 % (case 1); *Red*: 20 % (case 2); and *Blue*: 40 % (case 3) (Color figure online)

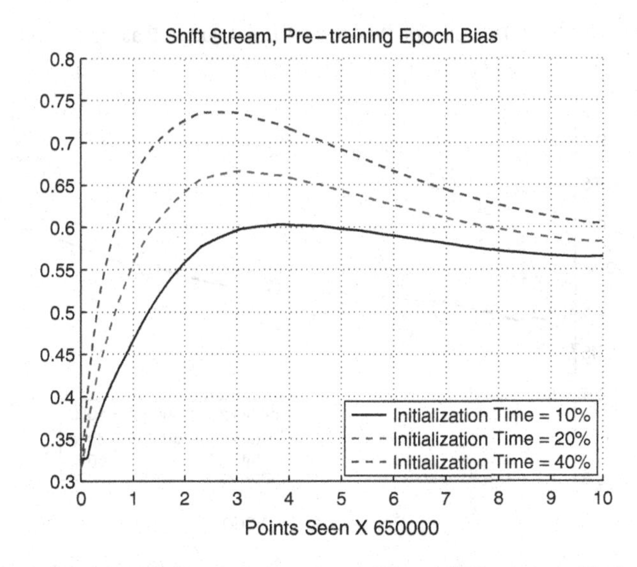

Fig. 18.7 Pre-training epoch bias experiment. *preqDR* on **shift** stream. *Black*: default 10 % (case 1); *Red*: 20 % (case 2); and *Blue*: 40 % (case 3) (Color figure online)

It appears that there is no lasting benefit to be gained from biasing more training epochs to the pre-training period. Under the **shift** dataset (Fig. 18.7), a significant regression back to the vicinity of 'case 1' detection rate appears by the end of the stream. In short, the benefit gained by introducing biases towards more pre-training time are lost over the remainder of the stream.[5]

Pre-training label budget bias experiment: Here we ask wether using more of the label budget during pre-training will provide the basis for better models during the remainder of the stream. Specifically, rather than providing more time to the pre-training period we provide a larger proportion of label budget to the pre-training period. Note that the overall label budget is intact, however more labels are requested during model construction and less thereafter. This leads to the following three configurations:

Case 1 The default case of uniform sampling through the stream, i.e. 10 % of label and training budget is requested during construction and 90 % after.

Case 2 20 % of label and training budget is requested during model construction and 80 % after.

Case 3 40 % of label and training budget is requested during model construction and 60 % after.

[5]Electricity demand and forest cover type datasets observed similar effects and therefore results are not explicitly reported.

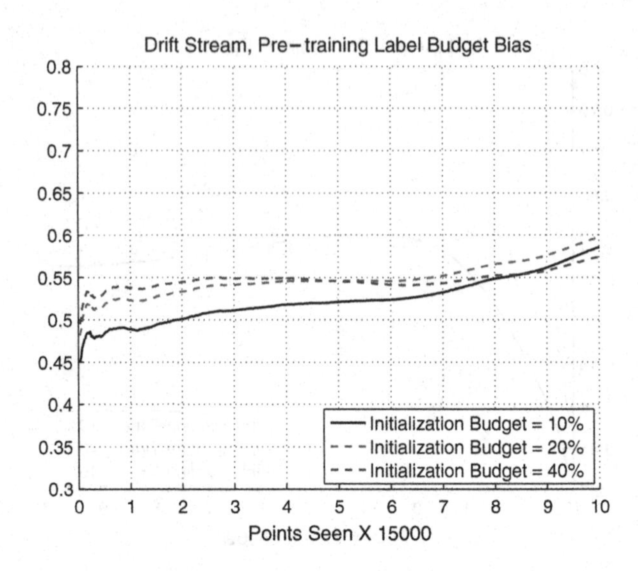

Fig. 18.8 Pre-training label budget bias experiment. *preqDR* on **drift** stream. *Black*: default 10 % (case 1); *Red*: 20 % (case 2); and *Blue*: 40 % (case 3) (Color figure online)

Figures 18.8 and 18.9 summarize the impact of pre-training label budget bias on detection rate for the concept **drift** and **shift** datasets respectively. In both cases the results are very similar to those observed under the bias to training epochs.[6] Any improvement relative to the drift data set (compare Fig. 18.8 to Fig. 18.6) being lost under the shift data set (compare Fig. 18.9 to Fig. 18.7). In short, no real benefit is observed in biasing more training generations or label budget towards the initial pre-training period.

DS Oversampling experiment: Sections 18.3 and 18.4.2 made the case for relaxing the relation between updates to the data subset ($DS(i)$) and performing a training epoch ($Gen(j)$, Fig. 18.1). Thus, for each update to the data subset rather than conduct a single generation, multiple generations might be performed. This does not change the label budget and does not represent a bias towards pre-training as it is performed throughout the whole stream. Three parameterizations are considered:

[6] Similar effects being observed for the electricity and forest cover type datasets.

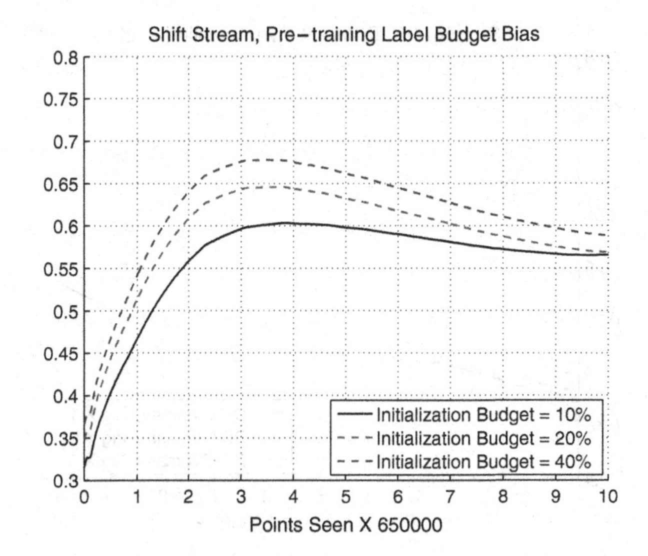

Fig. 18.9 Pre-training label budget bias experiment. *preqDR* on **shift** stream. *Black*: default 10 % (case 1); *Red*: 20 % (case 2); and Blue: 40 % (case 3) (Color figure online)

Case 1 The default case of uniform sampling through the stream.
Case 2 DS oversampling by a factor of 2.
Case 3 DS oversampling by a factor of 5.

Figures 18.10 and 18.11 illustrate the impact of oversampling in terms of incremental detection rate curves for concept **drift** and **shift** streams. Higher detection rates are now maintained throughout the stream. Indeed, the higher rate of oversampling appears to be preferable throughout, although further increases to the oversampling (a factor of 10) only had marginal effects compared to the case of oversampling to a factor of 5 (overlearning). Results for electricity and cover type were also positive and will be reported later when we compare with the Adaptive Naive Bayesian framework for streaming classification.

Monolithic vs. modular: In all the previous experiments StreamSBB was used in its original modular configuration, i.e. the number of symbionts per class labels were allowed to freely evolve. Previous benchmarking performed under a classical non-streaming setting of classification through supervised learning indicated that such open-ended evolution of modularity was particularly beneficial (Lichodzijewski and Heywood 2010). Other researchers have also noted that the open-ended evolution of modularity can be potentially beneficial in 'dynamic' environments (Sect. 18.2). In this experiment we compare StreamSBB with a modified version in which there can only be a single program per team per class. Although, still modular at some

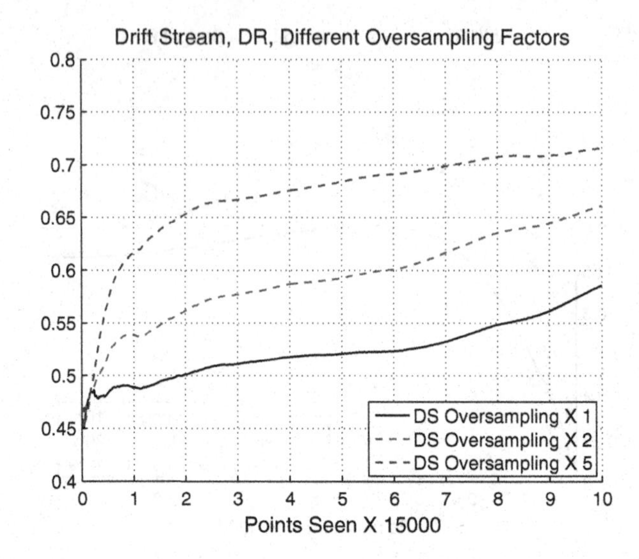

Fig. 18.10 Oversampling experiment. *preqDR* on **drift** stream. *Black*: default sampling; *Red*: ×2 oversampling; and *Blue*: ×5 oversampling (Color figure online)

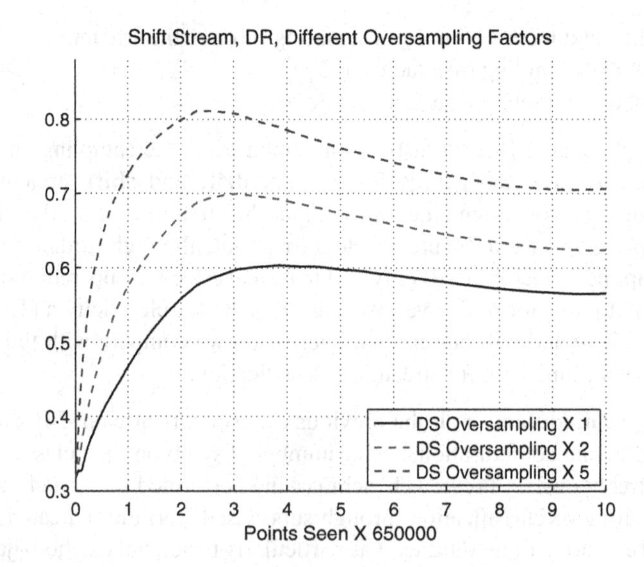

Fig. 18.11 Oversampling experiment. *preqDR* on **shift** stream. *Black*: default sampling; *Red*: ×2 oversampling; and *Blue*: ×5 oversampling (Color figure online)

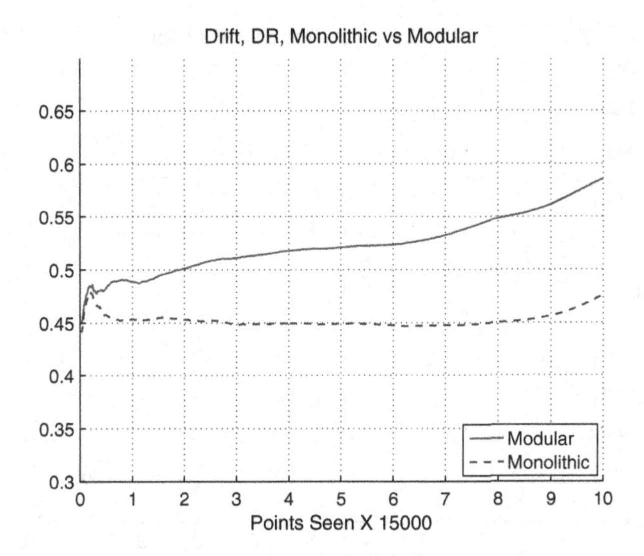

Fig. 18.12 Detection rate of concept **drift** stream under modular (*solid*) vs. monolithic (*dash*) configurations

level (there are as many programs as classes) we will refer to this as a monolithic classifier, and StreamSBB as a modular classifier.[7] The objective of this experiment is to quantify to what extent support for such open-ended evolution of modularity is beneficial under non-stationary streaming classification tasks.

Figures 18.12 and 18.13, summarize detection rate on the concept **drift** and **shift** streams respectively. There is a statistically significant difference in favour of assuming open-ended evolution of modularity (Student T-test p-value of 3.14×10^{-237} and 0 for concept drift and concept shift streams respectively under 0.01 significance level). Thus, modularity is synonymous with task decomposition, which under domains that undergo change potentially implies that only a subset of programs within a modular solution need to be revised when a change occurs. Conversely, under monolithic solutions, it is more difficult to explicitly delimit what variation operators modify, hence identifying the relevant parts of a program to modify becomes more difficult.

Under non-stationary streams, given that change takes place throughout the stream we can also review the development of age of the champion individual through the course of the stream. Note that the champion solution is the individual used to *provide* labels as the stream progresses, with identification of the champion performed relative to the current content of the data subset (Sect. 18.3.3). Naturally, each time the data subset is updated, the champion might change. Age is the number of training epochs for which an individual manages to exist.

[7]Other than the monolithic formulation of SBB being subject to the constraint that only one program may represent each class, the two implementations are the same.

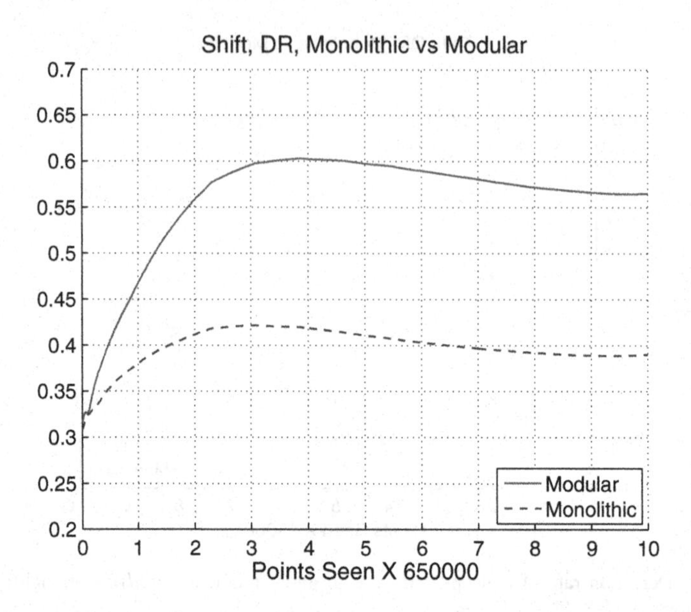

Fig. 18.13 Detection rate of concept **shift** stream under modular (*solid*) vs. monolithic (*dash*) configurations

Figures 18.14 and 18.15 illustrate the average age of the champion for concept **drift** and **shift** streams respectively. Both plots suggest that the average age of champion hosts of the modular configuration (red curve) remains lower throughout the stream. In effect, there is a higher rate of turn over of champion individuals when modularity is supported (implied by the lower age of champions). This reflects a stronger ability to react to change. Conversely, the much higher age of champions under the monolithic framework appears to indicate that the same champion has to be used for longer before better replacements are found. This has obvious decremental consequences for classifier performance.

StreamSBB vs. Adaptive Naive Bayes: Results for the second baseline classifier, the Adaptive Naive Bayesian (ANB) framework for streaming data classification under label budgets as implemented in the MOA toolkit (Sect. 18.4.3.2) are summarized in Figs. 18.16, 18.17, 18.18, 18.19. Detection rate of StreamSBB and ANB models with similar label budgets are compared against each other for the four datasets.

StreamSBB results are presented in terms of a set of curves illustrating the impact of assuming different DS oversampling rates, i.e. the number of training epochs performed per DS update. As noted in the earlier experiments this appears to be the most important design decision and has no impact on the label budget. In all cases the solid black curve is the incremental detection rate for ANB. Under the **drift** stream, ANB appears to take longer to develop an initial classifier. Thereafter, both models alternate until the end of stream where they settle on the same detection rate.

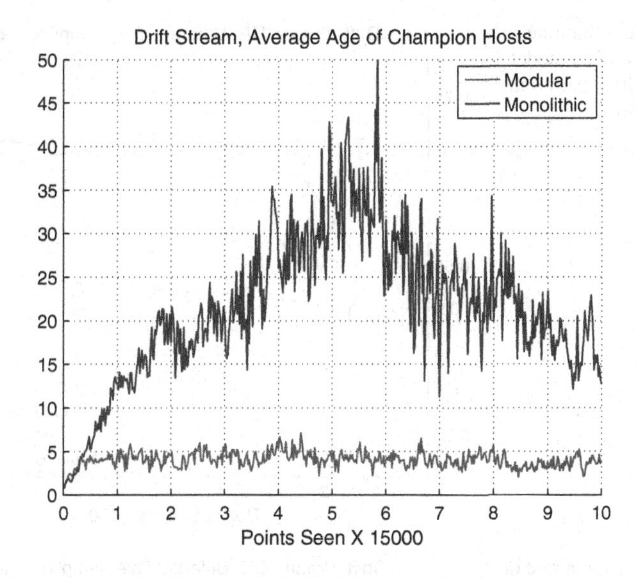

Fig. 18.14 Average age of the champion individual during evolutionary loop, concept **drift** stream

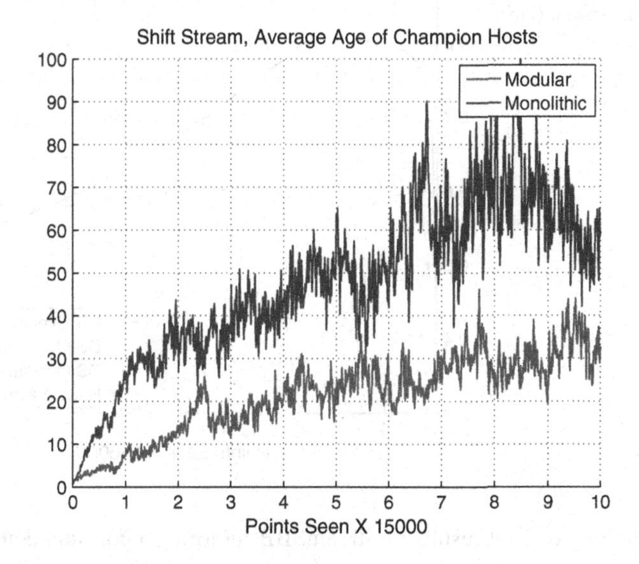

Fig. 18.15 Average age of the champion individual during evolutionary loop, concept **shift** stream

Conversely, ANB appears to over learn the initial configuration of the **shift** stream (corresponding to concept $C1$), and then lose 25 % of its initial detection rate (going from 80 % to 55 %) over the remaining 75 % of the stream. Conversely, StreamSBB experiences significantly less loss over the course of concept $C2$ being introduced during the last three quarters of the stream (Fig. 18.17).

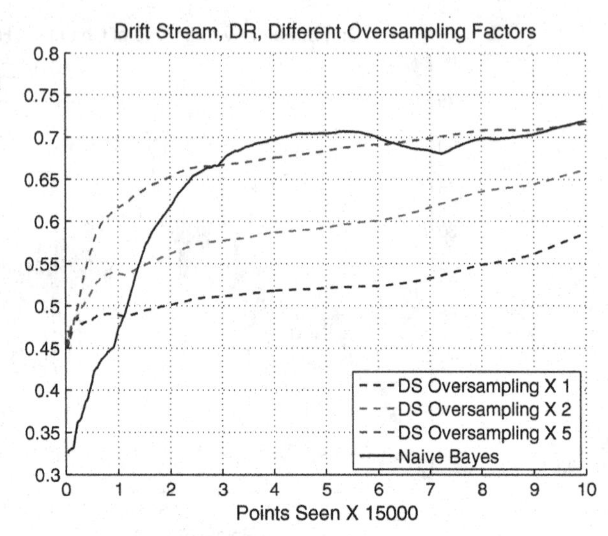

Fig. 18.16 Detection rate of StreamSBB vs. ANB model on concept **drift** stream (label budget = 6.6 %)

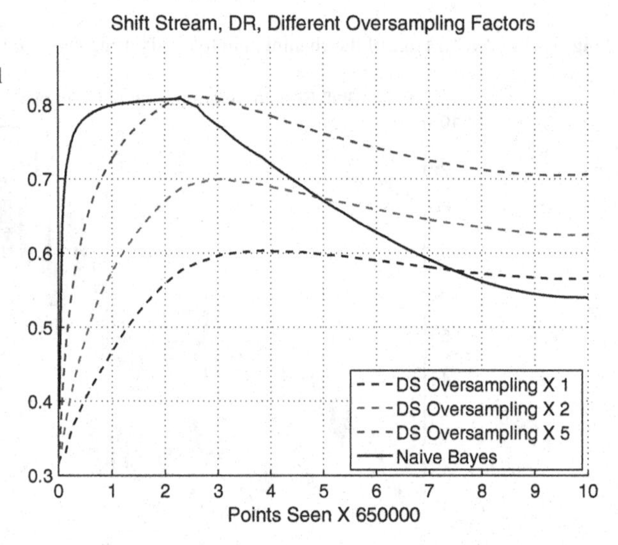

Fig. 18.17 Detection rate of StreamSBB vs. ANB model on concept **shift** stream (label budget = 0.3 %)

The **electricity** dataset results in StreamSBB returning a constant detection rate of 58–60 % throughout the stream, but never approaching the performance returned by ANB (Fig. 18.18). The behaviour under the **covertype** dataset is more interesting. During the first three intervals of the stream, both models undergo sharp changes in the detection rate (Fig. 18.19). A sudden drop then appears as the seventh class is encountered for the first time, and therefore all models miss-classify this class (see x-axis value $\approx$ 3). The ensuing gradual recovery of the detection rate undergoes a final jump in the last 20 000 exemplars of the sequence (last interval of the x-axis).

In summary ANB appears to have problems when there are sudden changes to the content of the stream (shift stream), whereas both algorithms are effective under

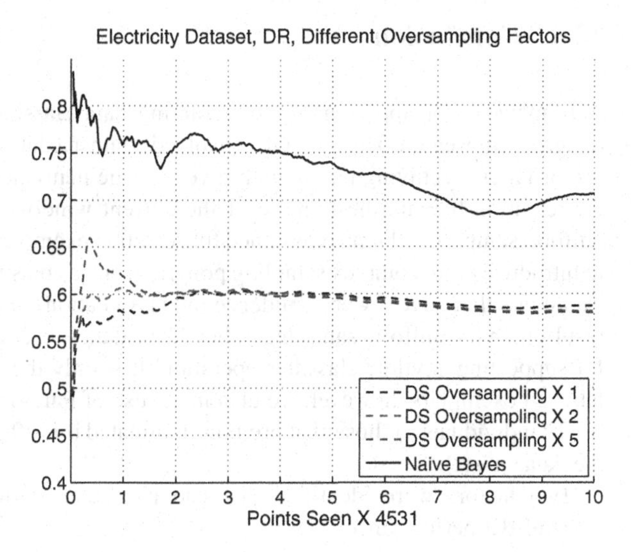

Fig. 18.18 Detection rate of StreamSBB vs. ANB model on **electricity** dataset (label budget = 22.1 %)

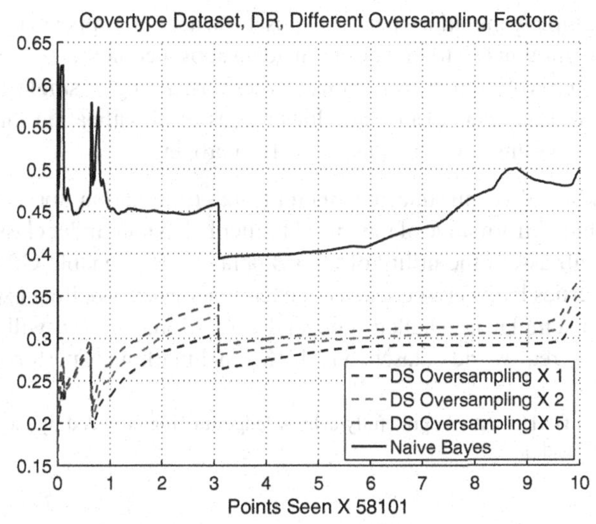

Fig. 18.19 Detection rate of StreamSBB vs. ANB model on **covertype** dataset (label budget = 3.4 %)

the drift stream. In both real-world datasets, ANB was more effective, however, StreamSBB might well benefit from the use of tapped delay lines when contracting models for such tasks (both models make classification decisions on the basis of a single exemplar). Further research indicates that StreamSBB does indeed benefit considerably from the use of a delay lines on real-world data sets (Vahdat et al. 2015).

18.6 Conclusion

A framework for applying GP to streaming data classification tasks under label budgets is presented. To do so, GP is evolved against a data subset. The subset makes use of Pareto archiving policy with diversity/age heuristics to prioritize exemplars for retention beyond the lifetime of the current window to the stream. A simple uniform sampling scheme is assumed for selecting exemplars for labelling. Attempts to introduce more complex sampling policies (such as biasing label requests towards exemplars that have lower confidence in classification) generally resulted in worse results than the uniform sampling policy.[8] The data subset was also used as the basis for supporting anytime classifier operation. It is only the data subset that contains labelled exemplars, however, we also make use of Pareto archiving to limit the set of GP individuals to those that are non-dominated (cf. effect of class imbalance on the data subset).

Two factors were identified that had particular significance with respect to StreamSBB performance:

- perform multiple generations per data subset—where this appears to improve the rate of adaptation in GP to updates to the data subset content.
- support for coevolution of programs—where assuming a single (monolithic) program per class resulted in much lower rates of classification than when the number of programs per class was an evolved property.

Benchmarking also introduced a incremental formulation for detection rate where this is more informative of the true classifier behaviour under class imbalance. Future work will assess the utility of tapped delay lines to enable GP to represent temporal properties between sequences of exemplars when labelling exemplar t. At present, each exemplar is labelled independently. Moreover, we will continue to extend the set of real-world datasets on which benchmarking is performed.

Acknowledgements The authors gratefully acknowledge funding provided by the NSERC CRD grant program (Canada).

References

A. Atwater and M. I. Heywood. Benchmarking Pareto archiving heuristics in the presence of concept drift: Diversity versus age. In *ACM Genetic and Evolutionary Computation Conference*, pages 885–892, 2013.

A. Atwater, M. I. Heywood, and A. N. Zincir-Heywood. GP under streaming data constraints: A case for Pareto archiving? In *ACM Genetic and Evolutionary Computation Conference*, pages 703–710, 2012.

K. Bache and M. Lichman. UCI machine learning repository, 2013.

[8]Results not shown for brevity.

M. Behdad and T. French. Online learning classifiers in dynamic environments with incomplete feedback. In *IEEE Congress on Evolutionary Computation*, pages 1786–1793, 2013.
A. Bifet. *Adaptive Stream Mining: Pattern Learning and Mining from Evolving Data Streams*, volume 207 of *Frontiers in Artificial Intelligence and Applications*. IOS Press, 2010.
A. Bifet and R. Gavalda. Learning from time-changing data with adaptive windowing. In *SIAM International Conference on Data Mining*, pages 443–448, 2007.
A. Bifet, I. Žliobaitė, B. Pfahringer, and G. Holmes. Pitfalls in benchmarking data stream classification and how to avoid them. In *Machine Learning and Knowledge Discovery in Databases*, volume 8188 of *LNCS*, pages 465–479, 2013.
T. Blackwell and J. Branke. Multiswarms, exclusion, and anti-convergence in dynamic environments. *IEEE Transactions on Evolutionary Computation*, 10(4):459–472, 2006.
G. Brown and L. I. Kuncheva. "Good" and "bad" diversity in majority vote ensembles. In *Multiple Classifier Systems*, volume 5997 of *LNCS*, pages 124–133, 2010.
T. Dasu, S. Krishnan, S. Venkatasubramanian, and K. Yi. An information-theoretic approach to detecting changes in multi-dimensional data streams. In *Proceedings of the Symposium on the Interface of Statistics*, 2006.
A. P. Dawid. Statistical theory: The prequential approach. *Journal of the Royal Statistical Society-A*, 147:278–292, 1984.
E. D. de Jong. A monotonic archive for pareto-coevolution. *Evolutionary Computation*, 15(1):61–94, 2007.
I. Dempsey, M. O'Neill, and A. Brabazon. *Foundations in Grammatical Evolution for Dynamic Environments*, volume 194 of *Studies in Computational Intelligence*. Springer, 2009.
G. Ditzler and R. Polikar. Hellinger distance based drift detection for non-stationary environments. In *IEEE Symposium on Computational Intelligence in Dynamic and Uncertain Environments*, pages 41–48, 2011.
J. A. Doucette, P. Lichodzijewski, and M. I. Heywood. Hierarchical task decomposition through symbiosis in reinforcement learning. In *ACM Genetic and Evolutionary Computation Conference*, pages 97–104, 2012a.
J. A. Doucette, A. R. McIntyre, P. Lichodzijewski, and M. I. Heywood. Symbiotic coevolutionary genetic programming: a benchmarking study under large attribute spaces. *Genetic Programming and Evolvable Machines*, 13(1), 2012b.
W. Fan, Y. Huang, H. Wang, and P. S. Yu. Active mining of data streams. In *Proceedings of SIAM International Conference on Data Mining*, pages 457–461, 2004.
G. Folino and G. Papuzzo. Handling different categories of concept drift in data streams using distributed GP. In *European Conference on Genetic Programming*, volume 6021 of *LNCS*, pages 74–85, 2010.
J. Gama. *Knowledge discovery from data streams*. CRC Press, 2010.
J. Gama. A survey on learning from data streams: Current and future trends. *Progress in Artificial Intelligence*, 1(1):45–55, 2012.
J. Gama, P. Medas, G. Castillo, and P. P. Rodrigues. Learning with drift detection. In *Advances in Artificial Intelligence*, volume 3171 of *LNCS*, pages 66–112, 2004.
J. Gama, R. Sebastião, and P. Rodrigues. On evaluating stream learning algorithms. *Machine Learning*, 90(3):317–346, 2013.
M. Harries. Splice-2 comparative evaluation: Electricity pricing. Technical report, University of New South Wales, 1999.
M. I. Heywood. Evolutionary model building under streaming data for classification tasks: opportunities and challenges. *Genetic Programming and Evolvable Machines*, 2015. DOI 10.1007/s10710-014-9236-y.
S. Huang and Y. Dong. An active learning system for mining time changing data streams. *Intelligent Data Analysis*, 11(4):401–419, 2007.
N. Kashtan, E. Noor, and U. Alon. Varying environments can speed up evolution. *Proceedings of the National Academy of Sciences*, 104(34):13713–13716, 2007.
D. Kifer, S. Ben-David, and J. Gehrke. Detecting change in data streams. In *Proceedings of the International Conference on Very Large Data Bases*, pages 180–191. Morgan Kaufmann, 2004.

C. Lanquillon. Information filtering in changing domains. In *Proceedings of the International Joint Conference on Artificial Intelligence*, pages 41–48, 1999.

P. Lichodzijewski and M. I. Heywood. Managing team-based problem solving with Symbiotic Bid-based Genetic Programming. In *ACM Genetic and Evolutionary Computation Conference*, pages 363–370, 2008.

P. Lichodzijewski and M. I. Heywood. Symbiosis, complexification and simplicity under GP. In *ACM Genetic and Evolutionary Computation Conference*, pages 853–860, 2010.

P. Lindstrom, B. MacNamee, and S. J. Delany. Handling concept drift in a text data stream constrained by high labelling cost. In *Proceedings of the International Florida Artificial Intelligence Research Society Conference*. AAAI, 2010.

P. Lindstrom, B. MacNamee, and S. J. Delany. Drift detection using uncertainty distribution divergence. *Evolutionary Intelligence*, 4(1):13–25, 2013.

L. L. Minku, A. P. White, and X. Yao. The impact of diversity on online ensemble learning in the presence of concept drift. *IEEE Transactions on Knowledge and Data Engineering*, 22(5):730–742, 2010.

M. Parter, N. Kashtan, and U. Alon. Facilitated variation: How evolution learns from past environments to generalize to new environments. *PLoS Computational Biology*, 4(11):e1000206, 2008.

J. Quinonero-Candela, M. Sugiyama, A. Schwaighofer, and N. D. Lawrence, editors. *Dataset shift in machine learning*. MIT Press, 2009.

R. Sebastio and J. Gama. Change detection in learning histograms from data streams. In *Proceedings of the Portuguese Conference on Artificial Intelligence*, volume 4874 of *LNCS*, pages 112–123. Springer, 2007.

R. Stapenhurst and G. Brown. Theoretical and empirical analysis of diversity in non-stationary learning. In *IEEE Symposium on Computational Intelligence in Dynamic and Uncertain Environments*, pages 25–32, 2011.

I. Žliobaitė, A. Bifet, B. Pfahringer, and G. Holmes. Active learning with evolving streaming data. In *Proceedings of the European Conference on Machine Learning and Knowledge Discovery in Databases*, pages 597–612. Springer, 2011.

I. Žliobaitė, A. Bifet, B. Pfahringer, and G. Holmes. Active learning with drifting streaming data. *IEEE Transactions on Neural Networks and Learning Systems*, 25(1):27–54, 2014.

A. Vahdat, A. Atwater, A. R. McIntyre, and M. I. Heywood. On the application of GP to streaming data classification tasks with label budgets. In *ACM Genetic and Evolutionary Computation Conference: ECBDL Workshop*, pages 1287–1294, 2014.

A. Vahdat, J. Morgan, A. R. McIntyre, M. I. Heywood, and A. N. Zincir-Heywood. Tapped delay lines for GP streaming data classification with label budgets. In *European Conference on Genetic Programming*, volume 9025 of *LNCS*. Springer, 2015.

P. Vorburger and A. Bernstein. Entropy-based concept shift detection. In *Proceedings of the Sixth International Conference on Data Mining*, pages 1113–1118, 2006.

G. P. Wagner and L. Altenberg. Complex adaptations and the evolution of evolvability. *Complexity*, 50(3):433–452, 1996.

X. Zhu, P. Zhang, X. Lin, and Y. Shi. Active learning from stream data using optimal weight classifier ensemble. *IEEE Transactions on Systems, Man, and Cybernetics – Part B*, 40(6):1607–1621, 2010.

Part III
Hybrid Approaches

Chapter 19
A New Evolutionary Approach to Geotechnical and Geo-Environmental Modelling

Mohammed S. Hussain, Alireza Ahangar-asr, Youliang Chen, and Akbar A. Javadi

19.1 Introduction

Introduced by Koza (1992) Genetic programming (GP) is a relatively new data mining method that solves problems in a systematic and domain-independent method. By following the principles of the evolutionary computation, GP learns to discover the appropriate mathematical models to fit a set of points. The 'fitness' of the solutions in the population is improved through successive generations. This automated induction of mathematical models of data using GP is commonly referred to as symbolic regression. In contract to the traditional numerical regressions, GP or more precisely symbolic regression does not need pre-specification of the regression structure by the users, where both the form of expression and its parameter values are found automatically. The nature of this automatic approach permits global

---

Electronic supplementary material The online version of this chapter (doi: 10.1007/978-3-319-20883-1_19) contains supplementary material, which is available to authorized users.

M.S. Hussain • A.A. Javadi (✉)
Department of Engineering, Computational Geomechanics Group, University of Exeter, North Park Road, Exeter EX4 4QF, UK
e-mail: msh218@ex.ac.uk; a.a.javadi@ex.ac.uk

A. Ahangar-asr
School of Computing, Science and Engineering, University of Salford, Newton Building, Peel Park Campus, Salford, Greater Manchester, M5 4WT, UK
e-mail: a.ahangarasr@salford.ac.uk

Y. Chen
Department of Civil Engineering, University of Shanghai for Science and Technology, 516 Jungong Road, Shanghai 200093, People's Republic of China
e-mail: chenyouliang2001@yahoo.com.cn

A.H. Gandomi et al. (eds.), *Handbook of Genetic Programming Applications*,
DOI 10.1007/978-3-319-20883-1_19

exploration of expressions and allows the user to have an insight into the relationship between input and output data.

In a last decade the genetic programming has been widely used in different fields of civil engineering such as structural and material engineering (e.g. Gandomi et al. 2009, 2010a, b, 2013; Gandomi and Alavi 2012b, 2013; Gandomi and Yun 2014); geotechnical and earthquake engineering (e.g. Yang et al. 2004; Johari et al. 2006; Narendra et al. 2006; Rezania and Javadi 2007; Baykasoğlu et al. 2008; Alavi et al. 2009, 2010, 2012, 2013a, b; Alavi and Gandomi 2011, 2013; Mousavi et al. 2011; Gandomi et al. 2011a, b; Gandomi and Alavi 2012a) and hydrology and water resources engineering (e.g. Dorado et al. 2003; Babovic and Keijzer 2006; Gaur and Deo 2008; Makkeasorn et al. 2008; Parasuraman and Elshorbagy 2008; Wang et al. 2009; Sreekanth and Datta 2011a). This popularity of GP is attributed to its success at discovering complex nonlinear spaces and its robustness in practical applications. Accordingly, a significant reduction in computational time, less uncertainty and also being better suited to the simulation/optimization framework using adaptive search space have been reported as positive potentials of GP (Sreekanth and Datta 2010, 2011b).

Despite the advantages, GP also suffers from certain limitations, as it tends to produce functions that grow in length over time (Davidson et al. 1999). In an attempt to overcome these limitations, Davidson et al. (1999) introduced a new regression method for creating polynomial models based on both numerical and symbolic terms. The structure of the polynomial regressions and the values for the constants in the expressions are simultaneously captured by GP and least squares optimisation respectively. Following the same principle Giustolisi and Savic (2006) developed Evolutionary Polynomial Regression EPR as a hybrid method that integrates the best features of numerical regression with the effectiveness of GP technique. Therefore EPR as a data driven technique, belongs to the family of GP strategies. EPR has been applied to model complicated civil engineering materials and systems (e.g. Javadi et al. 2012b; Ahangar-Asr et al. 2012, 2014; Faramarzi et al. 2013).

The overall goal of this study is to depict the computational capability of the EPR methodology in predicating the results of two complex geotechnical and geo-environmental problems. In the first problem, EPR is used for modelling of thermo-mechanical behaviour of unsaturated soils by development a model for volumetric strain. The EPR model is developed and evaluated based on results from test data and is validated using cases of data that had been kept unseen to the EPR during the modelling process. This allows investigating the generalisation capabilities of the developed model. The EPR model predictions are compared with experimental measurements to evaluate the model performance in predicting the volumetric behaviour of soils and the level of accuracy of the predictions. In second problem, the EPR is trained on a set of numerical inputs/outputs and is used to predict the total mass of salt released in the aquifer during the seawater intrusion problem. The developed EPR mode is then integrated with a multi objective optimisation tool to assess the efficiency of a hydraulic barrier proposed to control the inland encroachment of seawater into the aquifer. The potential capability of developed simulation–optimization methodology is compared with different schemes of direct linking of numerical model with the same optimization tool.

19.2 Evolutionary Polynomial Regression

Evolutionary polynomial regression EPR is a data-driven method based on evolutionary computing, aimed to search for polynomial structures representing a system. EPR integrates numerical and symbolic regression to perform evolutionary polynomial regression. The strategy uses polynomial structures to take advantage of their favourable mathematical properties. The key idea behind the EPR is to use evolutionary search for exponents of polynomial expressions by means of a genetic algorithm (GA) engine. This allows (1) easy computational implementation of the algorithm, (2) efficient search for an explicit expression, and (3) improved control of the complexity of the expression generated (Giustolisi and Savic 2006). A physical system, having an output y, dependent on a set of inputs **X** and parameters $\boldsymbol{\theta}$, can be mathematically formulated as:

$$y = F(\mathbf{X}, \boldsymbol{\theta}) \tag{19.1}$$

where F is a function in an m-dimensional space and m is the number of inputs. To avoid the problem of mathematical expressions growing rapidly in length with time, in EPR the evolutionary procedure is conducted in the way that it searches for the exponents of a polynomial function with a fixed maximum number of terms. During one execution it returns a number of expressions with increasing numbers of terms up to a limit set by the user, to allow the optimum number of terms to be selected. The general form of expression used in EPR can be presented as (Giustolisi and Savic 2006):

$$y = \sum_{j=1}^{m} F\left(\mathbf{X}, f\left(\mathbf{X}\right), a_j\right) + a_0 \tag{19.2}$$

where y is the estimated vector of output of the process; a_j is a constant; F is a function constructed by the process; **X** is the matrix of input variables; f is a function defined by the user; and m is the number of terms of the target expression. The first step in identification of the model structure is to transfer Eq. (19.2) into the following vector form:

$$Y_{N\times 1}(\boldsymbol{\theta}, \mathbf{Z}) = \left[\mathbf{I}_{N\times 1}\ \mathbf{Z}^{j}_{N\times m}\right] \times \left[a_0\ a_1\ \cdots\ a_m\right]^T = \mathbf{Z}_{N\times d} \times \boldsymbol{\theta}^T_{d\times 1} \tag{19.3}$$

where $\mathbf{Y}_{N\times 1}(\boldsymbol{\theta},\mathbf{Z})$ is the least squares estimate vector of the N target values; $\boldsymbol{\theta}_{d\times 1}$ is the vector of $d = m + 1$ parameters a_j and a_0 ($\boldsymbol{\theta}^T$ is the transposed vector); and $\mathbf{Z}_{N\times d}$ is a matrix formed by **I** (unitary vector) for bias a_o, and m vectors of variables $\mathbf{Z}^j$. For a fixed j, the variables $\mathbf{Z}^j$ are a product of the independent predictor vectors of inputs, $\mathbf{X} = \langle \mathbf{X}1\ \mathbf{X}2\ \ldots\ \mathbf{X}k \rangle$.

In general, EPR is a two-stage technique for constructing symbolic models. Initially, using standard genetic algorithm (GA), it searches for the best form of the function structure, i.e. a combination of vectors of independent inputs, **Xs** = 1:k, and secondly it performs a least squares regression to find the adjustable parameters, **θ**, for each combination of inputs. In this way a global search algorithm is implemented for both the best set of input combinations and related exponents simultaneously, according to the user-defined cost function (Giustolisi and Savic 2006). The adjustable parameters, a_j, are evaluated by means of the linear least squares (LS) method based on minimization of the sum of squared errors (SSE) as the cost function. The SSE function which is used to guide the search process towards the best fit model is as follows:

$$\mathbf{SSE} = \frac{\sum_{i=1}^{N} (y_a - y_p)^2}{N} \tag{19.4}$$

where y_a and y_p are the target experimental and the model prediction values respectively. The global search for the best form of the EPR equation is performed by means of a standard GA over the values in the user defined vector of exponents. The GA operates based on Darwinian evolution which begins with random creation of an initial population of solutions. Each parameter set in the population represents the individual's chromosomes. Each individual is assigned a fitness based on how well it performs in its environment. Through crossover and mutation operations, with the probabilities Pc and Pm respectively, the next generation is created. Fit individuals are selected for mating, whereas weak individuals die off. The mated parents create a child (offspring) with a chromosome set which is a mix of parents' chromosomes. In EPR integer GA coding with single point crossover is used to determine the location of the candidate exponents. The EPR process stops when the termination criterion, which can be either the maximum number of generations, the maximum number of terms in the target mathematical expression or a particular allowable error, is satisfied. A typical flow diagram for the EPR procedure is illustrated in Fig. 19.1.

19.3 Applications

19.3.1 Problem 1: Modelling Thermo-Mechanical Behaviour of Unsaturated Soils

Most of the research works on coupled thermal and mechanical behaviour of soils have focused on saturated soils. Khalili and Loret (2001) presented an alternative theory for heat and mass transport through deformable unsaturated porous media. They extended their previous work (Loret and Khalili 2000) on fully coupled isothermal flow and deformation in variably saturated porous media to include thermal coupling effects. Wu et al. (2004) presented a thermo-hydro-mechanical

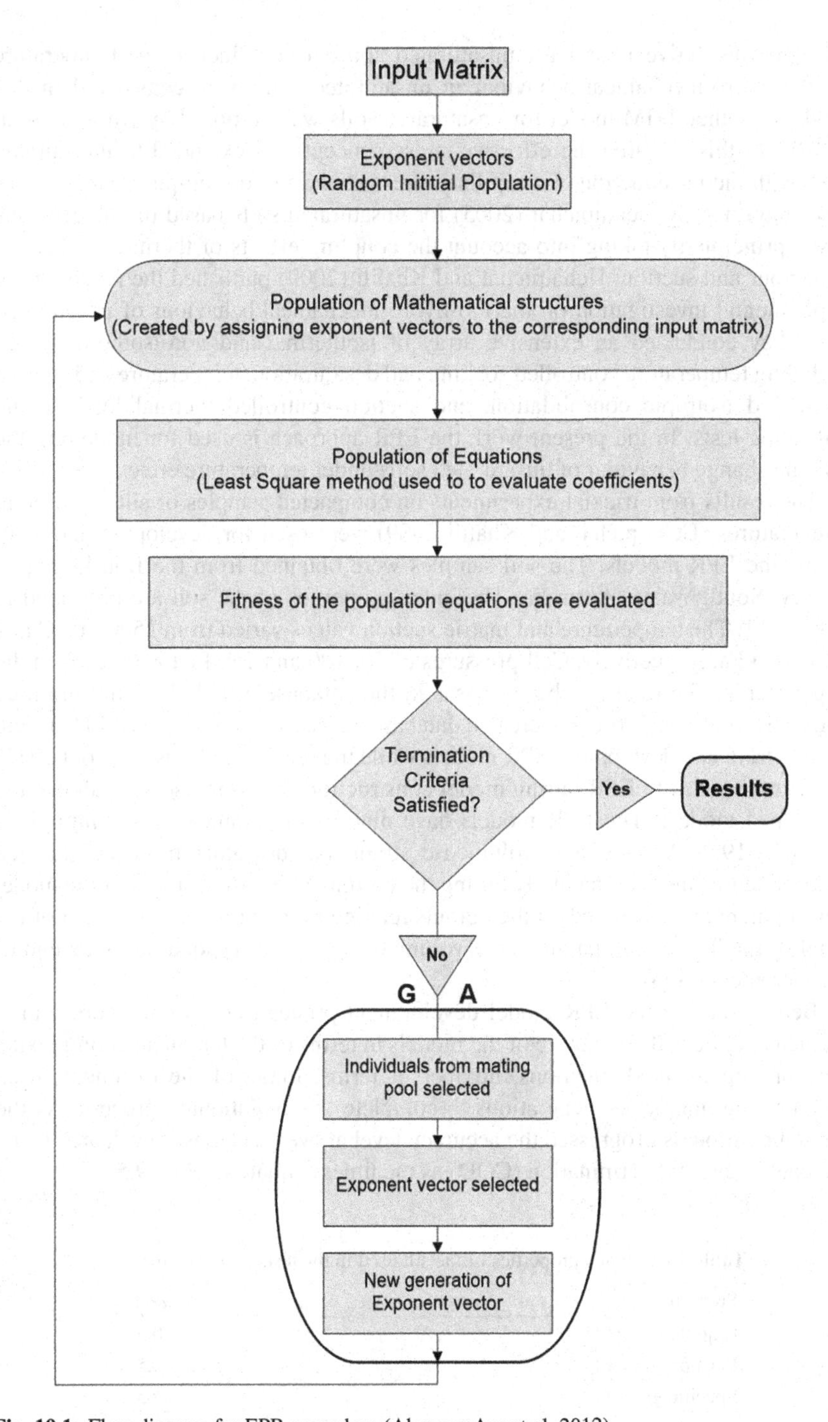

Fig. 19.1 Flow diagram for EPR procedure (Ahangar-Asr et al. 2012)

(THM) constitutive model for unsaturated soils. The influences of temperature on the hydro-mechanical behaviour in unsaturated soils were considered in this model. Another THM model for unsaturated soils was proposed by Dumont et al. (2010). In this research the effective stress concept was extended to unsaturated soils with the introduction of a capillary stress. A thermo-elastic-plastic model was also suggested by Uchaipichat (2005) for unsaturated soils based on the effective stress principle by taking into account the coupling effects of thermo-mechanical behaviour and suction. Uchaipichat and Khalili (2009) published the results of an experimental investigation on thermo-hydro-mechanical behaviour of unsaturated silt. They conducted an extensive array of isothermal and non-isothermal tests including temperature controlled soaking and desaturation, temperature and suction controlled isotropic consolidation, and suction controlled thermal loading and unloading tests. In the present work the EPR approach is used for modelling the volume change behaviour of unsaturated soils under temperature effects.

The results from triaxial experiments on compacted samples of silt at different temperatures (Uchaipichat and Khalili 2009) were used for developing and evaluating the EPR models. The soil samples were obtained from the Bourke region of New South Wales, Australia. The index properties of the soil are presented in Table 19.1. The temperature and matric suction values varied from 25 to 60 °C and 0 to 300 kPa, respectively. Cell pressures of 50, 100 and 150 kPa were used in the experiments. The total number of cases in the database was divided into training and testing datasets. From the created database 22 cases (approximately 80 %) were used to train and develop the EPR models while the remaining 5 cases (about 20 %) were kept unseen to EPR during model construction and were used to validate the developed models. The EPR models have nine input parameters as summarized in Table 19.2. Axial strain, volumetric strain and deviator stress are updated independently and incrementally during the training and testing stages of the model development process based on the outputs relating to the previous increment of the axial strain. The output parameter is volumetric strain corresponding to the end of the incremental step.

Before starting the EPR model development process, constraints were implemented to control the structure of the models in terms of the length and complexity, type of implemented functions, number of terms, range of the exponents used and also the number of generations to complete the evolutionary process. As the modelling process progressed the accuracy level at every stage was evaluated using the coefficient of determination (COD) as the fitness equation (Eq. 19.5).

Table 19.1 Index properties of the silt used in the tests (Bourke silt)

Properties	Values
Liquid limit (%)	20.5
Plastic limit (%)	14.5
Specific gravity	2.65
Air entry value (kPa)	18
Maximum dry unit weight from standard proctor test (kN/m^3)	18.8
Optimum moisture content from standard proctor test (%)	12.5

Table 19.2 Input and output parameters in the EPR models[a]

Parameters involved	Ranges of the parameter values	Unit
OCR (input)	1.3–4	–
P_{net} (input)	50–150	kPa
Su_i (input)	0–300	kPa
T (input)	25–60	°C
Sr_i (input)	0.3–1	–
ε_a (input)	0–25	%
q_i (input)	0–400	kPa
ε_{v_i} (input)	0–10	%
$\Delta\varepsilon_a$ (input)	0–2	%
$\varepsilon_{v_{i+1}}$ (output)	0–10	%

[a] OCR = overconsolidation ratio; P_{net} = mean net stress (kPa); Su_i = initial suction (kPa); T = temperature (°C); Sr_i = initial degree of saturation; ε_a = axial strain; q_i = deviator stress (kPa); ε_{vi} = volumetric strain; $\Delta\varepsilon_a$ = axial strain increment; $\varepsilon_{v_{i+1}}$ = volumetric strain corresponding to the next increment of axial strain

$$\mathrm{COD} = 1 - \frac{\sum_N (Y_a - Y_p)^2}{\sum_N \left(Y_a - \frac{1}{N}\sum_N Y_a\right)^2} \tag{19.5}$$

where Ya is the actual output value; Yp is the EPR predicted value and N is the number of data points on which the COD is computed. After completion of the modelling process, models were developed for volumetric strain (ε_v). From among the developed models some did not include all the defined parameters as inputs to the equations (the parameters that are known to affect the thermo-mechanical behaviour of soils) and hence were removed. The remaining models were considered and compared in terms of the robustness of the equations based on the COD, sensitivity analysis and also the level of complexity of the equations and the best model satisfying all these criteria was chosen as the final model. Equation (19.6) represent the selected EPR model for volumetric strain. The COD values of this EPR model are 99.99 and 99.86 % in training and testing datasets respectively.

$$\begin{aligned}\varepsilon_{v_{i+1}} = {} & \frac{1.06\mathrm{E}(-3)\, Sr_i q_i \Delta\varepsilon_a}{OCR\varepsilon_a} + 0.83\Delta\varepsilon_a + 0.98\varepsilon_{v_i} - 0.05\varepsilon_{v_i}\Delta\varepsilon_a - 4.44\mathrm{E}(-3)\, q_i\Delta\varepsilon_a \\ & + \frac{1.31\mathrm{E}(-7)\, Su_i^3 Sr_i^2 \varepsilon_a - 0.98T + 1.09\mathrm{E}(-3)\, T^3 - 9.15\mathrm{E}(-4)\, T^3 Sr_i}{T^2} \\ & + \frac{9.87\mathrm{E}(-7)\, Sr_i q_i^3 - 4.09}{P_{net}} + \frac{2.52\mathrm{E}(-4)\, {\varepsilon_{v_i}}^2 - 0.89 Sr_i^2 \Delta\varepsilon_a}{Sr_i} \\ & - 2.24\mathrm{E}(-4)\, q_i + 0.01 P_{net}\Delta\varepsilon_a + 0.1\end{aligned} \tag{19.6}$$

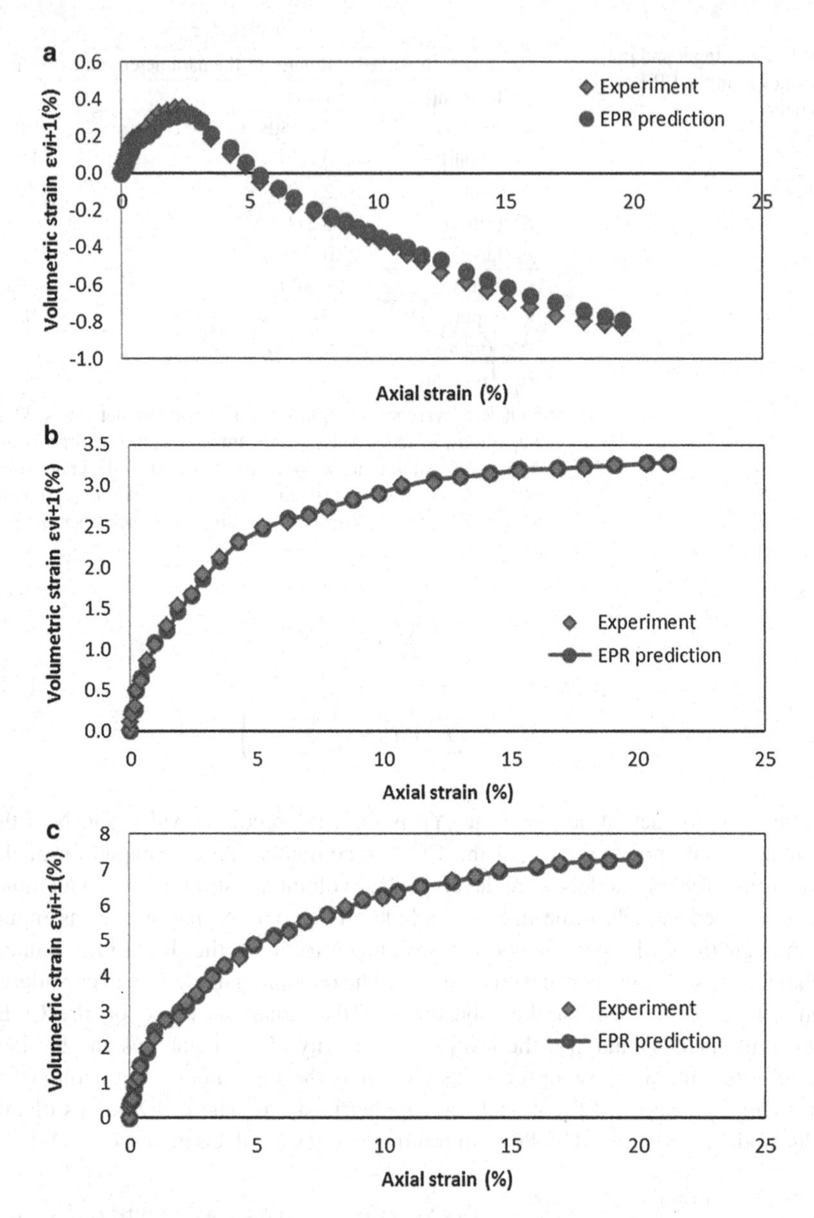

Fig. 19.2 Comparison between the EPR model predictions with experimental data for volumetric strain (**a**) (OCR = 4, mean net stress = 50 kPa, T = 25 °C); (**b**) (OCR = 2, mean net stress = 100 kPa, T = 40 °C); (**c**) (OCR = 1.33, mean net stress = 150 kPa, T = 60 °C)

Figure 19.2 shows the volumetric strain-axial strain curves predicted using the EPR model (Eq. 19.6) against the experimental results for the data used in training of the model. The results show the remarkable capabilities of the developed EPR model in capturing, predicting and also generalising the volume change behaviour of unsaturated soils considering temperature effects. The computational time to develop model using a Dual Core 2.20 GHz Intel Core i3 processor with 4.00 GB memory was less than 11 min.

The parameters chosen to be used in developing the model were also proven from the literature to be crucial contributors to the volume change behaviour of soils and have been widely implemented in previously developed models (Habibagahi and Bamdad 2003; Javadi et al. 2012b). An important feature of the proposed modelling approach in this work is the capability to phase any non-effective parameters out of the equations. However, exceptionally accurate prediction model developed using the considered parameters and without the need to add or remove any parameters to achieve the best results was another testimony to the fact that the contributing input parameters were correctly selected.

19.3.2 Problem 2: Management of Seawater Intrusion (Simulation–Optimization)

Seawater intrusion is a widespread environmental problem, particularly in arid and semi-arid coastal areas. Seawater intrusion is distinguished by the encroachment of saline water into zones previously occupied by fresh groundwater in coastal aquifer systems. Seawater intrusion problem is exacerbated by over-pumping which eventually can lead to other problems such as decrease of fresh water availability, human health and ecosystem damage (Patel and Shah 2008). Bruington (1972) and Todd (1974) list different methodologies that attempt to control seawater intrusion and to restore the quality of groundwater in aquifers. These include reduction of pumping rates, relocation of pumping wells, use of subsurface physical barriers, natural and artificial recharge, use of a line of injection wells (pressure barrier) along the coast and pumping of saline water (abstraction barrier) along the seacoast. The efficiencies of some of these methods have been investigated by integrating different simulation models (or meta models) with optimization tools to address long-term planning of groundwater management problems and to limit seawater intrusion (e.g. Das and Datta 1999; Bhattacharjya et al. 2007; Bhattacharjya and Datta 2009; Dhar and Datta 2009; Kourakos and Mantoglou 2009; Sreekanth and Datta 2010; Javadi et al. 2012a). This section involves the application of EPR as a metamodel to describe the response of the aquifer system under different pumping patterns. The model is trained and tested by data acquired from FE simulation. Thereafter, it is integrated with a multi-objective optimization algorithm to assess management scenario for controlling seawater intrusion. The results are compared with those obtained by direct integration of the numerical simulation model into the optimization model.

The study is conducted on a 2D cross section of an aquifer with 200 m length and 100 m depth (Fig. 19.3). The system was simulated using the finite element-based flow and solute transport model, SUTRA (Saturated-Unsaturated TRAansport) (Voss 1984). The main input data for the simulation model are given in Table 19.3.

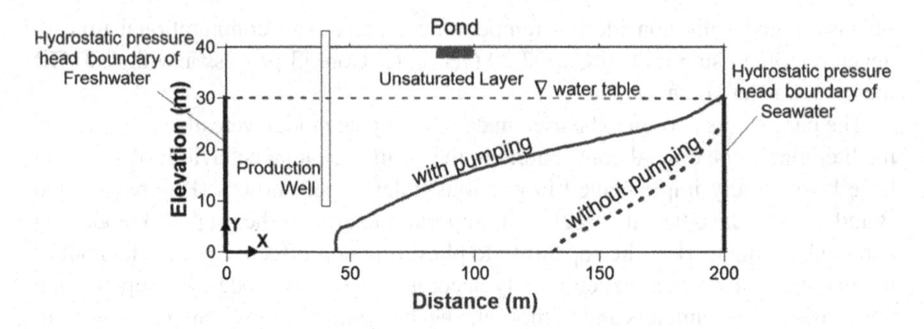

Fig. 19.3 Pre- and post-pumping distribution of salinity (0.5 isochlors)

Table 19.3 Parameters involved in the numerical simulation

Parameter	Description	Value
k	Permeability of saturated zone [m^2]	5.0×10^{-11}
	Permeability of unsaturated zone [m^2]	2.5×10^{-13}
ε	Porosity	0.3
μ	Fluid viscosity [kg/(m s)]	0.001
$\partial\rho/\partial C$	Change of fluid density with concentration [kg^2(seawater)/kg(dissolved solids) m^3]	700
C_f	Solute mass fraction of freshwater	0.0
C_{sea}	Solute mass fraction of seawater	0.0357
ρ_{sea}	Density of sea water [kg/m^3]	1025
ρ_o	Density of fresh water [kg/m^3]	1000
g	Gravitational acceleration [m/s^2]	9.8
$\alpha T/\alpha L$	Transverse/longitudinal dispersivity ratio	0.1
S_{res} [a]	residual saturation at immobile state of flow	0.23
α [a]	Fitting parameters [(m s^2)/kg]	2×10^{-4}
n [a]	Fitting parameters	1.3

[a]Parameters used in Van Genuchten (1980) model of unsaturated flow

To account for future demand, a production well pumping fresh water with constant rate of 26 m^3/day at a horizontal distance of 40 m from the inland boundary and depth of 30 m was incorporated in the model. The results of 50 % iso-concentration lines prior and after pumping (illustrated in Fig. 19.3) show that the system and the production well are threatened by seawater intrusion.

Therefore, to study optimal methods of control of seawater intrusion, the aquifer is subjected to a combined management scenario proposed by Javadi et al. (2012a). The management model is called ADR (Abstraction–Desalination–Recharge) and is based on continuous abstraction of brackish water near the coast, desalination of the abstracted brackish water and use of excess of the desalinated water as a source of artificial recharge while the rest of the desalinated water is used to meet part of the demands. The recharge was implemented by an artificial subsurface pond to allow the collected water to infiltrate into the aquifer.

Table 19.4 The values of the constant parameters involved in simulation–optimization model

Parameter	Description	Value	Unit
QR	Average rate of recharge by pond (calculated by SUTRA)	5.2	m^3/day
α_1	Cost of artificial recharge by pond	0.12	$/$m^3$
α_2	Cost of installation/drilling of well	200	($/m)
α_3	Unit cost for abstraction	0.42	$/$m^3$
α_4	Cost of treatment	0.6	$/$m^3$
α_5	Cost of construction of pond with (15 × 2 m dimensions)	350	$
α_6	Annual cost of maintenance and cleaning of pond	35	$
β	Market prices of desalinated water	1.5	$/$m^3$
Δt	Duration of application of the management strategy	10	years
r	Recovery ratio of the desalination plant	60	%

Coupling of (1) the numerical (FE) model and (2) the surrogate (EPR) model with a multi objective optimization model leads to the two different schemes used to investigate the efficiency of the control approach. The optimization model used was Non-dominated Sorting Genetic Algorithm (NSGAII) (Deb et al. 2002). Minimization of total mass of salinity in the aquifer (f_1) and minimization of the costs of construction and operation of the management process (f_2) are the two objectives considered in the simulation–optimization process. The objective functions and the set of implemented constraints used in direct coupling of FE model with GA (FE-GA) scheme are expressed mathematically as follows:

$$\min f_1 = \sum_{i=1}^{NN} C_i v_i \tag{19.7}$$

$$\begin{aligned} \min f_2 = {} & \alpha_1 QR\Delta t + \alpha_2 DA + (\alpha_3 + \alpha_4)\, QA\Delta t \\ & - \beta\,(r * QA - QR)\,\Delta t + \alpha_5 + \alpha_6 \end{aligned} \tag{19.8}$$

Subject to: $0.0 < QA(m^3/day) < 52$; $135 < XA(m) < 200$; $0 < YA(m) < 30$.

where f_1 and f_2 are objective functions, NN represents the total number of FE nodes in the domain, C_i is the solute concentration at node i, v_i is FE cell volume at node i, QA is the abstraction rate (m^3/day), and XA and YA are the spatial coordinates of the pumping well. DA is the depth of abstraction well (m). The definition and the corresponding values of other parameters (which were considered constant) are listed in Table 19.4. The values of the unit costs used were taken from literature (Javadi et al. 2012a).

In the EPR-GA scheme (the model based on coupling of the developed EPR model with GA), the EPR was first trained and tested to find the best models describing the values of first objective function (f_1) under different pumping patterns and it was then coupled with the optimization tool. The numerical simulation model

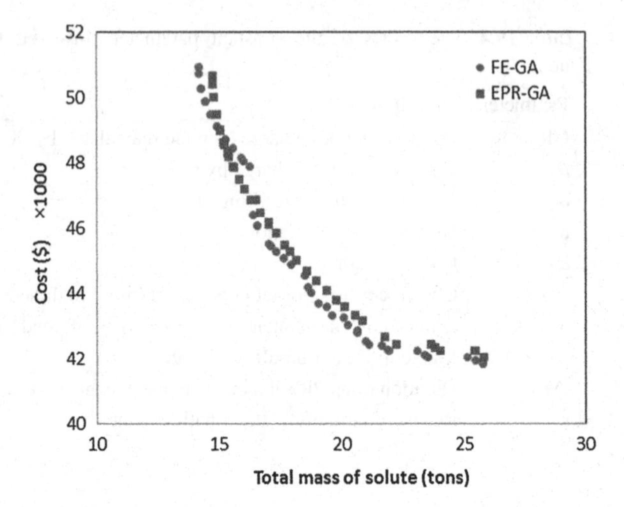

Fig. 19.4 Comparison of the optimal Pareto fronts obtained using FE-GA with the ones from EPR-GA for proposed management scenario

(SUTRA) was used to create the database including the response of the aquifer. The input parameters considered were XA, YA and QA of the abstraction well.

Using a uniform probability distribution a database of 500 data cases were randomly generated. Then, by multiple runs of the SUTRA code the outputs (total mass of salinity, f_1) corresponding to each set of data were calculated. 80 % of data were used to train the EPR model and the remaining 20 % were used for validation of the developed model. The best model (Eq. 19.9) in terms of having the optimum number of terms and complexity and also representing all contributing parameters with high levels of fitness (COD) values (94.76 % in trained and 94.27 % in tested datasets) was selected for prediction of total mass of solute (f_1):

$$\begin{aligned} f_1 = & -8461.72 + 1.14YA + 1787.86XA^{0.5} - 105.87XA \\ & + \left(1.10 * 10^{-1}\right) XA^2 - 40.92QA^{0.5} + \left(2.62 * 10^{-1}\right) YAQA^{0.5} \\ & + 6240.89QA^{0.5}XA^{-1} + \left(1.97 * 10^{-2}\right) QA^2 - 166.89YA^{0.5}QA^2XA^{-2} \end{aligned} \quad (19.9)$$

The input data of this equation were used concurrently to evaluate the second objective (cost) function (f_2) as well. Figure 19.4 shows the optimal results of the simulation–optimization process of the management model using both EPR-GA and FE-GA. It can be easily seen that the new non-dominated front obtained using the EPR-GA is very close to the one captured by FE-GA. The average time required to complete the analysis using EPR-GA (including the time to generate the database)

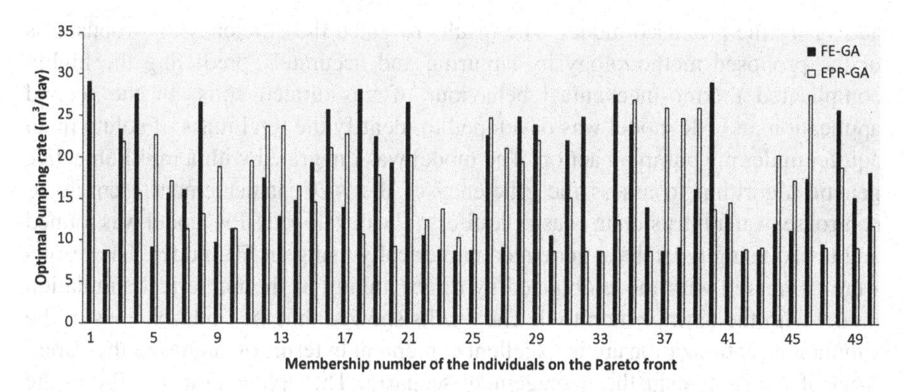

Fig. 19.5 Optimal pumping rates corresponding to optimal solutions on Pareto fronts

is less that 10 % of the time required by using FE-GA on an Intel(R) Core(TM) i7-2600 CPU @ 3.40 GHz with 16 GB RAM, which can be considered as a very significant difference.

Figure 19.5 shows the optimal pumping rates obtained through the simulation–optimization process. The obtained horizontal locations varied in the range 10–25 m from the coast and the depths between 36 and 38 m are proposed in both approaches as the optimal coordinates of designed abstraction well. Although, seawater intrusion is a highly nonlinear and density-dependent process controlled by natural hydrophysical and hydrogeological properties of the coastal system, other factors such as groundwater pumping also have important impacts on the overall hydrodynamic equilibrium of the process. The optimal controlling of the abstraction rates and the spatial distributions of the pumping wells can work against the encroached seawater to the extent that it has been introduced in literature as a specific strategy (known as abstraction barrier) to protect coastal aquifers (Pool and Carrera 2010; Sreekanth and Datta 2011a). Accordingly, and based on the management measure followed in this work, the pumping patterns (XA, YA and QA) of the designed abstraction well were considered as the only input parameters in the development of the EPR model (Eq. 19.9). However, the model cannot be generalized to other cases with different hydrogeological properties and with different control approaches unless relevant data is made available and the EPR is retrained.

19.4 Summary and Conclusion

Applications of EPR in two geotechnical and geo-environmental engineering systems were presented in this chapter. In first application an EPR model was developed to predict volumetric behaviour of unsaturated soils considering the temperature effects. An experimental dataset from the literature was used to develop and verify the proposed model. The results revealed the efficiency and robustness of the proposed methodology in capturing and accurately predicting the highly complicated thermo-mechanical behaviour of unsaturated soils. In the second application an EPR model was developed to identify the total mass of solute in an aquifer under the pumping action. The model was integrated with a multi objective genetic algorithm to assess the efficiency of the ADR management scenario to control seawater intrusion in coastal aquifers. The developed EPR model was trained and tested using a database generated numerically using an FE model. The results were compared with those obtained by direct linking of the numerical simulation model with the optimization tool. The results showed that the both schemes of the simulation–optimization are in excellent agreement in terms of capturing the Pareto front of the system in the management scenario. The application of EPR in the simulation–optimization framework resulted in significant reduction of the overall computational complexity and CPU time.

References

Ahangar-Asr A, Javadi AA, Johari A, Chen Y (2014) Lateral load bearing capacity modelling of piles in cohesive soils in undrained conditions: An intelligent evolutionary approach. Applied Soft Computing 24 (0):822–828. doi:http://dx.doi.org/10.1016/j.asoc.2014.07.027

Ahangar-Asr A, Johari A, Javadi AA (2012) An evolutionary approach to modelling the soil–water characteristic curve in unsaturated soils. Computers & Geosciences 43 (0):25–33. doi:http://dx.doi.org/10.1016/j.cageo.2012.02.021

Alavi A, Gandomi A (2013) Hybridizing Genetic Programming with Orthogonal Least Squares for Modeling of Soil Liquefaction. International Journal of Earthquake Engineering and Hazard Mitigation 1 (1):2–8

Alavi A, Gandomi A, Mollahasani A (2012) A Genetic Programming-Based Approach for the Performance Characteristics Assessment of Stabilized Soil. In: Chiong R, Weise T, Michalewicz Z (eds) Variants of Evolutionary Algorithms for Real-World Applications. Springer Berlin Heidelberg, pp 343–376. doi:10.1007/978-3-642-23424-8_11

Alavi A, Gandomi A, Nejad H, Mollahasani A, Rashed A (2013) Design equations for prediction of pressuremeter soil deformation moduli utilizing expression programming systems. Neural Comput & Applic 23 (6):1771–1786. doi:10.1007/s00521-012-1144-6

Alavi A, Gandomi A, Sahab M, Gandomi M (2010) Multi expression programming: a new approach to formulation of soil classification. Engineering with Computers 26 (2):111–118. doi:10.1007/s00366-009-0140-7

Alavi AH, Gandomi AH (2011) A robust data mining approach for formulation of geotechnical engineering systems. Engineering Computations 28 (3):242–274. doi:10.1108/02644401111118132

Alavi AH, Gandomi AH, Gandomi M, Sadat Hosseini SS (2009) Prediction of maximum dry density and optimum moisture content of stabilised soil using RBF neural networks. The IES Journal Part A: Civil & Structural Engineering 2 (2):98–106. doi:10.1080/19373260802659226

Alavi AH, Gandomi AH, Mollahasani A, Bazaz JB (2013) Linear and Tree-Based Genetic Programming for Solving Geotechnical Engineering Problems. In: Alavi X-SYHGTH (ed) Metaheuristics in Water, Geotechnical and Transport Engineering. Elsevier, Oxford, pp 289–310. doi:http://dx.doi.org/10.1016/B978-0-12-398296-4.00012-X

Babovic V, Keijzer M (2006) Rainfall-Runoff Modeling Based on Genetic Programming. In: Encyclopedia of Hydrological Sciences. John Wiley & Sons, Ltd. doi:10.1002/0470848944.hsa017

Baykasoğlu A, Güllü H, Çanakçı H, Özbakır L (2008) Prediction of compressive and tensile strength of limestone via genetic programming. Expert Systems with Applications 35 (1–2): 111–123. doi:http://dx.doi.org/10.1016/j.eswa.2007.06.006
Bhattacharjya RK, Datta B (2009) ANN-GA-based model for multiple objective management of coastal aquifers. J Water Resour Plann Manage 135 (5):314–322
Bhattacharjya RK, Datta B, Satish MG (2007) Artificial neural networks approximation of density dependent saltwater intrusion process in coastal aquifers. J Hydrol Eng 12 (3):273–282
Bruington AE (1972) Saltwater intrusion into aquifers1. JAWRA Journal of the American Water Resources Association 8 (1):150–160. doi:10.1111/j.1752-1688.1972.tb05104.x
Das A, Datta B (1999) Development of management models for sustainable use of coastal aquifers. J Irrig Drain Eng 125 (3):112–121
Davidson J, Savic DA, Walters GA (1999) Method for Identification of explicit polynomial formulae for the friction in turbulent pipe flow. Journal of Hydroinformatics 2 (1):115–126
Deb K, Pratap A, Agarwal S, Meyarivan T (2002) A fast and elitist multiobjective genetic algorithm: NSGA-II. IEEE Trans Evol Comput 6 (2):182–197. doi:10.1109/4235.996017
Dhar A, Datta B (2009) Saltwater intrusion management of coastal aquifers. I: Linked simulation-optimization. J Hydrol Eng 14 (12):1263–1272
Dorado J, RabuñAl JR, Pazos A, Rivero D, Santos A, Puertas J (2003) Prediction and modeling of the rainfall-runoff transformation of a typical urban basin using ann and gp. Applied Artificial Intelligence 17 (4):329–343. doi:10.1080/713827142
Dumont M, Taibi S, Fleureau J, Abou Bekr N, Saouab A (2010) Modelling the effect of temperature on unsaturated soil behaviour. Comptes Rendus Geoscience 342:892–900
Faramarzi A, Javadi AA, Ahangar-Asr A (2013) Numerical implementation of EPR-based material models in finite element analysis. Computers & Structures 118 (0):100–108. doi:10.1016/j.compstruc.2012.10.002
Gandomi A, Alavi A (2012) A new multi-gene genetic programming approach to non-linear system modeling. Part II: geotechnical and earthquake engineering problems. Neural Comput & Applic 21 (1):189–201. doi:10.1007/s00521-011-0735-y
Gandomi A, Alavi A (2012) A new multi-gene genetic programming approach to nonlinear system modeling. Part I: materials and structural engineering problems. Neural Comput & Applic 21 (1):171–187. doi:10.1007/s00521-011-0734-z
Gandomi A, Alavi A, Arjmandi P, Aghaeifar A, Seyednour R (2010) Genetic Programming and Orthogonal Least Squares: A Hybrid Approach to Modeling the Compressive Strength of CFRP-Confined Concrete Cylinders. Journal of Mechanics of Materials and Structures 5 (5):735–753. doi:10.2140/jomms.2010.5.735
Gandomi A, Alavi A, Sahab M (2010) New formulation for compressive strength of CFRP confined concrete cylinders using linear genetic programming. Mater Struct 43 (7):963–983. doi:10.1617/s11527-009-9559-y
Gandomi A, Alavi A, Yun G (2011) Formulation of uplift capacity of suction caissons using multi expression programming. KSCE J Civ Eng 15 (2):363–373. doi:10.1007/s12205-011-1117-9
Gandomi A, Yun G, Alavi A (2013) An evolutionary approach for modeling of shear strength of RC deep beams. Mater Struct 46 (12):2109–2119. doi:10.1617/s11527-013-0039-z
Gandomi AH, Alavi AH (2013) Expression Programming Techniques for Formulation of Structural Engineering Systems. In: Alavi AHG-SYTH (ed) Metaheuristic Applications in Structures and Infrastructures. Elsevier, Oxford, pp 439–455. doi:http://dx.doi.org/10.1016/B978-0-12-398364-0.00018-8
Gandomi AH, Alavi AH, Mousavi M, Tabatabaei SM (2011) A hybrid computational approach to derive new ground-motion prediction equations. Engineering Applications of Artificial Intelligence 24 (4):717–732. doi:http://dx.doi.org/10.1016/j.engappai.2011.01.005
Gandomi AH, Alavi, A. H., Kazemi, S., Alinia, M. M. (2009) Behavior appraisal of steel semi-rigid joints using Linear Genetic Programming. Journal of Constructional Steel Research 65 (8–9):1738–1750. doi:http://dx.doi.org/10.1016/j.jcsr.2009.04.010

Gandomi AH, Yun GJ (2014) Coupled SelfSim and genetic programming for non-linear material constitutive modelling. Inverse Problems in Science and Engineering: 1–19. doi:10.1080/17415977.2014.968149

Gaur S, Deo MC (2008) Real-time wave forecasting using genetic programming. Ocean Engineering 35 (11–12):1166–1172. doi:http://dx.doi.org/10.1016/j.oceaneng.2008.04.007

Giustolisi O, Savic DA (2006) A symbolic data-driven technique based on evolutionary polynomial regression. Journal of Hydroinformatics 8 (3):207–222. doi:10.2166/hydro.2006.020

Habibagahi G, Bamdad A (2003) A neural network framework for mechanical behavior of unsaturated soils. Canadian Geotechnical Journal 40 (3):684–693. doi:10.1139/t03-004

Javadi AA, Abd-Elhamid HF, Farmani R (2012) A simulation-optimization model to control seawater intrusion in coastal aquifers using abstraction/recharge wells. International Journal for Numerical and Analytical Methods in Geomechanics 36 (16):1757–1779. doi:10.1002/nag.1068

Javadi AA, Ahangar-Asr A, Johari A, Faramarzi A, Toll D (2012) Modelling stress–strain and volume change behaviour of unsaturated soils using an evolutionary based data mining technique, an incremental approach. Engineering Applications of Artificial Intelligence 25 (5):926–933. doi:http://dx.doi.org/10.1016/j.engappai.2012.03.006

Johari A, Habibagahi G, Ghahramani A (2006) Prediction of Soil–Water Characteristic Curve Using Genetic Programming. Journal of Geotechnical and Geoenvironmental Engineering 132 (5):661–665. doi:10.1061/(ASCE)1090-0241(2006)132:5(661)

Khalili N, Loret B (2001) An Elasto-plastic Model for Non-isothermal Analysis of Flow and Deformation in Unsaturated Porous Media: Formulation. International Journal of Solid and Structures 38:8305–8330

Kourakos G, Mantoglou A (2009) Pumping optimization of coastal aquifers based on evolutionary algorithms and surrogate modular neural network models. Adv Water Resour 32 (4):507–521

Koza JR (1992) Genetic programming: on the programming of computers by means of natural selection. MIT Press

Loret B, Khalili N (2000) A Three Phase Model for Unsaturated Soils. International Journal for Numerical and Analytical Methods in Geomechanics 24:893–927

Makkeasorn A, Chang NB, Zhou X (2008) Short-term streamflow forecasting with global climate change implications – A comparative study between genetic programming and neural network models. Journal of Hydrology 352 (3–4):336–354. doi:http://dx.doi.org/10.1016/j.jhydrol.2008.01.023

Mousavi S, Alavi A, Gandomi A, Mollahasani ALI (2011) Nonlinear genetic-based simulation of soil shear strength parameters. J Earth Syst Sci 120 (6):1001–1022. doi:10.1007/s12040-011-0119-9

Narendra BS, Sivapullaiah PV, Suresh S, Omkar SN (2006) Prediction of unconfined compressive strength of soft grounds using computational intelligence techniques: A comparative study. Computers and Geotechnics 33 (3):196–208. doi:http://dx.doi.org/10.1016/j.compgeo.2006.03.006

Parasuraman K, Elshorbagy A (2008) Toward improving the reliability of hydrologic prediction: Model structure uncertainty and its quantification using ensemble-based genetic programming framework. Water Resources Research 44 (12):W12406. doi:10.1029/2007WR006451

Patel AS, Shah DL (2008) Water management: Conservation, harvesting and artificial recharge. New age international (p) limited, publishers,

Pool M, Carrera J (2010) Dynamics of negative hydraulic barriers to prevent seawater intrusion. Hydrogeol J 18 (1):95–105. doi:10.1007/s10040-009-0516-1

Rezania M, Javadi AA (2007) A new genetic programming model for predicting settlement of shallow foundations. Canadian Geotechnical Journal 44 (12):1462–1473. doi:10.1139/T07-063

Sreekanth J, Datta B (2010) Multi-objective management of saltwater intrusion in coastal aquifers using genetic programming and modular neural network based surrogate models. J Hydrol 393 (3):245–256

Sreekanth J, Datta B (2011) Comparative Evaluation of Genetic Programming and Neural Network as Potential Surrogate Models for Coastal Aquifer Management. Water Resour Manage 25 (13):3201–3218. doi:10.1007/s11269-011-9852-8

Sreekanth J, Datta B (2011) Coupled simulation-optimization model for coastal aquifer management using genetic programming-based ensemble surrogate models and multiple-realization optimization. Water Resources Research 47 (4):W04516. doi:10.1029/2010WR009683

Todd DK (1974) Salt-Water Intrusion and Its Control. Journal (American Water Works Association) 66 (3):180–187. doi:10.2307/41266996

Uchaipichat A (2005) Experimental Investigation and Constitutive Modelling of Thermo-hydro-mechanical Coupling in Unsaturated Soils. The University of New South Wales, Sydney, New South Wales, Australia

Uchaipichat A, Khalili N (2009) Experimental investigation of thermo-hydro-mechanical behaviour of an unsaturated silt. Geotechnique 59 (4):339–353

Van Genuchten MT (1980) A closed-form equation for predicting the hydraulic conductivity of unsaturated soils. Soil Science Society of America Journal 44 (5):892–898. doi:10.2136/sssaj1980.03615995004400050002x

Voss CI (1984) A finite-element simulation model for saturated-unsaturated, fluid-density-dependent ground-water flow with energy transport or chemically-reactive single-species solute transport. U.S. Geol. Surv. (USGS), Water Resour. Invest.

Wang W-C, Chau K-W, Cheng C-T, Qiu L (2009) A comparison of performance of several artificial intelligence methods for forecasting monthly discharge time series. Journal of Hydrology 374 (3–4):294–306. doi:http://dx.doi.org/10.1016/j.jhydrol.2009.06.019

Wu W, Li X, Charlier R, Collin F (2004) A thermo-hydro-mechanical constitutive model and its numerical modelling for unsaturated soils. Computers and Geotechnics 31 (2):155–167. doi:http://dx.doi.org/10.1016/j.compgeo.2004.02.004

Yang C, Tham L, Feng X, Wang Y, Lee P (2004) Two-Stepped Evolutionary Algorithm and Its Application to Stability Analysis of Slopes. Journal of Computing in Civil Engineering 18 (2):145–153. doi:10.1061/(ASCE)0887-3801(2004)18:2(145)

Chapter 20
Application of GFA-MLR and G/PLS Techniques in QSAR/QSPR Studies with Application in Medicinal Chemistry and Predictive Toxicology

Partha Pratim Roy, Supratim Ray, and Kunal Roy

20.1 Introduction

The design of therapeutic molecules with desired properties and activities as well as ranking of the hazardous chemicals for screening of toxicity is a challenging task. Traditionally, drugs were developed by testing synthesized compounds in time-consuming multi-step processes which often required a trial-and-error procedure. The use of statistical models to predict biological and physicochemical properties started with linear regression models developed by Hansch in the 1960s (Hansch et al. 1962; Hansch and Fujita 1964) leading to the development of the discipline of Quantitative structure–activity relationship (QSAR), which is a part of the more general area quantitative structure-property relationship (QSPR). The target property to be modeled may also be a toxicity endpoint, the corresponding area being more specifically known as quantitative structure-toxicity relationship (QSTR). In QSAR/QSPR/QSTR modeling, a target activity/property/toxicity (response variable) is correlated with chemical structure information (descriptors) using appropriate statistical tools (Roy and Mitra 2011). Such methods have long been used in the context of drug design and predictive toxicology. The final

Electronic supplementary material The online version of this chapter (doi: 10.1007/978-3-319-20883-1_20) contains supplementary material, which is available to authorized users.

P.P. Roy
Institute of Pharmaceutical Sciences, Guru Ghasidas University, Bilaspur 495 009, India

S. Ray
Department of Pharmaceutical Sciences, Assam University, Silchar 788 011, India

K. Roy (✉)
Department of Pharmaceutical Technology, Jadavpur University, Kolkata 700 032, India
e-mail: kunalroy_in@yahoo.com; kroy@pharma.jdvu.ac.in

A.H. Gandomi et al. (eds.), *Handbook of Genetic Programming Applications*,
DOI 10.1007/978-3-319-20883-1_20

focus of the drug discovery process is to develop therapeutically active lead compounds as drug candidates. Among the high throughput screening techniques in drug discovery and toxicity screening, QSAR/QSPR/QSTR methods are practiced very often. This indicates deployment of computer aided molecular design methods in accelerating the drug development process (Blaney 1990; Bugg et al. 1993). QSAR/QSPR methods are also very useful for risk assessment of chemicals in the context of environmental toxicity. Predictive QSAR models reduce the need of animal experimentation for determination of chemical toxicity and can thus be used for regulatory purposes (Kar and Roy 2010).

The organization of economic co-operation and development (OECD) has recommended a set of guidelines for development and validation of QSAR models specifically for regulatory uses (Gramatica 2007). One of the guidelines states about an unambiguous algorithm for QSAR model development. Feature selection is one of the integral parts in the development of QSAR/QSPR models. Many applications are capable of generating hundreds or thousands of different molecular descriptors which are very large in number compared to number of compounds with biological activity/property to be modeled. Models developed using a large number of descriptors often suffer from loss of accuracy and/or the problem of overfitting. Another associated problem may be the redundancy of the descriptors, i.e., many descriptors may characterize the same feature of compounds. Standard regression models like multiple linear regression (MLR) and partial least squares (PLS) are routinely practised for QSAR/QSPR model development. Genetic algorithm has gained increased attention in the last two decades for providing available additional information not provided by standard regression techniques, even for data sets with many features. "*Genetic Algorithms in Molecular Modeling*" authored by Prof. J. Devillers (Devillers 1996) is the first book on the use of genetic algorithms in QSAR and drug design. It is also worth mentioning here that genetic programming, a branch of genetic algorithms, has found its application in drug design (Archetti et al. 2010; Pugazhenthi and Rajagopalan 2007), predictive toxicology (Harrigan et al. 2004) and chemical engineering problems (Bagheri et al. 2012a, b, 2014).

The increasing number of applications of GA in different fields since its inception from the Holland genetic algorithm (Holland 1975) indicates its effectiveness. In 1993, Forrest published a paper in *Science* on the mathematics of GA and its best use in several natural evolutionary systems (Forrest 1993). The use of genetic algorithm continued with different research groups in order to solve various problems (Maddox 1995). Different fields like digital image processing (Andrey and Tarroux 1994), scheduling problems and strategy planning (Cleveland and Smith 1989; Gabbert et al. 1991; Karr 1991), music composition (Horner and Goldberg 1991), criminology (Caldwell and Johnston 1991) and biology (Hightower et al. 1995; Jaeger et al. 1995) have been benefited from the applications of GA. Several published literatures indicate the application of genetic algorithms in chemistry and chemometrics (Hartke 1993; Chang and Lewis 1994; Rossi and Truhlar 1995). Computer aided molecular design is another field of application for GA for designing molecules with desired properties and activities (Venkatasubramanian et al. 1994a). GA is not limited only to the above mentioned applications, it has rather a broad spectrum applications in different regression methods. The calibration and selection of variables using large search domain (with n variables, 2^n-1 possible combinations) and many local optima makes GA as one of the suggested methods in feature selection (Wise et al. 1995, 1996). Hybridization of GA with other methods (MLR, PLS) greatly improves the results (Broadhurst et al. 1997). Increasing applications of genetic algorithms in the QSAR/QSPR/QSTR fields makes it a popular method of choice. In this context, this chapter is organized in the following manner. The first part includes a description of Genetic Function Approximation (GFA) by highlighting genetic algorithm and multivariate adaptive regression splines. The second part is about the fitness measures in GFA. The third section is a brief discussion on GFA-MLR and G/PLS techniques. In the final section, the applications of GFA-MLR and G/PLS in QSAR/QSPR studies are reviewed.

20.2 Genetic Function Approximation

The genetic function approximation algorithm is a combination of two different algorithms namely Holland's genetic algorithm and Friedman's multivariate adaptive regression spines (MARS) algorithm (Holland 1975; Friedman 1991).

20.2.1 Genetic Algorithm

Genetic algorithms couple the language of natural genetics and biological evolution. Adaptation to changing environment is essential for survival of each individual species. Hereditary transmission of genes to the offspring is a natural process of generation for survival of the fittest. The natural selection process was popularized by Charles Darwin (Darwin 1859). GA is the process where segments of genes of two individuals, i.e., parents are exchanged to produce two new individuals, i.e., children (Holland 1975). The driving force of a GA is recombination.

Usually a binary integer code (0, 1) of a definite length of vector of components or variables is assigned to each individual. For the continuation of genetic similarity, these individuals are linked to chromosomes and variables are regarded as genes. Therefore the chromosome consists of several genes or variables (Janikow and Michalewicz 1991; Mathias and Whitley 1994). The representation of chromosome may be like the following:

Chromosome 1: 1101100100
Chromosome 2: 1101111000

The chromosomes consist of several bits representing many characteristics, and encoding of the genes depends on the problem to be solved (Fraser 1957).

The following parameters influence the nature of chromosomes within a population of individuals:

- The competition among individuals for resources and mate.
- The successful individual will produce more offspring than poorly performed individuals
- A good individual will transmit good genes and produce better offspring than the parent.
- Therefore, each successive generation will be more adaptive to the environment.

The steps in genetic algorithm include the following: encoding mechanism; creation of a population of chromosomes; definition of a fitness function; genetic manipulation of the chromosomes.

To maintain equivalence to the natural selection, primarily individual solutions are randomly generated to form the initial individuals. Generally, increase in population size increases diversity. As a thumb rule, the population size depends on the nature of the problem to be solved, the representation used and the choice of the operators (Goldberg 1989a; Syswerda 1991).

The fitness function which allows the individuals of the population to be exposed to an evaluation function plays an important role in Darwinian evolution. The role of the fitness function is to assure the quality of individuals so that best individuals will receive the best fitness score (Goldberg 1989b). Another importance of the fitness function is prevention of domination of superior individuals in the selection process and to promote healthy competition between equal individuals. Different scaling functions, like linear scaling, the sigma truncation, the powerlaw scaling, the sigmoidal scaling and Gaussian scaling are applied along with the fitness function (Goldberg 1989b; Venkatasubramanian et al. 1994b).

Roulette Wheel Selection and tournament selection (Goldberg 1989b; Angeline 1995) procedure are employed in order to remove premature convergence in case of population with low average fitness values.

After selection of the best fitness function, the next operator is crossover followed by mutations for the formation of new chromosomes during the reproduction process. In this step, each parent chromosome shares some portion of their genes to create a child. Different literature indicates the details of different crossover techniques and their analysis (Schaffer et al. 1989; Syswerda 1989, 1993; Schaffer and Eshelman 1991; Spears 1993; Jones 1995).

The single-point crossover is the simplest form. In this, the crossover point is randomly selected and parts of chromosomes from two parents are exchanged from the randomly selected point to create two children (Fig. 20.1). Due to random selection of the crossover point, this crossover technique is characterized by high positional bias and low distributional bias (Eshelman et al. 1989).

The two point crossover technique (Fig. 20.2) is similar to the single point crossover in which two crossover points are randomly chosen and parts of the chromosomes between them are exchanged. The positional bias is reduced without affecting the distributional bias (Eshelman et al. 1989). Multipoint crossover is an extension of two point crossover just by increasing the number of crossover points which suffers both positional and distributional bias.

Mutation is the step followed by the crossover operation (Fig. 20.3). It introduces irregular and random alterations of genes in chromosomes (Spears 1993). It is an

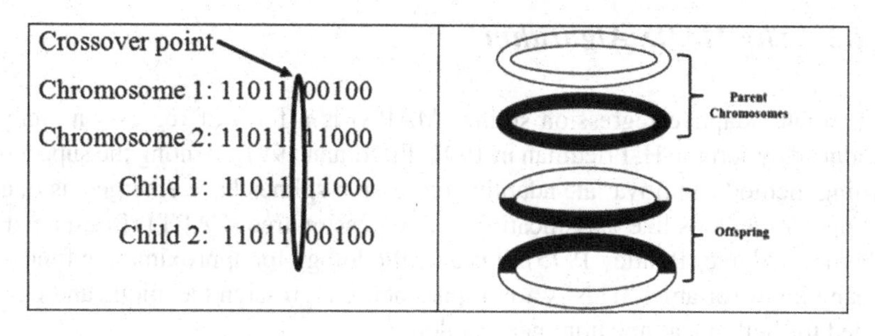

Fig. 20.1 Single point cross over

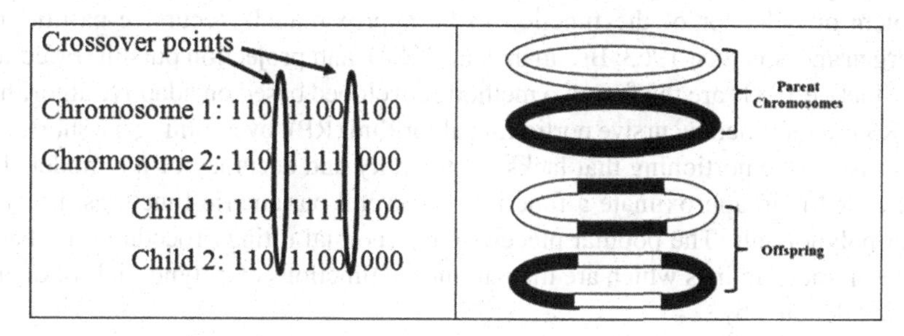

Fig. 20.2 Two point cross over

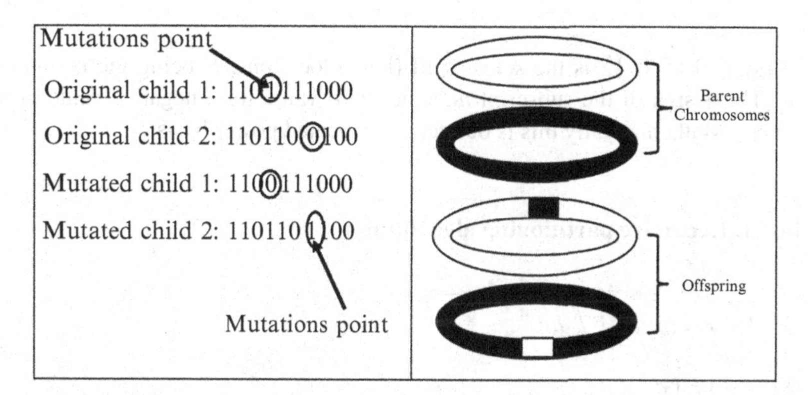

Fig. 20.3 Mutation

important genetic operator. The probability of mutation is user defined which can be kept constant or varied throughout the run of a genetic algorithm. Encoding and crossover techniques determine mutation probability (Bäck 1993; Tate and Smith 1993).

20.2.2 The MARS Algorithm

Multivariate adaptive regression spline (MARS) is a form of regression analysis introduced by Jerome H. Friedman in 1991 (Friedman 1991). Among the supervised learning methods, multivariate adaptive regression splines have emerged as one of the popular methods like classification and regression tree (CART) (Breiman et al. 1984) and *k-d* tree (Bentley 1975). It is a methodology for approximating functions of many input variables. This is a non-parametric regression technique and can be applied for both linear and nonlinear modeling.

In adaptive computation, the strategy is to adjust the behavior of particular problem or behavior of the function to be approximated. Recursive partitioning (Morgan and Sonquist 1963; Breiman et al. 1984), and projection pursuit (Friedman and Stuetzle 1981) are the first two methods developed based on adaptive algorithm. MARS is a modified recursive portioning algorithm (RP), overcoming the shortcoming of recursive portioning that lacks continuality and accuracy of the model. The idea of RP is to approximate a function by several parametric functions, i.e., low order polynomials. The popular piecewise polynomial fitting procedures are based on the splines. Splines which are the parametric functions of polynomials of degree of q are denoted by

$$\{(x - t_k)_+^q\}_1^K \tag{20.1}$$

In Eq. (20.1), $\{t_k\}_1^K$ is the set of split (knot) locations, K being the number of knots. The + sign of the subscript indicates zero value for a negative value of the argument. Mathematically this is defined as truncated power basis.

Box 1. Recursive partitioning algorithm

[1] $B_1(x) \leftarrow 1$
[2] *For* $M = 2$ *to* M_{max} do: $lof^* \leftarrow \infty$
[3] *For* $m = 1$ *to M-1 do*:
[4] *For* $\upsilon = 1$ *to n do*:
[5] *For* $t \in \{x_{\upsilon j} | B_m(x_j) > 0\}$
[6] $g \leftarrow \sum_{i \neq m} a_i B_i(x) + a_m B_m(x) H[+(x_\upsilon - t)] + a_M B_m(x) H[-(x_\upsilon - t)]$
[7] $lof \leftarrow min_{a1,\ldots,aM} LOF(g)$
[8] *if* $lof < lof^*$, *then* $lof^* \leftarrow lof$; $m^* \leftarrow m$; $v^* \leftarrow v$; $t^* \leftarrow t$ *end if*
[9] *end for*
[10] *end for*
[11] *end for*
[12] $B_M(x) \leftarrow B_{m^*}(x) H[-(x_{\upsilon^*} - t^*)]$
[13] $B_{m^*}(x) \leftarrow B_{m^*}(x) H[+(x_{\upsilon^*} - t^*)]$
[14] *end for*
[15] *end algorithm*

The RP algorithms (Box 1) are discontinuous because of the use of a step function H denoting a positive argument. Replacement of the step function with a continuous function produces a continuous model. In case of the recursive portioning algorithm, a special spline basis function is employed where $q = 0$.

The one-sided truncated power basis functions for representing qth order splines are represented as:

$$b_q(x - t) = (x - t)_+^q \tag{20.2}$$

In Eq. (20.2), t is the knot location, q is the order of the spline, and the subscript indicates the positive part of the argument. For $q > 0$, the spline approximation is continuous and has $q - 1$ continuous derivatives. A two-sided truncated power basis is a mixture of functions of the form:

$$b_q^{\pm}(x - t) = [\pm(x - t)]_+^q \tag{20.3}$$

The step functions appearing in RP are seen to be two-sided truncated power basis functions for $q = 0$ splines. The solution of discontinuity is application of the spline based function where the order of $q > 0$.

The second modification is related to the lack of good approximation for certain functions. After the first modification, a large number of variables can be involved. Therefore the final basis functions increase in the interaction order and result in complex function of higher order, which is indeed very difficult to approximate as in case of linear ones. The solution of the problem is to keep lower order parents rather deleting them resulting in availability of all basis functions for further splitting. Additionally, another splitting is applied on the parent but not on the child restricting MARS to increase its depth or addition of new factor to product.

Modification of the spline based system allows multiple splits on same predictors along with the path of binary tree allowing the final basis function carrying same variables in their product.

The last problem is to assign the value of q in MARS. The general idea is to use the value of 1 for q (Friedman 1991).

Generalization of recursive partitioning regression involves the following modifications to recursive partitioning algorithm:

(a) Replacing the step function $H[\pm(x - t)]$ (H being a step function indicating the positive argument) by a truncated power spline function. $[\pm(x - t)]_+^q$.
(b) Not removing the parent basis function B_{m^*} after it is split, thereby making it and both its daughters eligible for further splitting.
(c) Restricting the product associated with each basis function to factors involving distinct predictor variables.

Multivariate spline basis functions in the MARS algorithm (Box 2), after using two-sided truncated power basis functions instead of a step function, have now the following form:

$$B_m^{(q)}(x) = \prod_{k=1}^{K_m} \left[s_{km} \cdot \left(x_{v(k,m)} - t_{km}\right)\right]_+^q \tag{20.4}$$

where the quantity K_m is the number of splits that gave rise to basis function B_m, the quantity s_{km} takes the values ±1, υ(k,m) labels the predictor variables, and t_{km} represents values on the corresponding variables.

Box 2. The MARS algorithm (forward stepwise)

[1] $B_1(x) \leftarrow 1; M \leftarrow 2$
[2] *Loop until* $M > M_{max}$: $lof^* \leftarrow \infty$
[3] *For* $m = 1$ *to* $M\text{-}1$ *do:*
[4] *For* $\upsilon \notin \{\upsilon(k,m) | 1 \le k \le K_m\}$
[5] *For* $t \in \{x_{\upsilon j} | B_m(x_j) > 0\}$
[6] $g \leftarrow \sum_{i=1}^{M-1} a_i B_i(x) + a_M B_m(x)\left[+x_\upsilon - t)\right]_+ + a_{M=1} B_m(x)[-(x_\upsilon - t)]_+$
[7] $lof \leftarrow min_{a1,\ldots,aM+1} LOF(g)$
[8] *if* $lof < lof^*$, *then* $lof^* \leftarrow lof$; $m^* \leftarrow m$; $v^* \leftarrow v$; $t^* \leftarrow t$ *end if*
[9] *end for*
[10] *end for*
[11] *end for*
[12] $B_M(x) \leftarrow B_{m*}(x)\ [+ (x_{\upsilon*}\text{-}t^*)]_+$
[13] $B_{M+1}(x) \leftarrow B_{m*}(x)\ [\text{-}(x_{\upsilon*}\text{-}t^*)]_+$
[14] *end loop*
[15] *end algorithm*

In the algorithm above, truncated power basis functions ($q = 1$) are substituted for step functions H in 6th, 12th and 13th steps of the RP algorithm. The parent basis function is included in the modified model in 6th line and remains in the model through lines 12–13 (Box 2).

In the MARS algorithm, both forward and backward passes are used for the development of models. In case of the forward method, the model is built by adding basis functions in pair to models. The pair of basis functions selected in each step is based on a greedy algorithm which gives maximum reduction in sum-of-squares residual error. In the backward pass, the initial model is an *overfit* model. Pruning of the model is done in order to build a good general model. The basis functions are deleted according to their contribution and the least effective functions are deleted until the best supermodel is obtained. The cycles continue until the best possible balance of basis and variance is obtained.

20.2.3 GFA Spline in Regression Problem

As stated before, GFA can build models not only with linear polynomials but also with higher order polynomials (splines, gaussians, quadratic polynomials, and quadratic splines). By the application of spline-based terms, GFA can perform a

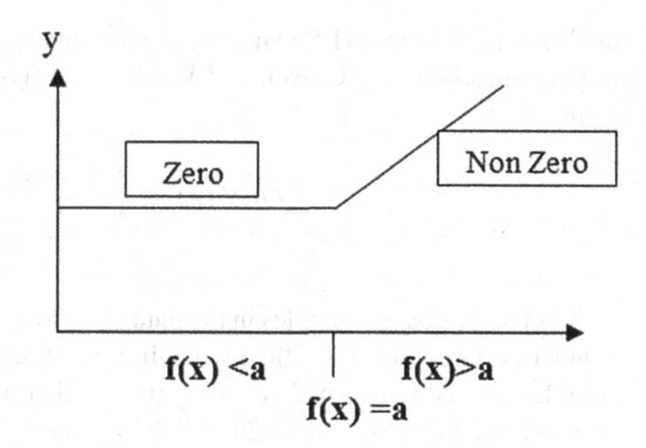

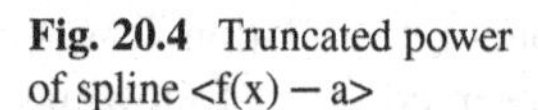
Fig. 20.4 Truncated power of spline <f(x) − a>

form of automatic outlier removal and classification. The splines used are truncated power splines and are denoted with angular brackets. For example, <f(x) – a> is equal to zero if the value of (f(x) – a) is negative, else it is equal to (f(x) – a). The constant 'a' is called the knot of the spline. A spline partitions the data samples into two classes, depending on the value of some feature. The value of the spline is zero for one of the classes and non-zero for the other classes (Fig. 20.4).

In addition, splines are applied either in range identification or outlier removal. Spline can identify a range of effect in case there are many members in the non-zero partition. If the members of the non-zero set are a few in number, then outliers are identified by splines.

The lack of fit (LOF) criterion optimizes the variables, knots and interaction criteria. In addition to variable selection, MARS also finds interactions between the variables and limits the optimal interactions. MARS can handle very complex high dimensional data. The MARS procedure is computationally intensive and expensive only for 20 features with more than 1000 input samples although it gives high levels of performance and is a competitor for many neural network techniques. Another limitation is its inability to discover models containing features that are good in groups but poor for individuals. David Rogers then incorporated genetic algorithm in MARS as a better search instead of exploring a large functional space in MARS by the incremental approach. Replacement of binary string by genetic approach led to G/SPLINE approach and it combinedly evolved into genetic function approximation approach (Rogers 1991, 1992).

20.2.4 Fitness Measures in GFA

The GFA algorithm builds multiple models rather than developing a single model. The appropriate fitness measure in GFA is Friedman's lack-of-fit (LOF) function (Friedman 1988). Generalized cross validation (GCV) is a criterion for appropriateness of a solution of a given problem λ. GCV uses a formula that approximates the least squares error (LSE) mainly determined by leave-one-out validation. It was initially developed by Craven and Wahba (1979) and represented in the following form:

$$GCV(\lambda) = \frac{1}{N}\frac{LSE}{\left[1 - \frac{C(\lambda)}{N}\right]^2} \tag{20.5}$$

N is the number of samples in the data set, and C (λ) is a complexity cost function which estimates the cost of the model. In case of regression splines where $\lambda = 0$, the complexity cost function C is simply the number of nonconstant basis function 'c' plus one and Eq. (20.5) takes the following form:

$$GCV(\lambda) = \frac{1}{N}\frac{LSE}{\left[1 - \frac{c+1}{N}\right]^2} \tag{20.6}$$

Finally, Friedman and Silverman (1989) added another penalty term to the basis function having the following form

$$LOF = \frac{1}{N}\frac{LSE}{\left[1 - \frac{c+d\times p}{N}\right]^2} \tag{20.7}$$

In the above equation c is the number of nonconstant basis functions, N is the number of samples in the data set, d is a smoothing factor to be set by the user, and p is the total number of parameters in the model. The value of d is assigned by the cross validation criterion. Based on the experiment, the following observations were suggested by Friedman and Silverman (1989):

(1) The actual accuracy of the modeling either in terms of expected squared error is fairly insensitive to the value of d in the range $2 < d < 4$. (2) However, the value of the LOF function for the final model does exhibit a moderate dependence on the choice of *d*. In fact, *d* is user defined. The overall process of GFA continues until the average LOF score of models in a population stops improving significantly. Usually for a population of 300 models, 3000–10,000 iterations are sufficient to achieve convergence (Rogers and Hopfinger, 1994). LOF and LSE are the fitness measures used in GFA-MLR and G/PLS model respectively.

20.2.5 GFA-MLR and G/PLS in Regression Studies

The general purpose of multiple regressions (the term was first used by Pearson 1908) is to learn more about the relationship between several independent or predictor variables and a dependent or criterion variable. Combination of genetic algorithm (GA) with multiple linear regression (MLR) results in increase in the predictivity and interpretability of models for QSAR and other applications like in spectroscopy (Broadhurst et al. 1997; Bangalore et al. 1996). GA is used as the optimal variable subset selection method and MLR is applied on the selected subset, and the process repeated until the best possible root-mean-square error of prediction (RMSEP) in a multiple linear regression (MLR) model is obtained (Allen 1971).

Partial Least Squares (PLS) regression is another recent technique that couples the features from principal component analysis and multiple regression and produces more robust result. PLS creates score vectors (also called latent vectors or components) by maximizing the covariance between different sets of variables. In case of the correlated X variables, there is a substantial risk of "overfitting", i.e., obtaining a well-fitted model, with little or no predictive capability (Wold 1995). G/PLS is derived from two QSAR calculation methods: GFA and partial least squares (PLS). The G/PLS algorithm uses GFA to select appropriate basis functions to be used in a model and PLS regression as the fitting technique to weigh the basis functions' relative contributions in the final model.

The advantages of incorporating PLS in GA are (Devillers 1996):

1. PLS can handle inter-correlated descriptors.
2. It gives a more robust model than MLR.
3. PLS scores and loadings provide information about the correlation structures of the variables and structural similarities/dissimilarities among the compounds.
4. PLS with the UNIPALA algorithm lowers the computational time for large data.
5. The GOPLE approach uses D-Optimal design methods to select the heavily loaded variables in the model which also partially removes the collinearity.
6. In PLS, there is a problem of "noise". GFA for subset selection filters the noise and improves the quality of models.

Hybridization of genetic algorithm with regression techniques is beneficial for its application in different fields. Other examples of similar hybrid evolutionary computation and regression analysis can also be found in the literature (Gandomi et al. 2010; Gandomi and Alavi 2013).

20.3 Application of Genetic Algorithm (GFA-MLR and G-PLS) in QSAR/QSPR Studies

GFA-MLR and G-PLS are frequently used by different research groups for the development of QSAR/QSPR models with application in medicinal chemistry and predictive toxicology. The genetic function approximation algorithm is derived from Rogers' G/SPLINES algorithm and allows a new way for construction of QSAR and QSPR models (Rogers 1991, 1992). The first work of GFA applied three data sets (Rogers and Hopfinger 1994). At first, QSAR models were developed with GFA on antimycin derivatives having antifilarial activity with an objective to illustrate the advantages and uses of multiple models. Then another set of compounds having acetylcholinesterase inhibitory activities was subjected to GFA analysis to illustrate the automatic partitioning behavior of spline based models in the QSAR study. Finally QSPR analysis was performed on some structurally diverse polymers to predict the glass transition temperatures and melt transition temperature with an objective to illustrate the applicability of the genetic analysis to QSPR problems. After this, a lot of studies related with QSAR/QSPR have been performed with the genetic function approximation algorithm, some of which are mentioned below. Note that this list is only representative and not exhaustive.

20.3.1 Application of Genetic Algorithm in QSAR Studies

The intrinsic ability of GFA for creation of multiple models makes it a preferable chemometric tool for QSAR study. GFA and G/PLS have been used in many applications of QSAR studies. The genetic method helps in appropriate feature selection leading to models with good predictive features. Some of the representative examples are cited here.

Modeling of a receptor surface provides information about drug receptor interactions. GFA was efficiently used in a receptor surface model for corticosteroid globulin binding data, because it allowed the construction of multiple probable models using the most valuable descriptors (Hahn and Rogers 1995). In another study (Klein and Hopfinger 1998), partial least squares along with GFA were used to construct QSAR models for a set of cationic-amphiphilic analogs having antiarrhythmic properties. For the same compounds, QSPR models were also developed for the interaction of the compounds with phospholipid membranes. The study revealed that spatial features of molecules along with partition coefficient play an important role in antiarrhythmic activity. The change in membrane transition temperature was also significant in the QSPR study (Klein and Hopfinger 1998). The importance of partial atomic charges on the ellipticine ring forming atoms for anticancer activity was identified by a GFA analysis of ellipticine analogs (Shi et al. 1998). A QSAR analysis (Kulkarni and Hopfinger 1999) was carried out for the intermolecular membrane solute interaction properties generated by molecular dynamics simulation along with intramolecular physicochemical properties of eye irritants (organic compounds) using GFA. The studies showed that the eye irritation by organic compounds was due to an increased aqueous solubility of compounds and its strength of binding to the membrane (Kulkarni and Hopfinger 1999). The importance of free energy force field terms like intramolecular vacuum energy of the unbound ligand, the intermolecular ligand receptor van der Waals interaction energy and the van der Waals energy of the bound ligand in regulation of glycogen metabolism in diabetes was identified using GFA-MLR and G/PLS techniques for a set of glucose analogue inhibitors of glycogen phosphorylase (Venkatarangan and Hopfinger 1999). In another study, cationic amphiphilic model compounds from a series of phenylpropylamine derivatives having antiarrhythmic activity were employed in a QSAR study using partial least squares and GFA techniques (Klein et al. 1999). Both intermolecular membrane-interaction descriptors generated from molecular dynamics simulation as well as intramolecular descriptors were used for development of the models. The studies suggest the importance of membrane-interaction descriptors regarding phase transition temperatures. The electrostatic properties of the compounds govern the calcium displacing activity at phosphatidylserine monolayers. The lipophilicity and molecular size of the compounds affect the antiarrhythmic activity (Klein et al. 1999). Terbinafine is one of the non-azole antifungal agents, which shows action by inhibition of the enzyme squalene epoxidase. A QSAR model was developed with GFA for a series of terbinafine analogues to explore that steric properties, and it was found that conformational rigidity of the side chains plays an important role for activity (Gokhale and Kulkarni 2000).

Multiple temperature molecular dynamics simulation along with G/PLS was used to develop a free energy force field 3D-QSAR model for ligand receptor binding (Santos-Filho et al. 2001). The receptor for the study was a specific mutant type of *Plasmodium falciparum* dihydrofolate reductase. The ligands were structurally diverse antifolate compounds. The developed models indicate some structural features of the compounds that are responsible for resistance of the enzyme (Santos-Filho et al. 2001). 3D-QSAR studies for 3-aryloxazolidin-2-one derivatives as antibacterial against *Staphylococcus aureus* were done using GFA. The study indicated the importance of electronic, spatial and thermodynamic factors of the molecules (Karki and Kulkarni 2001). A QSAR study using GFA was performed for some catechol and non-catechol derivatives having HIV-1 integrase inhibitory activity. From the study, it was found that for catechol derivatives, electronic, shape related and thermodynamic parameters where as for non-catechol derivatives spatial, structural and thermodynamic parameters played an important role in showing the activity (Makhija and Kulkarni 2002). The importance of size of the cluster on the predictive ability of a model was reported in a QSAR study of structurally diverse classes of compounds having HIV-1 integrase inhibition capacity using GFA as the chemometric tool. The best model suggested that the larger cluster of structural classes was better able to reproduce the biological activity (Yuan and Parrill 2002). Cytotoxic T cells are the integral part in adaptive immune response. The MHC class I molecules present antigenic peptides to cytotoxic T cells. QSAR models were developed with GFA and G/PLS algorithms to characterize interactions between bound peptides and MHC class I molecules (Davies et al. 2006). A GFA algorithm based QSAR model showed the importance of number of rotatable bonds, hydrogen-bonding properties and molecular connectivity descriptors in the binding affinity of arylpiperazines towards alpha-1 adrenoceptors (Maccari et al. 2006). Multiple temperature molecular dynamics simulation along with G/PLS was used to develop a free energy force field 3D-QSAR model for a ligand receptor binding process. This analysis was carried out for a set of p38-mitogen activated protein kinase inhibitors. The studies indicate the importance of van der Waals energy change upon binding and the electrostatic energy in the interaction of the ligands with the receptor (Romeiro et al. 2006). The antimicrobial and haemolytic activities of cyclic cationic peptide derived from protegrin-1 were analyzed in a QSAR model using the GFA algorithm. The models correlate antimicrobial potencies to peptide's charge and amphipathicity index, where as the haemolytic effect correlates with lipophilicity of residues forming the nonpolar face of the beta-hairpin (Frecer 2006). Another QSAR study of aryl heterocycle based thrombin inhibitors was performed using the GFA algorithm. The developed model had the capacity not only to predict the activity of new compounds but also to explain the important region in a molecule necessary for the activity (Deswal and Roy 2006). Inhibition of histone deacetylases is important in cancer therapy. A quantitative analysis using GFA showed that thermodynamic, shape and structural descriptors were important for inhibition of this enzyme (Wagh et al. 2006). Chalcones and flavonoids were studied as antitubercular agents. A QSAR modeling was done using GFA for better understanding the relationship between biological activity and structural features of chalcones and flavonoids (Sivakumar et al. 2007). A structure–activity relationship study of urea and thiourea derivatives of oxazolidinediones as antibacterial agents using GFA as the statistical tool showed that electron withdrawing groups at the ortho position of the phenyl ring enhances the activity of the compounds against various bacterial strain including clinical isolates and quality control organisms (Aaramadaka et al. 2007). Another QSAR study was performed for assessment of cyclooxygenase (COX-2 vs. COX-1) selectivity of nonsteroidal anti-inflammatory drugs from clinical practice using genetic function approximation. The study revealed the importance of thermodynamic, electronic, structural and shape parameters which can modulate the selectivity pattern to avoid side effects (Zambre et al. 2007). The *Pseudomonas aeruginosa* deacetylase LpxC [UDP-3-O-(R-3-hydroxymyristoyl)-GlcNAc deacetylase] inhibitory activity of dual PDE4-TNF alpha inhibitors was analyzed in a QSAR study using GFA. The developed models helped to validate potential leads for LpxC inhibition (Kadam et al. 2007). The pharmacokinetic parameters of oral fluoroquinolones were analysed in a QSPR model using the GFA algorithm. The study showed that a small volume, large polarizability and surface area of substituents at carbon number 7 contributed to large area under the curve for fluoroquinolones. It was also observed that large polarizability and small volume of substituents at N-1 contribute to long half life elimination (Cheng et al. 2007). Protein tyrosine phosphatase 1B is a negative regulator of the insulin receptor signaling system. Formylchromone derivatives show inhibitory action against the enzyme. A QSAR study was carried out with formylchromone derivatives using GFA and it was found that the inhibitory activities of these compounds depend on electronic, thermodynamic and shape related parameters (Sachan et al. 2007). Neuraminidase is one of the targets against the influenza virus. Thiourea analogs possess influenza virus neuraminidase inhibitory action. A QSAR study of these compounds was performed with spatial, topological, electronic, thermodynamic and electrotopological state (E-state) indices. GFA was used as the statistical tool for variable selection to generate the models. From the study, it was found that atom type log P and shadow indices descriptors had enormous contributions for the inhibitory activity (Nair and Sobhia 2008). Azole compounds including some commercial fungicides were subjected to a QSAR analysis for their binding affinity with cytochrome enzymes CYP 3A and CYP 2B. GFA and G/PLS were used as the chemometric tools (Roy and Roy 2008a). The models show that the binding affinity was related to topological, steric, electronic and spatial properties of the molecules. The spline based genetic models also indicate optimum range of different parameters (Roy and Roy 2008a). A QSAR study was carried out using GFA to determine non-specific chromosomal genotoxicity in relation to lipophilicity of compounds. It was found that the relation of polar surface to the total molecular surface plays an important role in the determination of the activity (Dorn et al. 2008). Genetic function approximation was used as the statistical tool in the QSAR studies for kinetic parameters of polycyclic aromatic hydrocarbon biotransformation by *Sphingomonas paucimobills* strain EPA505. The spatial descriptors present in the models were essential in explaining biotransformation kinetics (Dimitriou-Christidis et al. 2008). Aryl alkenyl amides/imines can be used as bacterial efflux pump inhibitors. GFA was used for variable selection to generate the QSAR model from several aryl alkenyl amides/imines derivatives. The models explain the important regions in the molecules necessary for activity (Nargotra et al. 2009a, b). N-aryl derivatives of amides and imides display varied inhibitory activity towards acetylcholinesterase and butyrylcholinesterase in Alzheimer's drug discovery. A QSAR model with these derivatives using GFA technique showed the importance of parameters like lipophilicity, connectivity, shape and dipole parameters in describing the bioactivity of the compounds (Solomon et al. 2009). GFA and G/PLS were used as the chemometric tools for QSAR and QAAR of a series of naphthalene and non-naphthalene derivatives possessing cytochrome P450 2A6 and 2A5 inhibitory activities. The studies show that the CYP2A5 and CYP2A6 inhibition activity of compounds is related to charge distribution, surface area, electronic, hydrophobic and spatial properties of the molecules (Roy and Roy 2009). The piperine analogs may be used as *Staphylococcus aureus* noradrenaline efflux pump inhibitors. The theoretical models of piperine analogs using GFA as the statistical tool showed that an increase in the exposed partial negative surface area increases the inhibitory activity of compounds against noradrenaline where as the area of the molecular shadow in the XY plane was inversely proportional to the inhibitory activity (Nargotra et al. 2009a, b). The QSAR study of aryl alkanol piperazine derivatives possessing antidepressant activity using GFA indicates the importance of various structural, spatial descriptors necessary for activity (Chen et al. 2009). A series of hetero aromatic tetrahydro-1,4-oxazine derivatives possessing antioxidant as well as squalene synthase inhibitory activities was subjected to a QSAR analysis using GFA and G/PLS techniques. The developed models suggest that the antioxidant activity was controlled by electrophilic nature of the molecules together with the charges on the phenolic hydrogen and the steric volume occupied by the molecules. For squalene synthase inhibitory activity, the charges on the hetero aromatic nucleus as well as the charge surface area of the molecules and their size governed the response (Roy et al. 2009). Protoporphyrinogen oxidase inhibitor 3H-pyrazolo[3,4-d][1,2,3]triazin-4-one derivatives are potential herbicides to protect agricultural products from unwanted weeds. A QSAR study using GFA and G/PLS suggest that for better activity, the molecules should have symmetrical shape in 3D space. Along with charged surface area, electrophilic and nucleophilic characters of the molecules also influence the activity (Roy and Paul 2010a). GFA was used for the development of a QSAR model for compounds showing N-type calcium channel blocking activity. The developed model identified the physicochemical features of the compounds relevant to N-type calcium channel blocking activity (Mungalpara et al. 2010). QSAR analyses of structurally diverse compounds possessing cytochrome 11B2 and 11B1 enzyme inhibitory activity using GFA and G/PLS indicate the importance of pyridinylnaphalene and pyridylmethylene-indane scaffolds with less polar and electrophilic substituents for optimum CYP11B2 inhibitory activity and CYP11B2/CYP11B1 selectivity (Roy and Roy 2010a). Another theoretical study using GFA was carried out for acetohydroxy acid synthase inhibitor sulfonylurea analogs. The developed models indicate the importance of bulky substitution, charged surface area, hydrogen bond acceptor parameters as well as the number of electronegative atom present in the molecules (Roy and Paul 2010b). Cytochrome 19 inhibitors are potential candidates for treatment of breast cancer. A QSAR study with GFA and G/PLS was carried out using molecular shape, spatial, electronic, structural and thermodynamic descriptors. The models obtained using the spline option showed better predictive capability (Roy and Roy 2010b). A QSAR study was performed for the free radical scavenging activity of flavone derivatives using GFA and G/PLS. The developed models indicate the importance of hydroxy and methoxy substituents present in the flavone moiety for the scavenging activity (Mitra et al. 2010). GFA and G/PLS were used for the development of QSAR models for androstendione derivatives possessing cytochrome 19 inhibitory activity. The developed models indicate the importance of spatial, structural and topological indices of different fragments (Roy and Roy 2010c).

9-Azido-noscapine and reduced 9-azido-noscapine were designed through a QSAR analysis of noscapinoids using GFA as the chemometric tool. The experimentally determined antitumor activities of the new compounds against human acute lymphoblastic leukemia cells were close to the predicted activity (Santoshi et al. 2011). Phosphodiesterase-4 enzyme inhibitors are used for various diseases like asthma, inflammation, rheumatoid arthritis, etc. A QSAR study was carried out using GFA for N-substituted cis-tetra and cis-hexahydrophthalazinone derivatives with potent anti-inflammatory activity. The analyses indicate that shape and structural descriptors strongly govern the phosphodiesterase-4 enzyme inhibition (Raichurkar et al. 2011). Another QSAR model for benzodithiazine derivatives was built using GFA to identify novel HIV-1 integrase inhibitors. Four benzodithiazine derivatives were identified as novel HIV-1 integrase inhibitors (Gupta et al. 2012). GFA technique was used for the development of QSAR models for 2-nitroimidazo-[2,1-b][1,3] oxazines as antitubercular agents. The GFA model with spatial, thermodynamic and topological descriptors appeared to be the best model (Ray and Roy 2012). The inhibitors of p53-HDM2 interaction are useful for treatment of wild type p53 tumors. A QSAR study was performed with GFA and G/PLS for 1,4-benzodiazepine-2,5-diones as HMD2 antagonist. The bioactivities of some new compounds were predicted with this model (Dai et al. 2012). Several QSAR models were developed with GFA and G/PLS techniques for lipid peroxidation inhibitory activity of cinnamic acid and caffeic acid derivatives. The model from GFA-spline techniques yielded most satisfactory results. The studies signify the importance of ketonic oxygen of the amide/acid fragment and the ethereal oxygen substituted on the parent phenyl ring of the molecules (Mitra et al. 2012). The interaction mechanism and binding properties of flavonoid-lysozyme were investigated in a QSAR model using GFA analysis. The result showed the importance of dipole moment, molecular refractivity, hydrogen-bond donor capacity of the molecules (Yang et al. 2012). Another computational study was carried out on novel 1,4-diazepane-2,5-dione derivatives with chymase inhibitory activity for the development of cardiovascular and anti-allergy agents with the GFA technique to construct 3D-QSAR models. The study explored the crucial molecular features contributing to the binding specificity (Arooj et al. 2012). Human leukotriene A4 hydrolase inhibitors play an important role in the treatment of inflammatory response exhibited through leukotriene B4. QSAR models were developed using GFA for compounds with leukotriene A4 hydrolase inhibitory activity (Thangapandian et al. 2013). A QSAR approach was employed for 1,6-dihydropyrimidine derivatives showing antifungal activity against *Candida albicans* (MTCC, 227). By using GFA as the statistical tool, it was observed that electron withdrawing substitution on N-phenyl acetamide ring of 1,6-dihydropyrimidine moiety lead to good activity (Rami et al. 2013).

20.3.2 Application of Genetic Algorithm in QSPR/QSTR Studies

GFA can also be utilized as a powerful tool in the analysis of property and toxicity of chemicals. A QSPR study was done to evaluate the release of flavor from i-carrageenan matrix using GFA as the statistical tool. The study showed that carrageenan polymers only modulate the interaction of aroma compounds with water molecules (Chana et al. 2006). The characteristic aspects of dielectric constants of conjugated organic compounds were elucidated by a QSPR study with the GFA technique. The dielectric constants of the organic compounds depend on the orientational correlations of the constituent molecules. Hydrogen bonding and π–π interaction affect the correlations (Lee et al. 2012). A QSPR method with GFA was used to predict the molecular diffusivity of structurally diverse classes of non-electrolyte organic compounds in air at 298.15 K and atmospheric pressure (Mirkhani et al. 2012). Quantitative structure-fate relationship study was employed using GFA and G/PLS for environmental toxicological fate prediction of diverse organic chemicals based on steady state compartmental chemical mass ratio. The models suggest that partition coefficient, degradation parameters, vapor pressure, diffusivity, spatial descriptors, thermodynamic descriptors and electrotopological descriptors are important for predicting chemical mass ratios (Pramanik and Roy 2013). Polybrominated diphenyl ethers are used as effective flame retardants. QSPR models using GFA have been developed for predicting toxic endpoints of these compounds in mammalian cells (Rawat and Bruce 2014). GFA and G/PLS techniques were also applied for a QSPR modeling of bioconcentration factor of diverse chemicals. The developed model can be used in the context of aquatic chemical toxicity management (Pramanik and Roy 2014).

20.3.3 Comparison of QSAR/QSPR Models Generated by Genetic Algorithm with Those Using Other Statistical Techniques

Several QSAR models were developed for camptothecin analogues having antitumor activity using different statistical methods including GFA. The model developed from GFA showed the best statistical quality. The results indicate the importance of partial atomic charges, interatomic distances that define the relative spatial disposition of three significant atoms such as hydrogen of the hydroxyl group, lactone carbonyl oxygen and the carbonyl oxygen of the camptothecin analogues (Fan et al. 2001). 3D-QSAR, comparative molecular field analysis (CoMFA) and comparative molecular similarity indices analysis (CoMSIA) were carried out for 1,4 dihydropyridine derivatives showing antitubercular activity. Both CoMFA and CoMSIA models based on multifit alignment showed better correlative and predictive properties than other models. The QSAR models with GFA suggest the importance of spatial properties and conformational flexibility of the side chain for antitubercular activity (Kharkar et al. 2002). The QSTR modeling of the acute toxicity of phenylsulfonyl carboxylates to *Vibrio fischeri* was performed with extended topochemical atom (ETA) indices using GFA as the statistical tool. The developed equation was better in statistical quality than that obtained previously using principal component analysis as the data processing step (Roy and Ghosh 2005). Thiazole and thiadiazole derivatives show potent and selective human adenosine A3 receptor antagonistic activity. QSAR models were developed using factor analysis followed by MLR (FA-MLR) and GFA-MLR techniques. The best two equations derived from GFA showed better predicted values than that found in case of the best equation derived from FA-MLR (Bhattacharya et al. 2005). A theoretical study was performed by linear free energy related model for 5-phenyl-1-phenylamino-1H-imidazole derivative possessing anti-HIV activity. Both FA-MLR and GFA were used as the chemometric tools. In this study, GFA produced the same best equation as obtained with FA-MLR. The study showed the structural and physicochemical contributions of the compounds for the cytotoxicity (Roy and Leonard 2005). Considering the potential of adenosine A3 receptor ligands for development of therapeutic agents, the A3 receptor antagonistic activity of 1,2,4-triazolo[4,3-a]quinoxalin-1-one derivatives was subjected to QSAR analysis using GFA and FA-MLR. Both the techniques led to the development of same equation. The best equation derived from G/PLS showed a little improvement in the explained variance. The results suggested that presence of electron withdrawing group at the para position of the phenyl ring would be favorable for the binding affinity (Bhattacharya and Roy 2005).

The assessment of toxicity of organic chemicals with respect to their potential hazardous effects on living systems is very important. A QSTR analysis was performed for acute toxicity of benzene derivatives to tadpoles using GFA and FA-MLR techniques (Roy and Ghosh 2006). The study suggested a parabolic dependence of the toxicity on molecular size. It was also observed that toxicity increases with a chloro substituent and decreases with methoxy, hydroxyl, carboxy and amino groups present in the molecules. The GFA models outperformed the FA-MLR models in terms of various validation metrics. A QSTR and toxicophore study of hERG K^+ channel blockers was performed using GFA and hypogen techniques respectively. Statistically significant QSTR models were developed from GFA. The hypogen model showed three important features like hydrophobic groups, ring aromatic group and hydrogen bond acceptor lipid group for the hERG K^+ channel blockers (Garg et al. 2008). The authors commented that the GFA derived 2D-QSTR model and the toxicophore model could be used in combination as a preliminary guidance for explaining hERG channel liabilities in early lead candidates. 1-Aryl-tetrahydroisoquinoline derivatives show anti HIV activity. A QSAR study was carried out for these types of compounds using GFA and stepwise regression analysis. The models from both techniques have similar predictive quality (Chen et al. 2008). Stepwise regression, partial least squares, GFA and G/PLS were used as the statistical tools in comparative QSAR studies for flavonoids having CYP1A2 inhibitory activity. The best model was obtained from G/PLS (Roy and Roy 2008b). Indolyl aryl sulfone derivatives are a class of novel HIV-1 non-nucleoside reverse transcriptase inhibitors. Linear and non-linear predictive QSAR models were developed using stepwise regression analysis, partial least squares, FA-MLR, GFA, G/PLS and artificial neural network. The model from GFA shows the best external predictive capacity (Roy and Mandal 2008). Tetrahydroimidazo[4,5,1-jk][1,4]benzodiazepine derivatives have reverse transcriptase inhibitory property. Various linear and nonlinear QSAR models were developed for these compounds. The best internal predictive ability of a model was obtained from the model developed with GFA (spline) (Mandal and Roy 2009). The antioxidant activities of hydroxybenzalacetones against lipid peroxidation induced by t-butyl hydroperoxide, gamma-irradiation and also their 1,1-diphenyl-2-picrylhydrazyl radical scavenging activity were modeled using different QSAR techniques such as stepwise regression, GFA and G/PLS. The best models for these responses are obtained from GFA and G/PLS (Mitra et al. 2009). A QSPR analysis on amino acid conjugates of jasmonic acid as defense signaling molecules was carried out using GFA and molecular field analysis. The models derived from both techniques showed high statistical quality (Li et al. 2009). The prediction of flashpoint of ester derivatives was determined by development of a QSPR model using GFA and adaptive neuro-fuzzy inference system (ANFIS) techniques. The results obtained showed the ability of the developed GFA and ANFIS models for prediction of flash point of esters (Khajeh and Modarress 2010).

The above findings suggest the efficient use of genetic algorithm for the development of best predictive QSAR/QSPR models. However, it is not true that GFA or genetic methods will always give the best results. For example, a comparative QSAR modeling was performed for CCR5 receptor binding affinity of substituted 1-(3,3-diphenylpropyl)-piperidinyl amides and ureas using various statistical techniques like stepwise MLR, FA-MLR, FA-PLS, PCRA, GFA-MLR and G/PLS. The study showed the importance of structural and physicochemical parameters towards the activity. However, the GFA derived models show high intercorrelation among predictor variables. The G/PLS model shows the lowest statistical quality among all types of models (Leonard and Roy 2006). In another study, GFA, enhanced replacement method and stepwise regression analysis techniques were used for development of QSAR models for bisphenylbenzimidazoles as inhibitors of HIV-1 reverse transcriptase. However, the model from enhanced replacement method showed better statistical quality in comparison to models obtained from GFA or stepwise regression analysis (Kumar and Tiwari 2013). The problem of intercorrelation in GFA derived models may be overcome by running PLS regression using GFA selected descriptors (Roy et al. 2015).

20.4 An Illustrative Example of GFA Model Development

Recently Roy and Popelier reported model development for the chromatographic lipophilicity parameter ($log\,k_0$) of ionic liquid cations with extended topochemical atom (ETA) and quantum topochemical molecular similarity (QTMS) descriptors (Roy and Popelier 2014). Experimental $log\,k_0$ values obtained from high performance liquid chromatography for 65 cations were considered for model development and validation. A total of thirty eight descriptors was computed. The ETA descriptors were calculated using the PaDEL-Descriptor openware (Yap 2011) and the QTMS descriptors were computed using the computer program MORPHY (Popelier 1996). The whole data set was divided into a training set and a test set (70 % and 30 % respectively of full data set size) using the Kennard–Stone algorithm (Kennard and Stone 1969). Initially MLR equations were developed from the training set compounds using various techniques of descriptor selection like stepwise selection, all possible subset regression, factor analysis and genetic function approximation. The stepwise selection was based on the F value ($F = 4.0$ or higher for inclusion and $F = 3.9$ or lower for exclusion). In case of all-possible-subset regression, a maximum cut-off inter-correlation of 0.7 and minimum R^2 of 0.9 were used. Factor analysis was performed to display multidimensional data in a space of lower dimensionality with minimum loss of information (i.e., explaining more than 95 % of the variance of the data matrix) and to extract the basic features behind the data, and using the factor scores important descriptors were selected for MLR model development. For GFA-MLR models, a spline option was used because it partitions the data set into groups having similar features and can account for non-linear behavior. Finally, partial least squares (PLS) regression was performed using the terms appearing in the MLR equations. PLS avoids the problem of inter-correlation among descriptors as present in MLR. For PLS model development, optimum number of latent variables was selected through leave-one-out (LOO) cross-validation. The developed models were validated internally on LOO validation and Y-randomization test and externally using test set prediction. Considering all statistical parameters, the PLS model developed using descriptors obtained from GFA (spline)-MLR was found to be the best one. The model is as follows:

$$\log k_0 = -0.580 + 0.484 < \sum \alpha - 4.633 > +10.289 < 0.125 - \Delta\varepsilon_A > +11.177 < G - 0.266 >$$
$$LVs = 2, R^2 = 0.974, Q^2_{LOO} = 0.968, Q^2_{ext} = 0.888 \quad (20.8)$$

In the above equation, $\Sigma\alpha$ and $\Delta\varepsilon_A$ are the ETA descriptors and G is a QTMS descriptor. LV indicates the number of latent variables selected, R^2 is the determination coefficient of the model, Q^2_{LOO} is the leave-one-out cross-validation metric and Q^2_{ext} is the external validation metric. The data set (training and test set composition with descriptor values and the response being modeled) for the model presented in Eq. (20.8) is separately provided with this chapter. The positive coefficient of the term $<\Sigma\alpha - 4.633>$ in the above model indicates that when the value of $\Sigma\alpha$ is higher than 4.633, it exerts a positive contribution to *log* k_0. The positive coefficient of the descriptor $<0.125 - \Delta\varepsilon_A>$ indicates that when the value of the descriptor $\Delta\varepsilon_A$ is lower than 0.125, it contributes positively to the *log* k_0. The positive coefficient of the descriptor $<G - 0.266>$ indicates that when the value of G is higher than 0.266, it has a positive contribution to *log* k_0. In this way, the developed model gives us an idea about different range values for different descriptors for their contribution to the response being modeled and thus helps the modeler to select or design samples with optimum ranges of the descriptor values for optimizing the endpoint.

20.5 Overview and Conclusion

The present chapter has attempted to explain the use of genetic algorithm along with multivariate adaptive regression splines in regression studies and the role of GFA and G/PLS in medicinal chemistry as well as in predictive toxicology. The initial discussion has been made on the algorithm of genetic function approximation followed by fitness measures in GFA. This is followed by a brief discussion on the use of GFA-spline techniques in regression studies. Finally the applications of GFA-MLR and G-PLS in QSAR/QSPR/QSTR studies have been reviewed. An increasing number of published papers on the use of genetic algorithm in regression based QSARs show effectiveness and validity of the genetic method as one of popular feature selecting algorithms. Another advantage of GA is that it is able to extract the most information rich combination of descriptors form a complex pool of data. Although it is computationally expensive but the facilities available now make it faster and computationally economic. Hybridization of GA with spines can handle non-linear character in the data. Use of GFA along with MLR and PLS increases predictive and interpretable nature of the models significantly. Drawbacks of MLR (collinearity) and PLS (noise problem) can be reduced to some extent by GA and the combined methods signify themselves to be valuable analysis tools in the case where the number of data points is much more than the samples. As a result, these methods infiltrate in different fields of chemistry, biology and chemometrics for solving different problems as addressed by the users. However, GFA models may still suffer from the problem of inter-correlation among descriptors which may be overcome by running PLS regression on the GFA selected descriptor combination. Again, G/PLS may sometimes show poor performance which may be overcome by thinning the initial pool of descriptors before applying the genetic method. We have already discussed the application of GFA-MLR and G/PLS in QSAR and QSPR as well as toxicity modeling. It is observed that in many cases, genetic methods can provide statistically more robust models in comparison to other conventional statistical techniques. LOF and LSE are the fitness measures for GFA and G/PLS respectively. It is also to be pointed out that the models developed are not assessed solely on these two parameters. Along with these two parameters (LOF, LSE), different internal and external validation metrics (Roy and Mitra 2011) are checked in order to validate the models before their possible application to a new set of data for ranking and prioritization of chemicals. Therefore, not only hybridization of the techniques but also combined use of different statistical parameters may yield robust and predictive models in different QSAR/QSPR and toxicity modeling studies.

References

Aaramadaka SK, Guha MK, Prabhu G et al (2007) Synthesis and evaluation of urea and thiourea derivatives of oxazolidinones as antibacterial agents. Chem Pharm Bull (Tokyo) 55: 236–240.

Allen DM (1971) Mean Square Error of Prediction as a Criterion for Selecting Variables. Technometrics 13: 469–475.

Andrey P, Tarroux P (1994) Unsupervised image segmentation using a distributed genetic algorithm. Pattern Recogn 27: 659–673.

Angeline PJ (1995) Evolution revolution: An introduction to the special track on genetic and evolutionary programming, IEEE Expert Intell Syst Appl 10 (June): 6–10.

Archetti F, Giordani I, Vanneschi L (2010) Genetic programming for QSAR investigation of docking energy. Appl Soft Comput J 10: 170–182.

Arooj M, Thangapandian S, John S et al (2012) Computational studies of novel chymase inhibitors against cardiovascular and allergic diseases: mechanism and inhibition. Chem Biol Drug Des 80: 862–875.

Bagheri M, Bagheri M, Gandomi AH, Golbriakh A (2012) Simple yet accurate prediction method for sublimation enthalpies of organic contaminants using their molecular structure. Thermochimica Acta 543: 96–106.

Bagheri M, Borhani TN, Gandomi AH, Manan ZA (2014) A simple modelling approach for prediction of standard state real gas entropy of pure materials. SAR QSAR Environ Res 25: 695–710.

Bagheri M, Gandomi AH, Bagheri M, Shahbaznezhad M (2012) Multi-expression programming based model for prediction of formation enthalpies of nitro-energetic materials. Expert Systems 30: 66–78.

Bäck T (1993) Optimal mutation rates in genetic search. In: Forrest S (ed) Proceedings of the Fifth International Conference on Genetic Algorithms, Morgan Kaufmann Publishers, San Mateo, California, p 2–8.

Bangalore AS, Shaffer RE, Small GW (1996) Genetic algorithm-based method for selecting wavelengths and model size for use with partial least-squares regression: application to near-infrared spectroscopy. Anal Chem 68: 4200–4212.

Bentley J (1975) Multidimensional Binary Search Trees used for Associative Searching. Commun ACM 18: 509–517.

Bhattacharya P, Leonard JT, Roy K (2005) Exploring QSAR of thiazole and thiadiazole derivatives as potent and selective human adenosine A3 receptor antagonist using FA and GFA techniques. Bioorg Med Chem 13: 1159–1165.

Bhattacharya P, Roy K (2005) QSAR of adenosine A3 receptor antagonist 1, 2, 4-triazolo [4, 3-a] quinoxalin-1-one derivatives using chemometric tools. Bioorg Med Chem Lett 15: 3737–3743.

Blaney F (1990) Molecular modelling in the pharmaceutical industry Chem. Indus XII: 791–794.

Breiman L, Friedman J, Olshen R et al (1984) Classification and Regression Trees. Wadsworth, Belmont, CA.

Broadhurst D, Goodacre R, Jones A et al (1997) Genetic algorithms as a method for variable selection in multiple linear regression and partial least squares regression, with applications to pyrolysis mass spectrometry. Anal Chim Acta 348: 71–86.

Bugg CE, Carson WM, Montgomery JA (1993) Drugs by design. Sci Am December: 60–66.

Caldwell C, Johnston VS (1991) Tracking a criminal suspect through 'face-space' with a genetic algorithm. In: Belew RK, Booker LB (eds) Proceedings of the Fourth International Conference on Genetic Algorithms, Morgan Kaufman n Publishers, San Mateo, California, p 416–421.

Cleveland GA and Smith SF (1989) Using genetic algorithms to schedule flow shop releases. In: Schaffer JD (ed) Proceedings of the Third International Conference on Genetic Algorithms, Morgan Kaufmann Publishers, San Mateo, California, p 160–169.

Chana A, Tromelin A, Andriot I et al (2006) Flavor release from i-carrageenan matrix: a quantitative structure property relationships approach. J Agric Food Chem 54: 3679–3685.

Chang G and Lewis M (1994) Using genetic algorithms for solving heavy-atom sites. Acta Cryst D50: 667–674.

Chen KX, Xie HY, Li ZG et al (2008) Quantitative structure-activity relationship studies on 1-aryl-tetrahydroisoquinoline analogs as active anti-HIV agents. Bioorg Med Chem Lett 18: 5381–5386.

Chen KX, Li ZG, Xie HY et al (2009) Quantitative structure-activity relationship analysis of aryl alkanol piperazine derivatives with antidepressant activities. Eur J Med Chem 44: 4367–4375.

Cheng D, Xu WR, Liu CX (2007) Relationship of quantitative structure and pharmacokinetics in fluoroquinolone antibacterials. World J Gastroenterol 13: 2496–2503.

Craven P and Wahba G (1979) Smoothing noisy data with spline functions. Numer Math 31: 377–403.

Dai Y, Chen N, Wang Q et al (2012) Docking analysis and multidimensional hybrid QSAR model of 1, 4-benzodiazepine-2, 5-diones as HM2 antagonists. Iran J Pharm Res 11: 807–830.

Darwin C (1859) On the origin of Species by Means of Natural selections, or the Preservation of Favoured Races in the Struggle for life. Nature 5 (121): 502.

Davies MN, Hattotuwagama CK, Moss DS et al (2006) Statistical deconvolution of enthalpic energetic contributions to MHC-peptide binding affinity. BMC Struct BIOL 6: 5.

Deswal S, Roy N (2006) Quantitative structure activity relationship studies of aryl heterocycle-based thrombin inhibitors. Eur J Med Chem 41:1339–1346.

Dimitriou-Christidis P, Autenrieth RL, Abraham MH (2008) Quantitative structure-activity relationships for kinetic parameters of polycyclic aromatic hydrocarbon biotransformation. Environ Toxicol Chem 27: 1496–1504.

Dorn SB, Degen GH, Bolt HM et al (2008) Some molecular descriptors for non-specific chromosomal genotoxicity based on hydrophobic interactions. Arch Toxicol 82: 333–338.

Devillers J (1996) Genetic Algorithms in Molecular Modeling, Elsevier Science & Technology Books.

Eshelman LJ, Caruana RA, Schaffer JD (1989) Biases in the crossover landscape. In: Schaffer JD (ed) Proceedings of the Third International Conference on Genetic Algorithms, Morgan Kaufmann Publishers, San Mateo, California, p 10–19.

Fan Y, Shi LM, Kohn KW et al (2001) Quantitative structure-antitumor activity relationships of camptothecin analogues: cluster analysis and genetic algorithm-based studies. J Med Chem 44: 3254–3263.

Forrest S (1993) Genetic algorithms: principles of natural selection applied to computation. Science 261:872–878.

Fraser A (1957) Simulation of genetic systems by automatic digital computers. I. Introduction. Aust J Biol Sci 10: 484–491.

Frecer V (2006) QSAR analysis of antimicrobial and haemolytic effects of cyclic cationic antimicrobial peptides derived from protegrin-1. Bioorg Med Chem 14: 6065–6074.

Friedman J (1988) Multivariate Adaptive Regression Splines, Technical Report No. 102, Laboratory for Computational Statistics, Department of Statistics, Stanford University, Stanford, CA, Nov 1988 (revised Aug 1990).

Friedman JH (1991) Multivariate adaptive regression splines. Annals of Statistics 19: 1–141.

Friedman JH and Silverman BW (1989) Flexible parsimonious smoothing and additive modeling. Technometrics 31: 3–39.

Friedman JH and Stuetzle W (1981) Projection pursuit regression. J Amer Statist Assoc 76: 817–823.

Gabbert PS, Markowicz BP, Brown DE et al (1991) A system for learning routes and schedules with genetic algorithms. In: Belew RK and Booker LB (eds) The Proceedings of the Fourth International Conference on Genetic Algorithms, Morgan Kaufmann Publishers, San Mateo, California, p 430–436.

Gandomi AH, Alavi AH (2013) Hybridizing genetic programming with orthogonal least squares for modeling of soil liquefaction. Int J Earthquake Engg Hazard Mitig 1: 2–8.

Gandomi AH, Alavi AH, Arjmandi P, Aghaeifar A, Seyednour R (2010) Genetic programming and orthogonal least squares: A hybrid approach to modeling the compressive strength of CFRP-confined concrete cylinders. J Mech Mater Struct 5: 735–753.

Garg D, Gandhi T, Gopi Mohan C (2008) Exploring QSTR and toxicophore of hERG K^+ channel blockers using GFA and HypoGen techniques. J Mol Graph Model 26: 966–976.

Goldberg DE (1989a) Sizing populations for serial and parallel genetic algorithms. In: Schaffer JD (ed) Proceedings of the Third International Conference on Genetic Algorithms, Morgan Kaufmann Publishers, San Mateo, California, p 70–79.

Goldberg DE (1989b) Genetic Algorithms in Search, Optimization & Machine Learning. Addison-Wesley Publishing Company, Reading, p 412.

Gokhale VM, Kulkarni VM (2000) Understanding the antifungal activity of terbinafine analogues using quantitative structure-activity relationship (QSAR) models. Bioorg Med Chem 8: 2487–2499.

Gramatica P (2007) Principles of QSAR models validation: internal and external. QSAR Comb Sci 26: 694–701.

Gupta P, Garg P, Roy N (2012) Identification of novel HIV-1 integrase inhibitors using shape-based screening, QSAR and docking approach. Chem Biol Drug Des 79: 835–849.

Hahn M, Rogers D (1995) Receptor surface models. 2. Application to quantitative structure-activity relationship studies. J Med Chem 38: 2091–2102.

Hansch C, Fujita T (1964) p-σ-π Analysis. A Method for the Correlation of Biological Activity and Chemical Structure. J Am Chem Soc 86: 1616–1626.

Hansch C, Maloney PP, Fujita T et al (1962) Correlation of Biological Activity of Phenoxyacetic Acids with Hammett Substituent Constants and Partition Coefficients. Nature 194: 178–180.

Harrigan GG, LaPlante RH, Cosma GN, Cockerell G, Goodacre R, Maddox JF, Luyendyk JP, Ganey PE, Roth RA (2004) Application of high-throughput Fourier-transform infrared spectroscopy in toxicology studies: Contribution to a study on the development of an animal model for idiosyncratic toxicity. Toxicol Lett 146: 197–205.

Hartke B (1993) Global geometry optimization of clusters using genetic algorithms. J Phys Chem 97: 9973–9976.

Hightower RR, Forrest S, Perelson AS (1995) The evolution of emergent organization in immune system gene libraries. In: Eshelman LJ (ed) Proceedings of the Sixth International Conference on Genetic Algorithms, Morgan Kaufmann Publishers, San Francisco, California, p 344–350.

Holland J (1975) Adaptation in Artificial and Natural Systems. University of Michigan Press, Ann Arbor, MI, 1975.

Horner A, Goldberg DE (1991) Genetic algorithms and computer-assisted n music composition. In: Belew RK, Booker LB (eds) Proceedings of the Fourth International Conference on Genetic Algorithms, Morgan Kaufmann Publishers, San Mateo, California, p 437–441.

Jaeger EP, Pevear DC, Felock PJ et al (1995) Genetic algorithm based method to design a primary screen for antirhinovirus agents. In: Reynold CH, Holloway MK, Cox HK (eds) Computer-Aided Molecular Design: Applications in Agrochemicals, Materials, and Pharmaceuticals ACS Symposium Series 589, American Chemical Society, Washington DC, p 139–155.

Janikow CZ, Michalewicz Z (1991) An Experimental Comparison of Binary and Floating Point Representations in Genetic Algorithms. In: Belew RK and Booker LB (eds) Proceedings of the Fourth International Conference on Genetic Algorithms, p 31–36.

Jones T (1995) Crossover, macromutation, and population-based search. In: Eshelman LJ (ed) Proceedings of the Sixth International Conference on Genetic Algorithms, Morgan Kaufmann Publishers, San Francisco, California, p 73–80.

Kadam RU, Garg D, Chavan A et al (2007) Evaluation of pseudomonas aeruginosa deacetylase LpxC inhibitory activity of dual PDE4-TNF alpha inhibitors: a multiscreening approach. J Chem Inf Model 47: 1188–1195.

Kar S, Roy K (2010) Predictive toxicology using QSAR : A perspective. J Indian Chem Soc 87: 1455–1515.

Karki RG, Kulkarni VM (2001) Three dimensional quantitative structure-activity relationship (3D-QSAR) of 3-aryloxazolidin-2-one antibacterials. Bioorg Med Chem 9:3153–3160.

Karr CL (1991) Air-injected hydrocyclone optimization via genetic algorithm. In: Davis L (ed) Handbook of Genetic Algorithms, Van Nostrand Reinhold, New York, p 222–236.

Kennard RW, Stone LA (1969) Computer aided design of experiments. Technometrics 11: 137–148.

Khajeh A, Modarress H (2010) QSPR prediction of flash point of esters by means of GFA and ANFIS. J Hazard Mater 179: 715–720.

Kharkar PS, Desai B, Gaveria H et al (2002) Three dimensional quantitative structure-activity relationship of 1, 4-dihydropyridines as antitubercular agents. J Med Chem 45: 4858–4867.

Klein CD, Klingmuller M, Schellinski C et al (1999) Synthesis, pharmacological and biophysical characterization and membrane-interaction QSAR analysis of cationic amphiphilic model compounds. J Med Chem 42:3874–3888.

Klein CD, Hopfinger AJ (1998) Pharmacological activity and membrane interactions of antiarrhythmics: 4D QSAR/QSPR analysis. Pharm Res 15:303–311.

Kulkarni AS, Hopfinger AJ (1999) Membrane-interaction QSAR analysis: application to the estimation of eye irritation by organic compounds. Pharm Res 16:1245–1253.

Kumar S, Tiwari M (2013) Variable selection based QSAR modeling on Bisphenylbenzimidazole as inhibitor of HIV-1 reverse transcriptase. Med Chem 9: 955–967.

Lee A, Kim D, Kim KH et al (2012) Elucidation of specific aspects of dielectric constants of conjugated organic compounds: a QSAR approach. J Mol Model 18: 251–256.

Leonard JT, Roy K (2006) Comparative QSAR modeling of CCR5 receptor binding affinity of substituted 1-(3, 3-diphenylpropyl)-piperidinyl amides and ureas. Bioorg Med Chem Lett 16: 4467–4474.

Li ZG, Chen KX, Xie HY et al (2009) Quantitative structure-property relationship studies on amino acid conjugates of jasmonic acid as defense signaling molecules. J Integr Plant Biol 51: 581–592.

Maddox J (1995) Genetics helping molecular dynamics. Nature 376: 209.

Maccari L, Magnani M, Strappaghetti G et al (2006) A genetic-function-approximation-based QSAR model for the affinity of arylpiperazines toward alpha 1 adrenoceptors. J Chem Inf Model 46: 1466–1478.

Mandal AS, Roy K (2009) Predictive QSAR modeling of HIV reverse transcriptase inhibitor TIBO derivatives. Eur J Med Chem 44: 1509–1524.

Makhija MT, Kulkarni VM (2002) QSAR of HIV-1 integrase inhibitors by genetic function approximation method. Bioorg Med Chem 10:1483–1497.

Mathias KE and Whitley LD (1994) Transforming the search space with Gray coding. Proceedings of the First IEEE Conference on Evolutionary Computation. In: IEEE World Congress on Computational Intelligence, Vol 1.

Mirkhani SA, Gharagheizi F, Sattari M (2012) A QSPR model for prediction of diffusion coefficient of non-electrolyte organic compounds in air at ambient condition. Chemosphere 86: 959–966.

Mitra I, Saha A, Roy K (2009) Quantitative structure-activity relationship modeling of antioxidant activities of hydroxybenzalacetones using quantum chemical, physicochemical and spatial descriptors. Chem Biol Drug Des 73: 526–536.

Mitra I, Saha A, Roy K (2010) Chemometric modeling of free radical scavenging activity of flavone derivatives. Eur J Med Chem 45: 5071–5079.

Mitra I, Saha A, Roy K (2012) In silico development. Validation and comparison of predictive QSAR models for lipid peroxidation inhibitory activity of cinnamic acid, caffeic acid derivatives using multiple chemometric and cheminformatics tools. J Mol Model 18: 3951–3967.

Morgan JN, Sonquist JA (1963) Problems in the analysis of survey data, and a proposal. J Amer Statist Assoc 58: 415–434.

Mungalpara J, Pandey A, Jain V et al (2010) Molecular modeling and QSAR analysis of some structurally diverse N-type calcium channel blockers. J Mol Model 16: 629–644.

Nargotra A, Sharma S, Koul JL et al (2009) Quantitative structure-activity relationship (QSAR) of piperine analogs for bacterial NorA efflux pump inhibitors. Eur J Med Chem 44: 4128–4135.

Nair PC, Sobhia ME (2008) Quantitative structure activity relationship studies on thiourea analogues as influenza virus neuraminidase inhibitors. Eur J Med Chem 43: 293–299.

Nargotra A, Koul S, Sharma S et al (2009) Quantitative structure-activity relationship (QSAR) of aryl alkenyl amides/imines for bacterial efflux pump inhibitors. Eur J Med Chem 44: 229–238.

Pearson K (1908) On the generalized probable error in multiple normal correlation. Biometrika, 6: 59–68.

Popelier PLA (1996) MORPHY, a program for an automated "atoms in molecules" analysis. Comput Phys Commun 93: 212–240.

Pramanik S, Roy K (2013) Environmental toxicological fate prediction of diverse organic chemicals based on steady-state compartmental chemical mass ratio using quantitative structure-fate relationship (QSFR) studies. Chemosphere 92: 600–607.

Pramanik S, Roy K (2014) Modeling bioconcentration factor (BCF) using mechanically interpretable descriptors computed from open source tool "PaDEL-Descriptor". Environ Sci Pollut Res Int 21: 2955–2965.

Pugazhenthi D, Rajagopalan SP (2007) Machine learning technique approaches in drug discovery, design and development. Inf Tech J 6: 718–724.

Raichurkar AV, Shah UA, Kulkarni VM (2011) 3D-QSAR of novel phosphodiesterase-4 inhibitors by genetic function approximation. Med Chem 7: 543–552.

Rami C, Patel L, Patel CN et al (2013) Synthesis, antifungal activity and QSAR studies of 1, 6-dihydropyrimidine derivatives. J Pharm Bioallied Sci 5: 277–289.

Rawat S, Bruce ED (2014) Designing quantitative structure activity relationships to predict specific toxic endpoints for polybrominated diphenyl ethers in mammalian cells. SAR QSAR Environ Res 16: 1–23.

Ray S, Roy PP (2012) A QSAR study of biphenyl analogues of 2-nitroimidazo-[2, 1-b] [1, 3]-oxazines as antitubercular agents using genetic function approximation. Med Chem 8: 717–726.

Rogers D (1991) G/SPLINES: A hybrid of Friedman's multivariate adaptive regression splines (MARS) algorithm with Holland's genetic algorithm. In: The proceedings of fourth international conference on genetic algorithm, San Diego, July 1991.

Rogers D (1992) Data analysis using G/SPLINES. In: Advances in neural processing systems 4, Kaufmann, San Mateo, CA, 1992.

Rogers D, Hopfinger AJ (1994) Application of genetic function approximation to quantitative structure-activity relationships and quantitative structure-property relationships. J Chem Inf Comput Sci 34:854–866.

Romeiro NC, Albuquerque MG, de Alencastro Rb et al. (2006) Free-energy force-field three-dimensional quantitative structure-activity relationship analysis of a set of p38-mitogen activated protein kinase inhibitors. J Mol Model 12:855–868.

Rossi I, Truhlar DG (1995) Parameterization of NDDO wave functions using genetic algorithms. An evolutionary approach to parameterizing potential energy surfaces and direct dynamics calculations for organic reactions. Chem Phys Lett 23: 231–236.

Roy K, Das RN, Popelier PLA (2015) Predictive QSAR modelling of algal toxicity of ionic liquids and its interspecies correlation with Daphnia toxicity. Environ Sci Pollut Res, 22: 6634–6641

Roy K, Ghosh G (2005) QSTR with extended topochemical atom indices. Part 5: Modeling of the acute toxicity of phenylsulfonyl carboxylates to *Vibrio fischeri* using genetic function approximation. Bioorg Med Chem 13: 1185–1194.

Roy K, Ghosh G (2006) QSTR with extended topochemical atom (ETA) indices. VI. Acute toxicity of benzene derivatives to tadpoles (*Rana japonica*). J Mol Model 12: 306–316.

Roy K, Leonard JT (2005) QSAR by LFER model of cytotoxicity data of anti-HIV 5-phenyl-1-phenylamino-1H-imidazole derivatives using principal component factor analysis and genetic function approximation. Bioorg Med Chem 13: 2967–2973.

Roy K, Mandal AS (2008) Development of linear and nonlinear predictive QSAR models and their external validation using molecular similarity principle for anti-HIV indolyl aryl sulfones. J Enzyme Inhib Med chem. 23: 980–995.

Roy K, Mitra I (2011) On various metrics used for validation of predictive QSAR models with applications in virtual screening and focused library design. Comb Chem High Throughput Screen 14: 450–474.

Roy K, Mitra I, Saha A (2009) Molecular shape analysis of antioxidant and squalene synthase inhibitory activities of aromatic tetrahydro-1, 4-oxazine derivatives. Chem Biol Drug Des 74: 507–516.

Roy K, Paul S (2010a) Docking and 3D QSAR studies of protoporphyrinogen oxidase inhibitor 3H-pyrazolo [3, 4-d][1, 2, 3]triazin-4-one derivatives. J Mol Model 16: 137–153.

Roy K, Paul S (2010b) Docking and 3D QSAR studies of acetohydroxy acid synthase inhibitor sulfonylurea derivatives. J Mol Model 16: 951–964.

Roy K, Popelier PLA (2014) Chemometric modeling of the chromatographic lipophilicity parameter $\log k_0$ of ionic liquid cations with ETA and QTMS descriptors. J Mol Liq 200: 223–228.

Roy K, Roy PP (2008a) Exploring QSARs for binding affinity of azoles with CYP2B and CYP3A enzymes using GFA and G/PLS techniques. Chem Biol Drug Des 71: 464–473.

Roy K, Roy PP (2008b) Comparative QSAR studies of CYP1A2 inhibitor flavonoids using 2D and 3D descriptors. Chem Biol Drug Des 72: 370–382.

Roy K, Roy PP (2009) Exploring QSAR and QAAR for inhibitors of cytochrome P450 2A6 and 2A5 enzymes using GFA and G/PLS techniques. Eur J Med Chem 44: 1941–1951.

Roy PP, Roy K (2010a) Exploring QSAR for CYP11B2 binding affinity and CYP11B2/CYP11B1 selectivity of diverse functional compounds using GFA and G/PLS techniques. J Enzyme Inhib Med Chem 25: 354–369.

Roy PP, Roy K (2010b) Docking and 3D QSAR studies of diverse classes of human aromatase (CYP19) inhibitors. J Mol Model 16: 1597–1616.
Roy PP, Roy K (2010c) Molecular docking and QSAR studies of aromatase inhibitor androstendione derivatives. J Pharm Pharmacol 62: 1717–1728.
Sachan N, Kadam SS, Kulkarni VM (2007) Human protein tyrosine phosphatase 1B inhibitors: QSAR by genetic function approximation. J Enzyme Inhib Med Chem 22: 267–276.
Santoshi S, Naik PK, Joshi HC (2011) Rational design of novel anti-microtubule agent (9-azido noscapine) from quantitative structure activity relationship (QSAR) evaluation of noscapinoids. J Biomol Screen 16: 1047–1058.
Santos-Filho OA, Mishra RK, Hopfinger AJ (2001) Free energy force field (FEFF) 3D-QSAR analysis of a set of plasmodium falciparum dihydrofolate reductase inhibitors. J Comput Aided Mol Des 15: 787–810.
Schaffer JD, Caruana RA, Eshelman LJ, et al (1989) A study of control parameters affecting online performance of genetic algorithms for function optimization. In: Schaffer JD (ed) Proceedings of the Third International Conference on Genetic Algorithms, Morgan Kaufmann Publishers, San Mateo, California, p 51–60
Schaffer JD, Eshelman LJ (1991) On crossover as an evolutionarily viable strategy. In: Belew RK, Booker LB (eds) Proceedings of the Fourth International Conference on Genetic Algorithms. Morgan Kaufmann Publishers, San Mateo, California, p 61–68.
Shi LM, Fan Y, Myers TG et al (1998) Mining the NCI anticancer drug discovery databases: genetic function approximation for the QSAR study of anticancer ellipticine analogues. J Chem Inf Comput Sci 38:189–199.
Sivakumar PM, Geeta Babu SK, Mukesh D (2007) QSAR studies on chalcones and flavonoids as anti-tuberculosis agents using genetic function approximation (GFA) method. Chem Pharm Bull (Tokyo) 55: 44–49.
Solomon KA, Sundararajan S, Abirami V (2009) QSAR studies on N-aryl derivative activity towards Alzheimer's disease. Molecules 14: 1448–1455.
Spears WM (1993) Crossover or mutation? In Whitley LD (ed) Foundations of Genetic Algorithms. 2, Morgan Kaufmann Publishers, San Mateo, California, p 221–237.
Syswerda G (1989) Uniform crossover in genetic algorithms. In: Schaffer JD (ed) Proceedings of the Third International Conference on Genetic Algorithms, Morgan Kaufmann Publishers, San Mateo, California, p 2–9.
Syswerda G (1991) Schedule optimization using genetic algorithms. In: Davis L (ed) Handbook of Genetic Algorithms, Van Nostrand Reinhold, New York, p 332–349.
Syswerda G (1993) Simulated crossover in genetic algorithms. In: Whitley LD (ed), Foundations of Genetic Algorithms. 2, Morgan Kaufmann Publishers, San Mateo, California, p 239–255.
Tate DM, Smith AE (1993) Expected allele coverage and the role of mutation in genetic algorithms. In: Forrest S (ed) Proceedings of the Fifth International Conference on Genetic Algorithms, Morgan Kaufmann Publishers, San Mateo, California, p 31–37.
Thangapandian S, John S, Son M et al (2013) Development of predictive quantitative structure-activity relationship model and its application in the discovery of human leukotriene A4 hydrolase inhibitors. Future Med Chem 5: 27–40.
Venkatasubramanian V, Chan K, Caruthers JM (1994) Computer-aided molecular design using genetic algorithms. Comput Chem Engng 18: 833–844.
Venkatasubramanian V, Chan K, Caruthers JM (1994) On the performance of genetic search for large-scale molecular design. Proc. PSE'94, 1001–1006.
Venkatarangan P, Hopfinger AJ (1999) Prediction of ligand-receptor binding thermodynamics by free energy force field three-dimensional quantitative structure-activity relationship analysis: applications to a set of glucose analogue inhibitors of glycogen phosphorylase. J Med Chem 42:2169–2179.
Wagh NK, Deokar HS, Juvale DC et al (2006) 3D-QSAR of histone deacetylase inhibitors as anticancer agents by genetic function approximation. Indian J Biochem Biophys 43: 360–371.

Wise BM, Gallagher NB, Eschbach PA et al (1995) Optimization of prediction error using genetic algorithms and continuum regression: determination of the reactivity of automobile emissions from FTIR spectra. Fourth Scand Symp. On Chemometrics (SSC4), Lund, June 1995.

Wise MB, Gallagher NB, Eschbach PA (1996) Application of a genetic algorithm to variable selection for PLS models. EUCHEM Conf, Gothenburg, June 1996.

Wold S (1995) Applications of Statistical Experimental Design and PLS Modeling in QSAR. In: Waterbeemd H (ed) Chemometric Methods in Molecular Design, VCH, Weinheim, Germany p 195–218.

Yang R, Yu L, Zeng H et al (2012) The interaction of flavonoid-lysozyme and the relationship between molecular structure of flavonoids and their binding activity to lysozyme. J Fluoresc 22: 1449–1459.

Yap CW (2011) PaDEL-Descriptor: An open source software to calculate molecular descriptors and fingerprints. J Comput Chem 32:1466–1474.

Yuan H, Parrill AL (2002) QSAR studies of HIV-1 integrase inhibition. Bioorg Med Chem 10: 4169–4183.

Zambre AP, Ganure AL, Shinde DB et al (2007) Perspective assessment of COX-1 and COX-2 selectivity of nonsteroidal anti-inflammatory drugs from clinical practice: Use of genetic function approximation. J Chem Inf Model 47: 635–643.

Chapter 21
Trading Volatility Using Highly Accurate Symbolic Regression

Michael F. Korns

21.1 Introduction

The discipline of Symbolic Regression (SR) has matured significantly in the last few years. There is at least one commercial package on the market for several years (http://www.rmltech.com/). There is now at least one well documented commercial symbolic regression package available for Mathematica (www.evolved-analytics.com). There is at least one very well done open source symbolic regression package available for free download (http://ccsl.mae.cornell.edu/eureqa).

In addition to our own ARC system (Korns 2013, 2014), currently used internally for massive financial data nonlinear regressions, there are a number of other mature symbolic regression packages currently used in industry including Smits et al. (2010) and Castillo et al. (2010). Plus there is an interesting work in progress by McConaghy et al. (2009).

Research efforts, directed at increasing the accuracy and dependability of Symbolic Regression (SR), have resulted in significant improvements in symbolic regression's range, accuracy, and dependability (Korns 2013, 2014). Previous research has also demonstrated the practicability of estimating corporate forward 12 month earnings, using advanced symbolic regression (Korns 2012a, b). In this paper we put these prior results and techniques together to select a 100 stock semi-passive index portfolio (*VEP100*), from the Value Line Timeliness (*Value Line*), which delivers consistent performance in both bull and bear decades.

We intend to produce our VEP100 buy list on a weekly basis using automated *ftmEPS* prediction involving the analysis of many securities, involving multiple training regressions each on hundreds of thousands of training examples. Plus the timeliness issue will require that our analytic tools be strong and thoroughly

M.F. Korns (✉)
The AIS Foundation, 2240 Village Walk Drive, Henderson, NV 89052, USA
e-mail: mkorns@korns.com

A.H. Gandomi et al. (eds.), *Handbook of Genetic Programming Applications*,
DOI 10.1007/978-3-319-20883-1_21

matured. Our new VEP100 semi-passive index fund should have great appeal to many high net worth clients, enjoy low management costs, and be easily acceptable to the compliance and regulatory authorities.

Valuation of securities via their forward 12 month price earnings ratio (*ftmPE*) is a very common securities valuation method in the industry. Obviously the *ftmPE* valuation depends heavily on the estimate of forward 12 month corporate earnings per share (*ftmEPS*). Obvious inputs to the *ftmEPS* prediction process are the past earnings time series plus one or more analyst predictions.

Valuation via *ftmEPS* is a necessary but not a sufficient attraction for a semi-passive index fund. So we will introduce the advantages of trading volatility. Our thesis will be that emotional trading patterns tend to make markets less efficient.

The efficient market hypothesis assumes rational trading patterns and equal and open access to information. Trading on insider information is illegal in most developed securities markets; but, *trading when others are emotional is unregulated.* In this paper we will develop a set of factors—all of which incorporate a measure of volatility indicating possible overly emotional trading patterns. The theme of our new VEP100 semi-passive index fund will be "*Buy value from those who are selling in a highly emotional state*".

Now would be a good time to provide an overview general introduction to symbolic regression as follows.

Symbolic Regression is an approach to general nonlinear regression which is the subject of many scholarly articles in the Genetic Programming community. A broad generalization of general nonlinear regression is embodied as the class of *Generalized Linear Models* (GLMs) as described in Nelder and Wedderburn (1972). A GLM is a linear combination of **I** basis functions $\mathbf{B_i}$; $i = 1,2, \ldots, I$, a dependent variable **y**, and an independent data point with **M** features $\mathbf{x} = <x_1, x_2, x_3, \ldots, x_m>$: such that

$$y = \gamma(x) = C_0 + \sum_{i=1}^{I} c_i B_i(x) + err \tag{21.1}$$

As a broad generalization, GLMs can represent any possible nonlinear formula. However the format of the GLM makes it amenable to existing linear regression theory and tools since the GLM model is linear on each of the basis functions $\mathbf{B_i}$.

For a given vector of dependent variables, Y, and a vector of independent data points, X, symbolic regression will search for a set of basis functions and coefficients which minimize *err*. In Koza (1992) the basis functions selected by symbolic regression will be formulas as in the following examples:

$$B_1 = x_3 \tag{21.2}$$

$$B_2 = x_1 + x_4 \tag{21.3}$$

$$B_3 = \mathrm{sqrt}(x_2) / \tan(x_5/4.56) \tag{21.4}$$

$$B_4 = \tanh(\cos(x_2 * .2) * \text{cube}(x_5 + \text{abs}(x_1))) \tag{21.5}$$

If we are minimizing the least squared error, ***LSE***, once a suitable set of basis functions {**B**} have been selected, we can discover the proper set of coefficients {**C**} deterministically using standard univariate or multivariate regression. The value of the GLM model is that one can use standard regression techniques and theory. Viewing the problem in this fashion, we gain an important insight. Symbolic regression does not add anything to the standard techniques of regression. The value added by symbolic regression lies in its abilities as a search technique: how quickly and how accurately can SR find an optimal set of basis functions {**B**}.

The immense size of the search space provides ample need for improved search techniques In standard Koza-style tree-based Genetic Programming (Koza 1992) the genome and the individual are the same Lisp s-expression which is usually illustrated as a tree. Of course the tree-view of an s-expression is a visual aid, since a Lisp s-expression is normally a list which is a special Lisp data structure. Without altering or restricting standard tree-based GP in any way, we can view the individuals not as trees but instead as s-expressions such as this depth 2 binary tree s-exp: (/ (+x_2 3.45) (*x_0 x_2)), or this depth 2 irregular tree s-exp: (/ (+x_2 3.45) 2.0).

In standard GP, applied to symbolic regression, the non-terminal nodes are all operators (implemented as Lisp function calls), and the terminal nodes are always either real number constants or features. The maximum depth of a GP individual is limited by the available computational resources; but, it is standard practice to limit the maximum depth of a GP individual to some manageable limit at the start of a symbolic regression run.

Given any selected maximum depth k, it is an easy process to construct a maximal binary tree s-expression U_k, which can be produced by the GP system without violating the selected maximum depth limit. As long as we are reminded that each f represents a function node while each t represents a terminal node, the construction algorithm is simple and recursive as follows.

$$U_0 : t$$

$$U_1 : (f\ t\ t)$$

$$U_2 : (f\ (f\ t\ t)\ (f\ t\ t))$$

$$U_3 : (f\ (f\ (f\ t\ t)\ (f\ t\ t))\ (f\ (f\ t\ t)\ (f\ t\ t)))$$

$$U_k : (f\ U_{k-1} U_{k-1})$$

Any basis function produced by the standard GP system will be represented by at least one element of U_k. In fact, U_k is isomorphic to the set of all possible basis functions generated by the standard GP system.

Given this formalism of the search space, it is easy to compute the size of the search space, and it is easy to see that the search space is huge even for rather simple basis functions. For our use in this chapter the function set will be the following functions: **F** = {+ − * / **abs sqrt square cube cos sin tan tan h log exp max min**ℵ} (where ℵ(a,b) = ℵ(a) = a). The terminal set is the features $\mathbf{x_0}$ thru $\mathbf{x_m}$ and the real constant **c**, which we shall consider to be 2^{64} in size. Where $|\mathbf{F}| = 17$, $\mathbf{M}=20$, and $\mathbf{k} = 0$, the search space is $S_0 = \mathbf{M} + 2^{64} = 20 + 2^{64} = 1.84 \times 10^{19}$. Where $\mathbf{k} = 1$, the search space is $S_1 = |\mathbf{F}| * S_0 * S_0 = 5.78 \times 10^{39}$. Where $\mathbf{k} = 2$, the search space grows to $S_2 = |\mathbf{F}| * S_1 * S_1 = 5.68 \times 10^{80}$. For $\mathbf{k} = 3$, the search space grows to $S_3 = |\mathbf{F}| * S_2 * S_2 = 5.5 \times 10^{162}$. Finally if we allow three basis functions $\mathbf{B} = 3$ for financial applications, then the final size of the search space is $S_3 * S_3 * S_3 = 5.5 \times 10^{486}$.

21.2 Methodology

Creating the weekly buy list for a modern semi-passive index fund requires many fully automated multiple regressions, all of which must be run in a timely fashion, and all of which must fit together seamlessly without human intervention. Our methodology is influenced by the practical issues of applying symbolic regression to the real world investment finance environment. First there is the issue that form of each symbolic regression must be preapproved by the regulatory authorities, the compliance officer, management, and clients. Second there is the issue of adapting symbolic regression to run in a real world financial application with massive amounts of data. Third there is the issue of modifying symbolic regression, as practiced in academia, to conform to the very difficult U.S. Securities Exchange Commission regulatory compliance environment.

Weekly preparation of our VEP100 semi-passive index fund buy list will require ∼1502 fully automated regressions (*as many as there are Value Line Timeliness stocks that week*). For each of the ∼1500 Value Line Timeliness stocks, a set of pre-approved earnings estimate inputs will be fed into a multiple linear regression for each stock, resulting in an interim forward 12 month earnings per share estimate for the stock. This will require ∼1500 regressions; but, they are relatively quick multiple linear regressions. Next, a set of preapproved earnings estimate inputs plus the interim *ftmEPS* estimate produced by the linear regressions will be input to a nonlinear weighted regression on all ∼1500 stocks. This expensive nonlinear weighted regression will produce a final *ftmEPS* estimate for each of the Value Line stocks. Finally, a set of preapproved z Score factor inputs plus the interim *ftmEPS* estimate produced by the linear regressions will be input to a nonlinear logistic regression on all ∼1500 stocks. This final very expensive nonlinear logistic regression will produce a final *expected forward 12 month total return* estimate for each of the Value Line stocks.

We use only statistical best practices out-of-sample testing methodology. For each regression, a matrix of independent variables will be constructed solely from the prior 10 years of historical data—520 weeks. No forward looking data will be allowed. This is very important because it will be the subject of detailed regulatory due diligence reviews. Then the preapproved regression model will be applied to produce the dependent variable.

For the forward estimation of corporate earnings, this paper uses an historical database of the Value Line stocks with daily price and volume data, weekly analyst estimates, and quarterly financial data from January 1990 to the December 2009. The data has been assembled from reports published at the time, so the database is highly representative of what information was realistically available at the point when trading decisions were actually made. No forward looking data is included in any historical point in the database.

From all of this historical data, 20 years (*1990 thru 2009*) have been used to produce the results shown in this research. This 24 year period includes a historically significant bull market decade followed by an equally historically significant bear market decade.

Multiple vendor sources have been used in assembling the data so that single vendor bias can be eliminated. The construction of this point in time database has focused on collecting weekly consolidated data tables, collected every Friday from January 3, 1986 to the present, representing detailed point in time input to this study and cover the Value Line stocks on a weekly basis. Each stock record contains daily price and volume data, weekly analyst estimates and rankings, plus quarterly financial data as reported. The primary focus is on gross and net revenues.

Our historical database contains 1050 weeks of data between January 1990 and December 2009. In a full training and testing protocol there is a separate symbolic regression run for each of these 1050 weeks. Each SR run consists of predicting the *ftmEPS* for each of the Value Line stocks available in that week, using the 520 prior weeks as the training data set for that week. A sliding training/testing window will be constructed to follow a strict statistical out-of-sample testing protocol.

For each of the 1050 weeks, the 520 prior weeks training examples will be extracted from records in the historical trailing 10 years behind the selected record BUT *not including any data from the selected week or ahead in time*. The training dependent variable will be extracted from the historical data record exactly 52 weeks forward in time from the selected record BUT *not including any data from the selected week or ahead in time*. Thus, as a practical observation, the training will not include any records in the first 52 weeks prior to the selected record—*because that would require a training dependent variable which was not available at the time*.

For each of the 1050 weeks, the testing samples will be extracted from records in the historical trailing 10 years behind the selected record *including all data from the selected week BUT not ahead in time*. The testing dependent variable will be extracted from the historical data record exactly 52 weeks forward in time from the selected record.

Each experimental protocol will produce approximately ~1500 linear regressions and 2 symbolic regression runs over an average of ~780,000 (*~1500 × 520*) records for each training run and for ~1500 records for each testing run. Ten hours will be allocated for training. Of course separate R-Square statistics will be produced for each experimental protocol. We will examine the R-Square statistics for evidence favoring the addition of swarm intelligence over the base line and for evidence favoring one swarm intelligence technique over another.

Finally we will need to adapt our methodology to conform to the rigorous United States Securities and Exchange Commission oversight and regulations on investment managers. The SEC mandates that every investment firm have a compliance officer. For any automated forward earnings prediction algorithm, *which would be used as the basis for later stock recommendations to external clients or internal portfolio managers*, the computer software code used in each prediction, the historical data used in each prediction, and each historical prediction itself, must be filed with the compliance officer in such form and manner so as to allow a surprise SEC compliance audit to reproduce each individual forward prediction exactly as it was at the original time of publication to external clients or internal portfolio managers.

Of course this means that we must provide a copy of all code, all data, and each forward prediction for each stock in each of the 1050 weeks, to our compliance officer. Once management accepts our symbolic regression system, we will also have to provide a copy of all forward predictions on an ongoing basis to the compliance officer.

Furthermore there is an additional challenge in meeting these SEC compliance details. The normal manner of operating GP, and symbolic regression systems in academia will not be acceptable in a real world compliance environment. Normally, in academia, we recognize that symbolic regression is a heuristic search process and so we perform multiple SR runs, each starting with a different random number seed. We then report based on a statistical analysis of results across multiple runs. This approach produces *different results* each time the SR system is run. In a real world compliance environment such practice would subject us to serious monetary fines and also to jail time.

The SEC compliance requirements are far from arbitrary. Once management accepts such an SR system, the weekly automated predictions will influence the flow of millions and even billions of dollars into one stock or another and the historical back testing results will be used to sell prospective external clients and internal portfolio managers on using the system's predictions going forward.

First the authorities want to make sure that as time goes forward, *in the event that the predictions begin to perform poorly*, we will not simply rerun the original predictions again and again, with a different random number seed, until we obtain better historical performance and then substitute the new better performing historical performance results in our sales material.

Second the authorities want to make sure that, *in the event our firm should own many shares of the subsequently poorly performing stock of "ABC" Corp*, that we do not simply rerun the current week's predictions again and again, with a different random number seed, until we obtain a higher ranking for "ABC" stock thus improperly influencing our external clients and internal portfolio managers to drive the price of "ABC" stock higher.

In order to meet SEC compliance regulations we have altered our symbolic regression system, used in this chapter across all experiments, to use a pseudo random number generator with a pre-specified starting seed. Multiple runs always produce *exactly the same results.*

21.3 Investing Strategies

Value investing (Graham and Dodd 2008) has produced several of the wealthiest investors in the world including Warren Buffet. Nevertheless, value investing has a host of competing strategies including momentum (Bernstein 2001) and hedging (Nicholas 2000).

One of the most difficult challenges in devising a securities investing strategy is the a priori identification of pending regime changes. For instance, momentum investing strategies were very profitable in the 1990s and not so profitable in the 2000s while value investing strategies were not so profitable in the 1990s but turned profitable in the 2000s. Long Short hedging strategies were profitable in the 1990s and early 2000s but collapsed dramatically in the late 2007 thru 2008 period. Knowing when to switch from Momentum to Value, Value to Hedging, and Hedging back to Value was critical for making consistent above average profits during the 20 year period from 1990 thru 2009.

The challenge becomes even more difficult when one adds the numerous technical and fundamental buy/sell triggers to currently popular active management investing strategies. Bollinger Bands, MACD, Earning Surprises, etc. all have complex and dramatic effects on the implementation of securities investing strategies, and all are vulnerable to regime changes. The question arises, "*Is there a simple securities investing strategy which is less vulnerable to regime changes than other strategies?*".

An idealized value investing hypothesis is put forward: "*Given perfect foresight, buying stocks with the best future earning yield* (***Future12MoEPS/CurrentPrice***) (*ftmEP*) *and holding for 12 months will produce above average securities investing returns*".

Of course the ideal hypothesis is *impossible to implement* because it requires perfect foresight which is, in the absence of time travel, unobtainable. Nevertheless the ideal hypothesis represents the theoretical upper limit on the profits realizable from a strategy of buying future net revenue cheaply; yet, the theoretical profits are so rich that one cannot help but ask the question, "*Are there revenue prediction models which will allow one to capture some portion of the profits from the ideal hypothesis?*".

The easiest revenue prediction model involves simply using the current year's trailing 12 month revenue as a proxy for future revenue.

Table 21.1 Returns for SP100 High ttmEP/ftmEP 100

Year	SP100 stocks	100ttmEP stocks	100 ftmEP stocks
1990	(6 %)	(17 %)	3 %
1991	24 %	40 %	111 %
1992	3 %	22 %	56 %
1993	8 %	9 %	46 %
1994	0 %	6 %	18 %
1995	36 %	22 %	49 %
1996	23 %	28 %	38 %
1997	28 %	27 %	51 %
1998	32 %	12 %	12 %
1999	31 %	38 %	22 %
2000	(13 %)	14 %	45 %
2001	(15 %)	11 %	56 %
2002	(24 %)	(15 %)	8 %
2003	24 %	52 %	67 %
2004	4 %	13 %	45 %
2005	(1 %)	17 %	43 %
2006	16 %	7 %	19 %
2007	3 %	(5 %)	20 %
2008	(37 %)	(28 %)	(17 %)
2009	19 %	43 %	120 %
CAGR%	6 %	14 %	37 %
Volatility	20 %	20 %	30 %
CAGR% 1990s	17 %	18 %	38 %
CAGR% 2000s	(4 %)	8 %	37 %

Note: Per annum total returns for each year

The data supports the conclusion that even using this current revenue proxy model buying the top one hundred stocks with the highest (***current12MoEPS/ currentPrice***) (*ttmEP*) and holding for 1 year produces above average securities investing profits, *as least for the Value Line stocks*, as shown in Table 21.1.

Nevertheless, buying a stock with high EP, *but whose future 12 month earnings will plummet bringing on bankruptcy*, is an obviously poor choice. So why is high EP investing so successful given that future 12 month earnings can vary significantly? Placing current earnings yield investing in this context puts a new spin on this standard *value investing* measure. In this context we are saying that current earnings yield (also known as high EP investing) works precisely to the extent that *current earnings are a reasonable predictor of future earnings*! In situations where current earnings are NOT a good predictor of future earnings, then current earnings yield investing loses its efficacy.

This agrees with our common sense understanding. For instance, given two stocks with the same high current earnings yield, where one will go bankrupt next year and the other will double its earnings next year; we would prefer the stock whose earnings will double. Implying that, *in the ideal*, current earnings are just a data point. We want to buy *future earnings* cheap!

Precisely because the per annum returns from this current revenue prediction model are far less than the returns achieved with perfect prescience, we must now look for more accurate methods of net revenue prediction.

21.3.1 Estimating Forward 12 Month EPS

Each week we will perform ~1500 linear regressions, one for each of the Value Line stocks. The preapproved linear regressions are expressed by the following Regression Query Language **RQL** (Korns 2013) expression:

$$\textbf{regress}\,(\mathbf{x0}, \mathbf{x1}, \mathbf{x2}, \mathbf{x3}, \mathbf{x4}, \mathbf{x5}, \mathbf{x6})\ \textbf{where}\ \{\}$$

For each of the ~1500 Value Line stocks in the current week, from each of the 520 trailing historical weeks for that stock (*see our methodology section above*) the following seven input (*independent*) variables will be collected:

1.	**Estimated12MoEPS**(x0)	Wall Street analysts 12Mo forward EPS estimate
2.	**Forward12MoEPS**(x1)	CurrentEPS + (CurrentEPS-Past1YrEPS)
3.	**Projected12MoEPS**(x2)	CurrentEPS + ((CurrentEPS-Past1QtrEPS) * 4)
4.	**EstimatedS12MoEPS**(x3)	(Wall Street analysts 12Mo forward SPS estimate) * CurrentMargin
5.	**ForwardS12MoEPS**(x4)	(CurrentSPS + (CurrentSPS-Past1YrSPS)) * CurrentMargin
6.	**ProjectedS12MoEPS**(x5)	(CurrentSPS + ((CurrentSPS-Past1QtrSPS) * 4)) * CurrentMargin
7.	**WeeksSinceLastReport**(x6)	Absolute count of weeks since last quarterly report

Each of the ~1500 linear regressions produces an *ftmEPS* estimate for each of the Value Line stocks for that week (**LRegress12MoEPS**). This regression output is then used as an input to a single preapproved nonlinear weighted regression on the following input variables:

The preapproved nonlinear weighted regression is expressed by the following Regression Query Language **RQL** (Korns 2013) expression:

1.	**Estimated12MoEPS** (*x*0)	Wall Street analysts 12Mo forward EPS estimate
2.	**Forward12MoEPS**(*x*1)	CurrentEPS + (CurrentEPS-Past1YrEPS)
3.	**Projected12MoEPS**(*x*2)	CurrentEPS + ((CurrentEPS-Past1QtrEPS) * 4)
4.	**EstimatedS12MoEPS**(*x*3)	(Wall Street analysts 12Mo forward SPS estimate) * CurrentMargin
5.	**ForwardS12MoEPS**(*x*4)	(CurrentSPS + (CurrentSPS-Past1YrSPS)) * CurrentMargin
6.	**ProjectedS12MoEPS**(*x*5)	(CurrentSPS + ((CurrentSPS-Past1QtrSPS) * 4)) * CurrentMargin
7.	**WeeksSinceLastReport**(*x*6)	Absolute count of weeks since last quarterly report
8.	**LRegress12MoEPS**(*x*7)	Result of the linear regression for the stock in question

$$\begin{aligned}&\mathbf{model}\big(\mathbf{c0} * \mathbf{f0}\,(\mathbf{x0}, \mathbf{v0}), \mathbf{c1} * \mathbf{f1}\,(\mathbf{x1}, \mathbf{v1}), \mathbf{c2} * \mathbf{f2}\,(\mathbf{x2}, \mathbf{v2}),\\ &\quad \mathbf{c3} * \mathbf{f3}\,(\mathbf{x3}, \mathbf{v3}), \mathbf{c4} * \mathbf{f4}\,(\mathbf{x4}, \mathbf{v4}), \mathbf{c5} * \mathbf{f5}\,(\mathbf{x5}, \mathbf{v5}),\\ &\quad \mathbf{c6} * \mathbf{f6}\,(\mathbf{x6}, \mathbf{v6}), \mathbf{c7} * \mathbf{f7}\,(\mathbf{x7}, \mathbf{v7})\big)\end{aligned}$$

$$\mathbf{where}\ \Big\{\mathbf{op}\,(\aleph, +, -, \mathbf{min}, \mathbf{max})$$

$$\mathbf{c0}\,(\mathbf{0.0}, \mathbf{1.0})\ \ \mathbf{c1}\,(\mathbf{0.0}, \mathbf{1.0})\ \ \mathbf{c2}\,(\mathbf{0.0}, \mathbf{1.0})\ \ \mathbf{c3}\,(\mathbf{0.0}, \mathbf{1.0})\ \ \mathbf{c4}\,(\mathbf{0.0}, \mathbf{1.0})$$

$$\mathbf{c5}\,(\mathbf{0.0}, \mathbf{1.0})\ \ \mathbf{c6}\,(\mathbf{0.0}, \mathbf{1.0})\ \ \mathbf{c7}\,(\mathbf{0.0}, \mathbf{1.0})\Big\}$$

This nonlinear weighted regression will achieve regulatory and client preapproval because it is so intuitive and so easy to explain. Let us start with the simplest case where the functions (**f0** thru **f7**) are all noops = ℵ, then the final result will always be like the following example:

$$\begin{aligned}\mathbf{NLREstimated12MoEPS}\,(y) = {} & \mathbf{.34} * \mathbf{x0} + \mathbf{.16} * \mathbf{x1} + \mathbf{.81} * \mathbf{x2} + \mathbf{.54} * \mathbf{x3}\\ & + \mathbf{.26} * \mathbf{x4} + \mathbf{.72} * \mathbf{x5} + \mathbf{.59} * \mathbf{x6} + \mathbf{.21} * \mathbf{x7}\end{aligned}$$

We have eight inputs in the form of dollar values for next year's estimated EPS. Our model simply assigns a weight (*0.0<= 1.0*) to each estimate—with the added benefit that, in the past 520 weeks for all ~1500 Value Line stocks, these weights have been the most successful in predicting next year's EPS values for the Value Line stocks. Now moving on to the case where one of more of the functions (**f0** thru **f7**) are other than noops, then the final result will always be something like the following example:

$$\begin{aligned}\textbf{NLREstimated12MoEPS}\ (y) = {} & \mathbf{.34 * x0 + .16 * x1 + .81 * max\,(x2, x0)} \\ & \mathbf{+\ .54 * x3 + .26 * x4 + .72 * x5 + .59 * x6} \\ & \mathbf{+\ .21 * x7}\end{aligned}$$

Again we have eight inputs in the form of dollar values for next year's estimated EPS. Our model simply assigns a weight (*0.0<= 1.0*) to each possible simple combination of those estimates—with the added benefit that, in the past 520 weeks, these weights and these combinations have been the most successful in predicting next year's EPS values for the SP100 stocks.

In all cases we are simply weighting simple estimates or simple combinations of estimates with combinations that will never get unruly or out of hand and with weights which will always remain safely between 0.0 and 1.0. For this intuitive nonlinear weighted regression, Regulatory and client preapproval will be easy to obtain.

21.3.2 Estimating Forward 12Mo Total Return

A close examination of the Efficient Market Hypothesis (*EMH*) shows that the expectation of rational investing decisions plays a significant role in the EMH conclusions in favor of passive index investing. Therefore, in addition to attempting to purchase cheap stocks (*via some estimate of future 12Mo earnings*), we would also like to purchase stocks from sellers whose decisions may not be as rational as the EMH might hope.

Normally each stock trades within its own average trading volume over the course of weeks and months. This trading volume can be expressed as a percent ***WeeksVolume* = (total number of shares traded today)/(total shares outstanding)**. For any given stock there will be periods of calm when weekly trading percent (*WeeksVolume*) is light compared to the its historical average, and periods of frenzy when the weekly trading percent (*WeeksVolume*) is very high compared to its historical average. Our assertion is that *when a trading frenzy is underway the buyer AND seller are less rational than on normal trading days.*

The following nine input factors (*each of which combines some measure of trading frenzy or intrinsic value or both*) will be converted to z Scores (Anderson et al. 2002) and are defined as follows:

First we see that **z Panic Level** is computed from the nonlinear regression future 12Mo EPS estimate divided by the week's closing price (*i.e. the estimated future EPS yield*) times the percent of outstanding shares traded that week (*Weeks Volume*) times the current week's trading percent as in comparison with the prior 52 weeks trading percent (*Volume52WeekRange*). This input will be high when the estimated future earnings yield is high (*the stock is cheap*), when a high percent of outstanding shares traded this week (*Weeks Volume*), and when this week's trading volume is on the high side compared to the previous 52 week trading history for this stock. This

1.	**zPanicLevel** ($x0$)	((NLRFuture12MoEPS/WeeksClose) * WeeksVolume * Volume52WeekRange))
2.	**zPriceMomentum**($x1$)	(Past52WeekReturn * WeeksVolume)
3.	**zDollarVolume**($x2$)	(WeeksVolume * Shares * WeeksClose)
4.	**zFutureEPSYield**($x3$)	(NLRFuture12MoEPS/WeeksClose)
5.	**zSalesAttractiveness**($x4$)	(Current12MoSPS/WeeksClose) * WeeksVolume
6.	**zCurrentValuation**($x5$)	(CurrentVPS/WeeksClose)
7.	**zValuationAttractiveness**($x6$)	(CurrentVPS/WeeksClose) * WeeksVolume
8.	**zWallStreetRank**($x7$)	Current Wall Street analysts ranking as a z Score
9.	**zFinancialRank**($x8$)	Current Wall Street financial ranking as a z Score

is a stock selling on much higher volume than normal with a very cheap future earnings yield. We use this input as a measure of panic on the seller's side. Since each of these inputs are z Scores, a high value for this input indicates that this stock is in a greater trading frenzy relative to other stocks this week.

Second we see that **z Price Momentum** is computed from the stock's past 52 week total return (*Past52WeekReturn*) times the week's trading volume (*Weeks Volume*). This input will be high for stocks with strong momentum selling on high trading volume. This is a popular stock, and we use this input as a measure of price momentum on the buyer's side. Since each of these inputs are z Scores, a high value for this input indicates that this stock enjoys greater momentum relative to other stocks this week.

Third we see that **z Dollar Volume** is an estimate of the total dollar value of the shares traded this week. This is a popular stock, and we use this input as a measure of relative dollar flow through this stock as opposed to other stocks this week. Since each of these inputs are z Scores, a high value for this input indicates that more dollars are flowing through this stock than other stocks this week.

Fourth we see that **z Future EPS Yield** is a measure of how cheap the future 12Mo EPS estimate divided by the week's closing price (*i.e. the estimated future EPS yield*) is compared to other stocks this week. This input will be high when the estimated future earnings yield is high (*the stock is cheap*). Since each of these inputs are z Scores, a high value for this input indicates that this stock is cheaper relative to other stocks this week.

Fifth we see that **z Sales Attractiveness** is computed from the current 12Mo SPS divided by the week's closing price (*i.e. the current sales yield*) times the percent of outstanding shares traded that week (*Weeks Volume*). This input will be high when the current sales yield is high (*the stock is cheap*), and when a high percent of outstanding shares traded this week (*Weeks Volume*. This is a stock selling on high volume with a very cheap current sales yield. We use this input as a measure of attraction on the buyer's side.

Sixth we see that **z Current Valuation** is a measure of the current enterprise value divided by the week's closing price (*i.e. the current VPS yield*). This input will be high when the current VPS yield is high (*the stock is cheap*). Since each of these inputs are z Scores, a high value for this input indicates that this stock is cheaper relative to other stocks this week.

Seventh we see that **z Valuation Attractiveness** is a measure of the current enterprise value divided by the week's closing price (*i.e. the current VPS yield*) times this week's trading volume (*Weeks Volume*). This input will be high when the current VPS yield is high (*the stock is cheap*), and trading volume is high. Since each of these inputs are z Scores, a high value for this input indicates that this stock is more attractive to buyers relative to other stocks this week.

Eighth we see that **z Wall Street Rank** is a measure of the current Wall Street analysts' rank for this stock. This input will be high when the Wall Street analysts' rank for this stock is high. Since each of these inputs are z Scores, a high value for this input indicates that this stock enjoys a higher Wall Street analyst ranking relative to other stocks this week.

Ninth we see that **z Financial Rank** is a measure of the current Wall Street analysts' financial rank for this stock. This input will be high when the Wall Street analysts' financial rank for this stock is high. Since each of these inputs are z Scores, a high value for this input indicates that this stock enjoys a higher Wall Street analyst financial ranking relative to other stocks this week.

The following single output factor (*what we train on*) will be converted to sigmoid score (Kleinbaum et al. 2010; Anderson et al. 2002) and is defined as follows:

1.	**sFuture12MoReturn** (*y*)	The actual Future 12Mo Total Return—as a sigmoid-score

Obviously we are not trying to predict actual future 12 month total return so much as we are trying to predict relative future 12 month total return. We don't really need to know actual future total 12 month returns. We only need to select the 100 Value Line stocks with the highest *relative* estimated future total 12 month return. This allows us the luxury of converting the output variable (**sFuture12MoReturn**) to a sigmoid factor, which allows us to perform a nonlinear logistic regression (Kleinbaum et al. 2010) of the following form.

$$\mathbf{logit}\big(\mathbf{f0\,(x0, v0)}, \mathbf{f1\,(x1, v1)}, \mathbf{f2\,(x2, v2)}, \mathbf{f3\,(x3, v3)}, \mathbf{f4\,(x4, v4)}, \mathbf{f5\,(x5, v5)}, \mathbf{f6\,(x6, v6)}, \mathbf{f7\,(x7, v7)}, \mathbf{f8\,(x8, v8)}\big)\ \mathbf{where}\ \{\mathbf{op}\,(\aleph, +, -, \mathbf{min}, \mathbf{max})\}$$

This simple and extremely intuitive nonlinear logistic regression will easily win regulatory and client preapproval. First of all this nonlinear regression will never produce unexpected or wild output. It will produce an orderly estimate for (**sFuture12MoReturn**) which will always lie between 0.0 and 1.0 for each stock. In essence, this nonlinear regression model will automatically rank each stock between 0.0 and 1.0 in terms of estimated future 12 month total return (*with 1.0 being the most desirable and 0.0 being the least desirable*). Let us start with the simplest case where the functions (**f0** thru **f8**) are all noops $= \aleph$, then the final result will always be like the following example:

$$\textbf{sFuture12MoReturn}(y) = \textbf{sigmoid}\big(\textbf{.34} * \textbf{x0} + \textbf{.16} * \textbf{x1} + \textbf{.81} * \textbf{x2} + \textbf{.54} * \textbf{x3} + \textbf{.26} * \textbf{x4} + \textbf{.72} * \textbf{x5} + \textbf{.59} * \textbf{x6} + \textbf{.21} * \textbf{x7} + \textbf{.91} * \textbf{x8}\big)$$

We have nine inputs in the form of z Scores for factors combining some measure of relative value and/or trading frenzy. Our model simply projects each weighted factor onto a relative ranking between 0.0 and 1.0—with the added benefit that, in the past 520 weeks, these weights have been the most successful in predicting next year's relative future 12Mo total return for the Value Line stocks.

Now moving on to the case where one of more of the functions (**f0** thru **f8**) are other than noops, then the final result will always be something like the following example:

$$\textbf{sFuture12MoReturn}(y) = \textbf{sigmoid}\big(\textbf{.34} * \textbf{x0} + \textbf{.16} * \textbf{x1} + \textbf{.81} * \textbf{max}\,(\textbf{x2}, \textbf{x0}) + \textbf{.54} * \textbf{x3} + \textbf{.26} * \textbf{x4} + \textbf{.72} * \textbf{x5} + \textbf{.59} * \textbf{x6} + \textbf{.21} * \textbf{x7} + \textbf{.91} * \textbf{x8}\big)$$

Again we have nine inputs in the form of z Scores for factors combining some measure of relative value and/or trading frenzy. Our model projects each weighted factor or simple combination of factors onto a relative ranking between 0.0 and 1.0—with the added benefit that, in the past 520 weeks, these weights and these combinations have been the most successful in predicting next year's relative future 12Mo total return for the Value Line stocks.

21.3.3 Historical Returns

Applying all of these tools, techniques, and factors to the task of creating our semi-passive VEP100 index fund, we perform our 1502 regression runs for the first week in each year from 1990 thru 2009. We select the 100 Value Line stocks with the highest **sFuture12MoReturn** values. And hold them for 1 year. We then compare the results to the SP100 passive index, buying the 100 stocks with the highest ttmEP, and buying the 100 stocks with the highest ftmEP and present the results in Table 21.2.

Our VEP100 semi-passive index produced a much higher compound annual growth rate (*CAGR%*) than the SP100 index and the 100 ttmEP method. However, it cannot compete with the ideal ftmEP method (*where one can see into the future*). Nevertheless the total return of our VEP100 semi-passive index is impressive and will definitely appeal to a wide range of high net worth clients.

So have we beaten the Efficient Market Hypothesis? With a little bit of humor I can answer with a definite **Yes** and **No**.

Table 21.2 Returns VEP 100

Year	SP100 stocks	100 ttmEP stocks	VEP100 index fund	100 ftmEP stocks
1990	(6 %)	(17 %)	(22 %)	3 %
1991	24 %	40 %	47 %	111 %
1992	3 %	22 %	33 %	56 %
1993	8 %	9 %	23 %	46 %
1994	0 %	6 %	0 %	18 %
1995	36 %	22 %	30 %	49 %
1996	23 %	28 %	24 %	38 %
1997	28 %	27 %	31 %	51 %
1998	32 %	12 %	0 %	12 %
1999	31 %	38 %	30 %	22 %
2000	(13 %)	14 %	10 %	45 %
2001	(15 %)	11 %	38 %	56 %
2002	(24 %)	(15 %)	(6 %)	8 %
2003	24 %	52 %	62 %	67 %
2004	4 %	13 %	30 %	45 %
2005	(1 %)	17 %	29 %	43 %
2006	16 %	7 %	8 %	19 %
2007	3 %	(5 %)	13 %	20 %
2008	(37 %)	(28 %)	(42 %)	(17 %)
2009	19 %	43 %	131 %	120 %
CAGR%	6 %	14 %	17 %	37 %
Volatility	20 %	20 %	30 %	30 %
CAGR% 1990s	17 %	18 %	18 %	38 %
CAGR% 2000s	(4 %)	8 %	20 %	37 %

Note: Per annum total returns for each year

Yes, because the VEP100 CAGR% of 17 % is a whopping 9 % per annum greater than the SP100! This is a significant amount which will be of interest to a large class of serious investors. Furthermore, the performance of the VEP100 is more consistent across bull and bear decades with a CAGR % of 18 % in the bullish 1990s and a CAGR% of 20 % in the bearish 2000s. Coupled with the transparent and intuitive methodology of the VEP100, there is definite added value here.

No, because the EMH does not actually claim that one cannot make higher profits than the indices. The EMH claims that one cannot increase returns without also increasing volatility, and this is exactly what happens with the VEP100 semi-passive index. Volatility increases from 20 % with the SP100 to 30 % with the VEP100. So in an important way, the VEP100 is a classic confirmation of the Efficient Market Hypothesis.

21.4 Summary

Advances in both the industrial strength and accuracy of Symbolic Regression packages can help overcome the resistance to SR in the investment finance industry. Management trust, regulatory approval, and client acceptance, are no longer the severe hurdles that they were in the past. Improvements in SR robustness, result invariance, demonstrable accuracy, and regression constraint languages, such as Regression Query Language **RQL** (Korns 2010, 2013, 2014), now support regulatory and client preapproval of important component SR processes.

In this research work, as series of cascade linear and nonlinear SR regressions are used to create a transparent semi-passive index fund with significantly higher returns, over the 1990–2009 two decade period, than its Standard &Poors 100 index benchmark. Because of its transparent and algorithmic nature, the new VEP100 semi-passive index fund could enjoy much lower costs than a standard active fund and yet enjoy attractive returns—costs similar in nature to the SP100 passive index fund.

Future research will focus on other semi-passive indices with performance tailored to various diverse client needs and requirements, and regulatory approval issues.

References

Graham, Benjamin, and David Dodd. 2008. Securities Analysis. New York, New York, USA. McGraw-Hill.

Korns, Michael F., 2010. Abstract Expression Grammar Symbolic Regression. In Riolo, Rick, L, Soule, Terrance, and Wortzel, Bill, editors, Genetic Programming Theory and Practice VIII, New York, New York, USA. Springer, pp. 109–128.

Guido Smits, Ekaterina Vladislavleva, and Mark Kotanchek 2010, Scalable Symbolic Regression by Continuous Evolution with Very Small Populations, in Riolo, Rick, L, Soule, Terrance, and Wortzel, Bill, editors, *Genetic Programming Theory and Practice VIII*, New York, New York, USA. Springer, pp. 147–160.

Flor Castillo, Arthur Kordon, and Carlos Villa 2010, Genetic Programming Transforms in Linear Regression Situations, in Riolo, Rick, L, Soule, Terrance, and Wortzel, Bill, editors, *Genetic Programming Theory and Practice VIII*, New York, New York, USA. Springer, pp. 175–194.

Trent McConaghy, Pieter Palmers, Gao Peng, Michiel Steyaert, Goerges Gielen 2009, Variation-Aware Analog Structural Synthesis: A Computational Intelligence Approach. New York, New York, USA. Springer.

J.A., Nelder, and R. W. Wedderburn, 1972, *Journal of the Royal Statistical Society, Series A, General*, 135:370–384.

John R Koza 1992, Genetic Programming: On the Programming of Computers by Means of Natural Selection. Cambridge Massachusetts, The MIT Press.

Bernstein, J., 2001. Momentum Stock Selection: Using The Momentum Method for Maximum Profits. New York, New York, McGraw Hill

Nicholas, J., 2000. Market-Neutral Investing: Long/Short Hedge Fund Strategies. New York, New York, Bloomberg Press.

Korns, Michael F, 2011b. Accuracy in Symbolic Regression. In Riolo, Rick, L, Soule, Terrance, and Wortzel, Bill, editors, Genetic Programming Theory and Practice IX, New York, New York, USA. Springer.
Korns, Michael F., 2012a. A Baseline Symbolic Regression Algorithm. In Soule, Terrance, and Wortzel, Bill, editors, Genetic Programming Theory and Practice X, New York, New York, USA. Springer.
Korns, Michael F., 2013. Extreme Accuracy in Symbolic Regression. In Soule, Terrance, and Wortzel, Bill, editors, Genetic Programming Theory and Practice XI, New York, New York, USA. Springer.
Korns, Michael F., 2014. Extremely Accurate Symbolic Regression for Large Feature Problems. In Soule, Terrance, and Wortzel, Bill, editors, Genetic Programming Theory and Practice XII, New York, New York, USA. Springer.
Korns, Michael F., 2012b. Predicting Corporate Forward 12 Month Earnings, 2012. Theory and New Applications of Swarm Intelligence, ISBN 978-953-51-0364-6, edited by Rafael Parpinelli and Heitor S. Lopes, InTech Academic Publishers.
Kleinbaum, David G., and Klein, Michael, 2010. Logistic Regression: A Self-Learning Text (Statistics for Biology and Health), ISBN 978–1441917416, New York, New York, USA. Springer.
Anderson, David R., Sweeney, Dennis J., and Williams, Thomas A, 2002. Essentials of Statistics for Business and Economics, ISBN 978–0324145809, Southwestern College Publishers.

Part IV
Tools

Chapter 22
GPTIPS 2: An Open-Source Software Platform for Symbolic Data Mining

Dominic P. Searson

22.1 Introduction

Genetic programming (GP; Koza 1992) is a biologically inspired machine learning method that evolves computer programs to perform a task. It does this by randomly generating a population of computer programs (usually represented by tree structures) and then breeding together the best performing trees to create a new population. Mimicking Darwinian evolution, this process is iterated until the population contains programs that solve the task well.

When building an empirical mathematical model of data acquired from a process or system, the process is known as symbolic data mining (SDM). SDM is an umbrella term to describe a variety of related activities including generating symbolic equations predicting a continuous valued response variable using input/predictor variables (symbolic regression); predicting the discrete category of a response variable using input variables (symbolic classification, e.g. see Espejo et al. 2010; Morrison et al. 2010) and generating equations that optimise some other criterion (symbolic optimisation, e.g. GPTIPS was used in this way to generate new chaotic attractors in Pan and Das 2014).

Symbolic regression is perhaps the most well known of these activities (it is closely related to classical regression modelling) and the most widely used. Hence, much of the functionality of GPTIPS is targeted at facilitating it. Unlike traditional regression analysis (in which the user must specify the structure of the model and then estimate the parameters from the data), symbolic regression automatically evolves both the structure and the parameters of the mathematical model from the data. This allows it to both select the inputs (features) of the model and capture non-linear behaviour.

D.P. Searson (✉)
School of Computing Science, Newcastle University, Newcastle, UK
e-mail: searson@gmail.com

A.H. Gandomi et al. (eds.), *Handbook of Genetic Programming Applications*,
DOI 10.1007/978-3-319-20883-1_22

Symbolic regression models are typically of the form:

$$\widehat{y} = f\left(x_1, \ldots, x_M\right) \tag{22.1}$$

where y is an output/response variable (the variable/property you are trying to predict), $\hat{y}$ is the model prediction of y and $x_1, \ldots, x_M$ are input/predictor variables (the variables/properties you know and want to use to predict y; they may or may not in fact be related to y) and f is a symbolic non-linear function (or a collection of non-linear functions). A typical simple symbolic regression model is:

$$\widehat{y} = 0.23x_1 + 0.33\left(x_1 - x_5\right) + 1.23{x_3}^2 - 3.34\,cos\left(x_1\right) + 0.22 \tag{22.2}$$

This model contains both linear and non-linear terms and the structure and parameterisation of these terms is automatically determined by the symbolic regression algorithm. Hence, it can be seen that symbolic regression provides a flexible—yet simple—approach to non-linear predictive modelling.

Additional advantages of symbolic regression are:

- It can automatically create compact, accurate equations to predict the behaviour of physical systems. This appeals to the notion of Occam's razor. In particular, the use of multigene GP (MGGP) within GPTIPS can exert a 'remarkable' degree of control of model complexity in comparison with standard GP (Gandomi and Alavi 2011).
- Unlike many soft-computing modelling methodologies—such as feed forward artificial neural networks or support vector machines (SVMs)—no specialised modelling software environment is required to deploy the trained symbolic models. And, because the symbolic models are simple constitutive equations, a non-modelling expert can easily and rapidly implement them in any modern computing language. Furthermore, the simplicity of the model form means they are more maintainable than typical black box predictive models.
- Examination of the evolved equations can often lead to human insight into the underlying physical processes or dynamics. In addition, the ability of a human user to understand the terms of a predictive equation can help instil trust in the model (Smits and Kotanchek 2004). It is hard to overstate the importance of user understanding and trust in predictive models, although this is not often discussed in the predictive modelling literature. In contrast, it is extremely difficult, if not impossible, to gain insight into a neural net model where the 'knowledge' about the data, system or process is encoded as network weights.
- Discovery of a population of models (rather than a single model as in the majority of other predictive modelling techniques). The evolved population can be regarded as a model library and usually contains diverse models of varying complexity and performance. This gives the user choice and the ability to gain understanding of the system being modelled by examination of the model library.

Note that the human related factors mentioned above, such as interpretation and deployment of models, are especially important when dealing with data obtained from highly multivariate non-linear systems of unknown structure (Smits and Kotanchek 2004) for which traditional analysis tends to be difficult or intractable.

Hence, symbolic regression (and symbolic data mining in general) has many features that make it an attractive basis for inducing simple, interpretable and deployable models from data where the 'true' underlying relationships are high dimensional and largely unknown. However, there has been a relative paucity of software that allows researchers to actually do symbolic data mining, and in many cases the existing software is either expensive, proprietary and closed source or requires a high degree of expertise in software configuration and machine learning to use it effectively.

GPTIPS (an acronym for Genetic Programming Toolbox for the Identification of Physical Systems) was written to reduce the technical barriers to using symbolic data mining and to help researchers, who are not necessarily experts in computing science or machine learning, to build and deploy symbolic models in their fields of research. It was also written to promote understanding of the model discovery mechanisms of MGGP and to allow researchers to add their own custom implementations of code to use MGGP in other non-regression contexts (e.g. Pan and Das 2014). To this end, it was written as a free (subject to the GNU public software license, GPL v3), open source project in MATLAB.

The use of MATLAB as the underlying platform confers the following benefits:

- Robust, trustable, fast and automatically multi-threaded implementations of many matrix and vector math algorithms (these are used extensively in GPTIPS).
- Widely taught at the undergraduate level and beyond at educational institutes around the world and hence is familiar (and site licensed) to a diverse array of students, researchers and other technical professionals. It is also heavily used in many commercial, technical and engineering environments.
- Supported, regularly updated and bug fixed and extremely well documented.
- Easy to use interface and interactive environment and supports the import and export of data in a wide variety of formats.
- A robust symbolic math engine (MuPAD) that is exceptionally useful for the post-run processing, simplification, visualisation and export of symbolic models in different formats using variable precision arithmetic.
- Runs on many OS platforms (i.e. Windows, Linux, Mac OSX) using the same code.
- Increasing emphasis on parallel computing (e.g. GPTIPS 2 has a parallel mode and can use unlimited multiple cores to evolve and evaluate new models), GPU computing, cloud computing and other so called 'big data' features such as memory-mapped variables.

This chapter is structured as follows: Sect. 22.2 provides a high level overview of GPTIPS and, in particular, the new features aimed at multigene regression model development in GPTIPS2. Section 22.3 is provided to review some different forms of symbolic regression in the context of classical regression analysis and describes the mechanisms of MGGP. Note that a basic tutorial level description of 'standard' GP is not provided here, as it is readily available elsewhere, e.g. (Poli et al. 2007).

Section 22.4 is used to demonstrate some of the features of GPTIPS 2, focusing on the visual analytics tools provided for the development of portable multigene symbolic regression models. Section 22.5 describes a new gene-centric approach to identifying and removing horizontal bloat in multigene regression models, with emphasis on the new visual analysis tool provided in GPTIPS to do this. Finally, the chapter ends with some concluding remarks in Sect. 22.6.

22.2 GPTIPS 2: Overview

GPTIPS (version 1) has become a widely used technology platform for symbolic data mining via MGGP. It is used by researchers globally and has been successfully deployed in dozens of application areas.[1]

GPTIPS using MGGP based regression has been shown to outperform existing soft-computing/machine learning methods such as neural networks, support vector machines etc. on many problem domains in terms of predictive performance and model simplicity. Examples include:

- Global solar irradiation prediction—MGGP was noted to give clearly better results than fuzzy logic and neural networks and the resulting equations were understandable by humans (Pan et al. 2013).
- The automated derivation of correlations governing the fundamental properties of the motion of particles in fluids, a key subject in powder technology, chemical and environmental engineering. The evolved models were significantly better (up to 70 %) than the existing empirical correlations (Barati et al. 2014).
- The reverse engineering of the structure of the interactions in biological transcription networks from time series data, attaining model accuracy of around 99 % (Floares and Luludachi 2014).
- The use of MGGP for the accurate modelling and analysis of data from complex geotechnical and earthquake engineering problems (Gandomi and Alavi 2011, 2012). It was noted that the evolved equations were highly accurate and 'particularly valuable for pre-design practices' (Gandomi and Alavi 2011).

The symbolic engine of GPTIPS, i.e. the mechanism whereby new equations are generated and improved over a number of iterations, is a variant of GP called multigene genetic programming (MGGP, e.g. see Searson 2002; Searson et al. 2007, 2010) which uses a modified GP algorithm to evolve data structures that contain multiple trees (genes). An example of a single tree representing a gene is shown in Fig. 22.1. This represents the equation $\sin(x_1) + \sin(3x_1)$. A typical GPTIPS multigene regression model consists of a weighted linear combination of genes such as these.

[1]A list of research literature using GPTIPS is maintained at https://sites.google.com/site/gptips4matlab/application-areas.

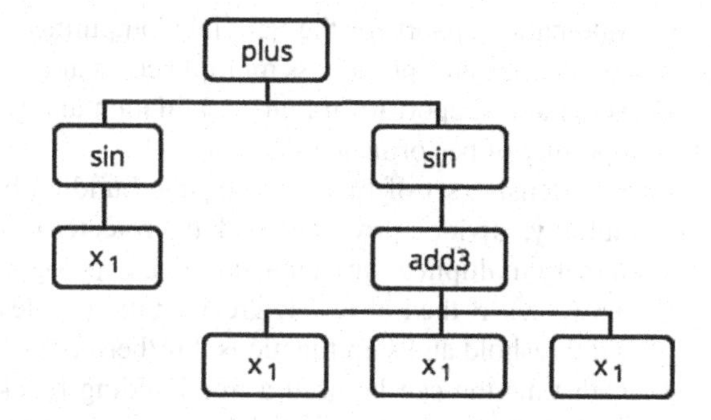

Fig. 22.1 Example of a tree (gene) representing the model term $\sin(x_1) + \sin(3x_1)$. This tree visualisation was created as a graphic within an HTML file using the GPTIPS 2 `drawtrees` function. The appearance of the trees is user customisable using simple CSS

GPTIPS is a generic tree based GP platform and has a pluggable architecture. This means that users can easily write their objective/fitness functions (e.g. for symbolic classification and symbolic optimisation) and plug them into GPTIPS without having to modify any GPTIPS code.

GPTIPS also has many features aimed specifically at developing multigene symbolic regression models. This combines the ability to evolve new equation model terms of MGGP with the power of classical linear least squares parameter estimation to optimally combine these model terms in order to minimise a prediction error metric over a data set. It is sometimes helpful to think of GPTIPS multigene regression models as pseudo-linear models in that they are linear combinations of low order non-linear transformations of the input variables. These transformations can be regarded as meta-variables in their own right.

Multigene symbolic regression has been shown to be able to evolve compact, accurate models and perform automatic feature selection even when there are more than 1500 input variables (Searson et al. 2010). It has been demonstrated that multigene symbolic regression can be more accurate and efficient than 'standard' GP for modelling nonlinear problems (e.g. see Gandomi and Alavi 2011, 2012).

22.2.1 GPTIPS Feature Overview

GPTIPS is mostly a command line driven modelling environment and it requires only a basic working knowledge of MATLAB. The user creates a simple configuration file where the data is loaded from file (or generated algorithmically within the configuration file) and configuration options set (numerous example configuration files and several example data sets are provided with GPTIPS). GPTIPS automatically generates default values for the majority of configuration options and these can be modified in the configuration file. Typical configuration options that the user sets are population size, maximum number of generations to run for, number of genes and tournament size. However, there are a large number of other run configuration options that the user can explore. In addition, GPTIPS 2 has the following features to support effective non-linear symbolic model development, analytics, export and deployment:

- Automatic support for the Parallel Computing Toolbox: fitness and complexity calculations are split across multiple cores allowing significant run speedup.
- Automatic support for training, validation and test data sets and comprehensive reporting of performance stats for each.
- An extensive set of functions for tree building blocks is provided: plus, minus, multiply, divide (protected and unprotected), add3 (ternary addition), mult3 (ternary multiplication), tanh, cos, sin, exp, $\log_{10}$, square, power, abs, cube, sqrt, exp $(-x)$, if-then-else, $-x$, greater than (>), less than (<), Gaussian ($\exp(x^2)$) and threshold and step functions. Furthermore—virtually any built in MATLAB math function can be used a tree building block function (sometimes a minor modification is required such as writing a wrapper function for the built in function). In general, it is very easy for users to define their own building block functions.
- Tight integration with MATLAB's MuPAD symbolic math engine to facilitate the post-run analysis, simplification and deployment of models.
- Run termination criteria. In addition to number of generations to run for, it is usually helpful to specify additional run termination criteria in order to avoid waste of computational effort. In GPTIPS, the maximum amount of time to run for (in seconds) can be set for each run as well as a target fitness. For example for multigene regression the target fitness can be set as model root mean squared error (RMSE) on the training data.
- Multiple independent runs where the populations are automatically merged after the completion of the runs. It is usually beneficial to allocate a relatively small amount of computational effort to each of multiple runs rather than to perform a single large run (e.g. 10 runs of 10 s each rather than a single run of 100 s). For example this 'multi-start' approach mitigates problems with the possible loss of model diversity over a run and with the GP algorithm getting stuck in local minima. In addition, GPTIPS 2 provides functionality such that final populations of separate runs may be manually merged by the user.
- Steady-state GP and fitness caching.
- Two measures of tree complexity: node count and expressional complexity (Smits and Kotanchek 2004). The latter is a more fine-grained measure of model complexity and is used to promote flatter trees over deep trees. This has significant benefits (albeit at extra computation cost) in evolving compact, low complexity models. For a single tree, expressional complexity is computed by summing together the node count of itself and all its possible *full* sub-trees (a leaf node is also considered a full sub-tree) as illustrated in (Smits and Kotanchek 2004). Hence, for two trees with the same node count, flatter and balanced trees have a lower expressional complexity than deeper ones. For instance, the tree shown in Fig. 22.2 has a total node count of 8 and contains 8 possible sub-trees. The sum of the node counts of the 8 possible full sub-trees gives, in this case, an expressional complexity of 23. For multigene individuals, the overall expressional complexity is computed as the simple sum of the expressional complexities of its constituent trees.

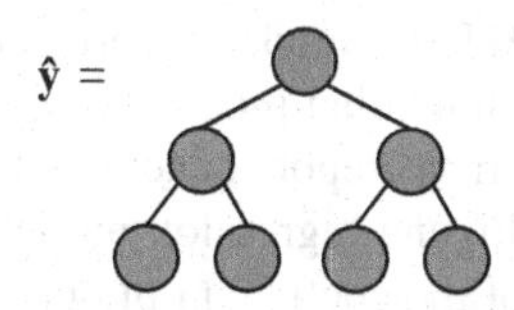

Fig. 22.2 Naïve symbolic regression. The prediction of the response data **y** is the unmodified output of a single tree that takes as its inputs one or more columns of the data matrix **X**

- Regular tournament selection (considers fitness only), Pareto tournament selection (considers fitness and model complexity) and lexicographic tournament selection (similar to regular tournament selection but always chooses the less complex model in the event of a fitness 'tie'). The user can set the probability of a particular tournament type occurring at every selection event (i.e. each time the GP algorithm selects an individual for crossover, mutation etc.). For example the user can set half of all selection events to be performed by regular tournament and half by Pareto tournament. Pareto tournaments of size P for two objectives are implemented using the $O(P^2)$ fast non-dominated sort algorithm described in (Deb et al. 2002).[2]
- Six different tree mutation operators.
- Interactive graphical population browser showing Pareto front individuals in terms of fitness (or for multigene regression models, the coefficient of determination R^2) and complexity on training, validation and test data sets. This facilitates the exploration of multigene regression models that are accurate but not overly complex and the identification of models that generalise well across data sets.
- A configurable multigene regression model filter object that enables the progressive refinement of populations according to model performance, model complexity and other user criteria (e.g. the presence of certain input variables in a model).
- Functions to export any symbolic regression model to (a) a symbolic math object (b) a standalone MATLAB file for use outside GPTIPS (c) snippets of optimised C code—which may be easily manually ported to other languages such as Java (d) an anonymous MATLAB function or function handle (e) an HTML formatted equation (f) a LaTeX formatted equation (g) a MATLAB data structure containing highly detailed information on the model as well as the individual gene predictions on training, test and validation data.
- Standalone (i.e. can be viewed in a web browser without the need for MATLAB) HTML model report generator. This enables a comprehensive performance and statistical analysis of any model in the population to be exported to HTML for later reference. The HTML report contains interactive graphical displays of model performance and model genotype and phenotype structure.

[2]Currently, the Pareto tournament implementation does not support more than two objectives.

- Customisable standalone HTML model report generator to visualise the tree structure(s) comprising an individual/model.
- Standalone HTML Pareto front report generator to allow the interactive visualisation of simplified multigene regression models in tabular format, sortable by performance (in terms of the coefficient of determination, i.e. model R^2) and model complexity.
- Regression Error Characteristic (REC; Bi and Bennett 2003) curves to allow simple graphical comparisons of the predictive performance of selected multigene regression models.

22.3 Multigene Symbolic Regression and MGGP: Overview and Mathematical Context

In this section, multigene symbolic regression is described in a mathematical context and compared with some other common symbolic regression methods as well as multiple linear regression (MLR). In addition, the mechanics of the MGGP algorithm are described, including a new, simplified high level crossover operator to expedite the exchange of genes between individuals during the simulated evolutionary process.

22.3.1 Multigene Symbolic Regression

22.3.1.1 Naïve Symbolic Regression

In early standard formulations of symbolic regression (which will be referred to as naïve symbolic regression) GP was often used to evolve a population of trees, each of which is interpreted *directly* as a symbolic mathematical equation that predicts a $(N \times 1)$ vector of outputs/responses $\mathbf{y}$ where N is the number of observations of the response variable y. The corresponding input matrix $\mathbf{X}$ is an $(N \times M)$ data matrix where M is the number of input variables. In general, only a subset of the M variables are 'selected' by GP to form the models. In naïve symbolic regression, the ith column of $\mathbf{X}$ comprises the N input values for the ith variable and is designated the input variable x_i. Figure 22.2 illustrates naïve symbolic regression.

Typically, the GP algorithm will attempt to minimise the sum of squared errors (SSE) between the observed response $\mathbf{y}$ and the predicted response $\hat{\mathbf{y}}$ (where the $(N \times 1)$ error vector $\mathbf{e}$ is $\mathbf{y} - \hat{\mathbf{y}}$) although other error measures are also frequently used, e.g. the mean squared error (MSE) and the root mean squared error (RMSE), the latter having the advantage that it is expressed in the units of the response variable y.

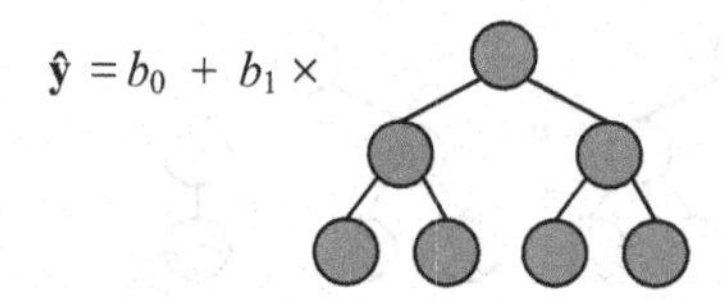

Fig. 22.3 Scaled symbolic regression. The prediction of the response data **y** is the vector output of single tree modified by a bias term b_0 and a scaling parameter b_1. These are determined by linear least squares

22.3.1.2 Scaled Symbolic Regression

To improve the efficacy of symbolic regression a bias (offset) term b_0 and a weighting/scaling term b_1 can be used to modify the tree output so that it fits **y** better. The values of these coefficients are determined by linear least squares and, for any valid tree, the prediction is guaranteed to be at least as good as the naïve prediction. It will almost always be better (the only case where it is not is the case $b_0 = 0$ and $b_1 = 1$). This method is essentially the same as scaled symbolic regression (Keijzer 2004) because the coefficients b_0 and b_1 translate and linearly scale the raw output of the tree in such a way as to minimise the prediction error of **y** as shown in Fig. 22.3.

Hence, the prediction of **y** is given by:

$$\hat{\mathbf{y}} = b_0 + b_1 \mathbf{t} \tag{22.3}$$

where **t** is the $(N \times 1)$ vector of outputs from the GP tree on the training data. This may also be written as:

$$\hat{\mathbf{y}} = \mathbf{Db} \tag{22.4}$$

where **b** is a (2×1) vector comprising the b_0 and b_1 coefficients and **D** is an $(N \times 2)$ matrix where the 1st column is a column of ones (this is used as a bias/offset input) and the 2nd column is the tree outputs **t**. The optimal linear least squares estimate (i.e. that which minimises the SSE $\mathbf{e}^T\mathbf{e}$) of **b** is computed from **y** and **D** using the well known least squares normal equation as shown in (22.5) where $\mathbf{D}^T$ is the matrix transpose of **D**. Note that the optimality of the estimate of **b** is only strictly true if a number of assumptions are met such as independence of the columns of **D** and normally distributed errors. In practice, these assumptions are rarely strictly met—but with the use of the Moore-Penrose pseudo-inverse (described in the following section)—the violations of these assumptions do not appear to prevent the practical development of effective symbolic regression models.

$$\mathbf{b} = \left(\mathbf{D}^T\mathbf{D}\right)^{-1}\mathbf{D}^T\mathbf{y} \tag{22.5}$$

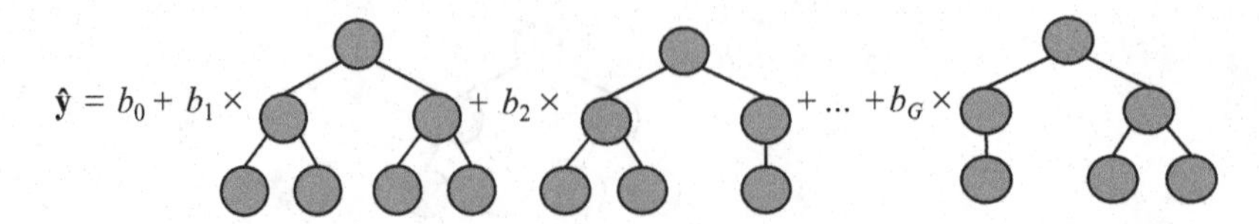

Fig. 22.4 Multigene symbolic regression. The prediction of the response data **y** is the vector output of G trees modified by bias term b_0 and scaling parameters $b_1, \ldots, b_G$

22.3.1.3 Multigene Symbolic Regression

A generalisation of the previous approach is to use G trees to predict the response data **y**. GPTIPS uses MGGP to evolve the trees comprising the additive model terms in each individual and this is referred to as multigene symbolic regression.

Again, there is an offset/bias coefficient b_0 and now the coefficients $b_1, b_2, \ldots, b_G$ are used for scaling the output of each tree/gene. A linear combination of scaled tree outputs can capture non-linear behaviour much more effectively than using scaled symbolic regression, in which one tree must capture all of the non-linear behaviour.

Moreover, by enforcing depth restricted trees and using other strategies such as Pareto tournaments and expressional complexity, this leads to the evolution of compact models that tend to have linearly separable terms and so lend themselves to automated post-run model simplification using symbolic math software. The structure of multigene symbolic regression models is illustrated in Fig. 22.4.

The prediction of the **y** training data is given by:

$$\widehat{\mathrm{y}} = b_0 + b_1\,\mathbf{t}_1 + \cdots + b_G\,\mathbf{t}_G \tag{22.6}$$

where $\mathbf{t}_i$ is the $(N \times 1)$ vector of outputs from the ith tree/gene comprising a multigene individual. Next, define $\mathbf{G}$ as a $(N \times (G+1))$ gene response matrix as follows in (22.7).

$$\mathbf{G} = [\mathbf{1}\mathbf{t}_1 \ldots \mathbf{t}_G] \tag{22.7}$$

where the $\mathbf{1}$ refers to a $(N \times 1)$ column of ones used as a bias/offset input.

Now (22.6) can be rewritten as:

$$\widehat{\mathrm{y}} = \mathbf{G}\mathbf{b} \tag{22.8}$$

The least squares estimate of the coefficients $b_0, b_1, b_2, \ldots, b_G$ formulated as a $((G+1) \times 1)$ vector can be computed from the training data as:

$$\mathbf{b} = \left(\mathbf{G}^{\mathrm{T}}\mathbf{G}\right)^{-1}\mathbf{G}^{\mathrm{T}}\mathbf{y} \tag{22.9}$$

In practice, the columns of the gene response matrix **G** may be collinear (e.g. due to duplicate genes in an individual, and so the Moore-Penrose pseudo-inverse (by means of the singular value decomposition; SVD) is used in (22.9) instead of the standard matrix inverse. Because this is computed for every individual in a GPTIPS population at each generation (except for cached individuals), the computation of the gene weighting coefficients represents a significant proportion of the computational expense of a run. In GPTIPS, the RMSE is then calculated from $\mathbf{e}^{\mathrm{T}}\mathbf{e}$ and is used as the fitness/objective function that is minimised by the MGGP algorithm.[3]

Compare this with classical MLR which is typically of the form:

$$\widehat{\mathbf{y}} = a_0 + a_1\mathbf{x}_1 + a_2\mathbf{x}_2 + \cdots + a_N\mathbf{x}_M \tag{22.10}$$

Here, the data/design matrix **X** is defined as:

$$\mathbf{X} = [\mathbf{1}\mathbf{x}_1 \ldots \mathbf{x}_M] \tag{22.11}$$

and this allows the least squares computation of the coefficients $a_0, a_1, \ldots a_M$ as:

$$\mathbf{a} = \left(\mathbf{X}^{\mathrm{T}}\mathbf{X}\right)^{-1}\mathbf{X}^{\mathrm{T}}\mathbf{y} \tag{22.12}$$

where **a** is a $((M+1) \times 1)$ vector containing the a coefficients.

This section described how a multigene individual can be interpreted as a linear-in-the-parameters regression model and how the model coefficients are computed using least squares. The following section outlines how MGGP actually generates and evolves the trees that the form the component genes of multigene regression models.

22.3.2 *Multigene Genetic Programming*

Here it is outlined how multigene individuals are created and then iteratively evolved by the MGGP algorithm. This algorithm is similar to a 'standard' GP algorithm except for modifications made to facilitate the crossover and mutation of multigene individuals. Note that—although GPTIPS uses MGGP primarily for symbolic regression—the algorithmic implementation of MGGP is *independent* of the interpretation of the multigene individuals as regression models. Multigene individuals can also be used in other contexts, e.g. classification trees (Morrison et al. 2010). In GPTIPS there is a clear modular separation of the MGGP code and the code that implements multigene regression. GPTIPS has a simple pluggable

[3]Although RMSE is the default fitness measure, this can be easily changed to, for example, MSE by a very minor edit to the file containing the default fitness function.

architecture in that it provides explicit code hooks to allow the addition of new code that interprets multigene individuals in a way of the user's choosing (the code for performing multigene regression is—by default—attached to these hooks). Note that MGGP also implicitly assumes that the specific ordering of genes in any individual is unimportant.

In the first generation of the MGGP algorithm, a population of random individuals is generated (it is currently not possible to seed the population with partial solutions). For each new individual, a tree representing each gene is randomly generated (subject to depth constraints) using the user's specified palette of building block functions and the available M input variables $x_1, \ldots, x_M$ as well as (optionally) ephemeral random constants (ERCs) which are generated in a range specified by the user (the default range is -10 to 10). In the first generation the MGGP algorithm attempts to maximise diversity by ensuring that no individuals contain duplicate genes. However, due to computational expense, this is not enforced for subsequent generations of evolved individuals.

Each individual is specified to contain (randomly) between 1 and $G_{\max}$ genes. G_{max} is a parameter set by the user. When using MGGP for regression, a high $G_{\max}$ may capture more non-linear behaviour but there is the risk of overfitting the training data and creating models that contain complex terms that contribute little or nothing to the model's predictive performance (horizontal bloat). This is discussed further in Sect. 22.5. Conversely, setting $G_{\max}$ to 1 is equivalent to performing scaled symbolic regression.

As in standard GP, at each generation individuals are selected probabilistically for breeding (using regular or Pareto tournaments or a mixture of both). Each tournament results in an individual being selected based on either its fitness or—for Pareto tournaments—its fitness and its complexity (the user can set this to be either the total node count of all the genes in an individual or the total expressional complexity of all the genes in an individual).

In MGGP, there are two types of crossover operators: high level crossover and the standard GP sub-tree crossover, which is referred to as low level crossover. The high level crossover operator is used as a probabilistically selected alternative to the ordinary low level crossover (in GPTIPS the default is that approximately a fifth of crossover events are high level crossovers).

When low level crossover is selected a gene is randomly chosen from each parent. These genes undergo GP sub-tree crossover with each other and the offspring genes replace the original genes in the parent models. The offspring are then copied into the new population.

When high level crossover is selected an individual may acquire whole genes—or have them deleted. This allows individuals to exchange one or more genes with another selected individual (subject to the $G_{\max}$ constraint).

In GPTIPS 2 the high level crossover operator described in (Searson 2002; Searson et al. 2007, 2010) has been simplified and is outlined below between a parent individual consisting of the three genes labelled (G1 G2 G3) and a parent individual consisting of the genes labelled (G4 G5 G6 G7) where (in this hypothetical case) $G_{\max} = 5$.

Parents (G1 **G2** G3)
(**G4** G5 G6 **G7**)

A crossover rate parameter *CR* (where $0 < CR < 1$) is defined. This is similar to the CR parameter used in differential evolution (DE, see Storn and Price 1997) and a uniform random number *r* between 0 and 1 is generated independently for each gene in the parents. If r is $\leq CR$ then the corresponding gene is moved to the other individual. The default value of *CR* in GPTIPS 2 is 0.5.

Hence, randomly selected genes (highlighted in boldface above) are exchanged resulting in two offspring in the next generation.

Offspring (G1 G3 **G4 G7**)
(G5 G6 **G2**)

This high level crossover mechanism is referred to as *rate based high level crossover* to distinguish it from the *two point high level crossover* mechanism in GPTIPS version 1 (which swapped contiguous sections of genes from individuals). Note that the rate based high level crossover mechanism results in new genes for both individuals as well as reducing the overall number of genes for one model and increasing the total number of genes for the other. If an exchange of genes results in either offspring containing more genes than the G_{max} constraint then genes are randomly deleted until the constraint is no longer violated.

22.4 Using GPTIPS

In this section it will be illustrated how GPTIPS 2 may be used to generate, analyse and export non-linear multigene regression models, both using command line tools and visual analytics tools and reports. The example screenshots in the figures contained in this section are taken from example runs from various data sets using configuration files and data that are provided with GPTIPS 2. The screenshots were obtained using MATLAB Release 2014b on OSX.

22.4.1 Running GPTIPS

As discussed in Sect. 22.2.1, the user creates a simple text configuration file that specifies some basic run parameters and either loads in the data to be modelled from file or algorithmically generates it. Any unspecified parameters are set to GPTIPS default values.

To run the configuration file (here called `configFileName.m`) the `rungp` function is used as follows:

```
gp = rungp(@configFileName)
```

where the @ symbol denotes a MATLAB function handle to the configuration file.

The GPTIPS run then begins. When it is complete—the population and all other relevant data is stored in the MATLAB 'struct' variable `gp`. This is used as a basis for all subsequent analyses.

22.4.2 *Exploratory Post Run Analyses*

GPTIPS provides a number of exploratory post-run interactive visualisation and analysis tools. For instance, a simple summary of any run can be generated using the `summary` function and an example is shown in Fig. 22.5.

For multigene symbolic regression this shows in the upper part of the chart—by default—the $\log_{10}$ value of the best RMSE (this is the error metric that GPTIPS attempts to minimise over the training data) achieved in the population over the generations of a run. The lower part of the chart shows the mean RMSE achieved in the population.

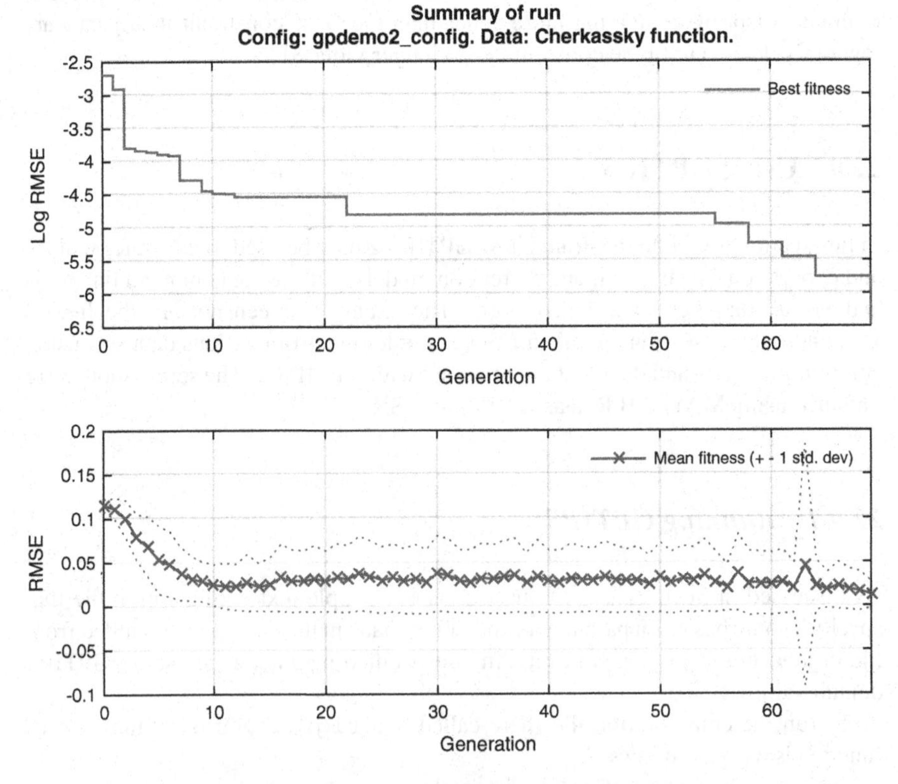

Fig. 22.5 An example of a run summary in GPTIPS. Generated using the `summary` function

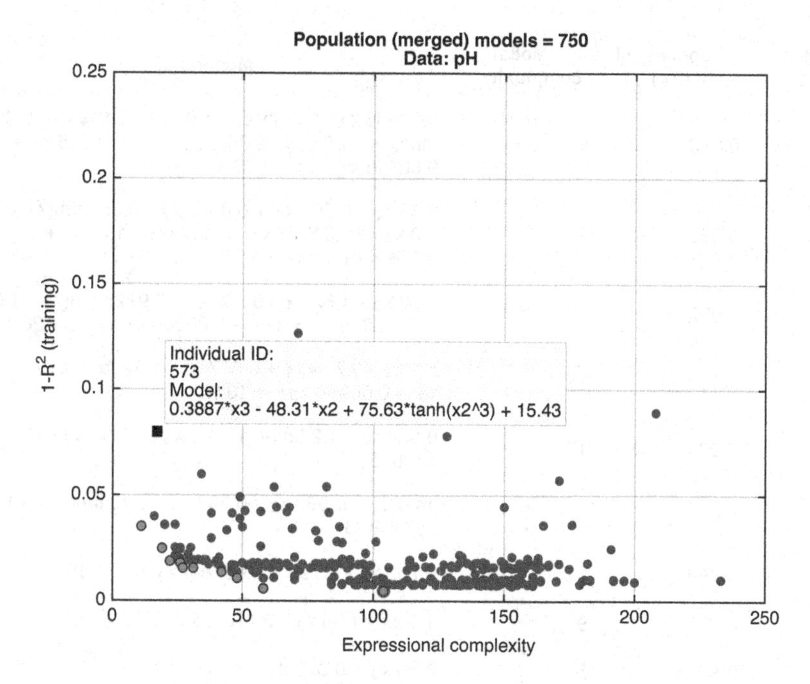

Fig. 22.6 Visually browsing a multigene regression model population. *Green dots* represent the Pareto front of models in terms of model performance $(1-R^2)$ and model complexity. *Blue dots* represent non-Pareto models. The *red circled dot* represents the best model in the population in terms of R^2 on the training data. Clicking on a *dot* shows a *yellow* popup containing the model ID and the simplified model equation. Generated using the `popbrowser` function

Other tools are intended to help the user to identify a model (or small set of models) that look promising and worthy of further investigation. One of the most useful visual analytic tools is the population browser. This interactive tool visually illustrates the entire population in terms of its predictive performance and model complexity characteristics. This is generated using the `popbrowser` function. An example of this is shown in Fig. 22.6. Each model is plotted as a dot with $(1-R^2)$ on the vertical axis and expressional complexity on the horizontal axis. The Pareto front models are highlighted in green and it is almost always these models that will be of the greatest interest to the user. In particular, the Pareto models in the lower left of the population (high R^2 and low complexity) are usually where a satisfactory solution may be found.

This visualisation may be used with the training, validation or test data sets. For example Fig. 22.6 was generated using:

```
popbrowser(gp,'train')
```

Another way of displaying information about Pareto front models in a population is by use of the `paretoreport` function. This creates a standalone HTML file—viewable in a web browser—that includes a table listing the simplified model equations along with the model performance and expressional complexity.

Model ID	Goodness of fit (R2)	Model complexity	Model
66	0.993	49	$0.00336 x_3 x_4 \tanh(x_1 - 1.0 x_3) - 0.134 x_4 - 2.89 \tanh(x_1 - 1.0 x_3) - 2.06e\text{-}4 x_3 x_4^2 - 0.845 x_1 + 0.00113 x_1 x_3 x_4 + 20.4$
13	0.993	37	$2.63 x_2 - 0.695 x_1 + 0.0679 x_3 - 3.35 \tanh(x_1 - 1.0 x_3) + 0.00391 x_3 x_4 \tanh(x_1 - 1.0 x_3) + 2.65e\text{-}4 x_1 x_3 x_4 + 15.0$
12	0.992	26	$6.53 x_2 - 1.62 x_1 - 0.262 x_3 - 0.972 \tanh(x_1 - 1.0 x_3) - 4.92e\text{-}4 x_3 x_4^2 + 0.00204 x_1 x_3 x_4 + 26.1$
5	0.989	18	$5.0 x_2 - 0.734 x_1 + 0.557 x_3 - 1.22 \tanh(x_1 - 1.0 x_3) - 0.00968 x_3^2 + 10.9$
27	0.983	17	$0.828 x_3 - 1.2 \tanh(x_1 - 1.0 x_3) - 7.0e\text{-}4 x_1 x_3 x_4 + 0.685$
7	0.974	12	$14.9 x_2 + 0.963 x_3 - 0.788 x_4 - 1.45 \tanh(x_1 - 1.0 x_3) + 10.3$
36	0.964	11	$0.752 x_4 - 1.79 x_1 - 1.57e\text{-}4 x_3 x_4^2 + 15.2$
49	0.953	9	$1.1 x_3 - 1.56 x_2 - 8.47e\text{-}4 x_1 x_3 x_4 - 1.55$
31	0.916	3	$8.86 x_2 + 0.372 x_3 - 3.29$

Fig. 22.7 Extract from a Pareto front model HTML report. GPTIPS 2 can generate a standalone interactive HTML report listing the multigene regression models on the Pareto front in terms of their simplified equation structure, expressional complexity and performance on the training data (R^2). The above table is sortable by clicking on the appropriate column header. Generated using the `paretoreport` function

The table is interactive and the models can be sorted by performance or complexity by clicking on the appropriate column header. An example of an extract from such a report is shown in Fig. 22.7. This report assists the user in rapidly identifying the most promising model or models to investigate in more detail.

It is also possible to filter populations according to various user criteria using the `gpmodelfilter` object. The output of this filter is another `gp` data structure which is functionally identical to the original (in the sense that any of the command line and visual analysis tools may be applied to it) except that models not fulfilling user criteria have been removed.

For example, if the user wants to only retain models that (a) have an R^2 greater than 0.8 (b) contain the input variables x_1 and x_2 and (c) do not contain the variable x_4 then the filter can be configured and executed as follows:

Create a new filter object `f`:

```
f = gpmodelfilter
```

Next set the user criteria, i.e. models must have R^2 (training data) greater or equal to 0.8:

```
f.minR2train = 0.8
```

Must include x_1 and x_2:

```
f.includeVars = [1 2]
```

Must exclude x4:

```
f.excludeVars = 4
```

Finally, apply the filter to the existing population structure `gp` to create a new one `gpf`:

```
gpf = f.applyFilter(gp)
```

At this point the user may apply the exploratory tools (e.g. `paretoreport`) to the refined population to zero in on models of interest fulfilling certain criteria. Other criteria that can be set include maximum expressional complexity, maximum and minimum number of variables and Pareto front (i.e. exclude all models not on the Pareto front).

22.4.3 Model Performance Analyses

Once a model (or set of models) has been identified using the tools described above, the detailed performance of the model can be assessed by use of the `runtree` function. This essentially re-runs the model on the training data (and validation and test data, if present) and generates a set of graphs including predicted vs actual y and scatterplots of predicted vs actual y. These graphs can be generated using the numeric model ID (e.g. from the `popbrowser` visualisation) as an input argument to `runtree` or by using keywords such as 'best' (best model on training data) and 'testbest' (best model on test data), e.g.

```
runtree(gp,'testbest')
```

This is a common design pattern across a large number of GPTIPS functions. An example of the scatterplots generated by `runtree` is shown in Fig. 22.8.

Additionally, for any model a standalone HTML report containing detailed tabulated run configuration, performance and structural (simplified model equations and trees structures) data may be generated using the `gpmodelreport` function. These reports contain interactive scatter charts similar to that in Fig. 22.8. The reports are fairly lengthy—however—and so are not illustrated here.

A way of comparing the performance of a small set of models simultaneously is to generate regression error characteristic (REC; Bi and Bennett 2003) curves using the `compareModelsREC` function. REC curves are similar to receiver operating characteristic curves (ROC) used to graphically depict the performance of classifiers on a data set. An example of REC curves generated using the `compareModelsREC` function is shown below in Fig. 22.9. The user can specify what curves to compare in the arguments to the function, e.g.

```
compareModelsREC(gp,[2 3 9], true)
```

where the final Boolean `true` argument indicates that the best model on the training data should also be plotted in addition to models 2, 3 and 9.

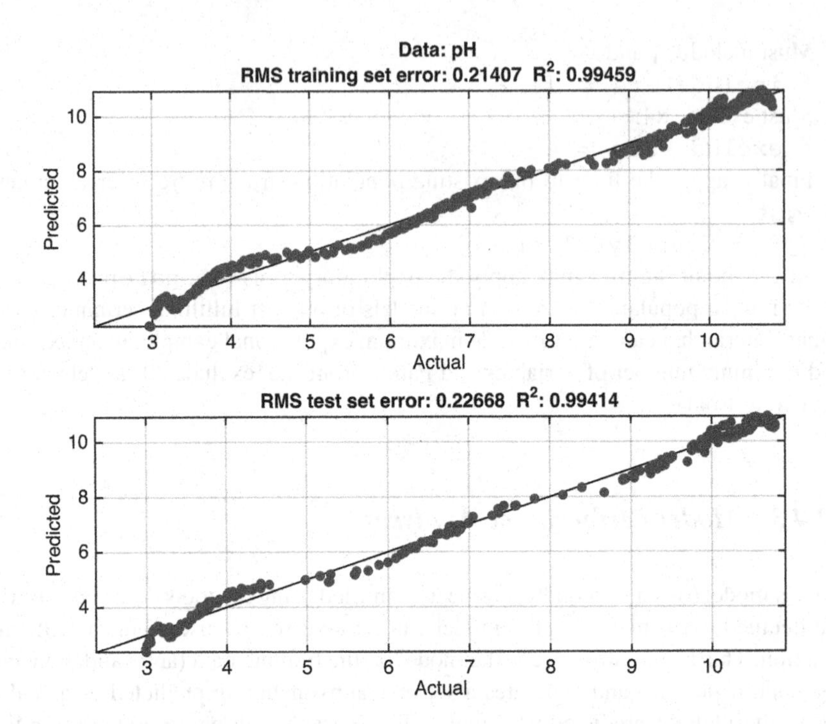

Fig. 22.8 Performance scatterplots on training and testing data sets for a selected multigene regression model. Generated by the `runtree` function

22.4.4 Model Conversion and Export

Finally, there is a variety of functions provided to convert and/or export models to different formats, e.g. to convert a model with numeric ID 5 to a standalone MATLAB M file called `model.m` then the `gpmodel2mfile` function may be used as follows:

```
gpmodel2mfile(gp,5,'model')
```

To convert a model to a symbolic math object, the `gpmodel2sym` function may be used in a similar way. A symbolic math object can then be converted to a string containing a snippet of C code using the `ccode` function.

22.5 Reducing Model Complexity Using Gene Analysis

22.5.1 Horizontal Model Bloat

GP frequently suffers from the phenomenon of 'bloat', i.e. the tendency to evolve trees that contain terms that confer little or no performance benefit, e.g. see (Luke and Panait 2006). In terms of model development this is related to the phenomenon

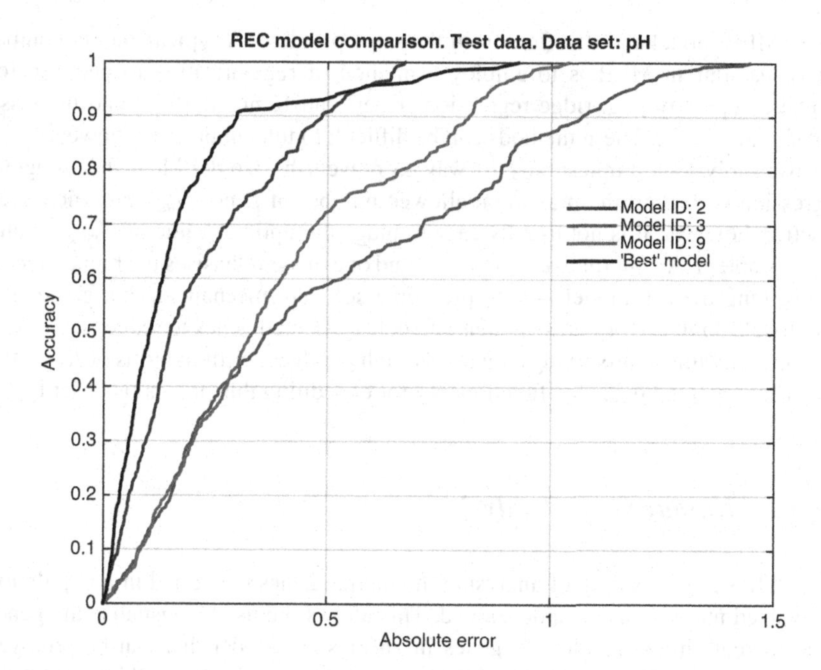

Fig. 22.9 Regression error characteristic (REC) curves. GPTIPS 2 allows the simple comparison between multigene regression models in terms of REC curves which are similar to receiver operating characteristic (ROC) curves for classifiers. The REC curves show the proportion of data points predicted (y axis) with an accuracy better than the corresponding point on the x axis. Hence, 'better' models lie to the upper left of the diagram. Generated using the `compareModelsREC` function

of overfitting. GPTIPS 2 contains a number of mechanisms intended to mitigate this. For instance: the use of fairly stringent restrictions on maximum tree depth (to ameliorate vertical bloat), the use of tree expressional complexity as a measure of model complexity (rather than a simple node count) to promote flatter trees over deeper ones during the simulated evolutionary process, the integration of the train-validate-test model development cycle, and the use of Pareto tournaments to select models that perform well (in terms of goodness of fit) and are not overly complex.

However, the use of multigene regression models in GPTIPS leads to another type of bloat that is referred to here as horizontal bloat. This is the tendency of multigene models to acquire genes that are either performance neutral (i.e. deliver no improvement in R^2 on the training data) or offer very small incremental performance improvements. Clearly—in the majority of practical applications—these terms are undesirable.

Horizontal bloat is the essentially the same behaviour exhibited by non-regularised MLR models, where it is well known that the addition of model terms leads to a monotonically increasing R^2 on training data even though the terms may not be meaningful (e.g. they are capturing noise) or allow the model to generalise well to testing or validation data sets. Multigene regression is a type of pseudo-linear MLR model and it suffers from the same problem. A typical way to combat this behaviour in MLR is to employ a method of regularisation to penalise for model complexity [e.g. ridge regression (Hoerl and Kennard 1970) and the lasso (Tibshirani 1996)]. These methods can be difficult to tune in practice, however.

Ostensibly, the simplest way to way to prevent horizontal bloat in multigene regression is to limit the maximum allowed number of genes G_{max} in a model. In practice, however, it is not usually easy to judge the optimal value of G_{max} for any given problem. An alternative approach—and one that emphasises the human factor in instilling trust in models—is to provide a software mechanism that guides the user to take high performance models and delete selected genes to reduce the model complexity whilst maintaining a relatively high goodness of fit in terms of R^2. In the following section GPTIPS 2 functionality for expediting this process is described.

22.5.2 Unique Gene Analysis

In GPTIPS 2, a new way of analysing the unique genes contained in a population of evolved models has been developed. This allows the user to visualise the genes in a population and to identify genes in an existing model that can be removed thus reducing model complexity whilst having only a relatively small impact on the model's predictive performance. The visualisation aspect (i.e. the ability to see the gene equation and the R^2 value if the gene were removed) is important because it allows the user to rapidly make an informed choice about which model terms to remove. Often this choice is based on problem domain knowledge of the system being modelled. For example, the user might want to delete a model term such as $\sin(1-x^3)$ because it is inconsistent with his or her knowledge about the underlying data or system. This gene-centric visualisation allows users to tailor evolved models to suit their own preferences and knowledge of the modelled data.

An additional benefit of being able to visualise the genes in a model is that it expedites the process of human understanding of the model and intuition into which model terms account for a high degree of predictive ability and which account for lower amounts.

After a GPTIPS run has been completed, the user can extract a MATLAB data structure containing all of the unique genes in a population using the `uniquegenes` function as indicated below:

```
genes = uniquegenes(gp)
```

This function does the following:

- Extracts every genotype i.e. tree encoded gene (gene weights are ignored) from each model in the population.
- Deletes duplicate genotypes.
- Converts the unique genotypes to symbolic math objects (phenotypes) and then analytically simplifies them using MATLAB's symbolic math engine (MuPAD).
- Deletes any duplicate symbolic math objects representing genes and assigns a numeric ID to the remaining unique gene objects.

Note that it is quite frequent that two different genotypes will, after conversion to symbolic math objects and automated analytic simplification, resolve to the same phenotype.

Next—to provide an interactive visualisation of the genes in the population and a selected model—the `genebrowser` function is used. In the example below, it is used on the model that performed best (in terms of R^2) on the training data.

```
genebrowser(gp,genes,'best')
```

Clicking on any blue bar shows a yellow popup containing the symbolic version of the gene and the reduction in R^2 that would result if that gene were to be removed from the model. Conversely, clicking on any orange bar in the lower axis does the same for genes that are not in the current model and shows the increase in R^2 that would be attained if that gene were added to the model (Fig. 22.10).

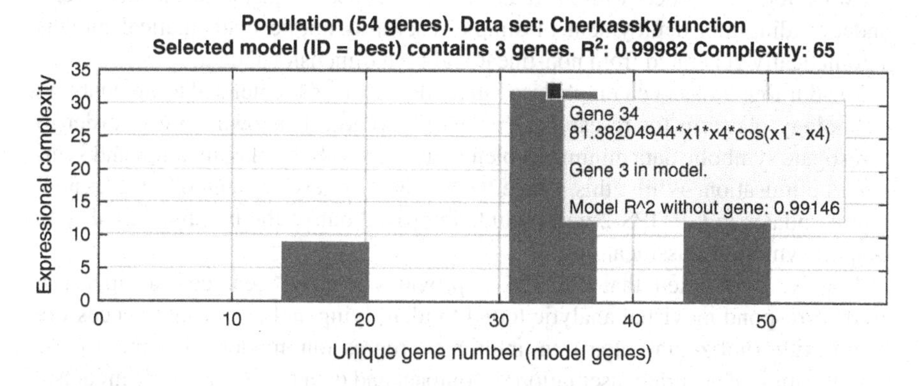

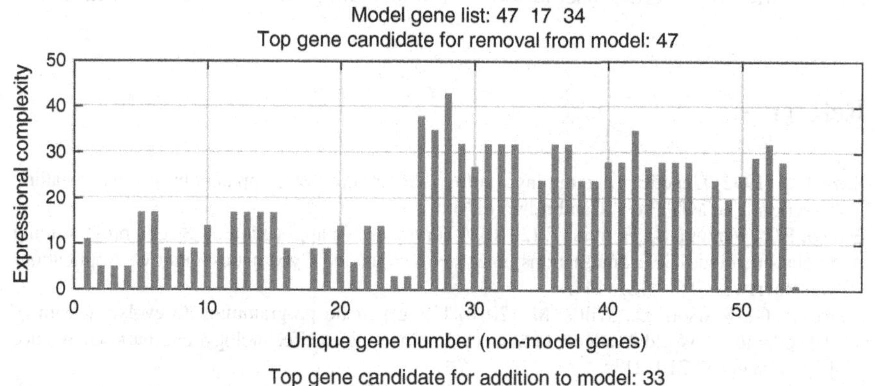

Fig. 22.10 Reducing model complexity using the `genebrowser` analysis tool. The *upper bar chart* shows the gene number and expressional complexity of genes comprising the selected model. The *lower bar chart* shows genes in the population but not in the selected model. Clicking on a *blue bar* representing a model gene reveals a popup containing the gene equation and the R^2 (on the training data) if that gene were removed from the model. Here it shows that the highlighted gene/model term $81.382x_1x_4\cos(x_1 - x_4)$ is a horizontal bloat term and could be removed from the model with a very minor decrease in R^2

Once the user has identified a suitable gene to be removed from the model, a new model without the gene can be generated using the `genes2gpmodel` function using the unique gene IDs as input arguments. The data structure returned from this function can be examined using the provided tools—as well as exported in various formats—in exactly the same way as any model contained within the population.

22.6 Conclusions

In this chapter GPTIPS 2, the latest version of the free open source software platform for symbolic data mining, has been described. It is emphasised that the software is aimed at non-experts in machine learning and computing science—and that the software tools provided within GPTIPS are intended to facilitate the discovery, understanding and deployment of simple, useful symbolic mathematical models automatically generated from non-linear and high dimensional data.

In addition, it has been emphasised that GPTIPS is also intended as an enabling technology platform for researchers who wish to add their own code in order to investigate symbolic data mining problems such as symbolic classification and symbolic optimisation. Whilst this article has focused largely on symbolic regression, future updates to GPTIPS 2 will include improved out-of-the-box functionality to support symbolic classification.

Finally, it is noted that GPTIPS 2 provides a novel gene-centric approach (and corresponding visual analytic tools) to identifying and removing unnecessary complexity (horizontal bloat) in multigene regression models, leading to the identification of accurate, user tailored, compact and data driven symbolic models.

References

Koza J.R. (1992) Genetic programming: on the programming of computers by means of natural selection, The MIT Press, Cambridge (MA).

Espejo, P.G., Ventura, S., Herrera, F. (2010) A survey on the application of genetic programming to classification, IEEE Transactions on Systems, Man and Cybernetics - Part C: Applications and Reviews, 40 (2), 121–144.

Morrison, G., Searson, D., Willis, M. (2010) Using genetic programming to evolve a team of data classifiers. World Academy of Science, Engineering and Technology, International Science Index 48, 4(12), 210–213.

Pan, I., Das, S. (2014) When Darwin meets Lorenz: Evolving new chaotic attractors through genetic programming. arXiv preprint arXiv:1409.7842.

Gandomi, A.H., Alavi, A.H. (2011) A new multi-gene genetic programming approach to non-linear system modeling. Part II: geotechnical and earthquake engineering problems, Neural Comput & Applic, 21(1), 171–187.

Smits, G.F., Kotanchek, M. (2004) Pareto-front exploitation in symbolic regression, Genetic Programming Theory and Practice II, 283–299.

Poli, R., Langdon, W.B., McPhee, N.F., Koza, J.R. (2007). Genetic programming: An introductory tutorial and a survey of techniques and applications. University of Essex, UK, Tech. Rep. CES-475.

Pan, I., Pandey, D.S., Das, S. (2013) Global solar irradiation prediction using a multi-gene genetic programming approach. Journal of Renewable and Sustainable Energy, 5(6), 063129.

Barati, R., Neyshabouri, S.A.A.S., Ahmadi, G. (2014) Development of empirical models with high accuracy for estimation of drag coefficient of flow around a smooth sphere: An evolutionary approach. Powder Technology, 257, 11–19.

Floares, A.G., Luludachi, I. (2014) Inferring transcription networks from data. Springer Handbook of Bio-/Neuroinformatics, Springer Berlin Heidelberg, 311–326.

Gandomi, A.H., Alavi, A.H. (2012) A new multi-gene genetic programming approach to nonlinear system modeling. Part I: materials and structural engineering problems. Neural Computing and Applications, 21(1), 171–187.

Searson, D.P. (2002) Non-linear PLS using genetic programming, PhD thesis, Newcastle University, UK.

Searson D.P., Willis M.J., Montague, G.A. (2007) Co-evolution of non-linear PLS model components, Journal of Chemometrics, 21 (12), 592–603.

Searson, D.P., Leahy, D.E., Willis, M.J. (2010) GPTIPS: an open source genetic programming toolbox for multigene symbolic regression, Proceedings of the International MultiConference of Engineers and Computer Scientists 2010 (IMECS 2010), Hong Kong, 17–19 March.

Deb, K., Pratap, A., Agarwal, S., Meyarivan, T.A.M.T (2002) A fast and elitist multiobjective genetic algorithm: NSGA-II. Evolutionary Computation, IEEE Transactions on, 6(2), 182–197.

Bi, J., Bennett, K.P. (2003) Regression error characteristic curves, Proceedings of the Twentieth International Conference on Machine Learning (ICML-2003), Washington DC, 43–50.

Keijzer, M. (2004) Scaled symbolic regression, Genetic Programming and Evolvable Machines, 5, 259–269.

Storn, R., Price, K. (1997) Differential evolution – a simple and efficient heuristic for global optimization over continuous spaces. Journal of global optimization, 11(4), 341–359.

Luke, S., Panait, L. (2006) A comparison of bloat control methods for genetic programming, Evol. Comput., 14(3), 309–344.

Hoerl, A. E., Kennard, R.W. (1970) Ridge regression: Biased estimation for nonorthogonal problems. Technometrics, 12(1), 55–67.

Tibshirani, R. (1996) Regression shrinkage and selection via the lasso. Journal of the Royal Statistical Society. Series B (Methodological), 267–288.

Chapter 23
eCrash: a Genetic Programming-Based Testing Tool for Object-Oriented Software

José Carlos Bregieiro Ribeiro, Ana Filipa Nogueira, Francisco Fernández de Vega, and Mário Alberto Zenha-Rela

23.1 Introduction

Modern software products typically contain millions of lines of code; precisely locating the source of errors can thus be very resource consuming. Most errors are introduced at the unit stage (Tassey 2002); Unit Testing is thus a key phase in projects that demand high quality and reliability, and plays a major role in the total testing efforts. Tools for automating Unit Testing improve this—largely informal and often human-dependant—process, and have a direct impact on the quality attributes of the implemented systems.

This paper details the architecture and functionalities of the *eCrash* tool, which has recently been deployed for public availability. *eCrash* implements a Test Data generation technique driven by an Evolutionary Algorithm (EA); the application of this type of algorithms to perform Software Testing activities is often Evolutionary Testing (ET) (Tonella 2004; Wappler and Wegener 2006) or Search-Based Test Case Generation (SBTCG) (McMinn 2004). In ET, meta-heuristic search techniques are used to select and produce high-quality Test Data. The search space is the input domain of the test object, and the problem is to find a set of Test Programs that satisfies a certain test criterion (e.g., achieving full structural coverage).

J.C.B. Ribeiro
Polytechnic Institute of Leiria, Morro do Lena, Alto do Vieiro, Leiria, Portugal
e-mail: jose.ribeiro@ipleiria.pt

A.F. Nogueira (✉) • M.A. Zenha-Rela
University of Coimbra, CISUC, DEI, 3030-290, Coimbra, Portugal
e-mail: afnog@dei.uc.pt; mzrela@dei.uc.pt

F.F. de Vega
University of Extremadura, C/ Sta Teresa de Jornet, 38, Mérida, Spain
e-mail: fcofdez@unex.es

A.H. Gandomi et al. (eds.), *Handbook of Genetic Programming Applications*,
DOI 10.1007/978-3-319-20883-1_23

eCrash employs Genetic Programming (GP) for evolving Test Data; it is arguably the most natural way to represent and evolve Object-Oriented (OO) programs, and its characteristics allow automating both the Test Object analysis and the Test Data generation processes. It is publicly available at:

http://sourceforge.net/p/ecrashtesting/

This paper is organised as follows. In Sect. 23.2 relevant concepts and terminology are introduced, and related work is reviewed. In Sect. 23.3, the *eCrash* tool is described in detail. Section 23.4 discusses *eCrash*'s current features and limitations, as well as topics for future work. Finally, in Sect. 23.5, some concluding considerations are presented.

23.2 Background and Related Work

Software Testing is the process of exercising an application to detect errors and to verify that it satisfies the specified requirements. When performing Unit Testing, the goal is to warrant the robustness of the smallest units—the Test Objects—by testing them in an isolated environment (Naik and Tripathy 2008). Unit Testing is performed by executing the Test Objects in different scenarios using relevant and interesting Test Programs (or Test Cases); a Test Set is said to be adequate with respect to a given criterion if the entirety of Test Cases in this set satisfies this criterion.

A Unit Test Case for OO software consists of a Method Call Sequence (MCS), which defines the test scenario (Harman et al. 2009). During Test Program execution, all participating objects are created and put into particular states through a series of method calls. Each Test Case focuses on the execution of one particular public method—the Method Under Test (MUT) that belongs to a specific Class Under Test (CUT). It is not always possible to test the operations of a class in isolation; testing a single method involves other classes, e.g., the classes that appear as parameter types in the method signatures of the CUT. Thus, the unit testing process includes the creation of auxiliary objects that will be used by the MUTs; the transitive set of classes which are relevant for testing a particular class is called the *test cluster* (Wappler and Wegener 2006).

In OO programs, an object stores its state in fields and exposes its behaviour through methods. Hiding internal state and requiring all interaction to be performed through an object's methods is known as data *encapsulation*—a fundamental principle of Object-Oriented programming. This principle is related to the *state problem* (McMinn and Holcombe 2003) which occurs with objects that exhibit state-like qualities by storing information in fields that are protected from external manipulation, and that can only be accessed through the public interface. The *state problem* is a major challenge when automating the test data generation process for unit testing of OO systems. Defining a test set that achieves full structural coverage may, in fact, involve the generation of complex and intricate test cases in order to define elaborate state scenarios, and requires the definition of carefully fine-tuned methodologies that promote the traversal of problematic structures and difficult control-flow paths. The metrics required for evaluating whether the test set is suitable can be collected by abstracting and modelling the programs' behaviours which are exhibited during execution. For this purpose, a Control-Flow Graph (CFG) is usually used to model the program representation (Vincenzi et al. 2006). Through *dynamic analysis*, which involves the execution and monitoring of the object under analysis, it is possible to observe which structural entities are traversed during the execution of unit test cases; it can be achieved by instrumenting the test object (Kinneer et al. 2006).

EAs use simulated evolution as a search strategy to evolve candidate solutions for a given problem, using operators inspired by genetics and natural selection. The application of EA to test data generation is often referred to in the literature as ET or SBTCG (Tonella 2004; McMinn 2004). The goal of ET problems is to find a set of test cases that satisfies a certain test criterion—such as full structural coverage of the test object. The test objective must be defined numerically; suitable fitness functions, which provide guidance to the search by telling how good each candidate solution is, must be defined (Harman 2007). The search space is the set of possible inputs to the test object—*the input domain*; in the particular case of OO programs, the input domain encompasses the parameters of the test object's public methods—including the implicit parameter (*this*). As such, the goal of the evolutionary search is to find Test Programs that define interesting state scenarios for the variables which will be passed, as arguments, in the call to the MUT.

GP (Koza 1992), in particular, is a specialization of Genetic Algorithm (GA) usually associated with the evolution of tree structures; it focuses on automatically creating computer programs by means of evolution, and is thus especially suited for representing and evolving test programs. The nodes of a GP tree are usually not typed—i.e., all the functions are able to accept every conceivable argument. Non-typed GP approaches are, however, unsuitable for representing test programs for OO software. Conversely, Strongly-Typed Genetic Programming (STGP) (Montana 1995) allows the definition of types for the variables, constants, arguments and returned values; the data type for each element must be specified beforehand in the Function Set. These specifications ensure that the initialization process and the various genetic operations only construct syntactically correct trees. Nevertheless, these syntactically correct and compilable MCS can originate *unfeasible* test cases. *Unfeasible* test cases abort prematurely during execution due to the occurrence of a runtime exception (Wappler and Wegener 2006); the exception prevents the evaluation of the individual (GP tree) because the final instruction of the MCS is never reached, i.e., the call to the MUT is not performed. Contrastingly, *feasible* test cases are successfully executed and terminate with the invocation of the MUT.

The first approach to the field of Object-Oriented Evolutionary Testing (OOET) was presented in (Tonella 2004); in this work, a technique for automatically generating input sequences for the structural Unit Testing of Java classes by means of GAs is proposed. Possible solutions are represented as chromosomes, which consist of the input values to use in test program execution; the creation of objects is also accounted for. A population of test programs is evolved in order to increase a measure of fitness accounting for their ability to satisfy a branch coverage criterion; new test programs are generated as long as there are targets to be covered or a maximum execution time is reached. The *eToc* framework software was implemented and made available as a result of this research.

Several OOET techniques as methodologies were proposed in the following years. However, the test data generation frameworks developed are seldom publicly available, with the only exceptions known to the authors being *eToc*,[1] *Testful*,[2] *EvoSuite*,[3] and (more recently) *eCrash*.

TestFul (Miraz et al. 2009; Baresi et al. 2010; Baresi and Miraz 2010) uses a holistic, incremental approach to the generation of test data for OO software, as the internal states reached with previous test programs are utilized as starting points to subsequent individuals. Strategies for enhancing the efficiency of the approach include local search (integrating the global evolutionary search in order to form a hybrid approach), seeding (providing an initial population so as to speed up the start of the evolutionary process) and fitness inheritance (replacing the evaluation of the fitness function by replacing the fitness of some individuals with and estimated fitness inherited from their parents). A multi-objective approach is used to combine coverage and length criteria. Test program quality is evaluated with a technique which merges black-box analysis (to evaluate the behaviours of tested classes and reward test programs accordingly) and white-box analysis (which utilizes coverage criteria).

EvoSuite (Fraser and Arcuri 2011a,b, 2013) applies a hybrid approach—integrating hybrid search, dynamic symbolic execution and testability transformation—for generating and optimizing test suites, while suggesting possible oracles by adding assertions that concisely summarize the current behaviour. It implements a "whole test suite" approach towards evolving test data, meaning that optimization is performed with respect to a coverage criterion, rather than individual coverage goals. The rationale for this methodology is related with the observation that coverage goals are not independent nor equally difficult, and are sometimes unfeasible; test suites are thus evolved with the aim of covering all coverage goals at the same time, while keeping the total size as small as possible. The *EvoSuite* tool has been utilized as a platform for research and experimentation. In Pavlov and Fraser (2012), a semi-automatic test generation approach based on EvoSuite is presented; a human tester is included in the test generation process, and given the opportunity to improve the current solution (an editor window is presented to the user with a preprocessed version of the current best individual) if and when the search stagnates, under the assumption that where the search algorithm struggles a human tester with domain knowledge can often produce solutions easily. In Fraser et al. (2013), the issue of primitive value (e.g., numbers and strings) optimization

[1] http://star.fbk.eu/etoc/.

[2] https://code.google.com/p/testful/.

[3] http://www.evosuite.org/.

is addressed by extending the global search applied in EvoSuite with local search on the individual statements of method sequences: at regular intervals, the search inspects the primitive variables and tries to improve them.

eCrash (Ribeiro 2010) employs STGP for evolving test programs. Its development supported a series of studies on defining strategies for addressing the challenges posed by the OO paradigm, which include: methodologies for systematizing both the Test Object analysis (Ribeiro et al. 2007a,b) and the Test Data generation (Ribeiro et al. 2007b; Ribeiro 2008) processes; introducing an Input Domain Reduction methodology, based on the concept of Purity Analysis, which allows the identification and removal of entries that are irrelevant to the search problem because they do not contribute to the definition of relevant test scenarios (Ribeiro et al. 2008, 2009); proposing an adaptive strategy for promoting the introduction of relevant instructions into the generated test cases by means of Mutation, which utilizes Adaptive EAs (Ribeiro et al. 2010); and defining an Object Reuse methodology for GP-based approaches to Evolutionary Testing, which allows that one object instance can be passed to multiple methods as an argument (or multiple times to the same method as arguments) and enables the generation of test programs that exercise structures of the software under test that would not be reachable otherwise (Ribeiro et al. 2010). Special attention is put on bridging and automating the static test object analysis and the iterative test data generation processes; the Function Set is computed automatically with basis on the Test Cluster, and the test programs are evolved iteratively solely with basis on Function Set information.

23.3 The *eCrash* Tool: Features, Architecture and Methodology

In this Section, the approach for automatic Test Data Generation employed by the *eCrash* framework is described. Figures 23.1 and 23.2 provide an overview of this framework and of the way in which its components interoperate. *eCrash* is composed of the following four core modules (Fig. 23.1):

- *Test Object Instrumentation (*TOI*) Module*— executes the tasks of building the CFG and instrumenting the Test Object.
- *Automatic Test Object Analysis (*ATOA*) Module*—performs the Test Object analysis; its main tasks are those of defining the Test Cluster, generating the Function Set and parameterising the Test Program generation process.
- *Test Program Generation (*TPG*) Module*—iteratively evolves potential solutions to the problem with basis on the GP paradigm.
- *Test Program Evaluation and Management (*TPEM*) Module*—synthesises, executes and evaluates Test Programs dynamically, and selects the Test Programs to be included into the Test Set.

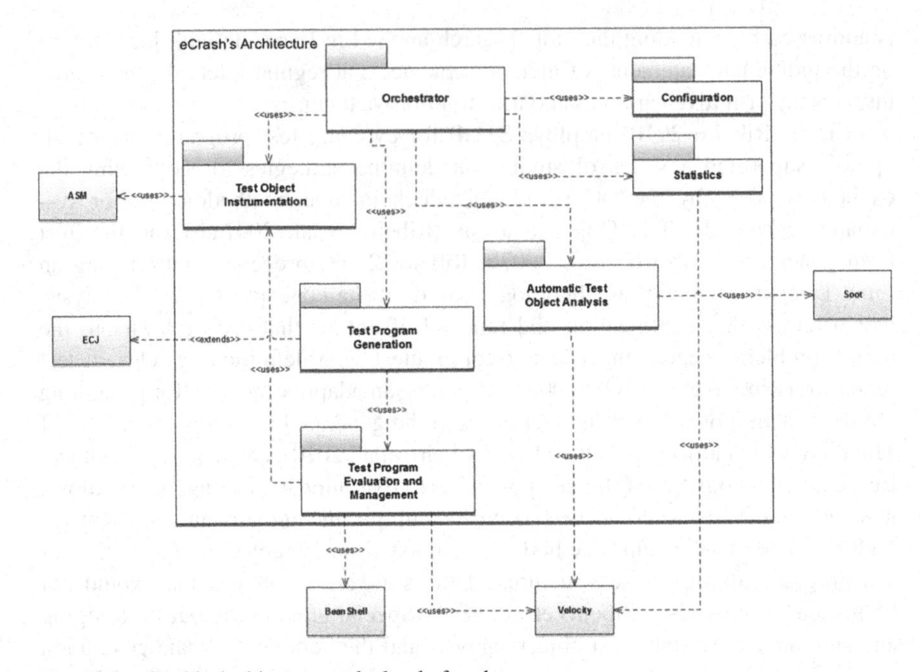

Fig. 23.1 eCrash's Architecture at the level of packages

Furthermore, there are three other modules that are critical to *eCrash*'s operation:

- *Statistics Module*—generates statistical data.
- *Configuration Module*—deals with the system's configurations.
- *Orchestration Module*—coordinates the different tasks and modules involved in the test data generation process.

The *eCrash*'s architecture (Fig. 23.1) includes third-party frameworks to provide part of the functionality, namely: a component for byte code instrumentation—the ASM library[4]; and a GP processing component—the Evolutionary Computation in Java (ECJ) framework (Luke 2013). ASM allows the CFG extraction from the classes under analysis, and the code instrumentation for structural coverage measurement purposes. ECJ provides the EA's infrastructure, i.e., it evolves and evaluates the genetic trees that represents each test case candidates.

Other third-party libraries were also included, namely: Soot,[5] Velocity[6] and BeanShell.[7] Soot is responsible for the Purity Analysis (Ribeiro et al. 2009)

[4]http://asm.ow2.org/.

[5]http://www.sable.mcgill.ca/soot/.

[6]http://velocity.apache.org/.

[7]http://www.beanshell.org/.

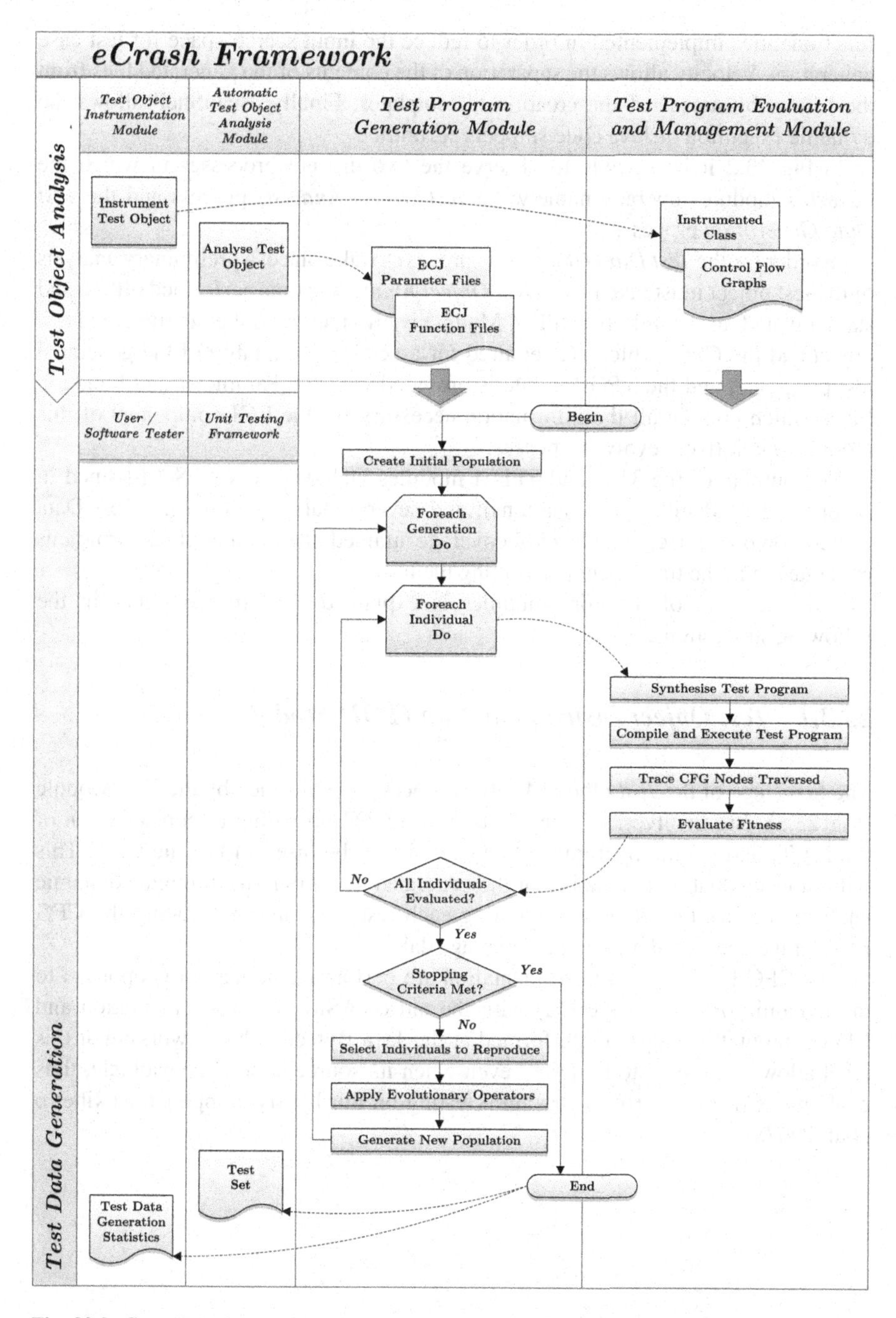

Fig. 23.2 Cross-Functional Diagram of the *eCrash* Framework (Ribeiro 2010)

functionalities implemented in order to reduce the input search space for test case generation. Velocity allows the separation of the contents of the generated files from the format by means of the creation of templates. Finally, BeanShell allows the dynamic execution of Java code snippets at runtime.

In Fig. 23.2 it is possible to observe the two distinct processes in which the *eCrash*'s modules operate, namely the *Test Object Analysis* process and the *Test Data Generation* process.

In order for the *Test Data Generation* process to take place, a preliminary analysis of the test object must take place. *Test Object Analysis* is thus performed offline and statically and, as a result, the TPEM Module is provided with the instrumented test object and the CFG, which are required for assessing the quality of the generated test programs; and the TPG Module is provided with the Parameter and Function Files, which contain all the information necessary for the ECJ component of this module to iteratively evolve test cases.

The outputs of the TPG and TPEM modules include: the Test Set (defined in accordance to JUnit's[8] specifications); and several statistics about the Test Data Generation process, e.g. the level of coverage attained, the number of test programs generated, and the time spent performing the task.

The operation of the core modules is explained with further detail in the following subsections.

23.3.1 Test Object Instrumentation (TOI) Module

The first stage of the *Test Object Analysis* process is performed by the TOI Module (Fig. 23.3), and involves: (1) the creation of a CFG providing a representation of the MUT; and (2) the instrumentation (i.e., the probe insertion) of the CUT. This will allow evaluating the quality of the generated test programs through a dynamic analysis process that requires executing each test program, and tracing the CFG nodes traversed in order to gather coverage data.

The CFG building and instrumentation are performed statically (as opposed to the dynamic process of the EA) with the aid of ASM. The instrumentation and CFG computation steps are performed at the Java Bytecode level; working at this level allows an object to be tested even when its source code is unavailable, thus enabling *eCrash* to perform structural testing on third-party components (Ribeiro et al. 2007a).

[8]http://junit.org/.

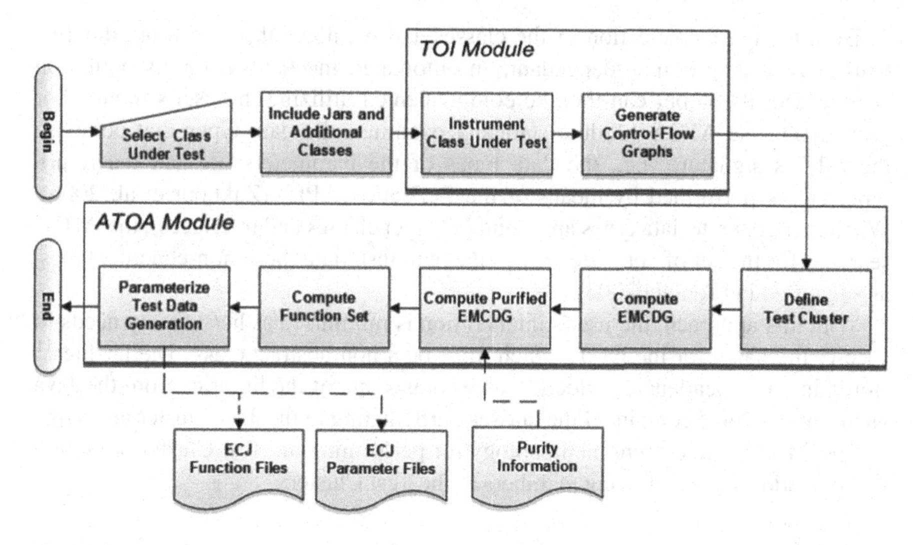

Fig. 23.3 Test Object Analysis (Adapted from Ribeiro 2010)

23.3.2 Automatic Test Object Analysis (ATOA) Module

The second stage of the *Test Object Analysis* process is performed by the ATOA Module (Fig. 23.3). It is in charge of generating all the data structures and files required by the *Test Data Generation process*, including: (1) the test cluster; (2) the ECJ Function Files; and (3) the ECJ Parameter Files. The test cluster gathers all the types, members and constant values that can be used to test a specific MUT, and its specification is essential to the generation of the ECJ's configuration files, which will subsequently be used by the GP algorithm to create suitable Test Data for the MUT.

23.3.2.1 Test Cluster Definition

The transitive set of classes which are relevant for testing a particular class is designated the Test Cluster for that class; its definition has great impact in the test data generated by the proposed evolutionary strategy, given that it contains the set of classes, methods, constructors and constants available for building the test cases. A call to the `search` method of the `Stack` class,[9] for example, requires both a `Stack` instance and an `Object` instance (the one to be searched) to be previously created. As such, both the `Object` and the `Stack` classes must be present in the Test Cluster.

[9]http://docs.oracle.com/javase/7/docs/api/java/util/Stack.html.

Even though the selection of the classes and members that constitute the Test Cluster is largely human-dependant, an automated mechanism for its definition is proposed; its output can then be complemented utilizing the user's inputs. For each MUT, the ATOA Module statically examines the data types that compose the MUT's signature, i.e., the data types of the parameters list and the return type; this is performed by means of the Reflection API[10] (Zakhour et al. 2006). When non-concrete data types are found (abstract classes and interfaces), the ATOA searches for the set of concrete classes that can instatiate those non-concrete types, as suggested in (Tonella 2004).

With this approach, the user's intervention is minimised as he/she only needs to specify the name for the CUT, which must be a non-abstract class. The jar files[11] stored in the dependencies folder(s), as well as some of the libraries from the Java environment, must contain all the classes participating in the Test Cluster analysis.

Specifically, the current methodology for performing the Test Cluster definition involves adding the following members to the Test Cluster:

- the CUT;
- all the public members (methods and constructors) of the CUT; these members form the set of MUT.
- all the data types (both reference and primitive) appearing as parameters in the CUT's members;
- the default constructors (i.e., constructors with no explicit parameters) of all the reference data types, if available;
- if the default constructors are not available for a data type, then the following attempts are executed (in order):
  - search for constructors with parameter types already included in the Test Cluster; if not available, then
  - search for methods (already included in the Test Cluster) that return an object of the data type; if not available, then
  - search for constructors that may add new reference data types to the Test Cluster.
- a default constant set for each of the primitive Java data types, which includes acceptable and boundary values, and is used to sample the search space in accordance to the methodology proposed in (Kropp and Siewiorek 1998); other values can be added by the user.
- *null* constants for the reference types included into the Test Cluster; if the CUT is listed as a parameter type then a *null* constant is added, otherwise it is not included.

[10]http://java.sun.com/javase/6/docs/api/java/lang/reflect/package-summary.html.

[11]Archive of Java classes or libraries.

An alternative methodology for specifying the Test Cluster is available; it consists of giving the user the possibility of providing the *eCrash* tool with a XML file containing a specification of the data types to be included.

After the Test Cluster is defined, the ATOA module proceeds to automatically generate the Function Set and, finally, all the configuration files required by ECJ to evolve test programs.

23.3.2.2 Function Set Generation

eCrash's testing methodology involves encoding and evolving candidate test programs as STGP trees. Each STGP tree must subscribe to a Function Set that defines the STGP nodes legally permitted in the tree.

The first task of the Function Set Generation process is that of modelling the call dependencies of the data types and members existing in the Test Cluster by employing an Extended Method Call Dependence Graph (EMCDG) (Wappler and Wegener 2006).

The *eCrash*'s Function Set generation process incorporates a technique for Input Domain Reduction used to remove irrelevant variables, with the goal of decreasing the number of distinct test programs that can be possibly created while searching for a particular test scenario. A strategy employing Purity Analysis (Ribeiro et al. 2009) was used to reduce the input domain of the search problem addressed by the *eCrash*'s approach. Parameter Purity Analysis is performed on the parameters of the methods included in the Test Cluster, and consists of annotating them with a label that identifies whether the parameter is read-only or is read-write. With basis on the resulting information, the EMCDG is pruned to remove irrelevant edges. Finally, the Function Set is derived from the purified EMCDG (by means of a process detailed in Ribeiro et al. (2009)), which in turn will be used to generate the ECJ configuration files.

23.3.2.3 ECJ Parameter and Function Files Generation

The evolution of potential solutions is performed by the TPG module, which is supported by the STGP mechanisms available in the ECJ framework. On the other hand, the ATOA module is responsible for automating the generation of the problem-specific ECJ configuration files, which are required for evolving the potential solutions: the Parameter and Function Files. The former parametrise the EA's configuration, whereas the latter are utilised to store the information contained by a particular Function Set entry and, consequently, the data which will be contained in the STGP trees' nodes. Therefore, the generation of these files is intrinsically based upon on the Function Set specification.

The set of Parameter Files comprises three distinct types of files:

- General Parameters—mostly encompass GP's configurations (e.g., termination criteria, population size, evolutionary operators, selection strategy, and tree builders).
- Test Object Specific Parameters—defined with basis on Function Set information; encode all problem-related configurations, and include the definition of the data types and GP node constraints.
- MUT Specific Parameters—define the root node of the STGP tree, which must necessarily be the MUT.

Function Files encode all the information that will be included into the Method Call Tree (MCT)' nodes, which will subsequently be used to decode the STGP tree into the Test Program (by means of the process conducted by the TPEM module). Relevant information includes the type of node (constructor, method, or constant), the member's parameters data types, and the return value data type.

23.3.3 Test Program Generation (TPG) Module

This Section details how the TPG module operates: it starts by describing the preparatory steps which precede the evolutionary run; and then, the iterative process by means of which candidate solutions to the problem are created.

23.3.3.1 Setting Up the Evolutionary Run

The main goal of each evolutionary run is to find a set of Test Programs that achieves full structural coverage of the CFG representing a particular MUT; as such, before commencing the evolutionary process, a list of the CFG nodes remaining—which initially includes all the CFG nodes—is created. At the beginning, all CFG nodes are assigned a predefined weight value (Ribeiro et al. 2009), and then the weights are re-evaluated in every generation.

Another preparatory step for the creation of the initial population of candidate solutions is that of initialising the constraints' selection probabilities. This particular parameter is related with the adaptive strategy implemented and described in Ribeiro et al. (2010), which promotes the introduction of relevant instructions into the generated Test Programs by means of Mutation.

23.3.3.2 Evolving Test Programs

Test Programs are evolved while there are CFG nodes left to be covered, or until a predefined number of generations is reached.

The initial population is composed by individuals created randomly by means of the selected tree building algorithm (e.g., Full and Grow); the seed for the random generator is also configurable by the user. Subsequent populations are formed by individuals originated from those existing in the preceding population, which may either be cloned directly or altered before being copied. Distinct probabilities of selecting the breeding operators may also be defined.

Each individual must be decoded into the corresponding Test Program, compiled and executed, so as to allow verifying if it should be included into the Test Set. Whenever a Test Case "hits" an unexercised CFG node, that node is removed from the list of remaining CFG nodes, and the Test Case is added to the Test Set.

The TPEM module is responsible for evaluating and providing feedback regarding the generated test programs; that feedback will determine whether there is the need to continue to evolve individuals—the evaluation process is described in more detail in the following subsections.

23.3.4 Test Program Evaluation and Management (TPEM) Module

Test program quality evaluation involves the execution of the generated test programs, with the intention of collecting trace information with which to derive coverage metrics. The TPEM module is the one responsible for: the process of transforming the individuals' genotypes—the MCT—into the phenotypes—the Test Programs; and the procedure of ascertaining the quality of test programs and computing the corresponding individuals' fitness.

23.3.4.1 Decoding Test Programs

Decoding an MCT into the Test Program for execution is a two step process involving: (1) the linearisation of the MCT, so as to obtain the MCS; and (2) the translation of the MCS into the Test Program. Therefore:

1. the MCS corresponds to the internal representation of the Test Program; specifically, it corresponds to the linearised MCT. Listing 1 depicts a MCS obtained from the linearisation of the MCT depicted in Fig. 23.4. Algorithm 2 details the depth-first transversal algorithm utilised to linearise the trees.
2. the Test Program corresponds to the syntactically correct, compilable, and executable version of the MCS, according to the Java programming language. Listing 2 contains the Java Test Program synthesised from the MCS shown in Listing 1.

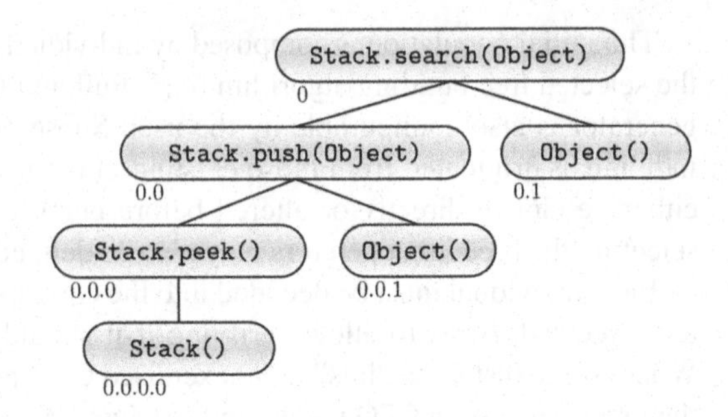

Fig. 23.4 Example Method Call Tree. The Method Under Test is the `search` method of the `Stack` class

Listing 1 Example MCS, resulting from the linearisation of the MCT depicted in Fig. 23.4.

```
1 0.0.0.0 Stack()
2 0.0.0 Stack.peek() [0.0.0.0 Stack]
3 0.0.1 Object()
4 0.0 Stack.push(Object) [0.0.0 Stack.peek(), 0.0.1 Object()]
5 0.1 Object()
6 0 Stack.search(Object) [0.0 Stack.push(Object), 0.1 Object()]
```

Algorithm 2: Algorithm for Method Call Tree linearisation.

Data: Method Call Tree
Result: Method Call Sequence

Global Variables:
Current Node ← Root Node;
Previous Method Information Object (MIO) ← null;
MCS ← empty sequence;

begin Function `linearizeMCT`(*Current Node*)
  if *Current Node* ≠ *Root Node* **then**
    Previous MIO ← get MIO from Parent Node of Current Node;
  Current MIO ← get MIO from Current Node;
  if *Previous* MIO ≠ *null* **then**
    add Current MIO to Parameter Providers List of Previous MIO;
  Child Nodes List ← get Child Nodes List from Current Node;
  foreach *Child Node* **in** *Child Nodes List* **do**
    call `linearizeMCT`(*Child Node*);
  add Current MIO to MCS;

Listing 2 Example Test Program, synthesised with basis on the MCS depicted in Listing 1.

```
1 Stack stack1 = new Stack();
2 stack1.peek();
3 Object object2 = new Object();
4 stack1.push(object2);
5 Object object3 = new Object();
6 stack1.search(object3);
```

Listing 2 depicts an example Test Program for Object-Oriented software; the MUT is the `search` method of the `Stack` class. In this program, instructions 1, 3 and 5 instantiate new objects, whereas instructions 2 and 4 aim to change the state of the `stack0` instance variable that will be used, as the implicit parameter, in the call to the MUT at instruction 6.

The Test Program's source-code synthesis is performed by translating MCS into Test Programs using the information contained in each MCS entry. Specifically, each MCS entry contains a MIO, which encloses the information contained in the Function Files, such as a method's name and class, parameter types and return type; and references to other MIOs providing the parameters (if any) for that method.

23.3.4.2 Evaluating Test Programs

The quality of a particular test program is related to the CFG nodes of the MUT which are the targets of the evolutionary search at the current stage of the search process. Test cases that exercise less explored CFG nodes and paths are favoured, with the objective of finding a set of test cases that achieves full structural coverage of the test object.

The issue of steering the search towards the traversal of interesting CFG nodes and paths was addressed by assigning weights to the CFG nodes; the higher the weight of a given node, the higher the cost of exercising it, and hence the higher the cost of traversing the corresponding control-flow path. The weights of the CFG nodes are re-evaluated at the beginning of every generation; nodes which have been recurrently traversed in previous generations and/or lead to uninteresting paths are penalised.

For *feasible* test cases, the fitness is computed with basis on their trace information; relevant trace information includes the "Hit List"—i.e., the set of traversed CFG nodes. This strategy causes the fitness of feasible test programs that exercise recurrently traversed structures to fluctuate throughout the search process. Frequently hit nodes will have their weight increased, thus worsening the fitness of the test cases that exercise them.

For *unfeasible* test cases, the fitness of the individual is calculated in terms of the distance between the runtime exception index (i.e., the position of the method call that threw the exception) and the MCS length. Also, an *unfeasible penalty constant* value is added to the final fitness value, so as to penalise unfeasibility. With this methodology, and depending on the value of the *unfeasible penalty constant* and on the fitness of feasible test cases, unfeasible test cases may be selected for breeding at certain points of the evolutionary search, thus favouring the diversity and complexity of MCSs.

The test program evaluation procedure utilized is described with detail in Ribeiro et al. (2008) and Ribeiro et al. (2009).

23.4 Discussion

Even though *eCrash* is still in a prototype development stage, it is applicable to a vast array of OO test objects. Preliminary empirical studies were conducted on container classes (e.g., Stack, Vector, and BitSet) in order to support the integration of research steps and establish the suitability of the proposed ET technique. Recent enhancements allowed performing experimental studies on larger, real-world software products; in (Nogueira et al. 2013) *eCrash* was utilised to generate structural test data for the Apache Ant project[12] release 1.8.4. In this context, *eCrash*'s performance was compared to that of two other well-known and established test data generation tools: the *Randoop* (Pacheco and Ernst 2007) random testing tool; and the *EvoSuite* ET tool. The custom-made test suite provided by the Apache Ant's distribution was also considered.

Results of recent experiments provided positive and solid indicators of the effectiveness and efficiency of the *eCrash* tool, as well as of its robustness and applicability to large and complex software products. Also, and most importantly, they allowed pinpointing several topics for future work. *eCrash* was unable to generate tests for some instance methods that enter infinite loops; and for some static methods in classes that are not able to provide instances of that data type (namely, when public constructors are not defined). Also, *eCrash* faced some difficulties when testing specific problematic methods that implement: instructions from the Reflection API; class loaders; input handlers; task and thread handlers; file and folder managers; managers for email and download; compilers; audio and image processors; and encapsulators of Unix commands. The difficulty in testing classes related to certain system's features and functionalities had, in fact, already been reported in the literature (Fraser and Arcuri 2012). Some Java's specific features which must also be addressed in future work include: the support for Generics; and enabling the testing of non-abstract classes.

Still, the positive feedback obtained from recent empirical results supported our decision to publish the *eCrash* tool for public usage. Currently, a prototype that can be executed via command-line is available, accompanied by tutorials which serve as a quick-start guide. We are actively working on developing an IDE-integrated version of the tool, which is expected to be more user-friendly and usable by Software Testers in a production environment and ET researchers alike.

23.5 Conclusions

eCrash is an automated test data generation tool for Object-Oriented software. This paper details the Evolutionary Testing methodology implemented by this

[12]http://ant.apache.org/.

framework, as well as its functionalities and architecture. Even though *eCrash* is currently on a prototype development stage, it has already been applied to software products of considerable complexity, and has recently been deployed for public availability.

eCrash's methodology for Test Data generation involves instrumenting the Test Object, executing it with the generated Test Cases, and tracing the structures traversed in order to derive coverage metrics. The primary goal is that of finding a set of Test Programs that achieves full structural coverage of the Test Object. Test Programs are encoded and evolved as STGP individuals.

This process is divided into two major moments: the Test Object Analysis phase and Test Data Generation phase. The *Test Object Analysis* phase includes the instrumentation of the CUT so as to allow the posterior coverage analysis task. Also, all the details regarding the CUT must be retrieved, such as the methods, dependencies and objects required to build interesting states for the class. Finally, the construction of specific configuration files takes place; these files will be provided as inputs for the ET algorithm. The *Test Data Generation* phase involves the creation and evaluation of Test Program candidates with the aid of the ECJ framework. This process outputs the generated test suite in JUnit compatible format.

Future work involves fixing some of the *eCrash*'s limitations identified in recent experiments, which focused on applying this tool to large and complex software products. It also entails the implementation of specific features that will allow *eCrash* to achieve better results in specific application scenarios, such as the support for the generation of test data for non-public members and abstract classes. Moreover, we expect to be able to use *eCrash* to gather information on several quality attributes of the software under test with basis on the characteristics of the automatically generated test suite.

eCrash is publicly available at:

http://sourceforge.net/p/ecrashtesting/

References

Baresi, L., Lanzi, P.L., Miraz, M.: Testful: An evolutionary test approach for java. In: Proceedings of the 2010 Third International Conference on Software Testing, Verification and Validation, ICST '10, pp. 185–194. IEEE Computer Society, Washington, DC, USA (2010). DOI 10.1109/ICST.2010.54. URL http://dx.doi.org/10.1109/ICST.2010.54

Baresi, L., Miraz, M.: Testful: automatic unit-test generation for java classes. In: Proceedings of the 32nd ACM/IEEE International Conference on Software Engineering - Volume 2, ICSE '10, pp. 281–284. ACM, New York, NY, USA (2010). DOI 10.1145/1810295.1810353. URL http://doi.acm.org/10.1145/1810295.1810353

Fraser, G., Arcuri, A.: Evolutionary generation of whole test suites. In: Proceedings of the 2011 11th International Conference on Quality Software, QSIC '11, pp. 31–40. IEEE Computer Society, Washington, DC, USA (2011). DOI 10.1109/QSIC.2011.19. URL http://dx.doi.org/10.1109/QSIC.2011.19

Fraser, G., Arcuri, A.: Evosuite: automatic test suite generation for object-oriented software. In: Proceedings of the 19th ACM SIGSOFT symposium and the 13th European conference on Foundations of software engineering, ESEC/FSE '11, pp. 416–419. ACM, New York, NY, USA (2011). DOI 10.1145/2025113.2025179. URL http://doi.acm.org/10.1145/2025113.2025179

Fraser, G., Arcuri, A.: Sound empirical evidence in software testing. In: 34th International Conference on Software Engineering, ICSE 2012, June 2–9, 2012, Zurich, Switzerland, pp. 178–188. IEEE (2012)

Fraser, G., Arcuri, A.: Whole test suite generation. IEEE Trans. Softw. Eng. **39**(2), 276–291 (2013). DOI 10.1109/TSE.2012.14. URL http://dx.doi.org/10.1109/TSE.2012.14

Fraser, G., Arcuri, A., McMinn, P.: Test suite generation with memetic algorithms. In: Proceeding of the fifteenth annual conference on Genetic and evolutionary computation conference, GECCO '13, pp. 1437–1444. ACM, New York, NY, USA (2013). DOI 10.1145/2463372.2463548. URL http://doi.acm.org/10.1145/2463372.2463548

Harman, M.: Automated test data generation using search based software engineering. In: AST '07: Proceedings of the Second International Workshop on Automation of Software Test, p. 2. IEEE Computer Society, Washington, DC, USA (2007). DOI http://dx.doi.org/10.1109/AST.2007.4

Harman, M., Mansouri, S.A., Zhang, Y.: Search based software engineering: A comprehensive analysis and review of trends techniques and applications. Tech. Rep. TR-09-03, Department of Computer Science, King's College London (2009). URL http://www.dcs.kcl.ac.uk/technical-reports/papers/TR-09-03.pdf

Kinneer, A., Dwyer, M., Rothermel, G.: Sofya: A flexible framework for development of dynamic program analysis for java software. Tech. Rep. TR-UNL-CSE-2006-0006, University of Nebraska (2006). URL http://sofya.unl.edu/

Koza, J.R.: Genetic Programming: On the Programming of Computers by Means of Natural Selection (Complex Adaptive Systems). The MIT Press (1992). URL http://www.amazon.ca/exec/obidos/redirect?tag=citeulike04-20&path=ASIN/0262111705

Kropp, N.P., Jr., P.J.K., Siewiorek, D.P.: Automated robustness testing of off-the-shelf software components. In: Symposium on Fault-Tolerant Computing, pp. 230–239 (1998). URL citeseer.ist.psu.edu/kropp98automated.html

Luke, S.: ECJ 21: A Java evolutionary computation library. http://cs.gmu.edu/~eclab/projects/ecj/ (2013)

McMinn, P.: Search-based software test data generation: A survey. Software Testing, Verification and Reliability **14**(2), 105–156 (2004). URL citeseer.ist.psu.edu/mcminn04searchbased.html

McMinn, P., Holcombe, M.: The state problem for evolutionary testing (2003). URL citeseer.ist.psu.edu/mcminn03state.html

Miraz, M., Lanzi, P.L., Baresi, L.: Testful: using a hybrid evolutionary algorithm for testing stateful systems. In: Proceedings of the 11th Annual conference on Genetic and evolutionary computation, GECCO '09, pp. 1947–1948. ACM, New York, NY, USA (2009). DOI 10.1145/1569901.1570252. URL http://doi.acm.org/10.1145/1569901.1570252

Montana, D.J.: Strongly typed genetic programming. Evolutionary Computation **3**(2), 199–230 (1995)

Naik, S., Tripathy, P.: Software Testing and Quality Assurance: Theory and Practice. Wiley (2008)

Nogueira, A.F., Ribeiro, J.C.B., de Vega, F.F., Zenha-Rela, M.A.: ecrash: An empirical study on the apache ant project. In: Proceedings of the 5th International Symposium on Search Based Software Engineering (SSBSE '13), vol. 8084. Springer, St. Petersburg, Russia (2013)

Pacheco, C., Ernst, M.D.: Randoop: feedback-directed random testing for java. In: OOPSLA '07: Companion to the 22nd ACM SIGPLAN conference on Object-oriented programming systems and applications companion, pp. 815–816. ACM, New York, NY, USA (2007). DOI http://doi.acm.org/10.1145/1297846.1297902

Pavlov, Y., Fraser, G.: Semi-automatic search-based test generation. In: Proceedings of the 2012 IEEE Fifth International Conference on Software Testing, Verification and Validation, ICST '12, pp. 777–784. IEEE Computer Society, Washington, DC, USA (2012). DOI 10.1109/ICST.2012.176. URL http://dx.doi.org/10.1109/ICST.2012.176

Ribeiro, J.C.B.: Search-based test case generation for object-oriented java software using strongly-typed genetic programming. In: GECCO '08: Proceedings of the 2008 GECCO Conference Companion on Genetic and Evolutionary Computation, pp. 1819–1822. ACM, New York, NY, USA (2008). DOI http://doi.acm.org/10.1145/1388969.1388979

Ribeiro, J.C.B.: Contributions for improving genetic programming-based approaches to the evolutionary testing of object-oriented software. Ph.D. thesis, Universidad de Extremadura, Espaã (2010)

Ribeiro, J.C.B., de Vega, F.F., Zenha-Rela, M.: Using dynamic analysis of java bytecode for evolutionary object-oriented unit testing. In: SBRC WTF 2007: Proceedings of the 8th Workshop on Testing and Fault Tolerance at the 25th Brazilian Symposium on Computer Networks and Distributed Systems, pp. 143–156. Brazilian Computer Society (SBC) (2007)

Ribeiro, J.C.B., Zenha-Rela, M., de Vega, F.F.: ecrash: a framework for performing evolutionary testing on third-party java components. In: CEDI JAEM'07: Proceedings of the I Jornadas sobre Algoritmos Evolutivos y Metaheuristicas at the II Congreso Español de Informática, pp. 137–144 (2007)

Ribeiro, J.C.B., Zenha-Rela, M., de Vega, F.F.: A strategy for evaluating feasible and unfeasible test cases for the evolutionary testing of object-oriented software. In: AST '08: Proceedings of the 3rd International Workshop on Automation of Software Test, pp. 85–92. ACM, New York, NY, USA (2008). DOI http://doi.acm.org/10.1145/1370042.1370061

Ribeiro, J.C.B., Zenha-Rela, M.A., Fernández de Vega, F.: Test case evaluation and input domain reduction strategies for the evolutionary testing of object-oriented software. Inf. Softw. Technol. **51**(11), 1534–1548 (2009). DOI http://dx.doi.org/10.1016/j.infsof.2009.06.009

Ribeiro, J.C.B., Zenha-Rela, M.A., de Vega, F.F.: Strongly-typed genetic programming and purity analysis: input domain reduction for evolutionary testing problems. In: GECCO '08: Proceedings of the 10th Annual Conference on Genetic and Evolutionary Computation, pp. 1783–1784. ACM, New York, NY, USA (2008). DOI http://doi.acm.org/10.1145/1389095.1389439

Ribeiro, J.C.B., Zenha-Rela, M.A., de Vega, F.F.: Adaptive evolutionary testing: an adaptive approach to search-based test case generation for object-oriented software. In: NICSO 2010 - International Workshop on Nature Inspired Cooperative Strategies for Optimization, Studies in Computational Intelligence. Springer (2010)

Ribeiro, J.C.B., Zenha-Rela, M.A., de Vega, F.F.: Enabling object reuse on genetic programming-based approaches to object-oriented evolutionary testing. In: EuroGP 2010 - 13th European Conference on Genetic Programming (to appear), Lecture Notes in Computer Science. Springer (2010)

Tassey, G.: The economic impacts of inadequate infrastructure for software testing. Tech. rep., National Institute of Standards and Technology (2002)

Tonella, P.: Evolutionary testing of classes. In: ISSTA '04: Proceedings of the 2004 ACM SIGSOFT international symposium on Software testing and analysis, pp. 119–128. ACM Press, New York, NY, USA (2004). DOI http://doi.acm.org/10.1145/1007512.1007528

Vincenzi, A.M.R., Delamaro, M.E., Maldonado, J.C., Wong, W.E.: Establishing structural testing criteria for java bytecode. Softw. Pract. Exper. **36**(14), 1513–1541 (2006). DOI http://dx.doi.org/10.1002/spe.v36:14

Wappler, S., Wegener, J.: Evolutionary unit testing of object-oriented software using a hybrid evolutionary algorithm. In: CEC'06: Proceedings of the 2006 IEEE Congress on Evolutionary Computation, pp. 851–858. IEEE (2006)

Wappler, S., Wegener, J.: Evolutionary unit testing of object-oriented software using strongly-typed genetic programming. In: GECCO '06: Proceedings of the 8th annual conference on Genetic and evolutionary computation, pp. 1925–1932. ACM Press, New York, NY, USA (2006). DOI http://doi.acm.org/10.1145/1143997.1144317

Zakhour, S., Hommel, S., Royal, J., Rabinovitch, I., Risser, T., Hoeber, M.: The Java Tutorial: A Short Course on the Basics, 4th Edition (Java Series), 4th edn. Prentice Hall PTR (2006). URL http://www.worldcat.org/isbn/0321334205

Printed in the United States
By Bookmasters